Engineering Mechanics of Solids and Structures

Engineering Mechanics of Solids and Structures is essential for self-study and reference, setting out the underlying mechanical concepts, methods, and theories applicable to a broad spectrum of structural and mechanical components—relevant to the analysis and design of a dome-like structure or a suspension bridge, the hull of a ship or a spacecraft, the mechanical component of an engine, or the power train of a vehicle. The principles are given a thorough mathematical treatment, which demonstrates the limitations of the approximations and provides the tools for more precise analyses, while less mathematics is needed to show the behavior of simple structural components.

This book covers both the Lagrangian and Newtonian approaches to the conditions of motion and equilibrium, offering the Lagrangean criteria of work and energy as the most effective means to formulate finite elements and other discrete approximations, which are amenable to digital computation. This gives advanced students in aerospace, mechanical, and civil engineering a solid foundation for the analysis and approximation of engineering problems, presented in a form that is both mathematically rigorous and physically meaningful.

Gerald Wempner held faculty positions at five major US state universities: Illinois, Arizona, Berkeley, Alabama, and Georgia Tech. He made seminal contributions to engineering mechanics, notably identifying shear locking in finite element formulations for thin shells, introducing a discrete analog of Kirchhoff's hypothesis, and developing the generalized arc method for nonlinear structural analysis. He passed away in 2024.

Abdul-Hamid Zureick is Professor of Structural Engineering, Mechanics, and Materials in the School of Civil and Environmental Engineering at Georgia Tech.

Engineering Mechanics of Solids and Structures

Gerald Wempner

Edited by
Abdul-Hamid Zureick

CRC Press
Taylor & Francis Group
Boca Raton London New York

CRC Press is an imprint of the
Taylor & Francis Group, an **informa** business

Front cover image: Gerald Wempner

First edition published 2026
by CRC Press
2385 NW Executive Center Drive, Suite 320, Boca Raton FL 33431

and by CRC Press
4 Park Square, Milton Park, Abingdon, Oxon, OX14 4RN

CRC Press is an imprint of Taylor & Francis Group, LLC

ISBN: 978-1-041-12570-9 (hbk)
ISBN: 978-1-041-12650-8 (pbk)
ISBN: 978-1-003-66586-1 (ebk)

DOI: 10.1201/9781003665861

Typeset in Times
by codeMantra

Contents

Foreword

The book *Engineering Mechanics of Solids and Structures* is the fifth and last book authored or co-authored by the late Professor Gerald (Gerry) Wempner. It covers a broad range of subjects that form the foundation of solid mechanics and structures. Each topic is presented with exceptional clarity and logical order, providing a lucid exposition that thoroughly outlines its fundamental principles, theories, assumptions, and limitations.

It is both an honor and a privilege to serve as an editor for this book, written by an admirable colleague and sincere friend who passed away at the age of 96 before completing his work. I first met Gerry in person in 1985 when I joined the faculty at Georgia Tech, having already known his published textbooks on solid mechanics and his breakthrough concept "the generalized arc-length method" published in 1971 in the *International Journal of Solid and Structures* as part of his paper entitled "Discrete Approximations related to Nonlinear Theories of Solids." I have since admired him not only as an impeccable scholar but also as a true gentleman and a sincere colleague.

Gerry Wempner grew up in a modest family. He found inspiration in schools, libraries, and the media of his time. Stories of his grandfather's stallion and his family's rural life in Minnesota and Wisconsin fostered a deep connection to animals, particularly horses, igniting his passion for artistic expression. He was pragmatic about life and understood during his youth that one must acquire some capabilities in a "meaningful" trade or profession to succeed in the adult world. This led him to pursue Mechanical Engineering under the G.I. Bill—a decision that shaped his academic journey, culminating in the publications of five textbooks and several groundbreaking research concepts and works in theoretical and applied mechanics. Noteworthy among his contributions are his 1968 paper, "New Concepts for Finite Elements of Shells," published in *Zeitschrift für Angewandte Mathematik und Mechanik (ZAMM),* which identified what is now known as "shear locking" and introduced the discrete analog of Kirchhoff's hypothesis. Also, his 1971 article "Discrete Approximations related to Nonlinear Theories of Solids" introduced a new method for effective numerical computations in nonlinear problems of structural mechanics, which attracted the attention of researchers and engineers. This concept was popularized in the late 1970s and in the 1980s under the name Riks-Wempner method and was implemented in major computational analysis tools that included ABAQUS, ANSYS, and NASTRAN.

In the classroom, Gerry was an exceptional professor who entered his lectures without a book or notes—armed only with chalks of different colors. He would derive every equation from first principles and illustrate the concepts on the blackboard with remarkable, artistically rendered drawings that greatly enhanced his students' understanding of the subject.

Gerry retired from Georgia Tech in the early 1990s, dividing his time between Atlanta and Plains, Montana, where he found a home that reconnected him with his roots, offering the tranquility of rural America and inspiring him to explore his passion for sculpture. While he was in Atlanta, he often visited me to share updates on his latest sculpture projects, discuss the future of higher education politics, and, of course, talk about his plans to publish a new book on the mechanics of solids. During the mid-2010 decade, Gerry underwent open heart surgery, and shortly thereafter, he was diagnosed with cancer. I maintained close contact with him throughout this challenging period and witnessed his extraordinary resilience. Despite his battle with cancer, he remained intellectually vibrant, often surrounded by a sharp pencil and loose papers filled with intricate mathematical expressions that elegantly captured the forces and deformations in solids and structures. When taking a break from the world of structural mechanics, Gerry turned his creativity toward sculpture. His meticulous works reflected both his precision and imagination. He generously donated many pieces to museums, including three remarkable sculptures gifted to the Bone Creek Museum of Agrarian Art in David City, Nebraska. These sculptures inspired the museum's 2017 exhibition, "The Great American Draft Horse," demonstrating Gerry's artistic vision and profound connection to agrarian themes.

Approximately two weeks prior to his passing, I contacted Gerry to check on him while he was in Plains, Montana. He shared with me some of the medical complications he had experienced during his hospital stay and said he didn't think he had much time left. He asked me to ensure that his final book manuscript—on the mechanics of solids and structures—which he had entrusted to me, would be published. Sadly, soon after that, Gerry passed away on December 30, 2024, marking the end of a celebrated career and a profound loss to the engineering mechanics community and the broader world of scholarship.

His pioneering contributions will continue to inspire and guide future generations of engineering students, researchers, and practitioners. He will be dearly missed. Gerry Wempner is survived by his two children, Paul and Susan, six grandchildren, and two great-grandchildren.

My heartfelt thanks go to Gerry's children, Paul and Susan, for their trust and encouragement in allowing me to edit their father's final work. I made no attempt to introduce any substantive changes to the original manuscript and took great care to preserve Gerry Wempner's original work and intent. It has been both an honor and a privilege to help preserve Gerry's legacy and share his enduring insights with the engineering community.

I would like to thank my Georgia Tech colleagues, Professors Arash Yavari and George Kardomateas, for their careful review of parts of the manuscript and for their thoughtful input and encouragement.

Last but not least, I would also like to extend my appreciation to Tony Moore and Andrew Carmount of the editorial team at CRC Press and Routledge for their valuable guidance and support throughout the publication process. Their guidance, professionalism, and dedication in bringing this book to completion are deeply appreciated.

Abdul-Hamid Zureick
Atlanta, October 15, 2025

Preface

The title of a technical book provides, at best, a general description of its contents. Here, we offer the prospective reader our version of the purpose, prerequisites, and utility of this book. The *Mechanics of Solids and Structures* is intended primarily for self-study and reference. Accordingly, it contains numerous illustrative examples to demonstrate and reinforce the underlying concepts, methods, and theories. The mathematical and mechanical prerequisites include trigonometry, algebra, calculus, vector analysis, and the mechanics of rigid bodies. Traditionally, textbooks on the mechanics of solids provide procedures and equations for specific problems, notably the elementary problems of beams and shafts. This text offers a full treatment of those topics, but they are presented as a means to exemplify the underlying concepts, methods, and theories. This approach reflects a contemporary perspective in the study and practice of engineering. In particular, the text focuses on the basic concepts that provide the foundations of modern methodologies, e.g., modeling via finite elements, which enables us to utilize the capabilities of modern digital computation. Although the mechanical behavior of a solid has its origins in the atomic structure, our treatment adopts the approximations of the continuum. The mathematical theories of the continuum remain the basis for most engineering analyses and designs. The underlying concepts and theories are applicable to a broad spectrum of structural and mechanical components; thus, they are equally relevant to the design of a dome-like structure or a suspension bridge, the hull of a ship or a spacecraft, the mechanical component of an engine, or the power train of a vehicle. Because the basic concepts of the continuous solid provide such powerful tools, our treatment emphasizes the foundations: Basic variables are given precise mathematical definitions and their physical/geometrical identities. This book places the most fundamental tools in the initial chapters. These early chapters include sections that offer very precise mathematical treatments; the precision helps identify the limitations of the approximations and provides the tools for more precise analyses, e.g., problems of severe deformations. Those sections are marked with an asterisk. The early chapters are followed by chapters that present the most elementary topics, those that require a modicum of mathematics, but demonstrate the behavior of simple structural components, e.g., beams and shafts. The mechanics of solids entails three distinct aspects: (1) Kinematics, which describes the geometrical aspects, displacements, and deformations; (2) dynamics/statics, which govern the state of motion/equilibrium; and (3) the behavior of the material, which ultimately relates the kinematical and dynamical variables, e.g., the deformations and forces. The kinematics and dynamics are introduced initially and independently; both topics are independent of the media. The definitions of essential quantities, e.g., strains and stresses, are introduced in simple geometric and physical terms. These are subsequently examined to reveal precise mathematical and physical interpretations. A distinctive feature of this text is the immediate introduction of the Lagrangean, as well as the Newtonian, approach to the conditions of motion/equilibrium. This is particularly relevant in our era of digital computation, since the Lagrangean criteria of work and energy are often the most effective means to formulate the discrete approximations, e.g., finite elements, which are amenable to digital computation. Energy criteria have also served in a perverse way to formulate continuous models to simulate the discrete behaviors of the real materials: A most notable example is the Griffith criteria of crack propagation [Section 4.20]. The consistent choices of associated strains and stresses [Section 3.16] depend on the equivalence of the work/energy density (per unit volume). In short, the work/energy criteria are crucial to the full understanding of our subject and the implementation of the various tools. The basic tools developed in Chapters 2–4 are applied to the simplest problems in Chapter 5, viz., simple extension, torsion, and bending of prismatic members. The results are the so-called "exact" solutions in the linear theory of elasticity. With due attention to Saint-Venant's principle [Section 5.2], these fundamental solutions are applied to systems of axially loaded members. The applications to trusses provide an entree to the practical usage of the fundamentals. Similarly, numerous examples demonstrate the practical usage of the solutions of simple torsion and bending. The analysis of beams is set apart in Chapter 6, since the theory incorporates specific approximations, differential equations, and the integration of those equations. The treatment includes the behavior of inelastic beams and buckling under axial loading. Chapter 7 introduces superposition and its application to combined loadings. Examples not only provide practical application of the prior solutions but also reveal the responses to various combined loadings. Chapter 8 provides an introduction to the mechanics of thin elastic plates. The concepts of the initial chapters are employed to describe extension and flexural deformations. The underlying hypothesis of Kirchhoff is developed, and a few rudimentary examples demonstrate the response to simple loadings. The approximation is extended to include a transverse shear; the latter is employed in Chapter 10 to produce a simple discrete approximation of the finite element. Chapter 9 introduces the basic concepts of virtual work and the principles of stationary and minimum potential; these are couched in the most general forms, i.e., they are not restricted to small deformations, nor to quadratic forms. The concept of minimum potential is cited as the basis for a stable conservative mechanical system. The Trefftz criteria for a critical state and the Koiter criteria for stability at the critical state are presented and illustrated. The latter is of much practical significance, since it identifies states of imminent collapse from those that merely limit load-carrying capacity. This chapter lays the foundations for many discrete approximations, i.e., finite elements, in Chapter 10. Chapter 10 introduces the concepts of the discrete approximations, i.e., the mechanisms by which the differential equations of the continuum theories are supplanted by the systems of difference (algebraic) equations. These concepts, together with the advent of the modern digital computer, have now opened the way to refine approximations of the solutions that had previously defied meaningful solutions. Here, our emphasis is on the mathematical formulations and the physical interpretations, e.g., the roles of the rigid and deformational modes of elements. Examples demonstrate the convergence of the difference equations to the differential equations. The overriding theme of this book is the presentation of concepts and methods with a view toward their applicability and limitations. Our intent is not only to present a handbook for the treatment of specific existing problems but also to provide a foundation for the solution/approximation of current and future problems. Hopefully, our presentations are mathematically correct and physically meaningful. On a personal note, the author must acknowledge his most influential teacher: His father, Paul C. Wempner, was a person with minimal formal education, but one who demonstrated the value of keen observation and concerted intellectual effort. In his words, "The mind is like a muscle. To develop, it requires intellectual exertion."

About the Author

Gerald Wempner, the late Emeritus Professor in the School of Civil and Environmental Engineering at the Georgia Institute of Technology in Atlanta, Georgia, began his career in 1953 with an appointment as an instructor of Theoretical and Applied Mechanics at the University of Illinois, Urbana-Champaign, where he received his Ph.D. in 1957. Apart from his short-term engineering and research activities at Stanford University, Lockheed Laboratories in Palo Alto, Ruhr University in Germany, and the University of Calgary in Canada, he held faculty positions at five major US universities. His teaching and research primarily focused on solid and structural mechanics, structural stability, plate and shell theories, as well as variational methods and approximations. Over his professional career, he authored and co-authored four books covering the mechanics of solids, structures, and shells. He passed away on December 30, 2024, at the age of 96.

1 Introduction

1.1 BRIEF HISTORICAL ACCOUNT

The "Mechanics of Solids" encompasses certain concepts (e.g., strain and stress) and mathematical descriptions of solid continua (e.g., elasticity), the formulations of the equations governing the mechanical behavior of solid bodies (e.g., differential equations and boundary values), simplifications for specific forms (e.g., beams, plates, and shells), and, finally, the approximations (e.g., finite elements), which simulate the continuous bodies via discrete variables. It is the latter that enables the engineer to utilize the enormous capabilities of the digital computer to perform the analysis and design of all manner of structures and machines.

Our subject has a long and storied history,[1] which can be traced to Galileo's book *Two New Sciences* (1638) on the mechanical properties of structural materials and on the strength of beams. Contemporary theories of elastic beams are attributed to eminent mathematicians, J. (Jacob) Bernoulli (1694) and L. Euler (1744). Our concepts of elastic behavior originated with R. Hooke.[2] The general theory of elastic bodies was established by A. L. Cauchy (1822). Important contributions were made by B. de Saint-Venant; among these are his theories of the torsion and flexure of beams (1847). It is significant that the latter was drawn from the general (three-dimensional) theory as opposed to the earlier studies of beams. In tracing the theories of plates and shells, one observes a similar dichotomy: Early formulations by A. L. Cauchy (1828) and S. D. Poisson (1829) were obtained by reductions from the three-dimensional equations of elasticity. The subsequent work of G. R. Kirchhoff (1876) followed his initial approximation of the flexural deformation, which immediately reduced the description to two dimensions. Kirchhoff's hypothesis was invoked by H. Aron (1874) and subsequently by A. E. H. Love (1888) to develop a general theory of Hookean shells. The foregoing overview serves only to highlight early "landmarks" in our subject. Equally important advances in general concepts of work and energy were set forth, most notably by J. (Jean) Bernoulli's principle of virtual work, L. Euler's development of the variational method (1744), and J. L. Lagrange's treatise (1788) on analytical mechanics. The latter are indispensable tools in contemporary simulation (e.g., discrete models via finite elements). Thus, without the developments of the eighteenth century, we could not readily exploit the powers of our modern electronic computers.

From a contemporary perspective, the foundations of our subject (mechanics of continuous solids) were in place at the turn of the century (1900). In particular, the general theory of elasticity was complete; Love's treatise (1892) left few significant voids. That theory and its evolution are described in Love's final edition (1927). Most concepts of inelasticity were also set forth in the nineteenth century: These include Tresca's yield condition (1864), Levy's flow relation (1870), and Maxwell's concept of viscoelasticity (1868). Still, general formulations and refinements followed: Theories of plasticity (time-independent inelasticity) are described by A. Nadai (1950, 1963), R. Hill (1950), and L. M. Kachanov (1971). We cite these few works only to provide some historical perspectives, but note that inelastic behavior is inherently more complicated than elastic. Hence, our knowledge of inelastic behavior and failure of solids remains incomplete.

As previously noted, the all-important concepts of work, energy, and the variational methods were established in the nineteenth century. Still, important developments followed: In particular, we note the generalized potential of H. C. Hu and K. Washizu (1955) and the complementary functional of E. Reissner (1950); both are useful in the contemporary formulations of finite elements.

DOI: 10.1201/9781003665861-1

The energy criteria of E. Trefftz (1933) and W. T. Koiter (1945) provide the means to assess the stability of modern structures, especially the critical states of thin shells. An exposition of these works can be found in the authors' previous book.[3]

1.2 RAISON D'ETRE

Every structure or machine is designed for a specific purpose, e.g., a bridge serves to carry traffic over an obstacle. Anything that impairs the ability of the structure to fulfill its intended purpose signals a failure. Most conditions of failure are predictable; however, unexpected circumstances and consequent failures occur. The failure of the Tacoma Narrows Bridge[4] in November 1940 was caused by unanticipated interactions. The bridge was dramatically destroyed by uncontrollable and violent oscillations induced by crosswinds; the phenomenon is known as aerodynamic flutter, which was not foreseen by the designers.

The effective design of any structural or mechanical component depends upon the engineer's ability to predict the responses under the anticipated conditions of service. Every solid body inevitably deforms under loads and changes of environment; failures are the consequences of excessive deformations or fracture. The means to predict those responses are founded upon the mechanics of solids. The underlying concepts, theories, and mathematical tools are the subjects of this book.

A review of the history of our subject indicates the diversity of problems, concepts, and methods that drew the attention of a broad spectrum of engineers, scientists, and mathematicians. It is noteworthy that the era from Galileo Galilei (1564–1642) to A. E. H. Love (1863–1940) saw no clear demarcation between those professions. Subsequently, the mechanics of solids has become increasingly fragmented; for example, studies of elasticity are often separated from those of inelasticity and from the more restrictive theories of beams, plates, and shells. Consequently, the student acquires a decidedly limited perspective of these topics. The dangers of such fragmentation are further amplified by the contemporary specialties of computational methods and programming. The specialist is seldom cognizant of the underlying theories and limitations. Therefore, the engineer is better served by a broader knowledge of the foundations and the limitations of his/her tools. A paramount goal of this book is a unified treatment that clearly sets forth the underlying concepts and identifies the limitations of every formulation, method, and result.

The foregoing historical sketch also reveals that the theoretical developments of our subject were established by the middle of the twentieth century. Coincidentally, at that time, the developments of solid-state physics (e.g., transistors and semiconductors) and integrated circuits led to the modern electronic computer. Previously, mechanicians and engineers had focused upon solving the differential and integral equations that govern the behavior of deformable bodies, structural, and mechanical components. The advent of the digital computer wrought a profound change. Quite abruptly, our attention turned to the discrete approximations of such bodies, e.g., assemblies of finite elements. Still, certain earlier solutions remain useful, e.g., the extension, flexure, and torsion of beams and shafts. Accordingly, our approach must recognize the merits of the earlier methods/results, but prepare to utilize these newfound capabilities. In the final chapter of this book, we introduce the concept of the finite element within the context of the prior theory of the continua. Of course, every topic in this book has been treated previously; topics in the earlier parts ("foundations" and "fundamental applications") are typically presented in the undergraduate textbooks on the mechanics of solids. Unfortunately, such introductory treatments are often superficial formulations for each specific topic as opposed to our unified presentation. The advanced topics on "plates," "work and energy methods," and "finite elements" are often presented in texts on those specific topics; again, these are treated in isolation from the essential foundations. Here, these topics are cast within the context of the basic concepts and solutions of the prior sections. This integration and unification is consistent with our overarching pedagogical philosophy: *Each subject is presented with emphasis on the foundations, with the requisite precision and with physical and geometrical interpretations.*

1.3 MECHANICAL AND MATHEMATICAL PERCEPTIONS

With a view toward an understanding of the construction and scope of this text, we provide the following overview of the mechanics of solid bodies: The resistance of a solid body to deformation and to fracture is dependent on the physical properties of the material and the geometry of the body. The former are manifestations of the discrete components of the micro/macro structure of the material, i.e., the atomic/molecular structure of a plastic, the crystalline structure of a metal, the filament/matrix structure of a modern composite or the cement/aggregate composition of concrete. Unfortunately, neither physics nor mathematics provides the means to incorporate specific micro/macro behavior into the analysis of actual mechanical/structural components. Consequently, the underlying mechanisms of deformation (both elastic and inelastic) and fracture are invested in the concepts and theories of continuous and cohesive media. These mathematical theories are drawn from experiment. Often criteria of work and energy provide the bridge between the actual discrete behavior and the continuous descriptions, as expressed by continuous mathematical functions. The underlying discrete constituents and their interactions are embedded in the useful, albeit approximate, descriptions as continuous media.

The mathematical descriptions of the continuous media lead to the mathematics of differential equations and boundary-value problems. These problems defy solutions in all but the simplest circumstances (the "fundamental applications") of our text. Engineers can effectively handle the elementary problems involving beams, plates, and shells of simple form by solving the boundary value problems. However, most real mechanical components are much too irregular in shape to allow such solutions; the reader might imagine the bounding surfaces of a connecting rod or crankshaft in a typical reciprocating engine. Ironically, but fortunately, the differential equations of the continuous body can be supplanted by algebraic (difference) equations. This requires approximations that effectively replace the continuous model of the solid body with an assembly of discrete "finite elements," one of the methods of contemporary engineering. Modern digital computers provide the capability to obtain numerical solutions of the very large systems of algebraic equations. Of course, the result is an approximation of the deformed (loaded) body in terms of numerical values at discrete sites, e.g., at the junctions of the elements. The foregoing account of the methodologies in the mechanics of solid bodies describes a curious, yet interesting, metamorphosis: the actual body of discrete constituents, e.g., the crystals that constitute a metal, is firstly approximated as a continuous body. Then, to obtain meaningful results, that continuous body is replaced by a discrete mathematical model.

The metamorphosis is depicted in Figure 1.1. Here, the roots of our subject lie in the discrete micro/macro structures of the actual solid body. That real body is supplanted by a continuous model that requires a transition, an approximation, at A. The continuous model leads to various theories, such as the Bernoulli theory of beams, and to the attendant mathematical equations, including differential equations. Our inability to solve the large number of boundary-value problems posed by these equations requires the subsequent transition B, wherein the continuous body is approximated by the discrete mathematical model. It is also noteworthy that both transitions from the continuous to the discrete models are accomplished by the balance of work and energy.

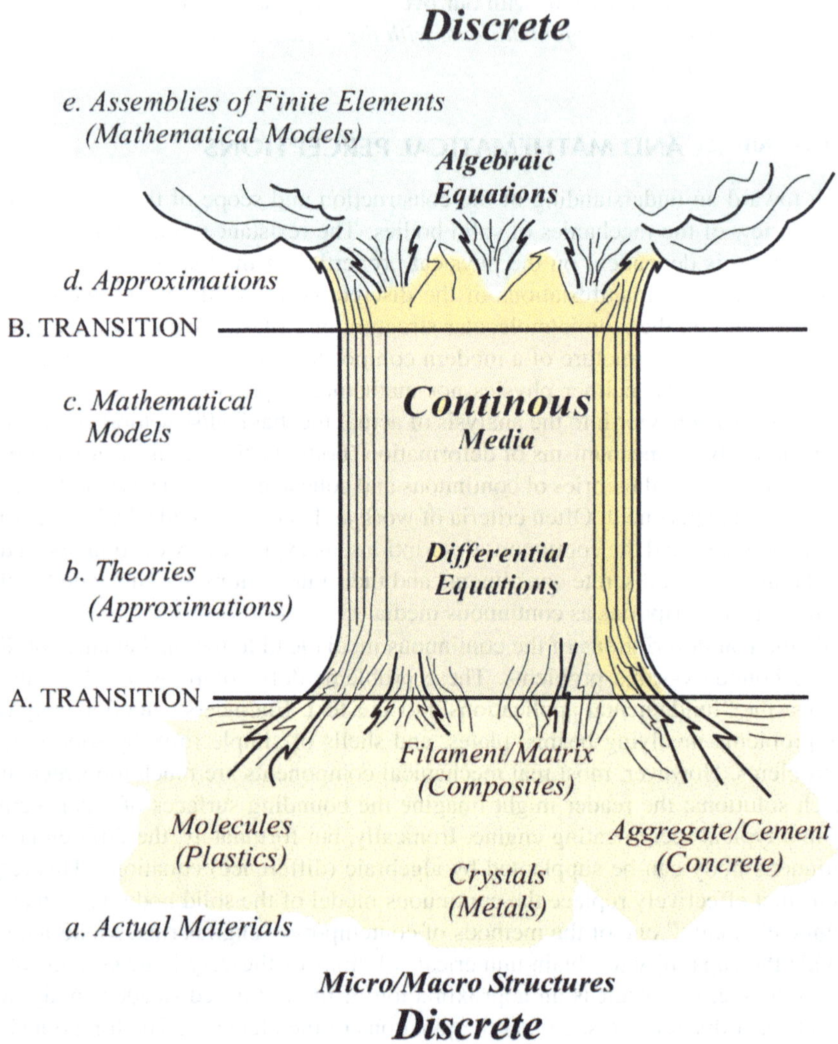

Figure 1.1

1.4 SCOPE

As previously noted, the behavior of the actual solid body rests with the discrete microscopic or macroscopic constituents, which constitute the body. A thorough study of the micro/macrostructures of materials is beyond the scope of this chapter, yet we must acknowledge their role in certain important phenomena. Specifically, we cite the role of the crystalline and atomic interactions that account for inelastic deformations and fracture. References for further study are provided.

Our goals are the practical means to describe the behavior of mechanical and structural members. The emphasis is placed on the effective methods that are founded on the concepts of a continuous medium, wherein the properties are described via continuous mathematical variables. Effective theories of continua (e.g., elasticity) have a long and storied history, dating to the works of such eminent scientists as Saint-Venant, Cauchy, and Euler. Those theories remain the basis for most current analyses and provide the basis for the contemporary approximations via discrete finite elements. These latter mathematical models/approximations have proven to be of great practical value with the advent of the modern digital computers.

Throughout the long history of our subject, the concepts, theories, and analyses of continuous media have been the foundations of structural and mechanical engineering. These remain a bulwark of engineering; they are the bases of most contemporary methods of approximation and computation and, specifically, the methods of finite elements. The mechanics of continuous deformable solids is the primary focus of this book; the final chapter provides a bridge to the contemporary methods of discrete approximation.

This book is divided into three parts; each has certain common attributes: The contents of Part I provide the common foundations for all subsequent parts. Chapters 2 and 3 present the kinematical and dynamical concepts, which are common to all theories of continuous deformable bodies. Chapter 4 describes the thermodynamic behavior of the materials; those descriptions are utilized in the subsequent parts, II and III. Part II presents basic applications; these are the simpler problems, which can be treated effectively with the mathematics of the continuum theories. As the name implies, the advanced topics are contained in Part III. Chapter 9 introduces the methods of work and energy, which are vital to the formulations of "finite elements" in Chapter 10.

1.4.1 PART I–FOUNDATIONS

Certain aspects of any theory of a continuous deformable body are common to all theories, whether the behavior is elastic or inelastic. Assuming only that the body is cohesive, the description of the deformation is applicable to all; a description of those geometrical changes is termed kinematics. Moreover, it is the most fundamental of all aspects. The kinematics of continua requires the definitions of essential geometrical quantities, e.g., the concept and definition of a strain. This most fundamental aspect of our subject is explored in Chapter 2.

A description of the forces, external and internal, and the criteria for equilibrium or motion are common to any analysis of deformable solid bodies. A description of the work expended by the forces is also common to any analysis. These aspects of our subject are presented in Chapter 3. We note that the forces and actions in question occur in the deformed body; hence, the concepts and analyses of Chapter 3 require the preceding kinematics of Chapter 2. The expenditure of work by forces, particularly the work of internal forces, requires the prior description of the deformation, as presented in Chapter 2.

We must also acknowledge the two alternative approaches to the equilibrium and/or motion: In the *Newtonian approach*, one views the actions of the force upon a free body and imposes Newton's "laws" of equilibrium or motion. In the *Lagrangean approach*, one examines the "virtual" work expended by the forces and requires the vanishing of that "virtual" work. Both apply to any mechanical system and, in particular, to the deformable bodies of any continuous medium. Chapter 3 provides the essential ingredients of both approaches. Both approaches are demonstrated in Section 1.6 of this introduction.

The kinematics of Chapter 2 and the kinetics of Chapter 3 provide a common foundation for all subsequent topics. The attributes of the media provide the links between the deformations (Chapter 2) and actions of forces (Chapter 3). Various properties and behaviors of solids have received the attention of many researchers and are the subjects of countless books. An extensive coverage is well beyond the scope of this book. Our presentation (Chapter 4) is limited to accepted theories, which have been applied effectively in engineering practice. The reader is alerted to the limitations, since all theories are but approximations of reality.

1.4.2 PART II–BASIC APPLICATIONS

Certain simple yet fundamental applications are contained in Chapter 5, e.g., flexure and torsion of bars; most are solutions that are consistent with the "exact" solutions in the theory of linear elasticity. With a view toward the practical application of such solutions, attention is placed upon the underlying assumptions and the consequent limitations. These fundamental applications provide an entree to the subsequent chapters that encompass the theory of beams (Chapter 6) and superposition (Chapter 7). These topics are essential tools of the structural engineer and included in his/her undergraduate studies. Chapter 8 provides an introduction to the mechanics of plates, which entails the solution of partial differential equations and associated boundary values.

1.4.3 PART III–ADVANCED TOPICS

As previously noted, the concept of virtual work provides an alternative approach to the kinetics of any mechanical system. Chapter 9 focuses on the concepts of virtual work, minimum potential, the variational method, and important applications to the stability of elastic structures. The methods of work and energy (Chapter 9) are essential to the formulation of finite elements (Chapter 10).

1.5 THE ELASTIC MATERIAL

A most ideal material is the perfectly elastic solid; it fully recovers an initial form after a deformation following the application and the removal of forces or temperatures, which caused the deformation. Work or energy (mechanical or thermal) expended upon the material during a deformation is stored as internal energy; all is available, and none is dissipated.

To exemplify the attributes of a simple elastic body, we consider the small rectangular block of Figure 1.2. Prior to the application of any force, it has dimensions Δx, Δy, and Δz, as shown in Figure 1.2a. The opposing forces, Δf and $-\Delta f$, are uniformly distributed and normal to the $y-z$ faces of the block in Figure 1.2b. As a consequence, the Δx dimension is extended to length Δx^*; typically the transverse dimensions, Δy and Δz, contract. The extension is the change:

$$\Delta e \equiv \Delta x^* - \Delta x$$

The proportionality of force and extension was proposed by Robert Hooke[5]; hence, the linear elastic behavior is termed Hookean. Such behavior might be expressed in the form:

$$\Delta f = k\Delta e$$

where k is a constant. Observe that k depends on the size of the block: Certainly the amount of extension Δe depends on the length Δx, and the magnitude of force depends on the size of the area ΔA. To describe the response of the material, rather than a specific block, we employ the concepts of *stress and strain*: The *normal stress* on the $y-z$ face is defined as

$$\sigma_x = \lim_{\Delta A \to 0} \frac{\Delta f}{\Delta A} \tag{1.1}$$

The *extensional strain* in the x direction is defined as

$$\varepsilon_x = \lim_{\Delta x \to 0} \frac{\Delta e}{\Delta x} \tag{1.2}$$

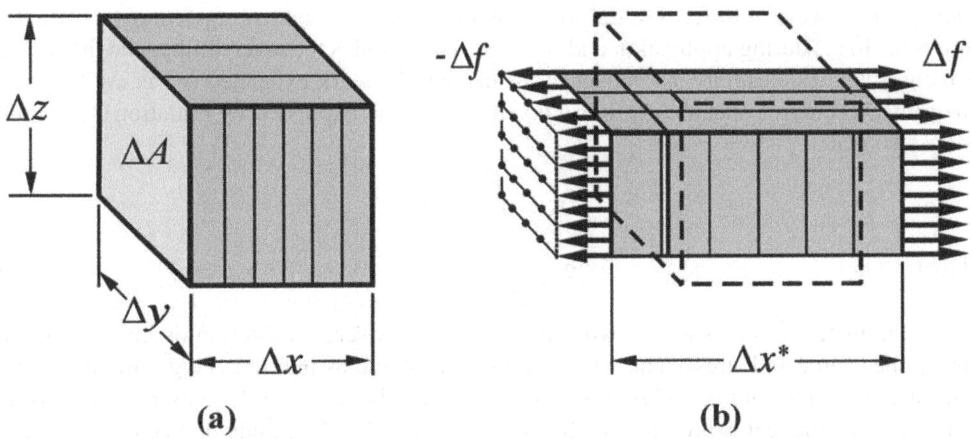

Figure 1.2

Now, the Hookean behavior of the *material* is expressed by the linear equation

$$\sigma_x = E_x \varepsilon_x \tag{1.3}$$

The constant E_x is the *modulus* of elasticity; unlike the constant k, E_x does not depend on the dimensions of the block. The modulus E_x is a *property* of the material; E_x has the dimensions of a stress. Several aspects of the foregoing equations must be noted: Firstly, the Equations (1.1) and (1.2) imply the *existence* of the limits; practically, this requires the continuity of the material. Our reference to the *constant* E_x implies a homogeneity; we note that such linearity is possible at each point of a non-homogeneous material, i.e., a modulus might be variable, $E_x = E_x(x, y, z)$. A comprehensive treatment of such attributes is provided in Chapter 4.

Let us now consider the work ΔW expended in the deformation of the block in Figure 1.2: During any incremental change of elongation, $\delta(\Delta\varepsilon)$, the work done by Δf is

$$\delta(\Delta W) = \Delta f \delta(\Delta e)$$

Since we have assumed that e (and ε_x) are continuous functions, one can replace the increment by the differential:

$$d\Delta W = \Delta f d(\Delta e)$$

The total work expended, as the block is deformed from the initial state (Figure 1.2a) to the deformed state (Figure 1.2b) is the integral:

$$\Delta W = \int_0^{\Delta e} \Delta f d(\Delta e) = \int_0^{\Delta e} k(\Delta e) d(\Delta e) = k \frac{(\Delta e)^2}{2}$$

Once again, we note that the work ΔW depends upon the size of the block. Employing the definitions (1.1) to (1.3), we have the *work per unit volume*

$$w = \lim_{\substack{\Delta x \to 0 \\ \Delta A \to 0}} \frac{\Delta W}{\Delta x \, \Delta A} = \frac{E_x \varepsilon_x^2}{2} \tag{1.4}$$

During the process of the *elastic* deformation, the force Δf remains proportional to Δe (and σ_x proportional to ε_x) during application and *removal*. This implies the reversibility, and the *recovery*, which characterizes elasticity. Moreover, it means that the work expended (w) is available in the form of internal energy; specifically, it is a *function* $w(\varepsilon_x)$, as expressed by Equation (1.4). Finally, we note also that

$$w = \int_0^{\varepsilon_x} \sigma_x d\varepsilon_x$$

and that

$$\sigma_x = \frac{dw}{d\varepsilon_x} = E_x \varepsilon_x \tag{1.5}$$

Since the material is elastic, the work expended is recovered during unloading, i.e., removal of the applied force and stress. The work expended is stored as internal energy. In view of (1.5), the density (per unit volume) of *internal energy* w, i.e., Equation (1.4), may be regarded as an alternative mathematical expression of Hookean behavior. The foregoing description of one ideal material is but a brief entree to the vast subject of material behavior. It is provided here only to enable an overview of the various facets of our subject and the alternative viewpoints; in the subsequent section, we provide an example to demonstrate the Newtonian and the Lagrangian approaches. The former utilizes the description (1.3); the latter employs (1.4).

1.6 INTRODUCTORY EXAMPLE

To exemplify the roles of kinematics, material behavior, and alternative formulations of the kinetics (Newtonian and Lagrangean), let us consider the plane truss depicted in Figure 1.3a. The three members are supposedly joined at their ends by ideal pins. Each is a so-called "two-force" member; each is subjected to opposing forces exerted by the pins at the ends. The members are pinned to fixed surfaces at B, C, and D; a constant static load F is applied at A. This simple structure might fail in various ways: Excessive displacement at A might be unacceptable, or excessive internal force (stress) in any member might cause permanent deformation or fracture. Any assessment requires a description of the material. For our example, let us assume the homogeneous Hookean behavior of the preceding Section 1.5, but recognize that such elastic behavior holds only within limits. Inelastic deformations or fracture always result at some state of excessive stress. As previously noted, the kinematics is common either to the Newtonian or to the Lagrangean approach. Accordingly, we begin with the geometrical consequences when the load F is applied and joint

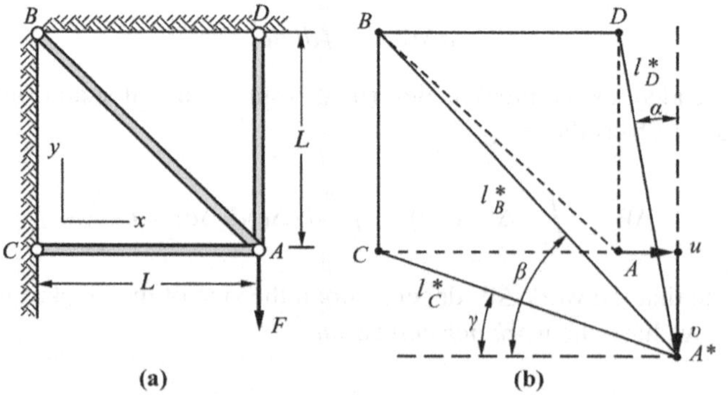

Figure 1.3

A displaces to A^* as shown in Figure 1.3b. Because the truss and loading are symmetrical about the x-y plane, the movement and deformation are also symmetrical about the x-y plane. The displacement of A is described by the two components u and v in directions x and y, respectively. We denote the elongation of a member by e_D, e_B, and e_C for members DA, BA, and CA, respectively. It is now a geometrical exercise to express the elongations in terms of the two displacement components (u,v). By the Pythagorean theorem,

$$\begin{aligned} e_D &\equiv D^*A - DA = \sqrt{(L+v)^2+u^2} - L \\ &\equiv l_D^* - L = L\sqrt{1+\frac{2v}{L}+\left(\frac{v}{L}\right)^2+\left(\frac{u}{L}\right)^2} - L \end{aligned} \tag{1.6a}$$

$$e_B \equiv l_B^* - l_B = L\sqrt{\left[1+\left(\frac{v}{L}\right)\right]^2+\left[1+\left(\frac{u}{L}\right)\right]^2} - \sqrt{2}L \tag{1.6b}$$

$$e_C \equiv l_C^* - L = L\sqrt{\left(\frac{v}{L}\right)^2+\left[1+\left(\frac{u}{L}\right)\right]^2} - L \tag{1.6c}$$

The uninitiated reader may question the need for these additional unknowns(e_D, e_B, e_C) and the three Equations (1.6a–c). We recall that these elongations are related to the forces acting upon the members; moreover, that relationship depends upon the material behavior. Indeed, our next step is the expression of the Hookean behavior.

We suppose that all members are composed of the same homogeneous material with the axial modulus E. Furthermore, we assume that the axial strains are uniform throughout a given member, though some local irregularities must exist near actual connections at the ends.[6] In accordance with (1.2),

$$\varepsilon_D = \frac{e_D}{L}, \quad \varepsilon_B = \frac{e_B}{\sqrt{2}L}, \quad \varepsilon_C = \frac{e_C}{L} \tag{1.7a,b,c}$$

The respective normal stresses $(\sigma_D, \sigma_B, \sigma_C)$ in cross-sections of the three members (DA, BA, CA) follow from (1.3):

$$\sigma_D = E\frac{e_D}{L}, \quad \sigma_B = E\frac{e_B}{\sqrt{2}L}, \quad \sigma_c = E\frac{e_C}{L} \tag{1.8a,b,c}$$

We assume that these stresses are uniformly distributed upon their respective cross-sections. For simplicity here, we suppose that the members have the same cross-sectional area A. Then, by (1.8a–c), (1.1), and (1.7a–c),

$$F_D = EA\varepsilon_D = \frac{EA}{L}e_D \tag{1.9a}$$

$$F_B = EA\varepsilon_B = \frac{EA}{\sqrt{2}L}e_B \tag{1.9b}$$

$$F_C = EA\varepsilon_C = \frac{EA}{L}e_C \tag{1.9c}$$

As an alternative to the stress-strain relations (1.3) and (1.8a–c), we have the internal energies; according to (1.4),

$$w_D = \frac{E\varepsilon_D^2}{2}, \quad w_B = \frac{E\varepsilon_B^2}{2}, \quad w_C = \frac{E\varepsilon_C^2}{2}$$

The total potential energy of each member follows in accordance with (1.7a–c):

$$W_D = ALw_D = \frac{EA}{2L}e_D^2 \tag{1.10a}$$

$$W_B = A\sqrt{2}Lw_B = \frac{EA}{2\sqrt{2}L}e_B^2 \tag{1.10b}$$

$$W_C = ALw_C = \frac{EA}{2L}e_c^2 \tag{1.10c}$$

In the manner of (1.5), we recover Equations (1.9a–c):

$$F_D = \frac{dW_D}{de_D} = \frac{EA}{\sqrt{L}}e_D \tag{1.11a}$$

$$F_B = \frac{dW_B}{de_B} = \frac{EA}{\sqrt{2}L}e_B \tag{1.11b}$$

$$F_C = \frac{dW_C}{de_C} = \frac{EA}{L}e_C \tag{1.11c}$$

Finally, we observe that the internal forces (F_D, F_B, F_C) are expressed in terms of the two components (u, v) by substituting the kinematic relations (1.6a–c) into (1.9a–c). Note that these are *nonlinear* in the displacements (u, v). Alternatively, the energies (1.10a–c) are expressed in terms of the two components (u, v) by substituting (1.6a–c) into (1.10a–c). With the former (F_D, F_B, F_C), we can formulate the governing equations of equilibrium via the Newtonian approach (i.e., by a *free-body*[7] diagram). With the latter (W_D, W_B, W_C), we can formulate the equations via the Lagrangean approach (i.e., by the vanishing of virtual work).

1.6.1 NEWTONIAN FORMULATION

The Newtonian approach requires a free-body diagram, Figure 1.4, which shows the forces acting upon the body. Then, we impose the two equations of equilibrium for this system of concurrent forces:

$$\sum F_x = -F_C\cos(\gamma) - F_B\cos(\beta) - F_D\sin(\alpha) = 0 \tag{1.12a}$$

$$\sum F_y = F_C\sin(\gamma) + F_B\sin(\beta) + F_D\cos(\alpha) - F = 0 \tag{1.12b}$$

where from Figure 1.3b

$$\begin{aligned} \sin\alpha &= \frac{u}{l_D^*}, \quad \cos\alpha = \frac{L+v}{l_D^*} \\ \sin\beta &= \frac{L+v}{l_B^*}, \quad \cos\beta = \frac{L+u}{l_B^*} \\ \sin\gamma &= \frac{v}{l_C^*}, \quad \cos\gamma = \frac{L+u}{l_C^*} \end{aligned} \tag{1.13}$$

As in the geometric relations (1.6a–c), the extended lengths (l_D^*, l_B^*, l_C^*) are readily expressed in terms of the (original) known lengths $(L, \sqrt{2}L, L)$ and the two displacements (u, v). The forces (F_D, F_B, F_C)) are also expressed in terms of the two displacements (u, v) by means of the constitutive Equations (1.9a–c) and the kinematic relations (1.6a–c). The successive substitutions render

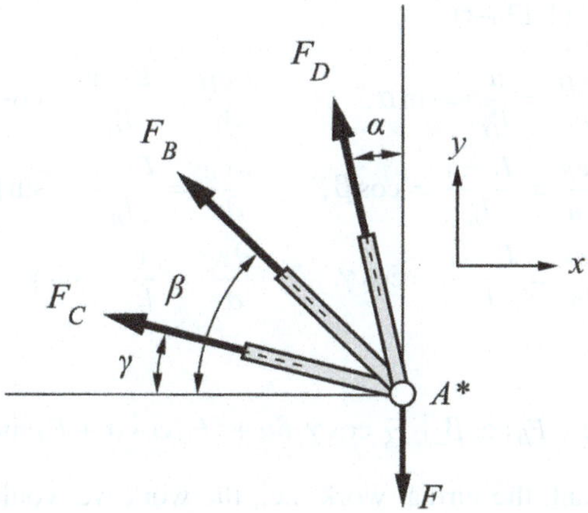

Figure 1.4

the two kinetic equations (1.12a,b) in terms of two unknowns (u, v), but incorporate the given dimensions (L, A) and the modulus E. The mechanical system has *two degrees-of-freedom*, i.e., the final configuration is determined by the two variables (u, v). We observe that the two Equations (1.12a,b) are nonlinear in those variables. Here, those nonlinearities stem entirely from the geometry of the deformed configuration; these arise in the kinematic relations (1.6a–c) and in the equilibrium equations (1.12a,b).

1.6.2 LAGRANGEAN FORMULATION

The Lagrangean approach does not require a free-body diagram. We obtain the equilibrium equations via the principle of *virtual work*. The virtual work can be defined as the work we would *expend* in an arbitrary admissible virtual displacement, i.e., an incremental displacement that does not violate any geometrical constraints. Here, our truss is constrained by the pins at A, B, C, and D; an inadmissible displacement is one which would disconnect those joints. The virtual work includes that expended against internal forces (F_D, F_B, F_C) and external force F. Here, we consider small incremental changes $(\delta u, \delta v)$ in (u, v); these are arbitrary, i.e., independent.

We have previously examined the work of internal forces (F_D, F_B, F_C). Because the members are elastic, the work expended upon those forces is manifested in the changes of their internal energies. Therefore, the virtual work upon those internal forces is

$$\begin{aligned}\delta W_I &= \delta W_D + \delta W_B + \delta W_C \\ &= \frac{dW_D}{de_D}\left(\frac{\partial e_D}{\partial u}\delta u + \frac{\partial e_D}{\delta v}\delta v\right) + \frac{dW_B}{de_B}\left(\frac{\partial e_B}{\partial u}\delta u + \frac{\partial e_B}{\delta v}\delta v\right) + \frac{dW_C}{de_C}\left(\frac{\partial e_C}{\partial u}\delta u + \frac{\partial e_C}{\delta v}\delta v\right)\end{aligned}$$

In view of (1.11a–c),

$$\delta W_I = F_D\left(\frac{\partial e_D}{\partial u}\delta u + \frac{\partial e_D}{\delta v}\delta v\right) + F_B\left(\frac{\partial e_B}{\partial u}\delta u + \frac{\partial e_B}{\delta v}\delta v\right) + F_C\left(\frac{\partial e_C}{\partial u}\delta u + \frac{\partial e_C}{\delta v}\delta v\right)$$

By Equations (1.6a–c) and (1.13a–f),

$$\frac{\partial e_D}{\partial u}=\frac{u}{l_D^*}=\sin\alpha,\qquad \frac{\partial e_D}{\partial v}=\frac{L+v}{l_D^*}=\cos\alpha$$

$$\frac{\partial e_B}{\partial u}=\frac{L+u}{l_B^*}=\cos\beta,\qquad \frac{\partial e_B}{\partial v}=\frac{L+v}{l_B^*}=\sin\beta$$

$$\frac{\partial e_c}{\partial u}=\frac{L+u}{l_C^*}=\cos\gamma,\qquad \frac{\partial e_C}{\partial v}=\frac{v}{l_C^*}=\sin\gamma$$

It follows that

$$\delta W_I=(F_D\sin\alpha+F_B\cos\beta+F_C\cos\gamma)\delta u+(F_D\cos\alpha+F_B\sin\beta+F_C\sin\gamma)\delta v$$

Since the load F is constant, the virtual work, i.e., the work we would expend as we lower the weight by δv, is

$$\delta W_e=-F\delta v$$

The total virtual work must vanish:

$$\delta W_{\text{tot}}=\delta W_I+\delta W_e=0$$

Now, we recall the term arbitrary displacement; $\delta W_{\text{tot}}=0$ if ($\delta u\neq 0$ and $\delta v=0$) or if ($\delta u=0$ and $\delta v\neq 0$). In short, the coefficients of δu and δv must both vanish in the work $W_(tot)$. It follows that

$$F_D\sin\alpha+F_B\cos\beta+F_C\cos\gamma=0,\quad F_D\cos\alpha+F_B\sin\beta+F_C\sin\gamma-F=0$$

These are precisely the Equations (1.12a,b) obtained via the Newtonian approach. Note that no *approximations* of the geometry are incorporated. The alternative approaches provide the same governing equations, irrespective of the magnitudes of deformations. Although we have presumed Hookean behavior, neither approach is restricted by the constitutive description.

1.6.3 SMALL DEFORMATIONS–LINEARITY

In most instances, the deformations of structural and mechanical systems are very small, often perceived only by careful measurements. In most cases, the components cannot suffer any *permanent* dimensional changes; strains and stresses must not exceed the limits of elastic behavior. Then, a typical metal part experiences strains O(10^{-3}). Consider the implications here: According to Equations (1.6a) and (1.7a),

$$\varepsilon_D=\frac{l_D^*-L}{L}=\left[1+\frac{2v}{L}+\left(\frac{v}{L}\right)^2+\left(\frac{u}{L}\right)^2\right]^{1/2}-1\tag{1.14}$$

If $\varepsilon_D=O(10^{-3})$, then both u/L and v/L are very small compared to unity. Then, products are negligible in the bracketed term of (1.14); the valid approximation follows:

$$\varepsilon_D\approx\left(1+\frac{2v}{L}\right)^{1/2}-1$$

By the binomial theorem,

$$(1\pm x)^n=1\pm nx+\frac{n(n-1)}{2!}x^2\pm\frac{n(n-1)(n-2)}{3!}x^3+\cdots,\quad (n>0)\tag{1.15a}$$

$$(1 \mp x)^{-n} = 1 \mp nx + \frac{n(n+1)}{2!}x^2 \mp \frac{n(n+1)(n+2)}{3!}x^3 + \cdots \quad (n > 0) \tag{1.15b}$$

Again, the higher powers of a small number ($x \ll 1$) can be neglected. Here, $x = 2v/L$ and $n = 1/2$, so that we have the first-order linear approximation.

$$\varepsilon_D \approx \frac{v}{L} - \frac{1}{2}\left(\frac{v}{L}\right)^2 + \cdots \approx \frac{v}{L} \tag{1.16a}$$

$$\varepsilon_B \approx \frac{u}{2L} + \frac{v}{2L} \tag{1.16b}$$

$$\varepsilon_C \approx \frac{u}{L} \tag{1.16c}$$

With such approximations in (1.7a–c) and (1.9a–c), the forces (F_D, F_B, F_C) are linear functions of the displacements (u, v). For these small displacements, the angles α and γ in Figure 1.4 are very small and $\beta \approx 45°$. Indeed, in terms of first order in the displacements,

$$\sin\alpha \approx \frac{u}{L}, \quad \sin\beta \approx \frac{1}{\sqrt{2}}(1 + \frac{v}{2L} - \frac{u}{2L}), \quad \sin\gamma \approx \frac{v}{L} \tag{1.17a}$$

$$\cos\alpha \approx 1, \quad \cos\beta \approx \frac{1}{\sqrt{2}}(1 + \frac{u}{2L} - \frac{v}{2L}), \quad \cos\gamma \approx 1 \tag{1.17b}$$

Although these are linear in the small numbers (u/L, v/L), the retention of those linear terms would result in quadratic terms in our final equations. Since these numbers are small [$(u/L) \ll 1$, $(v/L) \ll 1$], the consistent approximations follow:

$$\sin\alpha \approx \sin\gamma \approx 0, \quad \cos\alpha \approx \cos\gamma \approx 1, \quad \sin\beta \approx \cos\beta \approx \frac{1}{\sqrt{2}}$$

In other words, to achieve a *linear* approximation of our final equations, we *neglect* the deformation in our treatment of the *free-body* diagram (Figure 1.4). That is typical in most analyses of structural/mechanical systems. Very important exceptions must be noted; some examples follow:

Any analysis of structural instability must examine the kinetics (static or dynamic) of the deformed state.

Any analysis of severely deformed bodies, e.g., rubber-like elastic deformations or permanently deformed bodies as in the processes of stamping or extruding, must treat the deformed body.

It is instructive to obtain the linear approximations in a direct, though less rigorous, manner. Consider a blowup of the view at the joint A, as depicted in Figure 1.5. Because the displacement is relatively small, the rotation of member BA is nearly imperceptible. Indeed, a very small rotation about B does not stretch the bar. Taken in succession, the small rotation carries A to A' (nearly perpendicular to BA) and the elongation carries A' to A^* (nearly parallel to BA); hence, the first-order approximation of e_B is the projection of the displacement upon the line $B'A^*$ (parallel to BA in the first-order approximation), namely,

$$e_B \approx \frac{u}{\sqrt{2}} + \frac{v}{\sqrt{2}}$$

Then, (1.16b) follows:

$$\varepsilon_B = \frac{e_B}{\sqrt{2}L} \approx \frac{u}{2L} + \frac{v}{2L}$$

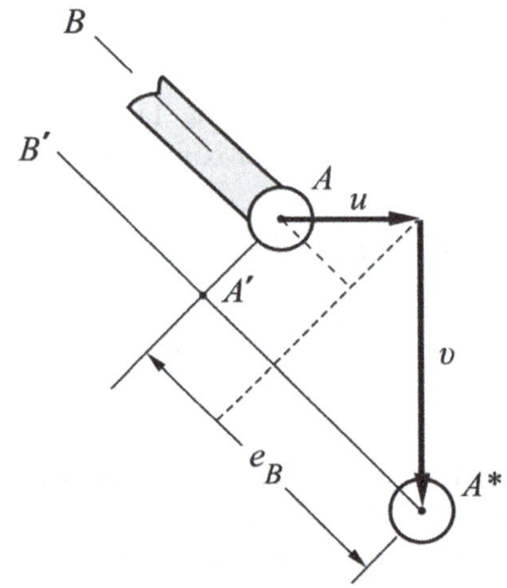

Figure 1.5

The foregoing example serves to illustrate essential aspects of our subject. Reviewing them, we see that internal and external forces are related by the conditions of *equilibrium* (1.12a,b), that elongations (changes of geometry) are related to displacement by *geometric* requirements (1.6a–c), and that the elongations are related to internal forces by the *resistive property of the material* (E) and the area (A) of the cross-section (1.9a–c). These three aspects of our subject are distinct, but each enters into the solution.

1.7 DISTRIBUTION OF FORCES

In the mechanics of *ideal* rigid bodies, it is possible to consider only certain attributes of a distributed force: If the resultant is a force, then only the line of action — and not the point of application — is needed. If the resultant is a couple, then the site of that action plays no role in the motion of the rigid body. In the mechanics of deformable bodies, we *cannot* take such liberties with the description of forces, external or internal. As an inviolable rule, one *cannot* alter the distribution of actual loading upon a deformable body. In our subsequent studies, we examine certain circumstances and effects that can be treated with imprecise descriptions of the loading (Section 5.2). Otherwise, *abide* by the rule! *Do not alter the loads!*. The following examples help to dispel any preconceived notions.

ILLUSTRATIVE EXAMPLE 1.1

Deformable blocks are depicted in Figure 1.6. Imagine two identical blocks of rubber bonded to the rigid base. In Figure 1.6a, a very concentrated force acts upon the left side; in Figure 1.6b, the same force acts along the same line, but it acts upon the right side. The difference in the deformations is evident; internal forces are also different.

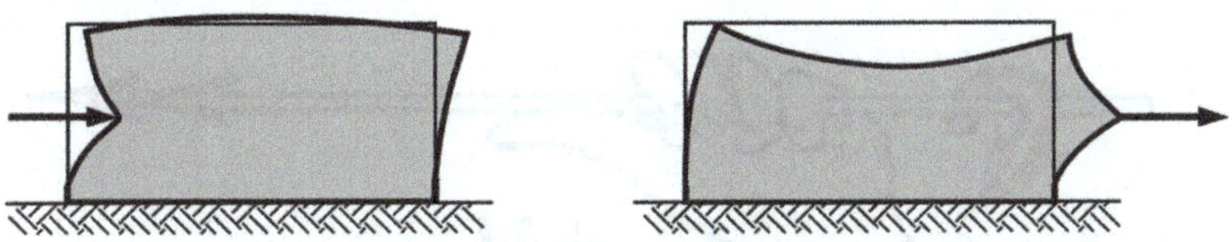

Figure 1.6

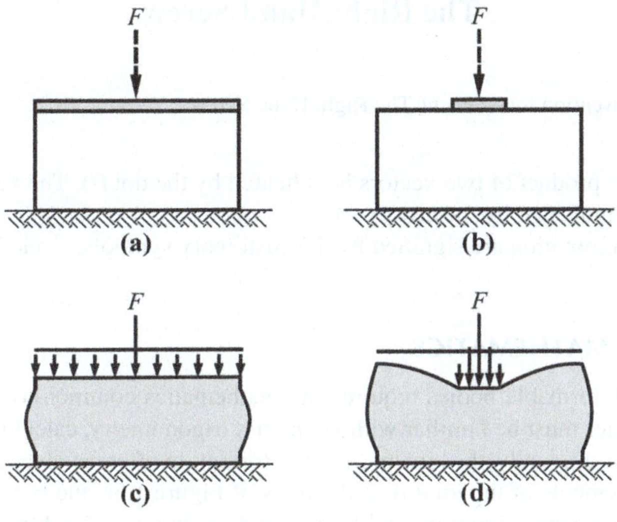

Figure 1.7

ILLUSTRATIVE EXAMPLE 1.2

The identical blocks in Figure 1.7 are subjected to the same resultant, a force F that acts along the centerline. In the case of Figure 1.7a, the force is applied upon a very rigid plate, which distributes that force over the upper surface. In the case of Figure 1.7b, the force is applied upon a very small plate, which then imposes a concentrated force on the block. Again, the difference (Figure 1.7c and d) in the deformations is evident.

1.8 SPECIFIC NOTATIONS

- A few specific notations are used throughout this book:
- Vectors are denoted by boldface fonts. Unit vectors are distinguished by a caret placed over the bold letter. Unit vectors in the directions of a fixed system of Cartesian coordinates (x, y, z) are denoted by the letters $\hat{\mathbf{i}}$, $\hat{\mathbf{j}}$, $\hat{\mathbf{k}}$.
- Vectors are represented graphically by a directed line segment with the direction indicated by an arrowhead.
- Forces are shown as vectors with the arrowhead indicating a point of application *when* the force acts at a specific site.
- Moments (or couples) are represented as vectors with double arrowheads. The direction of the turning action is indicated in accordance with the rule of the "right hand" as depicted in Figure 1.8.

Figure 1.8 Sign Convention for Moment-The Right Hand Screw

- The scalar (inner) product of two vectors is indicated by the dot $(\cdot)$. The vector (cross) product of two vectors is indicated by the cross $(\times)$.
- Integration and summation are signified by the customary symbols, $\int$ and $\sum$, respectively.

1.9 REQUISITE MATHEMATICS

The mechanics of deformable bodies requires the mathematics common to classical mechanics. Specifically, the reader must be familiar with geometry, trigonometry, calculus, and the rudiments of vector analysis. Additionally, the kinematics of deformation often involve relatively small quantities, as the displacements of the joint A in the truss of Figure 1.3a and b, where $(u/L) \ll 1$ and $(v/L) \ll 1$. Then, good approximations and much simplification are possible, but only with careful attention to the errors entailed. The following examples employ the relevant vector algebra.

ILLUSTRATIVE EXAMPLE 1.3

A quadrilateral has its vertices at the points $A(1,1,1)$, $B(2,3,2)$, $C(0,4,1)$, $D(-2,2,0)$. Six lines connect the four vertices. Which is the longest of the six lines? Which is most nearly parallel to the y axis? The length l_{IJ} of the line segment between two vertices I and J is given by

$$l_{IJ} = \sqrt{(x_J - x_I)^2 + (y_J - y_I)^2 + (z_J - z_I)^2}$$

where$(x_I, y_I, z_I)$and(x_J, y_J, z_J) are the Cartesian coordinates of vertices I and J. Thus,

$$l_{AB} = l_{BC} = \sqrt{6}, \quad l_{CD} = \sqrt{9}, \quad l_{DA} = \sqrt{11}, \quad l_{AC} = \sqrt{10}, \quad l_{BD} = \sqrt{21}$$

It follows that line (BD) is the longest of the six lines.
Let $\mathbf{r_I}$ and $\mathbf{r_J}$ be the position vectors of vertices I and J. The difference between those vectors

$$\mathbf{r_{IJ}} \equiv \mathbf{r_I} - \mathbf{r_J}$$

$\mathbf{r_{IJ}}$, is a vector with magnitude $|\mathbf{r_{IJ}}| \equiv l_{JI}$ in direction from vertex J to vertex I. If $\hat{\mathbf{j}}$ is the unit vector $(|\hat{\mathbf{j}}| = 1)$ in direction of the y axis, then the cosine of the angle θ between the line (JI) and the y axis is given by

$$\frac{\mathbf{r_{IJ}}}{l_{JI}} \cdot \hat{\mathbf{j}} = \frac{y_I - y_J}{l_{JI}} \equiv \cos\theta \tag{1.18}$$

If the line (JI) is parallel to the y axis, then $\cos\theta = \cos 0 = 1$. It follows that

$$\frac{y_J - y_I}{l_{IJ}} = \cos 0 = 1$$

By specializing Equation (1.18) for the line segments (AB),(BC),(CD),(DA), (AC), and (BD), respectively, it follows that line (AC) is most nearly parallel to the y axis.

ILLUSTRATIVE EXAMPLE 1.4

Let **a** and **b** be arbitrary vectors whose directions differ by the angle θ. Denote their sum by **c** and let φ be the angle between the directions of **c** and **b** (See Figure 1.9). Then, show that

$$c^2 = a^2 + b^2 + 2ab\cos\theta \text{and} \tan\varphi = \alpha\sin\theta/(b+\cos\theta)$$

where a, b, and c denote the magnitudes of vectors **a**, **b**, and **c**, respectively, i.e., $a \equiv |\mathbf{a}|$, $b \equiv |\mathbf{b}|$, and $c \equiv |\mathbf{c}|$.

$$c^2 = \mathbf{c}\cdot\mathbf{c} = (\mathbf{a}+\mathbf{b})\cdot(\mathbf{a}+\mathbf{b}) = \mathbf{a}\cdot\mathbf{a}+\mathbf{b}\cdot\mathbf{b}+2\mathbf{a}\cdot\mathbf{b} = a^2+b^2+2ab\cos\theta$$

The scalar product of vectors **c** and **b** assumes the form:

$$\mathbf{c}\cdot\mathbf{b} = c\cdot b\cdot\cos\varphi = (\mathbf{a}+\mathbf{b})\cdot\mathbf{b} = \mathbf{a}\cdot\mathbf{b}+\mathbf{b}\cdot\mathbf{b} = b^2+ab\cos\theta$$

It follows that

$$\cos\varphi = (b + a\cos\theta)/c \tag{1.19a}$$

Furthermore, the vector product of vector **c** with itself yields:

$$\mathbf{c}\times\mathbf{c} = \mathbf{c}\times(\mathbf{a}+\mathbf{b}) = \mathbf{c}\times\mathbf{a}+\mathbf{c}\times\mathbf{b} = (\mathbf{a}+\mathbf{b})\times\mathbf{a}+\mathbf{c}\times\mathbf{b} = \mathbf{b}\times\mathbf{a}+\mathbf{c}\times\mathbf{b} = (-ab\sin\theta+bc\sin\varphi)\hat{\mathbf{n}} = 0 \tag{1.19b}$$

where $\hat{\mathbf{n}}$ is a unit vector perpendicular to vectors **a**, **b**, and **c**. Since $\hat{\mathbf{n}} \neq 0$, it follows from Equation (1.196b):

$$\sin\varphi = \frac{a}{c}\sin\theta \tag{1.19c}$$

According to Equations (1.19a) and (1.19c),

$$\tan\varphi = \frac{\sin\varphi}{\cos\varphi} = \frac{a\sin\theta}{b+a\cos\theta}$$

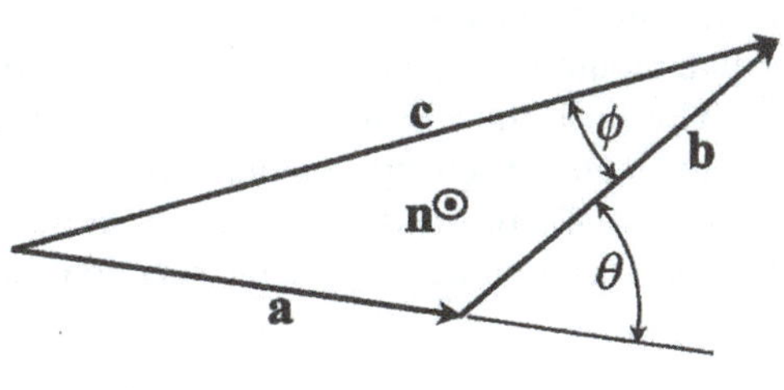

Figure 1.9

ILLUSTRATIVE EXAMPLE 1.5

A particle is acted upon by three forces; the corresponding force vectors ($\mathbf{F_D}, \mathbf{F_B}, \mathbf{F_C}$) are

$$\mathbf{F_A} = \frac{10}{\sqrt{59}}\left(1\hat{\mathbf{i}} + 3\hat{\mathbf{j}} + 7\hat{\mathbf{k}}\right) N$$

$$\mathbf{F_B} = \frac{30}{\sqrt{6}}\left(2\hat{\mathbf{i}} + 1\hat{\mathbf{j}} - 1\hat{\mathbf{k}}\right) N$$

$$\mathbf{F_C} = \frac{20}{\sqrt{9}}\left(-1\hat{\mathbf{i}} + 2\hat{\mathbf{j}} + 2\hat{\mathbf{k}}\right) N$$

where $\hat{\mathbf{i}}, \hat{\mathbf{j}}, \hat{\mathbf{k}}$ denote orthogonal unit vectors. The magnitudes follow:

$$F_A = 10N\ (2.25\text{lb}) \quad F_B = 30N(6.75lb) \quad F_C = 20N(4.50lb)$$

The resultant force $\mathbf{F}$ is obtained by summing the three force vectors:

$$\mathbf{F} = \mathbf{F_D} + \mathbf{F_B} + \mathbf{F_C} = (19.13\hat{\mathbf{i}} + 29.49\hat{\mathbf{j}} + 10.20\hat{\mathbf{k}})N$$

The magnitude of the resultant force is given by

$$|\mathbf{F}| = \sqrt{19.13^2 + 29.49^2 + 10.20^2} = 36.60\,N\ (8.23\ lb)$$

The direction cosines of the resultant force with respect to the positive directions of the x, y, and z axis (unit vectors $\hat{\mathbf{i}}$, $\hat{\mathbf{j}}$, $\hat{\mathbf{k}}$) follow:

$$\frac{\mathbf{F}}{|F|} \cdot \hat{\mathbf{i}} = \cos(F, x \text{ axis}) = 0.52$$

$$\frac{\mathbf{F}}{|F|} \cdot \hat{\mathbf{j}} = \cos(F, y \text{ axis}) = 0.81$$

$$\frac{\mathbf{F}}{|F|} \cdot \hat{\mathbf{k}} = \cos(F, z \text{ axis}) = 0.28$$

1.10 APPROXIMATION OF SMALL QUANTITIES

Often a significant feature of a deformation is the smallness of strains and also rotations; for instance, the rotations of the members in a truss are imperceptible without careful measurements. Indeed, with little forethought, we tend to examine the equilibrium with disregard for these rotations. Of course, real engineering problems do arise, wherein the strains and/or rotations are not inconsequential; rubber-like mountings, membranes, and thin flexible bodies demand our attention to geometric changes. However, the vast majority of engineering structures and machines exhibit only small changes in geometry. Then, approximations are admissible. These approximations always hinge on a comparison of terms and the neglect of those that are *relatively* small. To illustrate, let us return to the small rotation θ. Usually the functions can be expanded in a power series; for example,

$$\sin\theta = \theta - \frac{\theta^3}{3!} + \frac{\theta^5}{5!} - \cdots, \tag{1.20a}$$

For small values of θ, that is, $\theta \ll 1$

$$\sin\theta \approx 0 \tag{1.20b}$$

In many structural problems, a rotation $\theta = 0.1(6^\circ)$ is large; the error is about 0.2%. Similarly, for small values of θ

$$\cos\theta = 1 - \frac{\theta^2}{2!} + \frac{\theta^4}{4!} - \cdots \approx 1 \tag{1.21}$$

The reader is reminded that *mathematical analyses require* that angles be given in radians (arc/radius). Recall also the expansion by the binomial theorem (1.15a, b). Some specific examples follow ($x \ll 1$, $y \ll 1$)

$$(1 \pm x)^n \approx 1 \pm nx, \quad (n > 0) \tag{1.22a}$$

$$(1+x)^2 \approx 1 + 2x \tag{1.22b}$$

$$\sqrt{1+x} \approx 1 + \frac{1}{2}x \tag{1.22c}$$

$$(1 \pm x)^{-n} \approx 1 \mp nx, \quad (n > 0) \tag{1.22d}$$

$$(1 \pm x)^{-1} \approx 1 \mp x \tag{1.22e}$$

$$(1+x)(1+y) \approx 1 + x + y \tag{1.22f}$$

ILLUSTRATIVE EXAMPLE 1.6

Treat θ, γ, and ε as small relative to all other parameters (α, κ, p, q) Find first-order approximations for the following expressions:

a. $\sin(\alpha + \theta)$ **b.** $\dfrac{1}{\kappa + \varepsilon}$ **c.** $\sqrt{(1+\varepsilon)^2 + (1-\theta)^2}$ **d.** $\dfrac{(1+2\varepsilon)^2}{1-\gamma}$

e. $\dfrac{1}{(\kappa+\varepsilon)^{3/2}}$ **f.** $\cos^2\left(\dfrac{\pi}{4} - \theta\right)$ **g.** $[(1+\varepsilon)^p + \gamma]^q$ **h.** $\dfrac{(1+\varepsilon)^p + \gamma}{(1+\varepsilon)^q - \gamma}$

a. Considering the approximations (1.20b) and (1.21b), the trigonometric relation for the sum of two angles leads to:

$$\sin(\alpha + \theta) = \sin\alpha\cos\theta + \cos\alpha\sin\theta \approx \sin\alpha + \theta\cos\alpha$$

b. In view of Equation (1.22e):

$$\frac{1}{\kappa + \varepsilon} = \frac{1}{\kappa}\left(1 + \frac{\varepsilon}{\kappa}\right)^{-1} \approx \frac{1}{\kappa}\left(1 - \frac{\varepsilon}{\kappa}\right)$$

c. Neglecting the squares of quantities ε and θ and considering relationship (1.22c), the following first-order approximation is obtained:

$$\sqrt{(1+\varepsilon)^2 + (1-\theta)^2} \approx \sqrt{2}[1 + (\varepsilon - \theta)]^{1/2} \approx \sqrt{2}\left[1 + \frac{1}{2}(\varepsilon - \theta)\right]$$

d. Utilizing the approximations (1.22b) and (1.22e) and neglecting products of the small quantities, ε and γ, leads to the following expressions:

$$\frac{(1+2\varepsilon)^2}{1-\gamma} \approx (1+4\varepsilon)(1+\gamma) \approx 1+\gamma+4\varepsilon$$

e. In view of (1.22d):

$$\frac{1}{(\kappa+\varepsilon)^{3/2}} = \kappa^{-3/2}\left(1+\frac{\varepsilon}{\kappa}\right)^{-3/2} \approx \kappa^{-3/2}\left(1-\frac{3}{2}\frac{\varepsilon}{\kappa}\right)$$

f. According to the approximations (1.20a,b), we obtain from the trigonometric relationship

$$\cos 2\theta = \cos^2\theta - \sin^2\theta$$

the equation:

$$\cos^2\left(\frac{\pi}{4}-\theta\right) = \frac{1}{2}\left[1+\cos\left(\frac{\pi}{2}-2\theta\right)\right] = \frac{1}{2}(1+\sin 2\theta) \approx \frac{1}{2}+\theta$$

g. Utilizing the approximation (1.22a), we obtain:

$$[(1+\varepsilon)^p+\gamma]^q = [1+p\varepsilon+\gamma]^q = 1+q(p\varepsilon+\gamma)$$

h. In the same manner, taking into account (1.22e) and neglecting the squares of the small quantities ε and γ, the following first-order approximation is obtained:

$$\begin{aligned}\frac{(1+\varepsilon)^p+\gamma}{(1+\varepsilon)^q-\gamma} &\approx \frac{(1+p\varepsilon+\gamma)}{(1+q\varepsilon-\gamma)}\\ &\approx (1+p\varepsilon+\gamma)(1-q\varepsilon+\gamma)\\ &\approx 1+(p-q)\varepsilon+2\gamma\end{aligned}$$

ILLUSTRATIVE EXAMPLE 1.7

A gauge for measuring curvature is shown in Figure 1.10a. The two outer contacts are fixed to the frame [length $l = 2.50$ cm (0.984 in.)]. The central contact moves transversely; its displacement is recorded by a dial. The gauge is zeroed on a perfectly flat surface and then placed on the surface of a sphere with the gauge normal to the surface. If the dial indicates that the center was moved 1.00×10^{-3} cm (3.94×10^{-4} in.), what is the approximate radius of the sphere?

Denoting by l the distance of an outer contact to the center contact and by h the displacement of the center contact from the flat surface, we observe from Figure 1.10b that:

$$r^2 = l^2 + (r-h)^2$$

Considering that h is a small quantity compared to the length l, the following approximate value for the radius r is obtained:

$$r \approx \frac{l^2}{2h} = 3.13 \approx 10^3 \text{ cm } (1.23 \times 10^3 \text{ in.})$$

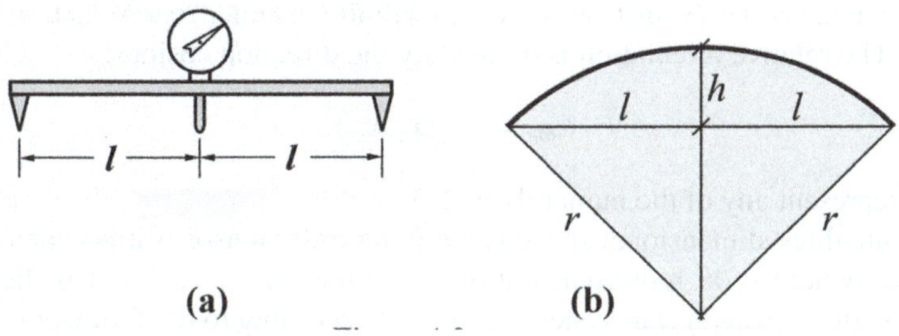

Figure 1.10

1.11 ACCURACY

For the sake of brevity, only nominal values are given in most problems. For example, a shaft might be described as one with a length of 900 mm (35.43 in.) and a diameter of 25.0 mm (0.98 in.). In actual practice, the length of a shaft might be held to tolerances of ± 0.25mm (± 0.01in.) and the diameter, to tolerances of ± 0.025 mm (± 0.001 in.). In such applications, one has accuracies of four decimal places. The length and thickness of a structural steel beam are less precisely specified, probably ± 1 mm (± 0.04 in.) and ± 0.25 mm (± 0.01 in.), respectively. Properties of materials are typically measured to accuracies of three decimal places. In short, the accuracy of the data is limited by the particular application, capabilities, and costs of machining and/or forming parts, and, ultimately, by cost-effective means of measurement and control. The reader must appreciate the practical significance of numerical values as they relate to the specific problem.

Often one loses accuracy in computation. For example, a wire is known to have a length $L = 2m$, ± 0.1 cm (78.74 in., ± 0.04 in.); the accuracy is ± 0.05% This wire is subjected to an axial load that causes it to elongate to a deformed length of $L^* = 200.4$ cm (78.89 in.); again, the ability to measure is limited by instrumentation to 0.1 cm (± 0.04 in.). Suppose that the important result is the change in length:

$$\Delta L = L^* - L = 200.4 - 200.0 = 0.4 \text{ cm } (0.157 \text{ in.} \approx 0.2 \text{ in.})$$

Both measurements are subject to an error of ±0.1 cm (±0.04 in.). Conceivably, the numerical value $\Delta L = 0.4$ cm could have an error of ±0.2 cm (0.078 in.), or 50%. Beware! In an era of electronic computation, we can easily punch 6, 8, or 10 keys; however, we must recognize the accuracy of our results, which are limited by practical considerations, as stated previously.

1.12 TENSORS

The analysis of deformable solids introduces important concepts to describe geometrical changes (deformations) and internal forces. These are the concepts of *strain* and *stress*, respectively. In the three-dimensional case, both are characterized by systems of nine components. *Mathematically* these systems have a remarkable similarity: The components can be referred to different systems of coordinates and are transformed from one system to another by the same linear equations. That common transformation identifies these systems as *tensors*. Here, we limit our attention to Cartesian tensors and the transformation, which transforms the components in one system of Cartesian coordinates to the components in another. To be more specific, let (x_1, x_2, x_3) and $(\bar{x}_1, \bar{x}_2, \bar{x}_3)$ denote the coordinates of two rectangular Cartesian systems.

The notations, i.e., $(\bar{x_1}, \bar{x_2}, \bar{x_3})$ instead of (x, y, z), admit a simpler representation of the tensorial transformation. The relative orientation is defined by the direction cosines:

$$\alpha_{nm} = \cos(x_n, \bar{x_m}),$$

where n and m represent any of the numerals, $1, 2, 3$.

An entity in the three-dimensional space is a second order tensor if it is characterized by nine components A_{nm}, where n, m represent any of the numerals, 1, 2, 3 and if these components transform from system (x_1, x_2, x_3) to system $(\bar{x_1}, \bar{x_2}, \bar{x_3})$ according to the following system of linear equations:

$$\bar{A}_{st} = \sum_{n=1,2,3} \sum_{m=1,2,3} \alpha_{ns} \alpha_{mt} A_{nm}$$

In matrix notation,

$$\left|\bar{A}_{st}\right| = |\alpha_{ns}|\,|\alpha_{mt}|\,|A_{nm}|$$

NOTES

1. I. Todhunter and K. Pearson. *A History of the Theory of Elasticity and of Strength of Materials from Galilei to the Present Time*, first edition, volume 1, 1886, volume 2, 1893. University Press, Cambridge, 1886–1893 (Reprinted: Dover Publications, New York, 1960); S. P. Timoshenko. *History of Strength of Materials*, first edition. McGraw-Hill Book Co., Inc., New York, 1953.
2. Hooke, Robert. Lectures de Potentia Restitutiva, Or of Spring Explaining the Power of Springing Bodies, Printed for John Martyn printer to the Royal Society, Bell in St. Paul's church-yard, 1678.
3. G. Wempner and D. Talaslidis. *Mechanics of Solids and Shells: Theories and Approximations*. CRC Press, Boca Raton, 2003.
4. This suspension bridge spanned the Tacoma Narrows, a strait between Tacoma and the Kitsap Peninsula in the state of Washington, U.S.
5. R. Hooke. *De Potentia Restitutiva*, (*ut tensio sic vis*). London: John Martyn, 1678.
6. Such irregularities and consequences are discussed in a later Section 5.2 (Saint-Venant's Principle).
7. It is important that the reader understands the precise definition of a *free-body* diagram: This is a diagram of a body, isolated from all adjoning parts of a structure or machine and that diagram displays all external forces acting upon that body.

2 Deformation

As demonstrated in the introduction (Chapter 1), every analysis of a deformable body embodies a description of the kinematic changes, which characterize the deformation. Such kinematic descriptions are entirely geometric and independent of any causes (forces, temperature, etc.) or the composition of the body (rubber, plastic, steel, etc.). The geometric analyses presuppose only that the medium is continuous and cohesive, as described in Section 2.1.

Because each elemental portion of a component (e.g., a connecting rod or a beam) can deform differently (e.g., one portion may stretch while another contracts), we must devise means to completely describe the deformation at each location. This is accomplished via the concept of strain, which is introduced in Section 2.3. The characteristics and the description of strains, special properties, and special states are explored in the subsequent sections. Section 2.24 illustrates the concepts via simple, yet important, problems.

The concept of strain is a very essential one. Every description of continuous media and all analyses of deformable bodies require this important tool. It is needed in our description of materials in Chapter 4 and in all subsequent analyses. Our treatment of this most basic subject is directed toward a thorough presentation of the fundamental concepts and an accurate development of the theory. Specifically, the various definitions of strain components and interpretations are emphasized; any approximations and limitations are noted.

2.1 PRELIMINARY CONSIDERATIONS

2.1.1 DEFORMATION

The term deformation denotes any change in shape and size of a body or any of its parts. The deformation may be caused by the application of external loads or by changes of temperature; it may be visible to the naked eye or detectable only by delicate instrumentation. Deformations always occur when a body is subjected to forces. A rubber band will extend to several times its original length when placed around a bundle of papers. In contrast, a person standing on a railroad rail will cause a slight, but unnoticeable, deformation of the rail. Another cause of deformation is a change in temperature. The operation of the bimetallic strip in a conventional thermostat is an example of temperature-induced deformation.

For the present, we disregard the causes of deformation and consider only the geometric problems associated with describing changes in size and shape of a body.

2.1.2 COHESIVE, CONTINUOUS MEDIA

To develop a reliable and workable theory of the behavior of deformable bodies, we assume that every body occupies all the space within its apparent surface boundaries. In other words, any arbitrarily small portion within boundaries is a part of the body. Such an idealization of the matter is called a *continuous medium.*

Materials, such as sand, differ from those such as wood, concrete, or metal. Adjacent particles in a box of sand may be separated. On the other hand, a line of particles in a bar of steel remains intact throughout a deformation of the bar. An idealized continuous medium is said to be cohesive if the continuity of a line of particles must persist during deformation. In Chapter 4, we address

DOI: 10.1201/9781003665861-2

the questions of rupture and fracture, which signify a failure of cohesion. In all other analyses, we assume that the media are *cohesive and continuous*.

2.2 DISPLACEMENT

A cylinder is shown in Figure 2.1. Each particle of material within this cylinder occupies some point in space. One such particle occupies the point labeled P. We call point P the initial position of the particle. If a system of applied forces stretches the cylinder, it will become longer and thinner, as shown in Figure 2.2. The particles of the cylinder change their positions in space. The particle of material that was originally at P now occupies a new position in space, denoted by P^*. The point P^* is called the final position of the particle. This particle, which was at P, has been displaced to P^*. Its displacement can be represented as a vector, $\mathbf{u}(P)$. The displacement vector $u(P)$ depends, of course, on the initial position of the particle. Notice that our attention is focused on a particle and the positions it occupies in space.

In Figure 2.1, Q is the initial position of another particle of the cylinder. When the cylinder is stretched, this particle is displaced to Q^*. Since the displacement depends on the initial position of this particle, the displacement vector is denoted by $\mathbf{u}(Q)$.

In the deformed body, the length of PQ is generally different from the undeformed length, P^*Q^*. The particles undergo relative changes in position. It is possible to displace every particle of a body so that no change in distance occurs between any pair of particles. Such displacements are very special and are called *rigid-body displacements*. In general, these displacements consist of a *translation* (all particles are moved the same distance in the same direction) and a *rotation* (all lines are rotated through the same angle about some fixed line). The following examples help clarify the concept of *rigid-body motion*:

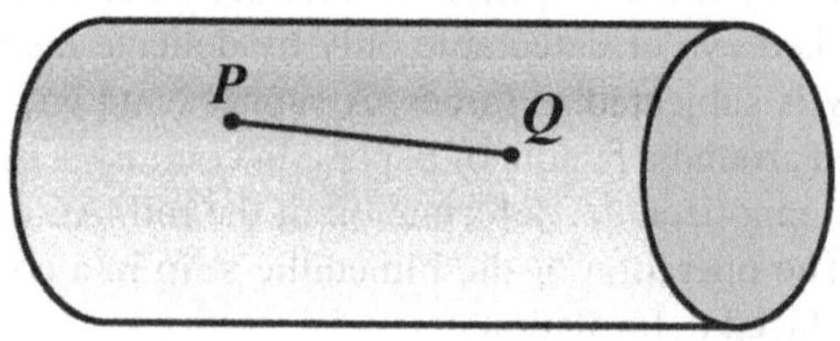

Figure 2.1

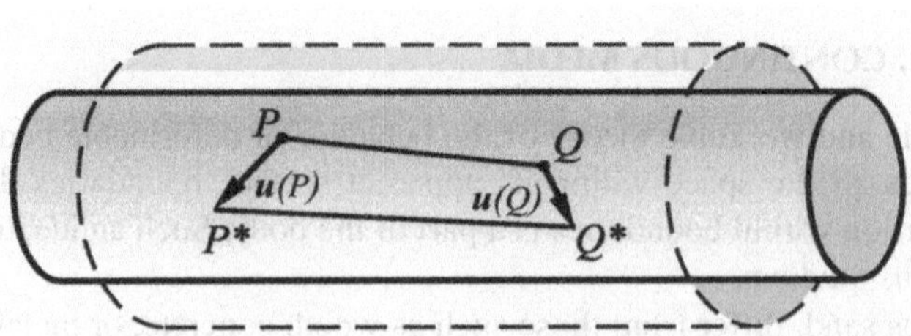

Figure 2.2

Imagine, for example, a piston moving in the cylinder of a stationary engine. As the piston moves from the top to the bottom of the cylinder, all the particles in the piston move parallel to the cylinder wall. The piston undergoes a translation.

On the other hand, a flywheel that turns about a fixed axle undergoes a rotation because, when the wheel turns, every radius turns through the same angle.

We are primarily concerned with displacements that change the *relative* positions of particles in a body, since deformation is a manifestation of such displacements.

2.3 DEFORMATION AND STRAIN

2.3.1 DEFORMATION OF A BODY AND ITS PARTS

Deformation is sensed by noting the overall change in the shape and size of a body. When a spherical balloon is inflated to a spherical form, the change in radius is sufficient to describe the balloon's alteration in size. If a solid circular cylinder is stretched so that it remains a circular cylinder, the changes in radius and length are enough to account for its final size. These observable effects are the manifestations of changes in shape and size that occur in every portion of the body.

2.3.2 STRAIN

Two forms of deformation can be observed. They are illustrated by the body of Figure 2.3a, which is deformed into the shape of Figure 2.3b. The straight lines PQ and PR of Figure 2.3a are cohesive, as chains of particles. They are, therefore, deformed to the continuous curves P^*Q^* and P^*R^* of Figure 2.3b. In general, the lengths of the lines are changed, and the angle between the lines has also changed. This suggests that we devise means to measure changes in the lengths of line segments and changes of angles between line segments. These changes, which occur during deformation, are generally not uniform throughout a body. Some portions of a line segment will stretch; others will contract. Short segments of the same length originating at the same point but oriented in different directions will experience different extensions. Similarly, the changes in angle between segments will vary with position and orientation. To account for such variations, it is necessary to define quantities that characterize these changes in the neighborhood of a point. A quantity that measures the changes of *length* is an *extensional strain*. A quantity that measures changes of *angle* is a *shear strain*. Notice that strains measure only the *relative* positions of neighboring particles; consequently, a strain is independent of any rigid-body motion.

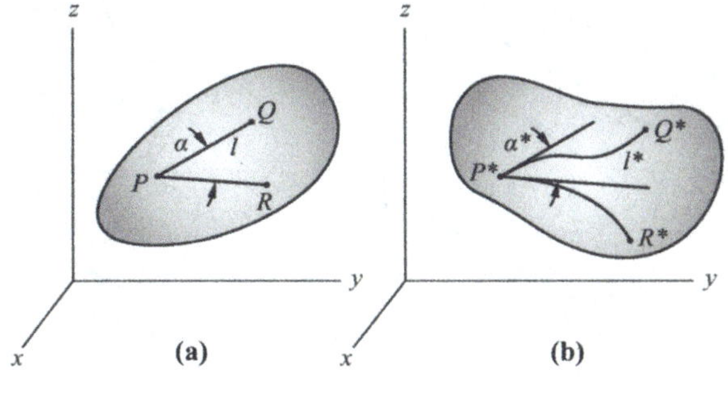

Figure 2.3 Strain

2.4 EXTENSIONAL STRAIN

Figure 2.4a shows a cohesive body before deformation and a straight line of particles PQ in the interior. After a deformation, the line PQ is displaced and deformed into the continuous curve P^*Q^* of Figure 2.4b. The length of P^*Q^* is generally different from the length of PQ. The initial line PQ of Figure 2.4a is shown subdivided into many small, but equal, segments by labeling intermediate particles P_1, P_2, ..., which are displaced to P_1^*, P_2^*, ..., on P^*Q^* of Figure 2.4b. Note that, in general, each segment undergoes a different change; for example, $P^*P_1^*$ is longer than PP_1, whereas $P_1^*P_2^*$ is shorter than P_1P_2. Evidently, the extension (or contraction) depends upon the location along the line and the length of the segment. To describe the extensional deformation, we require a measure at each point; we must compare the original and deformed lengths, but we must do so as the lengths shrink to the point in question. Such a measure is called an *extensional strain*. Note, too, that a line with different orientation will, in general, experience a different extensional strain. The cylinder of Figure 2.2 is a simple example: Lines in an axial direction are stretched, whereas lines in a radial direction are contracted. Any definition of extensional strain must characterize extension at the *specific point* and in the *specific direction*.

The length of one segment PP_1 in Figure 2.4a is Δs. The direction of the segment is identified by n. After the deformation, the particles at P and P_1 are at P^* and P_1^* in Figure 2.4b, and the length has changed to Δs^*. We *define*[1] an *extensional strain* $\varepsilon_n(P)$ at the point P in the direction n by the limit

$$\varepsilon_n(P) = \lim_{\Delta s \to 0} \frac{\Delta s^* - \Delta s}{\Delta s} \tag{2.1}$$

Note that the lengths Δs and Δs^* diminish as P_1 approaches P, because the body is continuous and cohesive. The extensional strain (2.1) is obtained as a ratio of two lengths, so it is a dimensionless quantity.

Let us examine the meaning of the notation for extensional strain. To specify $\varepsilon_n(P)$, we must (1) specify the point P, (2) specify the direction n, and (3) give the numerical value of $\varepsilon_n(P)$.

The quantity $\varepsilon_n(P)$ enables us to compute the approximate change in length of a short line segment in the body when the body is deformed. The particular segment originates at P in the undeformed body and is in the direction n. At P, in the direction n, we can choose a very short

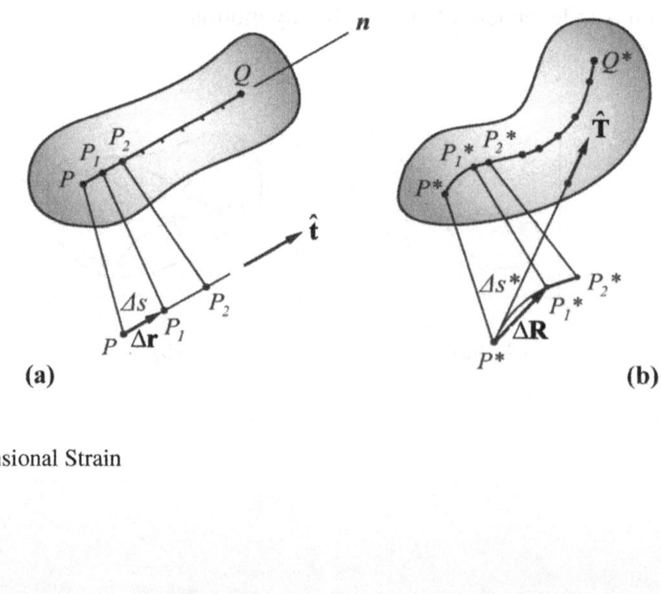

Figure 2.4 Extensional Strain

segment of length Δs. From equation (2.1), we have a good approximation for the final length of this segment:

$$\Delta s^* \doteq [1+\varepsilon_n(P)]\Delta s \tag{2.2}$$

If $\varepsilon_n(P)$ is positive, the segment of length Δs has extended; if $\varepsilon_n(P)$ is negative, this segment has shortened. It is also possible that $\varepsilon_n(P)$ is zero. If so, there is no change in length of the segment.

To determine the change in length of a line of particles of finite length, we need to know the extensional strain at every point along the line in the direction of the line. In Figure 2.4a, the line PQ has an initial length L, which has been divided into m equal segments of length Δs. After deformation, the line is deformed to $P^*\ Q^*$; each of the particles $P_1, P_2, \ldots,$ has been displaced to $P_1^*, P_2^*, \ldots$ In general, each segment has a new length: $P^*P_1^*$ has length Δs_1^*; $P_1^*P_2^*$ has length Δs_2^* and so on. Using equation (2.2) and denoting P by P_0, we have

$$\begin{aligned} \Delta s_1^* &\doteq [1+\varepsilon_n(P_0)]\Delta s \\ \Delta s_2^* &\doteq [1+\varepsilon_n(P_1)]\Delta s \\ \Delta s_3^* &\doteq [1+\varepsilon_n(P_2)]\Delta s \\ &\vdots \end{aligned} \tag{2.3}$$

If L^* is the length of P^*Q^*, then approximately

$$L^* \doteq \sum_{k=0}^{m-1}[1+\varepsilon_n(P_k)]\Delta s$$

As we pass to the limit ($m \to \infty$, $\Delta s \to 0$), $\varepsilon_n = \varepsilon_n(s)$, a function of distance along PQ; then we obtain

$$L^* = \int_0^L [1+\varepsilon_n(s)]ds = L + \int_0^L \varepsilon_n(s)ds \tag{2.4}$$

Thus, knowing $\varepsilon_n(s)$ at every point along PQ, equation (2.4) enables us to compute the exact value of the new length, P^*Q^*

ILLUSTRATIVE EXAMPLE 2.1

To illustrate equation (2.4), consider the straight wire shown in Figure 2.5a. Suppose that a nonuniform heating of the wire causes an extensional strain, ε_x , given by

$$\varepsilon_x = k\left(\frac{x}{L}\right)^2$$

We can determine the change in length of the wire by computing the change in length of the line of particles along the x axis. The new length L^* is obtained by placing the expression for ε_x in equation (2.4):

$$L^* = L + \int_0^L k\left(\frac{x}{L}\right)^2 ds = L + \frac{kL}{3}$$

Consequently, the change in length of the wire is

$$\Delta L = L^* - L = \frac{kL}{3}$$

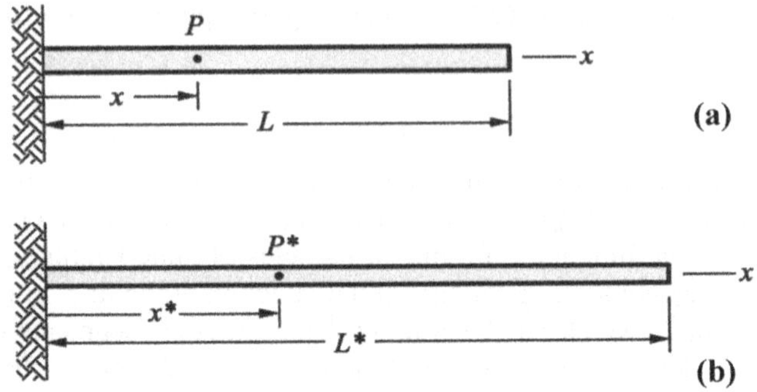

Figure 2.5

ILLUSTRATIVE EXAMPLE 2.2

In general, straight lines of a body are deformed to curves, and extensional strain is dependent upon the position. To illustrate, let us consider the hypothetical, but conceivable, situation of Figure 2.6. Here, the line that lies initially along the x axis is deformed to the parabolic curve $y = ax^2$, but no particle of the line displaces in the x direction. By the definition (2.1)

$$\varepsilon_x = \lim_{\Delta x \to 0} \frac{\Delta s^*}{\Delta x} - 1$$

Here, as depicted in Figure 2.6,

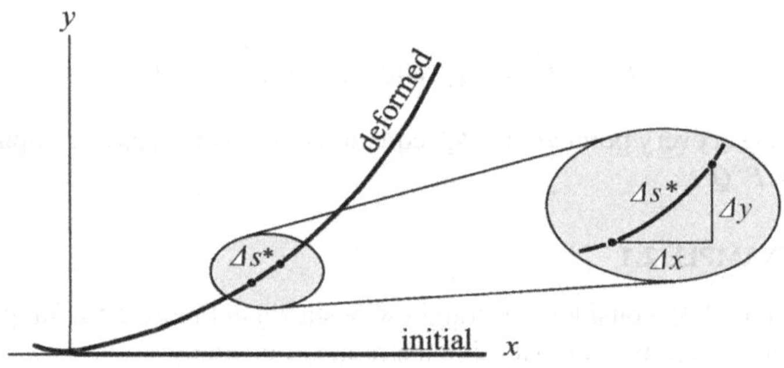

Figure 2.6

$$\frac{\Delta s^*}{\Delta x} = \frac{\sqrt{\Delta y^2 + \Delta x^2}}{\Delta x} = \sqrt{\left(\frac{\Delta y}{\Delta x}\right)^2 + 1}$$

Since

$$\lim_{\Delta x \to 0)} \frac{\Delta y}{\Delta x} = \frac{dy}{dx} = 2ax$$

it follows that

$$\varepsilon_x = \lim_{\Delta x \to 0} \frac{\Delta s^*}{\Delta x} - 1 = \sqrt{(2ax)^2 + 1} - 1$$

In this case, the extensional strain ε_x equals zero at the origin and increases with x.

If the initial length of the wire is 10 and if the wire terminates at ($x = 10$, $y = 2$), then $a = 1/50$ and

$$\varepsilon_{x-\max} = \left[\sqrt{(2ax)^2 + 1} - 1\right]_{x=10} = \sqrt{\left[(2)\left(\frac{1}{50}\right)(10)\right]^2 + 1} - 1 = 0.077$$

2.5 AVERAGE EXTENSIONAL STRAIN

Let the initial length of a segment be L and its final length be L^*. The ratio

$$\varepsilon = \frac{L^* - L}{L} \tag{2.5}$$

is called the *average extensional strain* of the line segment

ILLUSTRATIVE EXAMPLE 2.3

The two wires shown in Figure 2.7 are joined at P and to the fixed horizontal surface at B and C. The joint P is displaced to P^*; horizontal and vertical components of displacement are u and v, as shown. Let us obtain the exact expression for the average extensional strains of each wire in terms of u and v and then determine the linear approximation of each strain, valid if $u/L \ll 1$, $v/L \ll 1$. Finally, let us determine the error in the linear approximation when $u = v = 0.00100L$; for instance, if $L = 2$ m (78.74 in.), $u = v = 0.2$ cm (0.0787 in.).

In order to show the difference between exact and approximate values for the strains, eight-digit accuracy is used. However, any difference beyond three-place accuracy is meaningless because most material properties are not known beyond three places.

By means of the Pythagorean Theorem, the new lengths of the wires are

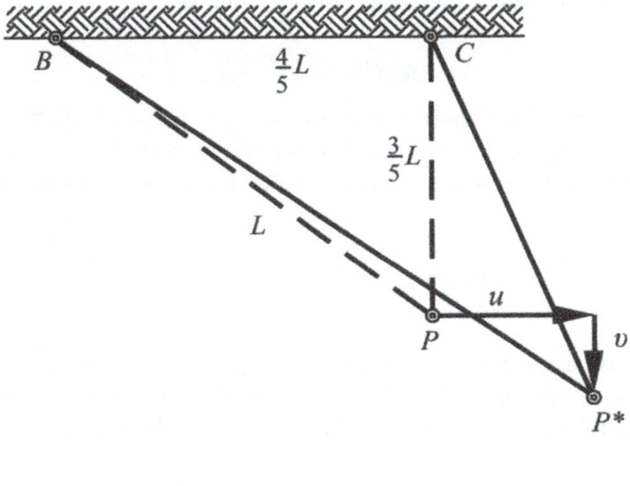

Figure 2.7

$$BP^* = \sqrt{\left(\frac{4}{5}L+u\right)^2 + \left(\frac{3}{5}L+v\right)^2} = L\sqrt{\left(\frac{4}{5}+\frac{u}{L}\right)^2 + \left(\frac{3}{5}+\frac{v}{L}\right)^2}$$

$$CP^* = L\sqrt{\left(\frac{u}{L}\right)^2 + \left(\frac{3}{5}+\frac{v}{L}\right)^2}$$

The extensional strains, ε_B and ε_C, of wires BP and CP, respectively, follow from the definition:

$$\varepsilon_B = \frac{BP^*}{L} - 1 = \left[\left(\frac{4}{5}+\frac{u}{L}\right)^2 + \left(\frac{3}{5}+\frac{v}{L}\right)^2\right]^{1/2} - 1 = 0.00140002$$

$$\varepsilon_C = \frac{5CP^*}{3L} - 1 = \frac{5}{3}\left[\left(\frac{u}{L}\right)^2 + \left(\frac{3}{5}+\frac{v}{L}\right)^2\right]^{1/2} - 1 = 0.00166805$$

By means of the binomial theorem, we obtain the linear approximation

$$\varepsilon_B \doteq \frac{4}{5}\frac{u}{L} + \frac{3}{5}\frac{v}{L} = 0.00140000$$

$$\varepsilon_B \doteq \frac{5}{3}\frac{v}{L} = 0.00166667$$

Practically, the linear approximations are as good as the "exact" results.

2.6 SHEAR STRAIN

In Figure 2.8a, two perpendicular directions n and t are shown at the point P. The particles along PQ deform into the curve P^*Q^*, and those along PR deform into P^*R^*, as shown in Figure 2.8b. The angles between the curves P^*R^* and P^*Q^* at P^* is φ^*_{nt} (in radians). The shear strain at P associated with the perpendicular directions n and t is defined by[2]

$$\gamma_{nt}(P) \equiv \pi/2 - \varphi^*_{nt} \tag{2.6}$$

If φ^*_{nt} is less than $\pi/2$, the shear strain is positive. If φ^*_{nt} is greater than $\pi/2$, the shear strain is negative. In order to specify $\gamma_{nt}(P)$, we must (1) identify the point P, (2) specify the two perpendicular directions at P, and (3) give a numerical value for the shear strain $\gamma_{nt}(P)$. We may also define $\gamma_{nt}(P)$ as

$$\gamma_{nt}(P) \equiv \pi/2 - \lim_{\substack{\Delta s_1 \to 0 \\ \Delta s_2 \to 0}} \angle R^*P^*Q^* \tag{2.7}$$

where $\angle R^*P^*Q^*$ is the angle between the chords P^*Q^* and P^*R^* in Figure 2.8b. Clearly, the limiting value of this angle is φ^*_{nt}. This alternative definition is sometimes useful for computing shear strains. If the shear strain is small $\gamma_{nt} \ll 1$, then we have the approximation:

$$\gamma_{nt} \equiv \pi/2 - \varphi^*_{nt} \doteq \cos\varphi^*_{nt} = \hat{\mathbf{T}} \cdot \hat{\mathbf{N}}$$

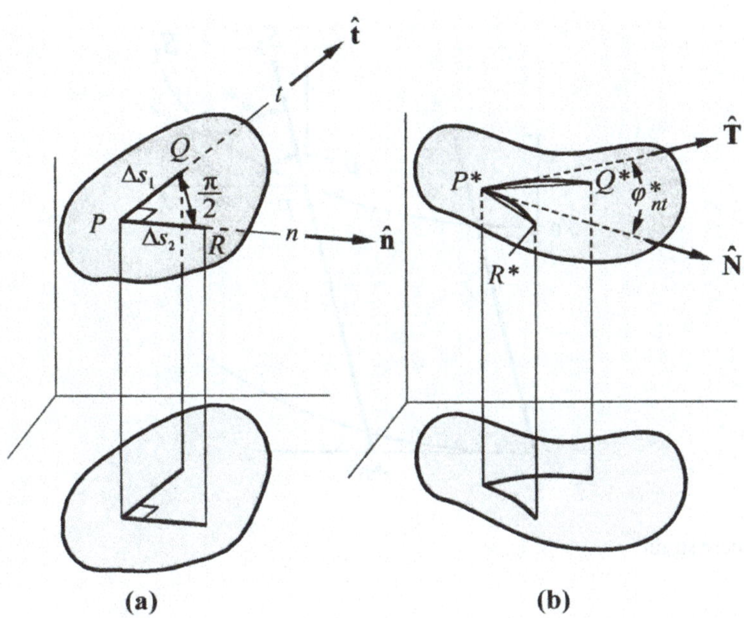

Figure 2.8

ILLUSTRATIVE EXAMPLE 2.4

A panel is deformed such that every horizontal line ($y = y_0 = constant$) is deformed into a similar parabola ($y - y_0 = x^2/400$) and every vertical line ($x = x_0$ =constant) is deformed into a similar straight line ($x - x_0 = y/40$) (see Figure 2.9). Obtain the exact and the approximate value for the shear strain $\gamma_{xy}(x, y)$ at point P. At P:

$$\begin{aligned}\gamma_{xy} &= \pi/2 - \varphi^* = \alpha + \beta \\ &= \arctan \left.\frac{dy}{dx}\right|_{s_1} + \arctan \left.\frac{dx}{dy}\right|_{s_2} = \arctan \left.\frac{x}{200}\right|_{s_1} + \arctan \left.\frac{1}{40}\right|_{s_2}\end{aligned} \tag{A}$$

For small angles we have the *approximation*

$$\gamma_{xy} \doteq \frac{x}{200} + \frac{1}{40} \tag{B}$$

Numerical example: Assume the panel is 25 cm (9.84 in.) long. Then, the exact equation (A) yields:

$$\gamma_{xy}(P) = \arctan \frac{25}{200} + \arctan \frac{1}{40} = 0.1493498$$

whereas the approximation (B) yields:

$$\gamma_{xy}(P) \doteq \frac{25}{200} + \frac{1}{40} = 0.1500000$$

Note that strains vary throughout most members. In this example, the shear strain γ_{xy} varies from 0.025 at the origin to 0.149 at point P.

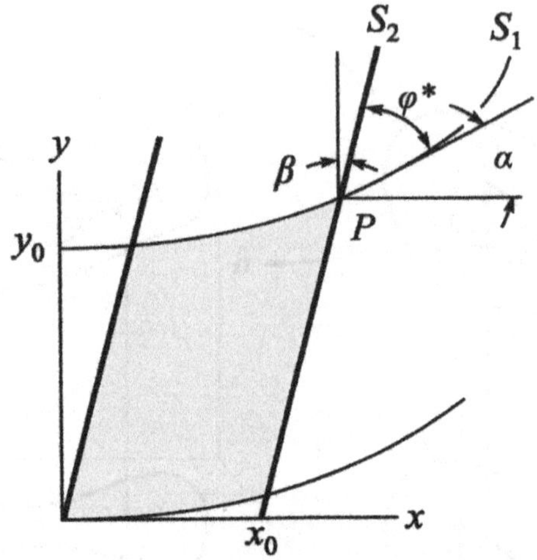

Figure 2.9 Shear strain

ILLUSTRATIVE EXAMPLE 2.5

The rectangular panel (4 m × 3 m) of Figure 2.10 is uniformly deformed such that the edges experience the extensional strains ε_x and ε_y and the diagonal OC experiences the extensional strain ε_c. If the extensional strains take values: $\varepsilon_x = -0.01$, $\varepsilon_y = 0.02$, and $\varepsilon_c = 0.02$, compute the shear strain γ_{xy}.

By definition,

$$O^*C^* = (1+\varepsilon_c)L, \qquad O^*A^* = B^*C^* = (1+\varepsilon_x)l, \qquad O^*B^* = A^*C^* = (1+\varepsilon_y)h \tag{A}$$

By the Cosine Law,

$$(O^*C^*)^2 = (O^*A^*)^2 + (A^*C^*)^2 - (O^*A^*)(A^*C^*)\cos\left(\frac{\pi}{2}+\gamma_{xy}\right) \tag{B}$$

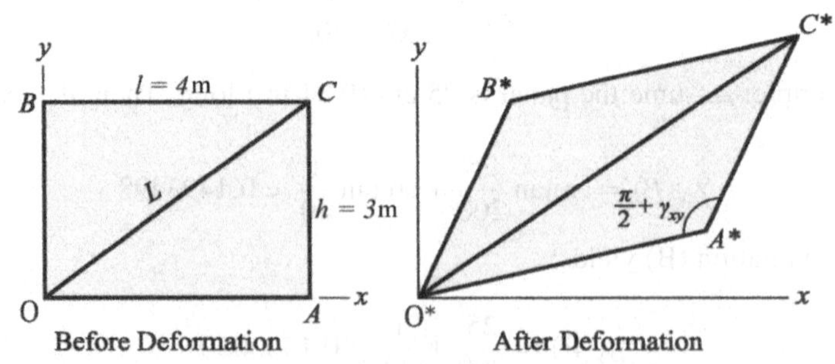

Figure 2.10

Substitution yields

$$(1+\varepsilon_c)^2 = (1+\varepsilon_x)^2\left(\frac{l}{L}\right)^2 + (1+\varepsilon_y)^2\left(\frac{h}{L}\right)^2 - 2(1+\varepsilon_x)(1+\varepsilon_y)\left(\frac{l}{L}\right)\left(\frac{h}{L}\right)\cos\left(\frac{\pi}{2}+\gamma_{xy}\right) \tag{C}$$

Since

$$\frac{l}{L}=0.8, \qquad \frac{h}{L}=0.6, \qquad \cos\left(\frac{\pi}{2}+\gamma_{xy}\right) = -\sin\gamma_{xy}$$

it follows that

$$(1.02)^2 = (0.99)^2(0.64) + (1.02)^2(0.36) + 2(0.99)(1.02)(0.48)\sin\gamma_{xy}$$

from which we obtain

$$\gamma_{xy} = 0.0398$$

Since changes are *small*, i.e., $\varepsilon_x \ll 1$, $\varepsilon_y \ll 1$ $\gamma_{xy} \ll 1$, we have the *approximations*

$$(1+\varepsilon)^2 \doteq 1+2\varepsilon, \qquad \sin\gamma_{xy} \doteq \gamma_{xy}$$

Therefore,

$$\varepsilon_c \doteq \varepsilon_x\left(\frac{l}{L}\right)^2 + \varepsilon_y\left(\frac{h}{L}\right)^2 + \gamma_{xy}\left(\frac{l}{L}\right)\left(\frac{h}{L}\right)$$

$$0.02 \doteq -(0.01)(0.64) + (0.02)(0.36) + (0.48)\gamma_{xy}$$

As a result, we obtain: $\gamma_{xy} \doteq 0.0400$.

2.7 ALTERNATIVE MEASURES OF STRAIN

2.7.1 EXTENSIONAL STRAIN OF CAUCHY-GREEN

Definitions of strain, such as the extensional strain of equation (2.1) or the shear strain of equation (2.6), are somewhat arbitrary. The practitioner uses the simplest one that meets his or her requirements. Another definition of extensional strain is sometimes more convenient; with the same basic quantities of Figure 2.4, we can define an alternative extensional strain $\tilde{\varepsilon}_n(P)$:[3]

$$\tilde{\varepsilon}_n(P) \equiv \lim_{\substack{\Delta s \to 0 \\ \Delta P_1 \to P}} \frac{1}{2}\left[\left(\frac{\Delta s^*}{\Delta s}\right)^2 - 1\right] \tag{2.8}$$

Instead of the approximation (2.2), we obtain

$$\Delta s^* \doteq \sqrt{1+2\tilde{\varepsilon}_n}\,\Delta s \tag{2.9}$$

Also, Equations (2.2) and (2.8) give, in the limit at P,

$$\tilde{\varepsilon}_n = \frac{1}{2}\left[(1+\varepsilon_n)^2 - 1\right] = \varepsilon_n + \frac{1}{2}\varepsilon_n^2 \tag{2.10}$$

This alternative measure $(\tilde{\varepsilon}_n)$ serves as well as the previous measure $(\varepsilon)_n$. If one is known, then the other can be computed by means of (2.10). Both vanish if there is no stretching or contracting–that is, if $\Delta s^* = \Delta s$. Note, too, that extensional strains are usually very small; in a typical member of a machine or structure, $\varepsilon_n = O(10^{-3})$. In such case, there is no practical difference between the values for (ε_n) and $(\tilde{\varepsilon}_n)$; the difference is usually too small to measure.

Likewise, we could adopt an alternative definition of a shear strain. With the angle φ^*_{nt} of Figure 2.8, we can define a shear strain

$$\tilde{\gamma}_{nt} \equiv (1+\varepsilon_n)(1+\varepsilon_t)\hat{\mathbf{N}}\cdot\hat{\mathbf{T}} \tag{2.11}$$

$$\equiv (1+\varepsilon_n)(1+\varepsilon_t)\sin\gamma_{nt} \doteq \gamma_{nt} \tag{2.12}$$

This, too, serves as well as the previous measure. Both vanish if the angle is unchanged–that is, if $\varphi^*_{nt} = \pi/2$. Again, the strain is often small $[O(10^{-3})]$; then the difference has no practical consequences.

The choice of definitions is a matter of practical and mathematical convenience. It is important, especially if the strains are large, to be consistent; any comparisons between theories and experiments must compare like quantities. Here, we introduce and use the earlier definitions, ε_n and γ_{nt} of Equations (2.1) and (2.6), because they are conceptually simple and yet are adequate for our needs. In a subsequent Section 2.13, we introduce alternative strains, the tensorial strains of Cauchy-Green, which offer mathematical advantages.

ILLUSTRATIVE EXAMPLE 2.6

Large strains in a typical structural or mechanical component might attain the values:

$$\varepsilon_n = \varepsilon_t = \gamma_{nt} = 4\times 10^{-3}$$

According to equations 2.10 and 2.11,

$$\tilde{\varepsilon}_n = \varepsilon_n + \frac{1}{2}\varepsilon_n^2 = 4.008\times 10^{-3}$$

$$\tilde{\gamma}_{nt} = (1+\varepsilon_n)(1+\varepsilon_t)\sin\gamma_{nt} = 4.032\times 10^{-3}$$

The differences are insignificant and immeasurable in practice

ILLUSTRATIVE EXAMPLE 2.7

A rubber-like material experiences an extensional strain $\varepsilon_x = \varepsilon_y = 0.400$ and shear strain $\gamma_{xy} = 0.300$. Compare these with the strains $\tilde{\varepsilon}_x$ and $\tilde{\gamma}_{xy}$.

Again according to equations 2.10 and 2.11:

$$\tilde{\varepsilon}_x = \varepsilon_x + \frac{1}{2}\varepsilon_x^2 = 0.4800,$$

$$\tilde{\gamma}_{xy} = (1+\varepsilon_x)(1+\varepsilon_y)\sin\gamma_{xy} = 0.5792$$

In this case, one must acknowledge the applicable definition, lest one incur errors of 20% and 93%.

2.7.2 ANOTHER DEFINITION OF EXTENSIONAL STRAIN

One may define extensional strain on the basis of final length; that is, instead of equation (2.1), one may write

$$\varepsilon_n'(P) = \lim_{\substack{P_1 \to P \\ \text{along } n}} \frac{\Delta s^* - \Delta s}{\Delta s^*}$$

Then, the difference in the values computed by this definition and by equation (2.1) is

$$\begin{aligned}\varepsilon_n(P) - \varepsilon_n'(P) &= \lim_{P_1 \to P} \left(\frac{\Delta s^*}{\Delta s} + \frac{\Delta s}{\Delta s^*} - 2 \right) \\ &= \lim_{P_1 \to P} \left(\frac{\Delta s^* - \Delta s}{\Delta s} \right) \lim_{P_1 \to P} \left(\frac{\Delta s^* - \Delta s}{\Delta s^*} \right) \\ &= \varepsilon_n(P)\varepsilon_n'(P)\end{aligned}$$

We note that this difference is a second-order term, and negligible when strains are small ($\varepsilon \ll 1$)

2.7.3 LOGARITHMIC OR NATURAL STRAIN

Another definition of extensional strain has been used to describe severe deformations. It is the so-called *logarithmic* or *natural* strain ε_n^l. An increment, or differential, of the logarithmic strain ($\delta\varepsilon_n^l$) is defined as the increment of current length $\delta(\Delta s')$ divided by the current length $\Delta s'$:

$$\delta\varepsilon_n^l \equiv \lim_{\Delta s' \to 0} \frac{\delta(\Delta s')}{\Delta s'}$$

The logarithmic strain is obtained by integrating the differential from the initial state ($\varepsilon_n^l = 0$, $\Delta s' = \Delta s$) to the final state (ε_n^l, $\Delta s' = \Delta s^*$):

$$\varepsilon_n^l = \ln(ds^*) - \ln(ds) = \ln \frac{ds^*}{ds} \tag{2.13}$$

If ε_n denotes the conventional strain defined by equation (2.1), then

$$\varepsilon_n^l = \ln(1 + \varepsilon_n) \tag{2.14}$$

Again, in structural applications, the differences are inconsequential. For example, if $\varepsilon_n = 0.01$, the difference is $\varepsilon_n^l - \varepsilon_n = -0.00005$.

2.8 STATE OF STRAIN

The deformation of a body can be completely described in terms of three extensional strains and three shear strains. To understand how this might be accomplished, imagine cutting the original body into a large number of very small blocks (see Figure 2.11). Any body can be subdivided into small blocks in numerous ways. The manner in which the body is subdivided determines the size and orientation of each of the small blocks. For example, the body could be subdivided into blocks whose edges are parallel to the rectangular coordinate axes (x, y, z). A few such blocks, as well as an enlarged typical block, are shown for the body in Figure 2.11. If these blocks are very

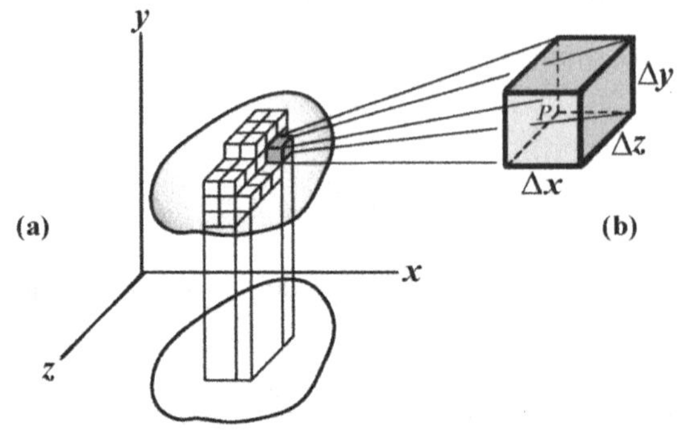

Figure 2.11 Discrete Elements

small, their deformed counterparts are approximately parallelepipeds. This result follows from the observation that very short line segments remain approximately straight and very small plane surfaces remain approximately plane after deformation.Thus, a small rectangular block becomes a parallelepiped, as illustrated in Figure 2.12.

Consider the process by which the rectangular block of Figure 2.12a can become the parallelepiped shown in Figure 2.12d

1. The sides *PB*, *PD*, and *PH* of the block in Figure 2.12a can be stretched until their lengths are equal to those of $P'B'$, $P'D'$, and $P'H'$, respectively, as in Figure 2.12b (three extensional strains).
2. The sides of the new block in Figure 2.12b can be skewed until the angles between the edges agree with those of the final state, as shown in Figure 2.12c (three shear strains).
3. The parallelepiped can be displaced as a rigid body until it is coincident with the final position of Figure 2.12d.

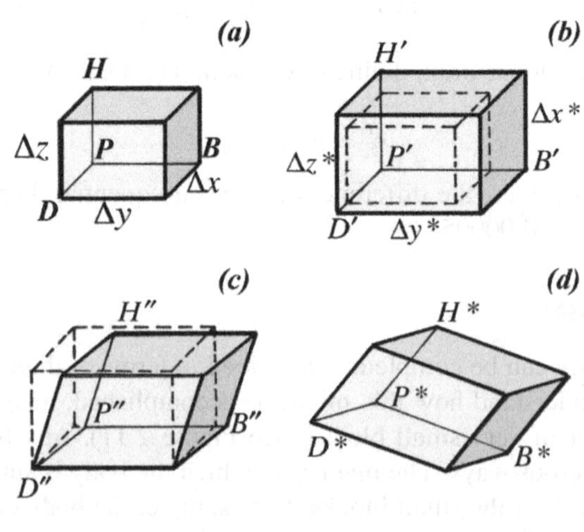

Figure 2.12 Extension, Shear and Displacement

This sequence of events can be classified as follows:

1. A change in size (extension or contraction) (a) to (b) (deformation)
2. A change in shape (shear or distortion) (b) to (c) (deformation)
3. A rigid-body translation and rotation (c) to (d) (rigid motion)

In any actual deformation, all these events would occur simultaneously. This breakdown is given only as an aid in understanding the process.

The deformation of the rectangular block is characterized by the changes in length of its edges and the changes in angle between the edges. Thus, the method we have used to describe the change in length of a line segment can be applied in turn to the edges of the block to determine their final lengths. If at P we know the extensional strains for the x, y, and z directions, then the change in size of the block is determined by these extensional strains. According to the notation introduced in Section 2.4, these normal strains are written as $\varepsilon_x(P)$, $\varepsilon_y(P)$, and $\varepsilon_z(P)$. The original block has dimensions Δx, Δy, and Δz. Hence, the new lengths of the sides in the deformed state are approximately

$$(1+\varepsilon_x)\Delta x, \qquad (1+\varepsilon_y)\Delta y, \qquad (1+\varepsilon_z)\Delta z$$

To describe the distortion or change in shape, we can introduce the shear strains between the pairs of directions (x and y), (y and z), and (z and x). The new angles between these edges are

$$\varphi^*_{xy} = \frac{\pi}{2} - \gamma_{xy}(P) \qquad \varphi^*_{yz} = \frac{\pi}{2} - \gamma_{yz}(P) \qquad \varphi^*_{zx} = \frac{\pi}{2} - \gamma_{zx}(P)$$

The deformation of the block is now completely determined because we have the lengths of the parallelepiped and the angles between the edges. If the six strains,

$$\varepsilon_x, \quad \varepsilon_y, \quad \varepsilon_z, \quad \gamma_{xy}, \quad \gamma_{yz}, \quad \gamma_{zx} \tag{2.15}$$

are known at every point throughout the body, then, by imagining the body to be composed of very small rectangular blocks with edges parallel to the x, y, and z coordinate axes, we can determine the shape and size of the deformed blocks. Because the change in size and shape of a body is the net result of all changes that occur in its elementary portions, the deformed state of the entire body is defined.[4]

There are other ways of subdividing a body into small blocks. The manner of subdivision is dictated by the coordinate system used. One additional example will serve to illustrate this. Suppose that cylindrical coordinates are used instead of Cartesian coordinates. If the body is subdivided into small blocks whose edges are along the coordinate lines, then, as we move throughout the body, these small blocks change their orientation. A few typical blocks and the orientation of these blocks are shown in Figure 2.13. A typical block is located by the point P, and its orientation is determined by the three edges that intersect at this point. In this case, one of the edges is in the direction of the radius, which is denoted by r. A second edge is in a direction denoted by θ; it is the direction of increasing θ. The third edge, perpendicular to the other two, is in the direction of the z axis. The changes in lengths of these edges can be characterized by the extensional strains associated with these directions. Thus, the extensional strains are $\varepsilon_r(P)$, $\varepsilon_\theta(P)$, and $\varepsilon_z(P)$. Although the small blocks obtained in this manner are not exactly rectangular, they are approximately so, and, as the subdivision becomes finer, the approximation gets better. We may take the lengths of the edges of the original block as Δr, $r\Delta\theta$, and Δz. The lengths of these edges in the deformed block are

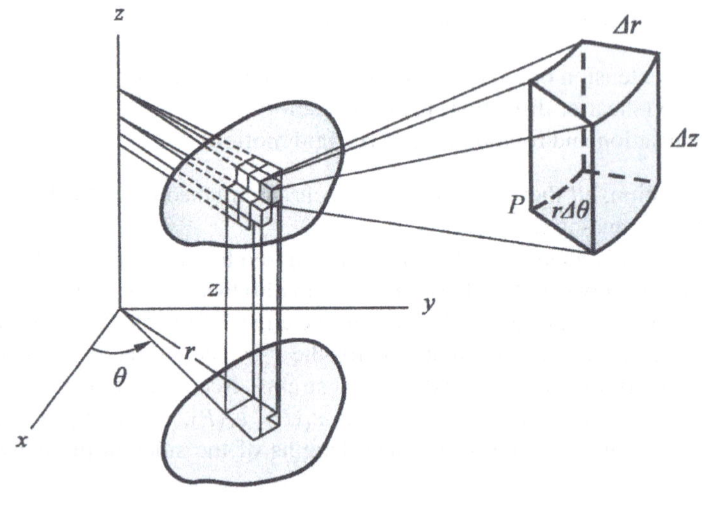

Figure 2.13 Element in Cylindrical Coordinates

$$(1+\varepsilon_r)\Delta r, \qquad (1+\varepsilon_\theta)r\Delta\theta, \qquad (1+\varepsilon_z)\Delta z$$

The new angles between these edges are

$$\phi^*_{r\theta} = \frac{\pi}{2} - \gamma_{r\theta}(P), \qquad \phi^*_{\theta z} = \frac{\pi}{2} - \gamma_{\theta z}(P), \qquad \phi^*_{zr} = \frac{\pi}{2} - \gamma_{zr}(P)$$

In this case, the six strains,

$$\varepsilon_r, \quad \varepsilon_\theta, \quad \varepsilon_z, \quad \gamma_{r\theta}, \quad \gamma_{\theta z}, \quad \gamma_{zr} \tag{2.16}$$

given throughout the body will determine the change in size and shape of every small block and, therefore, of the entire body.

Clearly, either set of strains, Equations (2.15) or (2.16), can be utilized to describe the deformation of a given body. In fact, if three extensional strains and three shear strains associated with three perpendicular directions are known throughout the body, the deformation of the body is completely described. The set of directions used to specify the strain is a matter of convenience and will vary from problem to problem. The coordinate lines need not be orthogonal; however, the description and mathematical formulations are more complicated.

Any set of six strains that describes the deformation in the neighborhood of a particle is said to determine the state of strain at the point.

2.9 SEVERITY OF DEFORMATION

Bodies subjected to forces or changes in temperature exhibit deformations of varying degree. A rubber band can be stretched several times its original length. Thin flat plates can be rolled into pipe. Boilers under internal pressure at elevated temperatures experience small changes in diameter and length. A rafter in a building bends slightly while supporting the roof above it. Circular shafts twist, but only slightly, when used to transmit torque.

In a great variety of situations, the deformation of a body is not too severe. Such small deformations are characterized by placing certain limits on the order of magnitude of strains. A strain, either extensional or shear, is considered small if it is small *compared to unity*– that is, $\varepsilon_n \ll 1$, $\gamma_{nt} \ll 1$. Most theories of deformable bodies, structures, and machines are founded on the assumption of small strains. This does not necessarily preclude large rotations and displacements.

Long, thin members, such as rods and thin sheets, can be rolled into a cylindrical form with very little stretching. A long, thin shaft can experience a large relative rotation of its opposite ends, even though the strains remain small. In general, thin bodies can experience large rotations and large deflections, though the strains are small throughout. On the other hand, large strains are always accompanied by large rotations.

To illustrate a circumstance of small strain accompanied by a large rotation, consider a thin straight strip of steel, such as a conventional hacksaw blade, as shown in Figure 2.14. The strip is clamped at one end; a transverse load is applied to the other. The load can cause a very large displacement of the loaded end, and the rotation might easily exceed 45°, yet the relative rotation of any two neighboring lines remains small. Hence, the shear strains and also the extensional strains are small throughout (so small that no permanent deformations result).

To appreciate the occurrence of large rotations, which usually accompany large strains, we need only consider the definition of a shear strain, which can be large only if a large rotation occurs (see Figure 2.10).

Most problems in this book involve only small strains. A few applications admit large rotations and, indeed, a later study of thin beams (Section 6.9) requires that we specifically include the relative rotations in our analysis.

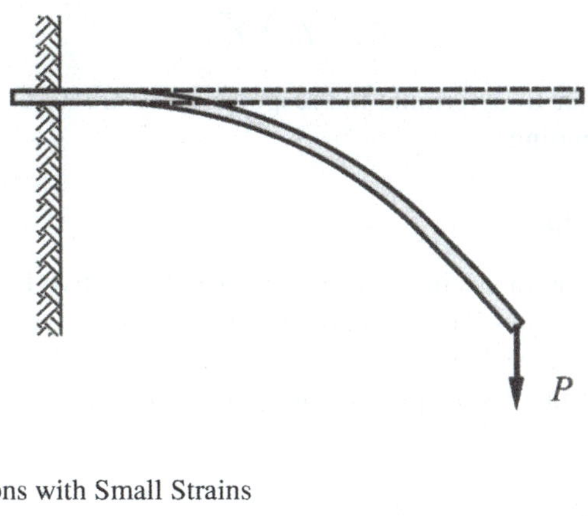

Figure 2.14 Large Rotations with Small Strains

2.10 SUPERPOSITION OF SMALL STRAINS

An important consequence of the assumption of small strain is that successive strains are additive. For example, suppose a short line segment Δx is subjected to an extensional strain ε_x' so that its length becomes $\Delta x^* = (1+\varepsilon_x')\,\Delta x$. If the element is then extended again, so that the additional strain is ε_x'', its new length is

$$\Delta x^{**} = (1+\varepsilon_x'')\Delta x^* = (1+\varepsilon_x'')(1+\varepsilon_x')\Delta x$$

If the strains are small compared to unity, only first-order terms are significant, so

$$\Delta x^{**} \doteq (1+\varepsilon_x' + \varepsilon_x'')\Delta x$$

Then, the result of these successive strains is the strain

$$\varepsilon_x = \lim_{\Delta x \to 0} \frac{\Delta x^{**} - \Delta x}{\Delta x} \doteq \varepsilon_x' + \varepsilon_x''$$

We therefore say that the successive strains may be superimposed. Similarly, the result of two successive shear strains γ_{xy}' and γ_{xy}''. which are small, is the net shear strain

$$\gamma_{xy} \doteq \gamma_{xy}' + \gamma_{xy}''$$

In general, suppose that

$$\varepsilon_x', \quad \varepsilon_y', \quad \varepsilon_z', \quad \gamma_{xy}', \quad \gamma_{yz}', \quad \gamma_{zx}'$$

describe a state of small strain throughout a body. If, in addition to this state, we impose a second state of small strain,

$$\varepsilon_x'', \quad \varepsilon_y'', \quad \varepsilon_z'', \quad \gamma_{xy}'', \quad \gamma_{yz}'', \quad \gamma_{zx}''$$

then, to first-order terms, the net effect is a state of strain given by

$$\begin{aligned} \varepsilon_x &= \varepsilon_x' + \varepsilon_x'', & \varepsilon_y &= \varepsilon_y' + \varepsilon_y'', & \varepsilon_z &= \varepsilon_z' + \varepsilon_z'' \\ \gamma_{xy} &= \gamma_{xy}' + \gamma_{xy}'', & \gamma_{yz} &= \gamma_{yz}' + \gamma_{yz}'', & \gamma_{zx} &= \gamma_{zx}' + \gamma_{zx}'' \end{aligned}$$

This result will be used later to determine the response of a structure subjected to simultaneous stretching, twisting, and bending.

ILLUSTRATIVE EXAMPLE 2.8

To illustrate the superposition of strain, let us reconsider Illustrative Example 2.3 of Section 2.5, wherein the displacement of point P has components u and v. We have seen that the linear approximation introduces a small error when $u = v = 0.00100L$. Suppose that the components u and then v were imposed in order. In the first displacement u, the strains are

$$\varepsilon_B' = \left[\left(\frac{4}{5} + \frac{u}{L}\right)^2 + \left(\frac{3}{5}\right)^2\right]^{1/2} - 1 = 0.00080$$

$$\varepsilon_C' = \frac{5}{3}\left[\left(\frac{u}{L}\right)^2 + \left(\frac{3}{5}\right)^2\right]^{1/2} - 1 \quad = 0.00000\ (0.000001)$$

In the second displacement v, the strains are

$$\varepsilon_B'' = \frac{\left[\left(\frac{4}{5} + \frac{u}{L}\right)^2 + \left(\frac{3}{5} + \frac{v}{L}\right)^2\right]^{1/2}}{\left[\left(\frac{4}{5} + \frac{u}{L}\right)^2 + \left(\frac{3}{5}\right)^2\right]^{1/2}} - 1 = 0.00060$$

$$\varepsilon_C'' = \frac{\left[\left(\frac{u}{L}\right)^2 + \left(\frac{3}{5} + \frac{v}{L}\right)^2\right]^{1/2}}{\left[\left(\frac{u}{L}\right)^2 + \left(\frac{3}{5}\right)^2\right]^{1/2}} - 1 = 0.00167$$

If we add these successive strains, we obtain

$$\varepsilon_B' + \varepsilon_B'' = 0.00140, \qquad \varepsilon_C' + \varepsilon_C'' = 0.00167$$

Of course, the latter is erroneous, but the errors are exceedingly small.

The linear approximations of the successive strains are

$$\varepsilon_B' \doteq \frac{4}{5}\frac{u}{L}, \qquad \varepsilon_C' \doteq 0$$

and

$$\varepsilon_B'' \doteq \frac{3}{5}\frac{v}{L}, \qquad \varepsilon_C' \doteq \frac{5}{3}\frac{v}{L}$$

The sums are the same as the previous result. In the context of first-order (linear) approximations, the successive strains are additive.

ILLUSTRATIVE EXAMPLE 2.9

Three square plates [each (25 cm × 25 cm) (9.84 in. × 9.84 in.)] are *uniformly* deformed as shown in Figure 2.15. Displacements (in centimeters, $d = .254$ cm) of the corners in the directions x and y, respectively, are indicated. Note that the displacements for plate c are the sum of the displacements for plates a and b.

Let us compute the strains ε_x^a, ε_x^b, ε_x^c , γ_{xy}^a, γ_{xy}^b and γ_{xy}^c, and then compare $\varepsilon_x^a + \varepsilon_x^b$ with ε_x^c and $\gamma_{xy}^a + \gamma_{xy}^b$ with γ_{xy}^c. For the extensional strains,

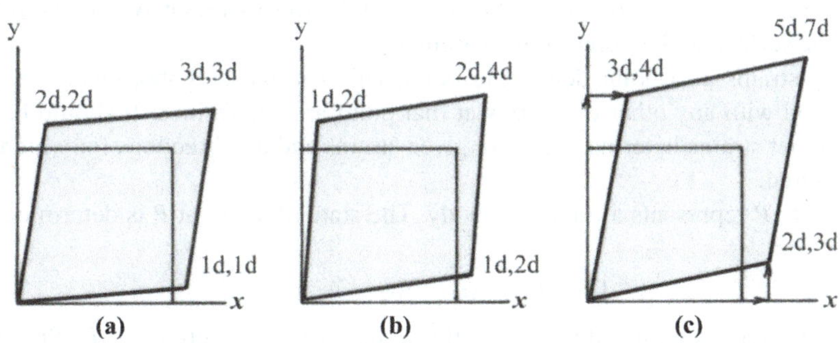

Figure 2.15 Superposition of Successive Strains

$$\varepsilon_x^a = \frac{[(25.00+0.254)^2+(0.254)^2]^{1/2}}{25.00} - 1 = 0.0102$$
$$\varepsilon_x^b = \frac{[(25.00+0.254)^2+(0.508)^2]^{1/2}}{25.00} - 1 = 0.0104$$
$$\varepsilon_x^c = \frac{[(25.00+0.508)^2+(0.762)^2]^{1/2}}{25.00} - 1 = 0.0208$$

Note that

$$\varepsilon_x^a + \varepsilon_x^b = 0.0206, \quad 1\% \text{ less than } \varepsilon_x^c!$$

For the shear strains,

$$\gamma_{xy}^a = \arctan\left(\frac{0.254}{25.00+0.254}\right) + \arctan\left(\frac{0.508}{25.00+0.508}\right) = 0.0300$$
$$\gamma_{xy}^b = \arctan\left(\frac{0.508}{25.00+0.254}\right) + \arctan\left(\frac{0.254}{25.00+0.508}\right) = 0.0301$$
$$\gamma_{xy}^c = \arctan\left(\frac{0.762}{25.00+0.508}\right) + \arctan\left(\frac{0.762}{25.00+1.016}\right) = 0.0591$$

Note that

$$\gamma_{xy}^a + \gamma_{xy}^b = 0.0601, \quad 1.7\% \text{ greater than } \gamma_{xy}^c!$$

Note too that these strains are much larger than those typically encountered in structural and mechanical components.

2.11 COMPLETENESS OF THE STATE OF STRAIN

As indicated in Section 2.8, the deformation of a body is completely described by a knowledge of six strains at each point throughout the body; for example, three extensional strains and three shear strains. These strains are commonly associated with three mutually perpendicular directions at each point. Occasionally, it is necessary to know an extensional strain or a shear strain at a point for directions other than those used to define the state of strain at the point. For example, in experimental work, extensional strains are measured in several directions at a point on the surface of a body. From these, the investigator is able to obtain the shear strain between two perpendicular directions in the surface at the point of measurement.

Because six strains at a point define the deformation in the neighborhood of the point, the strains associated with any other directions at that point are determined. It should be possible to express these other strains in terms of the six given strains and the directions for which these other strains are required.

In Figure 2.16, P represents a point in a body. The state of strain at P is determined by the six quantities

$$\varepsilon_x, \quad \varepsilon_y, \quad \varepsilon_z, \quad \gamma_{xy}, \quad \gamma_{yz}, \quad \gamma_{zx}$$

where x, y, and z represent any three mutually perpendicular directions at P. Suppose that we require the extensional strain at P in a direction n that is specified by direction angles α, β, γ or equivalently, by direction cosines l_n, m_n, n_n.

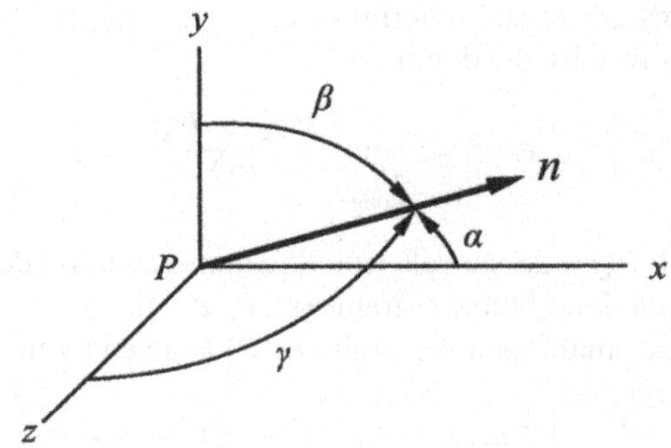

Figure 2.16

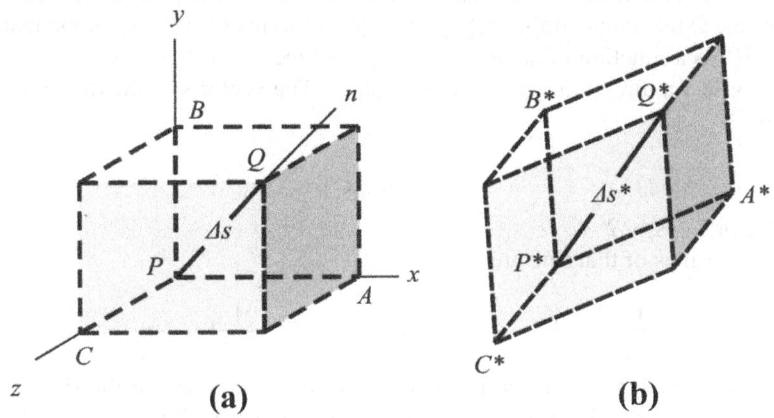

Figure 2.17

To obtain the required strain $\varepsilon_n(P)$, we might proceed as follows: We can select a short segment PQ of length Δs in the direction n, as shown in Figure 2.17a. The segment PQ forms the diagonal of the rectangular block with edges PA, PB, and PC in the directions x, y, and z, respectively. The lengths of the edges are given by the direction cosines:

$$PA = l_n \Delta s, \qquad PB = m_n \Delta s, \qquad PC = n_n \Delta s$$

When the body deforms, this block becomes a parallelepiped, as shown in Figure 2.17b Segment PQ becomes P^*Q^*, a diagonal of the parallelepiped. The lengths of the sides of the parallelepiped are

$$P^*A^* = l_n(1+\varepsilon_x)\Delta s, \qquad P^*B^* = m_n(1+\varepsilon_y)\Delta s, \qquad P^*C^* = n_n(1+\varepsilon_z)\Delta s$$

Also, the angles between adjacent edges are

$$\angle A^*P^*B^* = \frac{\pi}{2} - \gamma_{xy}, \qquad \angle A^*P^*C^* = \frac{\pi}{2} - \gamma_{xz}, \qquad \angle B^*P^*C^* = \frac{\pi}{2} - \gamma_{yz}$$

The parallelepiped is completely determined by the lengths of its edges and the angles between the edges. In particular, the length of the diagonal P^*Q^* can be expressed in terms of these six quantities, which are already expressed in terms of ε_x, ε_y, ε_z , γ_{xy}, γ_{yz}, γ_{zx}, l_n, m_n, n_n, and Δs. The extensional strain $\varepsilon_n(P)$ is given by the definition

$$\varepsilon_n(P) = \lim_{\substack{Q \to P \\ \text{along } n}} \frac{P^*Q^* - PQ}{PQ} \tag{2.17}$$

Note here that the length $PQ = \Delta s$ cancels from the numerator and denominator. The result, an expression for ε_n (P), is a function of the six strains (ε_x, ε_y, ε_z , γ_{xy}, γ_{yz}, γ_{zx}) and the direction cosines (l_n, m_n, n_n). If the strains are small, then the strain, $\varepsilon_n(P)$ is given by the linear approximation[5]

$$\varepsilon_n(P) \doteq l_n^2\varepsilon_x + l_n m_n \gamma_{xy} + l_n n_n \gamma_{xz} + m_n^2 \varepsilon_y + m_n n_n \gamma_{yz} + n_n^2 \varepsilon_z \tag{2.18}$$

Now, suppose that we require the shear strain for two directions n and t, which are initially perpendicular at P in Figure 2.18. The directions can be specified in various ways, but, in any case, we know the direction cosines l_n, m_n, n_n, and l_t, m_t, n_t. By the preceding arguments, we can always express the extensional strains $\varepsilon_n(P)$ and $\varepsilon_t(P)$ in terms of the six components (ε_x, ε_y, ε_z , γ_{xy}, γ_{yz}, γ_{zx}); $\varepsilon_n(P)$ is a function of l_n, m_n, n_n; $\varepsilon_t(P)$ is a function of l_t, m_t, n_t.

We can now select a direction s that bisects n and t. The vector **s** in the direction of s is given by the relation:

$$\mathbf{s} = \hat{\mathbf{n}} + \hat{\mathbf{t}}$$

and the unit vector $\hat{\mathbf{s}} = \mathbf{s}/\sqrt{2}$

The direction cosines of that line are

$$\frac{1}{\sqrt{2}}(l_n + l_t), \qquad \frac{1}{\sqrt{2}}(m_n + m_t), \qquad \frac{1}{\sqrt{2}}(n_n + n_t)$$

Again, the extensional strain of that line can be expressed in terms of the six components (ε_x, ε_y, ε_z , γ_{xy}, γ_{yz}, γ_{zx}) and the direction cosines (given previously). Now, consider the square with edges *PA* and *PB* along n and t and diagonal *PC* along s, as depicted in Figure 2.19a. This deforms to a parallelogram with edges P^*A^* and P^*B^* and diagonal P^*C^*. In Figure 2.19, we view *planar* views of both the original square *PACB* and the deformed parallelogram $P^*A^*C^*B^*$. If the original length of the diagonal is Δs, then the deformed lengths follow:

$$P^*C^* = (1+\varepsilon_s)\Delta s, \qquad P^*A^* = \frac{(1+\varepsilon_n)\Delta s}{\sqrt{2}}, \qquad P^*B^* = \frac{(1+\varepsilon_t)\Delta s}{\sqrt{2}}$$

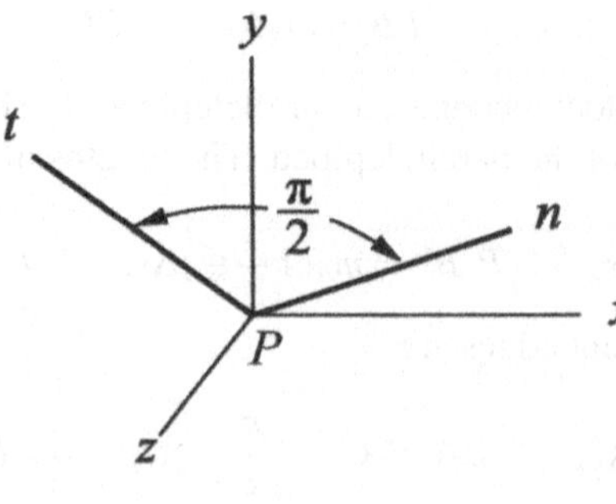

Figure 2.18

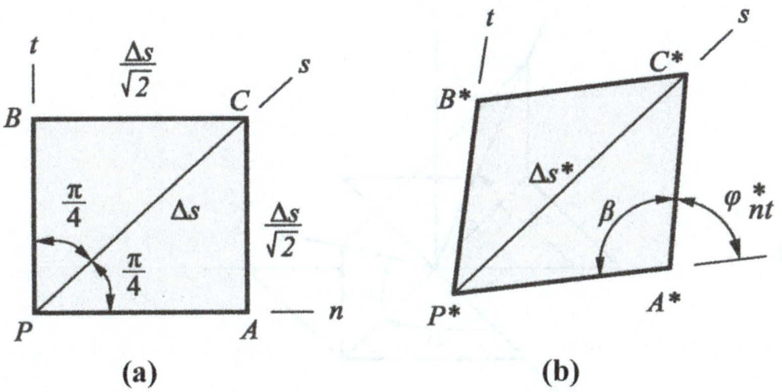

Figure 2.19

The length of each segment is known in terms of ε_n, ε_s , ε_t, and Δs, which, in turn, can be expressed in terms ε_x, ε_y, ε_z , γ_{xy}, γ_{yz}, γ_{zx}, l_n, m_n, n_n, l_t, m_t, n_t and Δs, as previously described. It is a geometric exercise (by the Cosine Law) to express the angle β in terms of these quantities; length Δs cancels. Note that $\gamma_{nt} = \beta - \pi/2$. It is thereby established that the shear strain for any directions (n,t) can be expressed in terms of the six given components, ε_x, ε_y, ε_z , γ_{xy}, γ_{yz}, γ_{zx}, and the direction cosines of those lines. If the strains are small, then the strain γ_{nt} is given by the approximation[6]

$$\begin{aligned}\frac{1}{2}\gamma_{nt}(P) \doteq\ & l_n l_t \varepsilon_x + \frac{1}{2} l_n m_t \gamma_{xy} + \frac{1}{2} l_n n_t \gamma_{xz} \\ & + \frac{1}{2} m_n l_t \gamma_{xy} + m_n m_t \varepsilon_y + \frac{1}{2} m_n n_t \gamma_{yz} \\ & + \frac{1}{2} n_n l_t \gamma_{xz} + \frac{1}{2} n_n m_t \gamma_{yz} + n_n n_t \varepsilon_z\end{aligned} \tag{2.19}$$

The reader will note the remarkable similarities between the linear approximations in (2.18) and (2.19). Both are quadratic forms in the direction cosines; of course, ε_n depends only on the direction n, whereas γ_{nt} depends on both directions n and t. The reader should note that both equations apply only to small strains, as defined by Equations (2.1) and (2.6); however, the displacements and rotations may be of arbitrary magnitude. The derivations of Equations (2.18) and (2.19) are given in the following section.

ILLUSTRATIVE EXAMPLE 2.10

To illustrate the transformation of strains, consider the specific orthogonal directions n and t shown in Figure 2.20. Axes n, t, and y lie in a common plane, which is inclined 45° to the x and z axes. From the geometry

$$l_n = n_n = \frac{1}{2}, \quad m_n = \frac{1}{\sqrt{2}}, \quad l_t = n_t = -\frac{1}{2}, \quad m_t = \frac{1}{\sqrt{2}}$$

the reader can verify the orthogonality—that is,

$$l_n l_t + m_n m_t + n_n n_t = 0$$

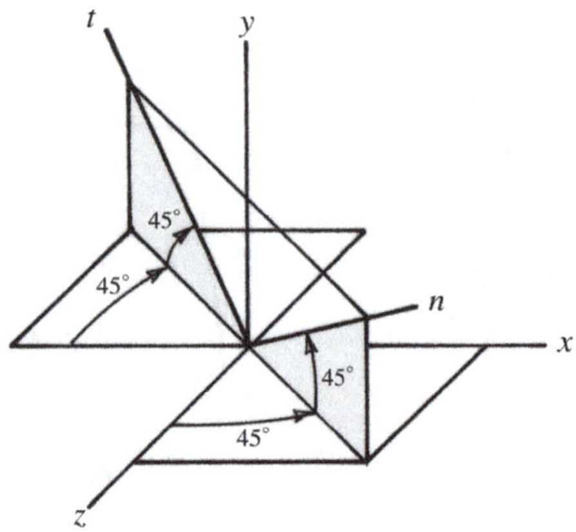

Figure 2.20

Now, suppose that the state of strain at the point is given by the six components ($\mu = 10^{-6}$)

$$\begin{pmatrix} \varepsilon_x & \gamma_{xy} & \gamma_{xz} \\ & \varepsilon_y & \gamma_{yz} \\ & & \varepsilon_z \end{pmatrix} = \begin{pmatrix} 100 & 60 & -80 \\ & -60 & 100 \\ & & 80 \end{pmatrix} \mu$$

By equation (2.18)

$$\begin{aligned} \varepsilon_n = & \left[\left(\frac{1}{2}\right)\left(\frac{1}{2}\right)(100) + \left(\frac{1}{2}\right)\left(\frac{1}{\sqrt{2}}\right)(60) + \left(\frac{1}{2}\right)\left(\frac{1}{2}\right)(-80) \right. \\ & \left. + \left(\frac{1}{\sqrt{2}}\right)\left(\frac{1}{\sqrt{2}}\right)(-60) + \left(\frac{1}{\sqrt{2}}\right)\left(\frac{1}{2}\right)(100) + \left(\frac{1}{2}\right)\left(\frac{1}{2}\right)(80) \right] \mu \\ = & \, 51.6\mu \end{aligned}$$

By equation (2.19)

$$\begin{aligned} \gamma_{nt} = & \left[(2)\left(\frac{1}{2}\right)\left(-\frac{1}{2}\right)(100) + \left(\frac{1}{2}\right)\left(\frac{1}{\sqrt{2}}\right)(60) + \left(\frac{1}{2}\right)\left(-\frac{1}{2}\right)(-80) \right. \\ & + \left(-\frac{1}{2}\right)\left(\frac{1}{\sqrt{2}}\right)(60) + (2)\left(\frac{1}{\sqrt{2}}\right)\left(\frac{1}{\sqrt{2}}\right)(-60) + \left(\frac{1}{\sqrt{2}}\right)\left(-\frac{1}{2}\right)(100) \\ & \left. + \left(\frac{1}{2}\right)\left(-\frac{1}{2}\right)(-80) + \left(\frac{1}{2}\right)\left(\frac{1}{\sqrt{2}}\right)(100) + (2)\left(\frac{1}{2}\right)\left(-\frac{1}{2}\right)(-80) \right] \mu \\ = & -110\mu \end{aligned}$$

2.12 TRANSFORMATION OF STRAIN COMPONENTS

In the preceding section, we argued that six components completely describe the state of strain at a point. Again, this means that the six components determine any other component at the point, an extensional strain of any line, or the shear strain between any two lines. Here, we derive the equations for that transformation.[7]

A particle of an undeformed body is at point P in Figure 2.21 and is located by the vector $\mathbf{r}$. Small changes of position accompany a change in each of the coordinates; for example, the change Δx carries the vector from P to A along the x line as $\mathbf{r}$ undergoes the change

$$\Delta \mathbf{r_x} = \frac{\partial \mathbf{r}}{\partial x}\Delta x = \hat{\mathbf{i}}\Delta x \tag{2.20}$$

Here, $\hat{\mathbf{i}}, \hat{\mathbf{j}}, \hat{\mathbf{k}}$ denote unit vectors in the directions of the rectangular lines x, y, and z, respectively. An arbitrary change in position carries the vector from P to Q; the change of position is given by the vector

$$\begin{aligned}\Delta \mathbf{r} &= \frac{\partial \mathbf{r}}{\partial x}\Delta x + \frac{\partial \mathbf{r}}{\partial y}\Delta y + \frac{\partial \mathbf{r}}{\partial z}\Delta z\\ &= \hat{\mathbf{i}}\Delta x + \hat{\mathbf{j}}\Delta y + \hat{\mathbf{k}}\Delta z\end{aligned}$$

Let Δn denote the length of this initial line segment PQ; $\Delta n = |\Delta \mathbf{r}|$. The notation n signifies the initial orientation of the line in the direction of the unit vector $\hat{\mathbf{n}}$:

$$\begin{aligned}\hat{\mathbf{n}} = \lim_{\Delta x \to 0}\frac{\Delta \mathbf{r}}{\Delta n} = \frac{\partial \mathbf{r}}{\partial n} &= \frac{\partial \mathbf{r}}{\partial x}\frac{\partial x}{\partial n} + \frac{\partial \mathbf{r}}{\partial y}\frac{\partial y}{\partial n} + \frac{\partial \mathbf{r}}{\partial z}\frac{\partial z}{\partial n}\\ &= \hat{\mathbf{i}}\frac{\partial x}{\partial n} + \hat{\mathbf{j}}\frac{\partial y}{\partial n} + \hat{\mathbf{k}}\frac{\partial z}{\partial n}\end{aligned}$$

where

$$\frac{\partial x}{\partial n}, \frac{\partial y}{\partial n}, \frac{\partial z}{\partial n} \equiv \lim_{\Delta n \to 0}\frac{\Delta x}{\Delta n}, \frac{\Delta y}{\Delta n}, \frac{\Delta z}{\Delta n}$$

These are the direction cosines (l_n, m_n, n_n):

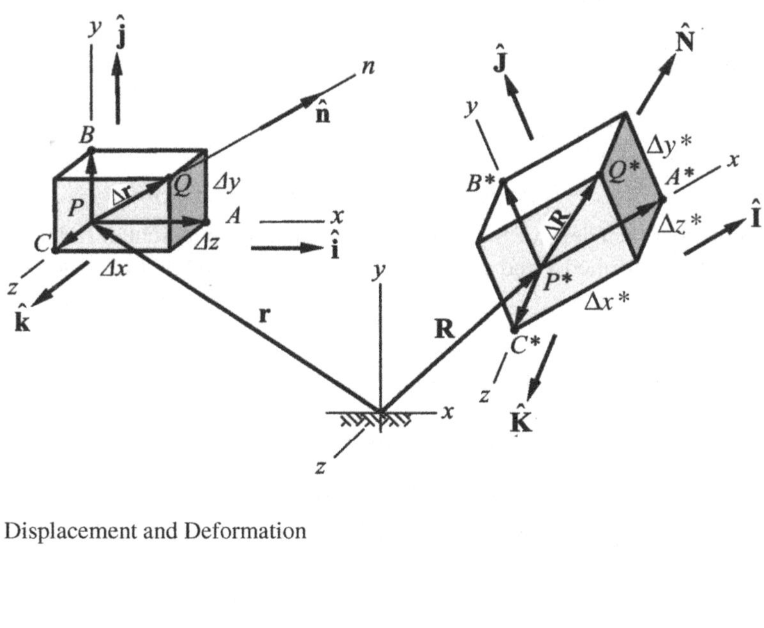

Figure 2.21 Displacement and Deformation

$$\hat{\mathbf{n}} = \frac{\partial \mathbf{r}}{\partial n} = l_n \hat{\mathbf{i}} + m_n \hat{\mathbf{j}} + n_n \hat{\mathbf{k}} \tag{2.21}$$

Upon deformation, the particle at P is transported to P^*, located now by the new vector $\mathbf{R}$. Particles at A, B, and C are moved to new locations A^*,B^*, and C^*. A very important feature is the deformation, which has *extended (or contracted)* each line and *sheared* the angles between the lines. The particle at A^* is now located relative to P^* by the vector

$$\Delta \mathbf{R}_x = \frac{\partial \mathbf{R}}{\partial x} \Delta x$$

Unlike equation (2.20), the vector $\partial \mathbf{R}/\partial x$ is *not* a unit vector. Here, the change $\Delta \mathbf{R_x}$ that accompanies the change Δx has the extended length P^*A^*. Indeed, by our earlier definition of an extensional strain,

$$\lim_{\Delta x \to 0} \left(\frac{P^*A^*}{PA} - 1 \right) \equiv \varepsilon_x$$

Stated otherwise, if $\hat{\mathbf{I}}$ is the unit vector in the new direction of the deformed x line, then

$$\lim_{\Delta x \to 0} \frac{\Delta \mathbf{R}_x}{\Delta x} \equiv \frac{\partial \mathbf{R}}{\partial x} = (1 + \varepsilon_x)\hat{\mathbf{I}}$$

Likewise,

$$\frac{\partial \mathbf{R}}{\partial y} = (1 + \varepsilon_y)\hat{\mathbf{J}}, \qquad \frac{\partial \mathbf{R}}{\partial z} = (1 + \varepsilon_z)\hat{\mathbf{K}}$$

where $\hat{\mathbf{I}}$, $\hat{\mathbf{J}}$ and $\hat{\mathbf{K}}$ are unit vectors along the displaced lines x, y, and z. Then, for the line P^*Q^*

$$\begin{aligned} \lim_{\Delta n \to 0} \frac{\Delta \mathbf{R}}{\Delta n} = \frac{\partial \mathbf{R}}{\partial n} &= \frac{\partial \mathbf{R}}{\partial x}\frac{\partial x}{\partial n} + \frac{\partial \mathbf{R}}{\partial y}\frac{\partial y}{\partial n} + \frac{\partial \mathbf{R}}{\partial z}\frac{\partial z}{\partial n} \\ \frac{\partial \mathbf{R}}{\partial n} &= l_n(1 + \varepsilon_x)\hat{\mathbf{I}} + m_n(1 + \varepsilon_y)\hat{\mathbf{J}} + n_n(1 + \varepsilon_z)\hat{\mathbf{K}} \end{aligned} \tag{2.22}$$

Just as the line PA is extended to P^*A^* by extensional strain ε_x, so the line PQ is extended to P^*Q^* by the extensional strain ε_n. If $\hat{\mathbf{N}}$ is the unit vector in the direction P^*Q^*, then

$$\begin{aligned} \frac{\partial \mathbf{R}}{\partial n} &= (1 + \varepsilon_n)\hat{\mathbf{N}} \\ &= l_n(1 + \varepsilon_x)\hat{\mathbf{I}} + m_n(1 + \varepsilon_y)\hat{\mathbf{J}} + n_n(1 + \varepsilon_z)\hat{\mathbf{K}} \end{aligned} \tag{2.23}$$

Now, we must note that the edges of the block are no longer perpendicular. Indeed, if φ_{xy} denotes the angle between the deformed x and y lines-that is, between P^*A^* and P^*B^*–then

$$\cos \varphi_{xy} = \hat{\mathbf{I}} \cdot \hat{\mathbf{J}}$$

By our earlier definition of the shear strain $\gamma_{xy}, \gamma_{yz}, \gamma_{zx}$,

$$\sin \gamma_{xy} = \cos \varphi_{xy} = \hat{\mathbf{I}} \cdot \hat{\mathbf{J}} \tag{2.24a}$$

$$\sin \gamma_{xz} = \hat{\mathbf{I}} \cdot \hat{\mathbf{K}} \quad \sin \gamma_{yz} = \hat{\mathbf{J}} \cdot \hat{\mathbf{K}} \tag{2.24b,c}$$

The extensional strain ε_n can now be evaluated by means of Equations (2.23) and (2.24a–c):

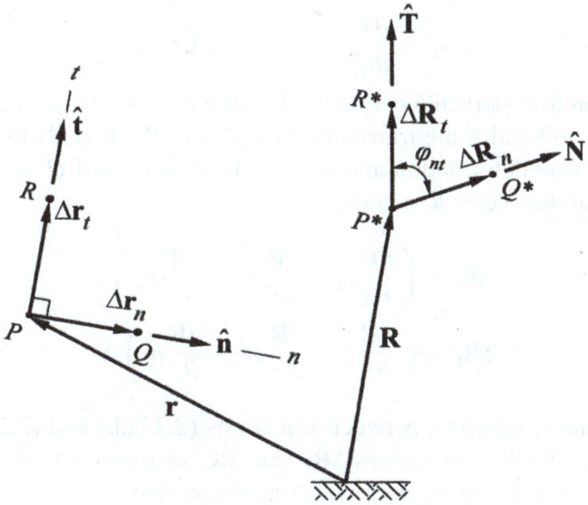

Figure 2.22

$$\begin{aligned}
\frac{\partial \mathbf{R}}{\partial n} \cdot \frac{\partial \mathbf{R}}{\partial n} &= (1+\varepsilon_n)^2 \\
&= l_n(1+\varepsilon_x)[l_n(1+\varepsilon_x) + m_n(1+\varepsilon_y)\sin\gamma_{xy} + n_n(1+\varepsilon_z)\sin\gamma_{xz}] \\
&\quad + m_n(1+\varepsilon_y)[l_n(1+\varepsilon_x)\sin\gamma_{xy} + m_n(1+\varepsilon_y) + n_n(1+\varepsilon_z)\sin\gamma_{yz}] \\
&\quad + n_n(1+\varepsilon_z)[l_n(1+\varepsilon_x)\sin\gamma_{xz} + m_n(1+\varepsilon_y)\sin\gamma_{yz} + n_n(1+\varepsilon_z)]
\end{aligned} \tag{2.25}$$

The result seems messy, but it does give the extensional strain ε_n *exactly* in terms of the six given components (ε_x, ε_y, ε_z , γ_{xy}, γ_{yz}, γ_{zx})

Before we undertake further examination of equation (2.25), let us derive the corresponding equation for the shear strain γ_{nt} in the arbitrary orthogonal directions n and t at point P of Figure 2.22. The unit vectors $\hat{\mathbf{n}}$ and $\hat{\mathbf{i}}$ are given in the forms

$$\hat{\mathbf{n}} = l_n\hat{\mathbf{i}} + m_n\hat{\mathbf{j}} + n_n\hat{\mathbf{k}}, \qquad \hat{\mathbf{t}} = l_t\hat{\mathbf{i}} + m_t\hat{\mathbf{j}} + n_t\hat{\mathbf{k}}$$

These satisfy the following conditions:

$$\hat{\mathbf{n}} \cdot \hat{\mathbf{n}} = l_n^2 + m_n^2 + n_n^2 = 1 \tag{2.26a}$$

$$\hat{\mathbf{t}} \cdot \hat{\mathbf{t}} = l_t^2 + m_t^2 + n_t^2 = 1 \tag{2.26b}$$

$$\hat{\mathbf{n}} \cdot \hat{\mathbf{t}} = l_n l_t + m_n m_t + n_n n_t = 0 \tag{2.26c}$$

Our attention focuses on the two orthogonal line elements PQ and PR. Each is defined by a relative change in the vector $\mathbf{r}$, namely,

$$\Delta\mathbf{r}_n = (l_n\hat{\mathbf{i}} + m_n\hat{\mathbf{i}} + n_n\hat{\mathbf{k}})\Delta n \tag{2.27a}$$

$$\Delta\mathbf{r}_t = (l_t\hat{\mathbf{i}} + m_t\hat{\mathbf{i}} + n_t\hat{\mathbf{k}})\Delta t \tag{2.27b}$$

Here, we call upon the earlier arguments, for example,

$$l_n = \frac{\partial x}{\partial n}, \qquad \Delta x = l_n \Delta n, \ldots$$

Recall that deformation transports the particle at P to the position P^* defined by the vector $\mathbf{R}$. The neighboring particles at Q and R are transported to Q^* and R^*. It is always the relative positions of particles that determine deformation and strain. The positions of Q^* and R^* relative to P^* are given by the changes in the vector R, namely,

$$\Delta \mathbf{R_n} = \left(\frac{\partial \mathbf{R}}{\partial x} l_n + \frac{\partial \mathbf{R}}{\partial y} m_n + \frac{\partial \mathbf{R}}{\partial z} n_n \right) \Delta n \tag{2.28a}$$

$$\Delta \mathbf{R_t} = \left(\frac{\partial \mathbf{R}}{\partial x} l_t + \frac{\partial \mathbf{R}}{\partial y} m_t + \frac{\partial \mathbf{R}}{\partial z} n_t \right) \Delta t \tag{2.28b}$$

The reader will note the similarities between Equations (2.27a,b) and (2.28a,b). More important are the differences; specifically, the vectors $\Delta \mathbf{R}_n$ and $\Delta \mathbf{R}_t$ are extended (or contracted) and are not, in general, orthogonal. Following the previous notation, we have

$$\frac{\partial \mathbf{R}}{\partial x} = (1+\varepsilon_x)\hat{\mathbf{I}}, \qquad \frac{\partial \mathbf{R}}{\partial y} = (1+\varepsilon_y)\hat{\mathbf{J}}, \qquad \frac{\partial \mathbf{R}}{\partial z} = (1+\varepsilon_z)\hat{\mathbf{K}} \tag{2.29a-c}$$

$$\frac{\partial \mathbf{R}}{\partial n} = (1+\varepsilon_n)\hat{\mathbf{N}}, \qquad \frac{\partial \mathbf{R}}{\partial t} = (1+\varepsilon_t)\hat{\mathbf{T}} \tag{2.29d,e}$$

Here, as before, $\hat{\mathbf{I}}$, $\hat{\mathbf{J}}$ and $\hat{\mathbf{K}}$ are unit vectors in directions of the deformed x, y, and z lines (not shown in Figure 2.22). $\hat{\mathbf{N}}$ and $\hat{\mathbf{T}}$ are unit vectors in the directions of the deformed n and t lines. From Equations (2.28a) and (2.29a–d) we obtain, as before, equation (2.23); from equation (2.28b), we also obtain a similar equation for the vector tangent to the deformed t line:

$$(1+\varepsilon_n)\hat{\mathbf{N}} = l_n(1+\varepsilon_x)\hat{\mathbf{I}} + m_n(1+\varepsilon_y)\hat{\mathbf{J}} + n_n(1+\varepsilon_z)\hat{\mathbf{K}} \tag{2.30a}$$

$$(1+\varepsilon_t)\hat{\mathbf{T}} = l_t(1+\varepsilon_x)\hat{\mathbf{I}} + m_t(1+\varepsilon_y)\hat{\mathbf{J}} + n_t(1+\varepsilon_z)\hat{\mathbf{K}} \tag{2.30b}$$

Now, the shear strain γ_{nt} can be obtained from the earlier definition:

$$\sin \gamma_{nt} = \cos \varphi_{nt} = \hat{\mathbf{N}} \cdot \hat{\mathbf{T}}$$

Recall also the products of Equations (2.24a-c). The product of the vectors of equations (2.30a, b) follows:

$$(1+\varepsilon_n)(1+\varepsilon_t)\hat{\mathbf{N}} \cdot \hat{\mathbf{T}} = (1+\varepsilon_n)(1+\varepsilon_t)\sin \gamma_{nt} \tag{2.31a}$$

$$\begin{aligned} &= l_n(1+\varepsilon_x)[l_t(1+\varepsilon_x) + m_t(1+\varepsilon_y)\sin\gamma_{xy} + n_t(1+\varepsilon_z)\sin\gamma_{xz}] \\ &\quad + m_n(1+\varepsilon_y)[l_t(1+\varepsilon_x)\sin\gamma_{xy} + m_t(1+\varepsilon_y) + n_t(1+\varepsilon_z)\sin\gamma_{yz}] \\ &\quad + n_n(1+\varepsilon_z)[l_t(1+\varepsilon_x)\sin\gamma_{xz} + m_t(1+\varepsilon_y)\sin\gamma_{yz} + n_t(1+\varepsilon_z)] \end{aligned} \tag{2.31b}$$

This result, like (2.25), is messy; it is a nonlinear equation, but it does give the shear strain γ_{nt} *exactly* in terms of the six given components (ε_x, ε_y, ε_z, γ_{xy}, γ_{yz}, γ_{zx}). Note that ε_t is obtained via (2.29e), as ε_n in (2.25).

Since our equations, (2.25) and (2.31 b), are so unwieldy, one can justly question their practical value: Firstly, we can adopt the pragmatic view that we intend to deal primarily with small strains; that is, the ε's and γ's are very small compared to unity. This is certainly true in the members of most structures and machines. Accordingly, we can expand both sides of (2.25) and (2.31b) in powers of these small quantities. We note the conditions (2.26a-c) and the approximation $\sin\gamma \doteq \gamma$ and retain only the linear terms in (2.25) and (2.31b); the linear approximations follow:

$$\varepsilon_n = l_n^2\varepsilon_x + l_n m_n \gamma_{xy} + l_n n_n \gamma_{xz} + m_n^2 \varepsilon_y + m_n n_n \gamma_{yz} + n_n^2 \varepsilon_z \tag{2.32}$$

$$\begin{aligned}\frac{1}{2}\gamma_{nt} \doteq{}& l_n l_t \varepsilon_x + l_n m_t \frac{1}{2}\gamma_{xy} + l_n n_t \frac{1}{2}\gamma_{xz} + m_n l_t \frac{1}{2}\gamma_{xy} + m_n m_t \varepsilon_y + m_n n_t \frac{1}{2}\gamma_{yz} \\ &+ n_n l_t \frac{1}{2}\gamma_{xz} + n_n m_t \frac{1}{2}\gamma_{yz} + n_n n_t \varepsilon_z\end{aligned} \tag{2.33}$$

The reader will note the similarities between Equations (2.32) and (2.33), the equations for ε_n and γ_{nt}. Both are linear in the components of strain; both are quadratic forms in the direction cosines of the initial lines n and t. Different versions of Equations (2.25) and (2.31b) are presented in Section 2.13; those versions require *no* approximations but employ alternative definitions of extensional and shear strain.

In summary: The exact equations for the transformation of the engineering strains (ε_x, ε_y, ε_z, γ_{xy}, γ_{yz}, γ_{zx}) are the nonlinear Equations (2.25) and (2.3lb). The linear approximations (2.32) and (2.33) are those cited previously [equation (2.18) and (2.19) in Section 2.11]. It is important to note that these forms are strictly limited to *small* engineering *strains* (ε's and γ's), but are not restricted to small *rotations*, nor small *displacements*.

ILLUSTRATIVE EXAMPLE 2.11

Suppose that the components of strain, in the rectangular system (x, y, z) at a point P are as follows ($\mu = 10^{-6}$):

$$\begin{pmatrix} \varepsilon_x & \gamma_{xy} & \gamma_{xz} \\ & \varepsilon_y & \gamma_{yz} \\ & & \varepsilon_z \end{pmatrix} = \begin{pmatrix} 80 & -60 & 40 \\ & -40 & 80 \\ & & 60 \end{pmatrix}\mu$$

Let us calculate the shear strain γ_{pr} in the orthogonal system (p, q, r) shown in Figure 2.23.

From geometrical considerations, it follows that

$$\alpha = 30°, \qquad ON = MP = \sqrt{3^2 + 4^2} = 5 = OP\sin 30° \Longrightarrow OP = 10$$

$$PQ = OP\tan 30° = \frac{10}{\sqrt{3}} \qquad MQ = MP\tan 30° = \frac{5}{\sqrt{3}}$$

$$OM = OP\cos 30° = 5\sqrt{3}$$

From Figure 2.23 and the preceding dimensions, we have the direction cosines of OP and PQ:

$$l_p = \frac{OR}{OP} = \frac{3}{10}, \qquad m_p = \frac{OS}{OP} = \frac{4}{10}, \qquad n_p = \frac{OM}{OP} = \frac{5\sqrt{3}}{10} = \cos 30°$$

$$l_q = \frac{RO}{QP} = -\frac{3\sqrt{3}}{10}, \qquad m_q = \frac{SO}{QP} = -\frac{4\sqrt{3}}{10}, \qquad n_q = \frac{MQ}{QP} = \frac{5}{10}$$

If $\hat{\mathbf{i}}$, $\hat{\mathbf{j}}$ and $\hat{\mathbf{k}}$ denote the *unit* vectors in the directions of the x, y, and z axes, respectively, and $\hat{\mathbf{r}}$ is the unit vector in the direction r, then

$$\hat{\mathbf{r}} = \hat{\mathbf{p}}\mathbf{x}\hat{\mathbf{q}} = \left(\frac{3}{10}\hat{\mathbf{i}} + \frac{4}{10}\hat{\mathbf{j}} + \frac{5\sqrt{3}}{10}\hat{\mathbf{k}}\right)\mathbf{x}\left(-\frac{3\sqrt{3}}{10}\hat{\mathbf{i}} - \frac{4\sqrt{3}}{10}\hat{\mathbf{j}} + \frac{5}{10}\hat{\mathbf{k}}\right) = \left(\frac{4}{5}\hat{\mathbf{i}} - \frac{3}{5}\hat{\mathbf{j}}\right)$$

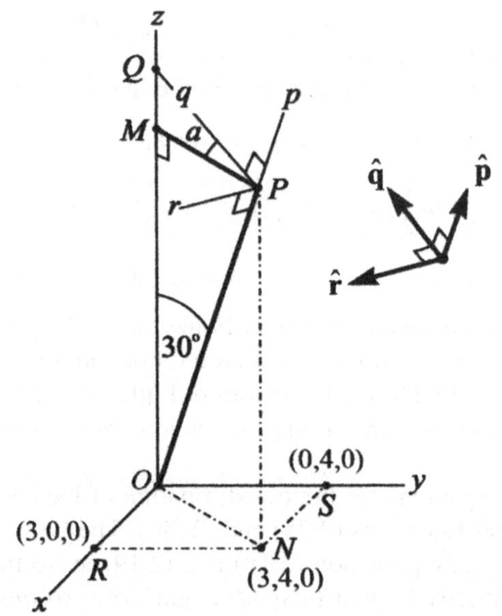

Figure 2.23

or

$$l_r = \frac{4}{5}, \qquad m_r = -\frac{3}{5}, \qquad n_r = 0$$

The reader can verify the following conditions:

$$l_p^2 + m_p^2 + n_p^2 = l_q^2 + m_q^2 + n_q^2 = l_r^2 + m_r^2 + n_r^2 = 1$$

$$l_p l_q + m_p m_q + n_p n_q = l_p l_r + m_p m_r + n_p n_r = l_q l_r + m_q m_r + n_q n_r = 0$$

Substituting the direction cosines into equation (2.33) we obtain:

$$\begin{aligned}\gamma_{pr} = &\left[(2)\left(\frac{3}{10}\right)\left(\frac{4}{5}\right)(80) + \left(\frac{3}{10}\right)\left(\frac{-3}{5}\right)\left(\frac{1}{2}\right)(-60)\right.\\ &+\left(\frac{4}{10}\right)\left(\frac{4}{5}\right)\left(\frac{1}{2}\right)(-60) + \left(\frac{4}{10}\right)\left(\frac{-3}{5}\right)(-40)\\ &\left.+\left(\frac{\sqrt{3}}{2}\right)\left(\frac{4}{5}\right)\left(\frac{1}{2}\right)(40) + \left(\frac{\sqrt{3}}{2}\right)\left(-\frac{3}{5}\right)\left(\frac{1}{2}\right)(80)\right]\mu\\ &= 35.3\mu\end{aligned}$$

2.13 TENSORIAL STRAINS OF CAUCHY-GREEN

In any treatment of large strains, one cannot employ the engineering strains and the linear equations of transformation, (2.32) and (2.33).[8] To use the nonlinear forms, (2.25) and (2.31b), would be cumbersome and, fortunately, unnecessary: It is far more expedient to employ the alternative strains, which are usually designated as the Cauchy[9] -Green[10] strains (see A. E . H. Love,[11] p. 59 for further historical remarks).

An extensional strain $\tilde{\varepsilon}_n$ in the n-direction (see Figure 2.21) is defined as follows:

$$\begin{aligned}\tilde{\varepsilon}_n &\equiv \lim_{\Delta n \to 0} \frac{1}{2}\left[\left(\frac{P^*Q^*}{PQ}\right)^2 - 1\right] \\ &= \frac{1}{2}\left[\frac{\partial \mathbf{R}}{\partial n} \cdot \frac{\partial \mathbf{R}}{\partial n} - 1\right] \\ &= \varepsilon_n + \frac{1}{2}\varepsilon_n^2\end{aligned} \tag{2.34}$$

The right side of equation (2.34) follows immediately from equation (2.23). This measure of strain ($\tilde{\varepsilon}_n$) serves as well as the other (ε_n); given ε_n, we can compute $\tilde{\varepsilon}_n$, and vice versa. Both strains vanish if the line is not extended or contracted.

A shear strain γ_{nt} in direction n and t (see Figure 2.22) is defined as follows:

$$\tilde{\gamma}_{nt} \equiv \frac{1}{2}\frac{\partial \mathbf{R}}{\partial n} \cdot \frac{\partial \mathbf{R}}{\partial t} \tag{2.35a}$$

$$= \frac{1}{2}(1+\varepsilon_n)(1+\varepsilon_t)\cos\varphi_{nt} \tag{2.35b}$$

$$= \frac{1}{2}(1+\varepsilon_n)(1+\varepsilon_t)\sin\gamma_{nt} \tag{2.35c}$$

Note that the tensorial component of equation (2.35) defers from that of equation 2.11 by the factor 1/2.

By the chain rule, we have

$$\frac{\partial \mathbf{R}}{\partial n} = \frac{\partial \mathbf{R}}{\partial x}\frac{\partial x}{\partial n} + \frac{\partial \mathbf{R}}{\partial y}\frac{\partial y}{\partial n} + \frac{\partial \mathbf{R}}{\partial z}\frac{\partial z}{\partial n} = \frac{\partial \mathbf{R}}{\partial x}l_n + \frac{\partial \mathbf{R}}{\partial y}m_n + \frac{\partial \mathbf{R}}{\partial z}n_n. \tag{2.36}$$

$$\frac{\partial \mathbf{R}}{\partial t} = \frac{\partial \mathbf{R}}{\partial x}\frac{\partial x}{\partial t} + \frac{\partial \mathbf{R}}{\partial y}\frac{\partial y}{\partial t} + \frac{\partial \mathbf{R}}{\partial z}\frac{\partial z}{\partial t} = \frac{\partial \mathbf{R}}{\partial x}l_t + \frac{\partial \mathbf{R}}{\partial y}m_t + \frac{\partial \mathbf{R}}{\partial z}n_t \tag{2.37}$$

By means of (2.36), (2.37), and the normality and orthogonality conditions (2.26a-c), equations (2.34) and (2.35) are expressed in terms of the Cauchy-Green components, viz., ($\tilde{\varepsilon}_x$, $\tilde{\varepsilon}_y$, $\tilde{\varepsilon}_z$, $\tilde{\gamma}_{xy}$, $\tilde{\gamma}_{yz}$, $\tilde{\gamma}_{zx}$):

$$\begin{aligned}\tilde{\varepsilon}_n = {} & l_n l_n \tilde{\varepsilon}_x + l_n m_n \tilde{\gamma}_{xy} + l_n n_n \tilde{\gamma}_{xz} \\ & + m_n l_n \tilde{\gamma}_{xy} + m_n m_n \tilde{\varepsilon}_y + m_n n_n \tilde{\gamma}_{yz} \\ & + n_n l_n \tilde{\gamma}_{xz} + n_n m_n \tilde{\gamma}_{yz} + n_n n_n \tilde{\varepsilon}_z\end{aligned} \tag{2.38}$$

$$\begin{aligned}\tilde{\gamma}_{nt} = {} & l_n l_t \tilde{\varepsilon}_x + l_n m_t \tilde{\gamma}_{xy} + l_n n_t \tilde{\gamma}_{xz} \\ & + m_n l_t \tilde{\gamma}_{xy} + m_n m_t \tilde{\varepsilon}_y + m_n n_t \tilde{\gamma}_{yz} \\ & + n_n l_t \tilde{\gamma}_{xz} + n_n m_t \tilde{\gamma}_{yz} + n_n n_t \tilde{\varepsilon}_z\end{aligned} \tag{2.39}$$

Note that these are linear equations, like (2.32) and (2.33); however, (2.38) and (2.39) entail no approximations.

Unlike the approximations (2.32) and (2.33), the tensorial strains in (2.38) and (2.39) need not be small; these equations are applicable to large strains, as may occur in rubberlike bodies. If the strains are small, then $\tilde{\varepsilon}_n \doteq \varepsilon_n$ and $\tilde{\gamma}_{nt} \doteq \gamma_{nt}/2$ and the exact equations, (2.38) and (2.39), are in full accord with the approximations, (2.32) and (2.33). In either case, the magnitudes of rigid displacements and rotations are irrelevant; only the *relative* movements of neighboring particles and lines are embodied in the strains. Finally, we note that the linear equations, (2.38) and (2.39). which transform components ($\tilde{\varepsilon}_x$, $\tilde{\varepsilon}_y$, $\tilde{\varepsilon}_z$, $\tilde{\gamma}_{xy}$, $\tilde{\gamma}_{yz}$, $\tilde{\gamma}_{zx}$) in one Cartesian system (x, y, z) to components $\tilde{\varepsilon}_n$, $\tilde{\varepsilon}_t$, $\tilde{\varepsilon}_s$, $\tilde{\gamma}_{nt}$, $\tilde{\gamma}_{ns}$, $\tilde{\gamma}_{ts}$) in another Cartesian system (n, t, s) constitute a tensorial transformation, i.e., these strain components comprise a Cartesian tensor. As mentioned at the beginning of this section, this particular strain tensor is associated with the works of Cauchy and Green.

2.14 PRINCIPAL DIRECTIONS; MAXIMUM AND MINIMUM VALUES OF STRAINS

The geometric relations, specifically Equations (2.18) and (2.19) [or the equivalent (2.32), (2.33) and (2.38), (2.39)], which enable us to obtain extensional and shear strains for arbitrarily oriented lines, also enable us to examine the state of strain. In particular, we can vary the orientation (l_n, m_n, n_n) of the direction n in equation (2.18) and ascertain the direction that provides the maximum or minimum extensional strain. We may vary the orientations (l_n, m_n, n_n; l_t, m_t, n_t) of directions n and t in equation (2.19) and determine directions of zero shear strain or the directions of maximum shear strain. Such investigation provides important information from a theoretical viewpoint and also from a practical viewpoint. The magnitude of an extensional strain or shear strain might limit the use of a material.

The mathematics is not unlike the search for the maximum of a function of one variable. To find the maximum (or minimum) of a smooth function $y(x)$, we must impose the condition of stationarity, $\mathrm{d}y/\mathrm{d}x = 0$. Here, the strain ε_n is a function of three variables (l_n, m_n, n_n) that are not independent ($l_n^2 + m_n^2 + n_n^2 = 1$). The strain γ_{nt} is a function of six variables (l_n, m_n, n_n; l_t, m_t, n_t) that are also interrelated. In view of the complexities, we omit the mathematical derivations and state the important results:

1. There exist, at every point of the continuous body, three *mutually perpendicular* directions of *zero shear strain*. These directions are called the *principal directions* of strain. The *extensional* strains in these three directions are called the *principal strains*. Figure 2.24 provides a graphic illustration. The edges of the small cube at P lie in the principal directions. After deformation, the particle at P has been transported to P^*. The cube is rectangular, since the shear strains of the edges are zero.
2. One principal strain is the *maximum extensional strain*, and one is the minimum *extensional* strain.
3. The directions of maximum shear strain bisect the principal directions of maximum and minimum extensional strain. In Figure 2.24, 1 and 2 are the directions of maximum and minimum extensional strain; the values are λ_1 and λ_2. The directions n and t, which initially bisect 1 and 2 at Point Q are the directions of maximum shear strain. The latter is given by the simple, but general, formula

$$\gamma_{\max}(= \gamma_{nt}) = |\lambda_1 - \lambda_2| \tag{2.40}$$

This formula is always valid for small strain ($\gamma_{nt} \ll 1$) but does not precisely give the angle $(\pi/2 - \varphi_{nt})$ when that angle is large.

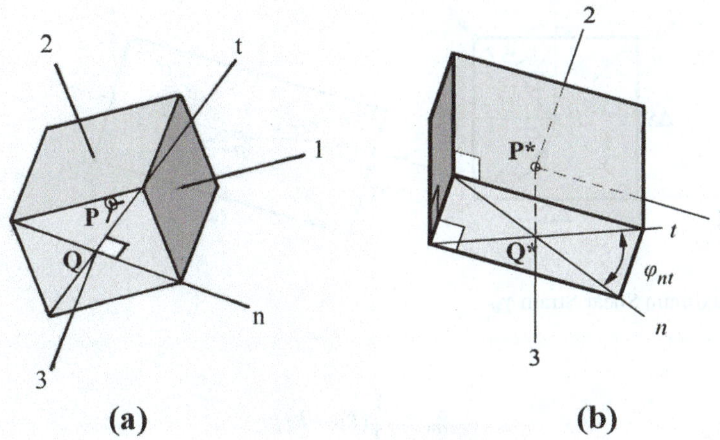

Figure 2.24

ILLUSTRATIVE EXAMPLE 2.12

The edges of the small square of Figure 2.25 lie in the principal directions. After deformation, the particle at P has been transported to P^* and the square assumes a rectangular form, since the shear strains of the edges are zero. If the maximum (ε_1) and minimum (ε_2) principal strains are known, evaluate the shear strain of the bisecting lines.

The lengths of the deformed edges, ΔS_1, ΔS_2, and the diagonal ΔS_n (see Figure 2.25) are given by:

$$\Delta S_1 = (1+\varepsilon_1)\Delta S, \qquad \Delta S_2 = (1+\varepsilon_2)\Delta S, \qquad \Delta S_n = \sqrt{2}(1+\varepsilon_n)\Delta S \tag{A}$$

By means of the Pythagoras Theorem, we obtain

$$\Delta S_n^2 = \Delta S_1^2 + \Delta S_2^2 = (1+\varepsilon_1)^2\Delta S^2 + (1+\varepsilon_2)^2\Delta S^2$$

By the Cosine Law,

$$(2\Delta S_2)^2 = \Delta S_n^2 + \Delta S_n^2 - 2\Delta S_n^2 cos\varphi_{nt} \tag{B}$$

Combining Equations (A) and (B), we obtain:

$$2(1+\varepsilon_n)^2\cos\varphi_{nt} = (1+\varepsilon_1)^2 - (1+\varepsilon_2)^2 = 2\varepsilon_1 - 2\varepsilon_2 + \varepsilon_1^2 - \varepsilon_2^2 \tag{C}$$

If the strains are small, $\varepsilon_n \ll 1$, $\varepsilon_1 \ll 1$, $\varepsilon_2 \ll 1$ $\gamma_{nt} \ll 1$, then equation (C) provides the approximation:

$$\cos\varphi_{nt} = \sin\gamma_{nt} \doteq \gamma_{nt} \doteq \varepsilon_1 - \varepsilon_2$$

In terms of the tensorial strains, $\tilde{\varepsilon}_n$, $\tilde{\gamma}_{nt}$, as defined by (2.34) and (2.35b,c) of Section 2.13, equation (C) assumes the form:

$$(1+\varepsilon_n)^2\cos\varphi_{nt} = (\tilde{\varepsilon}_1 - \tilde{\varepsilon}_2)$$

Since $\varepsilon_t = \varepsilon_n$, the left side is twice the tensorial strain of (2.35b); therefore,

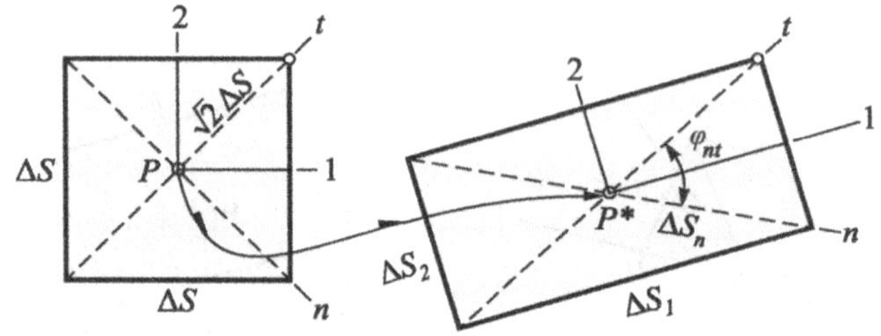

Figure 2.25 Maximum Shear Strain γ_{nt}

$$\tilde{\gamma}_{nt} = \frac{1}{2}(\tilde{\varepsilon}_1 - \tilde{\varepsilon}_2)$$

The latter is exact, i.e., it entails no approximations.

2.15 DETERMINATION OF PRINCIPAL STRAINS AND DIRECTIONS

The extensional strain for any given direction n is given by the equation (2.18),[12] a quadratic form in the direction cosines l, m, and n. The maximum or minimum values of ε_n occur in directions, which satisfy the stationary conditions $\partial\varepsilon_n/\partial l = \partial\varepsilon_n/\partial m = \partial\varepsilon_n/\partial n = 0$. In this instance, the variables l, m, and n are not independent, but must satisfy the normality condition, viz.,

$$l^2 + m^2 + n^2 = 1 \tag{2.41}$$

To accommodate the latter we can employ a Lagrange multiplier λ. Then we require the stationary values of the quadratic form:

$$\bar{\lambda} = \varepsilon_n - \lambda(l^2 + m^2 + n^2 - 1)$$

Since the added term is a constant, the extremal of the quadratic function $\bar{\lambda}$ occurs for the same directions as the strain ε_n. Hence, the directions of the stationary values of both are coincident. Accordingly, we require

$$\frac{\partial\bar{\lambda}}{\partial l} = \frac{\partial\bar{\lambda}}{\partial m} = \frac{\partial\bar{\lambda}}{\partial n} = 0$$

These conditions require the solution of the homogeneous equations:

$$(\varepsilon_x - \lambda)l + \frac{1}{2}\gamma_{xy}m + \frac{1}{2}\gamma_{xz}n = 0 \tag{2.42a}$$

$$\frac{1}{2}\gamma_{yx}l + (\varepsilon_y - \lambda)m + \frac{1}{2}\gamma_{yz}n = 0 \tag{2.42b}$$

$$\frac{1}{2}\gamma_{zx}l + \frac{1}{2}\gamma_{zy}l + (\varepsilon_z - \lambda)n = 0 \tag{2.42c}$$

These have a solution, if and only if, the determinant of the coefficients vanishes-that is, if

$$\begin{vmatrix} (\varepsilon_x - \lambda) & \frac{1}{2}\gamma_{xy} & \frac{1}{2}\gamma_{xz} \\ \frac{1}{2}\gamma_{yx} & (\varepsilon_y - \lambda) & \frac{1}{2}\gamma_{yz} \\ \frac{1}{2}\gamma_{zx} & \frac{1}{2}\gamma_{zy} & (\varepsilon_z - \lambda) \end{vmatrix} = 0 \tag{2.43a}$$

The latter is a cubic equation of the form:

$$\lambda^3 - I_1\lambda^2 + I_2\lambda - I_3 = 0 \tag{2.43b}$$

The labeling of the coefficients (I_1, I_2, I_3) is quite arbitrary. These are polynomials in the components (ε_x, ε_y, ε_z, γ_{xy}, γ_{yz}, γ_{zx}) , which follow from the expansion of the determinant; I_1 is linear, I_2 is quadratic, and I_3 is cubic. The roots of (2.43b) are the principal strains. When these have been computed, the direction for each is obtained by solving any two of the three linear Equations (2.42a–c) and the equation (2.41).

2.16 DILATATION

Dilatation measures the change of volume at a point; it might be called volumetric strain. Let ΔV denote the initial volume of an elemental region enclosing a point P and ΔV^* be the volume of the same material deformed and displaced to P^*. We require a measure of the change in the limit as the region shrinks to the point. We can adopt a definition analogous to the extensional strain (2.1) and define a dilatation e:

$$e \equiv \lim_{\Delta V \to 0} \frac{\Delta V^*}{\Delta V} - 1 \tag{2.44}$$

If we consider a small cube with faces in the principal directions, then the deformed counterpart is a rectangular parallelepiped. If the initial cube has edges of equal length $\Delta x = \Delta y = \Delta z$, then the deformed rectangle has edges of length $(1 + \varepsilon_x)\Delta x$, $(1 + \varepsilon_y)\,\Delta x$, and $(1 + \varepsilon_z)\,\Delta x$. In this case, the foregoing definition gives

$$\begin{aligned} e &= (1+\varepsilon_x)(1+\varepsilon_y)(1+\varepsilon_z) - 1 \\ e &= \varepsilon_x + \varepsilon_y + \varepsilon_z + \varepsilon_x\varepsilon_y + \varepsilon_x\varepsilon_z + \varepsilon_y\varepsilon_z + \varepsilon_x\varepsilon_y\varepsilon_z \end{aligned} \tag{2.45}$$

Of course, if the directions were not principal directions, then the cube would be deformed to a parallelepiped, and the dilatation would also depend on the components of shear strain. As previously noted, the geometry of such deformed parallelepipeds is fully determined by the six components ε_x, ε_y, ε_z, γ_{xy}, γ_{yz}, γ_{zx} and the initial length Δx. It is, therefore, a geometric exercise to obtain the expression for the dilatation e. Suffice it to note that the components of shear strain appear only in terms that are quadratic and higher degree. For small strains, the dilatation is given by the linear approximation

$$e \doteq \varepsilon_x + \varepsilon_y + \varepsilon_z \tag{2.46}$$

ILLUSTRATIVE EXAMPLE 2.13

If the principal strains according to the engineering definition take values:

$$\varepsilon_1 = 0.100, \qquad \varepsilon_2 = -0.040, \qquad \varepsilon_3 = 0.080$$

compute the maximum shear strain (i.e., the actual change of the right angle) γ_{max} and the dilatation e. Compare the results with the approximations of (2.40) and (2.46).

Lines in directions 1 and 2 (see Figure 2.26) form the initial square element and subsequent rectangular element. Diagonals OA and OB experience the largest shear strain (all changes are exaggerated). For the angle φ^* we obtain:

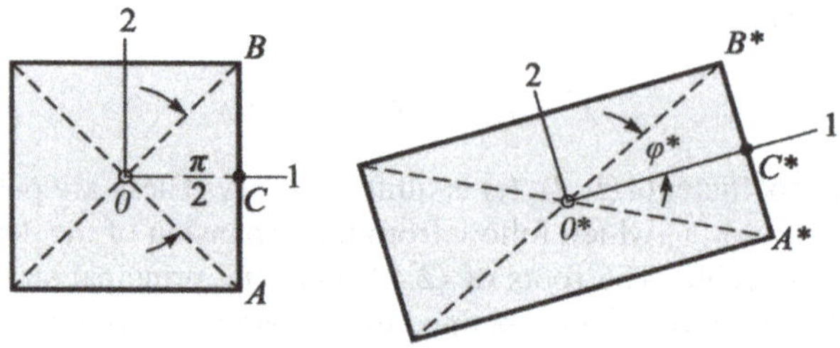

Figure 2.26

$$\varphi^* = \arctan\left(\frac{C^*B^*}{O^*C^*}\right) = \arctan\left(\frac{1+\varepsilon_2}{1+\varepsilon_1}\right)$$
$$= \arctan\left(\frac{0.960}{1.100}\right) = 0.7175 = 41.11^\circ$$

The maximum shear strain $\gamma_{\max}$ follows:

$$\gamma_{\max} = \frac{\pi}{2} - \angle A^*O^*B^* = \frac{\pi}{2} - 2\varphi^* = 0.1357 = 7.78^\circ$$

By approximation (2.40),

$$\gamma_{\max} \doteq |\varepsilon_1 - \varepsilon_2| = 0.140 = 8.02^\circ$$

According to the definitions (2.44) and (2.45), the dilatation e has the value:

$$e \equiv \frac{V^*}{V} - 1 = (1+\varepsilon_1)(1+\varepsilon_2)(1+\varepsilon_3) - 1 = 0.1405$$

By approximation (2.46),

$$e \doteq \varepsilon_1 + \varepsilon_2 + \varepsilon_3 = 0.1400$$

Note that these strains are *much larger* than those normally occurring in structures and machines. In practice, such strains would occur when a ductile metal is pressed or drawn as in the forming of a kettle or a casing.

2.17 SPECIAL STATES OF STRAIN

Several special states deserve our attention. The most special circumstance occurs if all three principal strains, the three roots of equation (2.43b), at a point are equal. It follows, in accordance with equation (2.40), that no shear strain occurs for any directions at that point. This situation is graphically illustrated by Figure 2.27. The small sphere at the point P is transported during the deformation to P^*. Lines may be rotated, but every radial line experiences the same extensional strain, and the angle between any two lines remains unchanged.

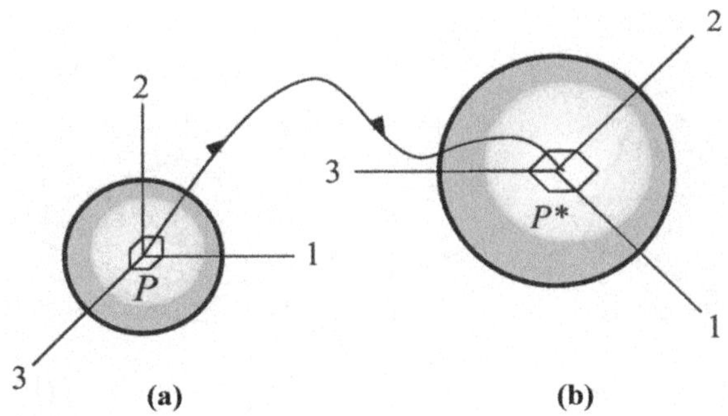

Figure 2.27 Simple Dilation

Another special circumstance arises if there is but one distinct principal direction of strain — equation (2.43b) has two equal roots. This situation is graphically illustrated in Figure 2.28, wherein a small circular cylinder at the point P is deformed to another circular cylinder at P^*. The axis lies in the direction of the one distinct principal strain (λ_1). Every radial line experiences the same extensional strain, i.e.,

$$\lambda_2 = \lambda_3 \neq \lambda_1$$

Then the angle between any two radial lines is unchanged. Note that the maximum shear strain occurs between lines that are inclined 45° to the axis.

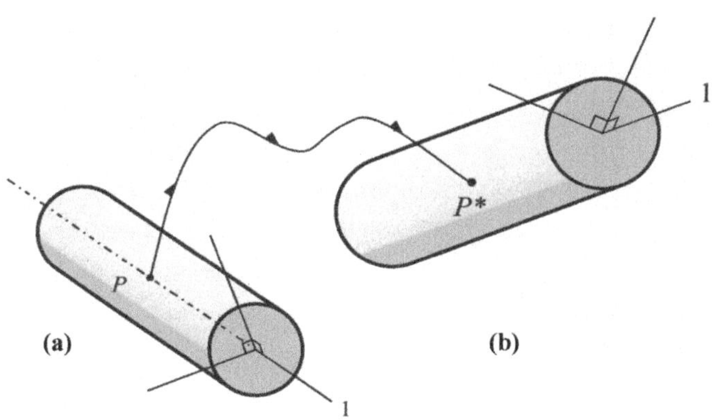

Figure 2.28 Two Equal Principal Strains

Another special state occurs if two principal strains are equal in magnitude but opposite in sign and the third is zero:

$$\lambda_1 = -\lambda_2, \qquad \lambda_3 = 0$$

By the preceding equations, the strain components for the directions that bisect the principal directions, 1 and 2, are all zero except the one shear strain. The situation is depicted in Figure 2.29; for clarity, it is viewed in the 1-2-plane. The maximum shear strain[13] is

$$\gamma_{nt} = \gamma_{\max} = 2\lambda_1$$

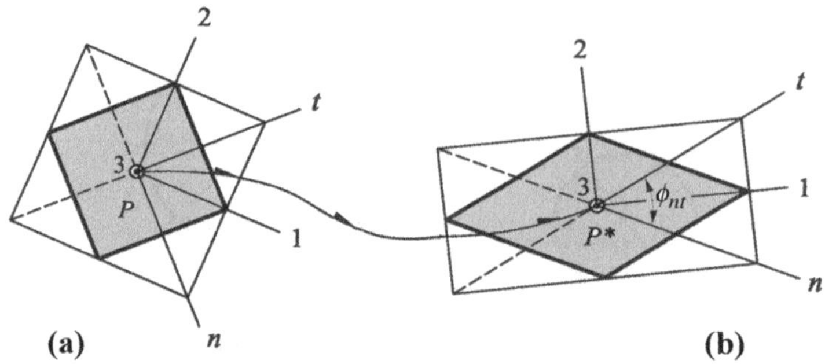

Figure 2.29 Maximum Shear Strain γ_{nt}

The state is *simple* shear. Relative to the rectangular directions n, t, and 3, γ_{nt} is the only nonzero strain. Note the simple shearing of the shaded element:

$$\gamma_{nt} = \pi/2 - \varphi_{nt}$$

2.18 STRAINS IN A SURFACE

Sometimes the engineer is concerned only with the deformations of a surface. Suppose that P in Figure 2.30a lies in a surface; x and y are orthogonal lines in the surface and z is normal to the surface.

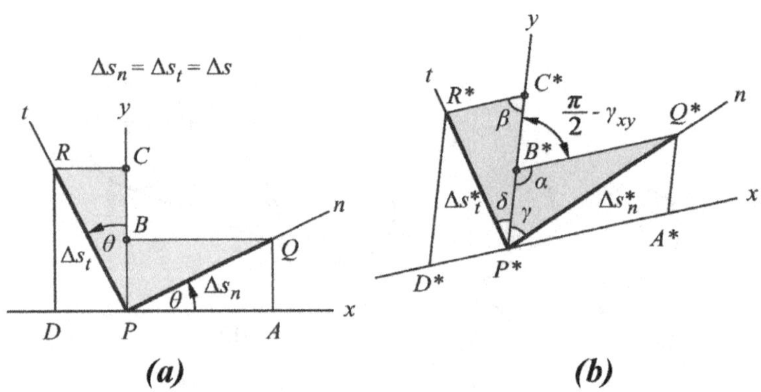

Figure 2.30

An arbitrary direction n in the surface is completely specified by one angle θ between the directions x and n. Likewise, direction t, also in the surface but perpendicular to n, is specified by the angle θ. We suppose that the components of strain ε_x, ε_y and γ_{xy} are known, and we require the extensional strain ε_n and shear strain γ_{nt}. The reader, who has assimilated the general transformation of Section 2.12 or the results of Section 2.11, will recognize this special case. Here, the direction cosines are

$$l_n = \cos\theta, \qquad m_n = \sin\theta, \qquad n_n = 0$$
$$l_t = -\sin\theta, \qquad m_t = \cos\theta, \qquad n_n = 0$$

When these are substituted in Equations (2.32) and (2.33) (or 2.18 and 2.19), we obtain

$$\varepsilon_n = \varepsilon_x \cos^2\theta + \gamma_{xy}\cos\theta\sin\theta + \varepsilon_y \sin^2\theta \tag{2.47}$$

$$\frac{1}{2}\gamma_{nt} = (\varepsilon_y - \varepsilon_x)\sin\theta\cos\theta + \frac{1}{2}\gamma_{xy}(\cos^2\theta - sin^2\theta) \tag{2.48}$$

As Equations (2.32) and (2.33), these are strictly applicable only if these strains are small compared to unity. If instead we employ the tensorial components of Section 2.13, viz., $\tilde{\varepsilon}_n$ and $\tilde{\gamma}_{nt}$ we have the *exact* equations.

$$\tilde{\varepsilon}_n = \tilde{\varepsilon}_x \cos^2\theta + 2\tilde{\gamma}_{xy}\sin\theta\cos\theta + \tilde{\varepsilon}_y \sin^2\theta \tag{2.49}$$

$$\tilde{\gamma}_{nt} = (\tilde{\varepsilon}_y - \tilde{\varepsilon}_x)\sin\theta\cos\theta + \tilde{\gamma}_{xy}(\cos^2\theta - \sin^2\theta) \tag{2.50}$$

These are applicable to finite strains; however, these strain components have different physical meanings (see Section 2.13).

It is often convenient to simplify Equations (2.47) and (2.48) by introducing the double angle formulas for the sine and cosine.

$$\sin 2\theta = 2\sin\theta\cos\theta, \qquad \cos 2\theta = \cos^2\theta - \sin^2\theta$$

When this is done, we obtain the following as alternative expression of Equations (2.47) and (2.48):

$$\varepsilon_n(P) = \frac{1}{2}(\varepsilon_x + \varepsilon_y) + \frac{1}{2}(\varepsilon_x - \varepsilon_y)\cos 2\theta + \frac{1}{2}\gamma_{xy}\sin 2\theta \tag{2.51}$$

$$\gamma_{nt}(P) = (\varepsilon_y - \varepsilon_x)\sin 2\theta + \gamma_{xy}\cos 2\theta \tag{2.52}$$

Again, we can introduce alternative definitions of the strains: Instead of ε_n and γ_{nt}, let us employ the Cauchy-Green strains. Then, in like manner, we obtain from (2.49) and (2.50):

$$\tilde{\varepsilon}_n(P) = \frac{1}{2}(\tilde{\varepsilon}_x + \tilde{\varepsilon}_y) + \frac{1}{2}(\tilde{\varepsilon}_x - \tilde{\varepsilon}_y)\cos 2\theta + \tilde{\gamma}_{xy}\sin 2\theta \tag{2.53}$$

$$\tilde{\gamma}_{nt}(P) = \frac{1}{2}(\tilde{\varepsilon}_y - \tilde{\varepsilon}_x)\sin 2\theta + \tilde{\gamma}_{xy}\cos 2\theta \tag{2.54}$$

Again, we remind the reader that the equations are only valid for small values of the engineering strains (ε's and γ's), but finite values of the Cauchy-Green strains ($\tilde{\varepsilon}'$s and $\tilde{\gamma}$'s).

Within a surface, we may seek the direction of maximum or minimum extension strain. The necessary condition follows from equation (2.51), namely,

$$\frac{d\varepsilon_n}{d\theta} = -(\varepsilon_y - \varepsilon_x)sin2\theta + \gamma_{xy}cos2\theta = 0 \tag{2.55a}$$

or, in another form,

$$\tan 2\theta = \frac{\gamma_{xy}}{(\varepsilon_x - \varepsilon_y)} \tag{2.55b}$$

Note that for any given real number, the function $\tan 2\theta$ assumes that value in the interval $0 \leq 2\theta \leq \pi$ and repeats the value at intervals of π; roots θ_1 and θ_2 differ by $\pi/2$. Stated otherwise, there are two orthogonal directions at each point of the surface that satisfy Equations (2.55a,b). One is the direction of maximum extensional strain and the other is the direction of minimum extensional strain. According to Equations (2.51) and (2.52), the shear strain vanishes in these directions *within the surface*. This means that at every point of the surface, we can find two orthogonal lines that remain orthogonal after the deformation. This circumstance is depicted in Figure 2.29. Here, a small square at a point P is transported and deformed to the rectangle at P^*. Lines 1 and 2 are the directions of maximum and minimum extensional strain-*within the surface*.

The directions of maximum (or minimum) shear strain follow from equation (2.52) and the condition

$$\frac{d\gamma_{nt}}{\partial\theta} = 2(\varepsilon_y - \varepsilon_x)\cos 2\theta - 2\gamma_{xy}\sin 2\theta = 0 \tag{2.56a}$$

or,

$$\tan 2\theta = -\frac{(\varepsilon_x - \varepsilon_y)}{\gamma_{xy}} \tag{2.56b}$$

It follows from Equations (2.55b) and (2.56b) that the directions (n and t) of maximum (or minimum) shear strain bisect the directions of maximum (or minimum) extensional strain. This is depicted in Figure 2.29. If $\varepsilon_{\max}$ and $\varepsilon_{\min}$ denote these extremal values of extensional strain, then the maximum shear strain follows:

$$\gamma_{\max} = |\varepsilon_{\max} - \varepsilon_{\min}| \tag{2.57}$$

Note that these directions in the surface *may not be the principal directions*. They are the principal directions if and only if the normal direction z is a principal direction. Note, too, that the foregoing equations apply only to small strain unless we employ Equations (2.49) and (2.50) which incorporate the tensorial components of Section 2.13.

It can be instructive to obtain Equations (2.47) and (2.48) from the plane geometry of the undeformed rectangles of Figure 2.30a and their deformed counterparts, the parallelograms of Figure 2.30b. To that end, we sketch the geometrical arguments:

Firstly, we recognize that the lines of finite length are generally deformed to curves. Here, we anticipate the infinitesimal elements, i.e., $\Delta s \to 0$, hence we justify the depiction of a "deformed parallelogram". Since directions n and t are orthogonal, we can take identical rectangles, *PAQB* and *PCRD*. Then $\Delta s_n = \Delta s_t = \Delta s$, and the orientations, i.e. angle θ, are known. Since the strains $(\varepsilon_x, \varepsilon_y, \gamma_{xy})$ of the initial system are known, the lengths of the deformed sides are also established, e.g.,

$$P^*A^* = B^*Q^* = PA(1+\varepsilon_x) = \Delta s(1+\varepsilon_x)\cos\theta$$

Likewise,

$$P^*B^* = A^*Q^* = PB(1+\varepsilon_y) = \Delta s(1+\varepsilon_y)\sin\theta$$

One angle of each shaded triangle is also known, e.g.,

$$\alpha = \pi/2 + \gamma_{xy}$$
$$\beta = \pi/2 - \gamma_{xy}$$

The geometry of each triangle is determined by the two sides and the included angle. The lengths, Δs_n^*; and Δs_t^*; are then expressed in terms of the initial length Δs (Cosine Law); division by that length (Δs) determines the extensional strain ε_n (the limit $\Delta s \to 0$ has been precluded). The strain follows: $\varepsilon_n = \Delta s_n^*/\Delta s - 1$. The angles γ and δ are also determined (Sine Law); these determine the shear strain $\gamma_{nt} = \pi/2 - \gamma - \delta$. The exact formulas are obtained according to the procedure outlined. The linear versions of Equations (2.47) and (2.48) are obtained by expanding in terms of *small strains* $\varepsilon_x, \varepsilon_y, \gamma_{xy}$ and retaining only the linear terms.

ILLUSTRATIVE EXAMPLE 2.14

Suppose that the strains in orthogonal (x, y) directions at a point on a surface are as follows:

$$\varepsilon_x = 700\mu, \qquad \varepsilon_y = -100\mu, \qquad \gamma_{xy} = 600\mu$$

where $\mu = 10^{-6}$. According to equation (2.55b), the directions of zero shear strain follow:

$$\tan 2\theta = \frac{\gamma_{xy}}{\varepsilon_x - \varepsilon_y} = \frac{600}{800} = \frac{3}{4}$$

$$\theta = 18.43^\circ, 108.43^\circ$$

$$\sin 2\theta = \pm\frac{3}{5}, \qquad \cos 2\theta = \pm\frac{4}{5}$$

According to equation (2.55a), these are the directions of extremal extensional strains. The latter are given by equation (2.51):

$$\varepsilon(18.43^\circ) = \frac{1}{2}(600)\mu + \frac{1}{2}(800)\frac{4}{5}\mu + \frac{1}{2}(600)\frac{3}{5}\mu = 800\mu$$
$$\varepsilon(108.43^\circ) = \frac{1}{2}(600)\mu - \frac{1}{2}(800)\frac{4}{5}\mu - \frac{1}{2}(600)\frac{3}{5}\mu = -200\mu.$$

In accordance with equation (2.56b), the extremal values of the shear strain bisect the directions of the maximum and minimum extensional strain; that is,

$$\tan 2\theta = -\frac{4}{3}$$

$$\theta = -26.56^\circ, \quad 63.44^\circ$$

The maximum shear is given by equation (2.52) [the absolute value, by (2.57)]:

$$\gamma_{\max} = -(800)\left(\frac{-4}{5}\right)\mu + 600\left(\frac{3}{5}\right)\mu$$
$$= 1000\mu$$

These values ($\varepsilon = 800\mu, -200\mu$) are only extremal values *within* the surface; they are principal values *only* if the orthogonal direction z is a principal direction. It follows too that the above $\gamma_{\max}$ need not be the maximum shear strain at this point.

ILLUSTRATIVE EXAMPLE 2.15

Lines a, b, and c lie in a surface with normal z at a point O (see Figure 2.31). The corresponding strains take values:

$$\varepsilon_a = 100\mu, \quad \varepsilon_b = 0, \quad \varepsilon_c = 80\mu, \quad \varepsilon_z = -70\mu, \quad \gamma_{cz} = 0, \quad \gamma_{bz} = 0$$

Calculate the strains ε_y and γ_{bc}, the principal directions, the principal strains, and the maximum shear strain.

Since the lines b and c are orthogonal, by means of equation (2.51) we obtain:

$$\varepsilon_y = \frac{1}{2}(\varepsilon_b + \varepsilon_c) + \frac{1}{2}(\varepsilon_b - \varepsilon_c)\cos(2\cdot 30^\circ) + \frac{1}{2}\gamma_{bc}\sin(2\cdot 30^\circ)$$

$$\varepsilon_y = 40\mu - 20\mu + \frac{\sqrt{3}}{4}\gamma_{bc} \qquad \text{(A)}$$

Here, the angle 30° is the angle between the axes b and y. Therefore,

$$\varepsilon_a = \frac{1}{2}(\varepsilon_b + \varepsilon_c) + \frac{1}{2}(\varepsilon_b - \varepsilon_c)\cos(-120^\circ) + \frac{1}{2}\gamma_{bc}\sin(-120^\circ)$$

$$100\mu = 40\mu + 20\mu - \frac{\sqrt{3}}{4}\gamma_{bc} \qquad \text{(B)}$$

Solving the two linear equations (A) and (B) for ε_y and γ_{bc}, yields:

$$\varepsilon_y = -20\mu, \qquad \gamma_{bc} = -\frac{160}{\sqrt{3}}\mu$$

Since $\gamma_{cz} = \gamma_{bz}$=0, z is a principal direction. The remaining two principal directions are determined by equation (2.52):

$$\gamma_{nt} = -(\varepsilon_b - \varepsilon_c)\sin 2\theta + \gamma_{bc} cos 2\theta = 0$$

$$\Longrightarrow \tan 2\theta = \frac{\gamma_{bc}}{\varepsilon_b - \varepsilon_c} = \frac{160}{80\sqrt{3}} = \frac{2}{\sqrt{3}}$$

$$\theta = 24.6^\circ, \quad 114.6^\circ$$

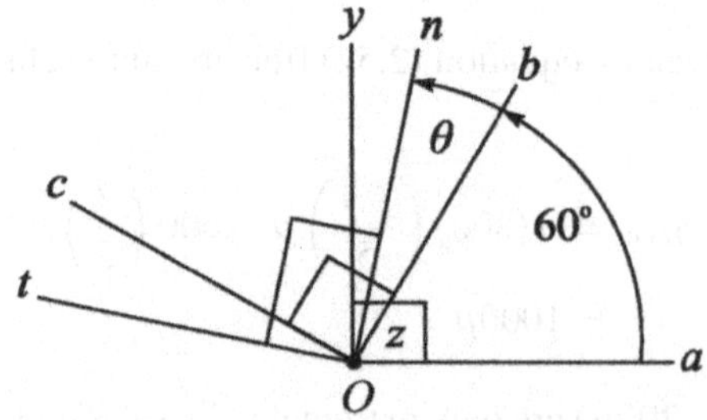

Figure 2.31

Since z is a principal direction, $\varepsilon_z = -70\mu$ is a principal strain.
The remaining two principal strains are determined by equation (2.51):

$$\begin{aligned}\varepsilon_n,\ \varepsilon_t &= \frac{1}{2}(\varepsilon_b+\varepsilon_c)+\frac{1}{2}(\varepsilon_b-\varepsilon_c)\cos(2\theta_{n,t})+\frac{1}{2}\gamma_{bc}\sin(2\theta_{n,t})\\ &= 40\mu-40\left(\pm\frac{\sqrt{3}}{\sqrt{7}}\right)\mu-\frac{80}{\sqrt{3}}\left(\pm\frac{2}{\sqrt{7}}\right)\mu\\ &= 21\mu,\quad 101\mu\end{aligned}$$

The maximum shear strain follows:

$$\gamma_{\max} = |\varepsilon_{\max}-\varepsilon_{\min}| = |101-(-70)|\mu = 171\mu$$

2.19 MOHR'S GRAPHICAL REPRESENTATION*

According to the preceding equations,[14,15] the components of strain (ε_n, γ_{nt}) for any orthogonal directions in a surface are given by Equations (2.51) and (2.52) [or, equivalently, (2.47) and (2.48)]. Also, according to the preceding arguments, there exist always two orthogonal directions (1 and 2) in which the shear strain vanishes. If the angle θ is measured from the direction 1, then $\gamma_{12}=0$ and Equations (2.51) and (2.52) assume the simpler forms

$$\varepsilon_n = \frac{\varepsilon_1+\varepsilon_2}{2}+\frac{\varepsilon_1-\varepsilon_2}{2}\cos 2\theta \tag{2.58}$$

$$-\frac{1}{2}\gamma_{nt} = \frac{\varepsilon_1-\varepsilon_2}{2}\sin 2\theta \tag{2.59}$$

Let us assume that the extremal values, ε_1 and ε_2, are known; for brevity, let

$$a = \frac{\varepsilon_1+\varepsilon_2}{2},\qquad b = \frac{\varepsilon_1-\varepsilon_2}{2} \tag{2.60}$$

Then Equations (2.58) and (2.59) take the abbreviated forms

$$\varepsilon_n = a+b\cos 2\theta \tag{2.61}$$

$$-\frac{1}{2}\gamma_{nt} = b\sin 2\theta \tag{2.62}$$

These equations are the parametric equations of a circle. Directions 1, 2, n, and t are shown in Figure 2.32a, and the plot of $-\gamma_{nt}/2$ versus ε_n is shown in Figure 2.32b. The labeled points correspond to the states 1, 2, n, and t. Observe that the shear strains corresponding to the points n and t have opposite signs. This is a consequence of the conventions: At any point, say n, the value is the shear strain between that line and the perpendicular line in the counterclockwise sense (n rotates toward t). Thus, at t the value is the shear strain between lines t and $-n$. If during deformation the angle (shear strain) between n and t decreases (positive shear), then the angle (shear strain) between t and $-n$ increases (negative shear) in the same amount (see Figure 2.32a and b).

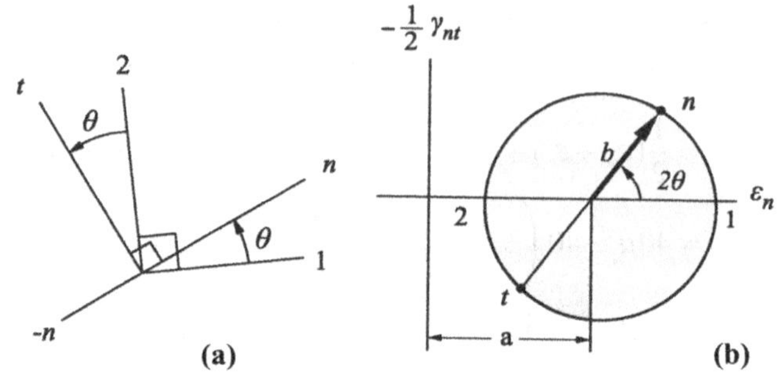

Figure 2.32 Mohr's Circle

2.20 MEASUREMENT OF STRAIN

Average extensional strains in the surface of a body can be determined in different ways. A widely used method is to bond a gage of thin wire to the surface of the body. The gage is actually a grid of continuous wire, wherein all segments lie in the direction of the extensional strain, as depicted in Figure 2.33. When the body deforms, the wire deforms with it; the electrical resistance of the wire then increases or decreases as the wire elongates or contracts. This change in resistance, as measured between terminals *A* and *B* is proportional to the change in length of the wire. Consequently, the average extensional strain can be measured by the change in resistance. When the length of the wire is very short, the average extensional strain of the wire will approach the point-wise extensional strain. These measurements, in conjunction with equation (2.47), or (2.51) can be used to determine the state of strain in the surface of a body.

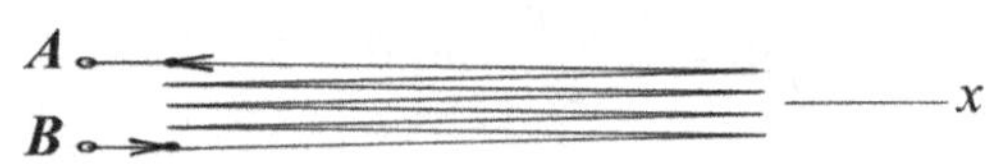

Figure 2.33 Wire Strain Gage

Suppose that the point *P* in Figure 2.34a is in the surface of a body and that *x* and *y* are any two perpendicular directions in the surface. If the extensional strain ε_a has been measured, it follows from equation (2.47) that

$$\varepsilon_a = \varepsilon_x \cos^2\theta_a + \varepsilon_y \sin^2\theta_a + \gamma_{xy}\sin\theta_a\cos\theta_a \tag{2.63a}$$

This is one equation in three unknowns, ε_x, ε_y, and γ_{xy}. Measurements of the extensional strains, ε_b and ε_c in two different directions, *b* and *c*, provide two similar equations:

$$\varepsilon_b = \varepsilon_x \cos^2\theta_b + \varepsilon_y \sin^2\theta_b + \gamma_{xy}\sin\theta_b\cos\theta_b \tag{2.63b}$$

$$\varepsilon_c = \varepsilon_x \cos^2\theta_c + \varepsilon_y \sin^2\theta_c + \gamma_{xy}\sin\theta_c\cos\theta_c \tag{2.63c}$$

The three equations provide a unique solution for the unknowns ε_x, ε_y and γ_{xy}; it is essential that the directions, *a*, *b*, and *c* are distinct. Of course, we can use equation (2.51) instead of (2.47). Much simplification results if measurements are made in orthogonal directions, *x* and *y*, and a third direction *a*. Then the solution of one equation determines the strain γ_{xy}. In any case, *three* measurements are enough to determine completely the state of strain *in the surface*, but, of course, no

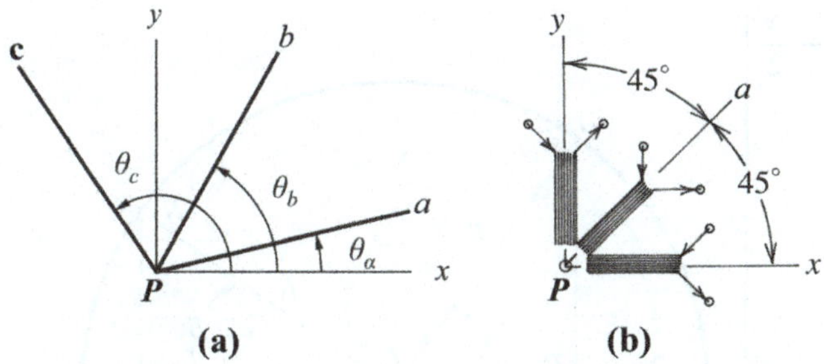

Figure 2.34

information about strains in the normal direction z is available. An array of three gages is required to provide the three extensional strains. Such assembly of three gages is a so-called "rosette", the $90°/45°$ rosette of Figure 2.34b provides the strains ε_x, ε_y and ε_a.

ILLUSTRATIVE EXAMPLE 2.16

The values of three extensional strains measured in directions x, a and y on a surface at a point P (see Figure 2.34b) are as follows:

$$\varepsilon_x = 100\mu \qquad \varepsilon_a = 30\mu, \qquad \varepsilon_y = 20\mu$$

Evaluate the shear strain γ_{xy}, the maximum extensional strain, and the maximum shear strain in the surface. By means of equation (2.51)

$$\varepsilon_a = \frac{1}{2}(\varepsilon_x + \varepsilon_y) + \frac{1}{2}(\varepsilon_x - \varepsilon_y)\cos 90° + \frac{1}{2}\gamma_{xy}\sin 90°$$

$$30\mu = 60\mu + \frac{1}{2}\gamma_{xy}, \Longrightarrow \gamma_{xy} = -60\mu$$

The orientations of the extremal values of extensional strain are determined by (2.55b):

$$\tan 2\theta = \frac{\gamma_{xy}}{\varepsilon_x - \varepsilon_y} = -\frac{60}{80}, \Longrightarrow \theta = -18.43°, 17.57°$$

The extremal values are given by equation (2.51)

$$\begin{aligned}
\varepsilon_{\text{max,min}} &= 60\mu + 40\mu\cos 2\theta - 30\mu\sin 2\theta \\
&= 60\mu + 40\mu\left(\pm\frac{4}{5}\right) - 30\mu\left(\mp\frac{3}{5}\right) \\
&= 60\mu \pm 32\mu \pm 18\mu \\
&= 110\mu, 10\mu
\end{aligned}$$

The maximum shear strain in the surface is given by (2.57)

$$\gamma_{\max} = \varepsilon_{\max} - \varepsilon_{\min} = 100\mu$$

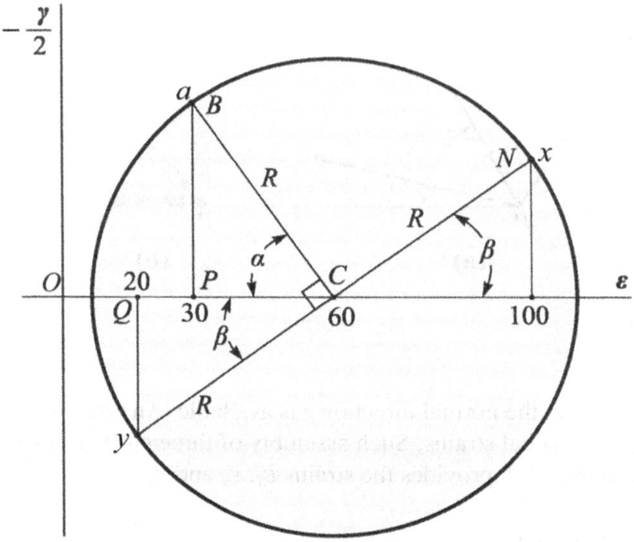

Figure 2.35 Mohr's Circle

The same problem can be treated via the geometry of Mohr's circle. On a Mohr's circle x and y are diametrically opposite and a lies at a point midway between x and y (see Figure 2.35.) Note that $|\gamma_{xy}| = |\gamma_{yx}|$. Therefore, the center must be at $\varepsilon = \frac{1}{2}(\varepsilon_x + \varepsilon_y)$. Since angles α and β are complementary,

$$\overline{PB} = \overline{CQ} = 40\mu$$

Also from geometry,

$$\overline{PC} = 60\mu - 30\mu = 30\mu$$

$$\tan\beta = \cot\alpha = \frac{\overline{PC}}{\overline{PB}} = \frac{3}{4}$$

$$R = \left[(\overline{PB})^2 + (\overline{PC})^2\right]^{1/2} = 50\mu$$

Therefore:

$$\begin{aligned}
\gamma_{xy} &= -2R\sin\beta = -100\frac{3}{5}\mu = -60\mu \\
\varepsilon_{\max} &= \overline{OC} + R = (60+50)\mu = 110\mu \\
\varepsilon_{\min} &= \overline{OC} - R = (60-50)\mu = 10\mu \\
\gamma_{\max} &= 2R = 100\mu
\end{aligned}$$

2.21 DECOMPOSITION OF THE ROTATION AND THE ENGINEERING STRAIN TENSOR[16]

Throughout our descriptions of motion and deformation, we have emphasized the distinction between the rigid motion (displacement and rotation) and the relative displacements, which account for strains. These can be decomposed into a sequence of mathematical transformations: The rigid rotation followed by the strain or, conversely, the strain followed by the rotation. The former is the more practical as the rotation is then described in the undeformed system of coordinates. Here, we intend only a conceptual, not a mathematical, description of that decomposition.

The actual rotation in that decomposition is the rotation of the principal lines of strain. It is also most readily perceived in those directions, since no *relative* rotations occur between the principal lines; recall that the shear strains vanish in those directions. Figure 2.36 depicts a rectangular block at P with edges in the initial directions of the principal lines.

In Figure 2.36 the orthogonal unit vectors ($\hat{\mathbf{i}}_1$, $\hat{\mathbf{i}}_2$, $\hat{\mathbf{i}}_3$) lie along the initial directions of the principal lines. A rigid rotation carries these to the positions of the orthogonal unit vectors ($\hat{\mathbf{I}}_1$, $\hat{\mathbf{I}}_2$, $\hat{\mathbf{I}}_3$) that lie in the directions of the deformed principal lines. As mentioned before, we denote the position of corner P by the vector $\mathbf{r}$ (from 0 to P) and the position of P^* by the vector $\mathbf{R}$ (from 0 to P^*). The initial lengths along the principal lines are denoted by (s_1, s_2, s_3). Then,

$$\hat{\mathbf{i}}_1 = \frac{\partial \mathbf{r}}{\partial s_1} \qquad \hat{\mathbf{i}}_2 = \frac{\partial \mathbf{r}}{\partial s_2} \qquad \hat{\mathbf{i}}_3 = \frac{\partial \mathbf{r}}{\partial s_3}$$

Along the deformed lines, we define the orthogonal triad ($\mathbf{G}_1$, $\mathbf{G}_2$, $\mathbf{G}_3$):

$$\mathbf{G}_1 \equiv \frac{\partial \mathbf{R}}{\partial s_1} \qquad \mathbf{G}_2 \equiv \frac{\partial \mathbf{R}}{\partial s_2} \qquad \mathbf{G}_3 \equiv \frac{\partial \mathbf{R}}{\partial s_3} \tag{2.64}$$

An increment $\Delta\mathbf{R}$ along the principal line s_1 does not have the magnitude of the increment Δs_1, but the magnitude of the strained line, say ΔS_1^*. Specifically,

$$\hat{\mathbf{I}}_1 \cdot \mathbf{G}_1 = \frac{\Delta S_1^*}{\Delta s_1} \qquad \hat{\mathbf{I}}_2 \cdot \mathbf{G}_2 = \frac{\Delta S_2^*}{\Delta s_2} \qquad \hat{\mathbf{I}}_3 \cdot \mathbf{G}_3 = \frac{\Delta S_3^*}{\Delta s_3}$$

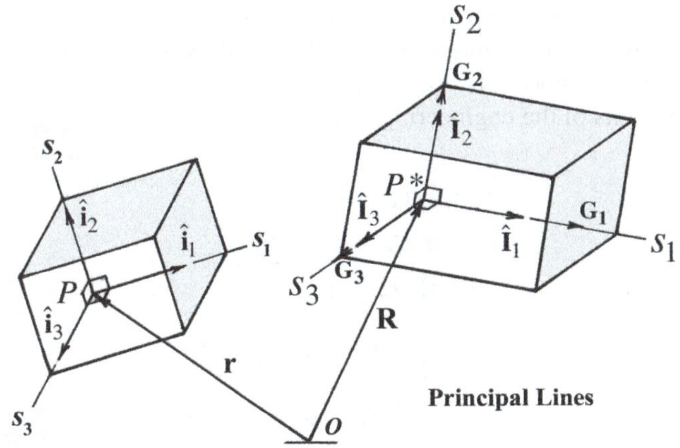

Figure 2.36 Motion and Deformation

The principal strains follow from the definition (2.1):

$$\varepsilon_1 = \hat{\mathbf{I}}_1 \cdot \mathbf{G}_1 - 1, \qquad \varepsilon_2 = \hat{\mathbf{I}}_2 \cdot \mathbf{G}_2 - 1, \qquad \varepsilon_3 = \hat{\mathbf{I}}_3 \cdot \mathbf{G}_3 - 1.$$

In a similar manner, we can perceive rectangular coordinate lines (x, y, z) originating at point P. Associated with those lines in Figure 2.37 is the triad of unit vectors ($\hat{\mathbf{i}}$, $\hat{\mathbf{j}}$, $\hat{\mathbf{k}}$). The particle at P is displaced to P^*. We imagine the triad ($\hat{\mathbf{i}}$, $\hat{\mathbf{j}}$, $\hat{\mathbf{k}}$) rigidly transported and rotated to an intermediate position ($\hat{\mathbf{I}}$, $\hat{\mathbf{J}}$, $\hat{\mathbf{K}}$) at P^*. The triad of vectors ($\mathbf{G}_x$, $\mathbf{G}_y$, and $\mathbf{G}_z$) are tangent to the deformed coordinate lines:

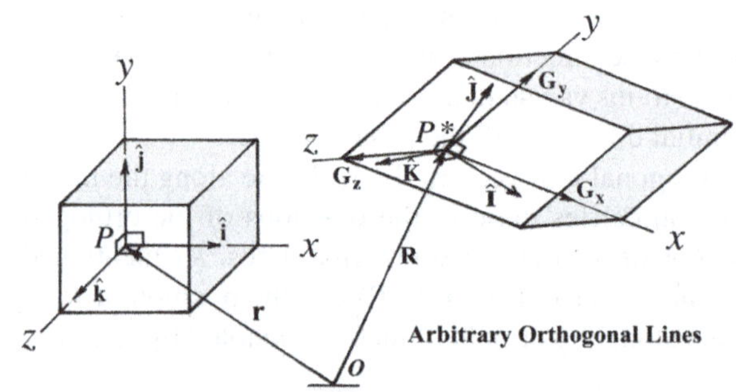

Figure 2.37 Motion and Deformation

$$\mathbf{G}_x \equiv \frac{\partial \mathbf{R}}{\partial x}, \qquad \mathbf{G}_y \equiv \frac{\partial \mathbf{R}}{\partial y} \qquad \mathbf{G}_z \equiv \frac{\partial \mathbf{R}}{\partial z}$$

These vectors are not orthogonal due to shear deformations; they are not unit vectors due to extensional strain.

The rotated position of the triad ($\hat{\mathbf{i}}$, $\hat{\mathbf{j}}$, $\hat{\mathbf{k}}$) is fixed by the following relations

$$\hat{\mathbf{I}} \cdot \mathbf{G}_y = \hat{\mathbf{J}} \cdot \mathbf{G}_x, \qquad \hat{\mathbf{I}} \cdot \mathbf{G}_z = \hat{\mathbf{K}} \cdot \mathbf{G}_x, \qquad \hat{\mathbf{J}} \cdot \mathbf{G}_z = \hat{\mathbf{K}} \cdot \mathbf{G}_y$$

The rotation of triad ($\hat{\mathbf{I}}$, $\hat{\mathbf{J}}$, $\hat{\mathbf{K}}$) is the rotation of triad ($\hat{\mathbf{i}}_1$, $\hat{\mathbf{i}}_2$, $\hat{\mathbf{i}}_3$). Since shear strains vanish in principal directions, each of the above vanishes if the lines (x, y, z) are the principal lines.

We now introduce tensorial components of strain that prove especially useful in analyses that entail large rotations. We adopt the terminology of Fraeijs de Veubeke,[17] i.e., "engineering strain," although these tensorial components are not entirely the same as those defined previously. We define the six components of the engineering strain tensor:

$$\bar{\varepsilon}_x \equiv \hat{\mathbf{I}} \cdot \mathbf{G}_x - 1 \tag{2.65a}$$

$$\bar{\varepsilon}_y \equiv \hat{\mathbf{J}} \cdot \mathbf{G}_y - 1 \tag{2.65b}$$

$$\bar{\varepsilon}_z \equiv \hat{\mathbf{K}} \cdot \mathbf{G}_z - 1 \tag{2.65c}$$

$$\bar{\gamma}_{xy} \equiv \frac{1}{2}(\hat{\mathbf{I}} \cdot \mathbf{G}_y + \hat{\mathbf{J}} \cdot \mathbf{G}_x) = \bar{\gamma}_{yx} \tag{2.65d}$$

$$\bar{\gamma}_{xz} \equiv \frac{1}{2}(\hat{\mathbf{I}} \cdot \mathbf{G}_z + \hat{\mathbf{K}} \cdot \mathbf{G}_x) = \bar{\gamma}_{zx} \tag{2.65e}$$

$$\bar{\gamma}_{yz} \equiv \frac{1}{2}(\hat{\mathbf{J}} \cdot \mathbf{G}_z + \hat{\mathbf{K}} \cdot \mathbf{G}_y) = \bar{\gamma}_{zy} \tag{2.65f}$$

In the principal directions, the extensional components are in accord with the customary definition, *only then*, we have

$$\bar{\varepsilon}_x = \varepsilon_x \qquad \bar{\varepsilon}_y = \varepsilon_y \qquad \bar{\varepsilon}_z = \varepsilon_z$$

This provides a justification for the terminology, i.e. "engineering tensorial components."

We must yet justify the adjective "tensorial"; this requires that the transformation to another system of orthogonal lines (n, t, s) follows the requisite rules, i.e., has the form of Equations (2.32) and (2.33). Two such systems are depicted in Figure 2.38 As before, we define

$$l_n \equiv \hat{\mathbf{i}} \cdot \hat{\mathbf{n}}, \qquad m_n \equiv \hat{\mathbf{j}} \cdot \hat{\mathbf{n}}, \qquad n_n \equiv \hat{\mathbf{k}} \cdot \hat{\mathbf{n}}$$
$$l_t \equiv \hat{\mathbf{i}} \cdot \hat{\mathbf{t}}, \qquad m_t \equiv \hat{\mathbf{j}} \cdot \hat{\mathbf{t}}, \qquad n_t \equiv \hat{\mathbf{k}} \cdot \hat{\mathbf{t}}$$

The tangent vectors $\mathbf{G}_n$ and $\mathbf{G}_t$ are given by the equations:

$$\mathbf{G}_n \equiv \frac{\partial \mathbf{R}}{\partial n} = \frac{\partial \mathbf{R}}{\partial x}\frac{\partial x}{\partial n} + \frac{\partial \mathbf{R}}{\partial y}\frac{\partial y}{\partial n} + \frac{\partial \mathbf{R}}{\partial z}\frac{\partial z}{\partial n} = l_n \mathbf{G}_x + m_n \mathbf{G}_y + n_n \mathbf{G}_z \tag{2.66a}$$

$$\mathbf{G}_t \equiv \frac{\partial \mathbf{R}}{\partial t} = l_t \mathbf{G}_x + m_t \mathbf{G}_y + n_t \mathbf{G}_z \tag{2.66b}$$

To be fully consistent, we introduce the *same* rigid rotation to carry the triad ($\hat{\mathbf{n}}$, $\hat{\mathbf{t}}$, $\hat{\mathbf{s}}$) to the position ($\hat{\mathbf{N}}$, $\hat{\mathbf{T}}$, $\hat{\mathbf{S}}$), viz., the rotation that carries ($\hat{\mathbf{i}}$, $\hat{\mathbf{j}}$, $\hat{\mathbf{k}}$) to ($\hat{\mathbf{I}}$, $\hat{\mathbf{J}}$, $\hat{\mathbf{K}}$). It follows that the rigidly rotated systems are related to the initial systems:

$$\hat{\mathbf{N}} = l_n \hat{\mathbf{I}} + m_n \hat{\mathbf{J}} + n_n \hat{\mathbf{K}}$$

$$\hat{\mathbf{T}} = l_t \hat{\mathbf{I}} + m_t \hat{\mathbf{J}} + n_t \hat{\mathbf{K}}$$

The equations of transformation follow:

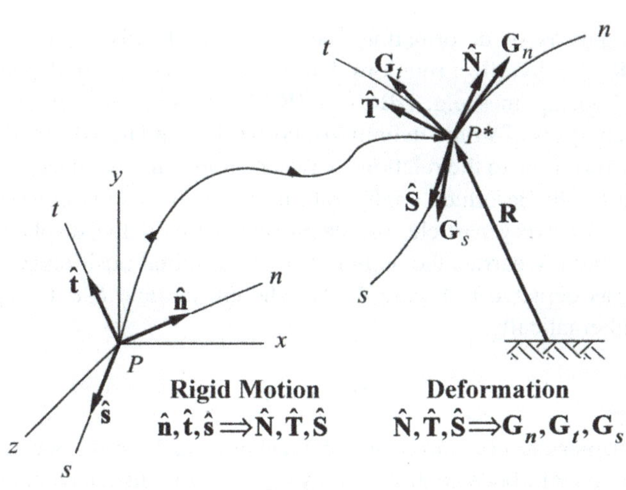

Figure 2.38

$$\begin{aligned}\bar{\varepsilon}_n &\equiv \hat{\mathbf{N}}\cdot\frac{\partial \mathbf{R}}{\partial n} - 1\\ &= l_n l_n \bar{\varepsilon}_x + l_n m_n \bar{\gamma}_{xy} + l_n n_n \bar{\gamma}_{xz}\\ &+ m_n l_n \bar{\gamma}_{xy} + m_n m_n \bar{\varepsilon}_y + m_n n_n \bar{\gamma}_{yz}\\ &+ n_n l_n \bar{\gamma}_{xz} + n_n m_n \bar{\gamma}_{yz} + n_n n_n \bar{\varepsilon}_z\end{aligned} \tag{2.67a}$$

$$\begin{aligned}\bar{\varepsilon}_{nt} &\equiv \frac{1}{2}\left(\hat{\mathbf{N}}\cdot\hat{\mathbf{G}}_t + \hat{\mathbf{T}}\cdot\hat{\mathbf{G}}_n\right)\\ &= l_n l_t \bar{\varepsilon}_x + l_n m_t \bar{\gamma}_{xy} + l_n n_t \bar{\gamma}_{xz}\\ &+ m_n l_t \bar{\gamma}_{xy} + m_n m_t \bar{\varepsilon}_y + m_n n_t \bar{\gamma}_{yz}\\ &+ n_n l_t \bar{\gamma}_{xz} + n_n m_t \bar{\gamma}_{yz} + n_n n_t \bar{\varepsilon}_z\end{aligned} \tag{2.67b}$$

Equations (2.67a,b) establish the tensorial character of these components. The six components ($\bar{\varepsilon}_x$,$\bar{\varepsilon}_y$, $\bar{\varepsilon}_z$, $\bar{\gamma}_{xy}$, $\bar{\gamma}_{xz}$, $\bar{\gamma}_{yz}$,) constitute a Cartesian tensor.

It is important to note that these strain components, as well as the rigid rotation, may have arbitrary magnitude. From a practical viewpoint, these are applicable to flexible bodies, e.g., springs, which exhibit large rotations, or rubber-like supports, which experience large strains.

The reader is reminded that the orientation of a rigid body can be fully established by three nonparallel lines, e.g., three orthogonal lines. At every point of a deformable body, there exist three orthogonal lines that remain orthogonal under a deformation, viz., the lines of principal strain (see Section 2.14). The rotation of those principal lines is also the rotation that carries triad ($\hat{\mathbf{n}}$, $\hat{\mathbf{t}}$, $\hat{\mathbf{s}}$) to ($\hat{\mathbf{N}}$, $\hat{\mathbf{T}}$, $\hat{\mathbf{S}}$) in Figure 2.38.[18] The tensorial components of Equations (2.67a,b) are especially useful in circumstances that entail large rotations because they are founded upon the decomposition of a rigid rotation and strain.

2.22 STRAIN-DISPLACEMENT RELATIONS

2.22.1 RELATIONS IN CARTESIAN COORDINATES

To this point, our analysis of deformation does not embody displacement.[19] The derivatives ($\partial\mathbf{R}/\partial x$, $\partial\mathbf{R}/\partial y$, $\partial\mathbf{R}/\partial z$) do reflect rotations, but our analysis of strain depends only on the relative rotations of neighboring lines, e.g., ($\partial\mathbf{R}/\partial x \cdot \partial\mathbf{R}/\partial y$); these relative rotations can be attributed to shear strains. All analyses of strain, definitions, properties, and transformations are independent of displacement. We turn now to the relations between strains and displacement.

As noted previously, the deformed configuration of a body is fully determined if the position (vector $\mathbf{R}$) is known for every particle, or, stated otherwise, if the displacement (vector $\mathbf{V}$) is known. The displacement $\mathbf{V}$ carries the particle from its original position (vector $\mathbf{r}$) to its current position (vector $\mathbf{R}$), as depicted in Figure 2.39a, wherein particle P is transported to P^* by the displacement $\mathbf{V}$; mathematically,

$$\mathbf{R} = \mathbf{r} + \mathbf{V} \tag{2.68}$$

The displacement, per se, need not cause deformation and strain, because the displacement might constitute only a rigid-body motion. For example, if the displacement of all particles were the same, then the body would be rigidly translated; then, in Figure 2.39a displacement of Q to Q^* would be the same as the displacement of P to P^* and line P^*Q^* would have the same length as

PQ. It is only the *relative* displacement to neighboring particles that can produce strains; we must compare the vector $\Delta\mathbf{R}$ with $\Delta\mathbf{r}$ (these determine the *relative* positions).

To determine the extensional strain ε_x, we consider a segment PT, initially parallel to the x axis at P, as shown in Figure 2.39b. The segment has initial length Δx and the initial position of T relative to P is

$$\Delta\mathbf{r} = \hat{\mathbf{i}}\Delta x$$

Following a motion, P is displaced to P^* and T to T^*, and the position of T^* relative to P^* is (approximately)

$$\Delta\mathbf{R_x} \doteq \frac{\partial\mathbf{R}}{\partial x}\Delta x$$

If Δx^* denotes the length of P^*T^*, then (approximately)

$$\begin{aligned}(\Delta x^*)^2 &\doteq \Delta\mathbf{R_x}\cdot\Delta\mathbf{R_x}\\ &\doteq \frac{\partial\mathbf{R}}{\partial x}\cdot\frac{\partial\mathbf{R}}{\partial x}\Delta x^2\end{aligned}$$

The extensional strain is obtained in accordance with definition (2.1):

$$\begin{aligned}\varepsilon_x &\equiv \lim_{\Delta x\to 0}\frac{\Delta x^* - \Delta x}{\Delta x}\\ &= \sqrt{\frac{\partial\mathbf{R}}{\partial x}\cdot\frac{\partial\mathbf{R}}{\partial x}} - 1\end{aligned} \tag{2.69}$$

Alternatively, with definition (2.8)

$$\tilde{\varepsilon}_x \equiv \lim_{\Delta x\to 0}\frac{1}{2}\left(\frac{\partial\mathbf{R}}{\partial x}\cdot\frac{\partial\mathbf{R}}{\partial x} - 1\right) \tag{2.70}$$

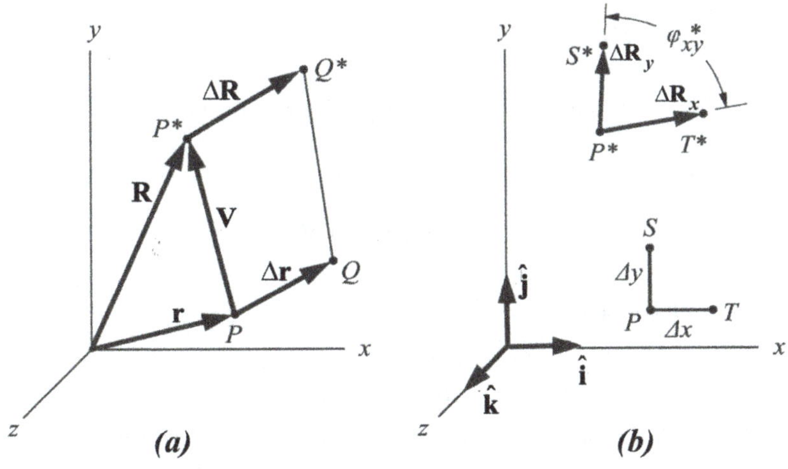

Figure 2.39 Displacement $\mathbf{V} = u\hat{\mathbf{i}} + v\hat{\mathbf{j}} + w\hat{\mathbf{k}}$

From equation (2.68) we have

$$\frac{\partial \mathbf{R}}{\partial x} = \frac{\partial}{\partial x}(\mathbf{r}+\mathbf{V}) = \hat{\mathbf{i}} + \frac{\partial \mathbf{V}}{\partial x} \tag{2.71}$$

The latter can be substituted into Equations (2.69) or (2.70) to obtain the desired strain in terms of the displacement. The strain $\tilde{\varepsilon}_x$ of (2.70) provides the equation:

$$\tilde{\varepsilon}_x \equiv \hat{\mathbf{i}} \cdot \frac{\partial \mathbf{V}}{\partial x} + \frac{1}{2}\frac{\partial \mathbf{V}}{\partial x} \cdot \frac{\partial \mathbf{V}}{\partial x} \tag{2.72}$$

Previously, we showed (see Equation 2.10) that the difference between the strains of Equations (2.69) and (2.70), between ε_x and $\tilde{\varepsilon}_x$, is inconsequential *if* strains are *small*. Then, equation (2.72) also provides a satisfactory approximation of the former (ε_x).

To determine the shear strain γ_{xy}, we must consider two lines, *PT* and *PS*, which are *initially* parallel to the x and y axes, respectively, as shown in Figure 2.39b. Following a motion, P is displaced to P^*; T and S are displaced to T^* and S^*. The positions of these particles, *relative* to P, are (approximately)

$$\Delta \mathbf{R_x} \doteq \frac{\partial \mathbf{R}}{\partial x}\Delta x, \qquad \Delta \mathbf{R_y} \doteq \frac{\partial \mathbf{R}}{\partial y}\Delta y$$

The shear strain γ_{xy} is given by definition (2.6); in the limit, $\Delta x \to 0$, $\Delta y \to 0$.

$$\gamma_{xy} = \frac{\pi}{2} - \varphi^*_{xy}$$

If ΔS_x and ΔS_y denote the lengths of $\Delta \mathbf{R}_x$ and $\Delta \mathbf{R}_y$, then

$$\cos \phi^*_{xy} = \lim_{\substack{\Delta x^* \to 0 \\ \Delta y^* \to 0}} \frac{\Delta \mathbf{R}_x \cdot \Delta \mathbf{R}_y}{\Delta x^* \cdot \Delta y^*} = \sin \gamma_{xy}$$

But $\Delta x^* \doteq (1+\varepsilon_x)\Delta x$, $\Delta y^* \doteq (1+\varepsilon_x)\Delta y$. Therefore,

$$\sin \gamma_{xy} \doteq \left(\frac{\Delta \mathbf{R}_x}{\Delta x} \cdot \frac{\Delta \mathbf{R}_y}{\Delta y}\right) / (1+\varepsilon_x)(1+\varepsilon_y)$$

In the limit, $\Delta x \to 0$, $\Delta y \to 0$,

$$\sin \gamma_{xy} = \left(\frac{\partial \mathbf{R}}{\partial x} \cdot \frac{\partial \mathbf{R}}{\partial y}\right) / (1+\varepsilon_x)(1+\varepsilon_y)$$

By definition (2.35a),

$$\tilde{\gamma}_{xy} = \frac{1}{2}\left(\frac{\partial \mathbf{R}}{\partial x} \cdot \frac{\partial \mathbf{R}}{\partial y}\right) \tag{2.73}$$

Just as $\partial \mathbf{R}/\partial x$ is given by equation (2.71),

$$\frac{\partial \mathbf{R}}{\partial y} = \hat{\mathbf{j}} + \frac{\partial \mathbf{V}}{\partial y} \tag{2.74}$$

Then, by substituting Equations (2.71) and (2.74) into equation (2.73), we obtain

$$\tilde{\gamma}_{xy} = \frac{1}{2}\left(\frac{\partial \mathbf{V}}{\partial x} \cdot \hat{\mathbf{j}} + \frac{\partial \mathbf{V}}{\partial y} \cdot \hat{\mathbf{i}}\right) + \frac{1}{2}\left(\frac{\partial \mathbf{V}}{\partial x} \cdot \frac{\partial \mathbf{V}}{\partial y}\right) \tag{2.75}$$

Previously, we expressed the strain $\tilde{\gamma}_{xy}$ in terms of the strain γ_{xy} (see Equation 2.35c). When the strains are small, $\gamma_{xy} \doteq 2\tilde{\gamma}_{xy}$. Then, equation (2.75) also provides a satisfactory approximation of the former $\gamma_{xy} \doteq 2\tilde{\gamma}_{xy}$.

The expression for $\tilde{\varepsilon}_y$ or $\tilde{\varepsilon}_z$ follows immediately from equation (2.72); we need only replace x by y or z and $\hat{\mathbf{i}}$ by $\hat{\mathbf{j}}$ or $\hat{\mathbf{k}}$. The expression for $\tilde{\gamma}_{yz}$ follows immediately from equation (2.75) we need only replace (x by y), (y by z), ($\hat{\mathbf{i}}$ by $\hat{\mathbf{j}}$), and ($\hat{\mathbf{j}}$ by $\hat{\mathbf{k}}$). In a similar manner, to obtain $\tilde{\gamma}_{xz}$ from the expression for $\tilde{\gamma}_{xy}$, we need to replace (y by z) and ($\hat{\mathbf{j}}$ by $\hat{\mathbf{k}}$).

The expressions for strains in terms of displacement can be elaborated in terms of components. The displacement can be expressed in terms of rectangular components:

$$\mathbf{V} = u\hat{\mathbf{i}} + v\hat{\mathbf{j}} + w\hat{\mathbf{k}}$$

The corresponding versions of Equations (2.72) and (2.75) follow:

$$\tilde{\varepsilon}_x = \frac{\partial u}{\partial x} + \frac{1}{2}\left(\frac{\partial u}{\partial x}\right)^2 + \frac{1}{2}\left(\frac{\partial v}{\partial x}\right)^2 + \frac{1}{2}\left(\frac{\partial w}{\partial x}\right)^2 \tag{2.76a}$$

$$\tilde{\gamma}_{xy} = \frac{1}{2}\left(\frac{\partial u}{\partial y} + \frac{\partial v}{\partial x}\right) + \frac{1}{2}\left(\frac{\partial u}{\partial x}\frac{\partial u}{\partial y}\right) + \frac{1}{2}\left(\frac{\partial v}{\partial x}\frac{\partial v}{\partial y}\right) + \frac{1}{2}\left(\frac{\partial w}{\partial x}\frac{\partial w}{\partial y}\right) \tag{2.76b}$$

Again, if the strains are small, then $\varepsilon_x \doteq \tilde{\varepsilon}_x$ and $\gamma_{xy} \doteq 2\tilde{\gamma}_{xy}$. The strains are usually small in the members of structures and machines. Then, these relations are valid irrespective of the magnitudes of displacements and rotations. If the rotations as well as the strains are everywhere small, we can delete the nonlinear terms of Equations (2.72), (2.75), (2.76a), and (2.76b)– that is, the products. The complete set of the *linear approximations* follows:

$$\varepsilon_x = \frac{\partial u}{\partial x}, \qquad \varepsilon_y = \frac{\partial v}{\partial y}, \qquad \varepsilon_z = \frac{\partial w}{\partial z} \tag{2.77a,b,c}$$

$$\gamma_{xy} = \frac{\partial u}{\partial y} + \frac{\partial v}{\partial x}, \qquad \gamma_{yz} = \frac{\partial v}{\partial z} + \frac{\partial w}{\partial y}, \qquad \gamma_{zx} = \frac{\partial w}{\partial x} + \frac{\partial u}{\partial z} \tag{2.77d,e,f}$$

Here, $\tilde{\varepsilon}_x \doteq \varepsilon_x$, $2\tilde{\gamma}_{xy} \doteq \gamma_{xy}$, etc. Note that the foregoing strain-displacement equations are valid *only* for Cartesian coordinates.

2.23 SMALL STRAINS AND MODERATE ROTATIONS

Most practical problems involve small strains; magnitude 0.01 would be large in most structural materials However, in problems of thin bodies, e.g. beams or plates, rotations might exceed $0.10\,(6^\circ)$. A typical blade for a hacksaw might easily bend 180°, yet be elastically deformed with small strains. Then terms of second order in a rotation are of the same order of magnitude as linear terms in the strains in the Equations (2.76).

Recall that a small shear strain, e.g. γ_{xy} , is essentially the change in angle between the two (x and y) lines, which were initially perpendicular in the undeformed state (see Figure 2.40)

$$\gamma_{xy} = \frac{1}{2}\left(\frac{\partial \mathbf{R}}{\partial x} \cdot \frac{\partial \mathbf{R}}{\partial y}\right) \doteq \frac{1}{2}\left(\frac{\partial v}{\partial x} + \frac{\partial u}{\partial y}\right)$$

Then the difference of those derivatives represents the average rotation (ω_z) about the z axis:

$$\omega_z \equiv \frac{1}{2}\left(\frac{\partial v}{\partial x} - \frac{\partial u}{\partial y}\right)$$

Then for small strains

$$\frac{\partial v}{\partial x} \doteq \gamma_{xy} + \omega_z \doteq \omega_z$$

Likewise

$$\frac{\partial u}{\partial y} \doteq \gamma_{xz} - \omega_z \doteq -\omega_z$$

Note that a rotation about the z axis does not move these lines from the $x-y$ plane. Note too that $\partial u/\partial x$ is of the order of ε_x (the linear term).

The rotation , ω_z as defined above, is revealed in Figure 2.40. Here we consider only a sequence of displacements in the x-y plane which moves the two segments, OP and OQ. Firstly we perceive a rigid translation $\mathbf{V_R}$ which transports the segments to O^*P^* and O^*Q^*. The rotations and deformations carry these segments to their final positions, $O^{**}P^{**}$ and O** Q**, respectively. These latter rotations are different ($\omega_u \neq \omega_v$) the difference is the shear deformation. Here we invoke the premise of small strains (ε_x, ε_y, and γ_{xy}) and moderate rotation (ω_z) about the z line. Since the shear strain γ_{xy} is perceived to be small, the rotations of both segments , $OP*$ and $OQ*$, are nearly the same; these are, respectively,

$$\omega_v \doteq \frac{\Delta v}{\Delta x}, \quad \omega_u \doteq -\frac{\Delta u}{\Delta y}$$

The average value follows:

$$\omega_z = \frac{1}{2}\left(\frac{\Delta v}{\Delta x} - \frac{\Delta u}{\Delta y}\right)$$

In the limit $(\Delta x \to 0, \Delta y \to 0)$

$$\omega_z = \frac{1}{2}\left(\frac{\partial v}{\partial x} - \frac{\partial u}{\partial y}\right)$$

Again, note that the difference in these rotations ($\omega_v + \omega_u$) provides our approximation of the small shear strain γ_{xy}. Substituting the foregoing approximations into Equations (2.76a,b), we obtain

$$\bar{\varepsilon}_x = \frac{\partial u}{\partial x} + \frac{1}{2}(\omega_z)^2 + \frac{1}{2}(\omega_y)^2 \tag{2.78a}$$

$$\gamma_{xy} = \frac{1}{2}\left(\frac{\partial u}{\partial y} + \frac{\partial v}{\partial x}\right) - \frac{1}{2}(\omega_x \omega_y) \tag{2.78b}$$

Corresponding expressions are obtained for the other components $(\varepsilon_x, \varepsilon_y, \gamma_{xz}, \gamma_{yz})$ by a permutation of the subscripts. Here we have neglected terms of order $\varepsilon_x^2, \gamma_{xy}, \omega_z$, etc.

The nonlinear terms of Equations (2.78a,b) are most likely to be relevant in analyses of thin beams, plates, and shells. For example, if w denotes the transverse displacement, then $\omega_y = -\partial w/\partial x$ and/or $\omega_x = \partial w/\partial y$ may be much larger than any component of strain.

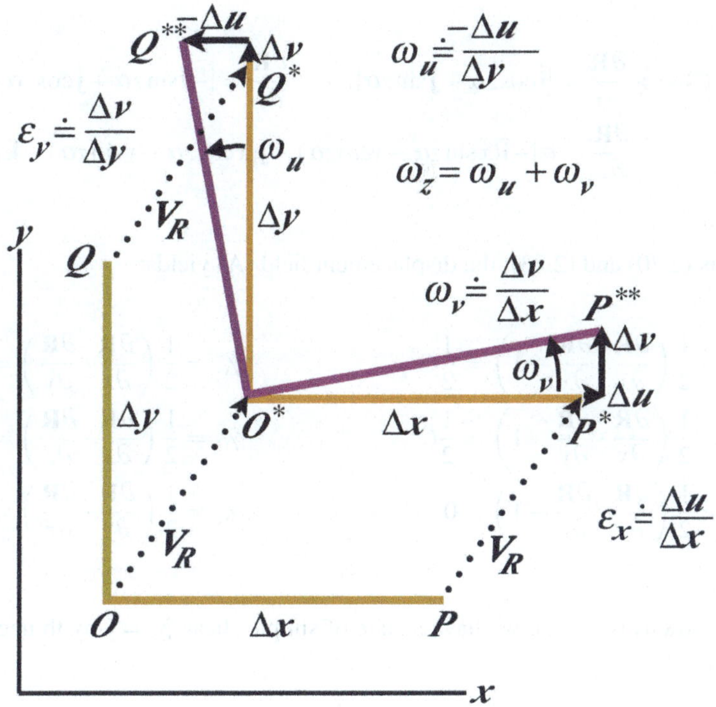

Figure 2.40 Small Strains with Moderate Rotations

ILLUSTRATIVE EXAMPLE 2.17

Calculate the strains for the following displacement fields **V** and the position vector **R**: for the three different examples:

$$\mathbf{V} = C(x\hat{\mathbf{j}} + y\hat{\mathbf{i}}) \tag{A}$$

$$\mathbf{V} = C(x\hat{\mathbf{j}} - y\hat{\mathbf{i}}) \tag{B}$$

$$\mathbf{R} = [\hat{\mathbf{i}}(x\cos z\alpha - y\sin z\alpha) + \hat{\mathbf{j}}(x\sin z\alpha + y\cos z\alpha) + z\hat{\mathbf{k}}] \tag{C}$$

In Equations (A) and (B), C denotes a constant and the physical meaning of α (rate of rotation of cross-sections along the axis x of a twisted circular shaft) in equation (C) is illustrated in Figure 2.52, where $\alpha = \partial\phi/\partial z$.

According to Equations (2.68), (2.71b), and (2.74), displacement fields (A) and (B) yield:

$$(A) \Longrightarrow \frac{\partial \mathbf{R}}{\partial x} = \hat{\mathbf{i}} + C\hat{\mathbf{j}}, \qquad \frac{\partial \mathbf{R}}{\partial y} = C\hat{\mathbf{i}} + \hat{\mathbf{j}}, \qquad \frac{\partial \mathbf{R}}{\partial z} = \hat{\mathbf{k}}$$

$$(B) \Longrightarrow \frac{\partial \mathbf{R}}{\partial x} = \hat{\mathbf{i}} + C\hat{\mathbf{j}}, \qquad \frac{\partial \mathbf{R}}{\partial y} = -C\hat{\mathbf{i}} + \hat{\mathbf{j}}, \qquad \frac{\partial \mathbf{R}}{\partial z} = \hat{\mathbf{k}}$$

For the position vector **R** of (C) we obtain:

$$(C) \Longrightarrow \frac{\partial \mathbf{R}}{\partial x} = [\hat{\mathbf{i}}\cos z\alpha + \hat{\mathbf{j}}\sin z\alpha], \qquad \frac{\partial \mathbf{R}}{\partial y} = [-\hat{\mathbf{i}}\sin z\alpha + \hat{\mathbf{j}}\cos z\alpha]$$

$$\frac{\partial \mathbf{R}}{\partial z} = \alpha[-\hat{\mathbf{i}}(x\sin z\alpha + ycosz\alpha) + \hat{\mathbf{j}}(x\cos z\alpha - y\sin z\alpha) + \hat{\mathbf{k}}]$$

By definitions (2.70) and (2.73), the displacement field (A) yields:

$$\tilde{\varepsilon}_x = \frac{1}{2}\left(\frac{\partial \mathbf{R}}{\partial x}\cdot\frac{\partial \mathbf{R}}{\partial x} - 1\right) = \frac{1}{2}C^2, \qquad \tilde{\gamma}_{xy} = \frac{1}{2}\left(\frac{\partial \mathbf{R}}{\partial x}\cdot\frac{\partial \mathbf{R}}{\partial y}\right) = C$$

$$\tilde{\varepsilon}_y = \frac{1}{2}\left(\frac{\partial \mathbf{R}}{\partial y}\cdot\frac{\partial \mathbf{R}}{\partial y} - 1\right) = \frac{1}{2}C^2, \qquad \tilde{\gamma}_{xz} = \frac{1}{2}\left(\frac{\partial \mathbf{R}}{\partial x}\cdot\frac{\partial \mathbf{R}}{\partial z}\right) = 0$$

$$\tilde{\varepsilon}_z = \frac{1}{2}\left(\frac{\partial \mathbf{R}}{\partial z}\cdot\frac{\partial \mathbf{R}}{\partial z} - 1\right) = 0, \qquad \tilde{\gamma}_{yz} = \frac{1}{2}\left(\frac{\partial \mathbf{R}}{\partial y}\cdot\frac{\partial \mathbf{R}}{\partial z}\right) = 0$$

In case of small strains (C $\ll$ 1), we have a state of simple shear $\tilde{\gamma}_{xy}$ = C with negligible uniform extensions, viz.,

$$\tilde{\varepsilon}_x = \tilde{\varepsilon}_y = \frac{1}{2}C^2 \approx 0$$

In a similar manner, we obtain for displacement field (B) the strains:

$$\tilde{\varepsilon}_x = \tilde{\varepsilon}_y = \frac{1}{2}C^2, \qquad \tilde{\varepsilon}_z = 0 \qquad \tilde{\gamma}_{xy} = \tilde{\gamma}_{xz} = \tilde{\gamma}_{yz} = 0$$

If strains are small ($C \ll 1$), all strains are negligible, and we have the simple rotation (see Figure 2.41)

Finally, displacement (C) leads to the following strains:

$$\tilde{\varepsilon}_x = \tilde{\varepsilon}_y = 0 \qquad \tilde{\varepsilon}_z = \frac{1}{2}\alpha^2(x^2 + y^2)$$

$$\tilde{\gamma}_{xy} = 0, \qquad \tilde{\gamma}_{xz} = -\frac{1}{2}\alpha y, \qquad \tilde{\gamma}_{yz} = \frac{1}{2}\alpha x$$

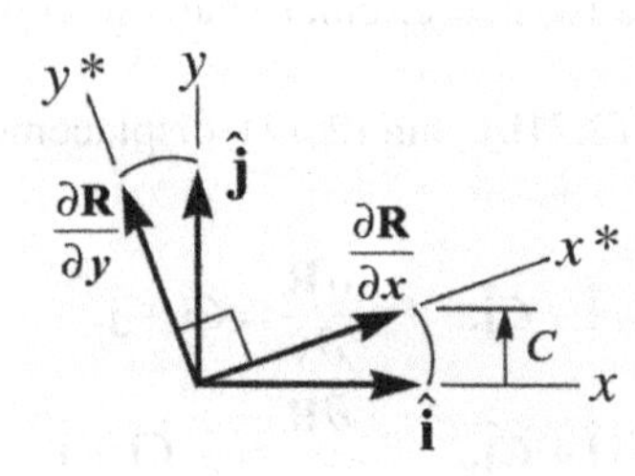

Figure 2.41

This is "simple torsion" about the z axis (see Illustrative Example 2.23). The extensional strain $\tilde{\varepsilon}_z$ occurs because the cross-sections are *not* displaced axially; therefore, an axial line (not on the axes, $x \neq 0$, $y \neq 0$) is stretched to a helical line.

Note: The rotation ($z\alpha$) of cross-sections increases with z and can be arbitrarily large while the strains may remain small as the rate of twist α.

2.24 RELATIONS IN CYLINDRICAL COORDINATES

A few problems of much significance have been solved in cylindrical coordinates. Those solutions require the strain components in that system.

Figure 2.42 depicts a segment *PACB* as delineated by the radial and circumferential lines (r, θ). The incremental vector $\Delta\mathbf{r}$ has components $\Delta r\mathbf{i}$ and $r\Delta\theta\mathbf{j}$, where $\mathbf{i}$ and $\mathbf{j}$ are orthogonal unit vectors in directions r and θ, respectively.

The particle at P is transported to P^* by the displacement $u\mathbf{i} + v\mathbf{j}$. The line PC (vector $\Delta\mathbf{r}$) is stretched to P^*C^* (vector $\Delta\mathbf{R}$) with components P^*A^* and P^*B^*. These components are stretched by the strains, ε_r and ε_θ.

Vectors $\hat{\mathbf{I}}$ and $\hat{\mathbf{J}}$ are unit vectors in directions of the deformed segments P^*A^* and P^*B^*, respectively. According to the engineering definition of extensional strain,

$$(1+\varepsilon_r)\hat{\mathbf{I}} = \frac{\partial \mathbf{R}}{\partial r}, \qquad (1+\varepsilon_\theta)\hat{\mathbf{J}} = \frac{1}{r}\frac{\partial \mathbf{R}}{\partial \theta} \tag{2.79a,b}$$

Owing to a shear strain, $\gamma_{r\theta} = \gamma_{\theta r}$, the vectors $\hat{\mathbf{I}}$ and $\hat{\mathbf{J}}$ are not orthogonal:

$$\hat{\mathbf{I}} \cdot \hat{\mathbf{J}} = \cos\phi = \sin\gamma_{r\theta}$$

In terms of displacements

$$\mathbf{R} = \hat{\mathbf{r}} + u\hat{\mathbf{i}} + v\hat{\mathbf{j}}$$

$$\frac{\partial \mathbf{R}}{\partial r} = \hat{\mathbf{i}} + \frac{\partial u}{\partial r}\hat{\mathbf{i}} + \frac{\partial v}{\partial r}\hat{\mathbf{j}} \tag{2.80a}$$

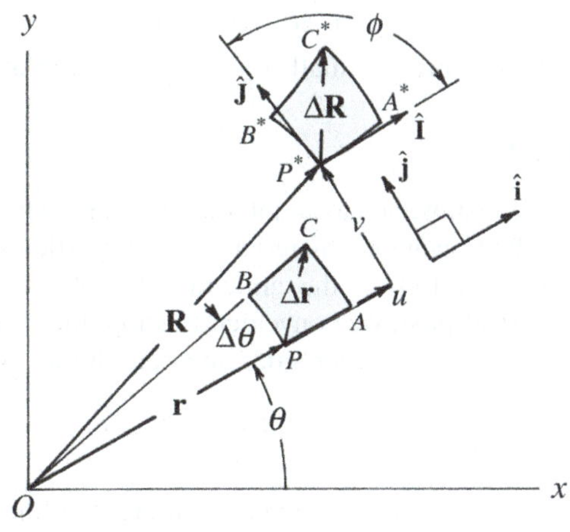

Figure 2.42 Cylindrical Coordinates

Note that vectors $\hat{\mathbf{i}}$ and $\hat{\mathbf{j}}$ turn with the angle θ:

$$\frac{\partial \hat{\mathbf{i}}}{\partial \theta} = \hat{\mathbf{j}}, \qquad \frac{\partial \hat{\mathbf{j}}}{\partial \theta} = -\hat{\mathbf{i}}$$

Therefore,

$$\frac{\partial \mathbf{R}}{r\partial \theta} = \hat{\mathbf{j}} + \frac{1}{r}\frac{\partial u}{\partial \theta}\hat{\mathbf{i}} + \frac{u}{r}\hat{\mathbf{j}} + \frac{1}{r}\frac{\partial v}{\partial \theta}\hat{\mathbf{j}} - \frac{v}{r}\hat{\mathbf{i}} \tag{2.80b}$$

We now form the inner products of (2.79a) and (2.80a), (2.79b) and (2.80b)

$$(1+\varepsilon_r)^2 = 1 + 2\frac{\partial u}{\partial r} + \left(\frac{\partial u}{\partial r}\right)^2 + \left(\frac{\partial v}{\partial r}\right)^2$$

$$(1+\varepsilon_\theta)^2 = 1 + 2\frac{u}{r} + \frac{2}{r}\frac{\partial v}{\partial \theta} + \frac{1}{r^2}\left(\frac{\partial u}{\partial \theta}\right)^2 + \left(\frac{u}{r}\right)^2$$

$$+2\frac{u}{r^2}\left(\frac{\partial v}{\partial \theta}\right) - \frac{2v}{r^2}\left(\frac{\partial u}{\partial \theta}\right) + \frac{1}{r^2}\left(\frac{\partial v}{\partial \theta}\right)^2 + \left(\frac{v}{r}\right)^2$$

If we accept the circumstance of small strains and small rotations, i.e., we neglected the products, then we obtain the approximations:

$$\varepsilon_r \doteq \frac{\partial u}{\partial r} \tag{2.81a}$$

$$\varepsilon_\theta \doteq \frac{1}{r}\frac{\partial v}{\partial \theta} + \frac{u}{r} \tag{2.81b}$$

The shear strain $\gamma_{r\theta}$ has the approximation given by the inner product:

$$\begin{aligned} \gamma_{r\theta} &\doteq \sin\gamma_{r\theta} = \hat{\mathbf{I}} \cdot \hat{\mathbf{J}} \\ &\doteq \frac{1}{r}\frac{\partial \mathbf{R}}{\partial r} \cdot \frac{\partial \mathbf{R}}{\partial \theta} \\ \gamma_{r\theta} &\doteq \frac{1}{r}\frac{\partial u}{\partial \theta} + \frac{\partial v}{\partial r} - \frac{v}{r} \end{aligned} \tag{2.81c}$$

This linear equation also holds only for small strains and small rotations.

ILLUSTRATIVE EXAMPLE 2.18

A thick-walled cylindrical pipe is subjected to an internal pressure, which causes radial expansion. If the pipe is composed of a homogeneous material, then all particles at any given radius r are displaced the same radial distance u to form the circle of radius R; this is depicted in Figure 2.43.

In an analysis of the cylindrical pipe, we conveniently adopt the cylindrical coordinates, r and θ. Then the extensional strain of the circumferential line (length l at radius r) is denoted

$$\varepsilon_\theta \equiv \frac{l^* - l}{l} = \frac{2\pi R - 2\pi r}{2\pi r} = \frac{u}{r}$$

We anticipate also a radial strain ε_r; e.g., as the circumferential line is stretched, we anticipate a contraction radially (ε_r), but that depends upon the properties of the material. We return to this problem later (Section 5.17).

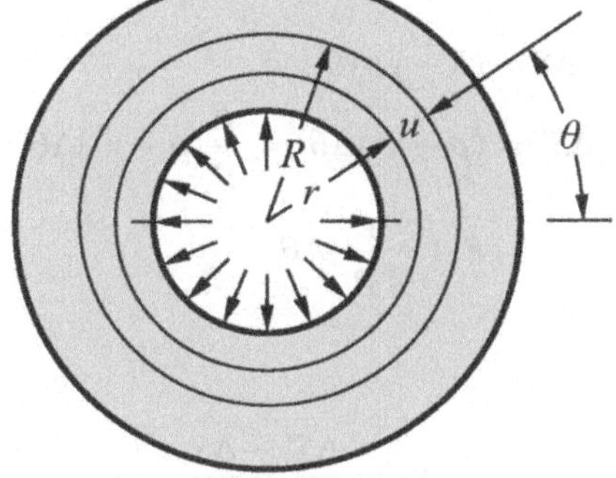

Figure 2.43 Cylindrical Pressure Vessel

ILLUSTRATIVE EXAMPLE 2.19

A segment of a curved beam is depicted in Figure 2.44a. The beam is symmetrical about the plane of the figure. It is of circular form and composed of a homogeneous material. It is subjected to the bending couples M $(-M)$ applied uniformly upon end plates. Under these conditions of symmetry, we assume that each cross-section remains plane and, furthermore, each similar segment undergoes similar deformation, as shown in Figure 2.44b.

We anticipate a compression and contraction of the innermost lines and the extension of the outermost. Therefore, an intermediate arc (a neutral line) remains unextended. We denote the initial radius to that line by R and the deformed radius by ρ, as depicted in Figure 2.44a and b, respectively. We examine a typical segment of the beam, which is subtended by angles $\Delta\theta$ and $\Delta\theta^*$, respectively. Our goal is an expression for the extensional strain ε_θ of an arbitrary line, which lies initially at a distance y from the neutral arc. The length ΔS_0 of the neutral line is unchanged; hence

$$\Delta S_0 = R\Delta\theta = \rho\Delta\theta^* \quad \text{or } \Delta\theta^* = \frac{R}{\rho}\Delta\theta$$

The lengths ΔS and ΔS^* are given by

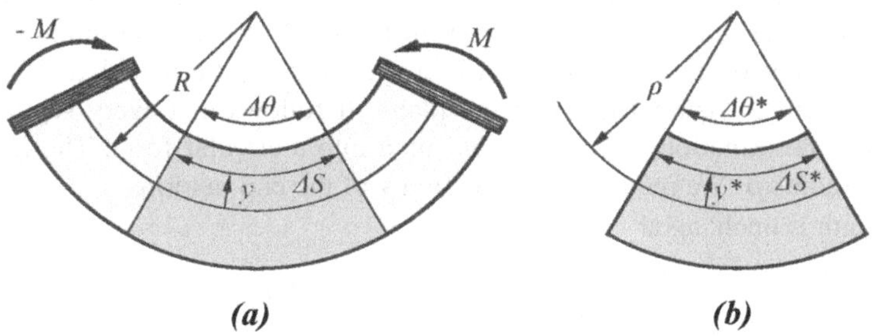

Figure 2.44 Simple Bending

$$\Delta S = (R - y)\Delta\theta, \tag{A}$$

$$\Delta S^* = (\rho - y^*)\Delta\theta^* = \frac{R}{\rho}(\rho - y^*)\Delta\theta$$
$$= R(1 - \frac{y^*}{\rho})\Delta\theta \tag{B}$$

By definition,

$$\varepsilon_\theta = \frac{\Delta S^* - \Delta S}{\Delta S} \tag{C}$$

It follows from (A), (B), and (C) that

$$\varepsilon_\theta = \left[\left(1 - \frac{y^*}{\rho}\right) \Big/ \left(1 - \frac{y}{R}\right)\right] - 1$$
$$= \left(\frac{y}{R} - \frac{y^*}{\rho}\right) \Big/ \left(1 - \frac{y}{R}\right) \tag{D}$$

The distance y^* differs from y by a mean value of the radial strain:

$$\overline{\varepsilon_r} = \frac{y^*}{y} - 1, \quad y^* = y(1 + \overline{\varepsilon_r})$$

In most practical circumstances, $\overline{\varepsilon_r} \ll 1$. Then, a very good approximation of (D) follows:

$$\varepsilon_\theta \doteq y\left(\frac{1}{R} - \frac{1}{\rho}\right) \Big/ \left(1 - \frac{y}{R}\right)$$

Our immediate goal is achieved. The distribution of internal force, the relation of the curvature $(1/\rho)$ to the applied couple M, and the location of the neutral line (R) depend upon the material. A complete solution is addressed in Section 5.17.

ILLUSTRATIVE EXAMPLE 2.20

Let us consider another very different circumstance, wherein a *thin* rod is deformed by the action of the couple applied at the free end, as shown in Figure 2.45. The rod initially encompasses a circular quadrant. The rod is composed of a homogeneous elastic material. Again, we suppose that the entire system is symmetric about the plane. It follows that every segment experiences similar action and deformation; the rod is deformed into a circular form. The initial radius R is deformed to the radius ρ. We require the deflection **v** of the end in terms of the curvature $1/\rho$.

Since the length is unchanged

$$\rho\theta^* = \frac{R\pi}{2}, \qquad \theta^* = \frac{R}{\rho}\frac{\pi}{2}$$

The displacement of the end is

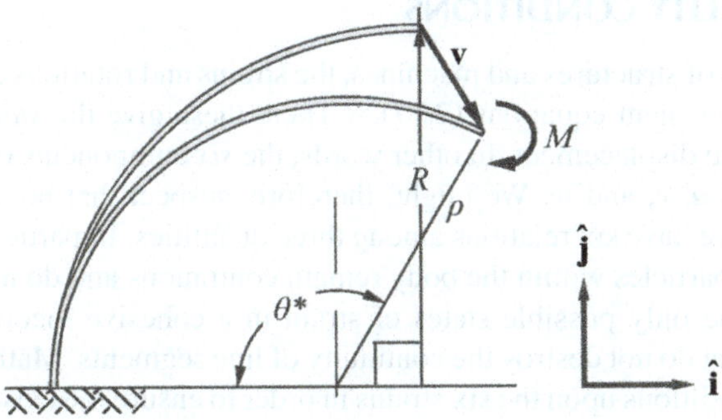

Figure 2.45

$$\begin{aligned}\mathbf{v} &= \hat{\mathbf{i}}(-\rho\cos\theta^* - R + \rho) + \hat{\mathbf{j}}(\rho\sin\theta^* - R) \\ &= -\hat{\mathbf{i}}\left[\rho\cos\left(\frac{R}{\rho}\frac{\pi}{2}\right) + R - \rho\right] + \hat{\mathbf{j}}\left[\rho\sin\left(\frac{R}{\rho}\frac{\pi}{2}\right) - R\right]\end{aligned}$$

Again, in practice, the displacement is often relatively small, i.e., $|\mathbf{v}| \ll R$; this also means that the difference $(1-\rho/R) \ll 1$. Let us define

$$\varepsilon \equiv \frac{1}{\rho}(R-\rho), \qquad \frac{R}{\rho} \equiv 1+\varepsilon$$

If $\varepsilon \ll 1$, then we have the approximations:

$$\begin{aligned}\cos\frac{\pi}{2}(1+\varepsilon) &= -\sin\frac{\pi}{2}\varepsilon \doteq -\frac{\pi}{2}\varepsilon \\ \sin\frac{\pi}{2}(1+\varepsilon) &= \cos\frac{\pi}{2}\varepsilon \doteq 1\end{aligned}$$

The approximation for *relatively* small displacement follows:

$$\begin{aligned}\mathbf{v} &\doteq \hat{\mathbf{i}}\left(\frac{\pi}{2}R - \frac{\pi}{2}\rho - R + \rho\right) + \hat{\mathbf{j}}(\rho - R) \\ &\doteq \hat{\mathbf{i}}\left(\frac{\pi}{2} - 1\right)(R-\rho) + \hat{\mathbf{j}}(\rho - R)\end{aligned}$$

In theories of curved rods and shells, one commonly employs the "change of curvature" κ:

$$\kappa \equiv \frac{1}{\rho} - \frac{1}{R}, \qquad R - \rho = R\rho\kappa \doteq R^2\kappa$$

Then,

$$\mathbf{v} \doteq \hat{\mathbf{i}}R^2\left(\frac{\pi}{2} - 1\right)\kappa - \hat{\mathbf{j}}R^2\kappa$$

If the rod is Hookean, then the kinematic variable κ is proportional to the bending couple. Later [Chapter 9], we apply an energy method to express displacement $\mathbf{v}$ in terms of couple M.

2.25 COMPATIBILITY CONDITIONS

In most practical cases of structures and machines, the strains and rotations are small enough to use the linear strain-displacement equations (2.77).[20] Then, these give the *state of strain* throughout the body in terms of the displacement. In other words, the *six* components of strain are determined by the *three* functions u, v, and w. We might, therefore, suspect that not any state of strain can exist in a body, since we have six relations among three quantities. In particular, it has always been assumed that lines of particles within the body remain continuous and do not overlap (see Section 2.1). Consequently, the only possible states of strain in a cohesive medium are those resulting from displacements that do not destroy the continuity of line segments. Mathematically, we need to establish sufficient conditions upon the six strains in order to ensure the existence of the continuous displacements. It is a simpler task to establish necessary conditions. To illustrate the latter, consider the three relations

$$\varepsilon_x = \frac{\partial u}{\partial x}, \qquad \varepsilon_y = \frac{\partial v}{\partial y}, \qquad \gamma_{xy} = \frac{\partial u}{\partial y} + \frac{\partial v}{\partial x}$$

Differentiating the first of these with respect to y twice, the second with respect to x twice, and the third with respect to y and x successively, we obtain

$$\frac{\partial^2 \varepsilon_x}{\partial y^2} = \frac{\partial^3 u}{\partial y^2 \partial x}, \qquad \frac{\partial^2 \varepsilon_y}{\partial x^2} = \frac{\partial^3 v}{\partial x^2 \partial y}, \qquad \frac{\partial^2 \gamma_{xy}}{\partial x \partial y} = \frac{\partial^3 u}{\partial x \partial y^2} + \frac{\partial^3 v}{\partial y \partial x^2}$$

Now, u and v must be such that the order of differentiation is interchangeable. Hence, it follows that

$$\frac{\partial^2 \varepsilon_x}{\partial y^2} + \frac{\partial^2 \varepsilon_y}{\partial x^2} = \frac{\partial^2 \gamma_{xy}}{\partial x \partial y}$$

Five other independent relations of this nature may also be derived. The entire set is as follows:

$$\begin{aligned}
&\frac{\partial^2 \varepsilon_x}{\partial y^2} + \frac{\partial^2 \varepsilon_y}{\partial x^2} = \frac{\partial^2 \gamma_{xy}}{\partial x \partial y} \qquad && 2\frac{\partial^2 \varepsilon_x}{\partial y \partial z} = \frac{\partial}{\partial x}\left(-\frac{\partial \gamma_{yz}}{\partial x} + \frac{\partial \gamma_{xz}}{\partial y} + \frac{\partial \gamma_{xy}}{\partial z}\right) \\
&\frac{\partial^2 \varepsilon_y}{\partial z^2} + \frac{\partial^2 \varepsilon_z}{\partial y^2} = \frac{\partial^2 \gamma_{yz}}{\partial y \partial z} \qquad && 2\frac{\partial^2 \varepsilon_y}{\partial z \partial x} = \frac{\partial}{\partial y}\left(-\frac{\partial \gamma_{zx}}{\partial y} + \frac{\partial \gamma_{yx}}{\partial z} + \frac{\partial \gamma_{yz}}{\partial x}\right) \\
&\frac{\partial^2 \varepsilon_z}{\partial x^2} + \frac{\partial^2 \varepsilon_x}{\partial z^2} = \frac{\partial^2 \gamma_{zx}}{\partial z \partial x} \qquad && 2\frac{\partial^2 \varepsilon_z}{\partial x \partial y} = \frac{\partial}{\partial z}\left(-\frac{\partial \gamma_{xy}}{\partial z} + \frac{\partial \gamma_{zy}}{\partial x} + \frac{\partial \gamma_{zx}}{\partial y}\right)
\end{aligned} \tag{2.82}$$

These six relations, called *compatibility conditions*, are necessary conditions imposed on the state of strain in order that no tears or folds occur in the body. It can also be demonstrated that these conditions are sufficient. Therefore, only those states of strain that satisfy these six partial differential equations are permissible states of strain in a cohesive medium. These compatibility relations complete our discussion of small strain and small rotation.

2.26 SOME ELEMENTARY STRAIN-DISPLACEMENT ANALYSES

Most fabricated structures consist of a number of interconnected members. The simplest types of members are all prismatic or cylindrical in shape. The distinguishing feature or characteristic of these simple members is the nature of the loads that they transmit or resist. Members of a pin-connected truss are subjected to loads that tend to stretch or compress them. A torque transmitted by means of a shaft tends to twist the shaft. The rafters of a building must support the load of the roof and thus tend to bend.

Individually, each of these members is a body. The complete deformational behavior of any of these bodies requires a knowledge of the displacements of all particles; however, some of this information is incidental. In a member of a pin-connected truss, for example, the overall extension or contraction can be treated without attention to localized effects.

Analyses of elementary members generally begin with some assumptions with regard to the general nature of the deformation of the member. These assumptions characterize the significant changes in the geometry of the member and are usually based upon the experience and observations of the engineer. The underlying assumptions are usually geometrical. Deformations, e.g. bending or twisting, are more perceptible than the distributions of internal stresses. Each of the following examples are founded upon an assumption, which provides a description of the displacement and, in turn, the consequent strain distribution.[21] Although intended to illustrate the strain-displacement relations, they provide the bases for subsequent analyses of rods, beams, and shafts.

ILLUSTRATIVE EXAMPLE 2.21

A variety of structural elements are cylindrical or prismatic in shape and are subjected to loads that tend to stretch or compress the member. For example, the pin-connected truss in Figure 2.46a consists of two members, AB and BC. When the load P is applied, each of these members is stretched.

Figure 2.46b shows a concrete cylinder supporting a bearing plate that carries a load P along the axis of the cylinder. Figure 2.46c shows a section of rail subjected to a rising temperature. As the rail tends to elongate, the adjacent portions restrain this expansion. Each of these members is a prismatic body whose geometric behavior can be described by the statements that (1) the axis of the cylinder remains straight after deformation and (2) any plane cross-section of particles remains plane and perpendicular to the axis after deformation. The latter (2) does not hold near connections and supports, e.g., joints A, B, and C of Figure 2.46 (see Section 5.2).

Suppose that the member of Figure 2.47a deforms according to these assumptions: A typical cross-section S at distance x from the left end remains plane and perpendicular to the axis, but displaces a distance $u(x)$ to the position S^* as depicted in Figure 2.47b. For convenience, we hold the left end fixed; $u(0) = 0$. The neighboring cross-section S_+ at distance $x + \Delta x$ displaces to the position S_+^* at distance $x + \Delta x + u(x + \Delta x)$ from the left end. The extensional strain ε_x of the line segment PQ is

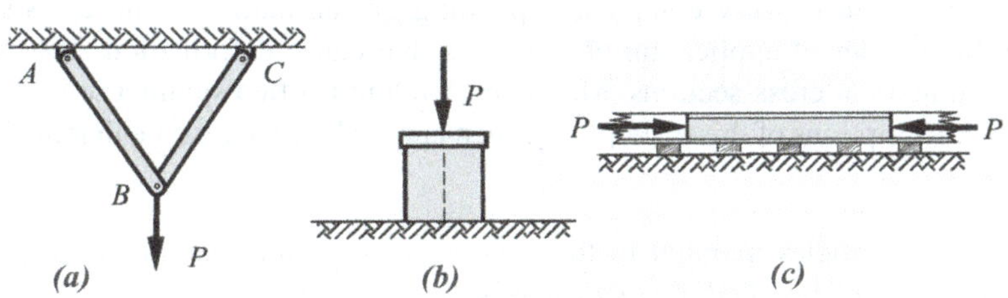

Figure 2.46

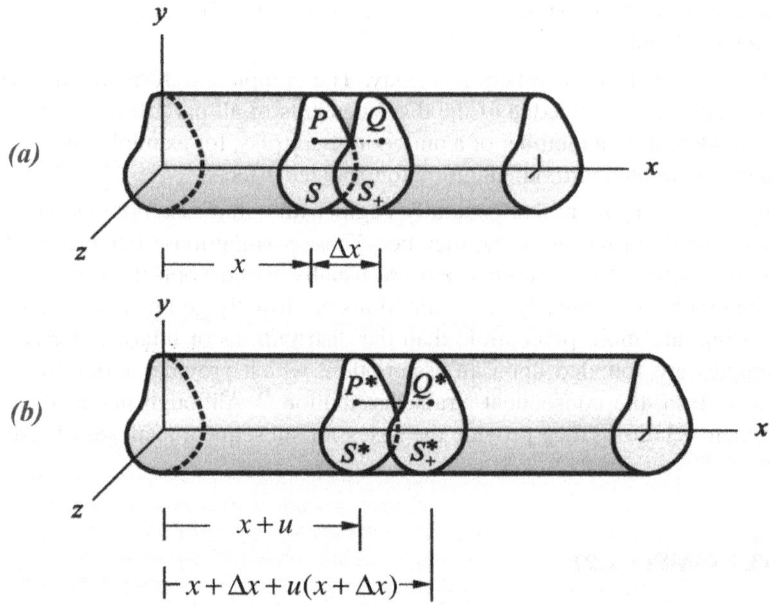

Figure 2.47 Simple Extension

$$\varepsilon_x(x) = \lim_{Q \to P} \frac{P^*Q^* - PQ}{PQ}$$

If we also assume that line P^*Q^* remains parallel to the x axis, then

$$\varepsilon_x(x) = \lim_{Q \to P} \frac{u(x+\Delta x) - u(x)}{\Delta x} = \frac{du}{dx}$$

By our assumptions, planes S^* and S_+^*. remain parallel; the site of P in S is arbitrary, so that $\varepsilon_x(x)$ is independent of coordinates y and z. Our assumptions provide no specific information about displacement in the y or z directions. Strain components ε_y, ε_z, γ_{xy}, γ_{xz} and γ_{yz} remain unknown.

ILLUSTRATIVE EXAMPLE 2.22

As a further example, let us consider the long straight *beams* of structures and machines; the beams carry loads transverse to their length. A floor joist in a building is a typical beam. The geometric behavior of a beam can be very complex, depending on the nature of the cross-section of the member and the manner of application of the loads that cause deformation. The most common beams have symmetrical cross-sections. Moreover, loading is often symmetric about the plane of symmetry. The assumptions of this example are then applicable. We assume that the beam deforms as follows:

1. One line of particles, parallel to the axis, remains in the plane and does not stretch or contract; that line is called the *neutral* axis.
2. Every cross-section perpendicular to the neutral axis remains perpendicular as the beam deforms.

These assumptions describe a plane *bending* of the beam and provide the basis of most analyses.

The straight beam of Figure 2.48a is symmetric about the x-y plane. For convenience, we take the x axis as the neutral axis. Bending carries the axis to the curve depicted in Figure 2.48b. The cross-section S is pierced by the axis at a point N; P identifies an arbitrary particle of section S. A neighboring cross-section S_+ is pierced by the axis at N_+; the point P_+ on S_+ identifies a segment PP_+ of a line parallel to the axis. Deformation carries section S to S^*, section S_+ to S_+^*, particles N to N^*, P to P^*, N_+ to N_+^*, and P_+ to P_+^*.

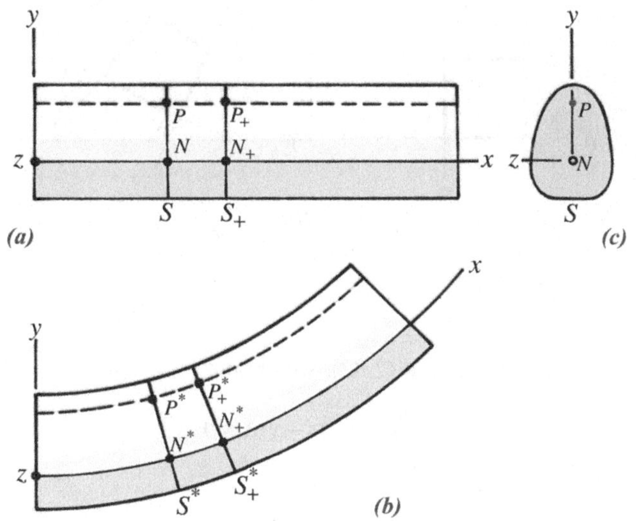

Figure 2.48 Bending of a Symmetrical Beam

Let us make the additional assumption that the deformation of the cross-section is negligible; we treat a cross-section as a rigid surface. This is tacitly implied in most analyses of beams and is usually justified. In the subsequent treatment of beams (Chapter 6), we examine this question and the extent of the cross-sectional deformations. With this additional constraint, all curves parallel to the axis are also deformed to plane curves. Specifically, the arbitrary segment $P^*P_+^*$ lies in a plane parallel to the x-y plane. It suffices then to view the deformation and the segments PP_+ and $P^*P_+^*$ in the two-dimensional view of Figure 2.49b. In the view of Figure 2.49b, the extended planes of S^* and S_+^* intersect at point C, the center of curvature of the deformed neutral axis x at the point N^*. We denote the radius of curvature $\rho \equiv \overline{CN^*}$. We recall that the dimensions of the section are presumably unchanged so that $\overline{P^*N^*} \doteq$ y, i.e., if P and P_+ are a distance y above the neutral axis before deformation, then P^* and P_+^* remain approximately the same distance from the neutral axis after deformation. The dihedral angle between planes S^* and S_+^* is denoted by $\Delta\theta$ in Figure 2.49b. Therefore,

$$\overline{P^*P_+^*} = \overline{CP^*}\Delta\theta \doteq (\rho - y)\Delta\theta \tag{2.83}$$

Since the neutral axis does not stretch or contract, $\overline{N^*N_+^*} = \Delta x$. Also, from the geometry of Figure 2.49b, $\overline{N^*N_+^*} = \rho\Delta\theta$ Therefore,

$$\Delta\theta = \Delta x/\rho \tag{2.84}$$

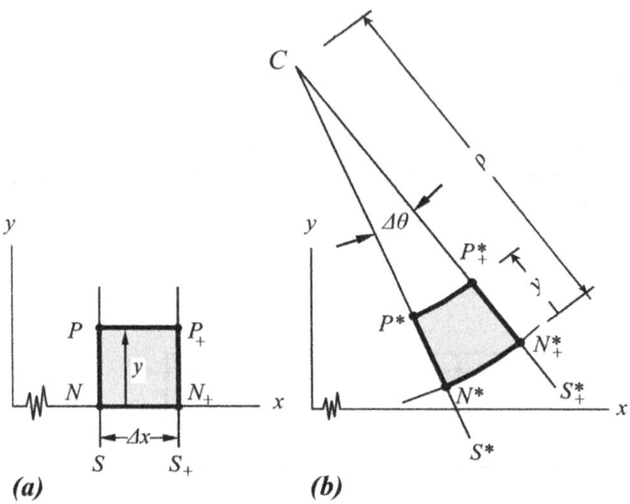

Figure 2.49

When equation (2.84) is substituted into (2.83), we obtain

$$\overline{P^*P_+^*} \doteq (\rho - y)\Delta x/\rho.$$

From the definition of extensional strain, we have

$$\varepsilon_x(x) = \lim_{P_+ \to P} \frac{\overline{P^*P_+^*} - \overline{PP_+}}{\overline{PP_+}} = -\frac{y}{\rho} \tag{2.85}$$

The extensional strain in the axial direction x at point P is linear in the distance y and proportional to the curvature $\kappa \equiv 1/\rho$ at the cross-section containing P; in general, $\rho = \rho(x)$. Our result is a basic kinematic relation in the elementary theories of beams.

ILLUSTRATIVE EXAMPLE 2.23

Let us consider the circular shaft shown in Figure 2.50. The shaft of Figure 2.50a is shown twisted in Figure 2.50b, i.e, one end of the shaft is rotated about the axis relative to the other. Such deformations result when a shaft is used to transmit torque. The drive shaft of an automobile, for example, behaves in this manner as it transmits power to the rear wheels. When a circular shaft is twisted slightly, the deformation can be described by the statements that (1) the axis of the cylinder remains straight after deformation and (2) all radii in a cross-section remain straight and rotate through the same angle about the axis of the cylinder. The circular shaft of Figure 2.50 illustrates a shaft that has been twisted according to these assumptions. A typical radius, OP, is shown in Figure 2.50a at a distance z from the rear end of the cylinder. For convenience, we shall assume that the back end of the cylinder is fixed. After deformation, the radius OP has rotated to OP^*, as shown in Figure 2.50b. Generally, the amount of rotation will depend on the location of the cross-section. We therefore denote it by $\varphi(z)$, where z is the distance of the cross-section from the rear end of the cylinder. Because we have assumed that the rear end of the cylinder is fixed, $\varphi(0) = 0$.

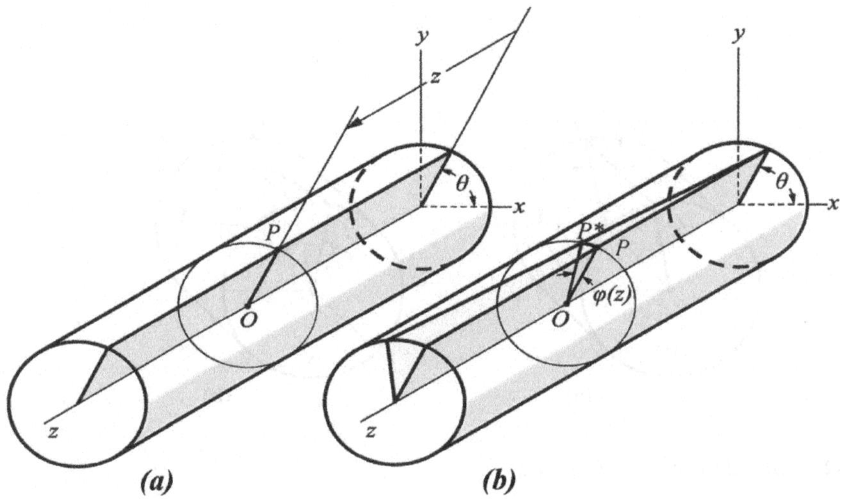

Figure 2.50 Torsion of a Circular Cylinder

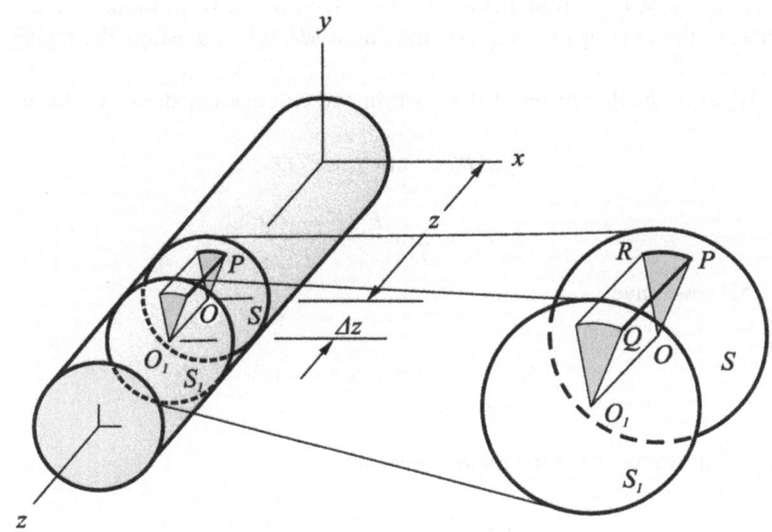

Figure 2.51 Undeformed Slice of the Cylinder

Figure 2.51 depicts a typical small slice of the cylinder before deformation. The slice is contained between the cross-sections Sand S_1, which are a distance Δz apart. When the cylinder twists, S_1 rotates relative to S, which causes a change in the angle QPR. To relate the shear strain $\gamma_{z\theta}(P)$ to the rotation $\varphi(z)$, we consider the edges PQ (in the axial direction) and PR (in the tangential direction). These edges are shown in Figure 2.52 before and after deformation.

When the cylinder twists, the radius OP, which initially makes an angle (θ) with the x axis, rotates through an angle $\varphi(z)$. Consequently, OP^* makes an angle $\theta + \varphi(z)$ with the x axis. On the other hand, the radius O_1Q, which initially makes an angle (θ) with the x axis, rotates into a

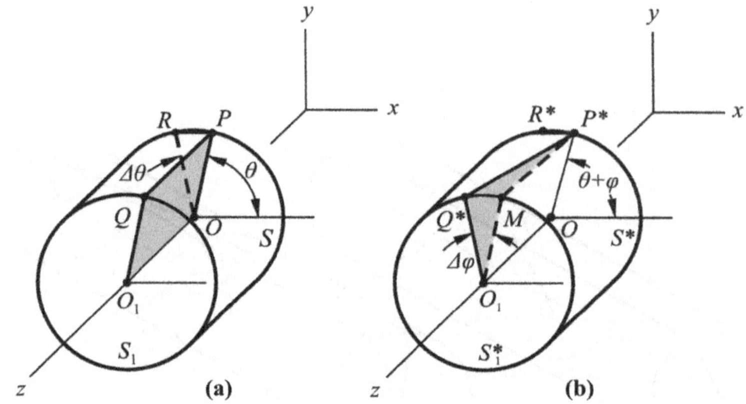

Figure 2.52 Twisted Slice of the Cylinder

position O_1Q^*, which makes an angle $\theta + \varphi(z+\Delta z)$ with the x axis. The angle is different because O_1Q is in a plane cross-section at a distance $(z+\Delta z)$ from the rear end of the cylinder. After deformation, the segment P^*R^* remains in the cross-section S^*. If, in Figure 2.52b, the line P^*M is drawn parallel to the axis of the cylinder, the angle MP^*Q^* represents the change in the right angle QPR.

Consequently, from the definition of shear strain and extensional strain, we have

$$\gamma_{z\theta}(P) = \lim_{Q\to P} \angle MP^*Q^* \tag{2.86a}$$

$$\varepsilon_z(P) = \lim_{Q\to P} \frac{P^*Q^* - PQ}{PQ} \tag{2.86b}$$

From Figure 2.52b, we have

$$MQ^* = r[\varphi(z+\Delta z) - \varphi(z)]$$
$$P^*M = \Delta z$$

Assuming a small strain, we have the approximation:

$$\angle MP^*Q^* \approx \frac{r[\varphi(z+\Delta z) - \varphi(z)]}{\Delta z}$$

Consequently,

$$\gamma_{z\theta}(P) = \lim_{Q\to P} \angle MP^*Q^* = \lim_{\Delta z\to 0} \frac{r[\varphi(z+\Delta z) - \varphi(z)]}{\Delta z} = r\frac{d\varphi}{dz} \tag{2.87}$$

In summary, we have related the shear strain associated with the axial and tangential directions at P to the rate of rotation of cross-sections along the axis of the cylinder.

Note that our derivation neglects any deformations of a cross-section; e.g., $OP^* = OP$. Axial extension or contraction is also neglected; e.g., $Q^*P^* \approx QP \approx MP^* \approx O_1O$. The latter implies the extensional strain ε_z in equation (2.86b) is negligible. These assumptions are implicit in most treatments and are usually justified. The same approximations would be inappropriate when treating the large strains of a rubber-like cylinder.

NOTES

1. Many alternative definitions are admissible. The definition adopted here is conventional in engineering; it is conceptually simple and adequate for most purposes.
2. Many alternative definitions are possible. The engineer adopts the one that serves most conveniently. Here, we give a definition that is conceptually simple, yet adequate.
3. This section represents a further development, which is not essential to subsequent applications
4. Upon further study (see Section 2.25), we find that these strains must satisfy certain differential equations to ensure the continuity of the body.
5. The mathematical derivation is provided in the following Section 2.12.
6. The mathematical derivation is provided in the following Section 2.12.
7. This rigorous development is not essential to the subsequent applications, but does emphasize the precise definitions and limitations.
8. This is a special topic, which is not essential to an understanding of subsequent developments.
9. A. L. Cauchy. Sur la condensation et la dilatation des corps solides. *Exercices de Mathematique*, 2:60–69, 1827.
10. G. Green. On the propagation of light in crystalline media. *Trans. Cambridge Phil. Soc.*, 7:121–140,1841.
11. A. E. H. Love. *A Treatise on the Mathematical Theory of Elasticity*, fourth edition. Cambridge University Press, Cambridge, 1927. Unabridged and unaltered republication of the fourth edition, Dover Publications, New York, 1944.
12. This is a special topic, which is not essential to an understanding of subsequent developments.
13. This formula is always valid for small strain ($\gamma_{nt} \ll 1$) but does not precisely give the angle ($\pi/2 - \varphi_{nt}$) when that angle is large.
14. The asterisk signifies a special topic, which is not essential to an understanding of subsequent developments.
15. In the past, Mohr's graphical representation was widely used as a tool to obtain strains and stresses. However, in contemporary practice, these transformations would be readily implemented via modern electronic facilities. Here, for pedagogical reasons, only a short description of the method is presented. The reader will find further details and examples in many traditional texts, e.g., the monographs by: R. G. Budynas. *Advanced Strength and, Applied Stress Analysis*, second edition. WCB/McGraw-Hill, Boston, 1999 (first edition: McGraw-Hill, New York, 1977). A. C. Ugural and S. K. Fenster. *Advanced Strength and Applied Elasticity*, fourth edition. Prentice Hall PTR, New Jersey, 2003 (second edition: Elsevier, New York, 1987).
16. The interested reader will find a complete mathematical account in the author's earlier book: "G. wempner and D. talaslidis. *mechanics of solids and shells: theories and approximations*. crc press, Boca Raton, 2003."
17. B. Fraeijs de Veubeke. A new variational principle for finite elastic displacements. *Int. J. Engrg. Sci.*, 10:745–763, 1972; B. Fraeijs de Veubeke. *A Course in Elasticity*. Springer-Verlag, New York, 1979. Translation from French.
18. G. Wempner and D. Talaslidis. *Mechanics of Solids and Shells: Theories and Approximations*. CRC Press, Boca Raton, 2003.
19. This rigorous development is not essential to the subsequent applications, but does emphasize the precise.
20. This is a special topic, which is not essential to an understanding of subsequent developments.
21. This is the initial step in the so-called inverse solutions in the theory of elasticity.

3 Internal Forces, Stresses, and Equilibrium

3.1 BRIEF HISTORICAL ACCOUNT

In our introductory chapter, we identified three distinct facets in the analysis of deformable bodies: kinematics, dynamics, and behavior of the material(s). To lay a firm foundation and to avoid misconceptions, the kinematic and dynamic aspects are treated independently: the former in Chapter 2, the latter in this chapter. Both are prerequisites to a description of the behavior of a material.

Here, the dynamics is viewed via the Newtonian and Lagrangian perspectives: Together, these viewpoints provide insight into the interactions of forces and the concept of stress; both yield the equations that govern equilibrium or motion. The reader has encountered the basic notions in previous studies of particles and "rigid" bodies via courses in physics and engineering. These ideas include the concepts of force, moment, and couple, the use of the free-body diagram, and the equations of equilibrium. Now, some extensions are needed to treat the mechanics of a *deformable* body.

Because each elemental portion of a body is subjected to different actions and undergoes different deformations, it is necessary to examine the actions upon internal surfaces. Hence, we pass surfaces *through* components of machines and structures and isolate the separate parts as free bodies. Those parts are subjected to forces that are internal to the entire component. Such *internal* forces require our special attention.

To describe accurately the effects of forces upon elements of *arbitrary* size, another concept is needed, that is, the concept of *stress*, which serves to define the intensity and the nature of force upon surfaces of arbitrary size and orientation and, specifically, surfaces of infinitesimal size. The definitions and analyses of *stress* and the *components* of *stress* are *essential* to our subsequent treatment of the behavior of materials and, ultimately, to our analysis of each deformable body.

3.2 CLASSIFICATION OF EXTERNAL FORCES

3.2.1 BODY AND SURFACE FORCES

The external forces that act on a body may be classified in different ways. For example, since every force is ultimately the action of one body on another, the force can arise in one of two ways. In one case, the force results from physical contact between two bodies. Thus, a book applies a force to a desk on which it rests. If the book is raised so that there is no physical contact between the book and the desk, then the force vanishes. Such forces are called contact forces. In the other case, one body exerts forces on another body even though there is no contact between them. The gravitational force called weight, which the earth applies to the book in the previous example, is such a force. The electromagnetic force that keeps a compass needle pointing north-south is another example. To distinguish these forces from contact forces, we call them *body forces*.

Since contact is not an essential element of body forces, it follows that all parts of the body are involved; that is, the force is distributed throughout the body. Contact forces, however, can act only on those portions of a body that are in direct contact with the body exerting the force; that is,

DOI: 10.1201/9781003665861-3

the force is distributed over a portion of the surface of the body. Hence, contact forces are often called *surface* forces or tractions. In many cases, a surface force is distributed over a relatively small region compared with the other dimensions of the body. It is sometimes possible to idealize such forces and treat them as if they acted at a point. Such an idealization is called a *concentrated* force. Also, if a surface force is distributed over a relatively narrow region, it might be idealized as a *line force* – that is, a force distributed along a line. Concentrated and line forces do not exist in reality, but are convenient idealizations. For instance, in some cases, it might be possible to regard the weight of a truck on a road as a concentrated force, while in other cases, the actual pressure distribution under the truck tires must be considered. The forces that a floor joist exerts on the floor it supports can generally be treated as a line force.

We depict a concentrated force with an arrow in the same way as we usually represent a vectorial quantity. The length is proportional to the magnitude; the direction is indicated by the arrowhead. The point of application of a concentrated force may be indicated by the point of the arrowhead. The simplified sketches of Figure 3.1 depict concentrated forces acting upon a plate (a), a beam (b), a cable (c), and a rod (d).

A line force is usually sketched as a line of arrows, as shown in the examples of Figure 3.2. The intensity of the line force at any point is proportional to the length of the arrow to that point. The units of line forces are [force]/[length]. To find the total force, we multiply the intensity by a length. Thus, for example, the total force acting on the bar in Figure 3.2a is pL, whereas the total force on the body in Figure 3.2b is $\int_0^L p(x)dx$. The total force distributed upon a surface is given by the integral over area A, e.g., $\int\int_A p(x,y)dA$. The resultant of a body force, such as a magnetic action, is an integral throughout the volume V, e.g, $\int\int\int_V p(x,y,z)dV$. Here, $p(x,y)$ and $p(x,y,z)$ are force/area and force/volume, respectively.

Body forces are sometimes represented by their resultant, although this practice is dangerous. For example, the weight of the cylinder shown in Figure 3.3a might be indicated by a single vector acting at the center of gravity; however, it is not a concentrated force. If we consider only half the cylinder, as shown in Figure 3.3b, the body force is only one-half of the previous resultant. The actual distribution must be considered when treating the deformable body.

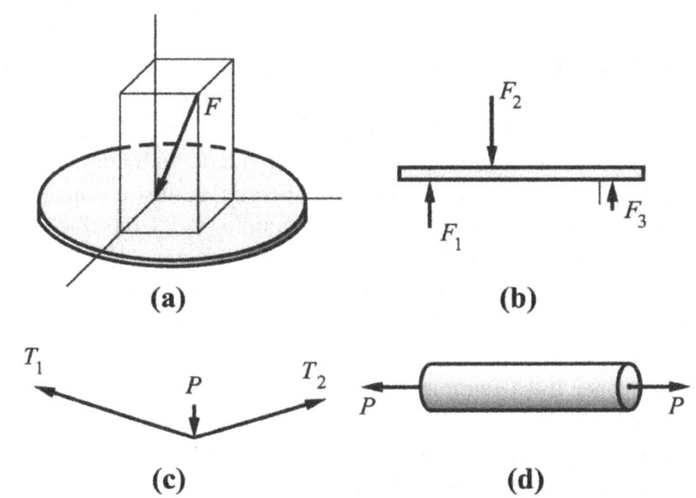

Figure 3.1 Concentrated Loads

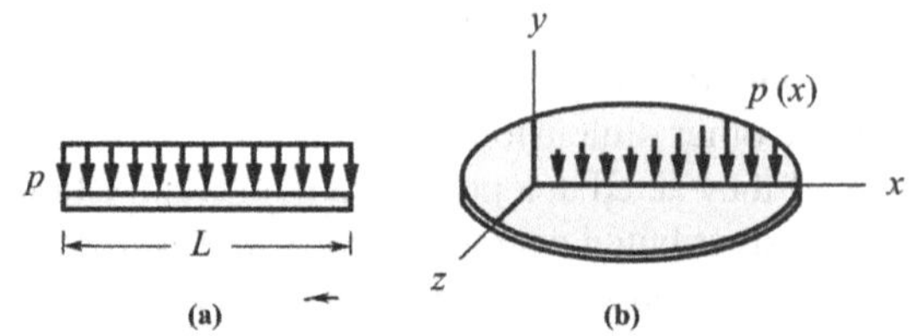

Figure 3.2

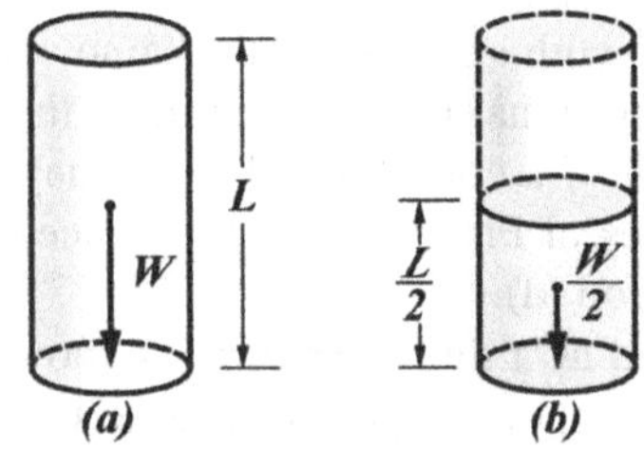

Figure 3.3

3.2.2 LOADS AND REACTIONS

Some of the external forces that act on a body arise from the connections that support the body. Let us return to the example of a book resting on a desk. The force that the desk exerts on the book is exactly enough to support it. If something else is placed on the top of the book, the supporting force of the desk adjusts so that the book remains where it is. As another example, let us consider a door on hinges. The supporting forces, which the hinges exert on the door, will allow the door to open or close but will restrict any other motion. The essential element of supporting forces is that they arise in order to restrict the displacement of the body that they support. Such supporting forces are called reactions. Any other external forces acting on a body are called *loads*. Thus, the weight of a person on a floor is a load on the floor, but the supporting forces exerted by the floor joists are reactions. The actual reactions (forces and/or couples) exerted by a support, and the restraints provided are often quite complicated. In many circumstances, approximations or idealizations are admissible. Some simple idealizations are tabulated in Figure 3.4 together with the motions, components of translation (→) and rotation (↠) that are or are not permitted, and the components of force (→) and couple (↠) that are provided. An inadmissible motion or action is indicated by a stroke (↛) or (↠̸). Notice that the restraint against a translation (↛) requires a force (→); a restraint against a rotation (↠̸) requires a couple (↠). Conversely, if the action (force or couple) is absent, then the corresponding motion occurs.

The effect of a rocker or roller (Case 4 in Figure 3.4) is to restrain displacement in one direction while leaving the body free to move in the perpendicular direction and/or rotate. This reaction is always a force perpendicular to the direction of the constrained motion. Such supports are common in bridge construction, since it is desirable to support the span while leaving it free to expand and contract with temperature changes.

A *clamped*, or *fixed*, support (Case 1 in Figure 3.4) does not permit displacement or rotation. These supports are required for overhanging structures. The reaction at a clamped support consists, in general, of a force to restrict displacement and a couple to restrict rotation.

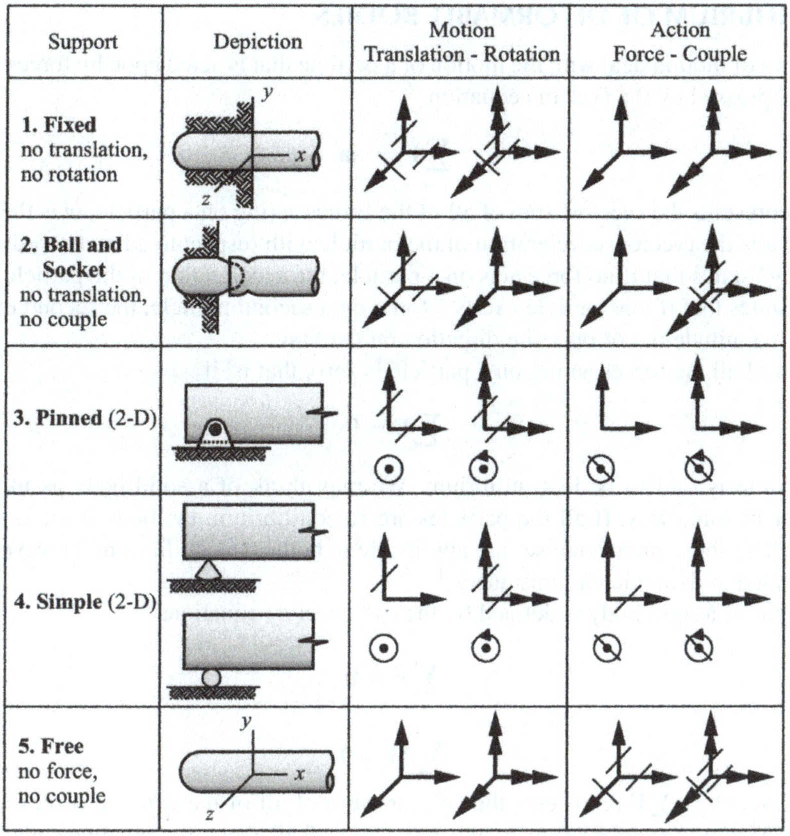

Figure 3.4 Idealized Supports

3.2.3 STATIC AND DYNAMIC LOADS

The effect of a load on a body might depend on how the load is applied. If a load is steady—that is, if the load does not change intensity, direction, or location with time—it is called a *static* load. Loads that change with time are called *dynamic* loads. Sometimes the changes in a dynamic load are so small or so gradual that the load may be treated as static. The weight of a building on its foundation as people enter and leave, and the fluid pressure at any instant at the bottom of a slowly filling tank is usually taken as static loads. Dynamic loads may be further subdivided: A truck being driven over a bridge applies a *moving* load to the structure; a connecting rod in a reciprocating engine is subjected to *repeated loading*; a hammer striking the head of a nail applies an *impact load.*

The different classifications of external forces discussed in this section are very useful. The terms should be treated as engineering definitions with specific meanings.

3.3 EQUILIBRIUM OF DEFORMABLE BODIES

Newton's laws of motion deal with the motion of a particle that is acted upon by forces. The second law can be expressed by the (vector) equation

$$\sum \mathbf{F} = m\mathbf{a} \tag{3.1}$$

where $\sum \mathbf{F}$ represents the (vector) sum of all of the forces acting on a particle, m is the mass of the particle, and $\mathbf{a}$ is the (vector) acceleration of the particle with respect to a fixed coordinate system. The first "law" states that if no force acts on a particle, the acceleration of the particle is zero. The third "law" states that if one particle exerts a force on a second particle, the second exerts a force of the same magnitude but of opposite direction on the first.

If the sum of all the forces acting on a particle is zero, that is, if

$$\sum \mathbf{F} = \mathbf{0} \tag{3.2}$$

then the particle is said to be in equilibrium. We may think of a solid body as many particles held together in some way. If all the particles are in equilibrium, the body itself is said to be in equilibrium. The subsequent analyses are applicable to bodies in equilibrium; however, results are readily extended to dynamic circumstances.[1]

Equilibrium of a rigid body is defined by the two (vector) equations

$$\sum \mathbf{F} = \mathbf{0} \tag{3.3a}$$

$$\sum \mathbf{M} = \mathbf{0} \tag{3.3b}$$

In Equations (3.3), $\sum \mathbf{F}$ represents the (vector) sum of all of the external forces acting on the body, and $\sum \mathbf{M}$ is the (vector) sum of the moments of all these forces about any point. For a deformable body in equilibrium, Equations (3.3a,b) also hold, not only for the body as a whole but also for any portion of it.

In any study of deformable bodies, it is necessary to examine the forces that act between adjacent parts of the body. With this in mind, consider the body depicted in Figure 3.5. It is acted upon

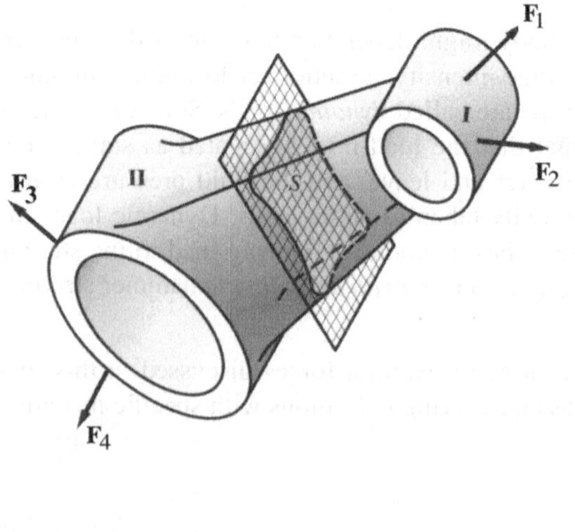

Figure 3.5

by the system of external forces $\mathbf{F}_1$, $\mathbf{F}_2$, $\mathbf{F}_3$, and $\mathbf{F}_4$. We suppose that the body is in equilibrium under the action of these forces. Imagine the body to be cut by a plane that divides it into two parts, I and II, with the common surface *S*. Let us examine one of these parts, for instance portion II, and the forces that act on it.

A free-body diagram of portion II is shown in Figure 3.6. The forces $\mathbf{F}_3$ and $\mathbf{F}_4$ are not the only forces that act on it. In addition, we also have the forces that the material of portion I exerts on the material in portion II. These forces are internal forces for the body as a whole but are external to portion II. Their distribution over the surface *S* is entirely unknown at present. We can, however, discuss the (vector) sum of these forces, which is denoted by $\mathbf{F}_{II}$ and the (vector) sum of the moments of these forces about a point *P* in *S*, which is denoted by $\mathbf{M}_{II}$. The force system consisting of the force $\mathbf{F}_{II}$ acting at *P* and the couple $\mathbf{M}_{II}$ is *statically equivalent* to the system of distributed forces on the surface *S*. This means that in any calculation involving only the equilibrium of portion II, the actual distribution of forces on *S* can be replaced by $\mathbf{F}_{II}$ acting at *P* and $\mathbf{M}_{II}$. We shall refer to $\mathbf{F}_{II}$ and $\mathbf{M}_{II}$ as the *net internal force system* on *S*.

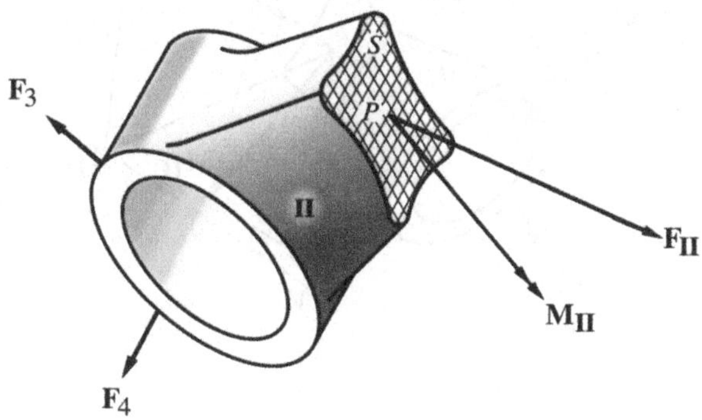

Figure 3.6

Since portion II is in equilibrium, the force system consisting of $\mathbf{F}_3$, $\mathbf{F}_4$, $\mathbf{F}_{II}$, and $\mathbf{M}_{II}$ must satisfy the equilibrium conditions, Equations (3.3a,b). From these equations, the net internal force system on *S* is

$$\mathbf{F}_{II} = -(\mathbf{F}_3 + \mathbf{F}_4), \quad \mathbf{M}_{II} = -(\mathbf{M}_3 + \mathbf{M}_4)$$

where $\mathbf{M}_3$ and $\mathbf{M}_4$ are the moments of the forces $\mathbf{F}_3$ and $\mathbf{F}_4$ *about* the point *P*. The point *P* is the assumed point of application of the net internal force $\mathbf{F}_{II}$: it plays a passive but important role in the use of Equation (3.3b) that cannot be overlooked. If we change the location of *P*, we will also, in general, change the value of $\mathbf{M}_{II}$. Consequently, we *must always identify the point P* in such representations of the net internal system. If we were to use a free-body diagram of portion I, as in Figure 3.7, instead of portion II, then the net internal force system on *S* would be $\mathbf{F}_1 + \mathbf{F}_2$ and $\mathbf{M}_1 + \mathbf{M}_2$. As a direct consequence of Newton's third law,

$$\mathbf{F}_1 + \mathbf{F}_2 = -\mathbf{F}_{II}, \quad \mathbf{M}_1 + \mathbf{M}_2 = -\mathbf{M}_{II}$$

Let us now focus on the actions in part II in Figure 3.8 where it is usually more convenient to use components of the net internal force and couple. One component is taken normal to the plane of the sectioning surface, and the other component is taken tangent to the plane. This decomposition is shown in Figure 3.8, where the axis *n* is normal to the plane of *S*.

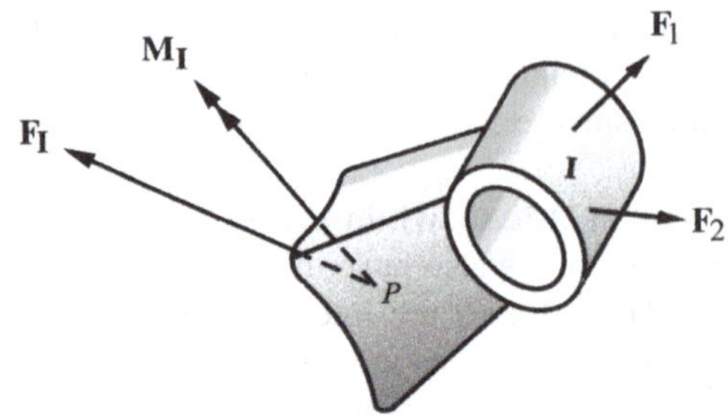

Figure 3.7

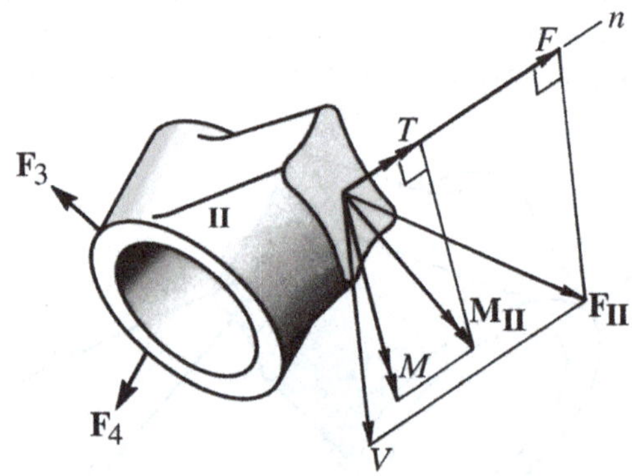

Figure 3.8

The force $\mathbf{F}_{II}$ is decomposed into a normal component F and a tangential component V. These are termed *normal force* (F) and the *shear force* (V), respectively. The couple $\mathbf{M}_{II}$ is similarly decomposed into a normal component T and a tangential component M. These are called the *twisting couple* (or torque T) and the *bending moment* (M), respectively.

The direction of the normal force F and twisting couple T is determined if the orientation of the sectioning plane is prescribed. However, the shear force V and bending couple M can have any direction in the plane of S. To be explicit, we can identify two directions tangent to surface S and decompose the shear force and bending couple into components in the chosen directions. Most conveniently, we identify orthogonal directions, s and t, tangent to the surface, as depicted in Figure 3.9. Here, V_s and V_t are the components of the shear force $\mathbf{V}$ in the s and t directions. Similarly, M_s and M_t are the components of the bending couple $\mathbf{M}$. Forces and couples are shown separately in Figure 3.9a and Figure 3.9b, respectively, simply to avoid confusion. These six components of the net internal force system can now be determined from the scalar form of the equilibrium conditions:

$$\begin{aligned} \sum F_s = 0, \quad \sum F_t = 0, \quad \sum F_n = 0 \\ \sum M_s = 0, \quad \sum M_t = 0, \quad \sum M_n = 0 \end{aligned} \tag{3.4}$$

Of course, if these equations are used, the components of the external forces (in this case $\mathbf{F}_3$ and $\mathbf{F}_4$) must be included. A normal force on a cross section arises when there is a tendency on the part of the external forces to compress or pull apart the two portions of the body on either side of the sectioning plane. A shear force comes about when the external forces tend to slide

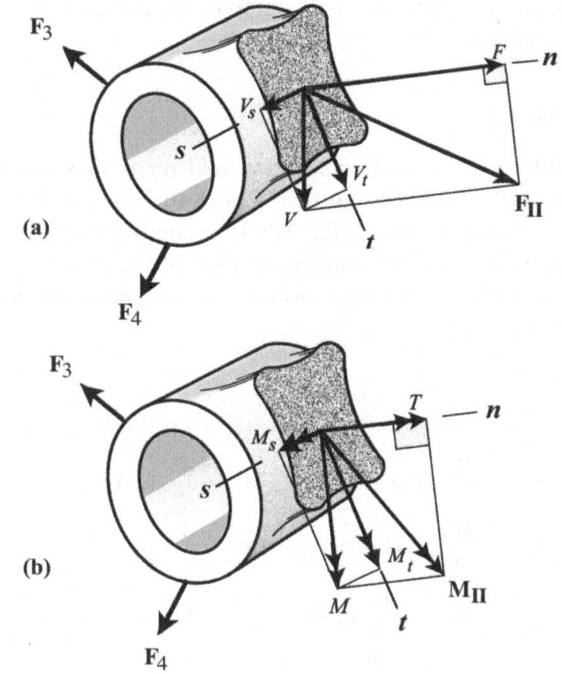

Figure 3.9

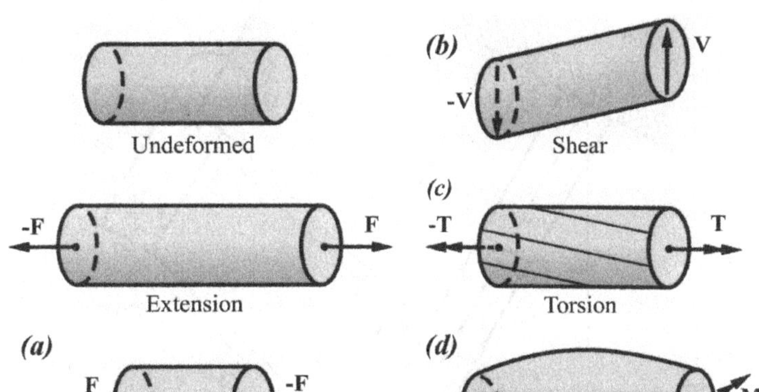

Figure 3.10

one portion over the other. When the external forces tend to twist one portion with respect to the other, a twisting couple arises. If the two portions are bent by the external forces, a bending couple results. These four situations are depicted in Figure 3.10, which depicts a cylinder stretched or compressed by a normal force F in (a), sheared by a force V in (b), twisted by a couple T in (c), and bent by a couple M (d).

To illustrate the application of these ideas, we compute the components of the net internal force system in the following examples.

ILLUSTRATIVE EXAMPLE 3.1

Consider a circular cylinder hanging under its own weight W, as shown in Figure 3.11a. The axis of the cylinder makes an angle θ with the vertical. Let us find the components of the net internal force system on a cross section one-third of the way from the free end of the cylinder.

In Figure 3.11b, a free-body diagram of the lower third of the cylinder is shown with an $(x\ y\ z)$ coordinate system indicated in the sectioning surface. The weight of this portion of the cylinder is $W/3$. Equilibrium equations yield

$$\sum F_x = 0,\ V - \frac{W}{3}\sin\theta = 0, \Longrightarrow \quad V = \frac{W}{3}\sin\theta$$

$$\sum F_y = 0,\ F - \frac{W}{3}\cos\theta = 0, \Longrightarrow \quad F = \frac{W}{3}\cos\theta$$

$$\sum M_{z(y=0)} = 0,\ M - \left(\frac{W}{3}\sin\theta\right)\left(\frac{L}{6}\right) = 0, \Longrightarrow \quad M = \frac{WL}{18}\sin\theta$$

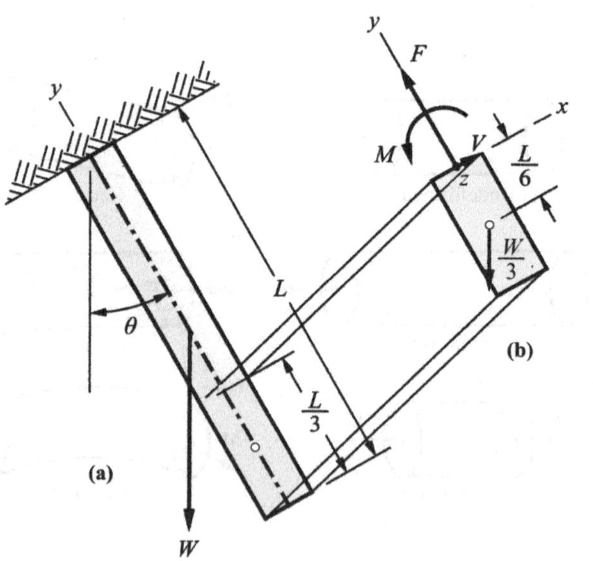

Figure 3.11

ILLUSTRATIVE EXAMPLE 3.2

Let us determine the shear force and bending couple on the center cross-section of the simply supported beam shown in Figure 3.12a.

Regardless of which half of the beam we take as a free body, an unknown reactive force at A or B will appear. We therefore consider a free-body diagram of the entire beam first (Figure 3.12b) to determine these forces:

$$\sum M_{z(x=0)} = 0$$
$$R_B(3.0) - (2250)(2.4) - (150 \times 3.0)(1.5) = 0$$
$$R_B = 2025\ \text{N}(455\ \text{lb})$$

$$\sum F_y = 0$$
$$R_A + R_B - 2250 - (150)(3.0) = 0$$
$$R_A = 675.0\ \text{N}(151.8\ \text{lb})$$

A free-body diagram of the left half of the beam is shown in Figure 3.12c, with the known value of R_A indicated. The equilibrium conditions for this portion enable us to determine V and M:

$$\sum F_y = 0; \quad V + 675 - (150 \times 1.5) = 0, \implies V = -450\ \text{N}(-101\ \text{lb})$$
$$\sum M_{z(x=1.5m)} = 0; \quad M - 675 \times 1.5 + (150 \times 1.5)(0.75) = 0, \implies M = 844\ \text{N} \times \text{m}(624\ \text{lb} \cdot \text{ft})$$

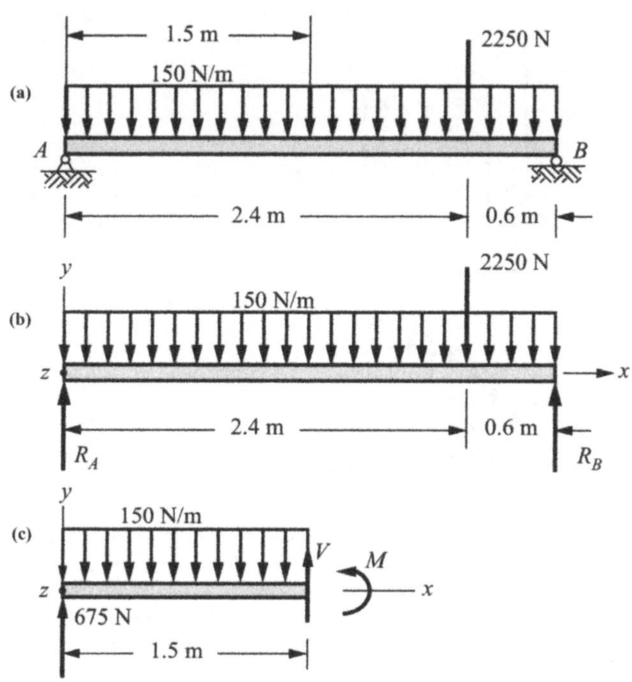

Figure 3.12

ILLUSTRATIVE EXAMPLE 3.3

The intensity of the lift force on the wing of the airplane shown in Figure 3.13 is given by

$$w_x = w_o \sin \frac{\pi x}{2L}$$

Determine the shear force and bending couple on the cross-section adjacent to the fuselage. The equilibrium conditions for the free-body diagram of Figure 3.13 yield the shear force V and bending moment M:

$$\sum F_y = 0$$

$$V - \int_0^L w_0 \sin \frac{\pi x}{2L} dx = V + w_0 \sin \frac{2L}{\pi} \left[\cos \frac{\pi x}{2L} \right]_0^L = 0$$

$$V = w_0 \frac{2L}{\pi}$$

$$\sum M_{z(x=L)} = 0:$$

$$M + \int_0^L w_0 (L - x) \sin \frac{\pi x}{2L} dx = M - w_0 \frac{2L^2}{\pi} \left[\cos \frac{\pi x}{2L} \right]_0^L - w_0 \left(\frac{2L}{\pi} \right)^2 \left[\sin \frac{\pi x}{2L} \right]_0^L + w_0 \frac{2L}{\pi} \left[x \cos \frac{\pi x}{2L} \right]_0^L = 0$$

$$M = w_0 \frac{2L^2}{\pi} \left(-1 + \frac{2}{\pi} \right)$$

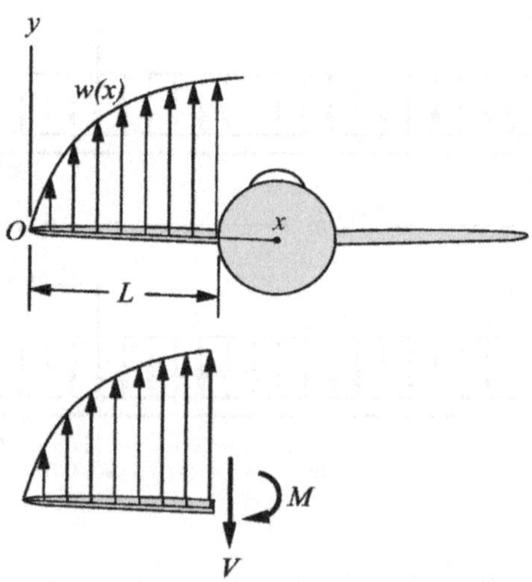

Figure 3.13

ILLUSTRATIVE EXAMPLE 3.4

The bent bar shown in Figure 3.14 consists of three straight portions joined consecutively at right angles. The first portion, OA, firmly clamped at its base O, has a length Y. The next portion, AB

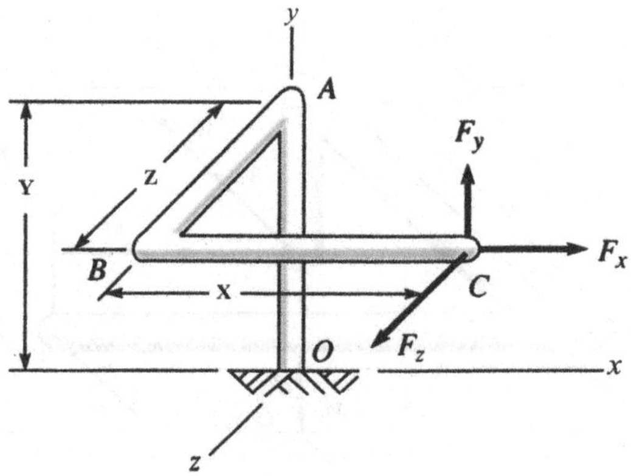

Figure 3.14

(parallel to the z-axis), of length Z, is connected to a portion BC of length X (parallel to the x-axis). At the free end C a force having components F_x, F_y, and F_z is applied. Compute the components of the net internal force system on the cross-sections at the center of each straight portion of the bar.

From the free-body diagrams shown in Figure 3.15a–c, we employ the equilibrium equations:

$$\sum F_x = 0, \quad \sum F_y = 0, \quad \sum F_x = 0$$
$$\sum M_x = 0, \quad \sum M_y = 0, \quad \sum M_z = 0$$

a) For the cross-section at the center of OA:

$$V_x = -F_x, \qquad F_n = F_y, \qquad V_z = F_z$$
$$M_x = F_y Z - \tfrac{1}{2} F_z Y, \qquad M_y = F_x Z - F_z X, \qquad M_z = F_y X - \tfrac{1}{2} F_x Y$$

b) For the cross-section at the center of AB:

$$V_x = -F_x, \qquad V_y = -F_y, \qquad F_n = F_z$$
$$M_x = \tfrac{1}{2} F_y Y, \qquad M_y = F_z X - \tfrac{1}{2} F_x Z, \qquad M_z = F_y X$$

c) For the cross-section at the center of BC:

$$F_n = F_x, \qquad V_y = -F_y, \qquad V_z = -F_z$$
$$M_x = 0, \qquad M_y = \tfrac{1}{2} F_z X, \qquad M_z = -\tfrac{1}{2} F_y X$$

Each segment of the bar is subject to bending couples: M_x and M_z on OA, M_x and M_y on AB, and M_y and M_z on BC. Segments OA and AB are subject to twisting couples M_y and M_z, respectively. All contribute to the deformation and to the displacement at C. Those deformations and the displacement are dependent on the dimensions and material properties, which are explored in Chapter 4. In Chapter 9, we return to this example to determine the deflection at C via an energy method.

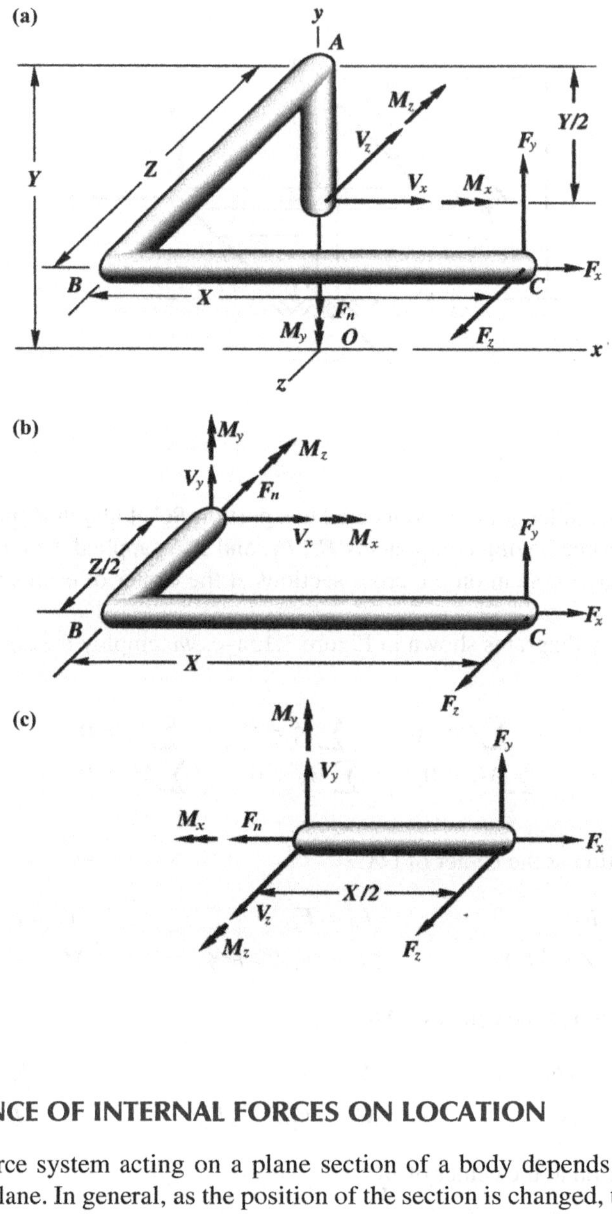

Figure 3.15

3.4 DEPENDENCE OF INTERNAL FORCES ON LOCATION

The net internal force system acting on a plane section of a body depends on the location and orientation of this plane. In general, as the position of the section is changed, the normal and shear forces, as well as the twisting and bending couples, change. These components of the net internal force system can be related to the variables that specify the location of the section. Some examples serve to illustrate these remarks.

ILLUSTRATIVE EXAMPLE 3.5

A cylindrical bar hanging under its own weight W is shown in Figure 3.16a.

We can specify the location of a cross section by its distance from the fixed end. Let us denote this distance as x. A free-body diagram of the portion of the cylinder below the cross-section at

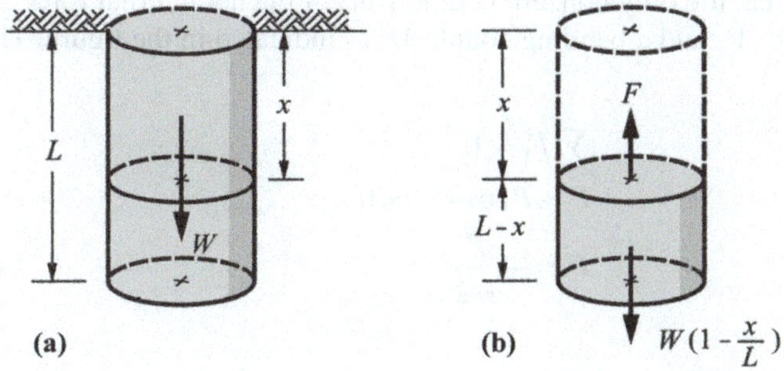

Figure 3.16

x is shown in Figure 3.16b. The net internal force system on this cross-section consists only of a normal force F. For equilibrium, all other components must be zero. The weight of the portion below the cross section at x is

$$W\frac{(L-x)}{L} = W\left(1-\frac{x}{L}\right)$$

For equilibrium of this portion, we must have

$$\sum F_x = 0$$
$$F - W\left(1-\frac{x}{L}\right) = 0, \Longrightarrow F = W\left(1-\frac{x}{L}\right)$$

This result can be illustrated graphically by a plot of the normal force F versus the coordinate x of the sectioning plane, as shown in Figure 3.17. This graph is called a *normal force diagram.*

Similar graphs can be made for other components of the net internal force system. A plot of the shear force versus x is called a *shear force diagram*, and a plot of the bending or twisting couple versus x is called a *bending couple diagram*, or a *twisting couple diagram.*

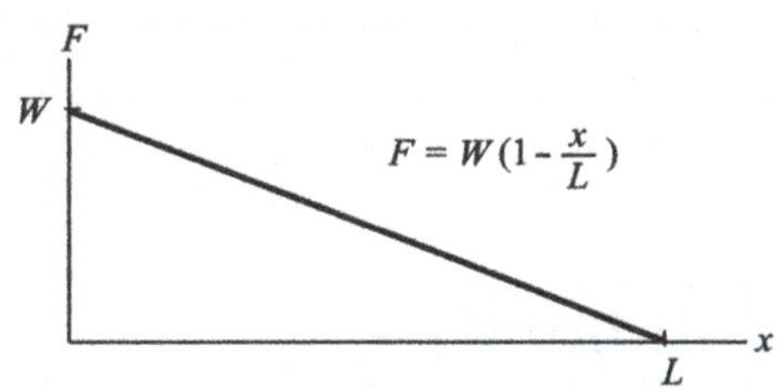

Figure 3.17

ILLUSTRATIVE EXAMPLE 3.6

Consider a cantilevered beam loaded as shown in Figure 3.18a. Again, an arbitrary cross-section of the beam can be located by the one coordinate x, the distance of this cross-section from the free end of the beam. To find the components of the net internal force system on this cross-section, we take, as a free body, the portion of the beam to the left of this cross-section (Figure 3.18b). Since the loading is coplanar, the only nonzero components of the net internal force system are a normal force F, a shear force V, and a bending couple M, as indicated in the figure. Then,

$$\sum F_x = 0$$
$$F + P\cos 45^\circ = 0$$
$$F = -\frac{P}{\sqrt{2}} \quad \text{(A)}$$

$$\sum F_y = 0$$
$$V - P\sin 45^\circ - wx = 0$$
$$V = \frac{P}{\sqrt{2}} + wx \quad \text{(B)}$$

$$\sum M_z = 0$$
$$M + (P\sin 45^\circ)x + (wx)\frac{x}{2} = 0$$
$$M = \frac{-Px}{\sqrt{2}} - \frac{wx^2}{2} \quad \text{(C)}$$

These equations are valid for every cross-section defined by a value of x between $x = 0$ and $x = L$, because the same free-body diagram, Figure 3.18b, would result for any x in this range. Therefore, these equations give the components of the net internal force system on every cross-section of the beam. They are plotted as the normal force diagram, the shear force diagram, and the bending couple diagram, in that order, in Figure 3.19. These diagrams give only the general shapes of the curves, since we do not have numerical values of P, L, and w.

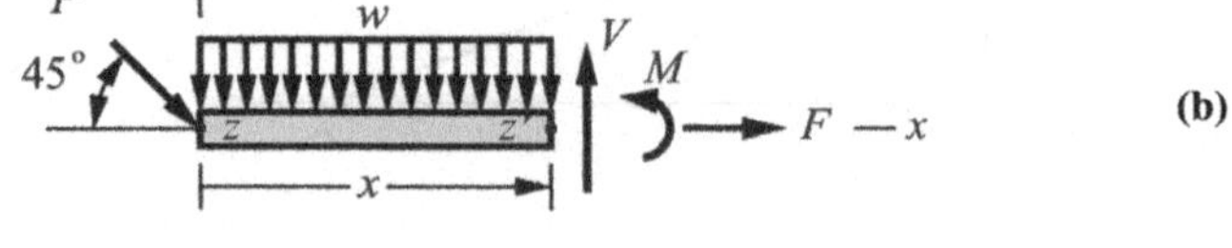

Figure 3.18

ILLUSTRATIVE EXAMPLE 3.7

Consider the beam shown in Figure 3.20a. An arbitrary cross-section can again be located by x, the distance of the cross-section from the pinned end of the beam. But when a free-body diagram is

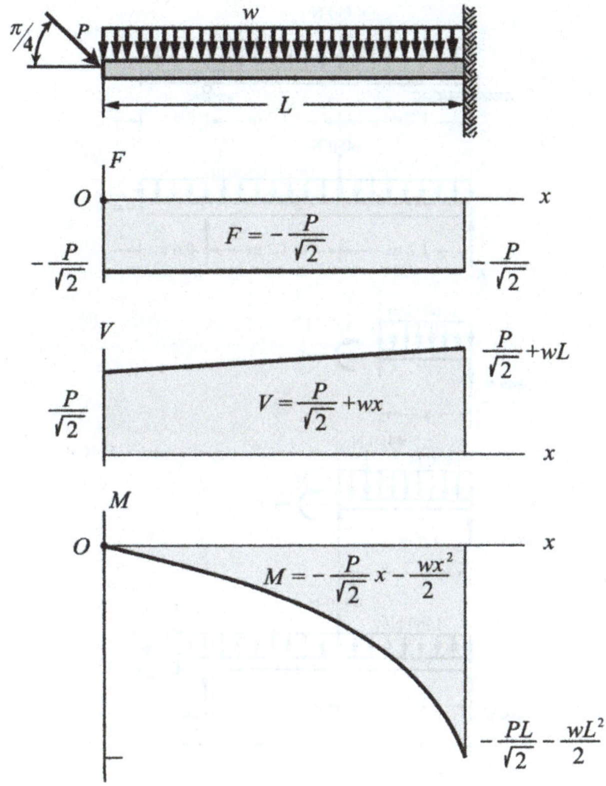

Figure 3.19

drawn of the portion to the left of a cross-section, three different situations arise. These are shown in Figure 3.20c–e depending on whether $0 < x < 1.2$ m, $1.2\text{ m} < x < 2.4$ m, or $2.4\text{ m} < x < 3.0$ m. Each of these free bodies is acted upon by a different system of external loading. Each must, therefore, be considered separately.

We first determine the reactions R_1 and R_2 from the equations of equilibrium applied to the beam as a whole. The free-body diagram is shown in Figure 3.20b.

$$\sum M_{z(x=0)} = 0 \Rightarrow R_2(2.4) - (1460 \times 3.0)(1.5) - (4450)(1.2) = 0 \Rightarrow R_2 = 4963\text{ N}(1115\text{ lb})$$
$$\sum F_y = 0 \Rightarrow R_1 + R_2 - 4450 - (1460)(3.0) = 0 \Rightarrow R_1 = 3867\text{ N}(869.5\text{ lb})$$

From Figure 3.20c, for $0 < x < 1.2$ m,

$$\sum F_y = 0; \Rightarrow V = (-3868 + 1460x)\text{N} \quad \text{(A)}$$
$$\sum M_{z(x=0)} = 0 \Rightarrow M = \left(3868x - 1460\frac{x^2}{2}\right)\text{ N}\cdot\text{m} \quad \text{(B)}$$

From Figure 3.20d, for $1.2m < x < 2.4$ m,

$$\sum F_y = 0 \Rightarrow V = (-3868 + 4450 + 1460x)\text{ N} = (582.5 + 1460x)\text{ N}) \quad \text{(C)}$$
$$\sum M_{z(x=0)} = 0 \Rightarrow M = [3868x - 4450(x - 1.2) - 730x^2]\text{ N}\cdot\text{m} \quad \text{(D)}$$

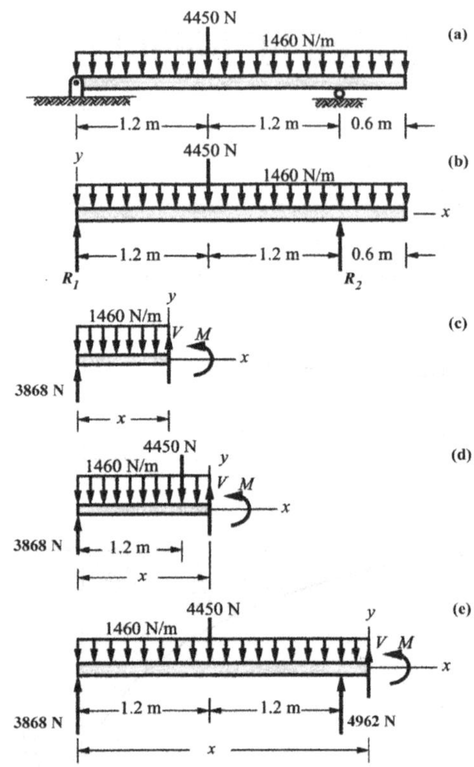

Figure 3.20

From Figure 3.20e, for $2.4m < x < 3.0$ m,

$$\sum F_y = 0 \Rightarrow V = (-3868x + 4450 - 4963 + 1460x)\ \text{N} = (-4380 + 1460x)\ \text{N} \qquad \text{(E)}$$

$$\sum M_{z(x=0)} = 0 \Rightarrow M = [3868x - 4450(x - 1.2) + 4963(x - 2.4) - 730x^2]\ \text{N}\cdot\text{m} \qquad \text{(F)}$$

$$M = (-6570 + 4380x - 730x^2)\ \text{N}\cdot\text{m}$$

Equations (A)–(F) can be plotted in the regions in which they are valid to give the shear force diagram and the bending couple diagram of Figure 3.21.

ILLUSTRATIVE EXAMPLE 3.8

As another example, let us consider the curved bar shown in Figure 3.22a. It is a quarter of a circular bar, fixed at one end, with a 500 N (112.4 lb) load applied to the free end perpendicular to the plane of the bar. In this case, we can locate an arbitrary cross-section by the angle θ shown in Figure 3.22. To find the components of the net internal force system on such a cross-section, we consider as a free body the portion of the bar shown in Figure 3.22b and Figure 3.22c. For equilibrium of this portion of the bar, we have

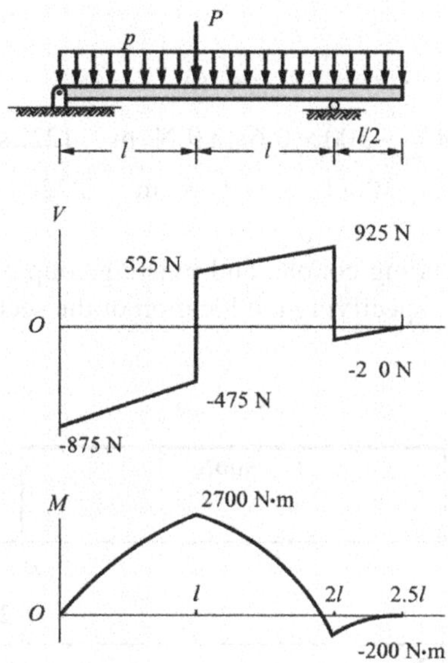

Figure 3.21

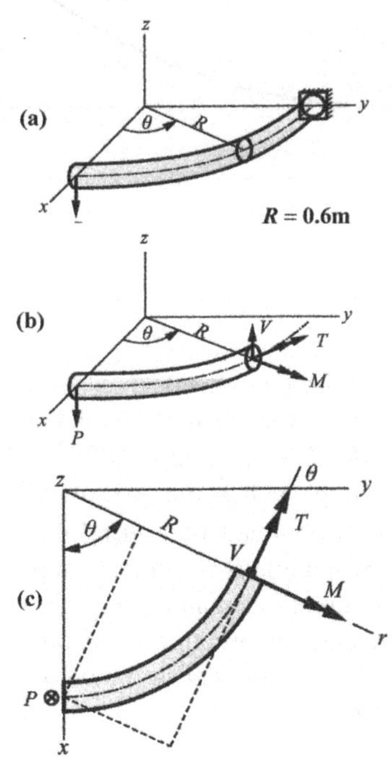

Figure 3.22

$$\sum F_z = 0; \quad V = 500\text{ N} \quad (1125\text{ lb})$$
$$\sum M_r = 0; \quad M = -500 \times 0.6 \sin\theta \text{ N}\cdot\text{m} \quad (222 \sin\theta \text{ lb}\cdot\text{ft})$$
$$\sum M_\theta = 0; \quad T = 300(1 - \cos\theta)\text{ N}\cdot\text{m} \quad [222(1 - \cos\theta)\text{ lb}\cdot\text{ft}]$$

The resulting shear force, bending couple, and twisting couple diagrams are shown in Figure 3.23. In this case, the coordinate specifying the location of the sectioning plane is an angle rather than a distance.

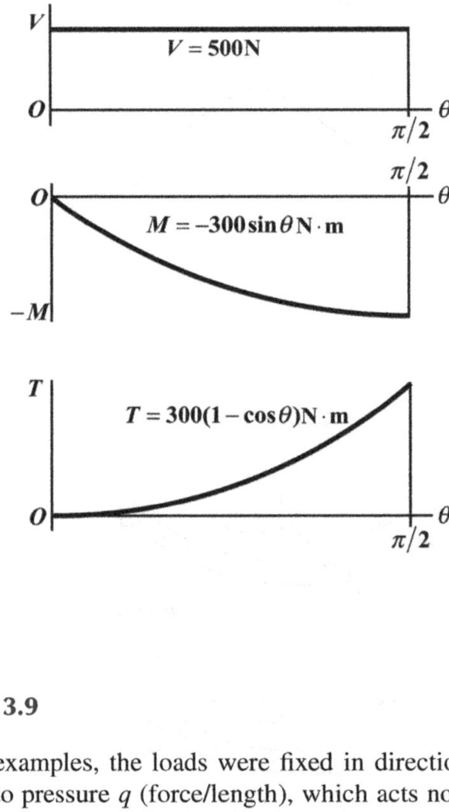

Figure 3.23

ILLUSTRATIVE EXAMPLE 3.9

In each of the preceding examples, the loads were fixed in direction. Now the circular ring of Figure 3.24a is subjected to pressure q (force/length), which acts normal to the surface; on each element of length $Rd\beta$, the force $qRd\beta$ is directed radially as shown. Often, domes of shell-like structures are subjected to such loading, such as the pressure of water upon a submerged hull.

The free body of Figure 3.24b is obtained by cutting the ring at an arbitrary angle θ. The external load upon a typical element at angle β is $qRd\beta$. Components of normal force F and shear force V act upon the section at A in directions n and t, respectively; additionally, a bending couple M acts upon the cross section. Note that the direction of the pressure changes with angle β, always normal to the surface of the ring. The component in direction n is $qRd\beta\ \sin(\theta - \beta)$, and the net component in direction n is the integral

$$\int_0^\theta qR\sin(\theta - \beta)d\beta = qR(1 - \cos\theta)$$

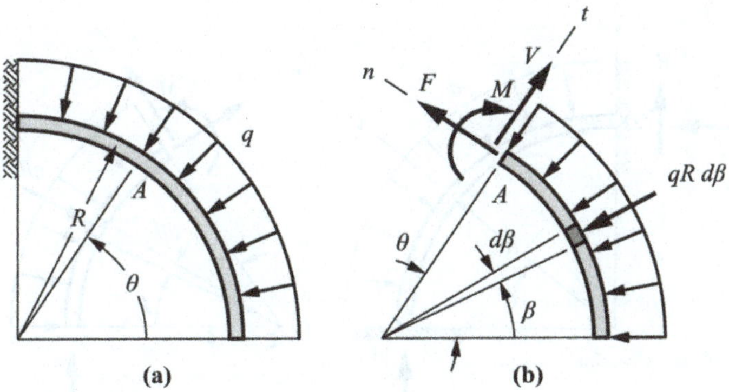

Figure 3.24

Equilibrium requires that

$$\sum F_n = F + \int_0^\theta qR\sin(\theta-\beta)d\beta = 0, F = -qR(1-cos\theta)$$

so that

$$F = -qR(1-\cos\beta)$$

Additionally

$$\sum F_t = V - \int_0^\theta qR\cos(\theta-\beta)\mathrm{d}\beta = 0, \qquad V = qR\sin\theta$$

$$\sum M_A = M + \int_0^\theta qR^2\sin(\theta-\beta)\mathrm{d}\beta = 0, \qquad M = -qR^2(1-\cos\theta)$$

The components of force F, V, and couple M can be plotted versus the variable θ, as in the previous example. It is most important, however, to note that we must duly account for the changing direction of the loading, as illustrated here.

ILLUSTRATIVE EXAMPLE 3.10

The cylindrical ring in Figure 3.25a is hinged at O and rests upon the smooth surface at A (neglecting friction at hinge O and surface A). The convex (outer) surface is subjected to a uniform pressure (normal to the surface) q (force/length). Express the normal force F, shear force V, and bending couple M on a cross section (M) in terms of the parameters (q, R) and position θ [$F(\theta)$, $V(\theta)$, $M(\theta)$].

The external load upon a typical element at angle β is $qR\mathrm{d}\beta$. The reactive forces R_1, R_2, and R_3 are determined from the equations of equilibrium applied to the circular ring as a whole:

$$\sum F_x = 0;\ R_2 = qR\int_0^{\pi/2}\cos\theta\mathrm{d}\theta = qR \tag{A}$$

$$\sum F_y = 0;\ R_1 + R_3 = qR\int_0^{\pi/2}\sin\theta\mathrm{d}\theta = qR \tag{B}$$

$$\sum M_z = 0;\ R_1\times R - R_2\times R = 0 \tag{C}$$

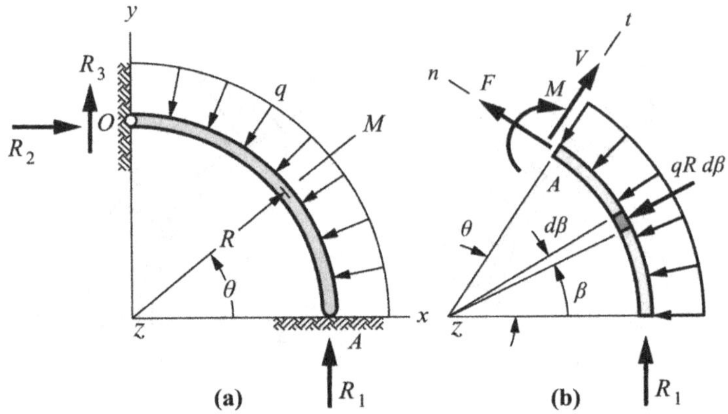

Figure 3.25

Equations (A)–(C) yield:

$$R_1 = qR, \qquad R_2 = qR, \qquad R_3 = 0$$

Components of normal force F and shear force V act upon the section at M in directions n and t, respectively; additionally, a bending couple M acts upon the cross section.

From the free-body diagram of Figure 3.25b we have:

$$\sum F_t = 0; \qquad V = -R_1 \sin\theta + qR\int_0^{\theta} \cos(\theta - \beta)\mathrm{d}\beta \tag{D}$$

$$\sum F_n = 0; \qquad F = -R_1 \cos\theta - qR\int_0^{\theta} \sin(\theta - \beta)\mathrm{d}\beta \tag{E}$$

$$\sum M_z = 0; \qquad M = R_1(R - Rcos\theta) - qR^2\int_0^{\theta} \sin(\theta - \beta)\mathrm{d}\beta \tag{F}$$

By considering that

$$\int_0^{\theta} \cos(\theta - \beta)\mathrm{d}\beta = \sin\theta, \qquad \int_0^{\theta} \sin(\theta - \beta)\mathrm{d}\beta = 1 - \cos\theta$$

we obtain from Equations (D)–(F)

$$F = -qR, \qquad V = 0, \qquad M = 0$$

ILLUSTRATIVE EXAMPLE 3.11

Repeat Illustrative Example 3.10 if the uniform pressure is replaced by a pressure that varies according to the equation $q = p(1 - \sin\theta)$. This corresponds to a hydrostatic pressure, wherein a fluid is contained with the surface at O to the depth A.

The equilibrium equations yield:

$$\sum F_x = 0; \qquad R_2 = pR\int_0^{\pi/2}(1-\sin\theta)\cos\theta d\theta \qquad \text{(A)}$$

$$\sum F_y = 0; \qquad R_1 + R_3 = pR\int_0^{\pi/2}(1-\sin\theta)\sin\theta d\theta \qquad \text{(B)}$$

$$\sum M_{z(O)} = 0; \; R_1 = pR\int_0^{\pi/2}(1-\sin\theta)\cos\theta d\theta \qquad \text{(C)}$$

Equations (A)–(C) lead to the following reactive forces:

$$R_1 = \frac{pR}{2}, \quad R_2 = \frac{pR}{2} \quad R_3 = \frac{pR}{2}\left(1-\frac{\pi}{2}\right)$$

Equilibrium of the free body in Figure 3.25b requires that

$$\sum F_t = 0; \quad V = R_1\sin\theta + pR\int_0^{\theta}(1-\sin\beta)\cos(\theta-\beta)d\beta \qquad \text{(D)}$$

$$\sum F_n = 0; \quad F = -R_1\cos\theta - pR\int_0^{\theta}(1-\sin\beta)\cos(\theta-\beta)d\beta \qquad \text{(E)}$$

$$\sum M_z = 0; \quad M = R_1(R - R\cos\theta) - pR^2\int_0^{\theta}(1-\sin\beta)\sin(\theta-\beta)d\beta \qquad \text{(F)}$$

By considering that

$$\int_0^{\theta}\sin\beta\cos(\theta-\beta)d\beta = \theta\frac{\sin\theta}{2}, \qquad \int_0^{\theta}\sin\beta\sin(\theta-\beta)d\beta = \frac{\sin\theta}{2} - \frac{\theta}{2}\cos\theta$$

we obtain from Equations (D)–(F):

$$F = \frac{pR}{2}(-2+\sin\theta+\cos\theta-\theta\cos\theta), \quad V = \frac{pR}{2}\sin\theta(1-\theta)$$

$$M = \frac{pR^2}{2}(-1+\sin\theta+\cos\theta-\theta\cos\theta)$$

ILLUSTRATIVE EXAMPLE 3.12

Suppose that the cylinders in the preceding Illustrative Examples (3.9–3.11) were clamped at O. Then a moment M_O must act at O to prevent rotation. It follows that the Equation (C) no longer holds; then, in each circumstance, there are four unknown reactions, R_1, R_2, R_3, and M_0. We are confronted with a problem that is statically indeterminate. A solution requires the consideration of the deformation, specifically, the relationship between the bending moment M and the deformation. Further analysis entails specific dimensions and properties of the material. The latter awaits our study of Chapter 4.

3.5 STRESS

The preceding sections were concerned with the net internal force system (the resultant) acting upon a plane section of a body. Such a resultant is sometimes said to be statically equivalent to the actual distribution of internal forces upon the surface. Here, it is very important to appreciate the implications of the term *statically equivalent*. The term means only that two *different* distributions of forces (or couples) have the same resultant. It means only that the systems have the same effect on a rigid body. Two different distributions are never equivalent in their effect on a *deformable* body. To appreciate this fact, consider a deformable block, say rubber, subjected to the two different distributions, as depicted in Figure 3.26. The resultants are the same, a central normal force; the effects are very different. In general, we must *never* replace one distribution by another and expect the same effects. The student of engineering *must* adopt this as an *inviolable* rule.

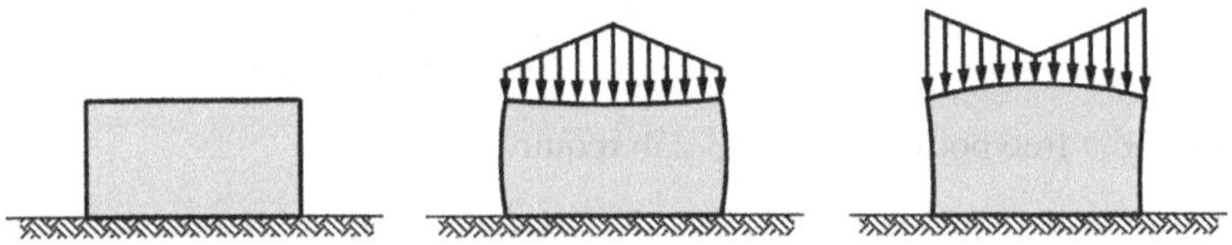

Figure 3.26

Since we wish eventually to relate the deformations of a body to the internal forces, the net internal force system alone will not adequately describe these internal forces. We need quantities that describe the internal forces in the immediate vicinity of a point. The necessary quantities are *stresses*, which are defined in the following:

Consider a member subjected to some external equilibrated loads, as depicted in Figure 3.27. The body may be divided by a surface *S*, as in the body of Figure 3.27. The surface divides the body into two free bodies, shown separated in Figure 3.28. Each part is subjected to interactions upon the intersection, actions, and reactions, respectively, as well as the external loads on that portion. We recognize, however, that representation by a resultant on a cross-section, say force and couple, is inadequate. To illustrate, suppose that each of the contiguous surfaces is subdivided into many small rectangular elements, as shown.

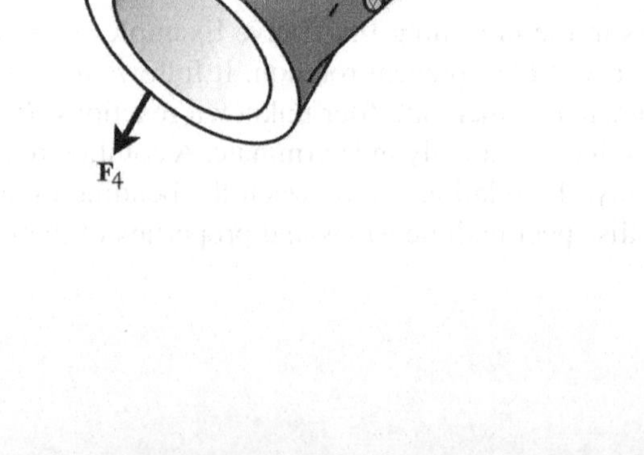

Figure 3.27

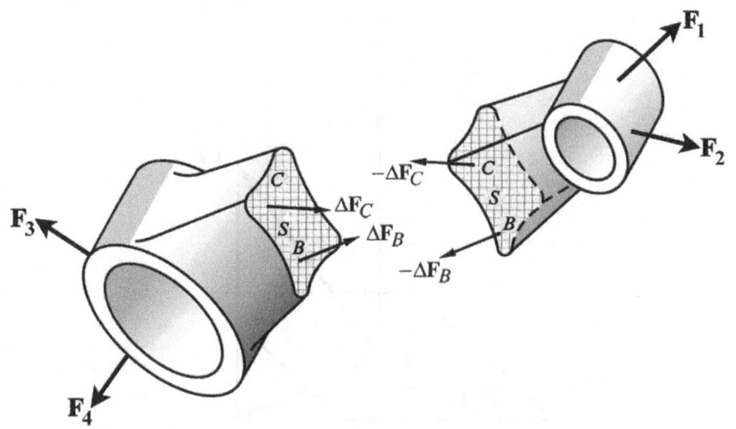

Figure 3.28

Each elemental area is subjected to a different force, $\Delta\mathbf{F}$, and the contiguous element of the separated body is subjected to the reaction, $-\Delta\mathbf{F}$, as in the elements at B and C. Moreover, the magnitude and direction of the force $\Delta\mathbf{F}$ change as the size of the element changes. An elemental surface at a point P is shown in Figure 3.29 with the force $\Delta\mathbf{F}$ that acts upon it. The area of the element is ΔA; the orientation of the element is established by the direction of the normal n. Again, $\Delta\mathbf{F}$ depends on the size of ΔA. We are concerned with the nature and intensity at the specific point P. Therefore, we define the *stress* (σ_n), a *vector*:

$$\sigma_n \equiv \lim_{\Delta A \to 0} \frac{\Delta\mathbf{F}}{\Delta A} \tag{3.5}$$

Several features must be noted:

The stress σ_n depends on the orientation of the surface; hence, the subscript n is needed. At the same location, the stress is generally different on surfaces of different orientation. A simple example illustrates this fact: Suppose that a thin bar is pulled by axial force. Upon a right section, we expect a large stress; upon a section parallel to the axis, we expect a very small (negligible) stress.

The stress presumes that $\Delta\mathbf{F}$ diminishes as ΔA diminishes. Mathematically, we presume that the limit exists. Practically, we assume that the quantity has physical significance; we are *assuming* that the media is *continuous*. Of course, on a *microscopic scale*, real materials are not continuous. Metals are composed of crystals, which are, in turn, composed of atoms; impurities, even spaces, exist between crystals. Other materials, such as wood and concrete, are more heterogeneous. This means that the concept of stress, as defined here, must be used with some caution; it may not suffice to describe phenomena that originate within a micro-structure. It is quite effective for the description of macroscopic phenomena and has proven most satisfactory for most analyses of structures and machines, particularly for the deformational behavior.

In practice, it is necessary to decompose the vector (σ_n) into components. Quite naturally, the vector can be decomposed into components normal to the surface and tangent to the surface; these rectangular components, denoted by σ_n and τ_n, are depicted in Figure 3.30.

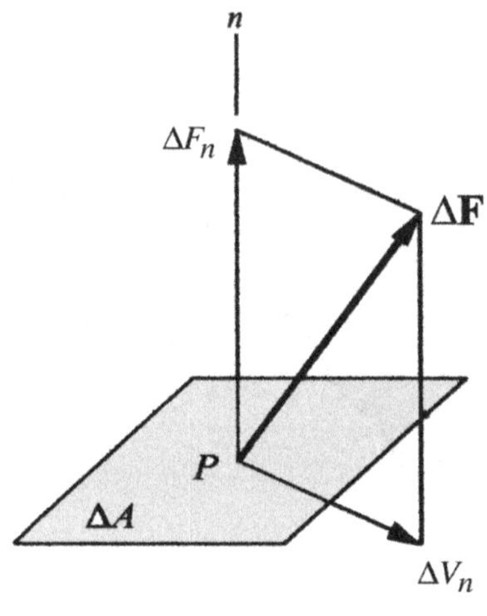

Figure 3.29

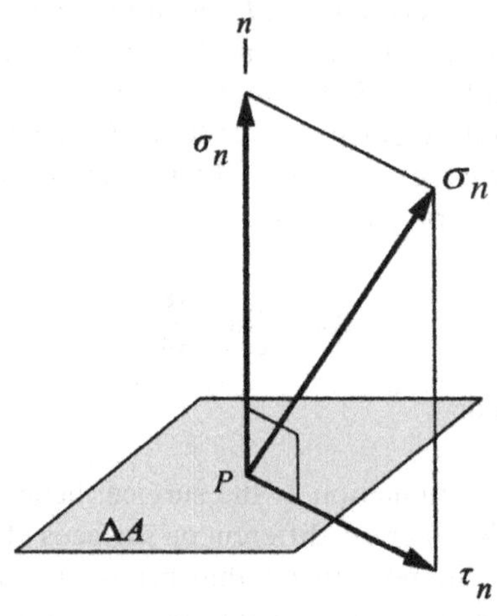

Figure 3.30

The component σ_n is the *normal* component of stress. The component τ_n is tangent to the surface; it is called the *shear* component because the action tends to shear the material (as scissors shear paper). If, at some point, we identify a direction n and give the normal component σ_n, the meaning is without ambiguity. If we give the shear component τ_n, the meaning is ambiguous, since that component could have any direction tangent to the surface. Physically, the consequences might be very different, depending upon the direction of the shear τ_n; the material might be more susceptible to shearing in certain directions, as wood shears more readily parallel to its grain. Accordingly, we select two perpendicular directions, such as s and t in Figure 3.31, and decompose the shear τ_n into the two rectangular components τ_{ns} and τ_{nt}. Note that two subscripts are needed; the first identifies the orientation of the surface, and the second identifies the direction.

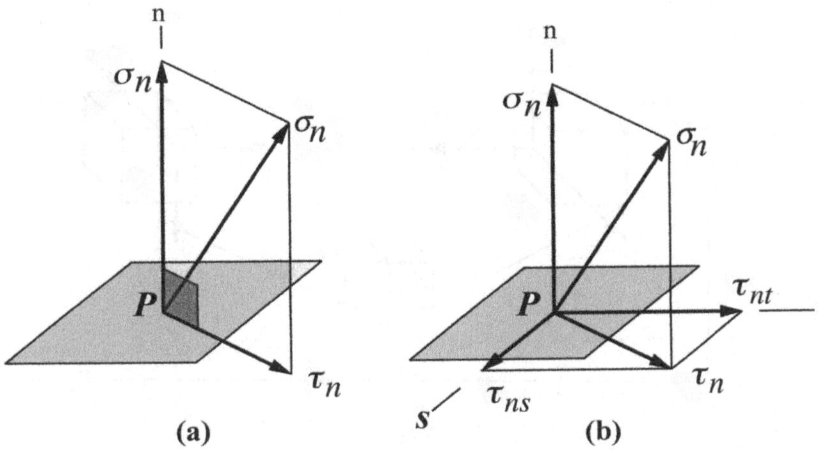

Figure 3.31

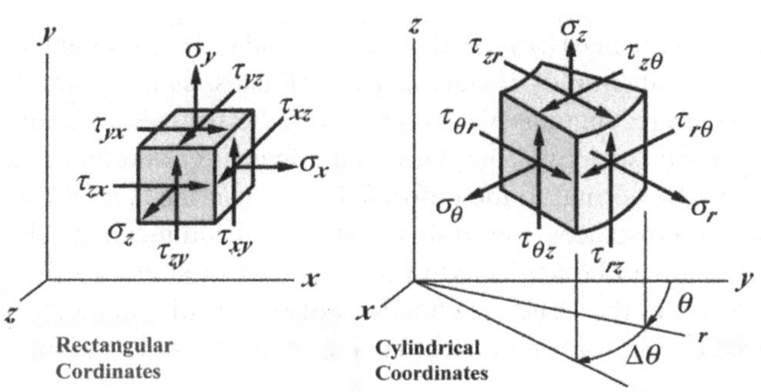

Figure 3.32

Notation: The symbols σ and τ are consistently used in this text to signify normal and shear components of stress. One subscript is needed to identify the surface on which the stress acts. The components of shear stress require a second subscript to indicate the direction of the action.

In any coordinate system, directions are indicated by the lines of the coordinates, a line along which only that coordinate changes. As examples, consider the rectangular coordinates (x, y, z) of Figure 3.32a and the polar coordinates (r, θ, z) of Figure 3.32b. The circular line (r =constant, z =constant) in Figure 3.32b is a θ line; the surface perpendicular to it is a θ surface, which is subject to the stress σ_θ with components σ_θ, $\tau_{\theta z}$, and $\tau_{\theta r}$.

Now consider the two adjoining rectangular blocks of Figure 3.33; the separated blocks are depicted at parts a and b of the figure. Components of stress on the x face of block A are labeled in accordance with the stated convention; the normal is the $+x$ direction. The actions upon the contiguous face of block B are in opposite directions, in accordance with Newton's law of action and opposing reaction. This opposite sense is signified, too, by the normal direction $(-x)$ and tangent directions $(-y$ and $-z)$.

A few words of caution are in order: In Chapter 2, we acknowledged the deformations, which cause elements to deform. An infinitesimal rectangular block is deformed to a parallelepiped;

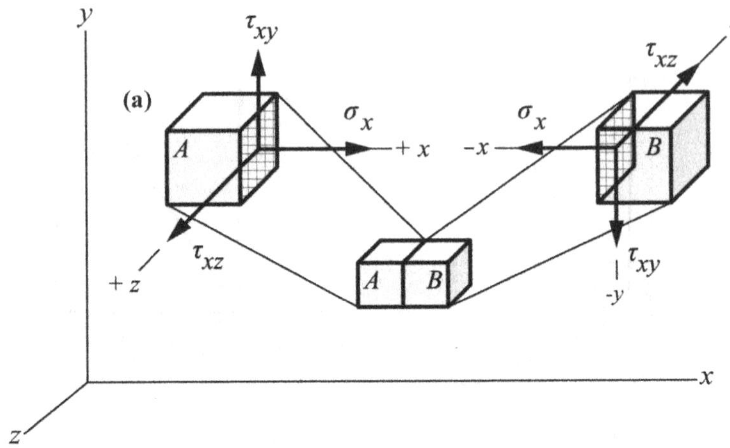

Figure 3.33

edges are extended (or contracted) by strains (ε_x, ε_y, ε_z) and right angles are altered by shear strains (γ_{xy}, γ_{xz}, γ_{yz}). Often we can employ rectangular coordinates, as in Figure 3.33, which subdivide the *undeformed* body into rectangular blocks. However, the *stresses* act upon the *deformed* blocks (see Section 3.15); strictly speaking, the deformed edges are nonorthogonal. This means that a component σ_x need not be normal to the deformed $(y - z)$ surface and τ_{xy} need not be tangent to that surface. In most circumstances, the strains of structural and mechanical elements are so small ($\varepsilon \ll 1$, $\gamma \ll 1$) that they can be neglected in the analyses of stresses. The reader must be wary when examining stresses in the large deformations of rubber-like materials or the large inelastic deformations incurred during forming processes, i.e., stamping or extruding.

ILLUSTRATIVE EXAMPLE 3.13

A stress vector acts on the surface of Figure 3.34 with a unit outward normal **n**, where

$$\sigma_{\mathbf{n}} = 60\hat{\boldsymbol{i}} + 30\hat{\boldsymbol{k}} \text{ ksi}, \qquad \hat{\boldsymbol{n}} = \left(\frac{1}{2}\hat{\boldsymbol{i}} + \frac{\sqrt{3}}{2}\hat{\boldsymbol{j}}\right)$$

Determine the components σ_n, τ_{nz}, and τ_{nb} where $\hat{\mathbf{b}}$ denotes the unit tangent vector given by

$$\hat{\boldsymbol{b}} = \hat{\boldsymbol{k}} \times \hat{\boldsymbol{n}} = -\frac{\sqrt{3}}{2}\hat{\boldsymbol{i}} + \frac{1}{2}\hat{\boldsymbol{j}}$$

The required components may be calculated from the relationships:

$$\sigma_n = \hat{\boldsymbol{n}} \cdot \boldsymbol{\sigma}_n = 30 \text{ ksi}$$
$$\tau_{nz} = \hat{\boldsymbol{k}} \cdot \boldsymbol{\sigma}_n = 30 \text{ ksi}$$
$$\tau_{nb} = \hat{\boldsymbol{b}} \cdot \boldsymbol{\sigma}_n = -30\sqrt{3} \text{ ksi}$$

As a numerical check, the following equation may be verified:

$$|\tau_{nb}| = \sqrt{\boldsymbol{\sigma}_n \cdot \boldsymbol{\sigma}_n - \sigma_{\boldsymbol{n}}^2 - \tau_{nz}^2} = \sqrt{(3600+900) - 900 - 900} = 30\sqrt{3}$$

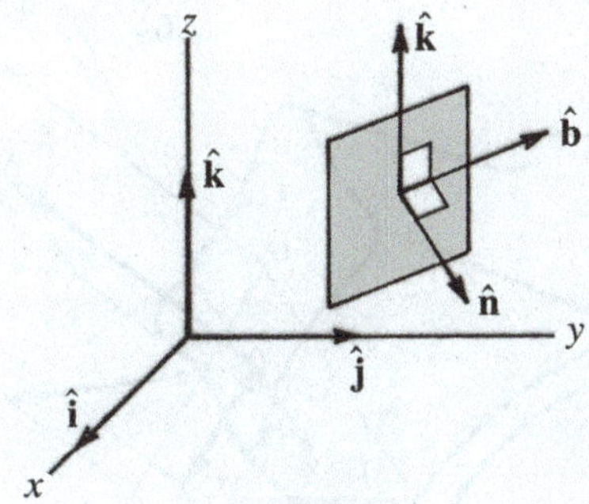

Figure 3.34

3.6 THE RELATIONSHIP BETWEEN STRESS AND THE NET INTERNAL FORCE SYSTEM

The actual distribution of internal forces on a sectioning surface is described by the stresses at each point of the surface. The net internal system is merely the resultant of the distributed stresses; it is fully determined by that distribution. We can express the resultant in terms of the stresses as follows.

Let ΔA be an element of area taken from a sectioning surface of a body, as shown in Figure 3.35. The rectangular coordinates (n, s, t) are normal (n) and tangent (s, t) to the section. Normal and shear components of stress are indicated on the enlarged element ΔA. $\Delta\mathbf{F}$ is the internal force acting on ΔA and ΔF, ΔV_s, and ΔV_t are the normal and two shear components of $\Delta\mathbf{F}$. Recall the definition

$$\sigma_n(P) \equiv \lim_{\substack{\Delta A \to 0 \\ P \text{ always in } \Delta A}} \frac{\Delta \mathbf{F}}{\Delta A}$$

If ΔA is small, we can write

$$\Delta\mathbf{F} = \boldsymbol{\sigma}_n \Delta A \tag{3.6}$$

where it is understood that the stress is evaluated for *some* point in ΔA. By a similar line of reasoning,

$$\Delta V_s = \tau_{ns} \Delta A \tag{3.7}$$

$$\Delta V_t = \tau_{nt} \Delta A \tag{3.8}$$

Furthermore, the moments of these forces about the s, t, and n axes at O (the centroid of S) are given by

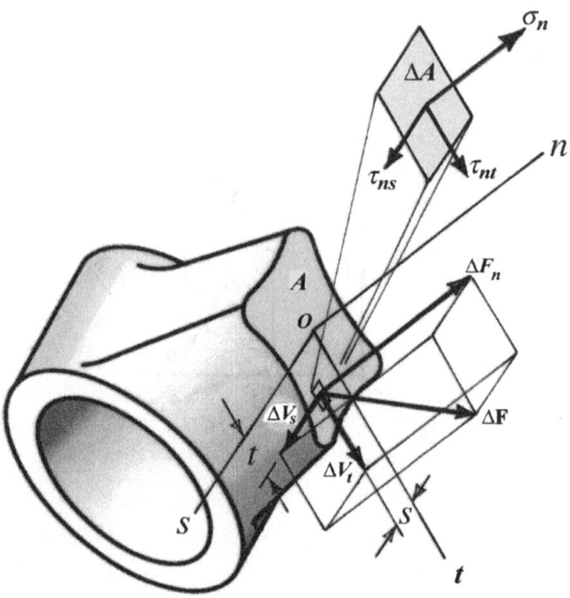

Figure 3.35

$$\Delta M_n \equiv \Delta T = s\Delta V_t - t\Delta V_s \approx (s\tau_{nt} - t\tau_{ns})\Delta A \tag{3.9}$$

$$\Delta M_s = t\Delta F_n \approx t\sigma_n\Delta A \tag{3.10}$$

$$\Delta M_t = -s\Delta F_n \approx -s\sigma_n\Delta A \tag{3.11}$$

Now, ΔA is a typical element of the sectioning surface S. To find the components of the net internal force system, we add the contributions from each element of the surface. In the limit, these sums become integrals. Thus, the components of the net internal force system can be expressed as integrals of the stresses. Normal stresses give rise to normal forces and bending couples; shear stresses give rise to shear forces and twisting couples. This is illustrated in the following examples.

ILLUSTRATIVE EXAMPLE 3.14

Suppose that the stresses on S of Figure 3.36 are known to be constant. The stresses on the small element of area ΔA give rise to the forces

$$\Delta F = \sigma_x \Delta A, \qquad \Delta V_y = \tau_{xy}\Delta A, \qquad \Delta V_z = \tau_{xz}\Delta A$$

These contributions to the net internal force system on S are added for every element of area A. In the limit,

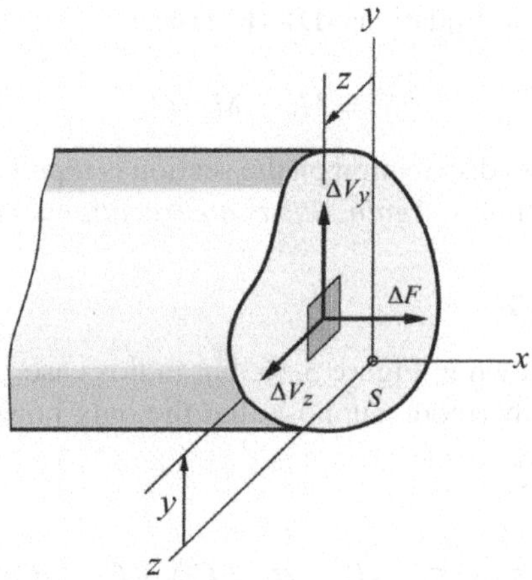

Figure 3.36

$$F = \iint_S \sigma_x dA \tag{A}$$

$$V_y = \iint_S \tau_{xy} dA \tag{B}$$

$$V_z = \iint_S \tau_{xz} dA \tag{C}$$

where F is the normal force on S and V_y and V_z are the components of the shear force. In this particular example, the stresses σ_x, τ_{xy}. and τ_{xz} are constant on S; therefore, they can be taken outside the integral signs in Equation (A, B, C), so that

$$F = \sigma_x A,\ \sigma_x = \frac{F}{A}; \quad V_y = \tau_{xy} A,\ \tau_{xy} = \frac{V_y}{A}; \quad V_z = \tau_{xz} A,\ \tau_{xz} = \frac{V_z}{A}$$

where A is the area of S. Note that the present example is an *exception*; usually, stresses are *variable*. The twisting couple T and the components of the bending couple, M_y and M_z, are also easily computed. In Figure 3.36 forces are shown acting on the element of area ΔA. The moments of these forces about the coordinate axes are

$$\Delta T = y\Delta V_z - z\Delta V_y = y\tau_{xz}\Delta A - z\tau_{xy}\Delta A, \quad \Delta M_y = z\Delta F = z\sigma_x \Delta A, \quad \Delta M_z = -y\Delta F = -y\sigma_x \Delta A$$

Summing these incremental moments over the entire area of S and again noting that the stresses are constant, we obtain

$$T = \tau_{xz}\iint_S y dA - \tau_{xy}\iint_S z dA, \quad M_y = \sigma_x \iint_S z dA, \quad M_z = -\sigma_x \iint_S y dA, \tag{D,E,F}$$

If the origin of coordinates is at the centroid of S, then

$$\iint_S y dA = \iint_S z dA = 0$$

Consequently, all the integrals in Equations (D)–(F) are zero:

$$T = M_y = M_z = 0$$

This means that a constant stress distribution on the section is represented by a force *at the centroid* with no bending or twisting couples. *Again, this is an exceptional circumstance.*

ILLUSTRATIVE EXAMPLE 3.15

Consider the same member shown in Figure 3.36, but in this case, suppose that the shear stresses τ_{xy} and τ_{xz} on S are zero. Furthermore, suppose that the only nonzero stress, σ_x, is linearly distributed on S. In equation form,

$$\tau_{xy} = \tau_{xz} = 0, \quad \sigma_x = C_1 y + C_2 z + C_3 \tag{A}$$

where C_1, C_2 and C_3 are, at most, functions of x. In Figure 3.36 only the normal force ΔF acts on the element of area ΔA, where

$$\Delta F = \sigma_x \Delta A = (C_1 y + C_2 z + C_3)\Delta A$$

Adding these increments of normal force over the entire area of S, we obtain

$$F = C_1 \iint_S y\mathrm{d}A + C_2 \iint_S z\mathrm{d}A + C_3 \iint_S \mathrm{d}A$$

The first two integrals are zero *if* y and z are measured from the centroid of S, and the last integral is simply the area A of S. Hence

$$F = C_3 A, \quad \text{or} \quad C_3 = \frac{F}{A} \tag{B}$$

The contributions of ΔF to the components of the bending couple are

$$\begin{aligned} \Delta M_y &= z\Delta F &&= z\sigma_x \Delta A &&= C_1 zy\Delta A + C_2 z^2 \Delta A + C_3 z\Delta A \\ \Delta M_z &= -y\Delta F &&= -y\sigma_x \Delta A &&= -C_1 y^2 \Delta A - C_2 yz\Delta A - C_3 y\Delta A \end{aligned}$$

Note that the third integral in each equation is again zero, because y and z are measured from the centroid of S. Then

$$M_y = C_1 \iint_S yz\mathrm{d}A + C_2 \iint_S z^2\mathrm{d}A \tag{C}$$

$$M_z = -C_1 \iint_S y^2\mathrm{d}A - C_2 \iint_S yz\mathrm{d}A \tag{D}$$

For convenience, set[2]

$$I_{yz} = \iint_S yz\mathrm{d}A, \quad I_z = \iint_S z^2\mathrm{d}A, \quad I_y = \iint_S y^2\mathrm{d}A \tag{E}$$

Then Equations (C) and (D) become

$$M_y = C_1 I_{yz} + C_2 I_z, \quad M_y = -C_1 I_y - C_2 I_{yz} \tag{F,G}$$

Solving Equations (F) and (G) for C_1 and C_2, we obtain

$$C_1 = -\frac{I_z M_z + I_{yz} M_y}{I_z I_y - I_{yz}^2}, \quad C_2 = \frac{I_y M_y + I_{yz} M_z}{I_z I_y - I_{yz}^2} \tag{H,I}$$

Then, upon substituting Equations (B), (H), and (I) into Equation (A), we obtain, finally,[3]

$$\sigma_x = -\frac{I_z M_z + I_{yz} M_y}{I_z I_y - I_{yz}^2} y + \frac{I_y M_y + I_{yz} M_z}{I_z I_y - I_{yz}^2} z + \frac{F}{A}$$

Since $\tau_{xy} = \tau_{xz} = 0$, the transverse shear forces and twisting couple vanish:

$$V_y = V_z = T = 0 \quad \text{on } S$$

ILLUSTRATIVE EXAMPLE 3.16

Consider the following stress distribution on the rectangular cross-section shown in Figure 3.37:

$$\sigma_x = 500 - 500(z/a) \text{ psi}, \quad \tau_{xy} = 0 \text{ psi}, \quad \tau_{xz} = 50 - 50(z/a)^2 \text{ psi}$$

Let us calculate the resultant, as a force F at the origin and a couple C, if $a = 1$ in.
Note: 1 lb ≡ 4.448 N, 1 lb·in. ≡ (1/8.86) N·m, 1 psi ≡ 6.895 kPa.
For the components (F_x, F_y, F_z of force **F** we obtain:

$$F_x = \iint_A \sigma_x dA = 4000 \text{ lb } (17.8 \text{ kN}) \qquad \text{(normal component)}$$

$$F_y = \iint_A \tau_{xy} dA = 0 \text{ lb } (0 \text{ kN}) \qquad \text{(shear component)}$$

$$F_z = \iint_A \tau_{xz} dA = 267 \text{ lb } (1.19 \text{ kN}) \qquad \text{(shear component)}$$

and for the components (C_x, C_y, C_z) of couple $\boldsymbol{C}$:

$$C_x = \iint_A (\tau_{xz} y - \tau_{xy} z) dA = 0 \text{ lb} \cdot \text{in. } (0 \text{ N} \cdot \text{m}) \qquad \text{(twisting component)}$$

$$C_y = \iint_A \sigma_x z dA = -1333 \text{ lb} \cdot \text{in. } (150.5 \text{ N} \cdot \text{m}) \qquad \text{(bending component)}$$

$$C_z = \iint_A -\sigma_x y dA = 0 \text{ lb} \cdot \text{in. } (0 \text{ N} \cdot \text{m}), \qquad \text{(bending component)}$$

It follows that:

$$\mathbf{F} = 4000\hat{\boldsymbol{i}} + 267\hat{\boldsymbol{k}} \text{ lb}, \qquad \boldsymbol{C} = -1333.3\hat{\boldsymbol{j}} \text{ lb} \cdot \text{in.}$$

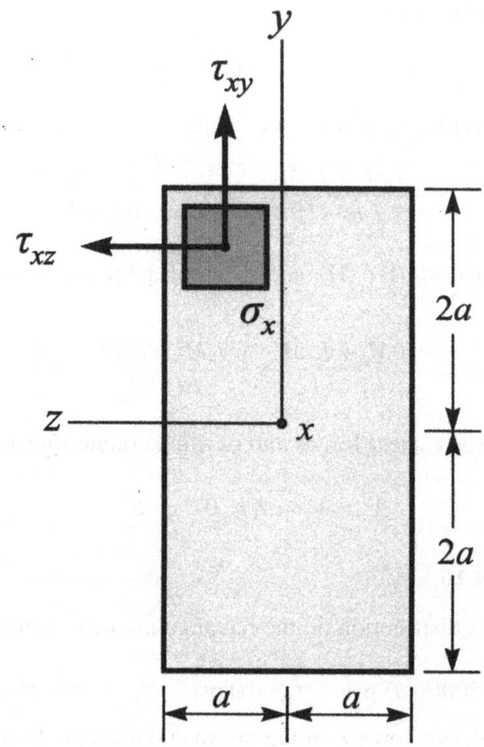

Figure 3.37

ILLUSTRATIVE EXAMPLE 3.17

Consider the circular cylinder shown in Figure 3.38. Let us adopt cylindrical coordinates r and θ in the cross section S, as indicated in Figure 3.39. This defines r and θ directions at each point of S. Suppose that the normal stress σ_x is zero throughout the cylinder and that the shear stress on S is such that τ_{xr} and $\tau_{x\theta}$ are symmetric about the x-axis. Specifically, τ_{xr} and $\tau_{x\theta}$ are independent of θ; i.e., $\tau_{xr} = \tau_{xr}(\mathrm{r})$, $\tau_{x\theta} = \tau_{x\theta}(\mathrm{r})$. Since $\sigma_x = 0$, the normal force $\mathbf{F}$ and bending couples, M_y and M_z, vanish.

On the element of area ΔA shown in Figure 3.40, the shear stresses τ_{xr} and $\tau_{x\theta}$ give rise to the force components

$$\Delta V_r = \tau_{xr}\Delta A, \qquad \Delta V_\theta = \tau_{x\theta}\Delta A$$

Since τ_{xr} is always directed radially, it contributes nothing to the twisting couple on S. The contribution to T from ΔV_θ is

$$\Delta T = r\Delta V_\theta = r\tau_{x\theta}\Delta A$$

In cylindrical coordinates, the element of area is $\Delta A = r\Delta r\Delta\theta$, so that

$$\Delta T = \tau_{x\theta} r^2 \Delta r \Delta\theta \tag{A}$$

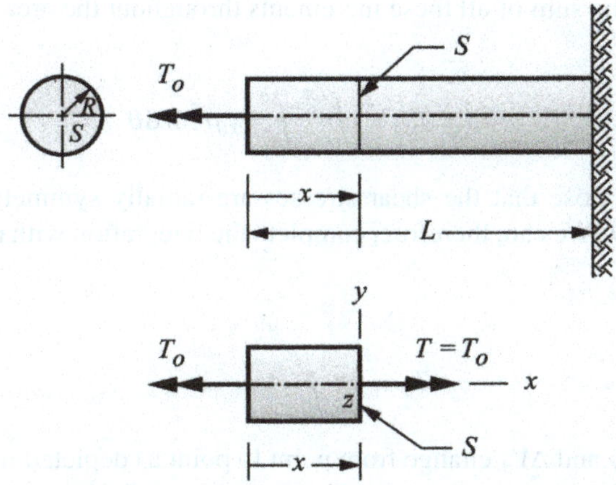

Figure 3.38

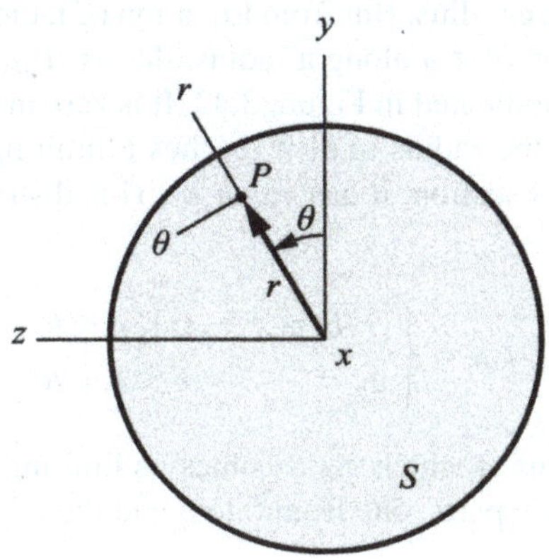

Figure 3.39

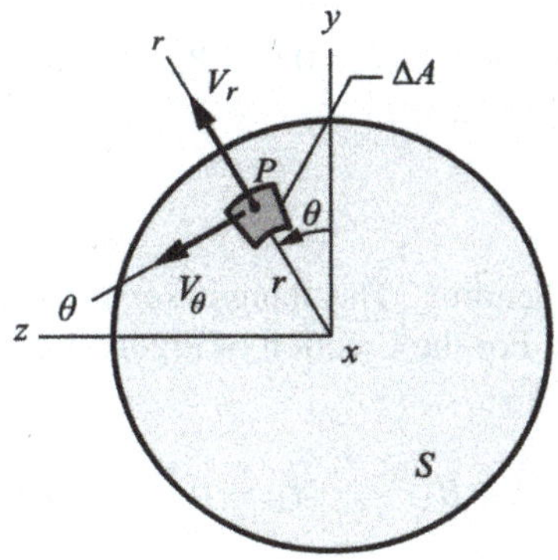

Figure 3.40

The twisting couple is the sum of all these increments throughout the area of S:

$$T(=T_0) = \int_0^{2\pi}\int_0^R \tau_{x\theta} r^2 \mathrm{d}r\mathrm{d}\theta \tag{B}$$

In our example, we suppose that the shear stresses are radially symmetric; that is, τ_{xr} and $\tau_{x\theta}$ depend on r but not on θ. We can, therefore, complete the integration with respect to θ in Equation (B) to obtain:

$$T_0 = 2\pi \int_0^R \tau_{x\theta} r^2 \mathrm{d}r \tag{C}$$

The directions of ΔV_r and ΔV_θ change from point to point as depicted in Figure 3.41. However, both τ_{xr} and $\tau_{x\theta}$ are axisymmetric. At each site (θ) in Figure 3.41, we note another site ($\theta + \pi$) where the contributions to both ΔV_y and ΔV_z precisely oppose those at the other (θ). Therefore, the forces V_y and V_z (the integrals) vanish. The cross-section is free of shear force, whatever the distribution of τ_{xr} and $\tau_{x\theta}$ along a radius. But from Equation (C), the value of T_0 can be determined only if we know the distribution of $\tau_{x\theta}$ along a radius, that is, $\tau_{x\theta}(\mathrm{r})$. For example, suppose that $\tau_{x\theta}$ varies along a diameter, as indicated in Figure 3.42. It is zero at the axis of the cylinder ($r = 0$) and increases in proportion to the radius until it reaches a limiting value τ_0 (say at $r = b < R$). Over the remainder of the cross section, it has value τ_0. This distribution of shear stress is given by the relation

$$\tau_{x\theta} = \begin{cases} (r/b)\tau_0, & 0 \le r \le b \\ \tau_0, & b \le r \le R \end{cases} \tag{D}$$

Here b is the (unspecified) radius at which $\tau_{x\theta}$ assumes its limiting value. To evaluate the integral (C), we split the integral into two parts, one from 0 to b and the other from b to R:

$$T_0 = 2\pi \int_0^R \tau_{x\theta} r^2 dr = 2\pi \int_0^b \frac{\tau_0}{b} r^3 dr + 2\pi \int_b^R \tau_0 r^2 dr = \frac{2\pi\tau_0}{3}\left(R^3 - \frac{b^4}{4}\right) \tag{E}$$

For this stress distribution, b must lie between 0 and R. According to Equation (E) T_0 must have a value between $(2\pi R^3 \tau_0)/3$ and $(\pi R^3 \tau_0)/2$.

ILLUSTRATIVE EXAMPLE 3.18

I-beams are designed to resist bending. The flanges serve to resist bending moment; the web resists the transverse shear force. For the I section of Figure 3.43, we might employ the following approximations:

$$\sigma_x = A + B\frac{y}{5}, \qquad \tau_{xz} = 0 \quad \text{(throughout)}$$

$$\tau_{xy} = \begin{cases} 0, & y > 4 & \text{(top flange)} \\ C, & 4 \ge y \ge -4 & \\ 0, & -4 > y & \text{(bottom flange)} \end{cases}$$

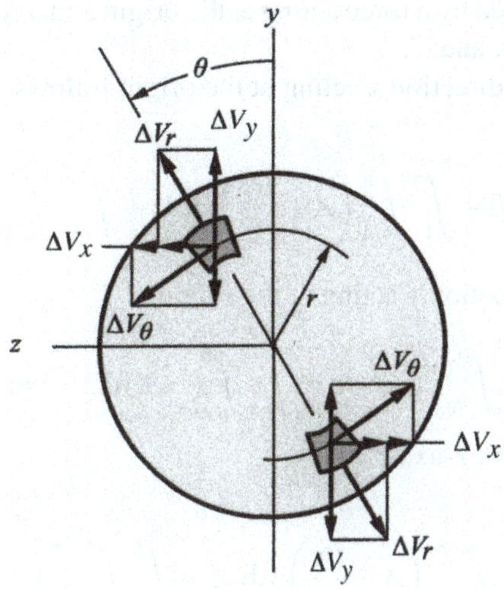

Figure 3.41

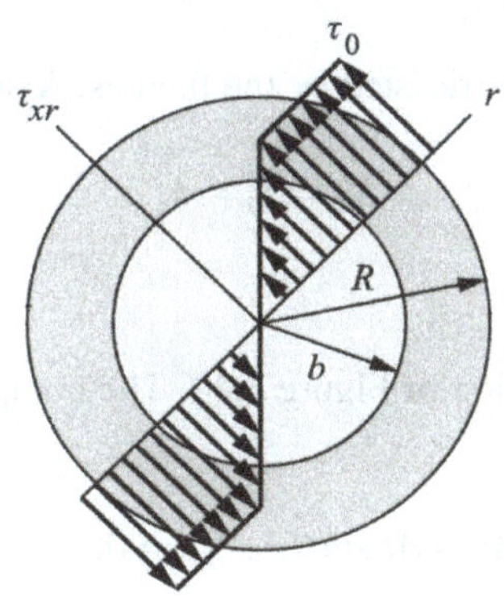

Figure 3.42

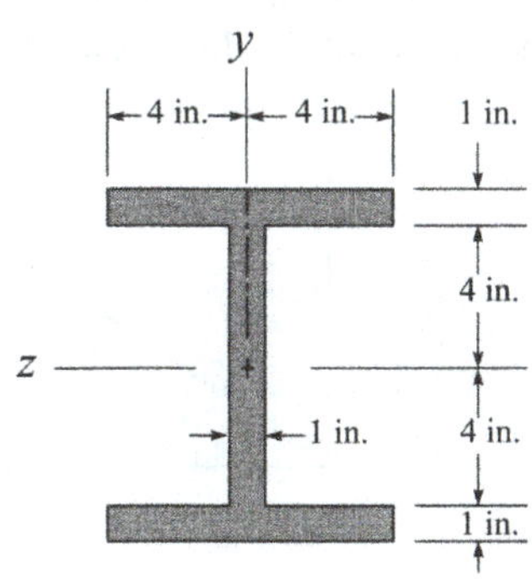

Figure 3.43

If the resultant is represented by a force acting at the origin and a couple, express all components in terms of the constants A, B, and C.

Component of force F_x in direction x acting at the origin follows:

$$F_x = \int_{-4}^{4}\int_{-0.5}^{0.5}\left(A+\frac{By}{5}\right)\mathrm{d}y\mathrm{d}z + \int_{4}^{5}\int_{-4}^{4}\left(A+\frac{By}{5}\right)\mathrm{d}y\mathrm{d}z + \int_{-5}^{-4}\int_{-4}^{4}\left(A+\frac{By}{5}\right)\mathrm{d}y\mathrm{d}z \quad \Longrightarrow F_x = 24A$$

Component of force V_y in direction y acting at the origin:

$$V_y = \int_{-4}^{4}\int_{-0.5}^{0.5}\tau_{xy}\mathrm{d}y\mathrm{d}z = \int_{-4}^{4}\int_{-0.5}^{0.5} C\mathrm{d}y\mathrm{d}z \quad \Longrightarrow V_y = 8C$$

Component of couple M_z about z-ax.is:

$$M_z = -\int_{-4}^{4}\int_{-0.5}^{0.5}\left(A+\frac{By}{5}\right)y\mathrm{d}y\mathrm{d}z - \int_{4}^{5}\int_{-4}^{4}\left(A+\frac{By}{5}\right)y\mathrm{d}y\mathrm{d}z$$
$$-\int_{-5}^{-4}\int_{-4}^{4}\left(A+\frac{By}{5}\right)y\mathrm{d}y\mathrm{d}z \quad \Longrightarrow M_z = -73.6B$$

Note that 88% of the moment M_z is resisted by the flanges. Also note that

$$V_z = T_x = M_y$$

ILLUSTRATIVE EXAMPLE 3.19

A shaft has the circular cross-section of Figure 3.44. The components of stress upon that section are

$$\sigma_z = Ay = Ar\sin\theta, \quad \tau_{zr} = 0, \quad \tau_{z\theta} = Br$$

The resultant can be represented by a force at the origin and a couple. We seek to express the components of each in terms of the constants a, A, and B.

For the components of force V_x, V_y, and F_z in directions x, y, and z, respectively, and for the components of moment M_x, M_y, and M_z about the directions x, y, and z, respectively, we obtain:

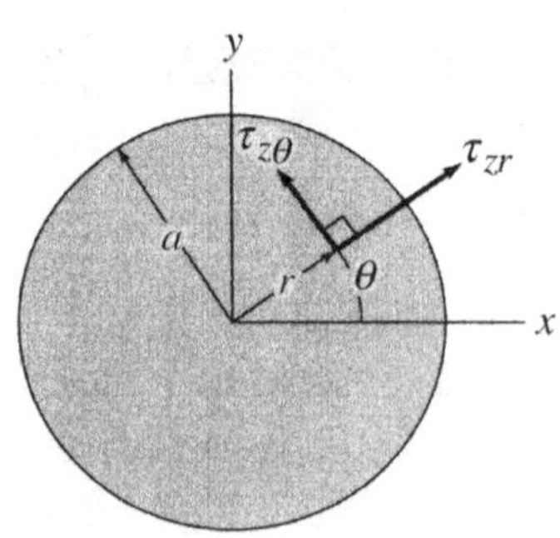

Figure 3.44

$$V_x = -\int_0^a \int_0^{2\pi} \tau_{z\theta} \sin\theta r dr d\theta = -\int_0^a \int_0^{2\pi} Br^2 \sin\theta dr d\theta = 0$$

$$V_y = \int_0^a \int_0^{2\pi} \tau_{z\theta} \cos\theta r dr d\theta = \int_0^a \int_0^{2\pi} Br^2 \cos\theta dr d\theta = 0$$

$$F_z = \int_0^a \int_0^{2\pi} \sigma_z r dr d\theta = \int_0^a \int_0^{2\pi} Ar^2 \sin\theta dr d\theta = 0$$

$$M_x = \int_0^a \int_0^{2\pi} \sigma_z y r dr d\theta = \int_0^a \int_0^{2\pi} Ar^3 \sin^2\theta dr d\theta = A\frac{a^4\pi}{4}$$

$$M_y = 0,$$

$$M_z = \int_0^a \int_0^{2\pi} \tau_{z\theta} r^2 dr d\theta = \int_0^a \int_0^{2\pi} Br^3 dr d\theta = B\frac{a^4\pi}{4}$$

3.7 STATE OF STRESS

The concept of stress was introduced to characterize the intensity of the internal forces in the vicinity of an interior point of a body. In addition to specifying the point at which a stress acts, we must specify the plane on which it acts. The stress on the plane can then be defined by a normal stress and two shear stresses. Thus, at every point of the body, a normal stress and two shear stresses act upon *every plane through the point*. The question now arises: How many stresses must be known at a point in order to determine the three stresses for every plane?

3.7.1 STRESSES ON PERPENDICULAR PLANES

A small rectangular block removed from the interior of a body is shown in Figure 3.45. On each of its six faces, the resultant of the internal forces is decomposed into three components.

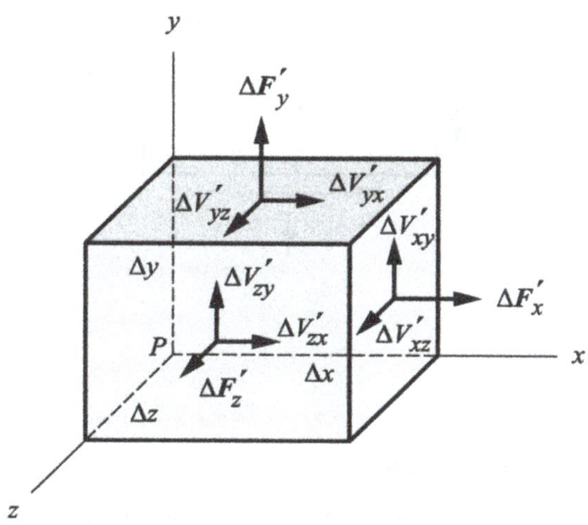

Figure 3.45

For clarity, this is shown only for the faces of the block visible in Figure 3.45. We have 18 distinct force components acting on the block. Each of these force components can be expressed in terms of an average stress. For example, the force components acting on the front face are

$$\Delta F_z' = \bar{\sigma}_z' \Delta x \Delta y$$
$$\Delta V_{zy}' = \bar{\tau}_{zy}' \Delta x \Delta y$$
$$\Delta V_{zx}' = \bar{\tau}_{zx}' \Delta x \Delta y$$

where bars are placed over average stresses to distinguish them from stresses at a point. Quantities with primes refer to the three faces visible in Figure 3.45, whereas quantities without primes refer to the three faces passing through the corner P.

If the dimensions Δx, Δy, and Δz are made smaller, the opposite faces get closer together. The average stresses with primes approach the values of the corresponding average stresses without primes. Furthermore, these average stresses become (in the limit) the stresses associated with the point P. Because we wish to examine only the stresses at point P, we must eventually perform this limiting process. Sacrificing mathematical rigor for simplicity, we regard the values of corresponding average stresses on opposite faces as equal to the values of the stresses at P. For example, the forces on the front and back faces are taken as

$$\Delta F_z \approx \Delta F_z' \doteq \sigma_z \Delta x \Delta y, \quad \Delta V_{zy} \approx \Delta V_{zy}' \approx \tau_{zy} \Delta x \Delta y$$
$$\Delta V_{zx} \approx \Delta V_{zx}' \approx \tau_{zx} \Delta x \Delta y \tag{3.12}$$

where σ_z, τ_{zy}, and τ_{zx} are evaluated at the point P. With these approximations, the actions on the element appear as noted in Figure 3.46. Components on the edge parallel to the z-axis are not shown.

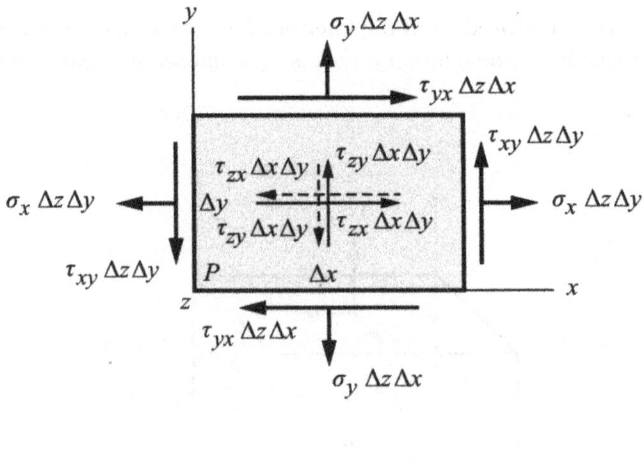

Figure 3.46

3.7.2 EQUILIBRIUM OF AN ELEMENT

Let us consider the equilibrium of the element: If we accept the approximations (3.12), then[4]

$$\sum F_x = 0, \quad \sum F_y = 0, \quad \sum F_z = 0$$

However, the sum of the moments is not identically zero. For example, from Figure 3.46 the sum of moments about the z axis (through point P) is

$$\sum M_z = (\tau_{xy}\Delta z\Delta y)\Delta x - (\tau_{yx}\Delta z\Delta x)\Delta y = (\tau_{xy} - \tau_{yx})\Delta x\Delta y\Delta z$$

For equilibrium, $\sum M_z = 0$. Upon dividing out the factor Δx, Δy, Δz, we conclude that

$$\tau_{xy} = \tau_{yx} \tag{3.13a}$$

Similarly, summing moments about the x and y axis, respectively, we conclude that

$$\tau_{zy} = \tau_{yz}, \quad \tau_{zx} = \tau_{xz} \tag{3.14b,c}$$

The number of distinct stresses needed to specify the internal forces acting on the elementary block is thereby reduced to six, namely,

$$\sigma_x, \quad \sigma_y, \quad \sigma_z, \quad \tau_{xy}, \quad \tau_{yz}, \quad \tau_{zx}$$

where τ_{xy} is written for both τ_{xy} and τ_{yx}, and so on.

We show in the next Section 3.8 that these six components evaluated at P are sufficient to determine the stresses on any plane through the point. In fact, the six components associated with any three non-co-planar surfaces at P suffice to determine the stresses on any other plane. The three normal stresses and the three (distinct) shear stresses associated with any three mutually perpendicular directions at P are said to define the *state of stress* at P.

3.8 STRESSES ON ANY SURFACE

Suppose that the six components in a rectangular system (x, y, z) are known at a point. Then the stresses on those surfaces are given by the vectors

$$\boldsymbol{\sigma_x} = \sigma_x\hat{\mathbf{i}} + \tau_{xy}\hat{\mathbf{j}} + \tau_{xz}\hat{\mathbf{k}} \tag{3.14a}$$
$$\boldsymbol{\sigma_y} = \tau_{yx}\hat{\mathbf{i}} + \sigma_y\hat{\mathbf{j}} + \tau_{yz}\hat{\mathbf{k}} \tag{3.14b}$$
$$\boldsymbol{\sigma_z} = \tau_{zx}\hat{\mathbf{i}} + \tau_{zy}\hat{\mathbf{j}} + \sigma_z\hat{\mathbf{k}} \tag{3.14c}$$

where $\hat{\mathbf{i}}$, $\hat{\mathbf{j}}$, and $\hat{\mathbf{k}}$ denote the unit vectors in the directions of the coordinates x, y, and z, respectively. Now suppose that we are interested in the stress upon a surface with some other orientation. The orientation of the surface in question can be identified by its unit normal $\hat{\mathbf{n}}$; this direction can be specified by the angles that the normal forms with axes x, y, z or, quite simply, by the direction cosines l_n, m_n, n_n. Then the unit normal has the form

$$\hat{\mathbf{n}} = l_n\hat{\mathbf{i}} + m_n\hat{\mathbf{j}} + n_n\hat{\mathbf{k}} \tag{3.15}$$

To determine the stress upon the surface, we suppose that a plane passes through the body with the normal $\hat{\boldsymbol{n}}$. The plane slices through the three rectangular planes (x, y, z) near the point P and together these planes enclose the small tetrahedron of Figure 3.47. Our inclined face has normal $\hat{\mathbf{n}}$; the rectangular faces have normals -$\hat{\mathbf{i}}$, -$\hat{\mathbf{j}}$, and -$\hat{\mathbf{k}}$. This small tetrahedron can be viewed as a free body, acted upon by the stress σ_n an on the inclined face and -$\boldsymbol{\sigma}_x$, -$\boldsymbol{\sigma}_y$ and -$\boldsymbol{\sigma}_z$ on the rectangular faces. The forces acting on the body of Figure 3.47 are the mean stresses multiplied by the respective areas. Those areas are denoted by ΔA_x, ΔA_y, and ΔA_z, where the subscripts signify the orientation of the face. Equilibrium requires that the sum of all forces vanish:

$$\sum \mathbf{F} = \boldsymbol{\sigma}_n \Delta A_n - \boldsymbol{\sigma}_x \Delta A_x - \boldsymbol{\sigma}_y \Delta A_y - \boldsymbol{\sigma}_z \Delta A_z = 0 \tag{3.16a}$$

or

$$\boldsymbol{\sigma}_n = \frac{\Delta A_x}{\Delta A_n}\boldsymbol{\sigma}_x + \frac{\Delta A_y}{\Delta A_n}\boldsymbol{\sigma}_y + \frac{\Delta A_z}{\Delta A_n}\boldsymbol{\sigma}_z \tag{3.16b}$$

When the elemental tetrahedron is shrunk to the point P, then the equation gives the stress at the point. The ratios of the areas depend on the orientation of the inclined face, or the direction of the normal $\hat{\mathbf{n}}$. These relations are simply

$$\frac{\Delta A_x}{\Delta A_n} = l_n, \quad \frac{\Delta A_y}{\Delta A_n} = m_n, \quad \frac{\Delta A_z}{\Delta A_n} n_n \tag{3.17}$$

Then the equation of equilibrium takes the conventional form:

$$\boldsymbol{\sigma}_n = l_n \boldsymbol{\sigma}_y + m_n \boldsymbol{\sigma}_y + n_n \boldsymbol{\sigma}_z \tag{3.18}$$

This result is most important: If the stresses are known for the three directions (x, y, z), in other words, the six components

$$\begin{matrix} \sigma_x & \tau_{xy} & \tau_{xz} \\ & \sigma_y & \tau_{yz} \\ & & \sigma_z \end{matrix} \tag{3.19}$$

then the stress is determined on any other surface in accordance with Equation (3.18). We need to specify only the orientation, direction n. The direction might be prescribed by the angles between the normal n and the directions x, y, z or by the direction cosines (l_n, m_n, n_n).

We might be interested in the normal component of $\boldsymbol{\sigma}_n$ i.e., σ_n. As example, we mention the case of a material that might be brittle and, therefore, susceptible to fracture caused by a normal

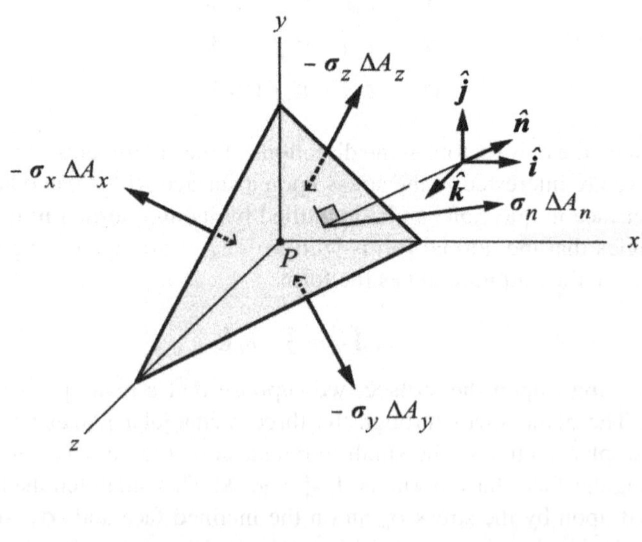

Figure 3.47

stress. In accordance with the definition, the normal component is the projection of the vector $\boldsymbol{\sigma}_n$ on the normal. This component is the scalar (dot) product of $\hat{\mathbf{n}}$ (3.15) and $\boldsymbol{\sigma_n}$ (3.18), where $\boldsymbol{\sigma}_x$, $\boldsymbol{\sigma}_y$, $\boldsymbol{\sigma}_z$ are given by 3.14:

$$\begin{aligned}\sigma_n = \hat{\mathbf{n}}\cdot\boldsymbol{\sigma}_n =& l_n l_n \sigma_x + m_n l_n \tau_{xy} + n_n l_n \tau_{xz}\\ &+ l_n m_n \tau_{yx} + m_n m_n \sigma_y + n_n m_n \tau_{yz}\\ & l_n n_n \tau_{zx} + m_n n_n \tau_{zy} + n_n n_n \sigma_z\end{aligned} \tag{3.20}$$

The component is depicted in Figure 3.48.

We might require a particular shear component of the stress $\boldsymbol{\sigma}_n$ (the material might be susceptible to shearing action in this particular direction along the surface). If so, we must prescribe the direction in question. Most conveniently, we could give the direction cosines (l_t, m_t, n_t) of the tangential direction t depicted in Figure 3.48. The unit tangent vector is then

$$\hat{\mathbf{t}} = l_t\hat{\mathbf{i}} + m_t\hat{\mathbf{j}} + n_t\hat{\mathbf{k}} \tag{3.21}$$

The required component on the surface (n) in the direction (t) is τ_{nt}, the projection of $\boldsymbol{\sigma}_n$ on the line t. It is also obtained as the scalar (dot) product of (3.18) and (3.21):

$$\begin{aligned}\tau_{nt} = \hat{\mathbf{t}}\cdot\boldsymbol{\sigma}_n =& l_t l_n \sigma_x + m_t l_n \tau_{xy} + n_t l_n \tau_{xz}\\ &+ l_t m_n \tau_{yx} + m_t m_n \sigma_y + n_t m_n \tau_{yz}\\ &+ l_t n_n \tau_{zx} + m_t n_n \tau_{zy} + n_t n_n \sigma_z\end{aligned} \tag{3.22}$$

In this manner, we can compute any component of stress upon any surface at a point if the six components are known. We say that the *state of stress* is determined by the six components.

It is noteworthy that the foregoing results and equations also hold if the body is in motion. A simple argument follows: Suppose that the element of Figure 3.47 were in motion with an acceleration **a**. The mass density is a number, say ρ, and the mass of the element is $\rho\Delta V$, where ΔV denotes the volume. Since the sum of the forces equals the product of the mass and acceleration,

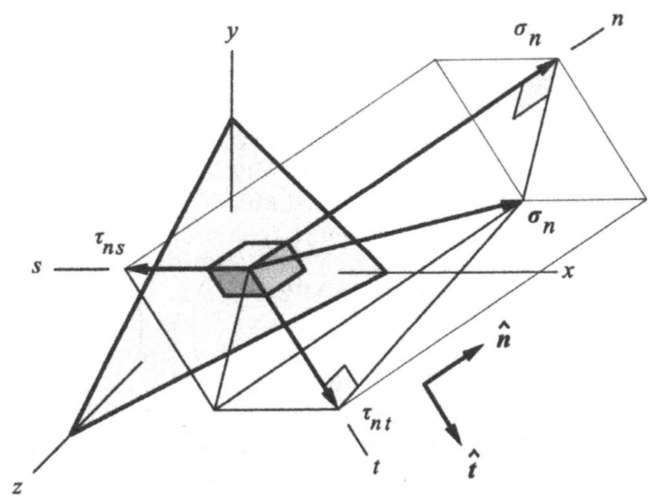

Figure 3.48

the right side of (3.16a) is not zero but is $\rho\mathbf{a}\Delta V$. Again, the equation is divided by the area ΔA_n and, upon passing to the limit $\Delta V/\Delta A_n \to 0$, we obtain the previous result (3.18).

Note that the six stress components (3.19), like the six strain components, compose a Cartesian tensor. Additional information on the mathematical significance of Equations (3.20) and (3.22) is given in Section 3.9.

The following example serves to illustrate the foregoing description of a state and the transformation from one rectangular system to another.

ILLUSTRATIVE EXAMPLE 3.20

A state of stress is described by six components; these need not be given with reference to the same rectangular system. Suppose that the four components (τ_{xy}, τ_{xz}, τ_{yz} and σ_z) are given in one system, (x, y, z), and the two components (σ_n, τ_{nt}) in another, (n, t, z), as shown in Figure 3.49. From the information, the definitions, and equilibrium, we can determine all components in either system; that is, we can fill in the blanks:

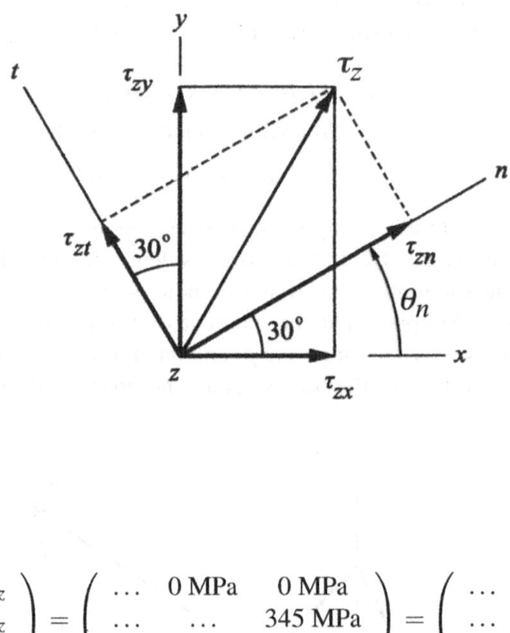

Figure 3.49

$$\begin{pmatrix} \sigma_x & \tau_{xy} & \tau_{xz} \\ \tau_{yx} & \sigma_y & \tau_{yz} \\ \tau_{zx} & \tau_{zy} & \sigma_z \end{pmatrix} = \begin{pmatrix} \dots & 0\text{ MPa} & 0\text{ MPa} \\ \dots & \dots & 345\text{ MPa} \\ \dots & \dots & 552\text{ MPa} \end{pmatrix} = \begin{pmatrix} \dots & 0\text{ ksi} & 0\text{ ksi} \\ \dots & \dots & 50\text{ ksi} \\ \dots & \dots & 80\text{ ksi} \end{pmatrix}$$

$$\begin{pmatrix} \sigma_n & \tau_{nt} & \tau_{nz} \\ \tau_{tn} & \sigma_t & \tau_{tz} \\ \tau_{zn} & \tau_{zt} & \sigma_z \end{pmatrix} = \begin{pmatrix} 414\text{ MPa} & 276\text{ MPa} & \dots \\ \dots & \dots & \dots \\ \dots & \dots & \dots \end{pmatrix} = \begin{pmatrix} 60\text{ ksi} & 40\text{ ksi} & \dots \\ \dots & \dots & \dots \\ \dots & \dots & \dots \end{pmatrix}$$

Of course, the relative orientations of the systems must be known. Here, the orthogonal directions n, t are also orthogonal to z and inclined to x and y, as shown in Figure 3.49.

We recall Equation (3.13) which enforce equilibrium of moment and assert the symmetry of the shear components with respect to their subscripts. Accordingly,

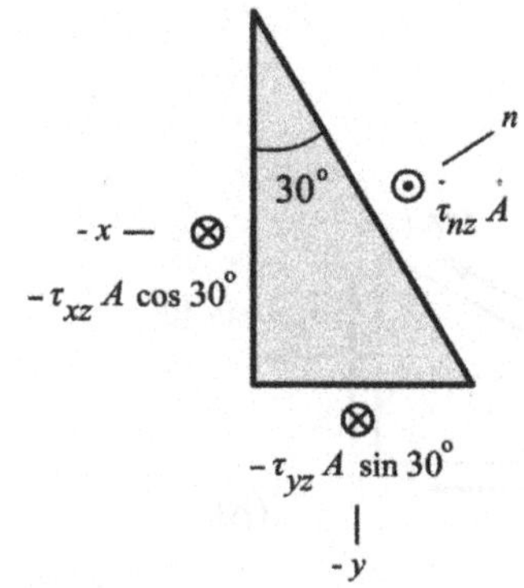

$$
\begin{aligned}
\tau_{yx} &= \tau_{xy} = 0 \\
\tau_{zx} &= \tau_{xz} = 0 \\
\tau_{yz} &= \tau_{zy} = 345 \text{ MPa}(50 \text{ ksi}) \\
\tau_{tn} &= \tau_{nt} = 276 \text{ MPa } (40 \text{ ksi})
\end{aligned}
$$

Figure 3.50

From the definitions, the shear stress upon the z surface is given by the two orthogonal components, τ_{zx} and τ_{zy}. This shear stress can be expressed as well by the components τ_{zn} and τ_{zt}. In other words, it is a matter of projecting the vector $\boldsymbol{\tau}_z$ in the directions n or t. From the figure we see that

$$
\begin{aligned}
\tau_{zt} &= -\tau_{zx} \sin 30° + \tau_{zy} \cos 30° = 345 \cos 30° = 299 \text{ MPa } (43 \text{ ksi}) \\
\tau_{zn} &= \tau_{zx} \cos 30° + \tau_{zy} \sin 30° \quad = 345 \sin 30° = 173 \text{ MPa } (25 \text{ ksi})
\end{aligned}
$$

Either result can be achieved by an appeal to equilibrium. The free body in Figure 3.50 is a wedge acted upon by the forces on the $-x$, $-y$, and n faces. For simplicity, we show only the relevant forces (stresses $\times$ areas); A denotes the area of the inclined n face. Equilibrium of the forces gives the preceding result for τ_{nz}. The same result is obtained by forming the scalar product $\hat{\boldsymbol{k}} \cdot \boldsymbol{\sigma}_n$. as in Equation (3.22); since $l_n = \cos\theta$, $m_n = \sin\theta$, and $n_n = 0$,

$$
\begin{aligned}
\tau_{nz} = \hat{\boldsymbol{k}} \cdot \boldsymbol{\sigma}_n &= \hat{\boldsymbol{k}} \cdot \boldsymbol{\sigma}_x \cos\theta + \hat{\boldsymbol{k}} \cdot \boldsymbol{\sigma}_y \sin\theta \\
&= \tau_{xz} \cos\theta + \tau_{yz} \sin\theta
\end{aligned}
$$

The remaining components, σ_y and σ_t, are also obtained by enforcing equilibrium. Each of the free bodies shown in Figure 3.51 is acted upon by one or more unknown actions ($A\sigma_x$, $A\sigma_y$, $A\sigma_t$). From the free-body diagram (F.B.D.) in Figure 3.51a, $\sum F_y = 0$ gives

$$
\sigma_y \sin 30° = 414 \sin 30° + 276 \cos 30°, \quad \sigma_y = 892 \text{ MPa } (129 \text{ ksi})
$$

and $\sum F_x = 0$ gives

$$
\sigma_x \cos 30° = 414 \cos 30° - 276 \sin 30°, \quad \sigma_x = 255 \text{ MPa } (37 \text{ ksi})
$$

From the free-body diagram (F.B.D.) in Figure 3.51(b), $\sum F_t = 0$ gives

$$
\sigma_t = \sigma_y \cos^2 30° + \sigma_x \sin^2 30° = 733 \text{ MPa } (106 \text{ ksi})
$$

The reader can verify that the latter component is the normal component of $\boldsymbol{\sigma}_t$. Indeed, the important message here is that transformations such as Equation (3.20) and (3.22) embody no more than the *definitions* and equilibrium.

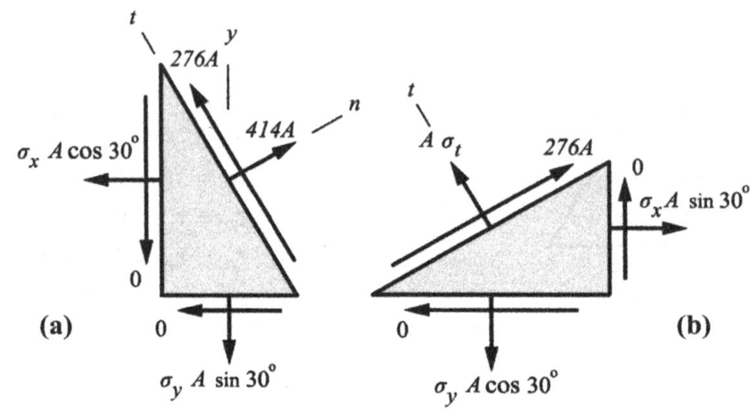

Figure 3.51

ILLUSTRATIVE EXAMPLE 3.21

The state of stress at a point is given by the six components of stress in the rectangular system (x, y, z):

$$\begin{pmatrix} \sigma_x & \tau_{xy} & \tau_{xz} \\ & \sigma_y & \tau_{yz} \\ & & \sigma_z \end{pmatrix} = \begin{pmatrix} 10 & 8 & -6 \\ & 4 & 10 \\ & & 8 \end{pmatrix} \text{MPa}$$

Evaluate the stress vector $\boldsymbol{\sigma}_n$ on a surface with normal $\hat{\boldsymbol{n}} = \frac{1}{2}\hat{\imath} + \frac{1}{2}\hat{\jmath} + \frac{1}{\sqrt{2}}\hat{\boldsymbol{k}}$: The components of $\boldsymbol{\sigma}_n$ in the direction of the normal and the shear components in the directions of the unit vectors:

$$\hat{\boldsymbol{a}} = -\frac{1}{2}\hat{\mathbf{i}} - \frac{1}{2}\hat{\mathbf{j}} + \frac{1}{\sqrt{2}}\hat{\boldsymbol{k}}, \quad \hat{\boldsymbol{b}} = \frac{1}{\sqrt{2}}\hat{\mathbf{i}} - \frac{1}{\sqrt{2}}\hat{\mathbf{j}}$$

As mentioned in Section 3.8, the *basic* tools are the *definitions and conditions of equilibrium*. By definition:

$$\boldsymbol{\sigma}_x = 10\hat{\mathbf{i}} + 8\hat{\jmath} - 6\hat{\mathbf{k}} \text{ MPa}$$
$$\boldsymbol{\sigma}_y = 8\hat{\mathbf{i}} + 4\hat{\jmath} + 10\hat{\mathbf{k}} \text{ MPa}$$
$$\boldsymbol{\sigma}_z = -6\hat{\mathbf{i}} + 10\hat{\jmath} + 8\hat{\mathbf{k}} \text{ MPa}$$

To determine the required stress upon the surface, the free-body diagram of Figure 3.52 is employed. The free body is acted upon by forces, which are the mean stresses multiplied by the respective areas. Those areas are denoted by ΔA_x, ΔA_y, ΔA_z, and ΔA_n where the subscripts signify the orientation of the face. Equilibrium requires that the sum of all forces vanish:

$$\sum \mathbf{F} = \boldsymbol{\sigma}_n \Delta A_n - \boldsymbol{\sigma}_x \Delta A_x - \boldsymbol{\sigma}_y \Delta A_y - \boldsymbol{\sigma}_z \Delta A_z = \mathbf{0}$$
$$\Longrightarrow \boldsymbol{\sigma}_n = \frac{\Delta A_x}{\Delta A_n}\boldsymbol{\sigma}_x + \frac{\Delta A_y}{\Delta A_n}\boldsymbol{\sigma}_y + \frac{\Delta A_z}{\Delta A_n}\boldsymbol{\sigma}_z$$

Here

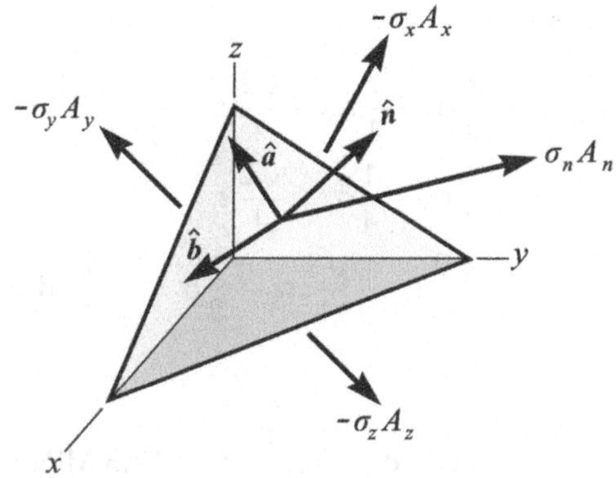

Figure 3.52

$$\frac{\Delta A_x}{\Delta A_n} = \hat{\mathbf{i}} \cdot \hat{\mathbf{n}} = \frac{1}{2}, \quad \frac{\Delta A_y}{\Delta A_n} = \hat{\mathbf{j}} \cdot \hat{\mathbf{n}} = \frac{1}{2}, \quad + \frac{\Delta A_z}{\Delta A_n} = \hat{\mathbf{k}} \cdot \hat{\mathbf{n}} = \frac{1}{\sqrt{2}}$$

Inserting these values into Equation (A), we obtain

$$\begin{aligned} \boldsymbol{\sigma}_n &= \frac{1}{2}\boldsymbol{\sigma}_x + \frac{1}{2}\boldsymbol{\sigma}_y + \frac{1}{\sqrt{2}}\boldsymbol{\sigma}_z \\ &= (9 - 3\sqrt{2})\hat{\mathbf{i}} + (6 + 5\sqrt{2})\hat{\mathbf{j}} + (2 + 4\sqrt{2})\hat{\mathbf{k}} \text{ MPa} \end{aligned}$$

The stress components follow from the definitions:

$$\begin{aligned} \sigma_n \equiv \hat{\boldsymbol{n}} \cdot \boldsymbol{\sigma}_n &= \frac{1}{2}(9 - 3\sqrt{2}) + \frac{1}{2}(6 + 5\sqrt{2}) + \frac{1}{2}(8 + 2\sqrt{2}) \\ &= \frac{1}{2}(23 + 4\sqrt{2}) = 14.3 \text{ MPa} \\ \tau_{na} \equiv \hat{\boldsymbol{a}} \cdot \boldsymbol{\sigma}_n &= -\frac{1}{2}(9 - 3\sqrt{2}) - \frac{1}{2}(6 + 5\sqrt{2}) + \frac{1}{2}(8 + 2\sqrt{2}) \\ &= -\frac{7}{2} = -3.5 \text{ MPa} \\ \tau_{nb} \equiv \hat{\boldsymbol{b}} \cdot \boldsymbol{\sigma}_n &= \frac{1}{2}(-6 + 9\sqrt{2}) - \frac{1}{2}(10 + 6\sqrt{2}) \\ &= -\frac{1}{2}(16 - 3\sqrt{2}) = -5.9 \text{ MPa} \end{aligned}$$

The above results may be verified by employing alternative relationships: For example, from the expressions for $\hat{\mathbf{a}}$ and $\hat{\boldsymbol{n}}$, we have

$$l_a = -\frac{1}{2}, \quad m_a = -\frac{1}{2}, \quad n_a = \frac{1}{\sqrt{2}}, \quad l_n = \frac{1}{2}, \quad m_n = \frac{1}{2}, \quad n_n = \frac{1}{\sqrt{2}}$$

Inserting these values into Equation (3.22) we obtain

$$\begin{aligned}\tau_{na} = & -\frac{1}{4}\sigma_x - \frac{1}{4}\tau_{xy} + \frac{\sqrt{2}}{4}\tau_{xz} \\ & -\frac{1}{4}\tau_{xy} - \frac{1}{4}\sigma_y + \frac{\sqrt{2}}{4}\tau_{yz} \\ & -\frac{\sqrt{2}}{4}\tau_{xz} - \frac{\sqrt{2}}{4}\tau_{yz} + \frac{1}{2}\sigma_z = -\frac{7}{2}\text{MPa}\end{aligned}$$

Also

$$|\sigma_n| = \sqrt{\boldsymbol{\sigma}_n \cdot \boldsymbol{\sigma}_n - \tau_{na}^2 - \tau_{nb}^2} = 14.3 \text{ MPa}$$

3.9 REMARK ON NOTATIONS AND TRANSFORMATIONS

To generalize the transformation of stress components presented in Section 3.8, the reader may recall a similar transformation of the components of strain, specifically Equation (2.18) and (2.19). Indeed, we need only replace the ε and $\gamma/2$ in Equations (2.18) and (2.19) by σ and τ, respectively, to obtain Equations (3.20) and (3.22). From a mathematical viewpoint, the six components of stress and the corresponding six components of strain transform as the components of symmetric second-order Cartesian tensors.

The mathematical jargon need not impress the engineer, but the recurrence of this linear transformation is noteworthy. This linear transformation of the six symmetrical components calls for an alternative system of notation. To this end, let

$$\begin{matrix}\sigma_x & \tau_{xy} & \tau_{xz} \\ & \sigma_y & \tau_{yz} \\ & & \sigma_z\end{matrix} \equiv \begin{matrix}\sigma_{xx} & \sigma_{xy} & \sigma_{xz} \\ & \sigma_{yy} & \sigma_{yz} \\ & & \sigma_{zz}\end{matrix}$$

Similarly set

$$\begin{matrix}\varepsilon_x & \gamma_{xy}/2 & \gamma_{xz}/2 \\ & \varepsilon_y & \gamma_{yz}/2 \\ & & \varepsilon_z\end{matrix} \equiv \begin{matrix}\varepsilon_{xx} & \varepsilon_{xy} & \varepsilon_{xz} \\ & \varepsilon_{yy} & \varepsilon_{yz} \\ & & \varepsilon_{zz}\end{matrix}$$

Again, these are merely alternative notations. The repeated subscripts signal a normal stress (e.g., σ_{xx}) or extensional strain (e.g., ε_{xx}), whereas the dissimilar subscripts signal a shear stress (e.g., $\sigma_{xy} = \tau_{xy}$) or a shear strain (e.g., $\varepsilon_{xy} = \gamma_{xy}/2$) . Then, we can write the equations of the linear transformation in the same form; to illustrate, Equations (2.18) and (2.19) assume the following forms:

$$\begin{aligned}\varepsilon_{nt} &= l_n l_n \varepsilon_{xx} + m_n l_n \varepsilon_{xy} + l_n n_n \varepsilon_{xz} \\ &\quad + m_n l_n \varepsilon_{yx} + m_n m_n \varepsilon_{yy} + m_n n_n \varepsilon_{yz} \\ &\quad + n_n l_n \varepsilon_{zx} + n_n m_n \varepsilon_{yz} + n_n n_n \varepsilon_{zz} \\ \varepsilon_{nt} &= l_n l_t \varepsilon_{xx} + l_n m_t \varepsilon_{xy} + l_n n_t \varepsilon_{xz} \\ &\quad + m_n l_t \varepsilon_{yx} + m_n m_t \varepsilon_{yy} + m_n n_t \varepsilon_{yz} \\ &\quad + n_n l_t \varepsilon_{zx} + n_n m_t \varepsilon_{yz} + n_n n_t \varepsilon_{zz}\end{aligned}$$

The same equations give the transformation of stress components; we need only replace ε by σ. If n and t are orthogonal to z, then $n_n = n_t = 0$, and we have the two-dimensional versions:

$$\begin{aligned}\varepsilon_{nn} &= l_n l_n \varepsilon_{xx} + l_n m_n \varepsilon_{xy} \\ &\quad + m_n l_n \varepsilon_{yx} + m_n m_n \varepsilon_{yy} \\ \varepsilon_{nt} &= l_n l_t \varepsilon_{xx} + l_n m_t \varepsilon_{xy} \\ &\quad + m_n l_t \varepsilon_{yx} + m_n m_t \varepsilon_{yy}\end{aligned}$$

The area integrals (I_y=I_{yy}, I_z=I_{zz}, I_{yz}= I_{zy} (See Illustrative Example 3.15, equation E) also transform via these equations.

3.10 PROPERTIES OF A STATE OF STRESS

In Section 3.8, we obtained the remarkable result expressed by Equation (3.18) [and Equations (3.20) and (3.22) as well] that the stress upon any surface can be expressed as a linear combination of the stresses on three surfaces. More specifically, the components of stress on the surface n can be expressed as a linear combination of the six components on the three rectangular faces (x, y, z). Much can be learned by examining these equations of transformation, (3.20) and (3.22). They are quadratic forms in the direction cosines. Specifically, the normal component σ_n is a quadratic function σ_n (l_n, m_n , n_n). We can ask: In what direction does the normal component attain the maximum value? This question might be of practical importance if a material is susceptible to failure under a tensile stress. Any smooth function attains a maximum (or minimum) only if the derivatives vanish; that is, $\partial\sigma_n/\partial l_n = \partial\sigma_n/\partial m_n = \partial\sigma_n/\partial n_n = 0$; the function σ_n is said to be *stationary*. Here, the mathematical exercise is complicated by the fact that the three variables are not independent; *specifically*, from (3.15)

$$\hat{\boldsymbol{n}} \cdot \hat{\boldsymbol{n}} = l_n^2 + m_n^2 + n_n^2 = 1 \tag{3.23}$$

Omitting the details, we give the required *stationary* criteria. The cosines (l_n, m_n, n_n) are to satisfy three linear equations:

$$\begin{aligned}(\sigma_x - \lambda) l_n + \tau_{xy} m_n + \tau_{xz} n_n &= 0 \\ \tau_{yx} l_n + (\sigma_y - \lambda) m_n + \tau_{yz} n_n &= 0 \\ \tau_{zx} l_n + \tau_{zy} m_n + (\sigma_z - \lambda) n_n &= 0\end{aligned} \tag{3.24}$$

Additionally, λ is to be determined by the condition that the determinant formed by the coefficients of the cosines l_n, m_n, n_n) in Equation (3.24) must vanish:

$$\begin{vmatrix}(\sigma_x - \lambda) & \tau_{xy} & \tau_{xz} \\ \tau_{yx} & (\sigma_y - \lambda) & \tau_{yz} \\ \tau_{zx} & \tau_{zy} & (\sigma_z - \lambda)\end{vmatrix} = 0 \tag{3.25}$$

The expansion of this determinant Equation (3.25) is a cubic equation in the unknown λ:

$$\lambda^3 - I_1 \lambda^2 + I_2 \lambda - I_3 = 0 \tag{3.26}$$

Here, the coefficients are functions of the *known* components. For example I_1 is the simple linear form:

$$I_1 = \sigma_x + \sigma_y + \sigma_z \tag{3.27}$$

I_2 and I_3 are, respectively, quadratic and cubic forms in the known components. To determine these special directions, one must first solve the cubic equation (3.26), which determines three real numbers (λ_1, λ_2, λ_3). For each number, we return to Equations (3.23) and (3.24) (any *two* of the latter) to determine three directions. These three directions are the principal directions of stress. The three numbers (λ_1, λ_2, λ_3) are the *normal stresses* for those directions. One is a maximum and one is a minimum.

We can also ask: Upon what surface is the shear stress a maximum? This too is a mathematical exercise. According to Equation (3.22), a component of shear τ_{nt} is a function of six cosines (l_n, m_n, n_n; l_t, m_t , n_t). One must find two directions that satisfy the stationary criteria, but the variables are not independent. Again, we bypass the mathematical exercise and give the remarkable results: The directions of maximum shear bisect the principal directions of maximum and minimum normal stress.

An important attribute of the quantities I_1, I_2 and I_3 in Equation (3.26): We note that the three roots of the cubic equation (3.26) are determined by the coefficients I_1, I_2 and I_3; those roots are the principal values. Those principal stresses cannot depend upon the choice of coordinates, and, therefore, the coefficients must also be independent of such choice. We say that these quantities are invariant (with respect to coordinate transformations). For example, if (x, y, z) and (x', y', z') are two different rectangular coordinates,

$$I_1 = \sigma_x + \sigma_y + \sigma_z = \sigma_{x'} + \sigma_{y'} + \sigma_{z'}$$

The expansion of the determinant equation (3.25) gives the explicit forms of the invariants I_2 and I_3; that expansion is left as an exercise.

3.11 PRINCIPAL DIRECTIONS; MAXIMUM NORMAL AND SHEAR STRESSES

Any state of stress is fully determined by six components, for example, σ_x, σ_y, σ_z, τ_{xy}, τ_{xz}, τ_{yz}. In other words, we can determine, from the six components, the stress upon any other surface at the point. It is a mathematical exercise (sketched in the preceding section) to determine various other properties. Specifically, we can ask the following:

1. Are there directions (surfaces) of zero shear stress?
2. In what direction is the normal stress a maximum?
3. In what direction is the shear stress a maximum?

The questions are interesting but are also of extreme practical importance. Ultimately the engineer must be concerned with the failure of members, machines, and structures; such failure often hinges on the magnitudes and directions of a normal or shear stress, depending upon the properties of a material. The very important answers follow:

1. At each point of a stressed body, there *exist three mutually perpendicular directions of zero shear stress*. In other words, at every point we can imagine a small rectangular element, as depicted in Figure 3.53; the faces of the element experience only normal stress (λ_1, λ_2, λ_3). These three perpendicular directions are called the *principal directions* of stress. The normal stresses are called the *principal stresses*.
2. One of the principal stresses is the *maximum normal* stress at the point. One of the principal stresses is the *minimum normal stress*.

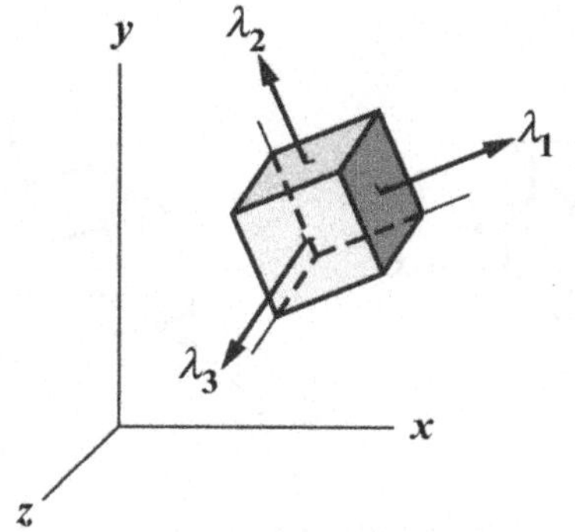

Figure 3.53

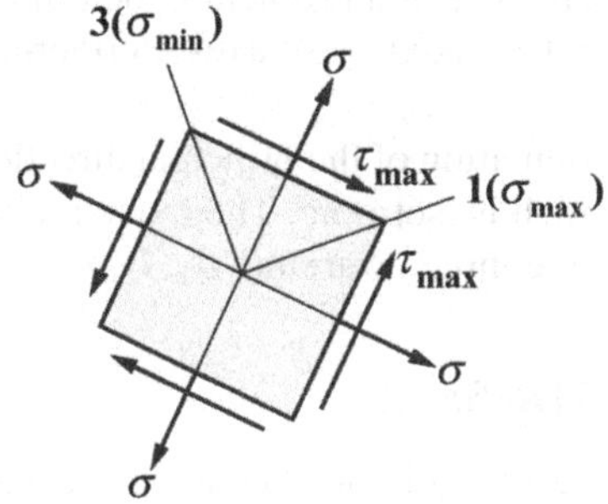

Figure 3.54

3. The *directions of maximum shear stress bisect the principal directions of maximum and minimum normal stress.* The directions of maximum shear stress are depicted in Figure 3.54. Here, principal directions 1 and 3 are the directions of maximum and minimum normal stress (λ_1 and λ_3); principal direction 2 is normal to the page. The maximum shear stress is always given by the simple, but useful, formula

$$\tau_{\max} = \frac{1}{2}\left|\sigma_{\max} - \sigma_{\min}\right| \tag{3.28}$$

   Note that the normal stress on these surfaces of maximum shear is not, in general, zero. Coincidentally, the normal stress on these surfaces is $\sigma = \frac{1}{2}\left|\sigma_{\max} + \sigma_{\min}\right|$.

Before leaving this discussion of state of stress at a point, it is important to reemphasize certain aspects of it. Suppose that a small rectangular parallelepiped is removed from a body, as shown in Figure 3.55. Whatever the orientation of this element, a knowledge of the six stresses acting on the faces of the element is sufficient to determine the stresses on any other plane. If we are able to pick an element whose edges are parallel to the principal directions, then we need only the three principal stresses to specify the state of stress. Thus, for the element shown in Figure 3.55a, there are- in general- six different stresses acting on the faces. However, in Figure 3.55b, the element is oriented so that the faces are principal planes, and there are, therefore, only three different stresses acting on the faces. The state of stress is also described fully by the three principal stresses *plus* three quantities, e.g. angles, which define the three principal directions. There is, however, one unfortunate feature connected with the use of principal stresses. Unless the principal directions

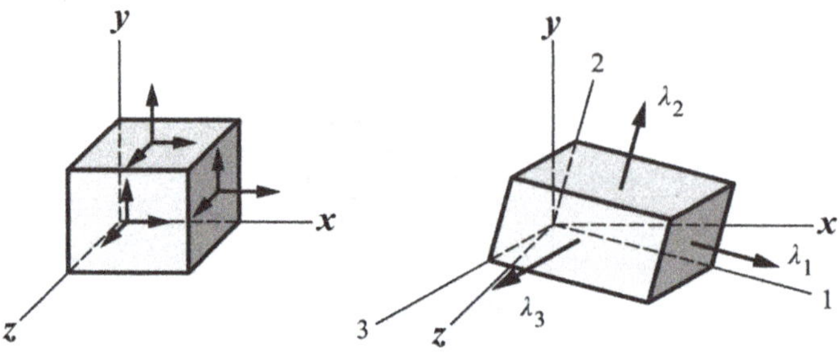

(a) (b)

Figure 3.55

can be determined easily, it is often more of a problem to determine the principal planes than it is to use the six stresses associated with a fixed set of axes. Furthermore, the principal directions can change from point to point.

In certain circumstances, the orientation of the principal directions is immaterial, e.g in describing the properties of a material which is isotropic. Then, one might describe the state by the three principal stresses or equivalently, the three invariants (I_1, I_2, I_3).

3.12 SPECIAL STATES OF STRESS

Several special states of stress deserve attention. The most special circumstance exists if all three principal stresses are equal (Equation 3.26 has but one root). Then, in accordance with equation 3.28, the maximum shear stress is zero; stated otherwise, *no* shear stress exists on *any* surface at such a point. Every direction may be called a principal direction. If one imagines a tiny sphere at the point, as in Figure 3.56a, then every elemental area of the surface experiences the same normal stress ($\sigma_n = T, \tau_n = 0$). This is called a *hydrostatic* state of stress. If we were submerged in water (hydro) at a great depth and if the water were quiescent (static), then every elemental surface of our bodies would experience the same pressure. Another special circumstance exists if only one principal stress is distinct: Two roots of (3.26) are equal. There is then only one distinct principal direction. We can imagine a tiny circular cylinder at such point, as depicted in Figure 3.56b. The axis lies in the one distinct principal direction; a right cross section experiences that distinct principal stress ($\sigma_n = T_1$). Every element of the cylindrical surface experiences only normal stress

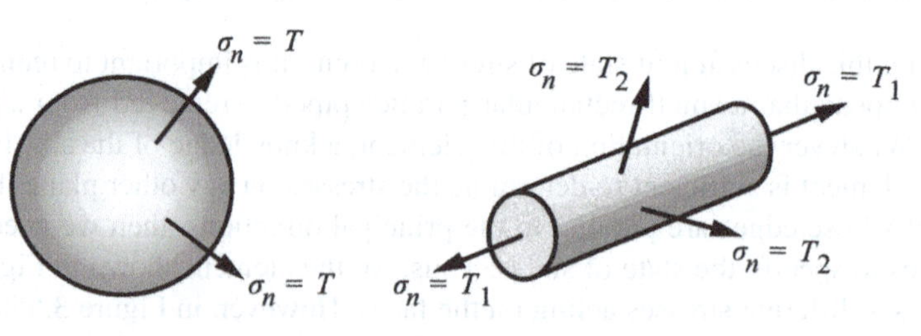

(a) (b)

Figure 3.56

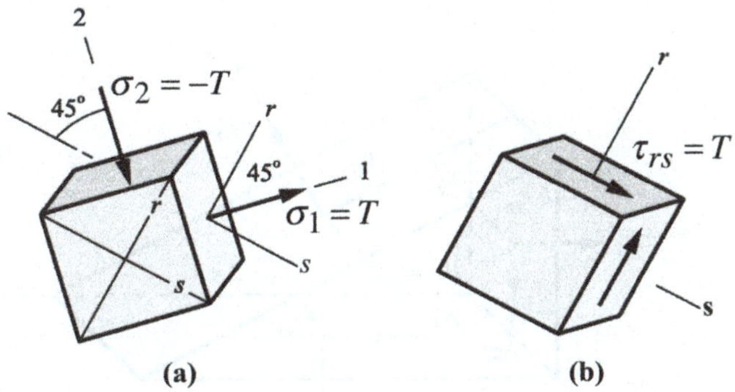

Figure 3.57

[$\sigma_n = T_2$, the repeated root of Equation (3.26)] ; every radial direction may be called a principal direction.

Another special state exists if the three principal stresses are $\sigma_1 = T$, $\sigma_2 = -T$, $\sigma_3 = 0$. This state is depicted in Figure 3.57a. The surfaces that bisect the two principal directions (1 and 2) experience only the shear stress ($\tau_{rs} = \tau_{sr}$ = T) as illustrated in Figure 3.57b. The transformation from the principal directions to the bisecting directions is given in Equations (3.20) and (3.22). Relative to the directions r and s this is seen as a state of simple shear stress.

3.13 STRESS ON A PLANE PARALLEL TO A COORDINATE LINE

Suppose that the direction n is normal to the z axis. Then a perpendicular plane cuts the xy plane to form the thin wedge shown in Figure 3.58a with the point P at one corner. This is a free-body diagram. The resultant of the internal forces on the inclined face has components ΔF_n, ΔV_{nt}, and ΔV_{nz} in the n (normal), t (tangential), and z (tangential) directions, respectively.

The wedge is redrawn in the planar view of Figure 3.58b. The force components on the front and rear faces of the wedge are not shown in either figure, because they cancel in the equations that equilibrate forces. The orientation of the inclined face is specified by θ, the angle between the x axis and n axis. From Figure 3.58b, the quantities Δx, Δy, Δs, and θ are related:

$$\Delta x = \Delta s \sin\theta, \quad \Delta y = \Delta s \cos\theta \tag{3.29}$$

According to the usual notations, σ_n denotes the normal stress on the inclined face, τ_{nt}, the shear stress in the t direction, and τ_{nz}, the shear stress in the z direction. The components of force on each of the faces may be expressed in terms of stresses and areas. Thus, on the inclined face,

$$\Delta F_n = \sigma_n \Delta s \Delta z, \quad \Delta V_{nt} = \tau_{nt} \Delta s \Delta z, \quad \Delta V_{nz} = \tau_{nz} \Delta s \Delta z$$

Also, by using equations 3.29. the force components on the other two faces of the wedge are

$$\Delta F_x = \sigma_x \Delta y \Delta z = \sigma_x \cos\theta \Delta s \Delta z \qquad \Delta F_y = \sigma_y \Delta x \Delta z = \sigma_y \sin\theta \Delta s \Delta z$$
$$\Delta V_{xy} = \tau_{xy} \Delta y \Delta z = \tau_{xy} \cos\theta \Delta s \Delta z \qquad \Delta V_{yx} = \tau_{yx} \Delta x \Delta z = \tau_{yx} \sin\theta \Delta s \Delta z$$
$$\Delta V_{xz} = \tau_{xz} \Delta y \Delta z = \tau_{xz} \cos\theta \Delta s \Delta z \qquad \Delta V_{yz} = \tau_{yz} \Delta x \Delta z = \tau_{yz} \sin\theta \Delta s \Delta z$$

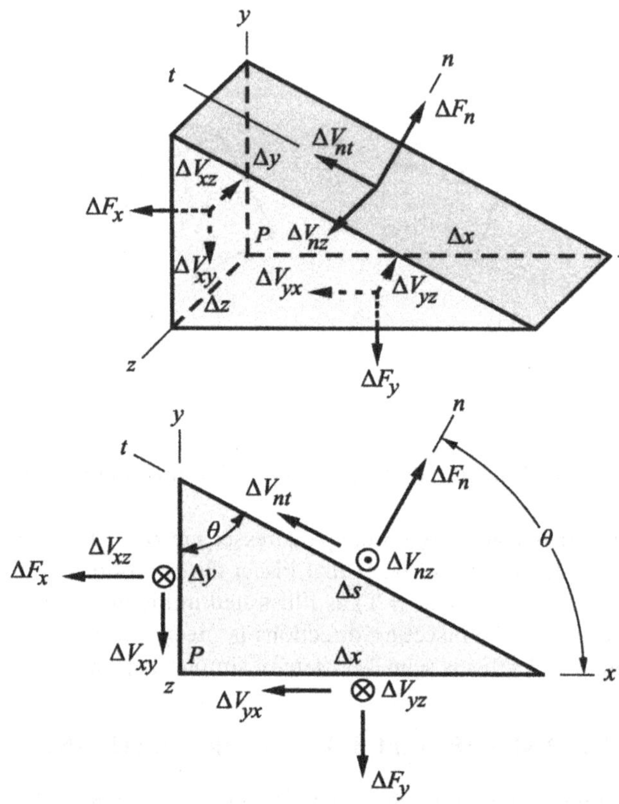

Figure 3.58

Here we have taken the liberty of replacing average stress values by the values at the point P. It is understood that the dimensions of the wedge are "infinitesimally" small, so that (in the limit) the inclined face passes through point P. Let us now write the equations for equilibrium of the forces. Moments were equilibrated previously to ascertain the equality $\tau_{xy} = \tau_{yx}$ For equilibrium, the sum of the forces in the n direction must vanish:

$$\Delta F_n - \Delta F_x \cos\theta - \Delta V_{xy} \sin\theta - \Delta F_y \sin\theta - \Delta V_{yx} \cos\theta = 0$$

or, in terms of stress components,

$$(\sigma_n - \sigma_x \cos^2\theta - \sigma_y sin^2\theta - 2\tau_{xy} \sin\theta \cos\theta)\Delta s \Delta z = 0 \tag{3.30}$$

Similarly, the sum of the forces in the t direction must also be zero:

$$\Delta V_{nt} + \Delta F_x \sin\theta - \Delta V_{xy} \cos\theta - \Delta F_y \cos\theta + \Delta V_{yx} \sin\theta = 0$$

or

$$[\tau_{nt} + (\sigma_x - \sigma_y)\sin\theta\cos\theta - \tau_{xy}(\cos^2\theta - \sin^2\theta)]\Delta s \Delta z = 0 \tag{3.31}$$

The common factor $\Delta s\, \Delta z$ can be divided out of Equations (3.30) and (3.31), so that

$$\sigma_n = \sigma_x \cos^2\theta + \sigma_y \sin^2\theta + 2\tau_{xy} \sin\theta \cos\theta \tag{3.32}$$

$$\tau_{nt} = (\sigma_y - \sigma_x)\sin\theta\cos\theta + \tau_{xy}(cos^2\theta - \sin^2\theta) \tag{3.33}$$

Note that Equations (3.32) and (3.33) are special forms of Equations (3.20) and (3.22) wherein

$$l_n = \cos\theta, \quad m_n = \sin\theta, \quad n_n = 0$$
$$l_t = -\sin\theta, \quad m_t = \cos\theta, \quad n_t = 0$$

These equations may be put in another form by substituting the trigonometric identities:

$$\cos^2\theta = \frac{1}{2} + \frac{1}{2}cos2\theta, \quad \sin^2\theta = \frac{1}{2} - \frac{1}{2}cos2\theta, \quad \sin\theta\cos\theta = \frac{1}{2}\sin 2\theta$$

Then Equations (3.32) and (3.33) become

$$\sigma_n = \frac{\sigma_x + \sigma_y}{2} + \frac{\sigma_x - \sigma_y}{2}\cos 2\theta + \tau_{xy}\sin 2\theta \tag{3.34}$$

$$\tau_{nt} = -\frac{\sigma_x - \sigma_y}{2}\sin 2\theta + \tau_{xy}\cos 2\theta \tag{3.35}$$

A third defining τ_{nz} is obtained by summing forces in the z direction:

$$\Delta V_{nz} - \Delta V_{xz} - \Delta yz = 0$$

which reduces to

$$\tau_{nz} = \tau_{xz}\cos\theta + \tau_{yz}\sin\theta \tag{3.36}$$

If the values of σ_x, σ_y, τ_{xy}, τ_{xz}, and τ_{yz} are known at a point, the stresses at that point on any other plane (parallel to the z axis) can be determined by this procedure. It is simply a matter of writing equilibrium equations for the forces acting on a properly chosen wedge. Note that only σ_x, σ_y, and τ_{xy} are required to compute σ_n and τ_{nt} for any perpendicular directions (n and t) in the xy plane.

Extreme care must be exercised in the use of Equations (3.34)–(3.36). If the angle θ is measured differently (say between the y and n axes) or if the t axis is opposite to that shown, the signs of the trigonometric functions appearing in these equations could be altered. For this reason, it is usually preferable to work directly from a free-body diagram of a wedge.

We can pose the following question: What direction n gives a maximum normal stress σ_n? By the necessary criterion, applied to Equation (3.34)

$$\frac{d\sigma_n}{d\theta} = -(\sigma_x - \sigma_y)\sin 2\theta + 2\tau_{xy}\cos 2\theta = 0 \tag{3.37a}$$

or

$$\tan 2\theta = \frac{2\tau_{xy}}{\sigma_x - \sigma_y} \tag{3.37b}$$

Any values (σ_x, σ_y, τ_{xy}) determine two angles (θ_1, θ_2), which differ by 90°. By comparison of Equations (3.37a) and (3.35), we see that the component τ_{nt} vanishes on the plane. Note that this is not necessarily a principal direction, since the component τ_{nz}, need not vanish. *If* we know also that the z direction is a principal direction, then $\tau_{zn} = \tau_{nz} = 0$, and the directions ($\theta_1$, θ_2) determined by Equation (3.37) are the remaining principal directions.

Equations (3.34) and (3.35) apply only if direction n is normal to axis z. Their applicability is, therefore, limited. They are useful primarily if one principal direction, z, is known a priori. Then the other two principal directions are determined by Equation (3.37); subsequently, the other two principal stresses are obtained from Equation (3.34), with the appropriate directions (θ_1, θ_2). The most common circumstance for such application occurs at a *free surface* of a member; free implies *free of stress* -in particular, free of shear stress. Such a free surface is a principal surface; the normal is a principal direction and the principal stress is zero. Then the other two principal directions are tangent to the surface and are determined in accordance with Equation (3.37).

ILLUSTRATIVE EXAMPLE 3.22

At the free surface of a body (direction z is normal to the surface, x and y are tangent) the components of stress are

$$\sigma_x = 100 \text{ ksi [689 MPa]}, \quad \tau_{xy} = 30 \text{ ksi [207 MPa]}, \quad \sigma_y = 20 \text{ ksi [138 MPa]}$$

Because the surface is free, that is, does not bear upon another body,

$$\sigma_z = \tau_{zx} = \tau_{zy} = 0$$

We wish to determine the maximum and minimum values of the normal stress and the maximum shear stress. Depending upon the material, these stresses could cause failure, fracture, or yielding.

Since $\tau_{zx} = \tau_{zy} = 0$, the z direction (normal to the surface) is one principal direction. It follows that the other two directions are tangent to the surface (in the xy plane). One principal stress is $\sigma_1 = \sigma_z = 0$. The other directions of vanishing shear—that is, principal directions—are determined by equation:

$$\tau_{nt} = 0$$

$$\tan 2\theta = \frac{2\tau_{xy}}{\sigma_x - \sigma_y}$$

We note that this also determines the directions in the xy plane that provide stationary values of normal stress in accordance with Equation (3.37). With the given numerical values,

$$\tan 2\theta = \frac{2 \times 30}{100 - 20} = \frac{3}{4}$$

Hence $2\theta_2 = 36.87°$ and $\theta_3 = 216.87°$, so $\theta_2 = 18.43°$ and $\theta_3 = 108.43°$. These give the other two principal directions. The corresponding normal stresses (principal stresses) follow according to Equation (3.34):

$$\sigma_{2,3} = \frac{1}{2}(120) + \frac{1}{2}(80)(\pm\frac{4}{5}) + (30)(\pm\frac{3}{5}) = 60 \pm \frac{160}{5} \pm \frac{90}{5} = 110 \text{ ksi or } 10 \text{ ksi [758 MPa or 9 MPa]}$$

The three principal stresses are

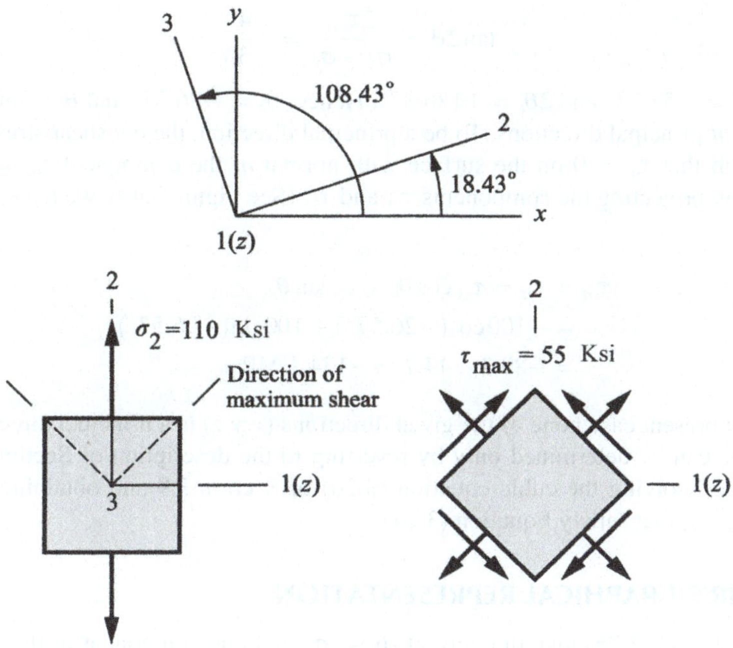

Figure 3.59

$$\sigma_1 = 0, \qquad \sigma_2 = 110 \text{ ksi [758 MPa]}, \qquad \sigma_3 = 10 \text{ ksi [69 MPa]}$$

Therefore, the maximum and minimum normal stresses are

$$\sigma_{max} = \sigma_2 = 110 \text{ ksi [758 MPa]}, \qquad \sigma_{min} = \sigma_1 = 0$$

In accordance with Equation (3.28), the maximum shear stress is

$$\tau_{max} = \frac{1}{2} |\sigma_{max} - \sigma_{min}| = 55 \text{ ksi[379] MPa}$$

Figure 3.59 shows the principal directions and the directions of maximum shear; the latter bisect the principal directions 1 and 2.

ILLUSTRATIVE EXAMPLE 3.23

Let us now examine the state of stress expressed by the following rectangular components:

$$\sigma_x = 200 \text{ MPa}, \qquad \tau_{xy} = -200 \text{ MPa}, \qquad \tau_{xz} = 100 \text{ MPa}$$
$$\sigma_y = -100 \text{ MPa}, \qquad \tau_{yz} = 100 \text{ MPa}, \qquad \sigma_z = 200 \text{ MPa}$$

Directions (n, t) perpendicular to z that bear no component τ_{nt} are obtained according to Equation (3.37): $\tau_{nt} = 0$ or

$$\tan 2\theta = \frac{2\tau_{xy}}{\sigma_x - \sigma_y} = -\frac{4}{3}$$

Therefore, $2\theta_n = -53.13^\circ$ and $2\theta_t = +126.87^\circ$. Hence, $\theta_n = -26.67^\circ$ and $\theta_t = +63.43^\circ$.

These are *not* principal directions. To be a principal direction, the *net* shear stress must vanish. It is not enough that $\tau_{nt} = 0$ on the surface with normal n; the component $\tau_{nz} = \tau_{zn}$ must also vanish. Here, by projecting the components τ_{zx} and τ_{zy} (See Figure 3.49), we have

$$\begin{aligned}\tau_{zn} = \tau_{nz} &= \tau_{zx}\cos\theta_n + \tau_{zy}\sin\theta_n \\ &= -100\cos(-26.57^\circ) + 100\sin(-26.57^\circ) \\ &= -89.4 - 44.7 = -134.1 \text{ MPa.}\end{aligned}$$

Note: In the present case none of the given directions (x, y, z) is a principal direction. The principal directions can be determined only by reverting to the description of Section 3.8 "Stresses on Any Surface," solving the cubic equation (3.26) of Section 3.9 and obtaining the directions (cosines l_n, m_n, n_n) that satisfy Equation (3.24).

3.14 MOHR'S GRAPHICAL REPRESENTATION

Equations (3.34) and (3.35) give the normal stress σ_n and one component of the shear stress τ_{nt} on a surface with normal n. These hold if and only if directions n and t are perpendicular to the direction z. None of the components on the z surface (σ_z, τ_{zx}, τ_{zy}) enter. As noted, we can always locate a direction n such that $\tau_{nt} = 0$. The direction follows in accordance with Equations (3.35) and (3.37 b). Now, let x' and y' denote these directions-that is, $\tau_{x'y'} = 0$ -and let θ denote the angle between x' and n. Then, in accordance with Equations (3.34) and (3.35),

$$\sigma_n = \frac{\sigma_{x'} + \sigma_{y'}}{2} + \frac{\sigma_{x'} - \sigma_{y'}}{2}\cos 2\theta \tag{3.38}$$

$$\tau_{nt} = -\frac{\sigma_{x'} + \sigma_{y'}}{2}\sin 2\theta \tag{3.39}$$

We suppose that the components ($\sigma_{x'}$, $\sigma_{y'}$) are known. For brevity. let

$$\frac{\sigma_{x'} + \sigma_{y'}}{2} \equiv a, \quad \frac{\sigma_{x'} - \sigma_{y'}}{2} \equiv b \tag{3.40a,b}$$

Then Equations (3.38) and (3.39) have the abbreviated forms

$$\sigma_n - a = b\cos 2\theta \tag{3.41}$$

$$-\tau_{nt} = b\sin 2\theta \tag{3.42}$$

These equations are the parametric equations of a circle, Mohr's circle; the center lies on the σ_n axis in Figure 3.60 at $\sigma_n = a$ and the radius is b. By squaring each of Equations (3.41) and (3.42) and adding the squares, we eliminate the parameter θ and obtain a familiar form:

$$(\sigma_n - a)^2 + \tau_{nt}^2 = b^2 \tag{3.43}$$

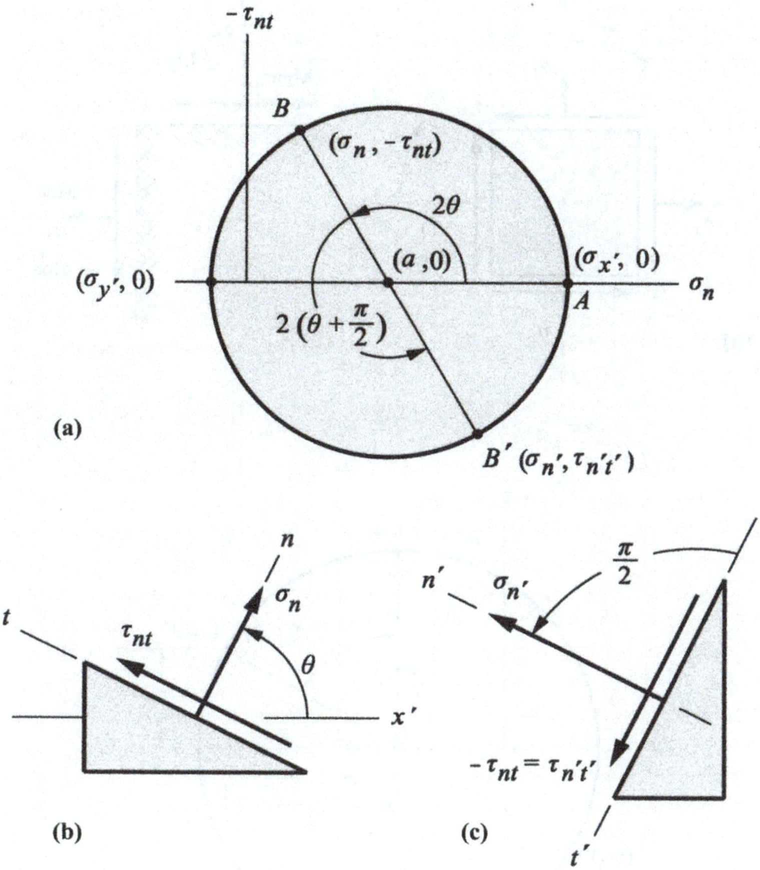

Figure 3.60

Note: To be entirely consistent with usual conventions, we plot $-\tau_{nt}$ versus σ_n. Then the radial line to the point B (σ_n, -τ_{nt}) on the circle forms the angle 2θ with the σ_n axis. Moreover, rotation from A ($\sigma_{x'}$,0), which represents the stress in the given direction x', to the point B($\sigma_{x'}$, -τ_{nt}), which represents the stress in direction n, has the sense of our original rotation (θ) from x' to n. Progressing around the circle in the counterclockwise sense through the angle $2\theta = \pi$ [through $\theta = \pi/2$ between n and n' in Figure 3.60b and c, we find that $\tau_{n't'} = -\tau_{nt}$. again consistent with the earlier findings ($\tau_{xy} = \tau_{yx}$). The graphical representation is an alternative to formulas (3.34) and (3.35). Consider the following example.

ILLUSTRATIVE EXAMPLE 3.24

Suppose that we are given the following components:

$$\begin{pmatrix} \sigma_x & \tau_{xy} & \tau_{xz} \\ & \sigma_y & \tau_{yz} \\ & & \sigma_z \end{pmatrix} = \begin{pmatrix} 8 & -3 & 2 \\ & -2 & 4 \\ & & 4 \end{pmatrix} \text{MPa}$$

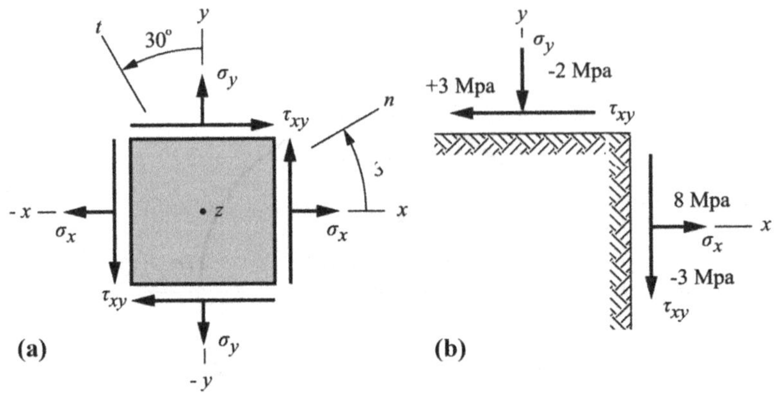

Figure 3.61

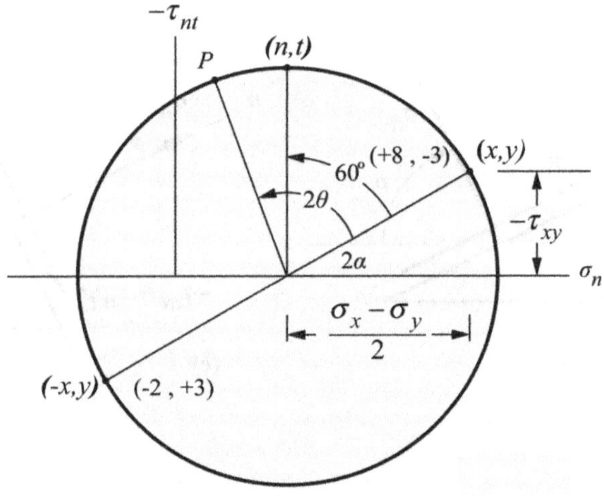

Figure 3.62

We require the components on a surface with normal n, which is perpendicular to direction z, as shown in Figure 3.61a. The components of stress are shown on surfaces with normals y and x in Figure 3.61b in accordance with the conventions used to derive equation (3.34), (3.35), and the plot of Mohr's circle. Since the rotation from x to y is an angle $\pi/2$, the corresponding points are diametrically opposed on Mohr's circle; these are labeled (x, y) and $(y, -x)$ in Figure 3.62. Note that $\tau_{xy} = -\tau_{y-x}$. These points are plotted, the diameter is drawn between them, and the circle is fully determined. The center is at $(3, 0)$ and the radius is

$$b = \sqrt{3^2 + 5^2} = \sqrt{34} \text{ MPa}$$

The angle 2α is

$$\tan^{-1}\frac{3}{5} = 30.96^{\circ}$$

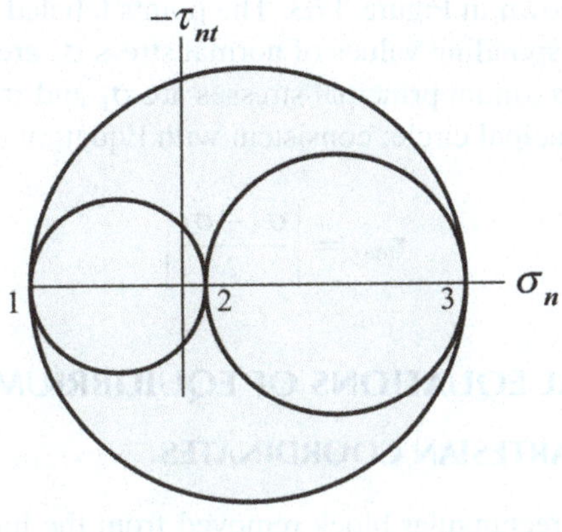

Figure 3.63

To reach the point corresponding to direction n, we must traverse the circle (counterclockwise) from the point labeled (x, y) through 60° (2 × 30°) to the point labeled (n, t). This lies near the top, where $\sigma_n = 3$ ksi and $\tau_{nt} = -5.8$ ksi. The component τ_{nz} cannot be obtained from the plot. We must revert to Equation (3.36). Suppose that the angle between x and n were some other value θ. Then the components (σ_n, τ_{nt}) lie at point P of Figure 3.62. Recall that the center lies at (σ_x + σ_y)/2; also, if the radius is denoted by b, then

$$\cos 2\alpha = \frac{\sigma_x - \sigma_y}{2b}, \qquad \sin 2\alpha = -\frac{\tau_{xy}}{b}$$

At point P,

$$\sigma_n = \frac{\sigma_x + \sigma_y}{2} + b\cos 2(\alpha + \theta) \tag{A}$$

$$-\tau_{nt} = b\sin 2(\alpha + \theta) \tag{B}$$

However,

$$\begin{aligned} \cos 2(\alpha + \theta) &= \cos 2\alpha \cos 2\theta - \sin 2\alpha \sin 2\theta \\ &= \frac{\sigma_x - \sigma_y}{2b}\cos 2\theta + \frac{\tau_{xy}}{b}\sin 2\theta \\ \sin 2(\alpha + \theta) &= \sin 2\alpha \cos 2\theta + \cos 2\alpha \sin 2\theta \\ &= \frac{\sigma_x - \sigma_y}{2b}\sin 2\theta - \frac{\tau_{xy}}{b}\cos 2\theta \end{aligned}$$

When we substitute the latter into Equations (A) and (B), we recover Equations (3.34) and (3.35). Again, Mohr's circle is a graphical representation of the equations, which can be reconstructed from the circle.

3.14.1 PRINCIPAL CIRCLES

We can plot one circle for all lines perpendicular to one principal direction, another for all lines perpendicular to the second principal direction, and another for all lines perpendicular to the third.

The three circles appear as shown in Figure 3.63. The points labeled 1, 2, and 3 correspond to the principal directions; the corresponding values of normal stress σ_n are the principal stresses σ_1, σ_2 and σ_3. The minimum and maximum principal stresses are σ_1 and σ_3. The maximum shear stress is the radius of the largest principal circle; consistent with Equation (3.28),

$$\tau_{\max} = \left| \frac{\sigma_3 - \sigma_1}{2} \right|$$

3.15 THE DIFFERENTIAL EQUATIONS OF EQUILIBRIUM

3.15.1 EQUATIONS IN CARTESIAN COORDINATES

Consider once again a small rectangular block removed from the interior of a body, as shown in Figure 3.64. We had previously neglected the small differences in the stresses on opposite faces of the block. That was justified on the basis that the block could be taken as small as we pleased and the faces, therefore, as close together as we pleased. Those small differences in the stresses on opposite faces are now our paramount concern.

Once again, primes are used for quantities on the three visible faces of the block in Figure 3.64. Quantities associated with the hidden faces are written without primes. The bars above the stresses denote average values over the appropriate faces. With this notation, we can write

$$\Delta \mathbf{F}_x' = \overline{\boldsymbol{\sigma}}_x' \Delta y \Delta z, \quad \Delta \mathbf{F}_x = \overline{\boldsymbol{\sigma}}_x \Delta y \Delta z \tag{3.44a,b}$$

Similar equations hold for the forces upon faces normal to the y and z lines.

We suppose too that the elemental block is acted upon by a body force ΔF, such as a weight or electromagnetic force. For equilibrium, the sum of the forces on all faces must vanish:

$$\Delta \mathbf{F}'_x - \Delta \mathbf{F}_x + \Delta \mathbf{F}'_y - \Delta \mathbf{F}_y + \Delta \mathbf{F}'_z - \Delta \mathbf{F}_z + \Delta \mathbf{F} = 0 \tag{3.45}$$

Expressing the forces in terms of the stresses, as in Equations (3.44a,b), and dividing by the volume ($\Delta x \Delta y \Delta z = \mathrm{dV}$) we obtain

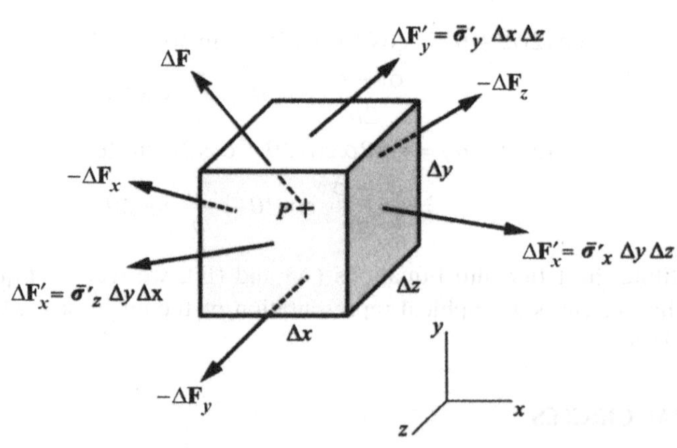

Figure 3.64

$$\frac{\overline{\boldsymbol{\sigma}}_x' - \overline{\boldsymbol{\sigma}}_x}{\Delta x} + \frac{\overline{\boldsymbol{\sigma}}_y' - \overline{\boldsymbol{\sigma}}_y}{\Delta y} + \frac{\overline{\boldsymbol{\sigma}}_z' - \overline{\boldsymbol{\sigma}}_z}{\Delta z} + \frac{\Delta \mathbf{F}}{\Delta v} = 0$$

In the limit ($\overline{\boldsymbol{\sigma}}_x' \to \overline{\boldsymbol{\sigma}}_x$, $\Delta x \to 0$, etc.) we obtain the differential equation of equilibrium:

$$\frac{\partial \boldsymbol{\sigma}_x}{\partial x} + \frac{\partial \boldsymbol{\sigma}_y}{\partial x} + \frac{\partial \boldsymbol{\sigma}_z}{\partial x} + \boldsymbol{f} = 0 \tag{3.46}$$

Here $\boldsymbol{f}$ denotes the force/unit volume. In accordance with expressions (3.14a–c) for the vectors $\boldsymbol{\sigma}_x$, $\boldsymbol{\sigma}_y$, and $\boldsymbol{\sigma}_z$, the one vectorial equation (3.46) represents three scalar equations. The three differential equations of equilibrium are then

$$\begin{aligned} \frac{\partial \sigma_x}{\partial x} + \frac{\partial \tau_{xy}}{\partial y} + \frac{\partial \tau_{xz}}{\partial z} + f_x = 0 \\ \frac{\partial \tau_{xy}}{\partial x} + \frac{\partial \sigma_y}{\partial y} + \frac{\partial \tau_{yz}}{\partial z} + f_y = 0 \\ \frac{\partial \tau_{xz}}{\partial x} + \frac{\partial \tau_{yz}}{\partial y} + \frac{\partial \sigma_z}{\partial z} + f_z = 0 \end{aligned} \tag{3.47}$$

(Note that we have written τ_{xy} for τ_{yx}, and so on.) These equations play a fundamental role in thorough studies of deformable bodies.

We must acknowledge a very important caveat: The stresses are acting upon the *deformed* surfaces in a loaded structural member. Typically, we have no a priori knowledge of the deformation. For example, we must delineate a rectangular system of lines and surfaces in a structural/mechanical component, *but* that system is deformed to different curvilinear lines and surfaces under applied loads (external forces and/or temperatures). A rigorous analysis would require that the governing differential equations be cast in the deformed curvilinear system. Fortunately, most problems entail only small strains and rotations; then we can pursue an analysis which neglects the deformation. The reader should be wary of circumstances which entail large rotations, as in flexible bodies or conditions of instability. We address these questions in Section 3.16.

3.15.2 EQUATIONS IN CYLINDRICAL COORDINATES

Figure 3.65 depicts an element delineated by the surfaces of the cylindrical coordinates (r, θ, z). The differential changes of force are shown acting upon the positive surfaces, i.e faces corresponding to the increasing values. A body force $\boldsymbol{f}$ (per unit volume) is also included.

The stresses on the respective faces (r, θ, z) follow:

$$\boldsymbol{\sigma}_r = \sigma_r \hat{\boldsymbol{i}} + \sigma_{r\theta} \hat{\boldsymbol{j}} + \sigma_{rz} \hat{\boldsymbol{k}} \tag{3.48a}$$

$$\boldsymbol{\sigma}_\theta = \sigma_{\theta r} \hat{\boldsymbol{i}} + \sigma_{\theta} \hat{\boldsymbol{j}} + \sigma_{\theta z} \hat{\boldsymbol{k}} \tag{3.48b}$$

$$\boldsymbol{\sigma}_z = \sigma_{zr} \hat{\boldsymbol{i}} + \sigma_{z\theta} \hat{\boldsymbol{j}} + \sigma_{z} \hat{\boldsymbol{k}} \tag{3.48c}$$

Equilibrium of the forces is expressed as follows:

$$\frac{\partial}{\partial r}(\boldsymbol{\sigma}_r r) + \frac{\partial}{\partial \theta}(\boldsymbol{\sigma}_\theta) + \frac{\partial}{\partial z}(\boldsymbol{\sigma}_z r) + \boldsymbol{f} = 0 \tag{3.49}$$

where we have divided by the differential $\mathrm{d}r\mathrm{d}\theta\mathrm{d}z$. When differentiating the vectors in Equations (3.48a–c), we observe that

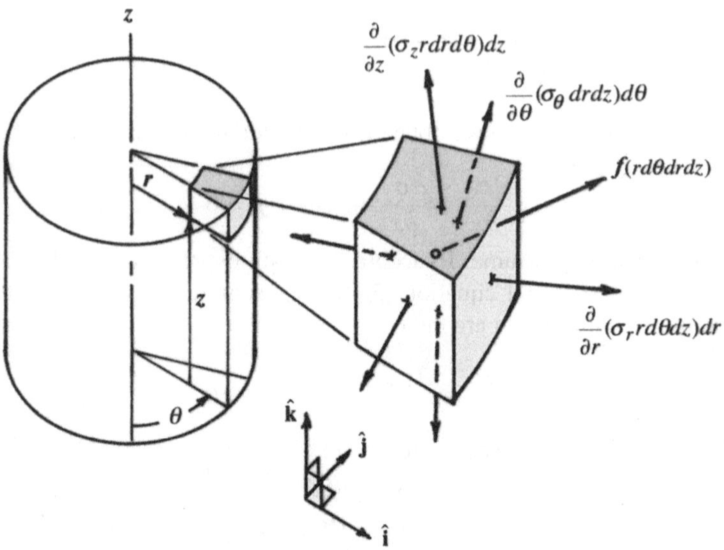

Figure 3.65

$$\frac{\partial \hat{\mathbf{i}}}{\partial \theta} = \hat{\mathbf{j}}, \qquad \frac{\partial \hat{\mathbf{j}}}{\partial \theta} = -\hat{\mathbf{i}}$$

Then Equation (3.49) results in the three components:

$$\frac{\partial \sigma_r}{\partial r} + \frac{1}{r}\frac{\partial \sigma_{\theta r}}{\partial \theta} + \frac{\partial \sigma_{zr}}{\partial z} + \frac{\sigma_r - \sigma_\theta}{r} + f_r = 0 \tag{3.50a}$$

$$\frac{\partial \sigma_{r\theta}}{\partial r} + \frac{1}{r}\frac{\partial \sigma_\theta}{\partial \theta} + \frac{\partial \sigma_{z\theta}}{\partial z} + \frac{2\sigma_{r\theta}}{r} + f_\theta = 0 \tag{3.50b}$$

$$\frac{\partial \sigma_{rz}}{\partial r} + \frac{1}{r}\frac{\partial \sigma_{\theta z}}{\partial \theta} + \frac{\partial \sigma_z}{\partial z} + \frac{\sigma_{rz}}{r} + f_z = 0 \tag{3.50c}$$

3.16 STRESSES UPON DEFORMED SURFACES

In Sections 2.13 and 2.21 we define tensorial components of strain which are applicable to large strains and rotations. We note, specifically, that orthogonal lines and surfaces in the undeformed body are, in general, not orthogonal in the deformed body. The deformation of surfaces requires that we distinguish between areas of the initial and deformed areas. Here we define stress per unit of the *initial* (deformed) area. In general, the occurrence of large strains requires precise definitions of stresses; these must embody the geometrical changes, strains and rotations which define the deformed state. Two versions are presented here; one complements the Cauchy-Green strains, the other complements the engineering strains as defined in Sections 2.13 and 2.21, respectively.

In Figure 3.66 we depict an elemental surface (x) orthogonal to an x line of the Cartesian system (x,y,z) at P. The elemental surface is transported and deformed to P^*; the positions are defined by vectors $\mathbf{r}(x,y,z)$ and $\mathbf{R}(x,y,z)$. As in Sections 2.13 and 2.21, we recognize the derivatives, vectors which are tangent to the deformed lines and adopt the definitions

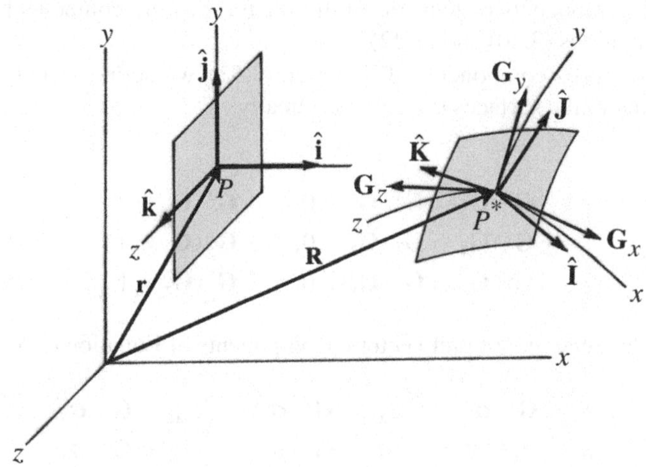

Figure 3.66

$$\mathbf{G}_x \equiv \frac{\partial \mathbf{R}}{\partial x}, \qquad \mathbf{G}_y \equiv \frac{\partial \mathbf{R}}{\partial y}, \qquad \mathbf{G}_z \equiv \frac{\partial \mathbf{R}}{\partial z} \tag{3.51a,b,c}$$

As in Section 2.21, we also introduce the *orthogonal* unit vectors $\hat{\mathbf{I}}$, $\hat{\mathbf{J}}$, $\hat{\mathbf{K}}$ which differ from the initial triad ($\hat{\mathbf{i}}$, $\hat{\mathbf{j}}$, $\hat{\mathbf{k}}$) by a rigid rotation such that

$$\hat{\mathbf{I}}\cdot\mathbf{G}_y = \hat{\mathbf{J}}\cdot\mathbf{G}_x, \qquad \hat{\mathbf{I}}\cdot\mathbf{G}_z = \hat{\mathbf{K}}\cdot\mathbf{G}_x \qquad \hat{\mathbf{J}}\cdot\mathbf{G}_z = \hat{\mathbf{K}}\cdot\mathbf{G}_y$$

The rotation is the rotation of the principal lines of strain. Observe that $\mathbf{G}_x$, $\mathbf{G}_y$ and $\mathbf{G}_z$ are orthogonal *only* when the lines (x, y, z) lie in the principal directions.

The x surface at P^* is subject to a stress $\boldsymbol{\sigma}_x$ per unit of initial area. We can represent that vector in two ways.

$$\boldsymbol{\sigma}_x \equiv t_x\hat{\mathbf{I}} + t_{xy}\hat{\mathbf{J}} + t_{xz}\hat{\mathbf{K}} \tag{3.52}$$

$$\boldsymbol{\sigma}_x \equiv s_x\mathbf{G}_x + s_{xy}\mathbf{G}_y + s_{xz}\mathbf{G}_z \tag{3.53}$$

These define two different systems of stress components. The former (3.52) was employed by Jaumann,[5] the latter (3.53) was described by Trefftz.[6] In like manner we define the components acting upon the deformed y and z surfaces; collectively, we have the nine components of each system:

$$\begin{matrix} t_x & t_{xy} & t_{xz} \\ t_{yx} & t_y & t_{yz} \\ t_{zx} & t_{zy} & t_z \end{matrix} \qquad\qquad \begin{matrix} s_x & s_{xy} & s_{xz} \\ s_{yx} & s_y & s_{yz} \\ s_{zx} & s_{zy} & s_z \end{matrix} \tag{3.54}$$

In the following section we show that these systems possess a symmetry, viz.

$$t_{xy} = t_{yx}, \qquad t_{xz} = t_{zx}, \qquad t_{yz} = t_{zy}$$

$$s_{xy} = s_{yx}, \qquad s_{xz} = s_{zx}, \qquad s_{yz} = s_{zy}$$

Moreover, both are components of tensors. Both transform as the components of Section 3.8 in accordance with Equations (3.20) and (3.22).

To determine the Trefftz components of Equation (3.53), we define a triad of vectors ($\mathbf{G}^x$, $\mathbf{G}^y$, $\mathbf{G}^z$) which are normal to the surfaces (x, y, z); specifically

$$\begin{aligned} &\mathbf{G}^x \cdot \mathbf{G}_y = \mathbf{G}^x \cdot \mathbf{G}_z = 0, \qquad \mathbf{G}^x \cdot \mathbf{G}_x = 1 \\ &\mathbf{G}^y \cdot \mathbf{G}_x = \mathbf{G}^y \cdot \mathbf{G}_z = 0, \qquad \mathbf{G}^y \cdot \mathbf{G}_y = 1 \\ &\mathbf{G}^z \cdot \mathbf{G}_x = \mathbf{G}^z \cdot \mathbf{G}_y = 0, \qquad \mathbf{G}^z \cdot \mathbf{G}_z = 1 \end{aligned}$$

Note that these are, in general, not unit vectors. Components of Equation (3.53) follow:

$$\begin{aligned} s_x &= \mathbf{G}^x \cdot \boldsymbol{\sigma}_x, & s_{xy} &= \mathbf{G}^y \cdot \boldsymbol{\sigma}_x, & s_{xz} &= \mathbf{G}^z \cdot \boldsymbol{\sigma}_x \\ s_{yx} &= \mathbf{G}^x \cdot \boldsymbol{\sigma}_y, & s_y &= \mathbf{G}^y \cdot \boldsymbol{\sigma}_y, & s_{yz} &= \mathbf{G}^z \cdot \boldsymbol{\sigma}_y \\ s_{zx} &= \mathbf{G}^x \cdot \boldsymbol{\sigma}_z, & s_{zy} &= \mathbf{G}^y \cdot \boldsymbol{\sigma}_z, & s_z &= \mathbf{G}^z \cdot \boldsymbol{\sigma}_z \end{aligned} \tag{3.55}$$

To determine the Jaumann components of Equation (3.52), we recall the orthogonal (rotated) triad ($\hat{\mathbf{I}}$, $\hat{\mathbf{J}}$, $\hat{\mathbf{K}}$) of Section 2.21: The components follow from Equation (3.52):

$$\begin{aligned} t_x &= \hat{\mathbf{I}} \cdot \boldsymbol{\sigma}_x, & t_{xy} &= \hat{\mathbf{J}} \cdot \boldsymbol{\sigma}_x, & t_{xz} &= \hat{\mathbf{K}} \cdot \boldsymbol{\sigma}_x \\ t_{yx} &= \hat{\mathbf{I}} \cdot \boldsymbol{\sigma}_y, & t_y &= \hat{\mathbf{J}} \cdot \boldsymbol{\sigma}_y, & t_{yz} &= \hat{\mathbf{K}} \cdot \boldsymbol{\sigma}_y \\ t_{zx} &= \hat{\mathbf{I}} \cdot \boldsymbol{\sigma}_z, & t_{zy} &= \hat{\mathbf{J}} \cdot \boldsymbol{\sigma}_z, & t_z &= \hat{\mathbf{K}} \cdot \boldsymbol{\sigma}_z \end{aligned} \tag{3.56}$$

3.17 VIRTUAL WORK UPON THE DEFORMED ELEMENT

Our previous views of stress were founded on the Newtonian perspective, viz. the balance of actions upon a free body. We now adopt the Lagrangean viewpoint wherein we consider the work expended in a virtual displacement of the element depicted in Figure 3.67.

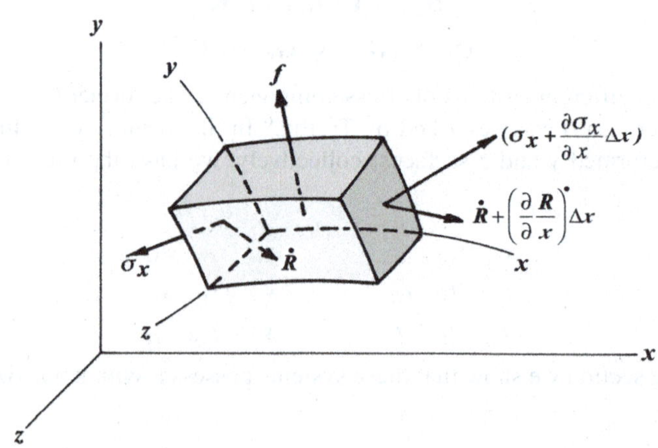

Figure 3.67

We denote the virtual displacement by $\dot{\mathbf{R}}$; We presume continuity of the derivatives such that

$$\frac{\partial \dot{\mathbf{R}}}{\partial x} = \left(\frac{\partial \mathbf{R}}{\partial x}\right)^{\bullet} \equiv \dot{\mathbf{G}}_x$$

In words, we can treat the derivative of the increment as the increment of the derivative. This is done is Figure (3.67).

The virtual work expended by the stress $\boldsymbol{\sigma}_x$ (per unit initial volume) follows:

$$\dot{w}_x = \frac{\partial \boldsymbol{\sigma}_x}{\partial x} \cdot \dot{\mathbf{R}} + \boldsymbol{\sigma}_x \cdot \dot{\mathbf{G}}_x$$

We obtain analogous expressions for the virtual work of the stresses ($\boldsymbol{\sigma}_y$, $\boldsymbol{\sigma}_z$) upon the y and z surfaces. Also, a body force expends work $\boldsymbol{f} \cdot \dot{\mathbf{R}}$ (per unit initial volume). Collectively the total virtual work follows:

$$\begin{aligned} \dot{w} = &\left(\frac{\partial \boldsymbol{\sigma}_x}{\partial x} + \frac{\partial \boldsymbol{\sigma}_y}{\partial y} + \frac{\partial \boldsymbol{\sigma}_z}{\partial z} + \boldsymbol{f}\right) \cdot \dot{\mathbf{R}} \\ &+ \boldsymbol{\sigma}_x \cdot \dot{\mathbf{G}}_x + \boldsymbol{\sigma}_y \cdot \dot{\mathbf{G}}_y + \boldsymbol{\sigma}_z \cdot \dot{\mathbf{G}}_z. \end{aligned} \tag{3.57}$$

The principle of virtual work asserts that the virtual work must vanish in a state of equilibrium for an *arbitrary* admissible displacement. An admissible displacement must satisfy the requisite continuity of the displacement and derivatives. The displacement $\dot{\mathbf{R}}$ represents a translation; it is independent of the gradients, $\dot{\mathbf{G}}_x$, $\dot{\mathbf{G}}_y$ and $\dot{\mathbf{G}}_z$. Consequently, the parenthetical expression must vanish; this requires the equilibrium condition:

$$\frac{\partial \boldsymbol{\sigma}_x}{\partial x} + \frac{\partial \boldsymbol{\sigma}_y}{\partial y} + \frac{\partial \boldsymbol{\sigma}_z}{\partial x} + \boldsymbol{f} = 0 \tag{3.58}$$

To examine the remaining terms of (3.57) we form the indicated products in accordance with Equation (3.52)

$$\begin{aligned} \dot{\Phi} \equiv\ & t_x \hat{\mathbf{I}} \cdot \dot{\mathbf{G}}_x + t_{xy} \hat{\mathbf{J}} \cdot \dot{\mathbf{G}}_x + t_{xz} \hat{\mathbf{K}} \cdot \dot{\mathbf{G}}_x \\ &+ t_{yx} \hat{\mathbf{I}} \cdot \dot{\mathbf{G}}_y + t_y \hat{\mathbf{J}} \cdot \dot{\mathbf{G}}_y + t_{yz} \hat{\mathbf{K}} \cdot \dot{\mathbf{G}}_y \\ &+ t_{zx} \hat{\mathbf{I}} \cdot \dot{\mathbf{G}}_z + t_{zy} \hat{\mathbf{J}} \cdot \dot{\mathbf{G}}_z + t_z \hat{\mathbf{K}} \cdot \dot{\mathbf{G}}_z \end{aligned} \tag{3.59}$$

To examine the quantity $\dot{\Phi}$ as defined by Equation (3.59), we recall the engineering strain component of Section 2.21. From the Equations (2.65a–f), we obtain

$$\dot{\bar{\varepsilon}}_x = \hat{\mathbf{I}} \cdot \dot{\mathbf{G}}_x, \qquad \dot{\bar{\gamma}}_{xy} = \frac{1}{2}(\hat{\mathbf{I}} \cdot \dot{\mathbf{G}}_y + \hat{\mathbf{J}} \cdot \dot{\mathbf{G}}_x)$$

Analogous results follow for the other components. We now define the following quantities:

$$\begin{aligned} \dot{\bar{\omega}}_x &\equiv \frac{1}{2}(\hat{\mathbf{K}} \cdot \dot{\mathbf{G}}_y - \hat{\mathbf{J}} \cdot \dot{\mathbf{G}}_z) \\ \dot{\bar{\omega}}_y &\equiv \frac{1}{2}(\hat{\mathbf{I}} \cdot \dot{\mathbf{G}}_z - \hat{\mathbf{K}} \cdot \dot{\mathbf{G}}_x) \\ \dot{\bar{\omega}}_z &\equiv \frac{1}{2}(\hat{\mathbf{J}} \cdot \dot{\mathbf{G}}_x - \hat{\mathbf{I}} \cdot \dot{\mathbf{G}}_y) \end{aligned} \tag{3.60}$$

With the foregoing definitions we obtain:

$$\hat{\mathbf{J}}\cdot\dot{\mathbf{G}}_x = \dot{\bar{\gamma}}_{xy} + \dot{\bar{\omega}}_z, \qquad \hat{\mathbf{I}}\cdot\dot{\mathbf{G}}_z = \dot{\bar{\gamma}}_{xz} + \dot{\bar{\omega}}_y$$
$$\hat{\mathbf{K}}\cdot\dot{\mathbf{G}}_x = \dot{\bar{\gamma}}_{xz} - \dot{\bar{\omega}}_y \qquad \hat{\mathbf{K}}\cdot\dot{\mathbf{G}}_y = \dot{\bar{\gamma}}_{xy} + \dot{\bar{\omega}}_x$$
$$\hat{\mathbf{I}}\cdot\dot{\mathbf{G}}_y = \dot{\bar{\gamma}}_{xy} - \dot{\bar{\omega}}_z, \qquad \hat{\mathbf{J}}\cdot\dot{\mathbf{G}}_z = \dot{\bar{\gamma}}_{xz} - \dot{\bar{\omega}}_x$$

In these notations Equation (3.59) assumes the form

$$\begin{aligned}\dot{\Phi} \equiv & t_x\dot{\bar{\varepsilon}}_x + t_{xy}\dot{\bar{\gamma}}_{xy} + t_{xz}\dot{\bar{\gamma}}_{xz} \\ & + t_{yx}\dot{\bar{\gamma}}_{yx} + t_y\dot{\bar{\varepsilon}}_y + t_{yz}\dot{\bar{\gamma}}_{yz} \\ & + t_{zx}\dot{\bar{\gamma}}_{zx} + t_{zy}\dot{\bar{\gamma}}_{zy} + t_z\dot{\bar{\varepsilon}}_z \\ & + (t_{xy} - t_{yx})\dot{\bar{\omega}}_z + (t_{zx} - t_{xz})\dot{\bar{\omega}}_y \\ & + (t_{yz} - t_{zy})\dot{\bar{\omega}}_x\end{aligned} \tag{3.61}$$

The work *expended in deformation* is represented by the products of the stress and strain components. That work is either stored in elastic deformations or dissipated (e.g. as heat) in inelastic deformations. The quantities $\dot{\bar{\omega}}_x$, $\dot{\bar{\omega}}_y$ and $\dot{\bar{\omega}}_z$ are incremental rigid rotations; for equilibrium (of moments) the parenthetical terms must vanish:

$$t_{yx} = t_{xy}, \qquad t_{xz} = t_{yz}, \qquad t_{xz} = t_{zx} \tag{3.62}$$

The foregoing analysis can be replicated when we employ the vector $\boldsymbol{\sigma}_x$ in the form (3.53). Instead of Equation (3.61) we obtain

$$\begin{aligned}\dot{\Phi} = & s_x\dot{\tilde{\varepsilon}}_x + s_{xy}\dot{\tilde{\gamma}}_{xy} + s_{xz}\dot{\tilde{\gamma}}_{xz} \\ & + s_{yx}\dot{\tilde{\gamma}}_{yx} + s_y\dot{\tilde{\varepsilon}}_y + s_{yz}\dot{\tilde{\gamma}}_{yz} \\ & + s_{zx}\dot{\tilde{\gamma}}_{zx} + s_{zy}\dot{\tilde{\gamma}}_{zy} + s_z\dot{\tilde{\gamma}}_z \\ & + (s_{xy} - s_{yx})\dot{\tilde{\omega}}_z + (s_{zx} - S_{xz})\dot{\tilde{\omega}}_y \\ & + (s_{yz} - s_{zy})\dot{\tilde{\omega}}_x\end{aligned} \tag{3.63}$$

Here as in (3.61), the incremental rotations are expressed as follows:

$$\begin{aligned}\dot{\tilde{\omega}}_x &= \frac{1}{2}(\mathbf{G}_z\cdot\dot{\mathbf{G}}_y - \mathbf{G}_y\cdot\dot{\mathbf{G}}_z) \\ \dot{\tilde{\omega}}_y &= \frac{1}{2}(\mathbf{G}_x\cdot\dot{\mathbf{G}}_z - \mathbf{G}_z\cdot\dot{\mathbf{G}}_x) \\ \dot{\tilde{\omega}}_z &= \frac{1}{2}(\mathbf{G}_y\cdot\dot{\mathbf{G}}_x - \mathbf{G}_x\cdot\dot{\mathbf{G}}_y)\end{aligned} \tag{3.64}$$

Since the virtual work vanishes under arbitrary rigid rotation, the parenthetical quantities of (3.63) must vanish, i.e.

$$s_{xy} = s_{yx}, \qquad s_{xz} = s_{yz}, \qquad s_{yz} = s_{zy} \tag{3.65}$$

The components of the Trefftz and Jaumann stresses constitute symmetric tensors.

The foregoing derivations establish another important feature of the alternative forms (3.52) and (3.53): In a precise treatment of large strains, the Jaumann stress of (3.52) requires the engineering strain components of (2.65); the Trefftz stress of (3.53) requires the Cauchy-Green strain components of Equations (2.34) and (2.35). The increments of work expenditure are the non-vanishing terms of (3.61) and (3.63), respectively.

NOTES

1. If necessary, the restriction can be removed at any point of the analysis by making use of D'Alembert's principle. For instance, if a particle is accelerating, we can define the so-called inertia force:

$$\mathbf{F}' = -m\mathbf{a}$$

Equation (3.1) can then be written

$$\sum \mathbf{F} + \mathbf{F}' = \mathbf{0}$$

which is now in the same form as Equation (3.2), the equation that defines equilibrium.
2. The reader will recognize these integrals as the so-called moments of the area of S with respect to the y and z axes in Figure 3.36.
3. This equation gives the distribution of normal stress upon an arbitrary cross-section of a Hookean beam in accordance with the Bernoulli hypothesis (See Chapter 6).
4. In Section 3.15 we acknowledge the differences, e.g. $\sigma_x' = \sigma_x + \Delta\sigma_x$, to obtain the differential equations of the equilibrium.
5. G. Jaumann, Physik der kontinuierlichen Medien. *Deutsche. Akad. Wiss.*, 95:41–562, 1918.
6. E. Trefftz, Zur Theorie der Stabilit́at des elastichen Gleighgewichts. *Z. Angew. Math. Mech.*, 13:160, 1933.

4 Thermodynamical Behavior of Materials

The response of any structure or machine to external loadings is related to the physical attributes of the materials that constitute the member. To appreciate the role of the material, the reader only needs to imagine applying an axial force upon a steel rod and then applying the same force upon a geometrically similar rod of rubber, and observing the consequent elongations of each. In this chapter, we explore the behavior of materials, some practical theories, and useful mathematical models to describe them. These formulations utilize the variables of strain and stress, as defined in Chapters 2 and 3. The former completely describes the nature and severity of the deformations at a *point* of a *continuous* medium. The latter describes the nature and the intensity of the forces at a *point* of a *continuous* medium.

Hence, any formulation in terms of these variables implies a *continuity* of the material and its properties. Of course, no real material is continuous on a microscopic scale; all have a micro (or macro)-structure, discrete irregularities, and imperfections. In short, none are strictly continuous. It follows that our mathematical models have limitations. Certain phenomena have their origins in the micro/macro structures of the material; then the concept of the continua and certain mathematical variables (e.g., strain and stress) may be inadequate. In the following, we must note such limitations. The mechanical behavior of most materials can be described in terms of strain and stress, but other variables, notably temperature and time, also play an important role. Here, we confine our attention to the thermodynamical models, which are expressed in terms of *strain, stress, temperature, and time*.

4.1 CAUSES AND EFFECTS

4.1.1 COMMON ATTRIBUTES

All solid bodies are affected in some way by their environment, by changes of environment, and by the application of external forces. The effects depend upon the size and shape of the body but also on the material that constitutes the body. Here, we want to characterize the attributes of the material; some of these are identified in the following.

A rubber band will stretch to several times its original length but recover when the forces are removed. It sustains no permanent deformation. It exhibits *elastic* behavior. A paper clip may be severely bent without breaking but does not recover its original shape. The clip is composed of a *ductile* material. In contrast, a piece of chalk bends only slightly, and elastically, before it fractures. The chalk exhibits a *brittle* behavior.

4.1.2 EFFECTS OF TEMPERATURE

Materials are also affected by temperature. One common effect is a change of volume, a *thermal expansion*. The response of a body to changes of temperature may also depend on the manner in which that change occurs: For example, if hot water is suddenly poured into a glass jar, the jar may fracture. An analysis would show that the fracture is a consequence of the nonuniform thermal expansion.

 DOI: 10.1201/9781003665861-4

Materials may exhibit quite different behaviors when subjected to changes in temperature: In particular, heating tends to soften many materials; their strength and resistance to deformation are reduced. Boiler tubes, which operate at high temperatures, may deform gradually over an extended period (*creep*) and eventually fail. Very low temperatures can have an entirely different effect: Many materials become brittle at low temperatures; their resistance to a suddenly applied load (*impact*) is greatly reduced.

4.1.3 EFFECTS OF TIME

The rate at which loads are applied and the duration of loading can play a major role in the response of certain materials: As previously noted, the sudden application of a load (*impact*) can cause fracture, when the slowly applied load does not.

Certain materials exhibit only elastic deformation upon initial loading, but continue to deform over an extended period of time, as a piece of taffy-candy deforms steadily under a slight, but constant force. Such gradual and continuous deformation is termed *creep*, and typically results in fracture.

Many repetitions of a load may cause fracture, though the same load is too slight to cause any damage when applied only a few times. The connecting rod in a reciprocating engine can fail in this manner after many hours of use. Such fractures are identified as *fatigue* failures.

4.1.4 CAUSE-EFFECT RELATIONSHIPS

Our immediate goal is the description of material via mathematical models, which reflect the *properties* of the material. The requisite relationships are expressed in terms of strain, stress, temperature, time, and the properties of the material. The specific properties must be incorporated in the *formulations* and relevant *parameters*, which are drawn from experimental observations.

The reader must be aware that other factors, such as radiation, corrosion, and moisture, can influence the behavior of a material. Here, the cause-effect relationships are expressed in terms of so-called *constitutive* equations, which reflect only the effects of the variables cited: *strain, stress, temperature* and *time*, and *parameters* which embody the properties of the specific material.

4.2 STRUCTURE OF MATERIALS

The material in any small portion of a body exhibits a structure composed of elementary constituents, which are arranged with varying degrees of regularity. The photographs in Figure 4.1 illustrate the elementary structures of four common materials: Figure 4.la is a cross-section of a typical concrete, which displays a random distribution of various aggregates (types of stone, rock, and sand) that are bonded by cement. Figure 4.ld is a photomicrograph of a filamentary composite in which the glass filaments are also randomly dispersed, but in a common direction; these are bonded in a matrix of epoxy. Figure 4.lb is a photomicrograph of a steel, which exhibits a crystalline structure. Individual crystals are oriented differently; they are often imperfect and interspersed with interstitial atoms, e.g., carbon and alloys. Finally, we note the random distribution of the molecular structure of the nylon in Figure 4.lf. In any of these materials, one can always anticipate voids and impurities. Of course, all properties of these materials are manifestations of their micro/macro structures.

The behavior and the properties of all materials are dependent upon the attributes of their elementary constituents and the interactions between them. For example, the behavior of a metal depends upon the arrangement of atoms within crystals, the bonds between those atoms, the size,

Figure 4.1 Micrographs-Courtesy of Russell Crutcher, Microlab Northwest

shape, distribution, and orientation of the various crystals, and the ever-present impurities. The behaviors of nonmetallic materials are likewise determined by their elementary structures. A complete understanding of the behavior of any material would require a study of the smallest structural components and the interactions between them. Ideally, such studies would lead to mathematical descriptions of their behavior and quantitative knowledge of the essential parameters. Because of the extreme complexities of their micro/macro-structures, that goal has been attained only in a few special cases (e.g., studies of single crystals). Such studies are still important, as they provide valuable insights into various phenomena.

Fortunately, one can ascertain certain properties from observations of tests upon specimens of a sufficient size to account for the random character of the distributions and orientations. Clearly, the size of the specimen depends on the material: The elemental structure of concrete has much greater scale than that of the steel; a specimen of concrete must be much larger than one of steel. In most cases, the testing of such specimens provides the essential information about elastic responses. Unfortunately, *inelastic* behaviors and, especially, the *fracture* of metals, are intimately related to the microstructure and to the *imperfections* within those structures. In any case, the identification of minute imperfections in every machine part is not a possibility with current technology. Consequently, the engineer must rely upon criteria which employ the variables of the continuum theories (e.g., strain and stress) and the information which has been gleaned from the observations of the full-scale testing.

4.3 RELEVANT TERMINOLOGY

4.3.1 HOMOGENEITY/HETEROGENEITY

The term homogeneous[1] signifies something that possesses uniformity. *A body is homogeneous if every similar portion possesses the same physical attributes.* In view of previous statements, it is unlikely that any materials of engineering are truly homogeneous. However, if we examine large enough portions of many materials, small irregularities and differences between the various portions become unimportant because of the random distributions and orientations of the small constituents. The lack of homogeneity is significant only on a very small scale. Where this is true, it is permissible to regard the material as homogeneous, meaning that any portion (of a certain minimum size) has properties that are essentially the same as every similar portion.

The extent to which a material may be regarded as homogeneous depends upon the size of the body and the effects being studied. A very small filament consisting of a few metal crystals cannot be treated as a homogeneous body; the same metal in a large structural component could be treated as homogeneous material. Also, certain phenomena have their origins in the small irregularities of the microstructure; that is particularly true of certain types of fracture such as the fatigue failures, which result from repeated loadings. Studies of such effects must take into consideration the actual *heterogeneous* structure of the material.

4.3.2 ISOTROPY/ANISOTROPY

All materials exhibit certain structures as evident in Figure 4.1. That structure may possess specific directional properties, as the composite of Figure 4.1c. A common example is wood, which is composed of parallel layers that form with each year of growth; appearance, properties, and behavior are dependent on the direction of the "grain"; it is much more difficult to cut (chop) across these layers than to split along the layers. Such preferential orientation of structure and properties signifies a form of *anisotropy*. A material, which exhibits no directional properties, is said to be *isotropic*.[2]

On some scale, nearly all materials are anisotropic: For example, the crystals within a metal possess certain directional properties. A single crystal usually has one or more planes of symmetry; its appearance and properties depend upon one's point of view. Noncrystalline materials also display anisotropic characteristics when examined on a small enough scale: In the process of formation, the molecules in rubber-like materials and plastics are often elongated in one or more directions. Though evidently anisotropic on the small scale, larger portions of these materials often appear and behave as isotropic; more importantly, they can be analyzed as isotropic materials. The isotropic behavior (on the larger scale) can be attributed to the random distribution and orientation of the various constituents.

The engineer often uses particular materials because of their unique anisotropic properties: Components of an aircraft might be fabricated of composites, which are composed of strong fibers of glass or carbon embedded in a matrix of epoxy or other plastic. Such composites are often lightweight, but possess great tensile strength in the direction of the fibers.

The reader should not confuse *homogeneity* and *isotropy*: A material may be homogeneous *and* anisotropic, viz., each similar portion has the same properties, but those properties exhibit directional attributes.

4.3.3 CONTINUUM = CONTINUOUS MEDIUM

In our analysis of deformable bodies, it is usually impossible to take full account of the complex structure of materials. We must, therefore, adopt a simple conceptual model of the medium, one which is amenable to mathematical analyses. That model is the *continuous medium*: The continuous medium has no voids, and all properties are continuously variable. The properties may vary from point to point, but gradually, so that they can be described by continuous mathematical functions. Of course, discrete discontinuities can be accommodated, but only by the implementation of the requisite mathematical tools. The concept of the continuum has been mentioned previously. Indeed, it is essential to our definitions of strain and stress; both incorporate the notion of continuity and the limit. They are valid descriptions of deformations and internal forces *only* in continuous media.

4.3.4 COHESION

The present study is primarily concerned with solid materials, so that adjacent portions of material adhere to one another except when fracture occurs. It is not possible for adjacent portions of the material to be separated or to pass through one another. Furthermore, if adjoining portions must adhere, lines and surfaces embedded within the body can never be destroyed, except when fracture occurs. A material of this kind is termed *cohesive*. Throughout this study of deformable bodies, we shall regard materials as *cohesive continuous media*.

4.3.5 RUPTURE/FRACTURE

The terms rupture and fracture have been used, almost interchangeably, to describe breakage. They derive from the Latin, *ruptura* and *fractura, break or to break* (as nouns and as verbs). The choice of terminology has been largely a matter of context. The term "rupture" was more prevalent in the past; today it is often employed in a broader context to indicate breakage, which may, or may not, be accompanied by some inelastic deformations. The term "fracture" is more often associated with breakage, which is accompanied by little inelastic deformation. It is interesting to note that the medical profession employs both terms; the former frequently signifies a failure of soft tissues, whereas the latter applies to the breakage of bones.

4.4 DETERMINATION OF MECHANICAL BEHAVIOR AND PROPERTIES

The engineer must acknowledge that the behavior and properties of all materials originate with the interactions of their elementary constituents, as described in Section 4.2. The complexities of those interactions compel us to adopt a phenomenological approach: The essential features of the observed behaviors are incorporated in formulas and criteria, which are amenable to the analyses of structures and machines. These must be drawn from experiments upon specimens of simple form, which admit to the requisite measurements under prescribed conditions and loads. The observations and measurements suggest the relationships between the variables (strain, stress, temperature, and time) and provide the essential parameters. Because experimentation is always limited, it is necessary to infer rather general relations (theories) from the observations and the measurements under the specific conditions of the tests. Consequently, the engineer must always be wary when a theory is applied in unusual circumstances. Here, we describe the most common tests, results, and properties. Subsequently, these are incorporated into theories, which have proven useful in many practical situations.

4.4.1 SIMPLE TENSION TEST

The simple tension test consists of the gradual application of an axial force to extend a prismatic bar. The cross section of the test specimen is uniform and usually of circular or rectangular shape. The important data to be recorded are the measurements of axial force and the corresponding length and width of the bar. The test is performed by gripping the specimen at both ends in the two heads of a testing device, as shown in Figure 4.2.

Two mechanisms can be employed to apply an axial force: Early machines extended the specimen via screw mechanisms; then, the applied force is measured via a balancing system, much as a conventional scale measures weights. Most contemporary machines apply the axial force via hydraulic systems, which measure the force via a pressure gauge. In either case, the extension of a prescribed *gauge length* is measured by an extensometer as shown in Figure 4.2. The preselected length is marked by points Q and R; that length is denoted here by L. Points Q and R are located far enough from the clamped ends that the portion QR is unaffected by local disturbances at the grips. Care must be taken to align the grips so that the bar experiences axial extension without bending.

As an illustration, consider the test of a typical structural metal (an aluminum or steel alloy). For simplicity, suppose that the specimen has a uniform circular cross section of diameter d. At regular intervals, during continuous stretching of the bar, recordings are made of the length of portion QR of the bar, the diameter of the bar, and the axial force. As the bar is stretched, the distance between Q and R increases. Let us denote the new length by L^* and the axial force at the

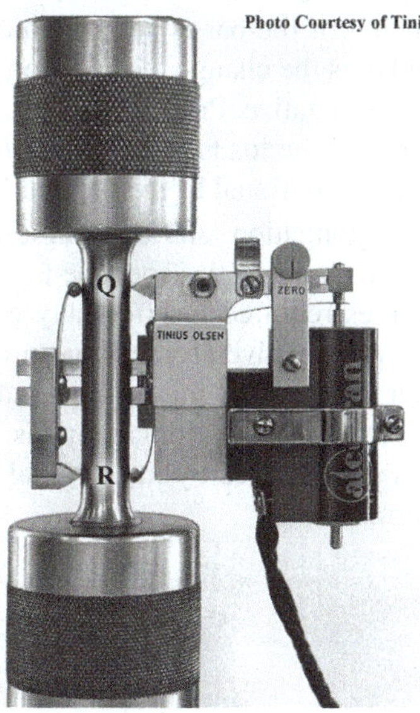

Figure 4.2 Tensile Specimen and Extensometer

same time by P. The increase in the gage length is $e = L^* - L$. From the data taken during the test, a plot of load P versus elongation e may be made. From sufficient data the continuous curve $OABCD$ of Figure 4.3 may be drawn.

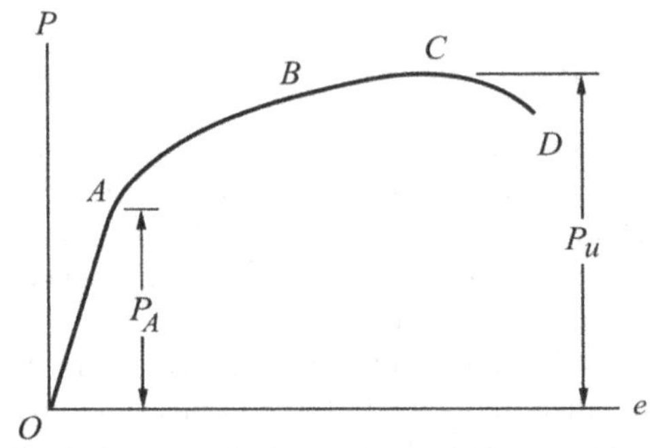

Figure 4.3 Load versus Elongation

Several important features of this diagram should be noted. The initial portion OA is approximately a straight line; that is, for loads less than P_A, the elongation e is proportional to the load P. Additional extension of the bar requires a greater load until a maximum force P_u (at point C) is attained. This maximum force P_u is termed the ultimate load; it is the greatest force that the specimen can resist. Subsequently, the bar elongates under the action of less and less force until it fractures (point D). The decrease in load prior to fracture is attributed to a phenomenon known as *necking*. Necking is a localized deformation that reduces the cross-sectional area near the location of the fracture, as shown in Figure 4.4. Necking begins to manifest itself when the load approaches the ultimate load P_u. At the time of fracture, there is an obvious and very localized contraction in diameter. Both necking and the resulting drop in load are pronounced only in ductile materials.

However, even a slight extension of the bar is accompanied by a reduction in diameter. If d^* denotes the reduced diameter and e' is the change of diameter, then $e' = d^* - d$. Note that because d^* is smaller than d, the change e' is negative. Prior to necking, the contraction e' is nearly constant along the gage length. Furthermore, if the loads are less than P_A, the contraction e' is proportional to the extension e, and hence also proportional to the applied load P.

The relationships among load, elongation, and transverse contraction depend not only on the material, but also on the size and shape of the specimen. The greater the cross-section of the specimen, the greater are the forces required to effect any changes. Also, a longer gage length will result in a greater elongation for any given force. Our aim is to describe the material behavior without reference to a particular body. With this in mind, the axial force P is divided by the original cross-sectional area, a constant A; the elongation e is divided by the original gage length, a constant L; and the contraction e' is divided by the original diameter, a constant d. This results in new variables: the average normal stress

$$\overline{\sigma} = \frac{P}{A} \tag{4.1}$$

average axial strain,

$$\overline{\varepsilon} = \frac{e}{L} = \frac{L^*}{L} - 1 \tag{4.2}$$

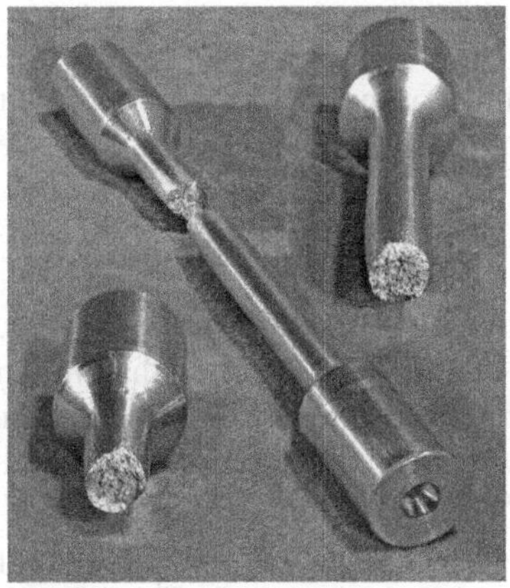

Figure 4.4 Ruptured Aluminum Specimen

the average transverse strain,

$$\bar{\varepsilon}' = \frac{e'}{d} = \frac{d^*}{d} - 1 \tag{4.3}$$

A plot of $\bar{\sigma}$ versus $\bar{\varepsilon}$ yields the curve *OABCD* of Figure 4.5. This curve is similar in every way to the curve *P-e* of Figure 4.3 because it is obtained from the *P-e* diagram by dividing the ordinate by the constant *A* and the corresponding abscissa by the constant *L*. Indeed, the only difference between the $\bar{\sigma} - \bar{\varepsilon}$ and *P-e* diagrams is a change of scale. Hence, all the previous remarks about the *P-e* diagram apply also to the $\bar{\sigma} - \bar{\varepsilon}$ diagram. However, the relationship of average stress to average strain is virtually independent of the size of the specimen. Only in the event of severe deformation (necking) is the $\bar{\sigma} - \bar{\varepsilon}$ relation greatly influenced by the dimensions of the specimen. For brevity in the following discussion, the quantities $\bar{\sigma}$, $\bar{\varepsilon}$ and $\bar{\varepsilon}'$ are referred to as the stress, strain, and transverse strain, respectively; however, these quantities are certain average values as defined by Equations (4.1)–(4.3).

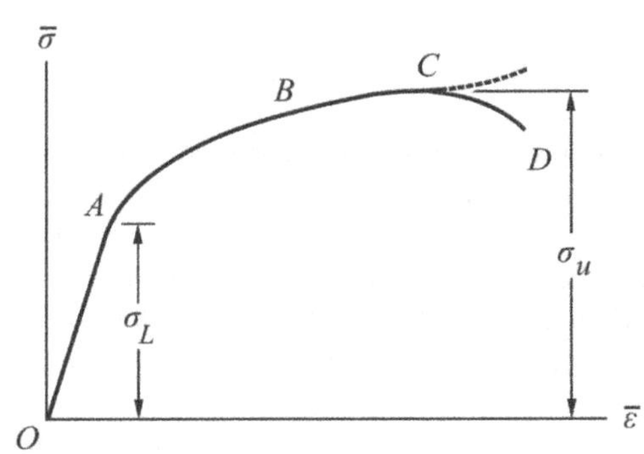

Figure 4.5 Stress versus Strain

4.4.2 MODULUS OF ELASTICITY AND POISSON'S RATIO

Let us examine the essential features of the $\overline{\sigma} - \overline{\varepsilon}$ diagram of Figure 4.5. The initial portion *OA* is approximately a straight line. For this initial portion, the $\overline{\sigma} - \overline{\varepsilon}$ relation is expressed by the linear equation

$$\overline{\sigma} = E\,\overline{\varepsilon} \tag{4.4}$$

where E is a constant of proportionality, the slope of the line *OA*. The constant E is a property of the material, called the *modulus of elasticity*. Note too that this stress is based on the initial areas. The "necking" which occurs at the onset of inelastic yielding causes a significant reduction in the area. Using the reduced area to compute the "true stress" provides a very different plot, as depicted by the dotted curve of Figure 4.5.

For the initial portion *OA*, the transverse contraction e' is proportional to the elongation e but has the opposite sign. Therefore, the transverse strain $\overline{\varepsilon}'$ is related to the axial strain $\overline{\varepsilon}$ by the linear equation

$$\overline{\varepsilon}' = -\nu\overline{\varepsilon} \tag{4.5}$$

where ν is a positive constant of proportionality, called *Poisson's ratio*. It follows from Equation (4.4) that $\overline{\varepsilon}'$ is also linearly related to the normal stress:

$$\overline{\varepsilon}' = -\frac{\nu}{E}\overline{\sigma} \tag{4.6}$$

Equations (4.4)–(4.6) apply only to the initial linear portion of the $\overline{\sigma} - \overline{\varepsilon}$ curve.

4.4.3 PROPORTIONAL LIMIT

When the stress $\overline{\sigma}$ attains a certain value, denoted by σ_L, the $\overline{\sigma} - \overline{\varepsilon}$ curve departs noticeably from the straight line *OA*. This deviation from the linear relation of Equation (4.4) is marked by the point *A* of Figure 4.5. The stress σ_L is termed the *proportional limit stress* (often simply the *proportional limit*).

4.4.4 ELASTIC LIMIT

So long as the applied force is sufficiently small, the specimen recovers its original size and shape upon removal of the force. This recovery of size and shape characterizes elastic behavior. The deformation, which disappears when the force is removed, is termed an *elastic deformation*. However, if the stress exceeds a certain value, σ_E (not shown), the specimen does not completely recover its size and shape when the force is removed. A permanent deformation is sustained. The stress σ_E is termed the *elastic limit*. The action, which results in a permanent deformation, is termed *yielding*.

The values σ_L and σ_E are not easily established. It is extremely difficult to determine accurately the stress at which the curve deviates from a straight line, and it is equally difficult to detect the earliest occurrence of permanent deformation. The precision with which these quantities are obtained is limited by the accuracy of the measuring devices. There is, in fact, good reason to doubt that any material exhibits a truly linear $\overline{\sigma} - \overline{\varepsilon}$ relation. Also, careful studies of deformation indicate that some permanent changes of structure accompany even the least application of force.

4.4.5 YIELD STRESS

For most metals, the proportional limit and elastic limit appear to be so close that, for practical purposes, a single value σ_0 ($\doteq \sigma_L \doteq \sigma_E$) characterizes the transition from linear as well as elastic behavior. Henceforth, we will refer to this stress σ_0 as the *yield stress*. The term *yielding* signifies the occurrence of permanent deformation. If the stress is less than the yield stress, then, for practical purposes, the observed behavior is elastic, and the stress-strain relation is linear in accordance with Equation (4.4). We say that this is *linearly elastic behavior*.

The linearly elastic behavior is also termed Hookean after Robert Hooke who first enunciated this proportionality in his anagram, viz *ceiiinosssttuv, id est, Ut Tensio sic vis.*[3]

4.4.6 ELASTIC AND PLASTIC BEHAVIOR

Stress exceeding the yield stress σ_0 will cause permanent deformation. Suppose, for example, that the specimen is continuously loaded to the point B of Figure 4.3 or 4.5 and then continuously unloaded. The $\overline{\sigma} - \overline{\varepsilon}$ diagram has been redrawn in Figures 4.6 and 4.7, respectively. In either figure, the unloading curve BO_p is practically a straight line, parallel to OA. (The deviation from linearity is exaggerated in the diagrams.) After the load is entirely removed, a permanent deformation remains; this is termed a plastic deformation. However, not all of the deformation caused by the loading is permanent. A part of the deformation is elastic (recoverable). In Figure 4.6, the plastic (permanent) strain is represented by OO_P, and the elastic strain, which is recovered during unloading, is represented by O_PO_B.

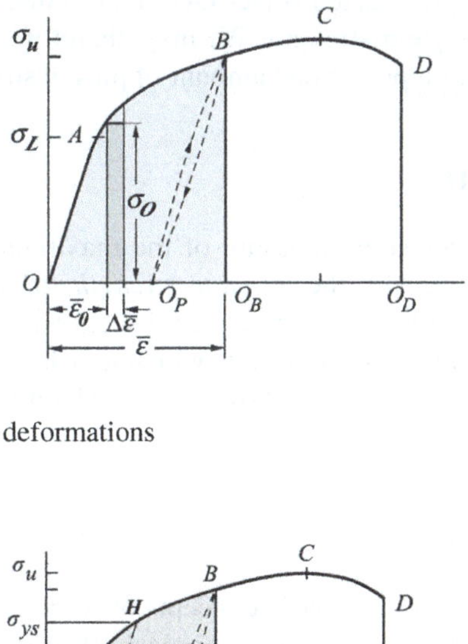

Figure 4.6 Elastic and Inelastic deformations

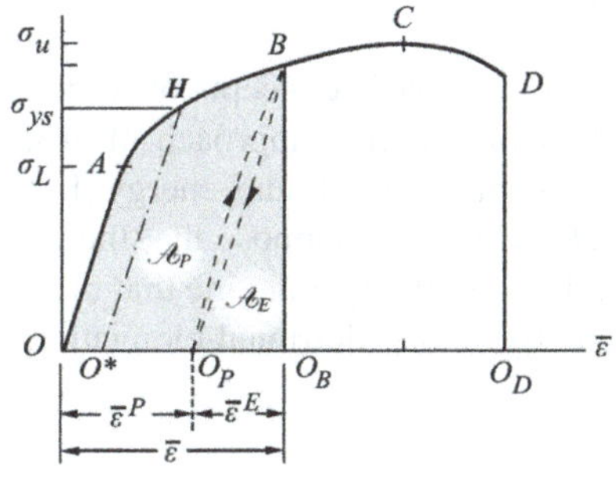

Figure 4.7 Dissipated (P) and stored Energy (E)

4.4.7 STRAIN HARDENING

Suppose that the specimen has been loaded and unloaded according to the path OBO_P. When the specimen is reloaded, the $\overline{\sigma}-\overline{\varepsilon}$ plot traces the indicated curve from O_P to B. This curve is again almost straight and parallel to OA. In other words, the $\overline{\sigma}-\overline{\varepsilon}$ relation is essentially linear. Furthermore, the behavior is essentially elastic, since unloading practically retraces the path to O_P. So long as the stress does not exceed the value at B, the behavior is approximately linear and elastic. Because of the previous plastic deformation (OO_P), a greater stress is required to initiate *additional* plastic deformation. The yield stress for the plastically deformed material is greater than that for the virgin material. We say that the material has *strain hardened.*

4.4.8 YIELD STRENGTH

As mentioned previously, the stress that initiates yielding is not well defined. For this reason, engineers sometimes use another quantity to characterize resistance to plastic deformation. This is the *yield strength* defined as follows:

From a preselected point O^* of the $\overline{\varepsilon}$ axis, a line O^*H is drawn parallel to the line OA, as shown in Figure 4.7. The stress determined by the intersection of this line and the $\overline{\sigma}-\overline{\varepsilon}$ curve is termed the yield strength denoted σ_{ys}. This value depends on the amount by which the point O^* is "offset" from the origin O; hence, the yield strength has a definite value only when the offset $OO^*$is specified.

According to the previous discussion, unloading from point H follows very nearly the straight line HO^* This means that the preselected offset OO^* represents, approximately, the permanent strain caused by loading to the yield strength. We may, therefore, interpret the yield strength as that average stress, which causes a prescribed amount of plastic strain in simple tension.

4.4.9 ULTIMATE STRENGTH

Quite often the engineer requires some measure of the maximum resistance of a material to a simple tensile force. This is given by the *ultimate strength*, σ_u in Figure 4.7. It is simply the ultimate force at point C divided by the original cross-sectional area of the specimen. We must interpret this with caution: Fracture occurs at a site which may have a much reduced cross-section. A more accurate strength is provided by the "true stress," which incorporates the reduced area at the site of the fracture.

4.4.10 WORK AND ENERGY

Throughout the previous discussion, the basic concepts of force and displacement have been employed to describe observed behavior. Another approach of great importance in the analyses of deformable bodies entails the concepts of work and energy. Let us therefore consider the work done in stretching the portion QR of the bar. Suppose that the portion QR, acted upon by a force P_0, has elongated the amount e_0. For simplicity, imagine that the point Q is held fixed, as shown in Figure 4.8. During an infinitesimally small additional elongation Δe, the work done by the force P_0 is

$$\Delta W = P_0 \Delta e$$

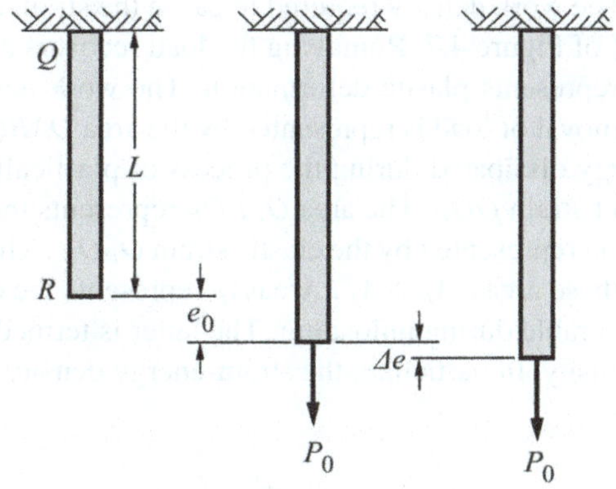

Figure 4.8

This small amount of work is represented by the shaded area in Figure 4.6. The total work done to effect an elongation e is then given by

$$W = \int_0^e P de \tag{4.7}$$

This integral is represented by the area $OABO_B$ of Figure 4.6. Work of amount W is required to cause an elongation of amount $e = OO_B$. Similarly, the total work done to fracture the bar is represented by the area $OABDO_D$ under the entire P-e curve of Figure 4.6.

Suppose now that Equation (4.7) is divided by the original area A and the original gauge length L. This yields

$$\frac{W}{AL} = \int_0^e \frac{P}{A}\frac{de}{L} = \int_0^{\overline{\varepsilon}} \overline{\sigma} d\overline{\varepsilon} \tag{4.8}$$

The product AL is simply the original volume; hence, the left side of Equation (4.8) is the work per unit volume. The right side is represented by the shaded area of Figure 4.7, the area $OABO_B$. For brevity, let us denote the work per unit volume by U, so that

$$U = \frac{W}{AL} = \int_0^{\overline{\varepsilon}} \overline{\sigma} d\overline{\varepsilon} \tag{4.9}$$

The quantity U is called the *work density*.

4.4.11 HYSTERESIS

Note that the trace of unloading, curve BO_p in Figure 4.7, differs from the path of reloading, O_PB. The differences are often very slight, but indicate that the work expended upon reloading exceeds the energy recovered during unloading. That difference is represented by the area enclosed in the loop BO_PB. The effect is called *hysteresis*.

4.4.12 TOUGHNESS

The total work density required to fracture the specimen is represented by the area $OABDO_D$ under the entire $\overline{\sigma} - \overline{\varepsilon}$ curve of Figure 4.7. This provides some measure of the ability of the material to resist fracture. It is termed the *toughness*.

Strain-Energy Density The work density required to cause the strain of amount $\overline{\varepsilon}$ is represented by the shaded area $OABO_B$ of Figure 4.7. Removing the load recovers a part of this strain (O_PO_B) and the other part (OO_P) represents plastic deformation. The work density expended during the cycle of application and removal of load is represented by the area $OABO_P$. This much of the work density represents the energy dissipated during the process of plastically deforming the material, which causes the permanent strain OO_P. The area O_PBO_B represents the work density required to cause the elastic deformation represented by the elastic strain O_PO_B . The total work is represented graphically by the sum of these areas, $A_P + A_E$. Area A_P represents the dissipated energy; area A_E represents the energy recoverable during unloading. The latter is termed *potential-energy*, *elastic-energy*, or *strain*-energy density. In particular, the strain-energy density when the stress $\overline{\sigma}$ reaches the proportional limit σ_L is

$$U_L = \frac{\sigma_L^2}{2E}$$

Since the proportional limit σ_L is generally very near the yield stress σ_0, the preceding energy density approximates the energy density required to initiate yielding in simple tension. It is called the *modulus of resilience* and is represented by the area under the linear portion (OA) of the $\overline{\sigma} - \overline{\varepsilon}$: curve of Figure 4.7.

4.4.13 DUCTILITY

Certain engineering applications demand material that can undergo severe permanent (plastic) deformation. The ability of materials to withstand plastic deformation is termed *ductility*. The processes of forging, stamping, and extrusion depend on the ductility of materials. Furthermore, ductility is essential if a structure is to absorb energy prior to fracture. Hence, the use of a ductile material provides a measure of safety in certain structural applications.

The plastic deformation incurred in the simple tension test is indicative of ductility. Measurements of the specimen before and after a test to fracture can be used to measure ductility. The *percent reduction* of the cross-sectional area and *percent elongation* of the gauge length are commonly accepted measures of ductility. Because of necking, both measures are dependent on the gauge length and, to some extent, on the size and shape of the material specimen. For this reason, test specifications require the use of a standard specimen.

4.4.14 STRUCTURAL STEEL

The preceding discussion of simple tension has revealed the essential characteristics of a typical ductile metal. Although the observed behavior is representative of many commonly used structural metals, certain other materials may exhibit quite different behavior. The behavior of common structural steel (hot-rolled low-carbon steel) is not entirely unlike the ductile metal just described. Its behavior under simple tension is illustrated by the diagram of Figure 4.9. Among the common engineering materials, it must be regarded as extremely ductile. The total plastic strain at fracture is more than one hundred times the elastic strain. The distinctive feature of structural steel is the portion BC of the $\overline{\sigma} - \overline{\varepsilon}$: diagram, which is very nearly a straight horizontal line. At point A the $\overline{\sigma} - \overline{\varepsilon}$ curve has a horizontal tangent. Additional plastic strain is accompanied by a decrease in stress. At point B, the curve again has a horizontal tangent. The stress corresponding to point A is known as the upper yield point, and the stress corresponding to point B is known as the lower yield point. The "flat" portion BC represents a plastic strain approximately twenty times the elastic

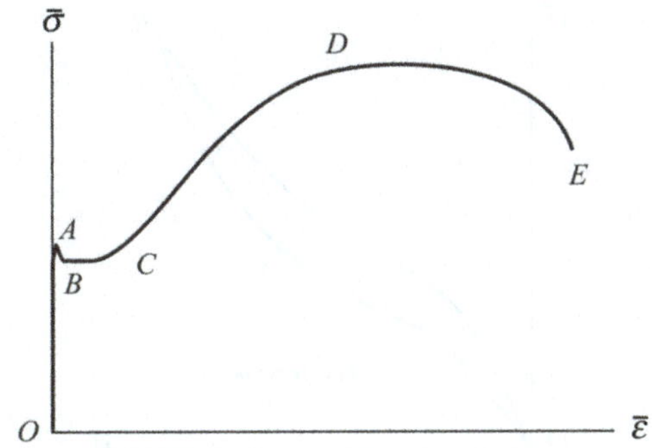

Figure 4.9

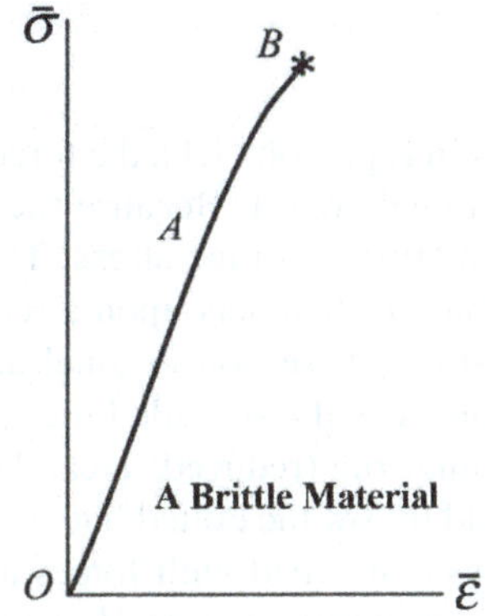

Figure 4.10

strain. In most other respects, the behavior of structural steel is similar to the behavior of the ductile metal described in the earlier parts of this section.

The behavior of a typical brittle material is illustrated by the diagram of Figure 4.10. Little plastic deformation occurs prior to fracture (point B). Furthermore, most brittle materials exhibit a linear relation almost to the point of fracture; that is, the entire curve is essentially a straight line.

4.4.15 RUBBER-LIKE MATERIAL

Not all materials display a linear relationship between the stress $\bar{\sigma}$ and the strain $\bar{\varepsilon}$. Many rubber-like materials exhibit a nonlinear $\bar{\sigma} - \bar{\varepsilon}$ relation, as illustrated by Figure 4.11. Although the $\bar{\sigma} - \bar{\varepsilon}$ curve is nonlinear, the body may return to its original size and shape when the applied forces are removed. Indeed, such materials exhibit very little tendency toward plastic deformation. If a specimen is loaded to point B and the load is removed, the curve for the unloading may not retrace the loading curve OAB. However, if the load is very slowly removed, the unloading curve BCO terminates at the origin O. Thus, there is no permanent deformation. If the unloading curve does not retrace the loading curve, the (shaded) area enclosed by the loading and unloading curves in Figure 4.11 represents dissipated energy. The cycle $OABCO$ is frequently termed a *hysteresis loop*.

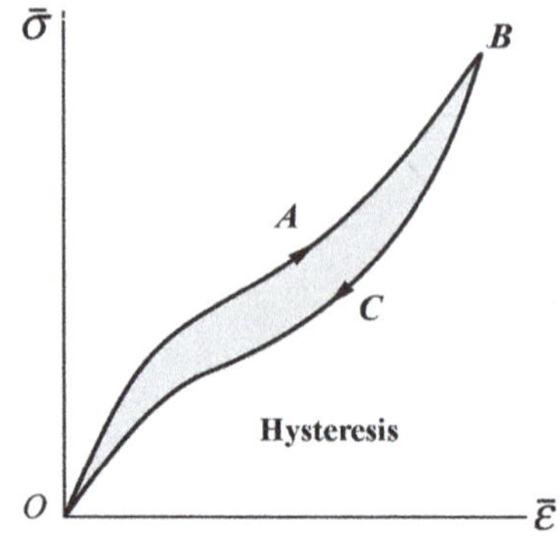

Figure 4.11

4.4.16 TRUE STRESS

In the definition of the average stress in Equation (4.1), the force P acting upon a transverse section is divided by the original cross-sectional area A. Because the cross-sectional area is reduced by deformation, the quantity $\overline{\sigma}$ is not a "true" average stress. This is particularly significant where necking occurs because the axial force P then acts upon a reduced area. The specimen deforms under less force, and the average stress (based on original area) actually decreases as necking progresses. A more meaningful quantity is the so-called true average stress obtained by dividing the axial force P by the corresponding true (reduced) area. The plot of true average stress for a typical test of a ductile material would follow the dotted line of Figure 4.5. Because, in the analyses of deformable bodies, we are always concerned with forces acting upon and within a deformed body, the discussions of Chapter 3 refer to true stresses. However, the deformations are not usually so severe that the distinction between original and deformed areas is important. For example, the difference between the curves of Figure 4.5 is significant only for very large strains. No distinction between true and conventional stress is required in most analyses of the members of structures and machines, but it is essential in analyses of severe inelastic deformations in forging and extrusion or the large elastic deformation of rubber-like materials.

4.4.17 SIMPLE COMPRESSION TEST

Basically, the simple compression test differs from the simple tension test only in the signs of the quantities that describe the causes and effects. An axial compressive force results in axial contraction of the test specimen. In accordance with previous sign conventions, the average stress $\overline{\sigma}$ and the average axial strain $\overline{\varepsilon}$ are negative. As a result of the applied compressive force, the test bar experiences positive transverse strains $\overline{\varepsilon}'$. The relationships between the quantities $\overline{\sigma}$ and $\overline{\varepsilon}$, and $\overline{\varepsilon}'$ are much like those observed in the tension test.

It is particularly informative to show the curves for tension and compression in a single diagram. Thus, in Figure 4.12, the curves $OABC$ and $OA'B'C'$ are obtained from a simple tension test and a simple compression test, respectively. For many common materials, these two curves are quite similar if the strain magnitudes are not much greater than the strain at initial yielding. A portion OA' of the curve $OA'B'C'$ is very nearly a straight line extension of AO. Accordingly, the linear relation of Equation (4.4) is, in most cases, a valid approximation for average stress magnitudes, which do not exceed the proportional limits in compression or in tension. In most cases, the proportional limit and the yield stress in compression have approximately the same magnitudes as those in tension.

For large magnitudes of average strain, the conventional tension and compression curves may be quite different. This is truer of some materials than of others. A very ductile material subjected to axial compression flattens into a thin disk. The applied load must, therefore, increase almost without limit as the deformation progresses. At the same time, the average strain is limited, since the change in length of the disk cannot exceed the original length.

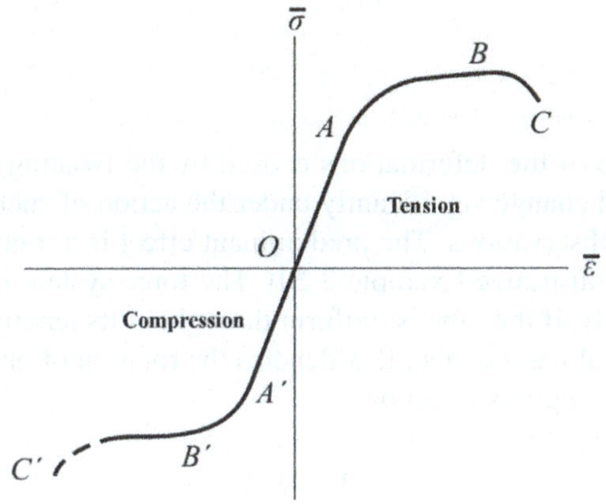

Figure 4.12

4.4.18 SIMPLE SHEAR TEST

The effects of shear forces are most readily studied by tests of thin circular cylindrical tubes subjected to twisting couples. Suitable fixtures at the ends maintain symmetry with respect to the tube axis. Under these conditions, it is reasonable to assume that the forces on any transverse section consist of shear forces distributed symmetrically about the axis. Any small element of the tube, which is oriented as shown in Figure 4.13a, is subject to the action of shear forces as shown in Figure 4.13b.

The average shear stress $\bar{\tau}$ is given by

$$\bar{\tau} = \frac{T}{2\pi\bar{R}t}$$

where T is the applied twisting couple, $\bar{R}$ is the mean radius of the tube, and t is the thickness of the wall. Since the tube is thin, we suppose that the shear stress on any element is approximately $\bar{\tau}$.

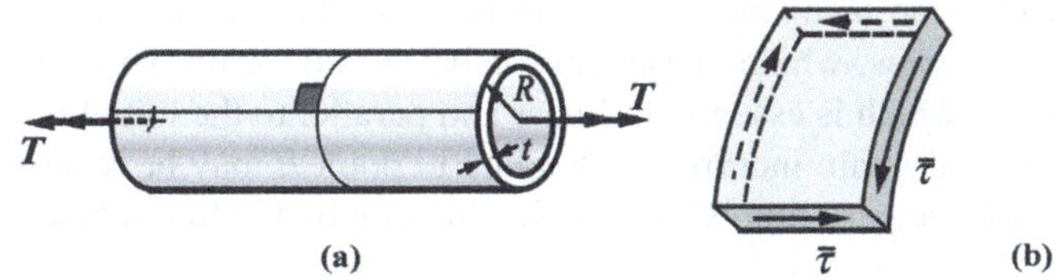

Figure 4.13

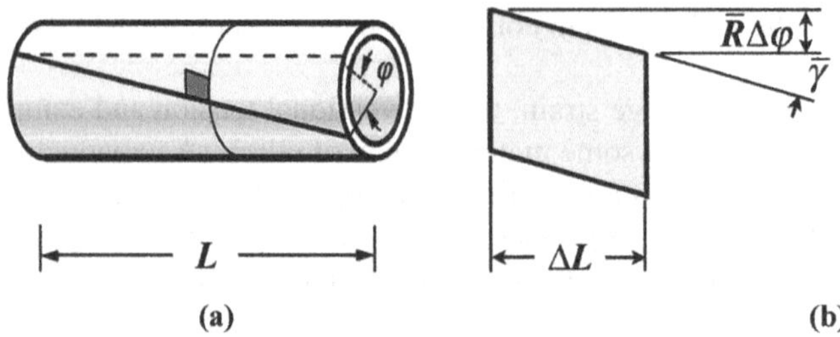

Figure 4.14

Let us consider the nature of the deformations caused by the twisting couple. It is unlikely that the radius or length would change significantly under the action of such external loading. That is verified by experimental observations. The predominant effect is a rotation of transverse sections (described in Chapter 2, Illustrative Example 2.20). The force system is identical on every transverse section. Consequently, if the tube is uniform throughout its length, the angle of twist θ (per unit of length) is constant along the tube. If ϕ denotes the rotation of one end relative to the other, the angle of twist per unit length is given by

$$\theta = \varphi / L$$

where L is the length of the tube. The deformation of a small element is then predominantly a shear strain, as illustrated in Figure 4.14. The average shear strain, according to Figure 4.14b, is

$$\bar{\gamma} = \arctan \bar{R}\frac{\Delta\varphi}{\Delta L} = \arctan \frac{\bar{R}\varphi}{L}$$

It is essentially the same for each similar element. Since the tube is thin, we suppose that the shear strain at any point is approximately $\bar{\gamma}$. In any event, the computations of average stresses and strains must suffice, as it is never possible to measure stresses and strains directly. In the torsion test, measurements of the rotation and the corresponding torque T are required. The additional measurements of L , R, and t will suggest that these dimensions do not change significantly.

Let us consider a typical ductile material, the same material that was examined in the tension and compression tests. During the gradual and continuous twisting of the cylinder, a plot of average shear stress $\bar{\tau}$ versus the average shear strain $\bar{\gamma}$ traces the curve $OABC$ shown in Figure 4.15. Reversing the sense of the applied torque would give a similar curve, $OA'B'C'$. If the material is isotropic, the two curves are identical in shape. The character of the shear stress-strain relation is very much like that of the tension-compression stress-strain relation. For all practical purposes, the portion $A'OA$ of the diagram is a straight line. Thus, for stress magnitudes within the shear proportional limits, $\bar{\tau} = G\bar{\gamma}$, where G is a constant of proportionality. This constant G, a property of the material, is called the *shear modulus* of elasticity. A shear yield stress τ_0 can be defined as the least average shear stress, which causes noticeable plastic strain. Unloading from point B to O_P traces a curve BO_P, which is essentially linear and parallel to the initial portion of the loading curve OA. Thus, the shear strain incurred by loading to point B may be regarded as composed of an elastic part and a plastic part, represented in the diagram by the line segments O_PO_B and OO_P, respectively.

The previous statements about work and energy densities in tension and compression are also applicable here. The work done (per unit volume) in causing a strain of amount OO_B is represented

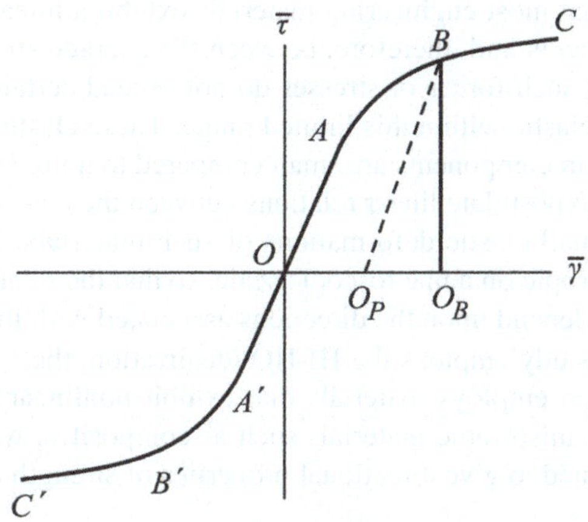

Figure 4.15

by the area OBO_B. The energy dissipated as a result of plastic deformation is represented by the area OBO_P, and the elastic (recoverable) strain energy by the area O_PBO_B. Plastic deformation alters the material properties, changes the subsequent yield stresses, and decreases ductility.

It should be noted that there is, in general, no irregularity in the deformation, such as the necking phenomenon observed in tension. However, if the tube is very thin, there is a possibility of localized buckling or collapse, which destroys the geometrical symmetry of the tube. Data obtained during such irregular deformations cannot serve to establish the relation between $\bar{\tau}$ and $\bar{\gamma}$.

Most ductile metals have a shear yield stress τ_0 that is approximately one-half of the tensile yield stress σ_0 obtained in tension or compression tests. In a simple tension test, the greatest average shear stress occurs on a plane inclined 45° to the plane of maximum normal stress. This maximum shear stress is one-half the average tensile stress. Hence, these tests suggest that yielding occurs in each case when the maximum shear stress for that case attains the value τ_0.

The average shear stress-strain diagrams for other materials may differ markedly from that of Figure 4.15. Brittle materials may be expected to exhibit an almost linear shear stress-strain relation to fracture; other materials may be quite nonlinear in their stress-strain relation.

4.5 LINEARLY ELASTIC (HOOKEAN) BEHAVIOR

The analysis of any engineering problem requires a suitable mathematical description of the physical situation. The formulation must be sufficiently simple mathematically to permit an analysis, yet realistic enough so that the solution will predict the response of the physical system with reasonable accuracy. As such, the mathematical description is an idealization. Our immediate attention focuses on the simplest idealization of solid materials, those that exhibit attributes of *elasticity, isotropy, and linearity* in the relations between strain, stress, and temperature. Linearly elastic materials are often termed Hookean after Robert Hooke,[4] who proposed the idealization. Since most of our applications also presume isotropy, we employ the acronym HI for Hookean Isotropic. When the HI material is also homogeneous—that is, the properties are constant throughout the body—we use the elaboration HI-HO, which represents: Hookean, Isotropic, and Homogeneous. The reader will be burdened with no other acronyms; HI and HI-HO materials are so prevalent in our introductory applications that such abbreviation seems justified.

Experiments indicate that most engineering materials exhibit a linear relation between the applied forces and displacements and, therefore, between the average stress and the corresponding average strains, so long as such forces or stresses do not exceed certain limits. Furthermore, the observed behavior is also elastic within this limited range. These elastic deformations are usually quite small, so that the strain components are small compared to unity ($\varepsilon, \gamma \ll 1$). In view of these observations, it is natural to postulate linear relations between the components of stress and strain in order to describe the small elastic deformations of such materials. Furthermore, many of the common materials are isotropic on a macroscopic scale, so that the relations between components of stress and strain do not depend upon the directions associated with the components.

Although much of our study employs the HI-HO idealization, the reader must be alert to limitations. The engineer often employs materials that exhibit nonlinear or inelastic behavior and intentionally designs with anisotropic materials such as composites, wherein filaments, or fibers of high strengths, are oriented to give directional properties of strength or stiffness.

4.6 SUPERPOSITION OF STRESSES

In the preceding study of strain (Section 2.10), we observed that small strains can be superposed. If two states of small strain are imposed, whether successively or simultaneously, they are additive; for example, if a body experiences two small strains, ε_x' and ε_x'', the practical result is a small strain,

$$\varepsilon_x \doteq \varepsilon_x' + \varepsilon_x'' \tag{4.10}$$

"Small" means small *compared* to unity; that is, $\varepsilon_x' \ll 1$, $\varepsilon_x'' \ll 1$, which justifies the neglect of higher-order terms.

Our study of stress follows our discussion of deformation, but clearly the imposition of stress (or temperature) causes deformation, and, in particular, the area of a surface is generally altered. For example, a simple stress $\overline{\sigma}_x' = F_x'/A'$ causes a change in area ($A' \to A''$) such that an additional force F_x'': results in a stress $\overline{\sigma}_x = (F_x' + F_x'')/A''$. Then, strictly speaking $\overline{\sigma}_x \neq \overline{\sigma}_x' + \overline{\sigma}_x''$. However, if the strains are small, then the dilation of the area is also small; specifically, if x, y, and z are principal directions, then

$$A'' = (1 + \overline{\varepsilon}_y')(1 + \overline{\varepsilon}_z')A'$$

Again, if the strains are small compared to unity, then $A'' \doteq A'$ and

$$\overline{\sigma}_x \doteq \overline{\sigma}_x' + \overline{\sigma}_x'' \tag{4.11}$$

The superposition of strains and stresses applies to arbitrary states provided that the former are small compared to unity. We now use superposition as a simple and direct way to synthesize the linear strain-stress-temperature equations for HI material.

4.7 HOOKEAN-ISOTROPIC (HI) BEHAVIOR

Consider the simple state of stress depicted in Figure 4.16. If the material is Hookean (a typical metal within the proportional limit), then the stress causes a proportional extensional strain:

$$\varepsilon_x' = \frac{1}{E}\sigma_x \tag{4.12}$$

The coefficient $(1/E)$ can only be inferred from the tension and/or compression tests; see Equation (4.4). Also, from such tests, we observe proportional strains (contractions) in the transverse directions (y and z):

$$\varepsilon_y' = \varepsilon_z' = -\frac{\nu}{E} \tag{4.13}$$

These are identical, since the material is *isotropic*. The coefficient $(-\nu/E)$ must be inferred from tests; see Equations (4.5) and (4.6). The constant E is the modulus of elasticity (also known as Young's modulus). It has the dimensions of stress (force/length^2). The constant ν, known as Poisson's ratio, is dimensionless. The modulus of elasticity can be determined by a simple tension or compression test. However, Poisson's ratio is not normally obtained in this way.

Now suppose that the same block of material is acted upon by the uniform stress σ_y only. Because the material is isotropic, it shows no preference for any direction. Consequently, the resulting extensional strains are

$$\varepsilon_y'' = \frac{1}{E}\sigma_y, \qquad \varepsilon_x'' = \varepsilon_z'' = -\frac{\nu}{E}\sigma_y$$

and if a normal stress σ_z acts alone, the resulting extensional strains are

$$\varepsilon_z''' = \frac{1}{E}\sigma_z, \qquad \varepsilon_x''' = \varepsilon_y''' = -\frac{\nu}{E}\sigma_z$$

The small strains caused by the combined application of the normal stresses, σ_x, σ_y, and σ_z, are obtained by superposition. For example, the resulting strain ε_x is the sum of the strains ε_x', ε_x'', and ε_x''' caused by each of the stress components. The effects of this combination of stress components are the strain components.

$$\varepsilon_x = \frac{1}{E}[\sigma_x - \nu(\sigma_y + \sigma_z)] \tag{4.14}$$

$$\varepsilon_y = \frac{1}{E}[\sigma_y - \nu(\sigma_z + \sigma_x)] \tag{4.15}$$

$$\varepsilon_z = \frac{1}{E}[\sigma_z - \nu(\sigma_x + \sigma_y)] \tag{4.16}$$

To this point we have considered only the relations between the normal stresses and extensional strains associated with the perpendicular directions x, y, and z. In the case of an isotropic material, the block of Figure 4.16 remains rectangular under the action of the normal stresses σ_x, σ_y, σ_z. That is, these normal stresses cause no shear strains γ_{xy}, γ_{yz}, γ_{zx}. This is a consequence of *isotropy* and is

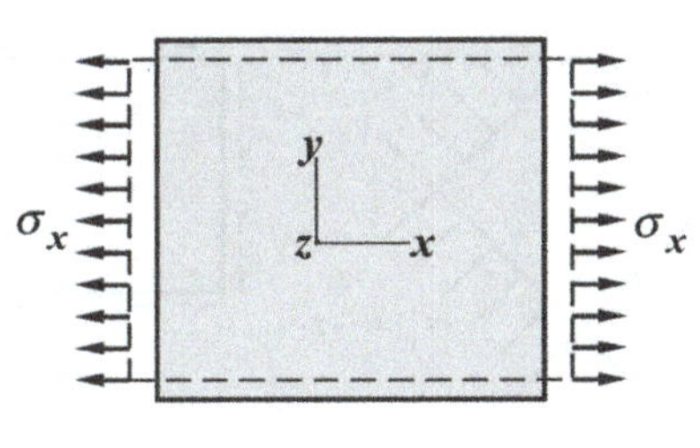

Figure 4.16

not limited to Hookean behavior. It implies that principal directions of stress ($\tau_{xy} = \tau_{xz} = \tau_{yz} = 0$) are also principal directions of strain ($\gamma_{xy} = \gamma_{xz} = \gamma_{yz} = 0$) and vice versa.

Consider now the state of stress depicted in Figure 4.17a; the only nonzero components are $\sigma'_x = \tau = -\sigma'_y$. By the equations of transformation (see Section 3.11, Figure 3.57), we obtain the components of stress in the rectangular system x, y, z, wherein x', y' are rotated 45° about z:

$$\begin{matrix} \sigma_x & \tau_{xy} & \tau_{xz} \\ & \sigma_y & \tau_{yz} \\ & & \sigma_z \end{matrix} = \begin{matrix} 0 & \tau & 0 \\ & 0 & 0 \\ & & 0 \end{matrix}$$

In other words, this is a state of simple shear, $\tau_{xy} = \tau_{yx} = \tau$, as depicted in Figure 4.17b.

By Equations (4.14)–(4.16), the state of strain for this HI material has the following components in the system (x', y', z'):

$$\varepsilon_{x'} = +\frac{(1+\nu)}{E}\tau, \qquad \varepsilon_{y'} = -\frac{(1+\nu)}{E}\tau$$
$$\varepsilon_{z'} = \gamma_{x'y'} = \gamma_{x'z'} = \gamma_{y'z'} = 0$$

The components in the system (x, y, z) are obtained by transformations (2.18) and (2.19) (see Section 2.11 and/or Section 2.18; see Figure 2.29):

$$\begin{gathered} \varepsilon_x = \varepsilon_y = \varepsilon_z = \gamma_{yz} = \gamma_{xz} = 0 \\ \gamma_{xy} = \frac{2(1+\nu)}{E}\tau = \frac{2(1+\nu)}{E}\tau_{xy} \end{gathered} \tag{4.17}$$

This is the state of simple shear strain as described in Section 2.17. One might infer that a simple state of shear stress τ_{xy}, as shown in Figure 4.17b, causes only the shear strain γ_{xy}. That conclusion is supported by experiments on HI materials, such as the torsional test of Figure 4.14. However, such experiments alone do not establish the relationship between the shear modulus G and the elastic moduli, E and ν. According to Equation (4.17), that relation follows:

$$G = \frac{E}{2(1+\nu)} \tag{4.18}$$

The modulus G might be obtained by the test of Figure 4.13 and the modulus E, by the tension test; then the Poisson's ratio can be determined by Equation (4.18). In any case, we note that the complete linear system of strain-stress equations for the HI material incorporates only two independent coefficients (mechanical properties).

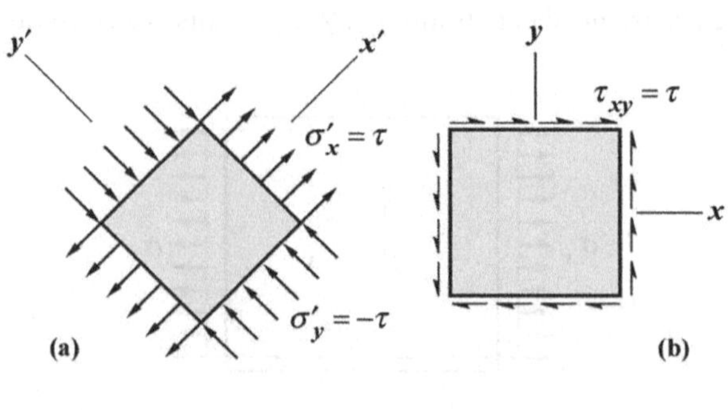

Figure 4.17

Since the behavior is isotropic, two equations like (4.17) apply to the other rectangular components, τ_{xz} and τ_{yz}. These and Equations (4.14)–(4.17) constitute the six strain-stress relations for the HI material. For convenience, these are collected here:

$$\varepsilon_x = \frac{1}{E}[\sigma_x - \nu(\sigma_y + \sigma_z)] \tag{4.19a}$$

$$\varepsilon_y = \frac{1}{E}[\sigma_y - \nu(\sigma_z + \sigma_x)] \tag{4.19b}$$

$$\varepsilon_z = \frac{1}{E}[\sigma_z - \nu(\sigma_x + \sigma_y)] \tag{4.19c}$$

$$\gamma_{xy} = \frac{1}{G}\tau_{xy}, \qquad \gamma_{yz} = \frac{1}{G}\tau_{yz}, \qquad \gamma_{zx} = \frac{1}{G}\tau_{zx} \tag{4.20a,b,c}$$

Often, it is necessary to compute the stress components from a knowledge of the strain components. For example, the strains may be determined by experimental means, using electric resistance strain gauges. If the material is linearly elastic, the stresses may be found by solving Equations (4.19) and (4.20) for the stresses in terms of strains. The solution follows:

$$\sigma_x = \frac{E}{(1-2\nu)(1+\nu)}[(1-\nu)\varepsilon_x + \nu(\varepsilon_y + \varepsilon_z)] \tag{4.21a}$$

$$\sigma_y = \frac{E}{(1-2\nu)(1+\nu)}[(1-\nu)\varepsilon_y + \nu(\varepsilon_z + \varepsilon_x)] \tag{4.21b}$$

$$\sigma_z = \frac{E}{(1-2\nu)(1+\nu)}[(1-\nu)\varepsilon_z + \nu(\varepsilon_x + \varepsilon_y)] \tag{4.21c}$$

$$\tau_{xy} = G\gamma_{xy}, \qquad \tau_{yz} = G\gamma_{yz}, \qquad \tau_{zx} = G\gamma_{zx} \tag{4.22a,b,c}$$

One additional property is occasionally employed in descriptions of HI materials; it is the coefficient of proportionality between a hydrostatic state of stress (see Section 3.11) and the dilatation (see Section 2.17). In the present case, strains are small so that the dilatation (or volumetric strain) is given by the linear approximation:

$$e = \varepsilon_x + \varepsilon_y + \varepsilon_z$$

It follows from (4.19) that

$$e = \frac{3(1-2\nu)}{E}\frac{(\sigma_x + \sigma_y + \sigma_z)}{3} \tag{4.23a}$$

or

$$\frac{(\sigma_x + \sigma_y + \sigma_z)}{3} = \frac{E}{3(1-2\nu)}e \tag{4.23b}$$

The quantity $(\sigma_x + \sigma_y + \sigma_z)/3$ is the *mean* normal stress, also termed the hydrostatic part of the state of stress. The coefficient of proportionality in Equation (4.23b) characterizes the resistance to dilatation (i.e., volumetric change). That coefficient, like E and G, has the dimensions of stress; like E and G, it is called a modulus, the *bulk modulus*:

$$K \equiv \frac{E}{3(1-2\nu)} \tag{4.24}$$

We recall that the modulus E and the Poisson's ratio are positive; also, the bulk modulus must be positive. Therefore, the denominator of (4.24) must remain positive; it follows that

$$0 < \nu < \frac{1}{2}$$

As Poisson's ratio approaches 1/2, the bulk modulus approaches infinity; stated otherwise, no amount of pressure can cause dilatation. When $\nu = 1/2$, the HI material is said to be *incompressible*.

One final word about HI behavior: Materials can be Hookean and elastic, yet non-homogeneous. For example, a composite might be formed with particles dispersed more densely in some portions; then the properties (E, G, K, ν) would be variable. However, most common materials (e.g., metals) are nearly homogeneous; then the properties are constant throughout. To avoid ambiguity, we identify those by the acronym HI-HO.

4.8 PLANE STRESS OF A HOOKEAN ISOTROPIC BODY

No stress acts upon a free surface (i.e., a surface not contacting another body). Therefore, the normal to a free surface identifies a principal direction of stress; moreover, that principal stress is zero. If z denotes the normal direction, then $\tau_{xz} = \tau_{yz} = \sigma_z = 0$. According to Equations (4.19) and (4.20),

$$\varepsilon_x = \frac{1}{E}(\sigma_x - \nu\sigma_y),\ \varepsilon_y = \frac{1}{E}(-\nu\sigma_x + \sigma_y),\ \varepsilon_z = -\frac{\nu}{E}(\sigma_x + \sigma_y) \qquad (4.25a,b,c)$$

$$\gamma_{xy} = \frac{1}{G}\tau_{xy}, \quad \gamma_{xz} = \gamma_{yz} = 0 \qquad (4.25d)$$

As previously noted, the normal direction is also a principal direction of strain. Note, too, that the extensional strain in the normal direction (ε_z) is not generally zero.

The inverse of the preceding equations gives the non-vanishing stresses:

$$\sigma_x = \frac{E}{(1-\nu^2)}(\varepsilon_x + \nu\varepsilon_y),\ \sigma_y = \frac{E}{(1-\nu^2)}(\nu\varepsilon_x + \varepsilon_y) \qquad (4.26a,b)$$

$$\tau_{xy} = G\gamma_{xy} \qquad (4.26c)$$

Since the extensional strains can be measured on the free surface (see Section 2.20), the strains for any directions in the surface (e.g., ε_x, ε_y, γ_{xy}) can be determined. Then, by means of the preceding equations, the nonzero stresses (σ_x, σ_y, τ_{xy}) can be determined and, by means of Equation (4.25c) the extensional strain (ε_z) can be found.

Because the normal to the free surface is a principal direction, the remaining two principal directions lie in the surface. Consequently, they can be determined by the special forms of Section (2.18).

Since the strains in a free surface are readily measured by electrical-resistance gauges (see Section 2.20) and since three components of stress are zero, one can determine the states of strain and stress if the body is known to be composed of the HI-HO material. An example serves to illustrate such computation.

ILLUSTRATIVE EXAMPLE 4.1

States of Strain and Stress at the Surface of a HI Body

The strain gauge (120° rosette) depicted in Figure 4.18 measures the extensional strains in the directions *a*, *b*, and *c* on the free surface of a steel member. The behavior is HI within the elastic limit; the modulus and Poisson ratio are $E = 30 \times 10^3$ ksi (approximately 210 MPa) and $\nu = 0.25$. The measured values are: $\varepsilon_a = 100 \times 10^{-5}, \quad \varepsilon_c = 40 \times 10^{-5}, \quad \varepsilon_b = -20 \times 10^{-5}$

We refer to the equations of transformations as given in Section 2.18 and let *x* denote the direction of *a*, *y* denote the orthogonal direction in the surface, and *z* denote the normal direction. Then

$$\varepsilon_b = \varepsilon_a \cos^2 120° + \varepsilon_y \sin^2 120° + \gamma_{xy} \sin 120° \cos 120°$$

$$\varepsilon_c = \varepsilon_a \cos^2 240° + \varepsilon_y \sin^2 240° + \gamma_{xy} \sin 240° \cos 240°$$

Substituting the numerical values, we have the two equations:

$$-20 \times 10^{-5} = 100 \times 10^{-5} \times \frac{1}{4} + \varepsilon_y \frac{3}{4} - \gamma_{xy} \frac{\sqrt{3}}{4}$$

$$40 \times 10^{-5} = 100 \times 10^{-5} \times \frac{1}{4} + \varepsilon_y \frac{3}{4} + \gamma_{xy} \frac{\sqrt{3}}{4}$$

The solution follows:

$$\varepsilon_y = -20 \times 10^{-5}, \quad \gamma_{xy} = 40\sqrt{3} \times 10^{-5}$$

Since the normal *z* is a principal direction of stress ($\sigma_z = \tau_{xz} = \tau_{yz} = 0$) and the material is isotropic, the normal *z* is also a principal direction of strain. The remaining principal directions lie in the surface; they are determined by Equation (2.55b):

$$\tan 2\theta = \frac{\gamma_{xy}}{(\varepsilon_x - \varepsilon_y)} = \frac{40\sqrt{3}}{100 + 20} = \frac{1}{\sqrt{3}}$$

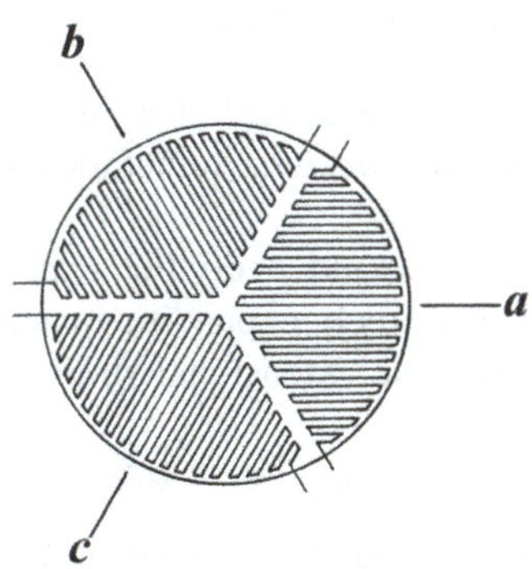

Figure 4.18

or

$$2\theta = 30^\circ, -150^\circ, \ \theta = 15^\circ, -75^\circ$$

The principal strains are then obtained by Equation (2.53):

$$\varepsilon_1(15^\circ) = (40 + 60\sqrt{3}) \times 10^{-5} = 144 \times 10^{-5}$$

$$\varepsilon_2(-75^\circ) = (40 - 60\sqrt{3}) \times 10^{-5} = -63.9 \times 10^{-5}$$

In the foregoing computations, we invoked only the isotropic property. The principal stresses now follow from Equations (4.26a,b):

$$\sigma_1 = \frac{E}{1-\nu^2}(\varepsilon_1 + \nu\varepsilon_2) = 41.0 \text{ ksi}$$

$$\sigma_2 = \frac{E}{1-\nu^2}(\varepsilon_2 + \nu\varepsilon_1) = -8.9 \text{ ksi}$$

The extensional strain in the third principal direction is given by Equation (4.25c)

$$\varepsilon_z = -\frac{\nu}{E}(\sigma_1 + \sigma_2) = -27 \times 10^{-5}$$

Often the maximum shear stress is important; by Equation (3.28)

$$\tau_{max} = \frac{1}{2}|\sigma_{max} - \sigma_{min}| = \frac{1}{2}(41.0 + 8.9) = 24.9 \text{ ksi}$$

Alternatively, the maximum shear strain is given by Equation (2.40):

$$\gamma_{max} = |\varepsilon_{max} - \varepsilon_{min}| = 208 \times 10^{-5}$$

Then, by Equation (4.22c),

$$\tau_{max} = G\gamma_{max} = \frac{E}{2(1+\nu)}\gamma_{max} = 24.9 \text{ ksi}$$

The latter computation serves only as an additional exercise and confirmation of our arithmetic.

4.9 EFFECTS OF TEMPERATURE

4.9.1 THERMAL EXPANSION

Almost all materials tend to expand when heated. The increase in volume that accompanies a rise in temperature is termed *thermal expansion.* Sometimes a temperature change has no other significant effect.

If a material is *isotropic* and free to expand, every line experiences the same extensional strain when the temperature is increased. The freely expanding material experiences no shear deformation. The simplest relation between these extensional strains and the temperature changes that cause them is the linear relations

$$\varepsilon_x = \varepsilon_y = \varepsilon_z = \alpha\Delta Q \tag{4.27}$$

where ΔQ denotes the temperature rise and α is a constant, the *coefficient of thermal expansion.* Experiments show that the linear equations are a satisfactory approximation for the solution of many engineering problems.

A uniform rise in the temperature of a free homogeneous body tends to cause a uniform expansion of every portion. Consequently, strains result, although there may be no internal forces (or stresses) present. Nonuniform temperature distributions tend to cause greater expansion in some portions than in others, which may cause high stress intensities in some parts of the body. Sudden localized heating, called *thermal shock*, causes severe nonuniform temperature distributions that may fracture brittle materials, e.g., a cold water-glass may fracture when abruptly filled with hot water.

The linear relations of equations (4.27) determine the strains that result from a temperature change ΔQ, *provided* that the material is free to expand. If the material is not free to expand, stresses arise. Consider, for example, the rectangular bar of Figure 4.19. At some initial temperature, Q_1 there are no stresses or strains present. The lateral surfaces are free, but the end faces are constrained by rigid surfaces. Suppose that the bar is uniformly heated to the temperature Q_2. The bar tends to expand but is constrained by the rigid walls; hence, normal stresses are developed on the end faces. For simplicity, we suppose that the conditions of stress and strain are uniform throughout the bar. If the bar were free, the effects of the temperature rise would be the extensional strains.

$$\varepsilon_x' = \varepsilon_y' = \varepsilon_z' = \alpha \Delta Q = \alpha (Q_2 - Q_1)$$

The presence of an axial stress $\sigma_x = -p$, a pressure exerted on the end faces by the constraining walls, would also cause deformation. For linearly elastic behavior, the extensional strains caused by this stress are

$$\varepsilon_x'' = -\frac{p}{E}, \qquad \varepsilon_z'' = \varepsilon_y'' = \left(\frac{\nu}{E}\right) p$$

When the bar is heated, both the effects of the rise in temperature, ΔQ, and the axial pressure, p, are present. Because the strains are small, these effects (strains) may be superimposed. Because of the constraining walls, the combined effect is zero net axial strain:

$$\varepsilon_x = \varepsilon_x' + \varepsilon_x'' = \alpha \Delta Q - \frac{p}{E} = 0 \tag{4.28}$$

Consequently, the axial stress induced by heating is

$$\sigma_x = -p = -E \alpha \Delta Q \tag{4.29}$$

The preceding example is a simple thermoelastic problem, in which strain-temperature and stress-strain relations are linear. A significant feature of the solution is the superposition of effects. Likewise, the most general problem of this type may be treated by adding the thermal strain $\alpha \Delta Q$ to

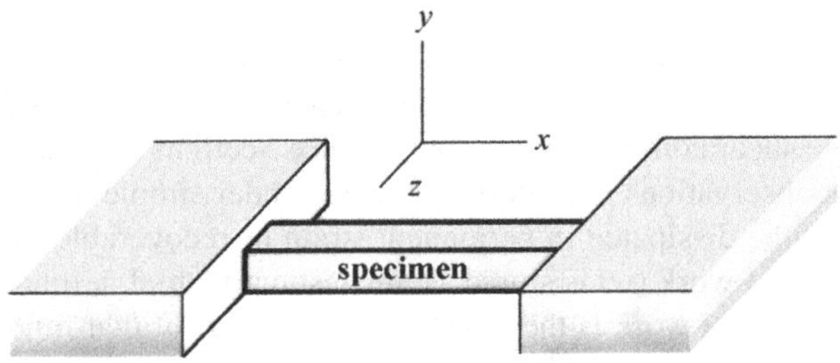

Figure 4.19

each one of Equations (4.19). Often these linear relations are adequate for the solution of thermal stress problems of HI bodies. For future reference, we collect complete strain-stress-temperature relations for the HI material:

$$\varepsilon_x = \frac{1}{E}(\sigma_x - \nu\sigma_y - \nu\sigma_z) + \alpha\Delta Q \tag{4.30a}$$

$$\varepsilon_y = \frac{1}{E}(-\nu\sigma_x + \sigma_y - \nu\sigma_z) + \alpha\Delta Q \tag{4.30b}$$

$$\varepsilon_z = \frac{1}{E}(-\nu\sigma_x - \nu\sigma_y + \sigma_z) + \alpha\Delta Q \tag{4.30c}$$

$$\gamma_{xy} = \frac{1}{G}\tau_{xy}, \quad \gamma_{xz} = \frac{1}{G}\tau_{xz}, \quad \gamma_{yz} = \frac{1}{G}\tau_{yz} \tag{4.30d}$$

4.9.2 EFFECTS OF TEMPERATURE ON MECHANICAL BEHAVIOR

Temperature also influences the mechanical behavior of materials. At high temperatures, most materials soften and lose strength, and metals corrode faster. Sustained elevated temperatures, as occur in boilers and gas turbines, can cause gradual deformations known as creep (see Section 4.14). Low temperatures cause some ductile materials to become brittle.

4.10 WORK AND ENERGY

From our earlier analyses of deformation (Chapter 2), we recall that a small rectangular element of the undeformed body is perceived as a parallelepiped after a deformation; our treatment is justified since we eventually pass to the limit (Δx, Δy, $\Delta z \to 0$). The initial rectangular element and the deformed parallelepiped are depicted in Figure 4.20. The vectors $\mathbf{G_x}$, $\mathbf{G_y}$, $\mathbf{G_z}$ are tangent to the *deformed* x, y, z lines, respectively (See Section 2.21, Equation 2.64).

In Section 3.16, we obtained general expressions for the work expended upon a deformed element (See Figure 3.67), as it undergoes a virtual displacement $\dot{\mathbf{R}}$. When the equations of motion or equilibrium are satisfied, that work is expended entirely upon the strains. In Section 3.15, we considered alternative measures of strain ($\tilde{\varepsilon}$, $\bar{\varepsilon}$) and the associated stresses (s, t). In either case, the virtual work assumes the following form:

$$\begin{aligned} \dot{w} = \sigma_x\dot{\varepsilon}_x + \tau_{xy}\dot{\gamma}_{xy} &+ \tau_{xz}\dot{\gamma}_{xz} \\ + \sigma_y\dot{\varepsilon}_y &+ \tau_{yz}\dot{\gamma}_{yz} \\ &+ \sigma_z\dot{\varepsilon}_z \end{aligned} \tag{4.31}$$

Here σ_n represents either of the normal components (s_n or t_n); τ_{nt} represents either of the shear components (s_{nt} or t_{nt}); ε_n represents either of the extensional components ($\tilde{\varepsilon}_n$, $\bar{\varepsilon}_n$) whereas $\gamma_{nt}/2$ represents either of the shear components ($\tilde{\gamma}_{nt}$ or $\bar{\gamma}_{nt}$) (See Sections 3.16, 2.13 and 2.21).

From our previous observations of material behavior under simple loadings, we know that the work expended (w) can be dissipated in permanent strain or recoverable as elastic energy. In the latter case, the incremental work ($\dot{w}$) is stored as an elastic potential, a functional (u) of the strain components. The incremental work is then the differential (du) of that function. The recoverable strain increments are also differentials. It follows that

$$\dot{w} = du(\varepsilon_x, \varepsilon_y, \varepsilon_z, \gamma_{xy}, \gamma_{xz}, \gamma_{yz})$$

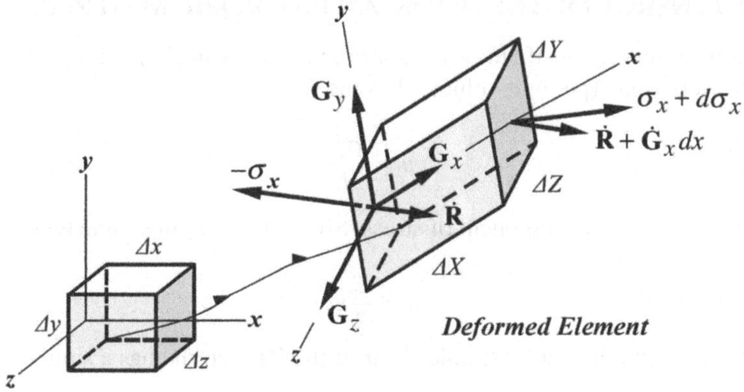

Figure 4.20

$$
\begin{aligned}
&= \frac{\partial u}{\partial \varepsilon_x} d\varepsilon_x + \frac{\partial u}{\partial \gamma_{xy}} d\gamma_{xy} + \frac{\partial u}{\partial \gamma_{xz}} d\gamma_{xz} \\
&\quad + \frac{\partial u}{\partial \varepsilon_y} d\varepsilon_y + \frac{\partial u}{\partial \gamma_{yz}} d\gamma_{yz} \\
&\quad + \frac{\partial u}{\partial \varepsilon_z} d\varepsilon_z
\end{aligned}
\tag{4.32}
$$

In the case of an elastic material, the two forms, (4.31) and (4.32), are equivalent for arbitrary differentials/increments of strain. Therefore, we must have

$$
\begin{aligned}
\sigma_x = \frac{\partial u}{\partial \varepsilon_x}, \tau_{xy} = \frac{\partial u}{\partial \gamma_{xy}}, \quad \tau_{xz} = \frac{\partial u}{\partial \gamma_{xz}} \\
\sigma_y = \frac{\partial u}{\partial \varepsilon_y}, \quad \tau_{yz} = \frac{\partial u}{\partial \gamma_{yz}}, \\
\sigma_z = \frac{\partial u}{\partial \varepsilon_z}
\end{aligned}
\tag{4.33}
$$

Equation (4.33) is a cornerstone in all subsequent analyses, which are founded upon principles of work and energy. These Equations (4.33) provide a general basis for the stress-strain relations of elastic materials in the absence of thermal changes. No limitations are imposed upon the magnitudes of the displacements nor rotations. The relations (4.33) can be applied to circumstances of large strains, if, and only if, the stress components are appropriately expressed in terms of the deformed areas and rotated directions (See Section 3.16).

Note: The potential u (the strain energy) can be expressed as a function of the components of the Cauchy-Green tensor ($\tilde{\varepsilon}$, $\tilde{\gamma}$) or as a function of the components of the engineering tensor ($\bar{\varepsilon}$, $\bar{\gamma}$). In the former case, the associated components of stress are those of the Trefftz tensor (s); in the latter case, the associated components of stress are those of the Jauman tensor (t); e.g. in simple tension.

$$du(\tilde{\varepsilon}_x) = s_x d\tilde{\varepsilon}_x, \quad du(\bar{\varepsilon}_x) = t_x d\bar{\varepsilon}_x$$

4.11 STRAIN ENERGY OF THE HOOKEAN ISOTROPIC MATERIAL

From our previous observations of Hookean isotropic behavior in simple tension (see Figure 4.8), we note that the work done (per unit volume) has the form

$$u = \frac{\sigma_x^2}{2E} \tag{4.34}$$

where σ_x is the only nonzero component of stress. Since $\sigma_x = E\varepsilon_x$ in simple tension, the Equation (4.34) has the alternative form

$$u = \frac{E}{2}\varepsilon_x^2 \tag{4.35}$$

In like manner, the work done under simple shear of the HI material has a similar form:

$$u = \frac{G}{2}\gamma_{xy}^2 \tag{4.36}$$

where γ_{xy} is the only nonzero component of strain. Forms (4.35) and (4.36) represent the strain energy (per unit volume) of the HI material in simple extension and in simple shear, respectively. These represent potentials of stored energy.

In general, the material of a deformed body experiences simultaneously the six components of strain. Then, an increment of work (per unit volume) of the material has the form of (4.31):

$$\begin{aligned} dw = {} & \sigma_x d\varepsilon_x + \sigma_y d\varepsilon_y + \sigma_z d\varepsilon_z \\ & + \tau_{xy} d\gamma_{xy} + \tau_{xz} d\gamma_{xz} + \tau_{yz} d\gamma_{yz} \end{aligned} \tag{4.37}$$

The deformed state is fully defined by the six components of strain. Consequently, the strain energy, in general, is a function of the six strain components[5]:

$$u(\varepsilon_x, \varepsilon_y, \varepsilon_z, \gamma_{xy}, \gamma_{xz}, \gamma_{yz})$$

The work done upon an elastic body is a conservative process; that work is manifested in the change of the function u. It is a "potential," and the incremental change is the differential of (4.32):

$$\begin{aligned} du = {} & \frac{\partial u}{\partial \varepsilon_x} d\varepsilon_x + \frac{\partial u}{\partial \varepsilon_y} d\varepsilon_y + \frac{\partial u}{\partial \varepsilon_z} d\varepsilon_z \\ & + \frac{\partial u}{\partial \gamma_{xy}} d\gamma_{xy} + \frac{\partial u}{\partial \gamma_{xz}} d\gamma_{xz} + \frac{\partial u}{\partial \gamma_{yz}} d\gamma_{yz} \end{aligned} \tag{4.38}$$

Now, $dw = du$ for arbitrary changes; for instance, the equality must hold if any one of the strains is changed. It follows that

$$\begin{aligned} \sigma_x &= \frac{\partial u}{\partial \varepsilon_x}, \quad \sigma_y = \frac{\partial u}{\partial \varepsilon_y}, \quad \sigma_z = \frac{\partial u}{\partial \varepsilon_z} \\ \tau_{xy} &= \frac{\partial u}{\partial \gamma_{xy}}, \quad \tau_{xz} = \frac{\partial u}{\partial \gamma_{xz}}, \quad \tau_{yz} = \frac{\partial u}{\partial \gamma_{yz}} \end{aligned} \tag{4.39}$$

If the material is Hookean, then each of the Equations (4.39) is a *linear* homogeneous form. Such form can follow only if the potential u is a homogeneous quadratic form in the strain components. If the material is *isotropic*, then the shear and extensional strains are uncoupled, according to Equations (4.21) and (4.22); also, isotropy demands that the three extensional strains enter in like manner and that the three shear strains enter in like manner. The form of u for the HI material must be [6]

$$u = \frac{1}{2}A(\varepsilon_x^2 + \varepsilon_y^2 + \varepsilon_z^2) + B(\varepsilon_x\varepsilon_y + \varepsilon_x\varepsilon_z + \varepsilon_y\varepsilon_z) + \frac{1}{2}C(\gamma_{xy}^2 + \gamma_{xz}^2 + \gamma_{yz}^2) \tag{4.40}$$

According to Equations (4.39) and (4.40),

$$\sigma_x = A\varepsilon_x + B(\varepsilon_y + \varepsilon_z), \quad \tau_{xy} = C\gamma_{xy}$$

Comparing these with Equations (4.21) and (4.22), we conclude that

$$A = \frac{(1-\nu)E}{(1-2\nu)(1+\nu)}, \quad B = \frac{\nu E}{(1-2\nu)(1+\nu)}, \quad C = G = \frac{E}{2(1+\nu)}$$

and

$$\begin{aligned} u = {} & \frac{E}{2(1-2\nu)(1+\nu)}[(1-\nu)(\varepsilon_x^2 + \varepsilon_y^2 + \varepsilon_z^2) + 2\nu(\varepsilon_x\varepsilon_y + \varepsilon_x\varepsilon_z + \varepsilon_y\varepsilon_z)] \\ & + \frac{G}{2}(\gamma_{xy}^2 + \gamma_{xz}^2 + \gamma_{yz}^2) \end{aligned} \tag{4.41a}$$

This is the general form of the potential for the HI material. In problems of thin bodies, beams, plates, or shells, the normal stresses in one direction (z) are negligible; then we can set $\sigma_z = \tau_{zx} = \tau_{xy} = 0$. It follows that

$$\varepsilon_z = -\frac{\nu}{1-\nu}(\varepsilon_x + \varepsilon_y), \; \gamma_{xz} = \gamma_{yz} = 0$$

Now the internal (strain) energy for this state of "plane stress" takes the form:

$$u = \frac{E}{2(1-\nu^2)}(\varepsilon_x^2 + \varepsilon_y^2 + 2\nu\varepsilon_x\varepsilon_y) + \frac{E}{4(1+\nu)}\gamma_{xy}^2 \tag{4.41b}$$

If only $\sigma_z = 0$ ($\tau_{xz} \neq 0, \tau_{yz} \neq 0$), then we have the additional "transverse" shear energy:

$$u_s = \frac{E}{4(1+\nu)}(\gamma_{xz}^2 + \gamma_{yz}^2)$$

4.12 HOOKEAN ANISOTROPIC AND ORTHOTROPIC BEHAVIOR

Many common materials, notably wood, exhibit directional properties.[7] Modern *composites* are often *designed* to possess specific anisotropy: Glass, graphite, or kevlar fibers are oriented in matrices of various plastics (e.g., epoxy) to provide greater strength and stiffness in the direction(s) of their orientation (s). Usually, the fibers are relatively small and are so uniformly dispersed that approximation as a continuum is acceptable. The mechanisms of failure are varied and complex; debonding of the constituents, crack propagation, and fracture all play roles. Subsequent to such failure, we can often describe the deformational behavior as linear and elastic, *but* anisotropic, that is, Hookean anisotropic (HA) with a generalized Hooke's law expressed in the form

$$\begin{bmatrix} \sigma_x \\ \sigma_y \\ \sigma_z \\ \tau_{yz} \\ \tau_{xz} \\ \tau_{xy} \end{bmatrix} = \begin{bmatrix} C_{11} & C_{12} & C_{13} & C_{14} & C_{15} & C_{16} \\ C_{12} & C_{22} & C_{23} & C_{24} & C_{25} & C_{26} \\ C_{13} & C_{23} & C_{33} & C_{34} & C_{35} & C_{36} \\ C_{14} & C_{24} & C_{34} & C_{44} & C_{45} & C_{46} \\ C_{15} & C_{25} & C_{35} & C_{45} & C_{55} & C_{56} \\ C_{16} & C_{26} & C_{36} & C_{46} & C_{56} & C_{66} \end{bmatrix} \begin{bmatrix} \varepsilon_x \\ \varepsilon_y \\ \varepsilon_z \\ \gamma_{yz} \\ \gamma_{xz} \\ \gamma_{xy} \end{bmatrix}$$

where C_{ij} $(i, j = 1, 2, \ldots, 6)$ are 21 independent coefficients.

In Section 4.10, we showed that elastic behavior is characterized by the existence of a strain-energy density (energy per unit volume), which (in the absence of thermal effects) is a function (u) of the six components of strain. Moreover, the stress-strain relations are derivable from that function according to Equation (4.33). In a theory of *small* (e.g. $\varepsilon \ll 1$) strain and *small* rotation, the components of stress (σ, τ) and strain (ε, γ) are approximately those of the linear theory $(\sigma_x, \tau_{xy}, \tau_{xz}, \sigma_y, \tau_{yz}, \sigma_z; \varepsilon_x, \frac{1}{2}\gamma_{xy}, \frac{1}{2}\gamma_{xz}, \varepsilon_y, \frac{1}{2}\gamma_{yz}, \varepsilon_z)$ (See Section 4.10). The energy function is a homogeneous quadratic form of the six strain components, given by:

$$
\begin{aligned}
u = \frac{1}{2}C_{11}\varepsilon_x^2 + C_{12}\varepsilon_x\varepsilon_y \quad &+ C_{13}\varepsilon_x\varepsilon_z \quad + \underline{C_{14}\varepsilon_x\gamma_{yz}} \quad + \underline{C_{15}\varepsilon_x\gamma_{xz}} + \underline{C_{16}\varepsilon_x\gamma_{xy}} \\
+ \frac{1}{2}C_{22}\varepsilon_y^2 \quad &+ C_{23}\varepsilon_y\varepsilon_z \quad + \underline{C_{24}\varepsilon_y\gamma_{yz}} \quad + \underline{C_{25}\varepsilon_y\gamma_{xz}} + \underline{C_{26}\varepsilon_y\gamma_{xy}} \\
&+ \frac{1}{2}C_{33}\varepsilon_z^2 \quad + \underline{C_{34}\varepsilon_z\gamma_{yz}} \quad + \underline{C_{35}\varepsilon_z\gamma_{xz}} + \underline{C_{36}\varepsilon_z\gamma_{xy}} \\
&\qquad + \frac{1}{2}C_{44}\gamma_{yz}^2 \quad + \underline{C_{45}\gamma_{yz}\gamma_{xz}} + \underline{C_{46}\gamma_{yz}\gamma_{xy}} \\
&\qquad\qquad + \frac{1}{2}C_{55}\gamma_{xz}^2 + \underline{C_{56}\gamma_{xz}\gamma_{xy}} \\
&\qquad\qquad\qquad + \frac{1}{2}C_{66}\gamma_{xy}^2
\end{aligned}
$$

The reader can verify that the 21 terms include all products; hence, the most general Hookean material is characterized by 21 elastic properties (C_{ij}). *If* the Hookean anisotropic (HA) material is also homogeneous (HA-HO), *then* these properties are constants. Special circumstances occur frequently in practice. Often a thin shell-like portion of an airframe, or a pressure vessel, is formed of laminations such that each layer, or laminate, contains parallel fibers. The fibers provide the strength and stiffness in the directions tangent to the shell; properties in the normal direction are distinctly different. If the fibers of successive laminations are orthogonal, then the material acquires the properties of *orthotropy*. Specifically, if the fibers lie in the orthogonal directions (x, y), then the material is said to be orthotropic with respect to the three directions (x, y, z). The physical properties of such a material are such that it appears (mechanically or structurally) the same when viewed from positive or negative positions on a line of orthotropy (x, y, or z). In other words, the mathematical description of the material (HA-HO) must be unchanged if x and $-x$ (y or $-y$, z or $-z$) are interchanged. We note that changing x to $-x$ means that the shear strains γ_{xy} and γ_{xz} change sign. The quadratic u would be changed if terms containing a multiplier γ_{xy} or γ_{xz} were present; such terms must be absent. The same argument applies to the other directions of orthotropy (y and z). It follows that every term, which contains one such multiplier (γ_{xy}, γ_{xz}, γ_{yz}) must vanish if the material is orthotropic with respect to those directions (x, y, z). The underlined terms in the quadratic u do not appear if the behavior is orthotropic. The orthotropic material is characterized by nine *elastic* coefficients (C_{11}, C_{12}, C_{13}, C_{22}, C_{23}, C_{33}, C_{44}, C_{55}, C_{66}).

The interested reader can explore other special forms of anisotropy in numerous books on elasticity.[8]

4.13 INELASTIC BEHAVIOR VERSUS NONLINEAR ELASTICITY

In general, nonlinear stress-strain relationships cause mathematical difficulties in the solution of engineering problems. The occurrence of plastic deformation introduces further difficulty in all but

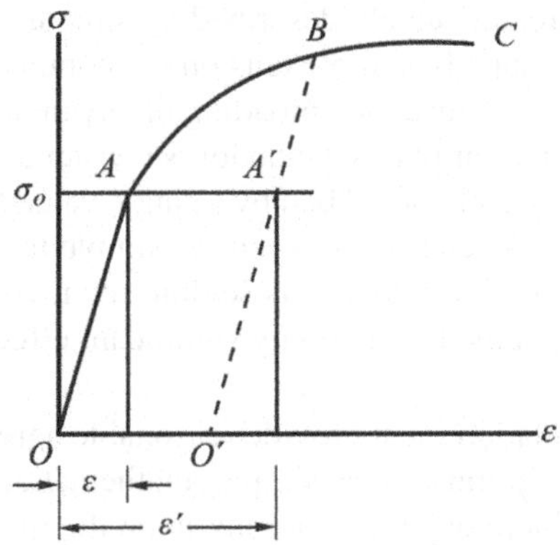

Figure 4.21

the simplest situations. In either case, the aforementioned principle of superposition is no longer applicable.

It is important to emphasize the essential distinction between elastic and inelastic deformations. The former can be described mathematically via stress-strain relations, linear or nonlinear. Those relations can be derived from a strain energy, a potential u, according to Equation (4.33). Work expended in the elastic deformations is stored as recoverable energy. Conversely, work spent upon inelastic deformations is dissipated; it is irrecoverable. The process is nonconservative. Moreover, every small increment of inelastic deformation alters the material. Consequently, most materials respond differently to each successive increment of inelastic strain. In general, inelastic behavior can only be described via relations between incremental strains, or rates, and stresses. Inelastic strains, which are independent of time, are often termed plastic strains; those which depend upon time, e.g. rate of occurrence, are identified as viscoelastic or viscoplastic strains. The phenomena of creep and relaxation are manifestations of viscoelasticity.

It is not our purpose here to formulate general stress-strain relations for nonlinearly elastic or plastic behaviors. However, it is well to understand the essential difference between nonlinear elasticity and plasticity. A simple example serves to illustrate this distinction. Suppose the stress-strain relation for a certain material in simple tension is that shown in Figure 4.21. If the material is perfectly elastic, its behavior is described by the curve *OABC*. In this case, the stress is uniquely determined by the strain (and vice versa), irrespective of the previous deformation (unloading retraces the loading curve). On the other hand, suppose that the material is not elastic. Then plastic deformation occurs during loading to *B*, and the unloading curve, BO', does not retrace the original loading curve. Consequently, the strain is not uniquely determined by the stress. For example, when the material is subjected to the tensile stress σ_0 shown in Figure 4.21, the extensional strain is ε if the loading has followed the curve *OA* and ε' if the loading has followed the curve $OABA'$. Thus, a plastic deformation depends upon the entire loading history. Of course, for combined states of stress, the loading history may be much more complex, and the stress-strain relations are further complicated.

In the special case of a simple stress, either tension, compression, or shear, there is no essential difference between elastic and inelastic behavior, *so long as there is no unloading*. For example, in Figure 4.21, if the material experiences no unloading, the stress-strain relation is represented in its entirety by the curve *OABC*. In this special circumstance only, the plastic stress-strain relation is expressed by the equation of a curve.

Many important problems are adequately described by simple states of stress. We will restrict our present discussion to these, namely, simple tension (or compression) and simple shear. Also, conditions of static loading seldom cause the unloading of any material element. Under these conditions, the mathematical treatment of plastic behavior is the same as nonlinearly elastic behavior, and the material behavior is adequately described by a single valued relation between the two variables, σ_x and ε_x (or τ_{xy} and γ_{xy}). A relation between the simple tensile (or compressive) stress σ_x and the transverse extensional strains ε_y and ε_z is seldom required. In the case of a simple shear stress τ_{xy}. It is usually safe to assume that the only significant effect is the associated shear strain γ_{xy}.

A satisfactory stress-strain relation must provide reasonable agreement with experimental data. At the same time, the relationship must be as simple, mathematically, as possible. The choice of a relation in any particular problem depends not only upon the resistive character of the material but also upon the mathematical complexity of the problem. Very elementary problems may permit refinements not possible in more complex situations. In the following discussion, we will consider but a few of the many possible stress-strain relations for the simple state of stress.

4.13.1 LINEARLY ELASTIC-IDEALLY PLASTIC MATERIAL

Figure 4.22a illustrates a stress-strain relation that is often used as an idealization for certain metals, notably low-carbon (structural) steel. The relation describes a linearly elastic behavior (line OA) to the yield stress σ_0, followed by plastic deformation at the constant stress σ_0. We refer to the behavior illustrated in Figure 4.22a as *ideally plastic or perfectly plastic*. It is a reasonable approximation for mild steel so long as the deformations are not too severe. For relatively large plastic strains (say 0.02), mild steel experiences strain hardening, and the actual $\sigma - \varepsilon$ curve rises.

Upon unloading, the behavior is linear and elastic; that is, unloading from point A' follows the straight line $A'O'$, which is parallel to OA. At the yield stress, the elastic strain is always

$$\varepsilon^E = \sigma_0/E$$

whereas the plastic strain ε_P can have any value.

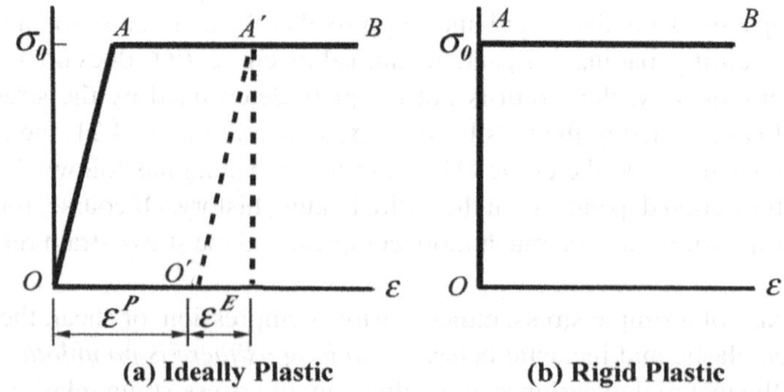

Figure 4.22

4.13.2 RIGID-PLASTIC MATERIAL

Sometimes plastic deformations are so great that elastic deformations are negligible. An idealization of this kind is referred to as a *rigid-plastic behavior*. It is applicable to processes of forming ductile materials, extrusion and drawing of wires and seamless tubes, or stamping parts for the bodies of automobiles. The rigid-plastic stress-strain relation is illustrated in Figure 4.22b. Such a material experiences no significant deformation until the stress reaches the yield point. If the material is rigid plastic, it can then experience unrestricted plastic deformation. However, a material element *within* a body does not necessarily experience unlimited strain when the yield stress is reached, because it may well be that the element is adjacent to material that has not yet yielded. If so, the deformation of the plastic element is constrained by the adjacent rigid material.

4.13.3 LINEARLY ELASTIC-LINEAR HARDENING MATERIAL

Many common metals experience strain hardening during plastic deformation. That is, the stress-strain curve has a positive (nonzero) slope. A simple approximation of this is illustrated in Figure 4.23a.

This stress-strain relation is linearly elastic (line OA) to the yield stress σ_0. Subsequent deformations are approximated by another linear relation, the straight line AB.Now suppose that unloading is also linear, so that the unloading curve from point A' traces the straight line $A'O'$, which is parallel to OA. Then the elastic strain is always

$$\varepsilon^E = \sigma/E$$

In addition, there is a plastic strain when the stress exceeds σ_0. This, too, is a linear function of the stress, so that

$$\varepsilon^P = \frac{\sigma - \sigma_0}{kE} \quad (\sigma > \sigma_0)$$

where k is a positive number ($0 < k < 1$). Hence, the actual strain (when $\sigma > \sigma_0$) is

$$\varepsilon = \varepsilon^E + \varepsilon^P = \frac{(1+k)\sigma}{kE} - \frac{\sigma_0}{kE} \quad (\sigma > \sigma_0) \tag{4.42}$$

This equation is the equation of the straight line AB of Figure 4.23a. We refer to this idealization as a *linear-hardening* material.

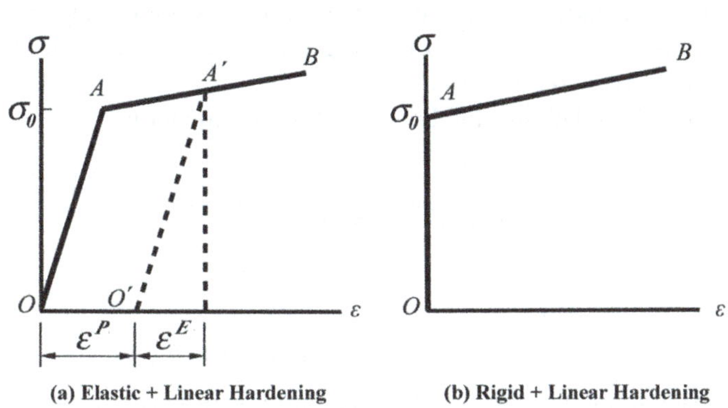

Figure 4.23

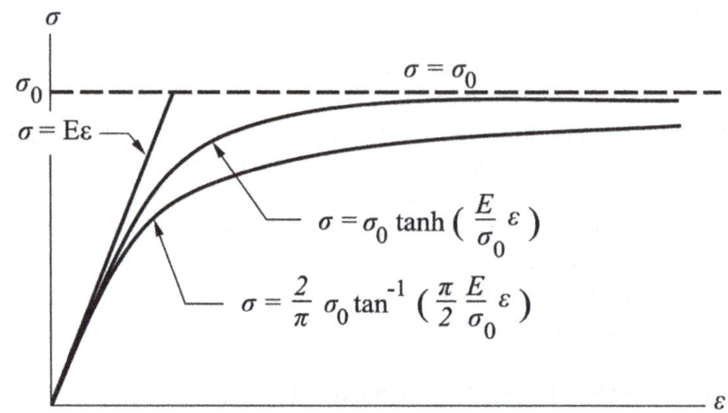

Figure 4.24

The ideal plastic behavior is a special case of the linear-hardening stress-strain relation. It corresponds to $k = 0$.

4.13.4 RIGID LINEAR-HARDENING MATERIAL

As before, if the elastic strain is negligible, the *rigid linear-hardening* idealization of Figure 4.23b may be justified. This stress-strain relation is expressed by

$$\varepsilon = 0 \quad (\sigma < \sigma_0)$$
$$\varepsilon = \varepsilon^P = \frac{\sigma - \sigma_0}{kE} \quad (\sigma > \sigma_0)$$

4.13.5 OTHER IDEALIZATIONS

In certain instances, a smooth curve provides a more useful description of a nonlinear stress-strain relation because a single mathematical relation can then be used to describe the entire curve. For example, two relations are shown in Figure 4.24. Both curves are closely related to the ideally plastic material. In both cases, the slope at the origin is the elastic modulus E, and both curves approach the horizontal asymptote $\sigma = \sigma_0$.

Another useful strain-stress relation is the Ramberg-Osgood equation:

$$\varepsilon = \frac{\sigma}{E}\left[1 + K\left(\frac{\sigma}{\sigma_0}\right)^n\right] \quad (n > 0) \tag{4.43}$$

It follows that

$$\frac{d\sigma}{d\varepsilon} = E_T \equiv \frac{E}{1 + (n+1)K(\sigma/\sigma_0)^n}$$

For all $n > 0$, the "tangent modulus" $E_T = E$ at the onset; that is, $\sigma = \varepsilon = 0$, as shown in Figure 4.25. Also, in the limit as $n \to \infty$, the relation approaches the ideally plastic approximation as depicted in Figure 4.25.

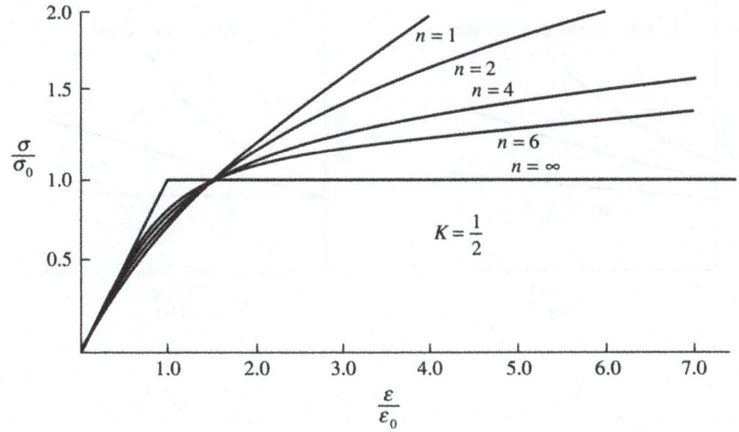

Figure 4.25

4.14 TIME-DEPENDENT EFFECTS

4.14.1 CREEP

In this section, we intend only to provide a general description of responses associated with the duration of loadings. The mechanisms of these time-independent effects are complex and depend on the micro-structures of the materials.

The term *creep* denotes a continuing deformation under constant load. Some materials (lead, pure aluminum, wood, plastics) exhibit considerable creep at ordinary temperatures. Most metals creep significantly only at elevated temperatures. For this reason, creep is a serious problem in high-temperature applications, such as boiler tubes or turbine blades. Often, creep deformations proceed very slowly, so that their effects become significant only after long periods of time.

In a standard creep test, the specimen is maintained under constant tensile load and at constant temperature while the strain and the corresponding time are recorded. The results of a creep test are shown in a plot of strain versus time. Typical curves are shown in Figure 4.26. The slope of a curve represents the creep rate (strain per unit time). The creep rate and time to fracture are important quantities derived from a test. In general, an increase in temperature or stress causes a greater creep rate and a reduction in the time to fracture. These effects are illustrated in Figure 4.26a and b.

In tests of long duration, the strain-time curve is nearly linear during most of the test; that is, the deformation proceeds at a steady rate. Immediately before fracture, the rate of deformation increases rapidly. In practice, creep may occur very slowly over long periods of time, often many years. The designer seeks to avoid excessive creep deformation and, above all, a high creep rate, which may lead to fracture. To do this the engineer must rely on extensive experimental data from creep tests, tests that are often of long duration. An alternative to overly conservative design is the common practice of periodically replacing parts to prevent significant creep damage.

4.14.2 RELAXATION

In certain instances the load required to maintain a given deformation decreases with time. For example, if a bar were extended and then the length held fixed, the tensile stress might diminish as

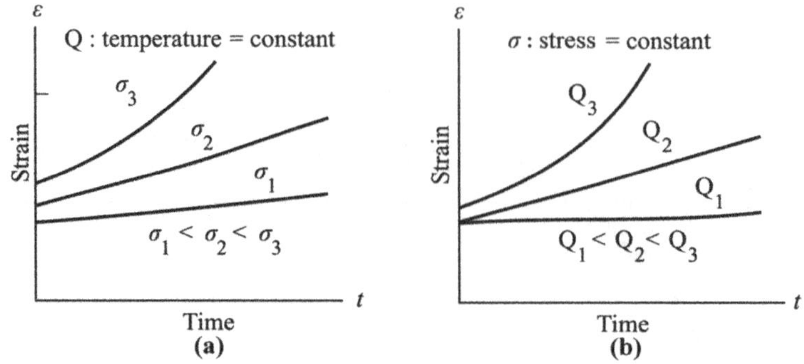

Figure 4.26

shown in Figure 4.27. This behavior is known as *relaxation*. Typically, materials that exhibit creep also exhibit relaxation; the phenomena are interrelated. In one test (creep) the stress is maintained constant, and in the other (relaxation) the strain is held constant. In most situations, both effects are present and both are dependent on temperature.

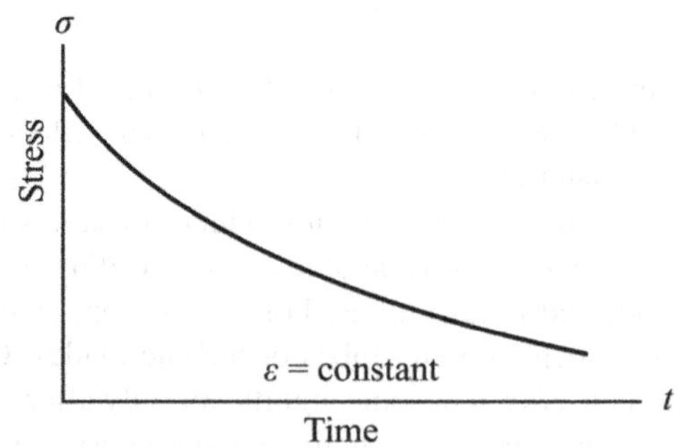

Figure 4.27

4.14.3 IMPACT LOADS

In many cases, materials respond to rapidly applied loads differently than they do to slowly applied loads. The complete explanation for this is not yet available but is thought to have its origins in certain microscopic irregularities of the material. However, the effects of rapid loading may be assessed, at least qualitatively, by testing materials under impact loads. Since it is virtually impossible to determine the magnitudes of very rapidly applied forces, measurements are made of the energy imparted to the specimen. The energy required to fracture a specimen in a standardized test is then used as a measure of resistance to impact loading.

Embrittlement of materials at low temperatures is vividly illustrated by impact tests. Tests are conducted at various temperatures, and energy-to-fracture versus temperature is plotted. The curve shown in Figure 4.28 is typical of some metals. In this case, a drastic reduction in impact resistance occurs over a very narrow temperature range.

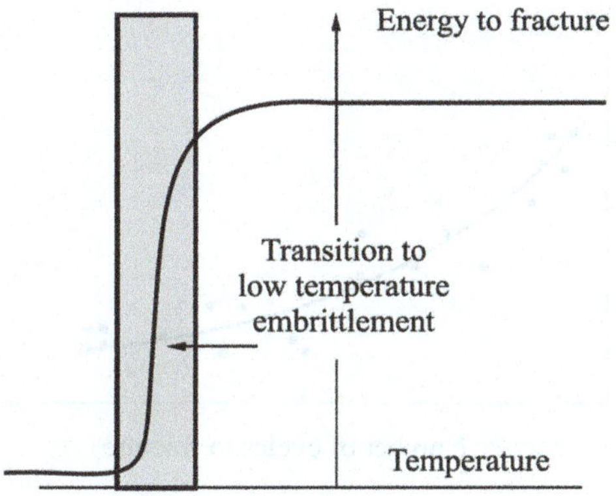

Figure 4.28

4.14.4 REPEATED LOADING-FATIGUE

Many repetitions of load can cause fracture, although the stress is much less than the ultimate strength of the material under static conditions. Such fractures are termed *fatigue failures*. They are a common concern in metal parts of reciprocating or rotating machinery. Fatigue failures are especially troublesome because they occur without warning and are often unpredictable. Typically, there is little evidence of inelastic deformation at the site of a fatigue fracture; it appears much as a brittle fracture.

A fatigue fracture originates with a minute crack at a point of high stress (See Section 4.19). Often, the crack starts at a small surface scratch, a toolmark, a notch or keyway, a corroded region or an imperfection in the material, usually a metal. The origins of a fatigue crack, the progressive damage and propagation, are microstructural phenomena. Here, we describe the phenomena as observed in simple testing procedures. The interested reader can find specific information in numerous texts, which are devoted to this subject.[9]

In the usual fatigue test, a specimen is subjected to alternating tensile and compressive stresses (of equal magnitude). The specimen is either subjected to this stress reversal until fracture, or the test is terminated after a preselected number of cycles. Many tests at varying stress levels are required for complete data. The result of each test is plotted on a diagram, with the magnitude of the alternating stress as the ordinate and the number of cycles to fracture as the abscissa. Usually, the plotted points are widely scattered, as illustrated in Figure 4.29. A curve drawn through these points represents only mean values of all experimental data. Such a curve is termed an S-N curve (Stress versus Number of cycles).

4.14.5 FATIGUE LIMIT

The *S-N* curve for some materials approaches a horizontal line for a very large number of cycles. In this event, the material is said to have a fatigue limit, a *stress magnitude below which most specimens will withstand an indefinite number of repetitions*. Certain materials, notably steels, have a fatigue limit. Most nonferrous metals and plastics have *S-N* curves similar to the aluminum alloy of Figure 4.30. These materials may fracture at very low stresses if the stress is repeated often enough.

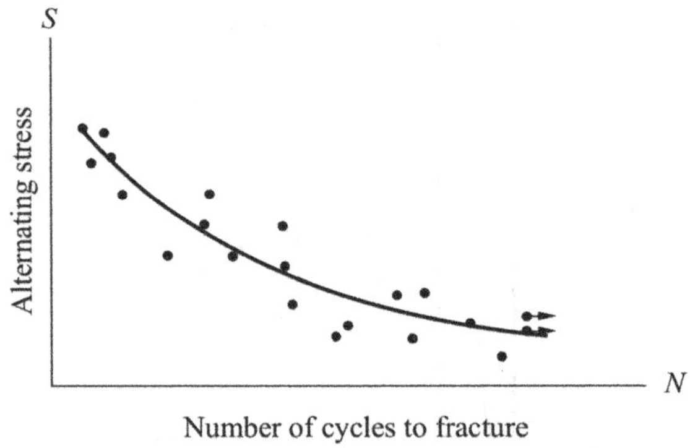

Figure 4.29

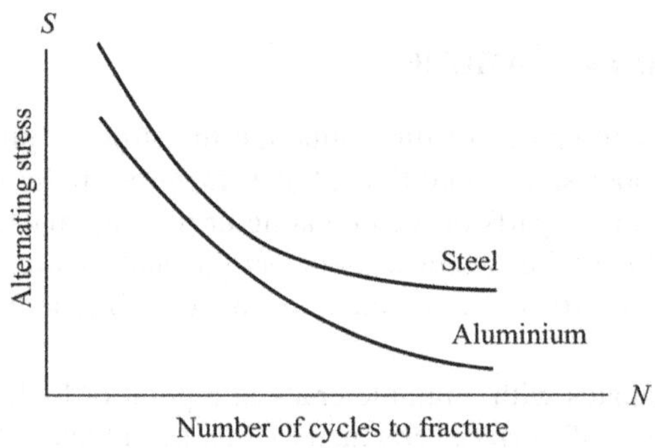

Figure 4.30

4.14.6 FATIGUE STRENGTH

The number of repetitions of stress before fatigue fracture is termed the *fatigue life*. The *fatigue strength* is defined in terms of a specified fatigue life. For any preselected fatigue life, the *S-N* diagram determines a stress magnitude, the fatigue strength of the material (for the prescribed life). Because of the unpredictability of fatigue failure, we can only say that the material will *probably* withstand this alternating stress for its prescribed fatigue life.

4.14.7 EFFECT OF MEAN STRESS

The presence of a static load, in addition to a repeated load, accelerates the growth of fatigue cracks. This effect is evaluated by tests in which alternating stress is superimposed on a static (mean) stress. The effect of various mean stress levels is illustrated by the *S-N* curves of Figure 4.31. In general, the fatigue strength or the fatigue life is reduced. For a given fatigue life, the influence of mean stress can be shown graphically by plotting the magnitude of the alternating stress versus the mean stress as in Figure 4.32. This graphical representation is known as a *Soderberg diagram*. As the mean stress level approaches the ultimate strength, the magnitude of the alternating stress diminishes to zero. At the opposite extreme, as the mean stress level approaches zero, the magnitude of the alternating stress approaches the fatigue strength for alternating stress alone.

In the absence of sufficient data, a reasonable approximation (usually conservative) is obtained by constructing a straight line through these two points.

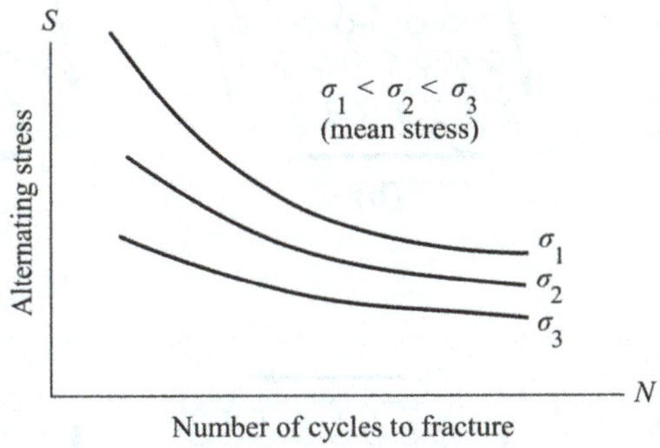

Figure 4.31

In view of the wide scatter in experimental data, it is not possible to predict fatigue strength accurately, nor is a stress below the fatigue limit a guarantee against fatigue failure. We can, at best, assess the probability of failure at a given stress. Furthermore, in practical applications, it is difficult and often impossible to predict the magnitudes and frequencies of repeated loads. For example, the loads on the axle of an automobile depend on the road conditions, as well as the speed and the dynamic response of the loaded vehicle. The prediction of fatigue failure is a statistical matter. The designer can only utilize experimental data to reduce the probability of failure to an acceptable limit.

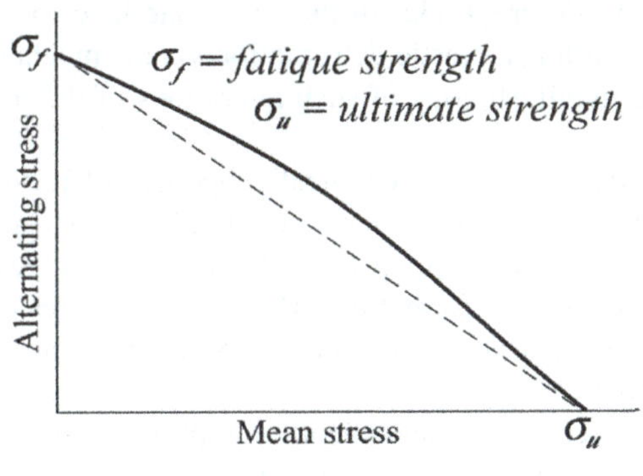

Figure 4.32

4.15 MECHANISM OF DEFORMATION IN METALS

To understand the mechanical behavior of materials, it is necessary to examine the behavior of basic constituents. In metals, the individual crystal is a basic structural element. Each crystal is composed of atoms which are bound together by interacting forces of an electrodynamic nature. In a perfectly formed crystal, the atoms are arranged in a regular pattern termed the *crystal lattice*. The simplified (two-dimensional) arrangement of Figure 4.33a will serve for purposes of illustration. During the application of force, we expect the crystal to deform as shown, for example,

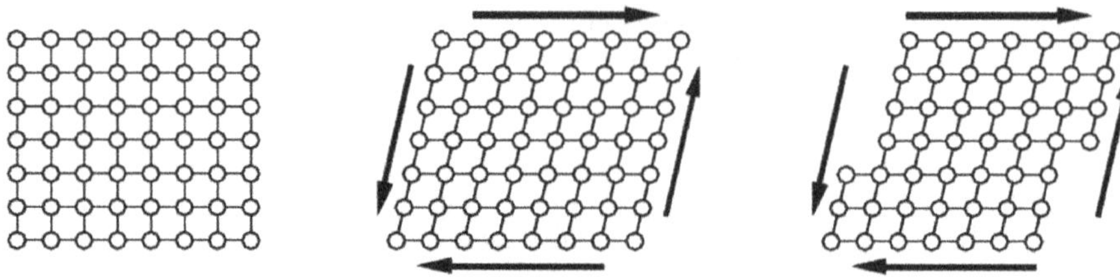

Figure 4.33

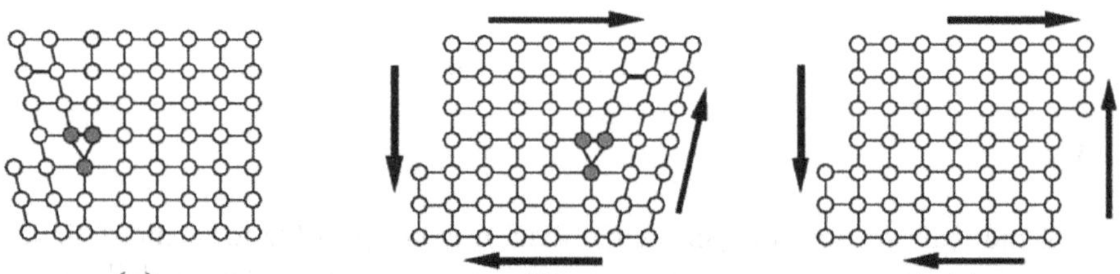

Figure 4.34

in Figure 4.33b. In this event, the pattern is slightly altered, but the original configuration is restored upon removal of the forces. This is an elastic deformation of the crystal. On the other hand, sufficient force may cause a permanent disarrangement of the crystal lattice. Such a permanent disfiguration is illustrated by Figure 4.33c. In this case, one layer of atoms *slips* over the adjacent layer. Thus, the crystal suffers a plastic deformation via the mechanism of slip. The observed deformations of metals are manifestations of such alterations of the many crystals that compose them.

It is reasonable to suppose that the mechanical properties of a metal might be obtained by accurate analyses of the forces and displacements in individual crystals. To a limited extent, this is true. The elastic moduli of metals have been predicted with a fair degree of accuracy by analyses of interatomic forces that resist elastic deformations. However, estimates of the yield strength, which are based on slip in perfect crystals, are far in excess of observed values. Apparently, yielding does not occur in quite this manner.

The strength of metals is very much dependent on imperfections in crystals. Of particular importance are disarrangements of the crystal lattice, known as *dislocations*. One type of dislocation (edge dislocation) is illustrated by Figure 4.34a. The forces necessary to cause a movement of the dislocation are much less than those required to effect slip in the perfect crystal. It is necessary only to propagate the dislocation as shown in Figure 4.34b and c.

By contrast, the shear force in Figure 4.33 must simultaneously overcome the bonds between many atoms in order to cause slip in the perfect crystal. This mechanism has been likened to the movement of a wrinkle in a rug, shown in Figure 4.35. The rug is easily displaced along the floor by moving a wrinkle from one end to the other, whereas much greater force is needed to slide the entire rug. Each metal crystal has a certain inherent resistance to plastic deformation, depending on the type of crystal lattice and the strength of interatomic bonds. In actual metals, many dislocations and impurities are present in each crystal. These impurities may take the form of small particles or foreign atoms within crystals. During deformation, dislocations move, join together, and even generate new dislocations. As the plastic deformation progresses, the dislocations may interfere with one another and so impede further movement. Foreign particles (or atoms) and crystal boundaries also block the movement of dislocations. As the motions of the dislocations are arrested, the resistance to plastic flow increases. The metal is said to *strain harden*. As the temperature increases, the mobility of atoms increases. Foreign atoms tend to diffuse, and dislocations tend to migrate around barriers. The obstacles, which impede dislocation movement, are dissipated by heating.

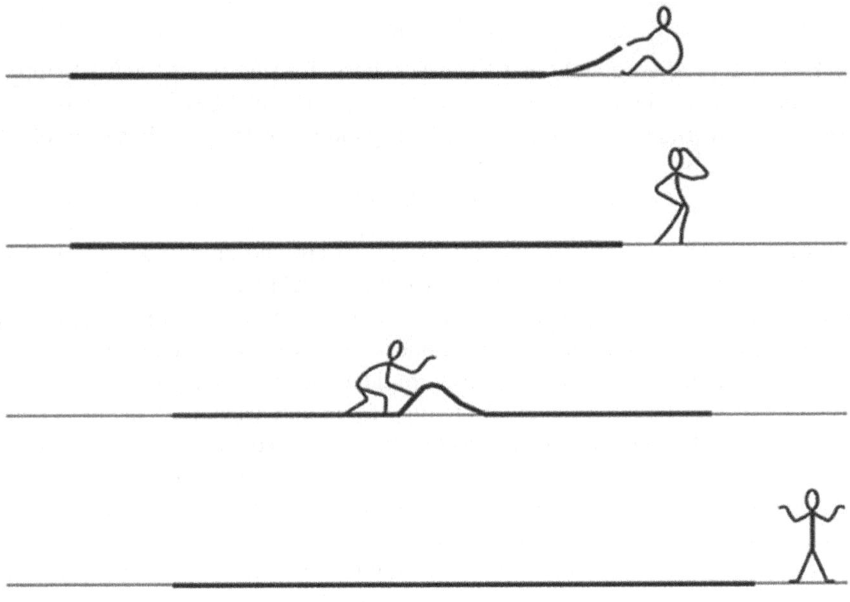

Figure 4.35

Thus, at elevated temperatures, the metal may continue to deform plastically under constant force. This is the phenomenon called *creep*. If, on the other hand, a metal is cooled, the activity of atoms is decreased. The movement of dislocations becomes more difficult. Thus, we observe that the normally ductile metals exhibit brittle behavior in very cold environments. Dislocations, as well as other crystal defects, play a major role in the mechanism of fracture. Since the movement of dislocations is effected by relatively little force, repeated loading can cause the movement of dislocations, even though stresses are well below the apparent yield stress. It would therefore appear that fatigue fractures originate with the movement of dislocations. Indeed, it appears that the elastic constants are the only mechanical properties that are not greatly influenced by these minute defects in the crystallographic structure.

4.16 YIELDING AND FRACTURE CRITERIA

The strength of a material is characterized by its resistance to plastic deformation and/or its resistance to fracture. Unfortunately, the mechanisms of yielding and fracture are very complex. These events originate and propagate at the microscopic, or macroscopic, levels: For example, the yielding and fracture of metals are interactions of the atomic and crystalline structures; moreover, these mechanisms involve imperfections, which are often imperceptible. Consequently, our attention is focused on the phenomenological descriptions and theories.

Previously, we noted the resistance of a material to simple tension. Its behavior is predominantly elastic as long as the stress does not exceed the yield stress. Greater stress is accompanied by plastic deformation. Thus, for simple tension, a mathematical statement of the condition for incipient plastic flow is

$$\sigma = \sigma_0 \tag{4.44}$$

If the tensile stress reaches the ultimate strength σ_μ, fracture is imminent. The condition for fracture is thus

$$\sigma = \sigma_\mu \tag{4.45}$$

Equations (4.44) and (4.45) are yield and fracture criteria for this simple state of stress. Neither criterion need be expressed in terms of stress but may equally well be written in terms of strain or even in terms of the work density required to effect yielding or fracture. For example, Equation (4.44) may be replaced by

$$\varepsilon = \varepsilon_0, \quad U = U_0$$

where ε_0 and U_0 denote, respectively, the extensional strain and energy density at the onset of plastic flow (in simple tension), i.e. the criterion for initial yielding.

In some machine parts or precision instruments, any yielding of the material constitutes failure. If plastic deformation does not destroy the usefulness of a structure, fracture may lead to failure. In the case of brittle materials, failure implies complete fracture. In any event, Equations (4.44) and (4.45) can serve as criteria to predict failure only for simple tension. In the most general case, the material is subjected to the combined action of six stresses (σ_x, σ_y, σ_z, τ_{xy}, τ_{yz}, τ_{zx}). Since the simultaneous action of all stresses will affect yielding and fracture in some manner, more general failure criteria are needed. They must take into account all possible combinations of the stresses. But it is virtually impossible to establish the criteria by testing materials under the many possible states of stress. For example, there is no means available for subjecting a material to a prescribed triaxial tension. Consequently, a *theory of failure* is needed.

A theory of failure sets forth the quantity (or quantities) that determine failure. The simplest theories ascribe failure to a single quantity. Failure is then supposed to occur when the specified quantity attains a certain critical value, which may be obtained from a simple test such as the tension test. The mathematical expression of the conditions for failure is termed the *failure criterion*. Clearly, we must distinguish between theories pertaining to yielding and those relating to fracture. If the occurrence of plastic deformation constitutes failure, an appropriate theory establishes a yield criterion. If failure is defined by fracture,the fracture criterion is established by a theory of fracture. In either case, the criterion must apply to an arbitrary state of stress. In particular, Equations (4.44) and (4.45) must be, respectively, special forms (the forms that apply to simple tension) of the general criteria for yielding and fracture.

Possibly, the simplest point of view is to regard these criteria as limiting conditions imposed on the stresses. In other words, we may regard the general yield and fracture criteria as generalizations of Equations (4.44) and (4.45). Then, from the mathematical point of view, the concepts of yield and fracture criteria are quite similar. This does not mean that the physical events associated with the two phenomena are alike, but rather that the matter of formulating the criteria is similar.

4.17 YIELD CRITERIA

Previous references to yielding pertained to simple states of stress. Then, the criterion for yielding was the stipulation of a yield stress, either σ_0 in tension or τ_0 in shear. In general, a material is subject to six components of stress; all might affect the yielding. To appreciate the problem, suppose that the material were subject only to the two components σ_x and σ_y. Imagine a test wherein different combinations of these stresses are imposed; in each case, the values at the onset of yielding are noted. Then, we might plot these values, σ_x° and σ_y°, as shown in Figure 4.36.

If the material is isotropic, then the curve intersects the respective axes at $\sigma_x = \sigma_y = \sigma_t$, the tensile yield stress, and at $\sigma_x = \sigma_y = -\sigma_c$, the compressive yield stress. In this relatively simple case, many tedious tests are needed. The resulting curve of Figure 4.36 can be approximated by an equation $f(\sigma_x, \sigma_y) = 0$, which constitutes the yield criterion. In general, a yield criterion must take the form:

$$f(\sigma_x, \sigma_y, \sigma_z, \tau_{xy}, \tau_{yz}, \tau_{zx}) = 0$$

Clearly, this is a formidable experimental and theoretical task. Fortunately, the criterion is simplified by certain attributes of common metals. Specifically, most are essentially *isotropic*. We note that any state of stress is determined by the principal stresses (σ_1, σ_2, σ_3) *and* the orientation of the principal directions (three angles would suffice). If the material is isotropic, then the orientation is irrelevant and the essential attributes of the state are the three principal stresses (σ_1, σ_2,σ_3). Now, we can imagine tests that impose various combinations of the principal stresses and note the values that initiate yielding. Such experiments would provide the data to plot a surface in the three-dimensional space (σ_1, σ_2, σ_3). That *yield* surface must enclose the origin. Progressing from the origin and following a path of loading, that is, changing the stresses (σ_1, σ_2,σ_3)- we would observe elastic behavior until our path strikes the predetermined yield surface. That surface can be approximated by a function

$$f(\sigma_1, \sigma_2, \sigma_3) = 0$$

Still, this is a hypothetical experiment that remains a formidable physical reality. Further observations and insights are needed to obtain a useful criterion.

One such observation follows from the extensive experiments of P.W. Bridgeman,[10] who conducted tests of materials under hydrostatic states. These led to the conclusion that hydrostatic

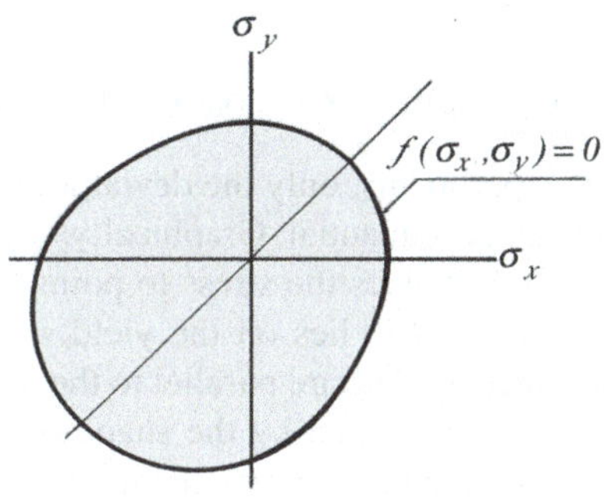

Figure 4.36

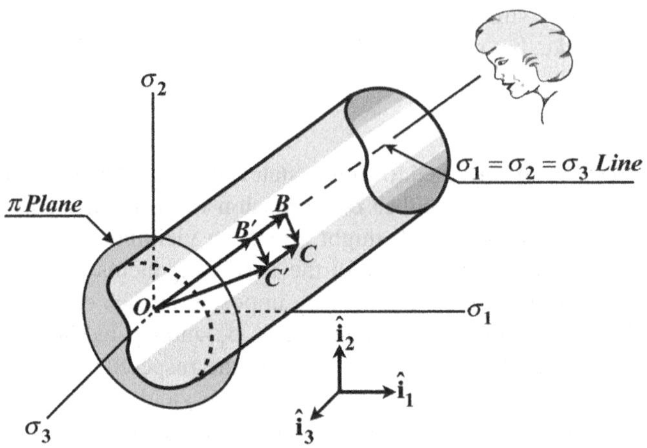

Figure 4.37

stress has little effect on yielding. Recall that a simple hydrostatic stress implies $\sigma_1 = \sigma_2 = \sigma_3$. That equation is represented by the line of Figure 4.37, which forms equal angles with the axes, σ_1, σ_2, and σ_3. If we adopt the hypothesis that a hydrostatic state does not influence yielding, then the yield surface must enclose the line $\sigma_1 = \sigma_2 = \sigma_3$.

In the three-dimensional space of Figure 4.37 (σ_1, σ_2, σ_3), any state of stress is represented by a vector:

$$\sigma = \sigma_1\hat{\mathbf{i}}_1 + \sigma_1\hat{\mathbf{i}}_2 + \sigma_3\hat{\mathbf{i}}_3 \tag{4.46}$$

If that stress reaches the surface for example, at point C' yielding is initiated. Now the vector to C', namely, OC', can be decomposed into a hydrostatic part OB' along the line $\sigma_1 = \sigma_2 = \sigma_3$ and an orthogonal component $B'C'$. The latter is called the deviatoric part. The components of the hydrostatic part are equal; they are the mean of the normal components, in this case, the mean of the principal stresses:

$$\sigma_m = \frac{\sigma_1 + \sigma_2 + \sigma_3}{3}$$

The deviatoric components (the components of vector $B'C'$ are obtained by subtracting these mean values (the components of the hydrostatic part) from the components of stress. Accordingly, the deviator $B'C'$ is the vector

$$\sigma' = (\sigma_1 - \sigma_m)\hat{\mathbf{i}}_1 + (\sigma_2 - \sigma_m)\hat{\mathbf{i}}_2 + (\sigma_3 - \sigma_m)\hat{\mathbf{i}}_3 \tag{4.47}$$

By the hypothesis, supported by experiments, only the deviator affects the yielding. Then any additional hydrostatic component is inconsequential. Graphically, if point C' lies on the yield surface in Figure 4.37. then the addition of $B'B$ takes the stress to point C *also* on the yield surface. The straight line $C'C$, parallel and equal to $B'B$, lies on the yield surface. In other words, the yield surface is cylindrical; generators, such as $C'C$, are parallel to the axis $OB'B$. The yield surface, and hence the function $f(\sigma_1, \sigma_2, \sigma_3)$, is determined by the shape of the cross section. A right cross section of the cylindrical surface is defined by a curve, the intersection of the surface and the plane (the π-plane) perpendicular to the line $\sigma_1 = \sigma_2 = \sigma_3$. The viewer in Figure 4.37 from a position on the line $\sigma_1 = \sigma_2 = \sigma_3$ sees that intersection on the "π plane." This is the yield curve as seen in

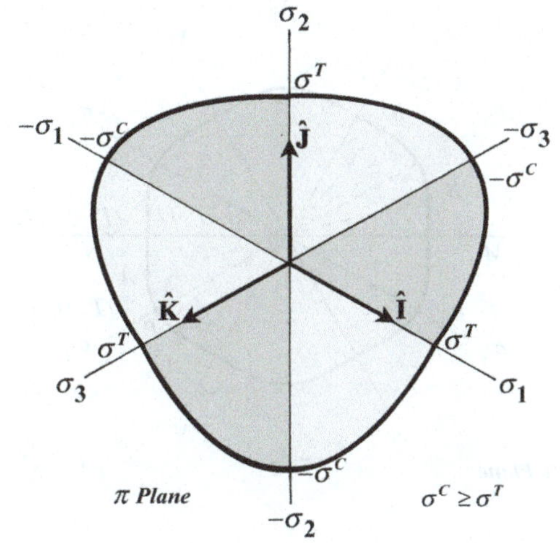

Figure 4.38

Figure 4.38; axes σ_1, σ_2 and σ_3 appear similarly foreshortened in this view. The *nonorthogonal* vectors $(\hat{\mathbf{I}}, \hat{\mathbf{J}}, \hat{\mathbf{K}})$ in the figure are expressed in terms of the orthogonal triad $(\hat{\mathbf{i}}, \hat{\mathbf{j}}, \hat{\mathbf{k}})$ as follows.

$$\hat{\mathbf{I}} = \frac{1}{\sqrt{6}}(2\hat{\mathbf{i}} - \hat{\mathbf{j}} - \hat{\mathbf{k}}),\ \hat{\mathbf{J}} = \frac{1}{\sqrt{6}}(-\hat{\mathbf{i}} + 2\hat{\mathbf{j}} - \hat{\mathbf{k}}),\ \hat{\mathbf{K}} = \frac{1}{\sqrt{6}}(-\hat{\mathbf{i}} - \hat{\mathbf{j}} + 2\hat{\mathbf{k}})$$

Since the material is presumed *isotropic*, the curve crosses each axis at σ^T, the tensile yield stress, and at -σ^C, the compressive yield stress. Because of the isotropy the elastic regions within the curve are fully defined by the 60° segments shown shaded. *If* the material yields at the same magnitudes of stress in simple tension and compression, then $\sigma^T = \sigma^C$.

If we presume that the material is isotropic and also has the same magnitudes of yield stress in tension and compression,$\sigma^T = \sigma^C$ then the yield curve has certain additional symmetries, which are shown in Figure 4.39. If point *A* at (*a*, *b*, *c*) is on the curve, then point *A'* at (*a*, *c*, *b*) is also on the curve since this constitutes only an interchange in the roles of σ_2 and σ_3, which has no effect on the isotropic materials. In other words, the curve is symmetrical about the σ_1 line and likewise about the σ_2 and σ_3 lines. If the points *A* at (*a*, *b*, *c*) and A' at (*a*, *c*, *b*) are on the curve, then the points *B* at $(-a, -b, -c)$ and B' at $(-a, -c, b)$ are also, since the yielding is affected the same if the sign of the stress is reversed. We see that the curve is also symmetrical about line *MN* and likewise about the lines $M'N'$ and $M''N''$. Indeed, the entire curve is therefore determined by one 30° segment, for example, the segment *PQ*. Note that a simple tensile test ($\sigma_1 = T$, $\sigma_2 = \sigma_3 = 0$) determines point *P*;a simple shear test ($\sigma_1 = T, \sigma_2 = 0$, and $\sigma_3 = T$) determines point *Q*. The most prevalent and simple criteria presume isotropy and similar response upon reversal of sign; for example, tension and compression are equally effective. These are the Tresca and Mises criteria.

4.17.1 TRESCA THEORY

In a paper published in 1864, H. Tresca[11] proposed that metals yield when the maximum shear stress attains a certain critical value; that is, if the shear stress at a point on any plane reaches a specified value, the material experiences plastic deformation at that point. Because the theory takes no account of the orientation of that plane, it is strictly applicable only to isotropic materials.

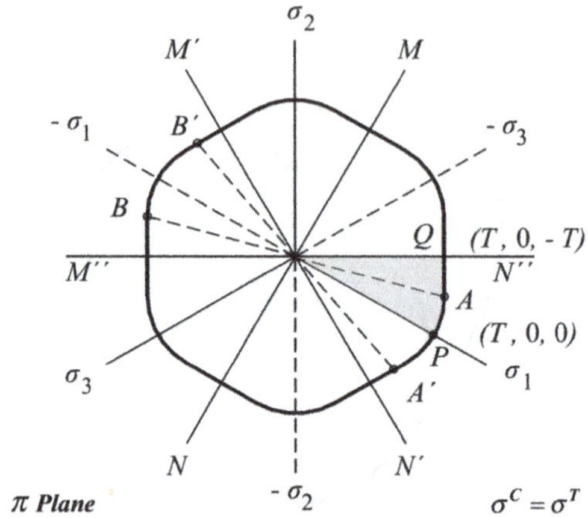

Figure 4.39

Let us consider a mathematical expression of the Tresca theory. We recall from Chapter 3 that the maximum shear stress at a point is simply half the difference between the maximum and minimum principal stresses (see Equation 3.28). Then, the criterion is expressed by one of six equations, depending on the relative magnitudes of the principal stresses:

$$\frac{1}{2}(\sigma_1 - \sigma_3) = \pm\tau_m \tag{4.48a}$$

$$\frac{1}{2}(\sigma_1 - \sigma_2) = \pm\tau_m \tag{4.48b}$$

$$\frac{1}{2}(\sigma_2 - \sigma_3) = \pm\tau_m \tag{4.48c}$$

where τ_m is the limiting value of the shear stress. This might be determined by a test in simple shear; then $\tau_m = \tau_0$, the yield stress in shear. However, in practice, the value τ_m is more readily determined by a test in simple tension, say $\sigma = \sigma_1$ ($\sigma_2 = \sigma_3 = 0$). If $\sigma = \sigma_0$ denotes the tensile stress at the onset of yielding, then by Equation (4.48),

$$\tau_m = \frac{\sigma_0}{2} \tag{4.49}$$

The Tresca criterion of maximum shear stress is expressed by equation 4.48. These are the equations of six planes, parallel to the line $\sigma_1 = \sigma_2 = \sigma_3$, which form a hexagonal cylinder. They intersect the π-plane to form the regular hexagon depicted in Figure 4.40. The criterion of Tresca presents difficulties because the six equations in (4.48a–c) are needed to express it generally. It is conveniently employed only when the relative magnitudes of the principal stresses are known. Then one of the six equations suffices. Geometrically, only one plane (one line in Figure 4.40) of the yield surface is then needed. A yield criterion that avoids the difficulty was proposed by R. von Mises in 1913.

Yielding according to the Tresca criterion is supported by an observation of the tension test on a specimen of hot-rolled steel: The advent of yielding is accompanied by the flaking of the brittle "mill scale" along the 45° lines; these lines of plastic deformation are known as "L*ü*ders' lines",[12]

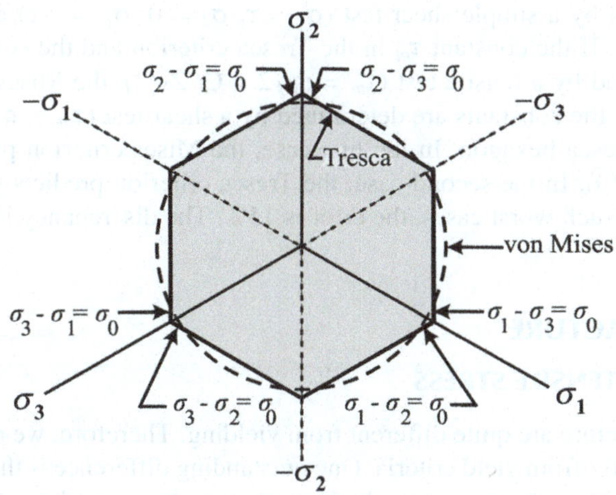

Figure 4.40

4.17.2 VON MISES THEORY

The von Mises[13] yield criterion replaces the six equations of the Tresca criterion by a single equation. In terms of principal stresses, the von Mises criterion follows

$$(\sigma_1 - \sigma_m)^2 + (\sigma_2 - \sigma_m)^2 + (\sigma_3 - \sigma_m)^2 = C^2/3 \tag{4.50}$$

Here C is an experimentally determined constant. In terms of the geometric representation, the von Mises criterion replaces the hexagonal cylinder of the Tresca criterion by a circular cylinder. Its axis is also the line $\sigma_1 = \sigma_2 = \sigma_3$.

Differences between the Tresca and von Mises criteria are evident from the curves of Figure 4.40. However, these curves may be brought into reasonable agreement by proper choice of the constants C and τ_m. From a practical point of view, the important difference lies in the mathematical expression of the criteria. Quite often the choice of a yield criterion is a matter of mathematical convenience or necessity in the solution of a given problem. Either the Tresca or von Mises criterion will provide a suitable approximation for most metals. Yielding of some materials conforms more nearly to the Tresca criterion; in others, yielding is more accurately described by the von Mises criterion. The von Mises criterion also has a physical interpretation. If the strain energy of an isotropic Hookean material (see Equation 4.41) is expressed in terms of the components of stress, it is a quadratic form; specifically,

$$u = \frac{(1+\nu)}{2E}[(\sigma_1 - \sigma_m)^2 + (\sigma_2 - \sigma_m)^2 + (\sigma_3 - \sigma_m)^2] + \frac{1}{2K}\sigma_m^2 \tag{4.51}$$

In this form, we can interpret the final term as energy associated with the dilatation and the hydrostatic part (σ_m) of the stress. The initial term is associated with distortion; it depends only on the deviatoric part of the stress (see Equation 4.47). Accordingly, the criterion has been interpreted[14] to mean that yielding is initiated when the distortional energy attains a specific limit $[(1+\nu)C^2/6E]$. In view of Equations (4.47) and (4.50), the constant C has also a geometric interpretation: The radius of the Mises circle is $R = \sqrt{\sigma' \cdot \sigma'} = C/\sqrt{3}$. Furthermore, if the constant is determined by a simple tensile test ($\sigma_1 = \sigma$, $\sigma_2 = \sigma_3 = 0$) and the yield stress is $\sigma = \sigma_0$, then $C = \sqrt{2}\sigma_0$. If the constant is determined by a simple shear test ($\sigma_1 = \tau$, $\sigma_2 = 0$, $\sigma_3 = -\tau$) and the yield stress is $\tau = \tau_0$, then $C = \sqrt{6}\tau_0$. If the constant τ_m in the Tresca criterion and the constant C in the Mises criterion are determined by a tensile test ($\tau_m = \sigma_0/2 = C/\, 2\sqrt{2}$), the Mises circle circumscribes the Tresca hexagon. If the constants are determined by a shear test ($\tau_m = \tau_0 = C/\sqrt{6}$), the Mises circle inscribes the Tresca hexagon. In the first case, the Mises criterion predicts a higher yield stress in shear ($2\tau_0/\sqrt{3}$). In the second case, the Tresca criterion predicts greater yield stress in tension ($2\sigma_T/\sqrt{3}$). In such worst cases, the error is 15%. The discrepancy is much reduced by an intermediate position.

4.18 BRITTLE FRACTURE

4.18.1 MAXIMUM TENSILE STRESS

The phenomena of fracture are quite different from yielding. Therefore, we must expect the mathematical criteria to differ from yield criteria. One outstanding difference is the effect of hydrostatic pressure; experiments have shown that the hydrostatic state has a marked effect on the resistance of certain materials to fracture. Very great pressures can even result in the plastic flow of otherwise brittle materials; the material becomes more ductile and resistant to fracture. On the other hand, fracture is much affected by tensile stress. Noting this, early investigators proposed a maximum normal stress theory of fracture. According to this theory, fracture occurs whenever the maximum (principal) stress attains a critical value, usually taken as the ultimate strength. Quite simply, fracture is to occur when

$$\sigma_1 \quad \text{or} \quad \sigma_2 \quad \text{or} \quad \sigma_3 = Y_t$$

Since the theory takes no account of the orientation of the principal directions, it is strictly applicable only to isotropic materials. It is the simplest of fracture criteria and applies primarily to brittle materials, which are essentially unaffected by the hydrostatic state.

Fracture of some brittle materials can be predicted by this criterion. A simple experiment provides some evidence: The torsion of a cylindrical stick of chalk is a classroom illustration. The twisted stick exhibits a helical surface of fracture; the surface is that of maximum normal stress.

4.18.2 CRITERION OF COULOMB-MOHR

Important structural materials are essentially brittle, exhibit little resistance to tensile stress, but are also much affected by the hydrostatic state. Concrete is a prominent example: Often engineers discount any tensile strength, but utilize the much greater resistance to compressive stresses. The tensile stresses in a concrete beam are sustained by steel "rebars" (see Illustrative Example 5.29). A. Nadai[15] describes the failures of certain cast metals, rocks, and granular media, which have little resistance to tensile stress; these are relatively brittle materials, which appear to fail along surfaces of "slip" under the action of shear stresses. In such materials, one can imagine that the normal stress on such surfaces might increase the cohesion, i.e., resistance to slip.

Mohr[16] proposed failure/fracture criteria, which embody the notion of the shearing action, but also the influence of the normal stress upon the slip surfaces. A version of Mohr's theory had been proposed earlier by the French scientist Coulomb: The Coulomb-Mohr criterion is expressed by a linear relation between the maximum shear stress and the normal stress upon the surface of the shear stress: If the principal stresses satisfy $\sigma_3 \geq \sigma_2 \geq \sigma_1$, then the criterion takes the form:

$$(\sigma_3 - \sigma_1) = C_1 + C_2(\sigma_3 + \sigma_1) \tag{4.52}$$

This is the equation of a plane in the space (σ_1, σ_2, σ_3). Six similar equations apply, as follows:

$$\pm(\sigma_3 - \sigma_1) = C_1 + C_2(\sigma_3 + \sigma_1), \qquad \sigma_3 \gtrless \sigma_2 \gtrless \sigma_1 \tag{4.53a}$$

$$\pm(\sigma_2 - \sigma_1) = C_1 + C_2(\sigma_2 + \sigma_1), \qquad \sigma_2 \gtrless \sigma_3 \gtrless \sigma_1 \tag{4.53b}$$

$$\pm(\sigma_3 - \sigma_2) = C_1 + C_2(\sigma_3 + \sigma_2), \qquad \sigma_3 \gtrless \sigma_1 \gtrless \sigma_2 \tag{4.53c}$$

The reader will note the similarities between these equations and Equation (4.48) of the Tresca yield criterion. In like manner, the planes defined by the Equation (4.53) cut the π plane (perpendicular to the line $\sigma_1 = \sigma_2 = \sigma_3$ at $\sigma_1 = \sigma_2 = \sigma_3$-= 0) in the form of a hexagon as depicted in Figure 4.41. Unlike the Tresca cylinder, the Coulomb-Mohr criterion forms a pyramid depicted in Figure 4.42, i.e., the tensile stresses are limited by the apex at a point on the line $\sigma_1 = \sigma_2 = \sigma_3$.

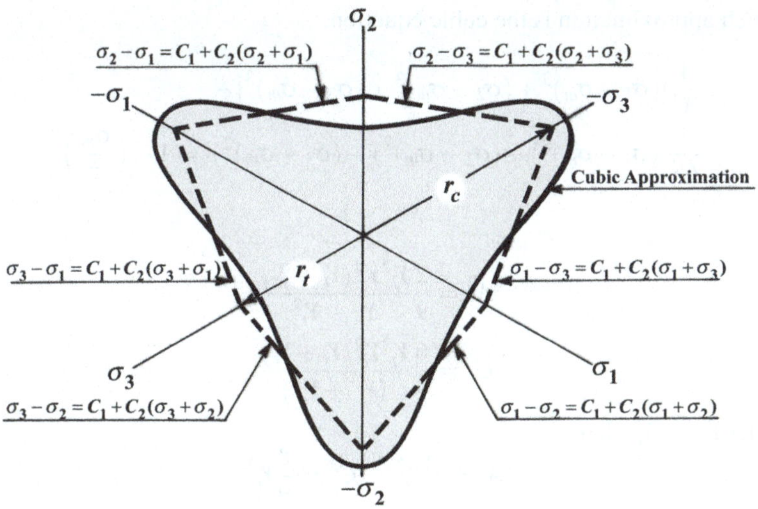

Figure 4.41

It remains to impart physical meaning to the constants, C_1 and C_2, in the Equation (4.53) and to provide additional interpretations. Two tests suffice to ascertain the constants: From a simple compression test, $\sigma_1 = -Y_c$, $\sigma_2 = \sigma_3 = 0$, we obtain

$$Y_c = C_1 - C_2 Y_c$$

From a simple tension test, $\sigma_1 = Y_t$, $\sigma_2 = \sigma_3 = 0$, we obtain

$$Y_t = C_1 + C_2 Y_t$$

It follows that

$$C_1 = \frac{2Y_c Y_t}{Y_c + Y_t}, \quad C_2 = \frac{Y_t - Y_c}{Y_c + Y_t} \tag{4.54}$$

We note that $C_2 \leq 0$, since $Y_c \geq Y_t$, for the materials in question. One can also obtain the lengths r_c and r_t of Figure 4.41.

$$r_c = \sqrt{\frac{2}{3}}\frac{2Y_c Y_t}{Y_c + 2Y_t}, \quad r_t = \sqrt{\frac{2}{3}}\frac{2Y_c Y_t}{Y_t + 2Y_c}$$

The pyramid/envelope of Figure 4.42 has the apex on the axis $\sigma_1 = \sigma_2 = \sigma_3 = Y_t/(1 - Y_t/Y_c)$. In the special case in which the magnitudes of the tensile and compressive stresses are the same, i.e. $Y_c = Y_t$, then the Coulomb-Mohr criterion degenerates to the Tresca criterion. Then $r_c = r_t$ in Figure 4.41.

It is interesting to reflect on the implications of the criterion: As expected, the tensile strength is limited and dependent on the ratio Y_t/Y_c. For example, concrete is typically weak in tension, but relatively strong in compression, e.g. $Y_c \approx O(10Y_1$. A cast metal might exhibit a ratio $Y_t/Y_c = 1/2$; then the failure of the metal under a tri-axial (hydrostatic) state of stress occurs at $\sigma_1 = \sigma_2 = \sigma_3 = 2Y_t$.

The criterion of Coulomb-Mohr poses a computational impediment: Like the Tresca yield criterion, the condition is represented by the equations of six planes. To facilitate implementation, we can use an approximation by a continuous function much as the Mises' criterion can replace the Tresca. Such approximation is the cubic equation.

$$\begin{aligned}&\frac{1}{A^3}[(\sigma_1-\sigma_m)^3+(\sigma_2-\sigma_m)^3+(\sigma_3-\sigma_m)^3]+\\&\frac{1}{C^2}[(\sigma_1-\sigma_m)^2)+(\sigma_2-\sigma_m)^2)+(\sigma_3-\sigma_m)^2)]=1-\left(\frac{\sigma_m}{\overline{\sigma}}\right)^n\end{aligned} \tag{4.55}$$

where

$$A^3 = \frac{2}{9}\frac{Y_t^2 Y_c^2 (Y_t + Y_c)}{Y_c^2 - Y_t^2}$$

$$C^2 = \frac{6}{9}\frac{Y_t^2 Y_c^2 (Y_t + Y_c)}{(Y_t^3 + Y_c^3)}$$

To illustrate, let $Y_c = 2Y_t$; then

$$A^3 = \frac{8}{9}Y_t^3, \quad C^2 = \frac{8}{9}Y_t^2$$

The intersection with the π plane is the solid curve of Figure 4.41. The surface assumes a conical form, similar to the pyramid of Figure 4.42. We need only set approximate values for the constant $\overline{\sigma}$ and exponent n of Equation (4.55); those values must derive from experiments.

Our treatment of the criteria for failure of brittle materials is very limited. Many alternatives exist, but they are beyond the scope of our introduction. Much insight is provided by the early classic of A. Nadai.[17]

4.19 STRESS CONCENTRATION

Most failures originate at the site of greatest stress; these are often localized and designated as *stress concentrations*. Depending upon the material and conditions of loading, such localized states of stress often initiate failure, the yielding of a ductile material, or the fracture of a brittle material. The latter often begins as a minute crack, then propagates with subsequent loading. Criteria for the propagation of cracks derive from the conditions, the state, and the intensity of the stress at the site of the concentration.

Several important analyses of linear elasticity focused attention on stress concentrations and led to subsequent investigations of fracture mechanics: One important result was provided by Inglis' solution [18] for the stress in a plate with an elliptic hole: The plate of Figure 4.43 is subjected to uniform tensile stress, $\sigma_y = \sigma$, at a great distance from the hole (theoretically $\sigma_y \to \sigma$ in the limit as $\pm x \to \infty$, $\pm y \to \infty$); from the practical viewpoint, the actual size of a small hole is irrelevant.

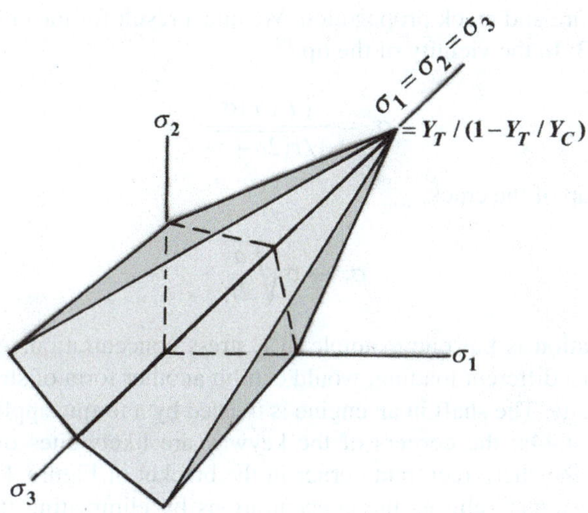

Figure 4.42

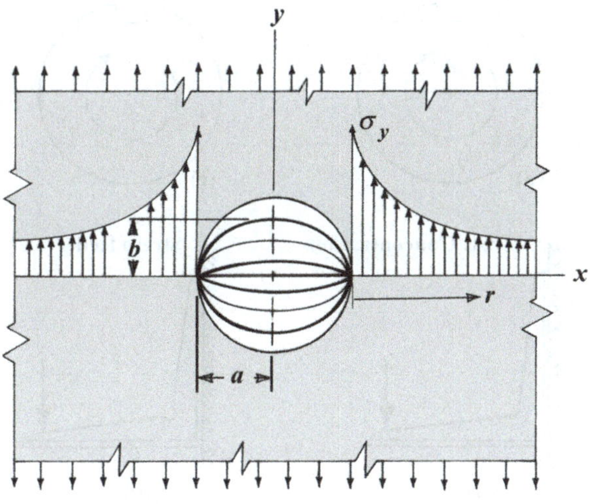

Figure 4.43

The important result is the stress at the ends of the major axis (x = $\pm a$), i.e., at the edge of the hole:

$$\sigma_y(a,0) = (1+2a/b)\sigma$$

As the ellipse grows narrow, $b/a \to 0$, $\sigma_y(a,0) \to \infty$. Of course, no material can resist an unlimited stress. Indeed, one can rightly argue that our definition of stress ceases to have meaning, as the region shrinks to the microscopic scale. On that scale, we see an assembly of atoms and imperfections, but certainly not the continuity inherent in our definition of strain and stress. Events at the microscopic level cause some form of damage, yielding, or fracture from a microscopic viewpoint.

Other results of mathematical analyses of linear elasticity are also incorporated into contemporary criteria of fracture and crack propagation. We cite a result for the cracked plate ($b = 0$), as depicted in Figure 4.43: In the vicinity of the tip [19]

$$\sigma_y = \frac{(a+r)\sigma}{\sqrt{r(2a+r)}} \tag{4.56a}$$

As $r/a \to 0$, near the tip of the crack,

$$\sigma_y \to \sigma\sqrt{\frac{a}{2r}} \tag{4.56b}$$

The foregoing situation is but one example of a stress concentration. A plate with a similar crack, but subjected to a different loading, would exhibit another form of stress concentration and a different stress intensity. The shaft in an engine is twisted by a torque applied through a keyway, as depicted in Figure 4.44a; the corners of the keyway are likely sites of stress concentration and failure. Likewise, the sharp reentrant corner in the bracket of Figure 4.44b is a site of stress concentrations. The designer relieves the concentrations by eliminating these sharp corners as depicted in Figure 4.44c and d. In short, any sharp notch, scratch, or re-entrant corner is a potential site of stress concentration.

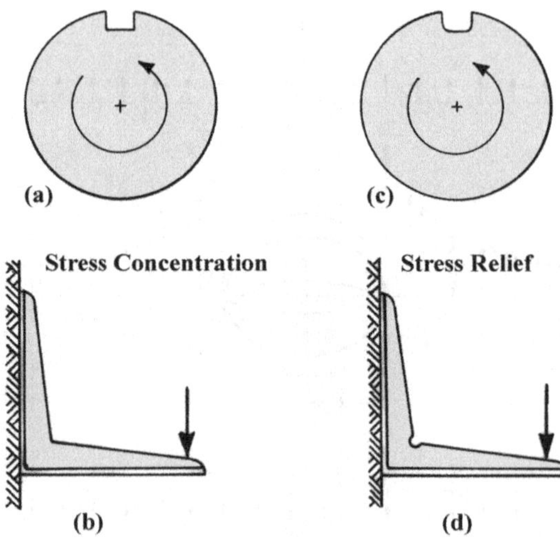

Figure 4.44

4.20 FRACTURE MECHANICS

A preceding section 4.18 presents but two of the many criteria for fracture. They are expressed in terms of stresses and, therefore, they embody the presumption of a continuous medium. Unfortunately, most fractures are initiated on a micro scale; they are attributed to various forms of imperfection or damage, which might involve the dislocations in crystalline materials or a progressive loss of interatomic bonds in nonmetallic materials. Unfortunately, such phenomena are not readily amenable to the tools of continuum mechanics, e.g., stresses and strains. The contemporary topic of *fracture mechanics* establishes criteria for the progressive extension of an existing microcrack; it provides the means to join the physics of the material to the mechanics of the continuum model.

Throughout the history of mechanics (see H. P. Rossmanith[20]) researchers have struggled with various inexplicable occurrences of fracture and the inadequacies of the prevailing criteria. For example, we have noted the unpredictability of fatigue failures (Section 4.14). Investigators have also observed the inexplicable consequences of other circumstances, e.g., effects of the size and the surface conditions of specimens; the earlier criteria of fracture take no account of such factors. In part, the difficulties rest with the dichotomy between the physics of the microstructural events and the desire to describe these phenomena via the mechanics of continua. Specifically, investigators sought theories to predict the occurrence and the propagation of these microcracks, but also to cast the criteria in the macroworld of the continuum. An important step was achieved by A. A. Griffith[21]: Griffith was motivated by his observations of fractures which occurred in glass; stresses were much less than predicted by the prevailing criteria and less than the anticipated limits of Hookean behavior. He focused his attention on the propagation of an existing crack and the energy required to overcome the cohesive forces between the contiguous surfaces at the tip. Subsequently, the Griffith theory was extended to the fracture of materials, notably steel, which exhibit inelastic deformations in a small zone around the crack tip. G. R. Irwin[22,23] and E. Orowan[24] amended the Griffith theory to include the additional work expended upon the inelastic deformations. In the following, we present only the underlying concepts and prominent features of those theories.

4.20.1 GRIFFITH THEORY

To present the underlying features of the Griffith criterion, we consider the plate with a central crack, as depicted in Figure 4.45. Realistically, the specimen is finite in extent; it occupies a region, $x, y = \pm l$, where the length l is much greater than the thickness t and much greater than the length ($2a$) of the crack. At the distant edges $y = \pm l$ the specimen is subject to the uniform stress $\sigma_y = \sigma$; the lateral surfaces $\sigma_z = \pm t/2$ are free. Let $\dot{a}$ denote the rate of crack growth; the area grows at the rate $\dot{A} = 2t\dot{a}$.

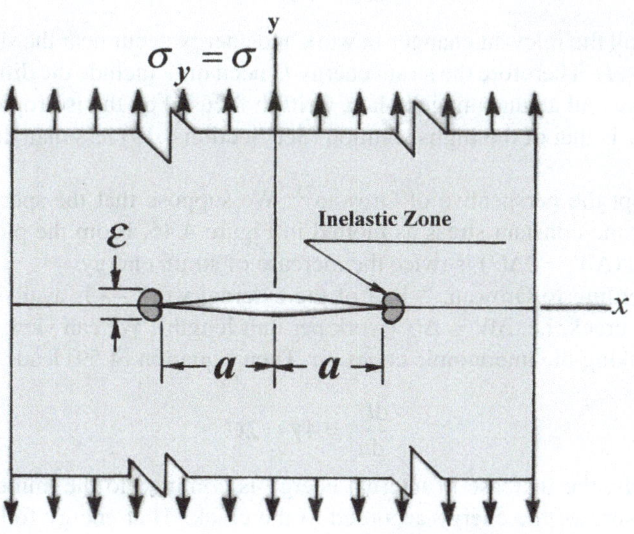

Figure 4.45

Any deformation of our specimen is accompanied by a corresponding change in the elastic potential U, the integral of the density u of Equation (4.41). The kinetic energy K may also increase; then the total mechanical energy is the sum: $U + K$. Energy supplied to the body consists of the work (W) expended by the stress ($\sigma_y = \sigma$) upon the edges ($y = \pm l$) and the heat supplied (Q); together the rate of energy input is the sum: $\dot{W} + \dot{Q}$. Our interpretation of Griffith's argument follows: According to the first principle of thermodynamics, the energy supplied to the system must appear as potential and kinetic energy or as dissipated energy. The dissipation is expressed by the words of Griffith: "It is known that, in the formation of a crack [increase in area] in a body composed of molecules, which attract one another, work must be done against the cohesive forces of the molecules on either side of the crack." If the work expended upon those forces (per unit area) is denoted by γ then the rate of dissipation is $2\gamma\dot{A}$ (upon the two surfaces). Our balance of work and energy follows:

$$(\dot{W}+\dot{Q})-(\dot{U}+\dot{K})=\frac{2\gamma\dot{A}}{t} \tag{4.57}$$

All work and energy are expressed per unit thickness (t). In recognition of Griffith, the work 2γ is assigned the letter:

$$G \equiv 2\gamma \tag{4.58}$$

G is usually called the "energy release rate." Griffith referred to the "surface tension T of the material" ($T = \gamma/t$). In the absence of heat transfer, $\dot{Q} = 0$. Additionally, we assume kinetic energy is not changing; this has been termed "controlled" or "stable" growth. Then Equation (4.57) assumes the form:

$$\dot{W}-\dot{U}=\frac{2\gamma\dot{A}}{t}=4\gamma\dot{a}$$

Propagation implies $\dot{a} > 0$; the process requires

$$\frac{\mathrm{d}W}{\mathrm{d}a}-\frac{\mathrm{d}U}{\mathrm{d}a}=4\gamma \tag{4.59}$$

We must note that all the relevant changes in work and energy occur near the site of the small crack in a large plate ($a \ll l$). Therefore the strain energy U need only include the difference between that for the cracked plate and an undamaged plate. Griffith focused on the isotropic Hookean material; hence the energy U is that of the Inglis solution (See Section 4.19) less than that of an undamaged plate.

Now let us adopt the perspective of Orowan[25]: We suppose that the specimen is slightly extended (Δv) under the constant stress as plotted in Figure 4.46. From the plot, we conclude that the work expended ($\Delta W = 2\Delta U$) is twice the increase of strain energy.

Therefore, according to Orowan, "–half of the external work — is available"– to enforce the propagation of the crack, i.e. $\Delta W = \Delta U$ (work per unit length). We can view that as dissipated in the process of breaking the interatomic cohesion. Then Equation (4.59) leads to the result:

$$\frac{\mathrm{d}U}{\mathrm{d}a}=4\gamma=2G \tag{4.60a}$$

As noted previously, the increase in internal energy is confined to the immediate neighborhood of the tip and consists of the energy accorded to the crack. That energy follows from the Inglis solution[26]:

$$U=\frac{\pi a^2\sigma^2}{E}$$

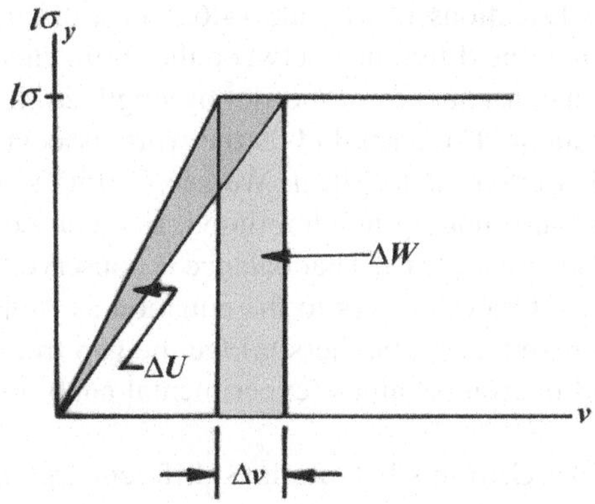

Figure 4.46

Then Equation (4.60a) provides the result:

$$\frac{dU}{da} = \frac{2\pi a\sigma^2}{E} = 4\gamma \tag{4.60b}$$

$$\sigma = \sqrt{\frac{2E\gamma}{\pi a}} \tag{4.60c}$$

In his original work (1921) Griffith adopted a different perspective. He observed that the potential (P) of the system consists of the strain energy (the increase due to the crack) and the potential of the "tension" T: $-4aT$. Here we adopt a different sign; we view the latter potential as energy available by virtue of the bonds. As those bonds are broken, as a increases, that potential decreases. Then the total potential (per unit thickness) is

$$P = \frac{\pi a^2\sigma^2}{E} - 4aTt \tag{4.61}$$

Again the strain energy is the difference between that of the cracked plate (Inglis solution) and that of the uniform plate. From Griffith's perspective, there is no external work expended. His condition for propagation is the stationary condition for the total potential:

$$\frac{dP}{da} = \frac{2\pi a\sigma^2}{E} - 4Tt = 0 \tag{4.62a}$$

or

$$\sigma = \sqrt{\frac{2ETt}{\pi a}} \tag{4.62b}$$

This is the previous result (4.60b) apart from the different interpretation of parameters:

$\gamma =$ force per unit length $T =$ force per unit area

Others have introduced the parameter G of (4.58):

$$G = 2\gamma = 2Tt \tag{4.63}$$

In any case, the parameters are properties of the material and the particular stress concentration.

One might also view the Equations (4.61) and (4.62) in a different light. The energy $U = \pi a^2 \sigma^2 / E$ is an elastic potential, the difference between the strain energy of the cracked plate and that of the solid plate; this is a consequence of the slot of length $2a$. It is calculated via our theory of the elastic (HI-HO) continuum. The term $4aT$ is the work associated with breaking the inter-atomic "tension" T along that crack of length a. We see Griffith's result as a balance between the energy (U) created in the surrounding neighborhood, as a consequence of the crack, and the work expended to produce that crack ($4aT$). That balance is conserved as the crack advances, only if Equation (4.62a) is satisfied. Orawan refers to the equation as "a necessary condition of completely brittle fracture". It is here that researchers bridge the gap from our theory of a continuum ((mathematical analyses) and micromechanics (experimental analyses) via the balance of energy and work.

Orowan obtains a result like Griffith's but employs different arguments and offers other interpretations. He introduces a "molecular strength":

$$\sigma_m = \sqrt{\frac{2E\alpha}{b}}$$

Here b denotes the space between two atomic planes and α is the force per unit length. He acknowledges a circular region at the root of the crack, a region of inelastic deformation such as those depicted in Figure 4.45. He cites an analysis by Neuber[27] which provides an estimate of the inelastic zone with radius b. The stress at the root of the notch of radius b was given by Westergaard[28]

$$\sigma_m = 2\sigma\sqrt{\frac{a}{b}}$$

Orowan asserts that the diameter of the inelastic region "is of the order of magnitude of the interatomic spacing b." Therefore,

$$\sigma = \sqrt{\frac{E\alpha}{2a}}$$

This differs from (4.60c) and (4.62b) by a factor $2/\sqrt{\pi}$ and the interpretation of the parameter α which is related to "the work of fracture per unit of the cross-sectional area". In each case, we see that propagation ultimately is dependent on a property (γ, T or α) of the material and its microstructure.

4.20.2 IRWIN/OROWAN THEORY

The occurrence of abrupt fracture, with little evidence of inelastic deformations, is normally associated with brittle materials, e.g., glass. However, similar fractures do occur also in relatively ductile materials. These are initiated at sites of imperfections; initially minute cracks can abruptly propagate and cause fracture. Such failures are promoted by prevailing conditions, e.g., extreme cold, or repeated loading, e.g., fatigue failures.

The abrupt brittle-like fracture of low-carbon steel received widespread attention in the wake of World War II: The United States had embarked upon the mass production of the so-called "liberty ships." The hulls of these vessels were formed of steel plates, joined by welds, which replaced the earlier riveted assemblies. Unexpectedly, many broke apart as a consequence of abrupt fractures. The immediate question was the differences between the earlier (riveted) and subsequent (welded) construction. Two factors were noted: The material at the weld was embrittled by the welding process and, more important, the rivet-holes had served to arrest the propagation of cracks; a smooth hole at the tip of a notch or crack suppresses the stress concentration (see Figure 4.47).

Investigators undertook extensive research, especially in the U.S. governmental facilities: The topic of "fracture mechanics" was born. The mechanics of Griffith's theory and its applicability to the ductile metals was particularly relevant. Irwin (1948)[29] and Orowan (1950,[30] 1955) recognized that crack propagation in metals is accompanied by inelastic deformations near the tip. They observed that the region of the inelastic behavior is localized, but the work expended in these inelastic deformations far exceeds the work expended to break the interatomic bonds, as described by Griffith. These observations led them to revisit and to modify Griffith's theory. Both investigators also noted that the inelastic zone at the tip is relatively small, $\varepsilon \ll a$ in Figure 4.45. Irwin appealed to this feature in his analysis (1957),[31] which acknowledges the stress associated with the HI-HO behavior, wherein the stress near the tip approaches (See Section 4.19, Equation 4.56b):

$$\sigma_y \to \sigma\sqrt{\frac{a}{2r}}$$

Irwin introduces a "force component" $\mathfrak{g}$ which is related to the crack propagation. He notes that $\dot{a}\,\mathfrak{g}$ is the "fixed grip" loss of energy from the strain-energy field (of a specimen) as the crack extends the amount $\dot{a}$; $\mathfrak{g}$ is comparable to the term $2T$ in Griffith's theory. Finally, Irwin (1957) proposed also the inclusion of a second parameter, an "optional added stress in the x direction $-\sigma_o x$,"... "the only stress acting at the edges of the crack." In his conclusion, he states: "The most significant of these parameters, the intensity factor, is $\sqrt{E\mathfrak{g}/\pi}$ for plane stress where $\mathfrak{g}$ is the force tending to cause crack extension."

In the preceding paragraphs, we have merely reviewed pioneering works by Griffith, Irwin, and Orowan. Their perspectives and interpretations differ; however, there is a common thread: Each calls upon a balance of work and energy. The work expended to break the interatomic bands is provided by the release of strain-energy in the surrounding continuous medium. The actual forces are incorporated in a parameter (T, G, or $\mathfrak{g}$). In the contemporary literature, these parameters are replaced by the critical "stress intensity factor" K_1:

$$K_1 = \sqrt{2E\gamma} = \sqrt{EG} = \sqrt{ETt} = \sqrt{E\mathfrak{g}}$$

Subscript 1 identifies the geometry of the concentration, viz the slit. Returning to Equation (4.60c), we have the critical stress:

$$\sigma = \frac{K_1}{\sqrt{\pi a}}$$

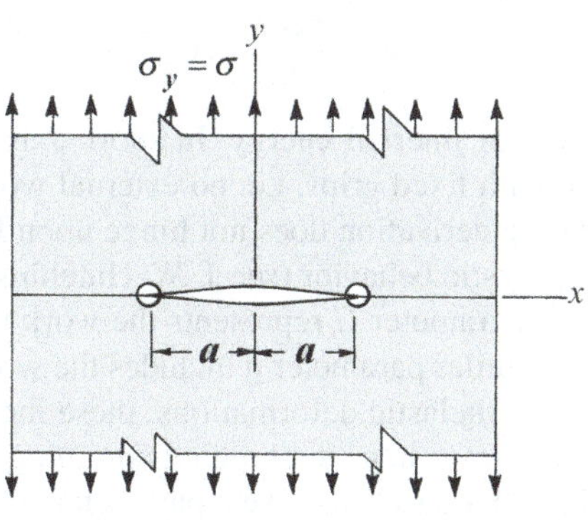

Figure 4.47

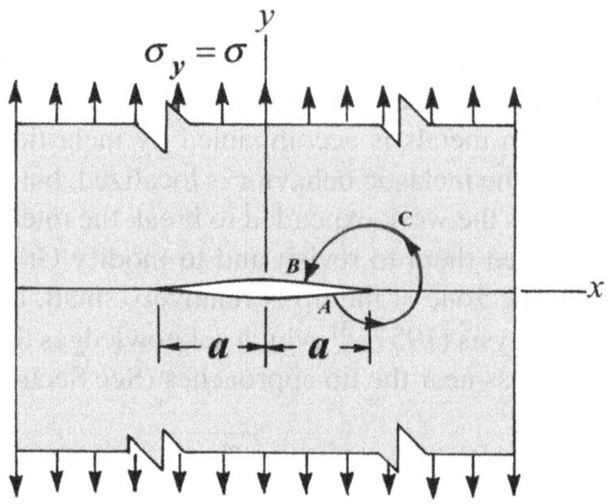

Figure 4.48

The factor K_I is determined by tests on a specimen at the specific material and geometrical configuration.

The Irwin-Orowan Theory modifies the Griffith Theory by identifying the essential parameter, the energy release rate G with the stress intensity factor K_I.[32] In the light of the Irwin-Orowan Theory, investigations focused upon the determination of the factor $G = K_I^2/E$. A relevant result was given by J. R. Rice[33] (1968). His investigations produced an integral about a contour (C in Figure 4.48) from one face to the other (A to B), which is independent of the path.

The so-called J integral follows:

$$J = \int_A^B \left(u\mathrm{d}y - T\frac{\partial v}{\mathrm{d}x} ds \right)$$

Here, T denotes the traction at C upon the enclosed region, u the strain energy density, and v the displacement. In the limit, as the contour shrinks about the tip,

$$J \to \oint u\mathrm{d}y$$

This is a "measure of the strain on the notch tip." Additionally, Rice showed that

$$J = \frac{\partial \Delta U}{\partial a} = G$$

Here, ΔU denotes the change of internal energy. In short, J is the energy release rate (G in equation 4.60a) for a specimen with fixed grips, i.e. no external work is expended upon the body. It is especially noteworthy that the derivation does not hinge upon Hookean elasticity. Hence, the result is applicable to nonlinear elastic behavior (see J. W. Hutchinson.[34])

In the Griffith theory, the one parameter G represents the work to overcome the atomic bonds. In the Griffith/Orowan theory, a similar parameter $\mathfrak{g}$ includes the work to rupture the atomic bonds but also the work dissipated in the inelastic deformations; these include slip within, and between, crystals and microfissures, which occur in a small region adjoining the tip. Both ($\mathfrak{g}$ and G) appear similarly in the formulation. In either case, we have a parameter, a *fracture toughness*

$$G_c = G_c(K_n) \tag{4.65}$$

The parameter G_c is a property of the material, but also depends on the state of stress, specifically, the relevant stress intensity factor K_n. We denote the latter by the symbol K_n to signify the numerous states, which might act near the crack (c.f. E. E. Gdoutos.[35] p. 16, Lemaitre and Chaboche[36] p. 454 and p. 468)

Our treatment of fracture mechanics is necessarily brief, deliberately lacking specific formulas, and devoid of numerical values. It is intended only to illuminate the underlying concepts and inherent difficulties, specifically, the dichotomy between the microstructural mechanism and our macro-tools of analysis. An engineer, who intends to predict and avoid crack propagation, also requires specific criteria and relevant properties (e.g., "fracture toughness"). To that end, we suggest a few sources: The book by Gdoutos provides lucid accounts of the phenomena; the book by Lemaitre and Chaboche offers much insight into the microstructural behaviors (metals, polymers, granular materials, and wood).

4.21 COMPOSITES

The term "composite" denotes a union of two (or more) distinctly different materials; these constituents play different roles to achieve a greater strength in the composition. Composites have been used for centuries in various forms: Bricks of clay have been reinforced with plant fibers. The plaster, used to finish walls, has been reinforced with animal hair. These fibrous materials provided tensile strength to the matrices (clays and plaster) which possess compressive strength but little resistance to tension. Engineers have long introduced steel rebars to resist the tensile stresses in concrete beams, steel mesh in slabs; concrete is yet another material with little tensile strength.

The contemporary focus on composites arises from newfound technologies to manufacture strong fibers and also in the materials which form the matrices. The latter are thermosetting materials, e.g. epoxy, which serve to bond the fibers. Very strong fibers have derived from methods which enable the production of very thin (diameters of 5–25 microns = 0.0002–0.001 in.), but extremely strong filaments of glass, carbon, and aramid. The excellent combination of high strength and low weight has made these fibers a preferred choice for reinforcement in aircraft, automotive, and railcar structures. Composites are also desirable in non-structural applications in which the designer strives to minimize weight. Examples of the latter are the molded housings of an appliance or instrument, or the interior panels of a vehicle. In these latter applications the composite might consist of a random dispersion of short fibers. Then the composite appears and behaves as a homogeneous and isotropic medium at the macro level. The mechanical and thermal behavior is more predictable via analyses and experimentation, as a continuum. Composites for primary structural components incorporate long filaments which are oriented to maximize the strength of the composite. For example, we think of the thin shell-like components of an aircraft wing. The filaments/fibers are laid in distinct plies; plies are oriented differently in successive layers. Orientations are based upon the anticipated loads and consequent stresses. Composites constituted of short fibers of random orientation can be treated as both homogeneous and isotropic. In most cases, the responses are linear. We then require the moduli E and G. If nonlinear, we require the effective moduli that govern successive increments of the nonlinear relations, so-called *tangent moduli*. A full treatment of this topic is beyond the scope of this book. For detailed discussions of composite constituents, manufacturing, mechanics, analysis, and design, readers are referred to ASM Handbook Volume 21—Composites.[37]

NOTES

1. Homogeneus, homogenus: from Greek *homogenes*, from *homn- hom-* (one and the same : alike : similar as opposed to *heter-*) + *-genes* (from *genos* kind, race).
2. From Greek *isotropos*, from *isos* (equal as opposed to *an-isos*) + *-tropos* (turn, way, manner, style).
3. R. Hooke, *De potentia restitutvia*. London: John Martyn, 1678; See I. Todhunter and K. Pearson. *Elasticity and Strength of Materials*. Dover Publication, New York, 1960.
4. R. Hooke. *De Potentia Restitutiva* . London: John Martyn, 1678.
5. Thermal effects are excluded here. Otherwise, the function would also depend on thermal variables.
6. Terms of the form $\gamma_{xy}\gamma_{xz}$ are precluded by the isotropy and Equations (4.39) such as $\tau_{xy} = \partial u/\partial \gamma_{xy}$.
7. This is a special topic, which is not essential to an understanding of subsequent applications.
8. See, for instance, A. C. H. Love. *A Treatise on the Mathematical Theory of Elasticity*, fourth edition. Dover Publications, New York, 1944. Unabridged and unaltered republication of the fourth (1927) edition, Cambridge University Press; A. C. Green and J. C. Adkins. *Large Elastic Deformations and Nonlinear Continuum Mechanics*, first edition. Oxford University Press, London, 1960; G. A. Wempner. *Mechanics of Solids with Applications to Thin Bodies*, first edition. Sijthoff & Noordhoff, Alphen aan den Rijn, The Netherlands 1981. Originally published: McGraw-Hill, Inc., New York, 1973.
9. J. Lemaitre and J.-L. Chaboche. *Mechanics of Solid Materials*. Cambridge University Press, Cambridge, 1990. Originally published by Dunod, Paris, 1985.
10. P. W. Bridgman, *The Physics of High Pressure*. Macmillan, New York, 1931.
11. H. Tresca. Memoire sur l' ecoulement des corps solides soumis *à* les fortes pressions. *Compt. Rend. Acad. Sci. Paris*, 59:754, 1864.
12. W. Lüders, Über die Äußerung der Elasticität an stahlartigen Eisenstäben, und über eine beim Biegen solcher Stäbe beobachtete Molecularbewegung. *Polytechnisches Journal*, Band 155, pp. 18–22, 1860.
13. R. von Mises. Mechanik der festen Körper im plastisch-deformablen Zustand. *Nachr. Akad. Wiss. Göttingen, Math.-Phys. Kl. I,* 582–592, 1913.
14. H. Hencky, Zur Theorie plastischer Deformationen und der hierdurch im Material herforgerufenen Nachspannungen. *Z. angew. Math. Mech.*, 4:323–334, 1924.
15. A. Nadai. *Theory of Flow and Fracture of Solids*, volume 1 of *Engineering Societies Monographs*, second edition. McGraw-Hill, New York, 1950.
16. O. Mohr. *Abhandlungen aus dem Gebiete der technischen Mechanik*, (2nd Edition), Berlin, Germany: Wilhelm Ernst & Sohn, 1914.
17. A. Nadai, *Theory of Flow and Fracture of Solids*, Vol. 1. McGraw-Hill, New York, 1949.
18. C. E. Inglis. Stresses in a plate due to the presence of cracks and sharp corners. *Trans. lnst. Naval Architects*, 55:219–241, 1913.
19. H. M. Westegaard, Bearing pressures and cracks, *J. Appl. Mech. Trans. ASME*, 6:A49–A53, 1939.
20. H. P. Rossmanith (Editor). *Fracture Research in Retrospect*, pp. 37–94. A. A. Balkema, Rotterdam, 1997.
21. A. A. Griffith. The phenomena of rupture and flow in solids. *Phil. Trans. Roy. Soc. London*, A221:l63–198, 1921.
22. G. R. Irwin. Fracture dynamics. In *Fracturing of Metals*, pp. 147–166. 1947 ASM Symposium (Trans. ASM 40A), Cleveland, 1948.
23. G. R. Irwin. Analysis of stresses and strains near the end of a crack traversing a plate. *J. Appl. Mech.,ASME*, 24(3):361–364, 1957.
24. E. Orowan. Energy criteria of fracture. *Welding J. Res. Supplement*, 34(3):142–145, 1955.
25. E. Orowan, Energy criteria of fracture. *Welding J. Res. Suppl.*, 34(3):157–160, 1955.
26. A. A.Griffith. The Phenomena of Rupture and Flow in Solids. *Phil Trans. Roy. Soc. A*, 221: 163–198,1921.
    See also: A. J. M. Spencer, *Int J. Eng. Sci.*, 3:441–449, 1965
27. H. Neubar. *Kerbspannungslehre*, p. 142. Springer Verlag, Berlin, 1937.
28. H. M. Westergaad. Bearing pressures and cracks.*Trans. ASME*, 61:A-49, 1939.
29. G. R. Irwin. Fracture dynamics. *Fracturing of Metals*, ASM, Cleveland, pp. 147–166, 1948.

30. E. Orowan. Fundamentals of Brittle Behavior in Metals. In W. M. Murray (Ed.), *Fatigue and Fracture of Metals*. Symposium MIT. Wiley, 1952.
31. G. R. Irwin. Analysis of stresses and strains near the end of a crack traversing a plate. *J. Appl. Mech.*, 24(3): 361–364, 1957.
32. The factor K_1 applies only to the plane problem as described (Figure 4.45); hence, we identify this particular factor with the subscript "1".
33. J. R. Rice. A Path Independent Integral and the Approximate Analysis of Strain Concentration by Notches and Cracks. *J. Appl. Mech.*, 35(2):379–386, 1968.
34. J. W. Hutchinson, Singular Behaviour at the End of a Tensile Crack in a Hardening Material. *J. Mech. Phys. Solids*, 16(1):13–31, 1968.
35. E. E. Gdoutos. *Fracture Mechanics*, Kluwer, Dordrecht, Netherlands, 1993.
36. J. Lemaitre and J.-L. Chaboche. *Mechanics of Solid Mat.* Cambridge University Press, Cambridge, 1990.
37. ASM Handbook Volume 21—Composites (2002), ASM International, Material Park, Ohio.

5 Fundamental Applications

Our basis for the mechanics and the analyses of solid bodies is analogous to a three-legged stool. Each of the three legs must support a part of the load. Here, the three legs are constructed in Chapters 2–4.

Deformation: The basic mathematical tools are entirely geometrical. The *new* tool is strain, which is needed to describe deformation at a point.

Force, stress, and equilibrium: The basic mathematical tools are the equations of equilibrium. The *new* tool is stress, which is needed to describe internal forces at a point.

Behavior of materials: The mathematical models of continuous media relate the dynamic variables (e.g., stresses), kinematic variables (e.g., strains), and temperature. Additional theories and concepts of failure delimit the models of continua.

The purpose of this chapter is twofold: Most importantly, we develop the methods of utilizing the tools to obtain solutions. Secondarily, we obtain solutions to some fundamental problems of engineering importance; most notably, we obtain the classical solutions to the extension of prismatic bars, the torsion of circular cylindrical shafts, and the simple bending of symmetrical beams. Each of these is characterized as an *inverse* solution: In each, we assume a state of deformation. A state of stress follows via the relations of Hookean Isotropic and Homogeneous (HI-HO) elasticity. That stress distribution is subsequently shown to satisfy the conditions for equilibrium. Hence, these solutions are *exact* in the context of *linear* (HI-HO) elasticity.

5.1 FUNDAMENTAL PROBLEMS

The reader might ask what distinguishes our fundamental problems from others, as contained in subsequent chapters. Each of these fundamental problems is distinguished by simplicity of geometrical form, physical attributes, and loading. For example, we examine the torsion of a shaft, which is geometrically and physically symmetrical about an axis; it is subjected only to opposing couples upon the ends. The practical significance of such simplicity is a simple description of the deformation, state of strain, and stress, which are quite plausible (and in some instances, logical). Moreover, in all instances, our results are in agreement with so-called exact solutions that derive from advanced theories of elasticity and plasticity. Of course, the ultimate justification for an engineer's solution rests on the agreement between predictions and actual observations or the results of more elaborate and precise theories.

Although the fundamental problems, as cited, admit simple analyses, they also have practical importance. By invoking the Saint-Venant principle (Section 5.2), the solution of simple extension can be applied to assemblies of axially loaded members (e.g., the trusses of Section 5.6); likewise, the solution of simple torsion can be applied to the assemblies of shafts (Section 5.9). In such applications, the student is forewarned of limitations that are attributed to the local effects of the actual loadings. The extent of such effects is described in the following section.

Some examples are intended to provoke the reader's sense of engineering practice and to accept a degree of approximation by modifying the simple solutions to accommodate minor variations in form, properties, or loading (e.g., a bar or shaft may be *slightly* tapered). These provide a bridge

DOI: 10.1201/9781003665861-5

between theoretical analysis and engineering practice with their attendant approximations and limitations.

5.2 SAINT-VENANT'S PRINCIPLE

More often than not, the loads applied to structural members are contact forces; that is, the external forces are applied to the member by physical contact with other bodies. Often these forces are transmitted to a member through a connecting device, such as the bolts or rivets that fasten members together. In such cases, the actual distribution of the applied forces is so complicated that an accurate description of the load distribution is difficult. Fortunately, it is sometimes possible to assess the net effect of the loading without regard for the precise manner of load distribution. In most practical situations, the resultant of the distributed forces may be all that is needed.

As an illustration, consider the situations shown in Figure 5.1. Identical rectangular bars have a square gridwork of lines drawn on their exteriors. These bars are subjected to axial loads applied near the ends. The manner of application of the applied forces is quite different in the two cases shown, but the resultant load is the same: an axial force acting along the centroid of the cross sections. Near the ends (in the vicinity of the applied loads), the deformations are quite different; severe distortions of the small squares are evident in both cases. However, the observed deformations are essentially the same throughout the central portion of these two bars; here the squares are deformed into rectangles. Apparently, the only significant differences in behavior are localized effects near the ends, where the loads are applied. If these members are long and slender, the overall stretching of the two bars is approximately the same.

The essential idea demonstrated by this example was elucidated by Barre de Saint-Venant in 1855.[1] In essence, Saint-Venant's principle states that *two different distributions of force acting on the same portion of a body have essentially the same effects on parts of the body that are sufficiently far from the region of application, provided that these force distributions have the same resultant.* This principle applies not only to axially loaded members but also to all types of loading.

According to Saint-Venant's principle, it is often possible to assess the net effect of a loading without regard for the manner of load application. In the preceding example, the extension of the bar is essentially independent of the manner in which the load is applied. This statement is true because the load is applied over a relatively small portion of the member. A change in the load distribution has a localized effect. When these localized effects are of no consequence, the precise manner of load application is unimportant.

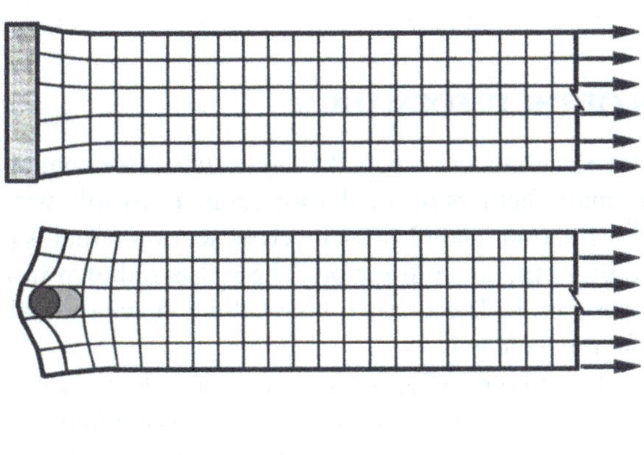

Figure 5.1

The principle of Saint-Venant is employed in a great variety of problems in which loads act upon relatively small portions of a body (essentially concentrated loads). These problems include many practical situations, especially those involving thin flexible bodies, such as bars, shafts, beams, springs, plates, and shells. Throughout the remainder of this chapter, the principle is tacitly employed unless specific details of loading are given. However, the reader should bear in mind that there are instances in which localized deformations and stresses are of the greatest importance. Fracture and fatigue failures are manifestations of local disturbances. Clearly, we cannot overlook local stress distributions when examining phenomena of this kind.

5.3 AXIAL EXTENSION AND LOADING

Often structures are composed of slender members that are subjected to axial loads, for example, the trusses that form the primary structures of most bridges. The members of a truss can be treated as prismatic bars or rods subjected to axial forces (usually applied at the ends of the members). As an example, we consider a situation of this kind. The rod shown in Figure 5.2a is subjected to an axial force P applied at the right end. For simplicity, consider the rod to be held fixed at the left end.

Let us consider a free-body diagram of a portion of this rod. The portion to the right of a cross section S is shown in Figure 5.2b. For equilibrium, the net internal force system on the cross section is an axial force F, collinear and equal to the applied load P.

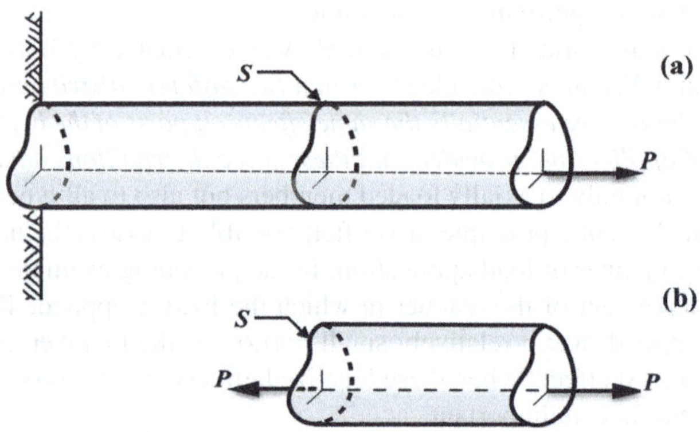

Figure 5.2

5.3.1 DESCRIPTION OF THE DEFORMATION

Because of the nature of the loading, we expect the rod to stretch (or contract) in the axial direction. We will assume that the main features of the deformation are as follows: (1) The axis of the rod remains straight after deformation, and (2) a plane cross section remains plane and perpendicular to the axis, displacing only in the axial direction. The rod is redrawn in Figure 5.3 as it appears both before and after deformation. The x-axis is taken along the centroidal axis of the rod, and the origin of coordinates is at the fixed end.

In the undeformed rod of Figure 5.3a, a cross section is located by the coordinate x and a nearby cross section is located at $x+\Delta x$. After the axial load is applied, these planes are displaced, as shown in Figure 5.3b. Since these cross sections remain plane and parallel, their deformed

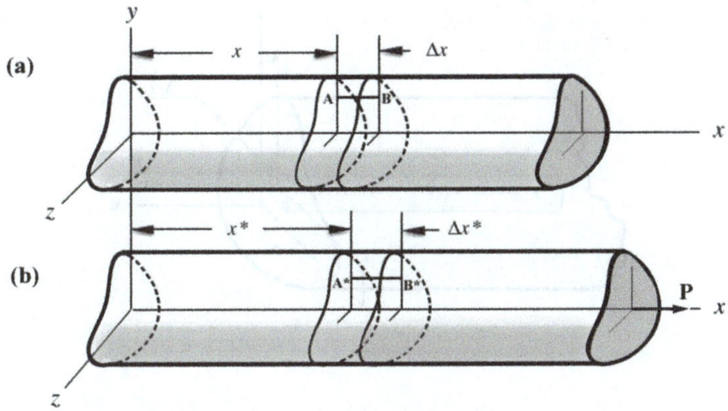

Figure 5.3

positions can be located by the coordinates $x^* + \Delta x^*$. If AB is any line segment between these cross sections, in Figure 5.3a, and parallel to the axis, its length must be Δx. When the rod is deformed, the new length A^*B^* is Δx^*. The extensional strain is

$$\varepsilon_x = \lim_{x \to 0} \frac{\Delta x^* - \Delta x}{\Delta x} \tag{5.1a}$$

On any cross section, the extension strain ε_x is constant; that is, it is independent of y and z. It might, of course, depend on the location of the cross section; that is, ε_x might depend on x but not on y or z. The extension (or contraction) of the rod can also be obtained, according to Equation (5.la): For the slice shown in Figure 5.3, the elongation is

$$\Delta e = \Delta x^* - \Delta x \doteq \varepsilon_x \Delta x \tag{5.1b}$$

The total elongation, e, of a member of length L is obtained as a summation and, in the limit ($\Delta x \to 0$), by the integral

$$e = \int_0^L \varepsilon_x dx \tag{5.1c}$$

From the foregoing description, we can say nothing about the extensional strains ε_z and ε_y; some additional information or approximation is needed. With this in mind, let us examine the nature of the internal forces.

5.3.2 STATICS

Consider a segment of the rod as drawn in Figure 5.4. Because of the normal force P, we anticipate a normal force ΔP on the elemental area ΔA. This, of course, will have to come from a normal stress σ_x. On the other hand, if we look at the narrow rectangular prism, we might suppose that the normal stresses on its lateral faces are inconsequential, since the rod is itself a slender prism whose lateral surfaces are free of applied stress. Assuming that this is the case, we take $\sigma_y = \sigma_z = 0$ throughout the rod.

On the element of area ΔA in the cross section S shown in Figure 5.4, the normal stress σ_x produces the increment of normal force

$$\Delta P = \sigma_x \Delta A$$

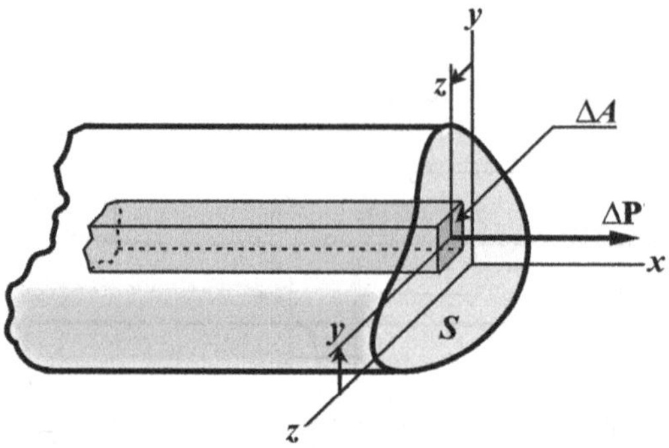

Figure 5.4

Then, over the entire cross section, the total normal force is

$$P = \iint_S \sigma_x dA \tag{5.2}$$

The loading is an *axial force*; that is, the resultant is a force P acting along the x-axis. Then, with respect to the $x-$ axis, $M_x = 0$. The moment of P about the y-axis is

$$M_y = \iint_S \sigma_x z dA = 0 \tag{5.3a}$$

Similarly,

$$M_z = -\iint_S \sigma_x y dA = 0 \tag{5.3b}$$

Equilibrium of the member is ensured by Equations (5.2) and (5.3). To complete a solution, we require a description of the material.

5.3.3 LINEARLY ELASTIC MATERIAL

Suppose that the bar is composed of a linearly elastic and isotropic (HI = Hookean Isotropic) material. Then, see (Section 4.7),

$$\begin{aligned} \varepsilon_x &= \frac{1}{E}(\sigma_x - \nu\sigma_y - \nu\sigma_z) \\ \varepsilon_x &= \frac{1}{E}(\sigma_y - \nu\sigma_z - \nu\sigma_x) \\ \varepsilon_z &= \frac{1}{E}(\sigma_z - \nu\sigma_x - \nu\sigma_y) \end{aligned} \tag{5.4a}$$

As previously noted, the lateral surfaces of the slender rod are stress-free. Therefore, we assume that σ_y and σ_z are negligible and Equation (5.4a) are simplified accordingly:

$$\varepsilon_x = \frac{\sigma_x}{E}, \quad \varepsilon_y = \varepsilon_z = -\nu\frac{\sigma_x}{E} \tag{5.4b}$$

Let us now limit our analysis to *homogenous* (HI-HO = Hookean, Isotropic, and Homogenous bars). Then E is constant and σ_x, as ε_x is independent of y and z.

With the *kinematic* equation (5.1), the equilibrium conditions (5.2) and (5.3), and the *hookean* relations (5.4), we can implement the solution.

Since σ_x is constant, the couples, M_y and M_z of Equation (5.3) vanish *if*

$$\iint_S z dA = \iint_S y dA = 0$$

In other words, the *origin* must be at the *centroid* of the cross section; the force P acts at the centroid. Then, by Equation (5.2)

$$\sigma_x = \frac{P}{A} \tag{5.5}$$

By Equations (5.4b) and (5.5)

$$\varepsilon_x = \frac{P}{EA}, \quad \varepsilon_y = \varepsilon_z = -\nu \frac{P}{EA} \tag{5.6}$$

By means of (5.lc), (5.4b), and (5.6), the elongation of the member follows:

$$e = \frac{PL}{EA} \tag{5.7}$$

Equations (5.4b)–(5.7) relate the most significant quantities: the applied load P, the normal stress σ_x, extensional strain ε_x, and total elongation e. Every cross section is acted upon by the same constant normal stress, and every cross section is subjected to the same constant extensional strain. Presumably, this applies to the end faces as well as to all the intermediate cross sections, which implies that a very special support is required in order that the rod behaves exactly as described.[2] Fortunately, Saint-Venant's principle enables us to apply this solution to a wide variety of problems. The manner in which the loads and reactions are applied does not significantly affect the *overall* deformation if the rod is slender. A similar statement can be made about the stresses in portions of the rod not near the ends.

In summary, we have assumed a deformation which is consistent with the constant extensional strains $(\varepsilon_x, \varepsilon_y, \varepsilon_z)$ of Equation (5.4b), the constant stress of equation (5.5), and our assumptions, $\sigma_y = \sigma_z = 0$. It is easy to conjecture that the remaining shear components vanish: i.e., $\gamma_{xy} = \gamma_{xz} = \gamma_{yz} = 0$ and $\tau_{xy} = \tau_{xz} = \tau_{yz} = 0$. Then the strain and stress distributions are fully prescribed. The interested reader can return to the compatibility conditions (Section 2.23) and equilibrium conditions (Section 3.14) to verify our *inverse* solution.

5.4 AXIAL EXTENSION OF HOOKEAN RODS

The preceding description of the slender bar subjected to axial loading is not limited to a Hookean material. In other circumstances, the axis of the bar remains straight, and cross sections remain perpendicular to it. For any homogeneous material, the extensional strains $(\varepsilon_x, \varepsilon_y, \varepsilon_z)$ are constant on a cross section. Further analysis requires a description of that material.

5.4.1 LINEAR-HARDENING MATERIAL

Suppose the rod is made of a linear-hardening material. The normal stress σ_x is related to the normal strain ε_x as in Figure 5.5a. (This stress-strain curve presupposes that $\sigma_y = \sigma_z = 0$.) increases in σ_x beyond σ_0 are accompanied by plastic deformation (see Section 4.13). If we assume that the bar is not unloaded, the material behavior can be expressed by

$$\varepsilon_x = \frac{\sigma_x}{E} \qquad \sigma_x \leq \sigma_0 \tag{5.8}$$

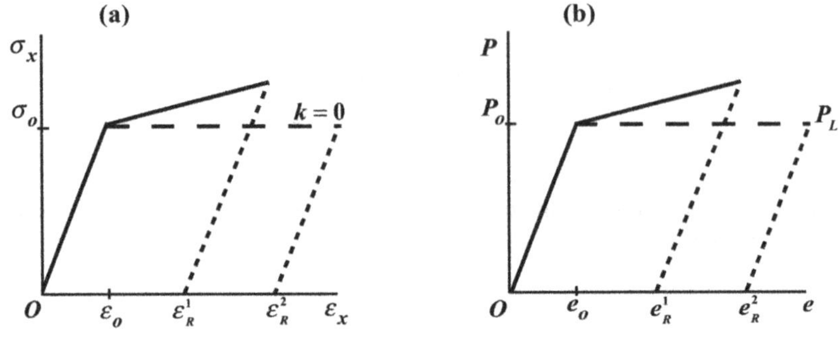

Figure 5.5

$$\varepsilon_x = \frac{\sigma_0}{E} + \frac{\sigma_x - \sigma_0}{kE} \qquad \sigma_x \geq \sigma_0 \tag{5.9}$$

where k is a positive number ($0 < k \leq 1$). Because ε_x is constant on any cross section, σ_x must also be constant on any cross section. It follows that the net internal force system consists of the normal force.

$$P = \sigma_x A \tag{5.10}$$

where A is the cross-sectional area.

Two situations can arise. If P is less than $P_0 = \sigma_0 A$, the behavior is linearly elastic, and the substitution of Equation (5.10) into Equation (5.8) gives

$$\varepsilon_x = \frac{P}{EA}, \quad P \leq P_0 \equiv A\sigma_0 \tag{5.11a}$$

On the other hand, if P exceeds P_0, some plastic deformation occurs, and the substitution of Equation (5.10) into Equation (5.9) yields

$$\varepsilon_x = \frac{P_0}{EA} + \frac{P - P_0}{kEA}, \quad P \geq P_0 \tag{5.11b}$$

As before, the load P must lie along the centroidal axis of the bar, since a constant normal stress cannot give rise to bending couples.

The total elongation of the bar is given by

$$e = \int_0^L \varepsilon_x dx$$

so that

$$e = \frac{PL}{EA} \quad P \leq P_0 \tag{5.12a}$$

$$e = \frac{P_0 L}{EA} + \frac{(P - P_0)L}{kEA} \quad P \geq P_0 \tag{5.12b}$$

Consider the extension of the bar as the load P is gradually increased from zero. This is defined by Equations (5.12a,b), which are plotted in Figure 5.5b. Initially, the behavior is linear and elastic; this is the same as the earlier example of this section. When the load P exceeds the value $P_0 = \sigma_0 A$, the material yields, and a permanent deformation is sustained by the bar. If the load were subsequently removed, unloading would not retrace the loading path, but would follow the dotted linear path, as shown in Figure 5.5b.

5.4.2 IDEALLY PLASTIC MATERIAL

A very important special case occurs when the constant k approaches zero. (The material in this case is ideally plastic, and the load-elongation curve becomes flat, as indicated by the dotted line in Figure 5.5. When the applied load attains the value $P_0 = A\sigma_0$, the bar undergoes unrestricted plastic deformation. This constitutes plastic collapse; because the bar cannot sustain a greater load, P_0 is termed the *limit load.*

5.5 APPLICATIONS OF AXIAL LOADING

A rigorous analysis of a deformable body requires a study of the deformation and equilibrium of each infinitesimal element of the body. This involves a study of the stress, strain, and displacement differentials from point to point throughout the body. Mathematically, such a treatment leads to the consideration of partial differential equations. To avoid these complexities, our current analyses are based on assumptions concerning the significant features of the deformation. These analyses combine kinematic concepts (strain), material behavior (stress-strain laws), and dynamic relations (equilibrium equations) to obtain load-stress and load-deformation relations. The procedure follows the same general pattern for all the examples in this chapter. Let us review this procedure as it was applied to the axial extension of prismatic bars.

The assumptions concerning the significant features of the deformation are used to deduce how the strains vary in the member. This information will generally be incomplete. For example, in the axial extension problem, the assumptions that the axis remains straight and cross sections remain perpendicular to it lead to the conclusion that the extensional strain on a cross section is constant. These assumptions contain no information about the other strains in the rod.

The material is described by relations between the strain and the stress distribution. Often the incompleteness of the strain description necessitates the introduction of other assumptions or simplifications before these stress-strain relations can be used. In the example of axial extension, it was necessary to introduce the further assumption that the transverse normal stresses (σ_y and σ_z) vanish in order to use the stress-strain relations considered. It then followed that the normal stress on a cross section is constant.

Equilibrium conditions enable us to express the internal force system on a sectioning surface in terms of the load. Since the stress distribution on a cross-section is known, the stresses can be related to the load. Thus, we found that the constant normal stress σ_x gave rise to a normal force equal to the load P, from which it followed that $\sigma_x = P/A$. Additionally, the constant stress σ_x requires that the load P act along the centroidal axis, such that $M_y = M_z = 0$.

We return to the stress-strain relations to express the strains in terms of the external load. The significant features of the deformation are then related to the load, since they can be determined from a knowledge of the strains. In the example of the axial extension we found, for elastic materials, $\varepsilon_x = \sigma_x/E = P/EA$. The significant feature of the deformation is the total elongation of the rod, which is related to the extension strain ε_x by $e = \int_0^L \varepsilon_x dx$. Hence, for the linearly elastic materials, $e = PL/EA$.

ILLUSTRATIVE EXAMPLE 5.1

A structural steel channel is initially 3.0 m long. It is stretched 0.80 cm. The steel can be treated as an ideally plastic material ($k = 0$ in Figure 5.5).

$$E = 200 \times 10^3 \text{ MPa and } \sigma_0 = 300 \text{ MPa}$$

We require the permanent elongation.

At the onset of yielding the elongation is entirely elastic, i.e., recoverable:

$$e_E = \frac{\sigma_0 L}{E} = \frac{300}{200} \times 10^{-3} \times 300 \text{ cm} = 0.45 \text{ cm}$$

However, the bar has been extended by 0.80 cm. Upon removal of the load, the channel recovers 0.45 cm; the load-elongation plot traces the linear unloading relation, i.e., $\Delta e = \sigma_x / E$, as depicted in Figure 5.5b. The permanent elongation is the difference:

$$e_R = (0.80 - 0.45) = 0.35 \text{ cm}$$

ILLUSTRATIVE EXAMPLE 5.2

Suppose that the cylindrical bar of Figure 5.2 were composed of a homogeneous material with the stress-strain relation:

$$\sigma_x = \sigma_0 \frac{E\varepsilon_x}{\sigma_0 + E\varepsilon_x}$$

The preceding kinematic functions are applicable if the bar is subjected to an axial centroidal force P. Therefore,

$$\varepsilon_x = \frac{e}{L}, \quad \sigma_x = \frac{P}{A}$$

When these are substituted into the above relation, the elongation follows:

$$e = \frac{\dfrac{PL}{EA}}{1 - \dfrac{P}{\sigma_0 A}}$$

This bar also exhibits a limit load; as $P \to \sigma_0 A$, the bar undergoes unconstrained elongation. Note that the initial response is Hookean: i.e.

$$\lim_{x \to 0} \frac{d\sigma_x}{d\varepsilon_x} \to E$$

ILLUSTRATIVE EXAMPLE 5.3

The predominant features of the simple extension (or contraction) occur under many circumstances besides the direct application of a central force. To illustrate this, consider the mechanical assembly of the bolt B, which extends through the cylindrical tube T as shown in the section of Figure 5.6a.

Suppose, for simplicity, that the washers beneath the head and nut are stiff enough so that we can assume that they remain plane as the nut is advanced, the bolt is extended, and the tube is compressed. To be specific, suppose that the bolt is $\frac{1}{4}$ in. (nominal diameter) with 24 threads per inch and that it is made of steel ($E_B = 30 \times 10^6$ lb/in^2, $\alpha_B = 6.5 \times 10^{-6}/^\circ F$). The tube has inner

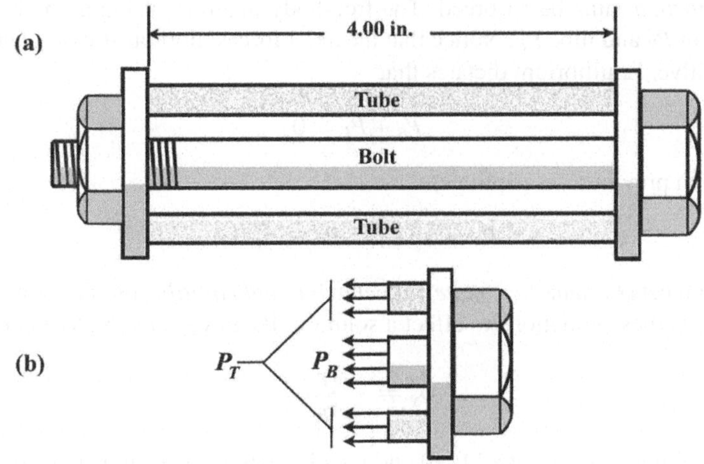

Figure 5.6

and outer diameters $d = 0.600$ in. and $D = 0.800$ in. and is made of aluminum alloy ($E_T = 10 \times 10^6$ lb/in^2, $\alpha_T = 13.5 \times 10^{-6}/°F$. At room temperature ($70°F$) the nut is snug and is then advanced one-quarter turn ($\frac{1}{96}$ in.). The assembly is then also heated to $470°F$. Our problem is to determine the mean normal stresses on cross sections of the bolt and the tube. The problem is a simple problem from an engineer's viewpoint, but one that serves to illustrate the essential features in the solution of a problem of deformable solids. Again, we must recognize that local effects occur near the ends and especially at the threads; yet simple extension provides a valid overall description in accordance with Saint-Venant's principle.

Firstly, let us consider the *kinematics*: Since the nut is advanced one turn, the effective length of the bolt (between the "rigid" washers) is reduced by $\frac{1}{96}$ in. The essential geometric constraint can be stated simply: The final effective length of the bolt and the tube must be the same (distance between the "rigid" washers). If L denotes the initial length of both, L_B^* and L_T^* are the final lengths, and e_B^*; and e_T^* are the elongations of bolt and tube, respectively, then

$$L - \frac{1}{96} + e_B = L_B^* = L_T^* = L + e_T \tag{5.13a}$$

or

$$e_B - e_T = \frac{1}{96} \tag{5.13b}$$

Secondly, the description of both steel and aluminum under the circumstances is *Hookean, isotropic, and homogeneous* (HI-HO). Moreover, our previous description of the axial extension suggests that the transverse normal stresses can be neglected in both components:

$$\sigma_y = \sigma_z = 0$$

Then the strain-stress-temperature relations for each are similar. If superscripts B and T signify steel bolt and aluminum tube, respectively, then

$$\varepsilon_x^B = \frac{\sigma_x^B}{E_B} + \alpha_B \Delta T, \quad \varepsilon_x^T = \frac{\sigma_x^T}{E_T} + \alpha_T \Delta T \tag{5.14a,b}$$

Finally, *equilibrium* must be enforced. The free-body diagram of Figure 5.6b shows the axial force upon the bolt P_B and tube P_T. Notice that the axial forces, normal stresses, and "elongations" are assumed positive. Equilibrium dictates that

$$P_B + P_T = 0 \tag{5.15}$$

In accordance with previous descriptions,

$$P_B = \sigma_x^B A_B, \quad P_T = \sigma_x^T A_T \tag{5.16}$$

Each of the three facets (*kinematics, material behavior, and equilibrium*) has now been addressed. It remains to utilize these equations to effect a solution. By Equations (5.15) and (5.16),

$$\sigma_x^B = -\frac{A_T}{A_B}\sigma_x^T \tag{5.17}$$

Then, σ_x^B can be eliminated from (5.14a,b) and the elongations determined, as follows

$$e_B = \int_0^L \varepsilon_x^B dx = L\left(\frac{\sigma_x^B}{E_B} + \alpha_B \Delta T\right) = \left(-L\frac{A_T}{A_B}\frac{\sigma_x^T}{E_B} + \alpha_B \Delta T\right)$$

Similarly,

$$e_T = L\left(\frac{\sigma_x^T}{E_T} + \alpha_T \Delta T\right)$$

Finally, the latter are substituted into the geometric constraint (5.13b):

$$-L\left(\frac{1}{E_T} + \frac{A_T}{A_B E_B}\right)\sigma_x^T = +\frac{1}{96} + L(\alpha_T - \alpha_B)\Delta T$$

With the foregoing values ($L = 4.00$ in.), we obtain

$$\sigma_x^T = -21{,}700\ \text{lb/in}^2, \quad \sigma_x^B = +97{,}000\ \text{lb/in}^2$$

The axial force upon the bolt and tube is 4770 lb. All numerical values are given to three decimal places of accuracy because dimensions and properties are seldom more accurate in most practical circumstances.

ILLUSTRATIVE EXAMPLE 5.4

A member is composed of many thin strips of Hookean materials with varying moduli. These are bonded together to form a laminated bar with a rectangular cross-section; the modulus of the composite varies linearly: $E = E_0\left(1 + k\frac{x}{D}\right)$ (see Figure 5.7).

The bar is clamped at the upper end; a load P is applied axially upon the rigid plate affixed to the lower end. We are asked to determine the position $\bar{x}$ such that the plate remains perpendicular to the axis as the load P is applied. Since laminae are extended equally,

$$\varepsilon_x(x) = \text{constant} \equiv \bar{\varepsilon}$$

Then the force P is the integral

$$P = \int_0^D E_o\left(1 + k\frac{x}{D}\right)\bar{\varepsilon}_x dx = E_0 D\bar{\varepsilon}\left(1 + \frac{k}{2}\right)$$

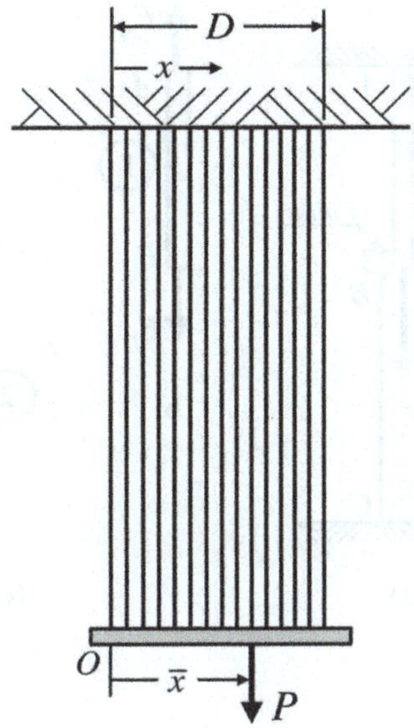

Figure 5.7

Since the resultant must be that force acting at distance $\bar{x}$, the moment with respect to the edge O follows:

$$\bar{x}P = \int_0^D \sigma_x x\mathrm{dx} = \int_0^D E_0\left(x + k\frac{x^2}{D}\right)\bar{\varepsilon}\mathrm{dx}$$

$$\bar{x}E_0 D\bar{\varepsilon}\left(1 + \frac{k}{2}\right) = E_0 D^2\left(\frac{1}{2} + \frac{k}{3}\right)\bar{\varepsilon}$$

$$\bar{x} = D\frac{(3+2k)}{(6+3k)}$$

Note: If $k = 0$, the bar is homogeneous and $\bar{x} = \frac{D}{2}$. If $k = 1$, i.e., the modulus varies from E_0 to $2E_0$, then $\bar{x} = \frac{5}{9}D$

ILLUSTRATIVE EXAMPLE 5.5: RESIDUAL STRESS

The stepped bar of Figure 5.8a is composed of a material that is linear elastic (HI-HO) to the yield stress:

$$\sigma_0 = \pm 40\ \text{ksi}\ (\pm 276\ \text{MPa})$$

$$E = 30 \times 10^3\ \text{ksi}$$

Subsequently, the material behaves as ideally plastic (see Figure 5.5 with $k = 0$).
The segments (1 and 2) have the following dimensions:

$$l_1 = 8\ \text{in}\ (20\ \text{cm}),\ l_2 = 10\ \text{in}\ (25\ \text{cm})$$

$$A_1 = 0.12\ \text{in}^2\ (0.77\ \text{cm}^2),\ A_2 = 0.20\ \text{in}^2\ (1.30\ \text{cm}^2)$$

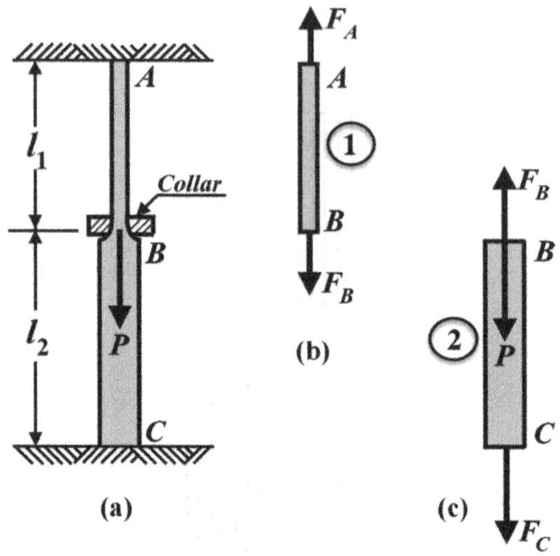

Figure 5.8

An axial load P is applied via the rigid collar upon the shoulder at B. The load is distributed on the collar such that the resultant force acts on the centroidal axis. We seek the relation between that load P and the displacement e_1 of the collar.

From the free bodies depicted in Figure 5.8b and c, we obtain the equilibrium conditions:

$$F_A - F_B = 0, \quad F_B - P - F_C = 0 \tag{5.18a,b}$$

The forces (F_A,F_B, and F_e) are *statically indeterminate*. We require the geometric constraint upon the deformations, the extensions, e_1 and e_2, and the linearly elastic description of the material. The constraints at the fixed ends require that

$$e_1 + e_2 = 0 \tag{5.19}$$

The linearly elastic description is embodied in the relations:

$$e_1 = \frac{F_A l_1}{EA_1}, \quad e_2 = \frac{F_C l_2}{EA_2} \tag{5.20a,b}$$

The substitution of Equation (5.20) into (5.19) provides the third equation in the three forces (F_A, F_B, and F_C):

$$\frac{F_A l_1}{A_1} + \frac{F_C l_2}{A_2} = 0 \tag{5.21}$$

The simultaneous solution of the Equations (5.18a,b) and (5.21) follows:

$$F_A = F_B = \frac{A_1 l_2 P}{A_1 l_2 + A_2 l_1}, \quad F_C = -\frac{A_2 l_1 P}{A_1 l_2 + A_2 l_1} \tag{5.22a,b}$$

The stresses in each segment are

$$\sigma_1 = \frac{F_A}{A_1} = \frac{l_2 P}{A_1 l_2 + A_2 l_1}, \quad \sigma_2 = \frac{F_C}{A_2} = -\frac{l_1 P}{A_1 l_2 + A_2 l_1} \tag{5.23a,b}$$

The displacement of the collar follows from Equations (5.20a) and (5.22a):

$$e_1 = \frac{l_1 l_2 P}{E\,(A_1 l_2 + A_2 l_1)}$$

The latter holds provided that no inelastic deformation occurs. Since the yield stresses in tension and compression have the same magnitudes, yielding is initiated in member 1 according to Equation (5.23a):

$$\sigma_1 = 40\text{ ksi}, \quad F_A^\circ = 4.8\text{ kip}, \quad P = 11.2\text{ kip}$$

Subsequently, the behavior of member 1 follows the ideally plastic description, viz. the stress and force in that member remain constant. As the load P is increased, only the force upon segment 2 increases, but only to the yield stress, $\sigma_2 = -40$ ksi. Then

$$F_C^\circ = -40A_2 = -8.0\text{ kip}$$

Force P then attains the limit load:

$$P = P_L = 12.8\text{ kip}$$

A trace of the load-elongation history is depicted in Figure 5.9. At the state A, the displacement of the shoulder is

$$e_1^\circ = \frac{\sigma_0 l_1}{E} = 0.0107\text{ in}$$

At the limit state B,

$$e_1 = -e_2^\circ = \frac{\sigma_0 l_2}{E} = 0.0133\text{ in}$$

Now suppose that the load P_L is removed after some inelastic deformation of bar 1. For simplicity, let us remove the load just as bar 2 is about to yield, i.e., at state B on the plot of Figure 5.9. Both segments respond elastically during the unloading (see Figure 5.5). The unloading traces the dotted line BO' of Figure 5.9.

Let us denote the changes in the forces, stresses, strains, and elongations by ΔF, $\Delta\sigma$, ΔE, and Δe, respectively. The equilibrium and kinematic conditions follow:

$$\Delta F_B - \Delta F_C = \Delta P \tag{5.24}$$

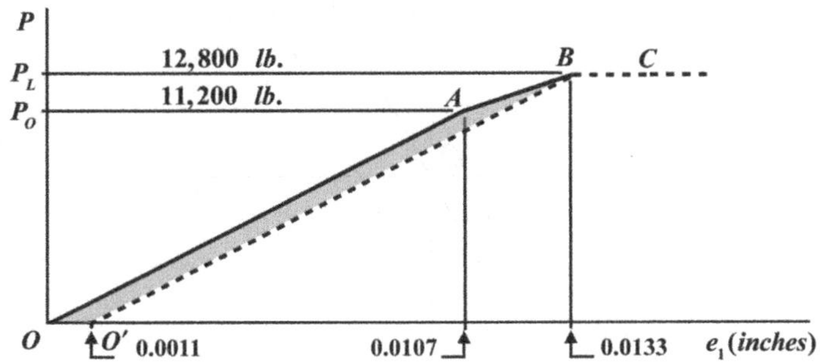

Figure 5.9

$$\Delta e_1 + \Delta e_2 = 0 \tag{5.25}$$

Again, the elastic response asserts that,

$$\Delta e_1 = \frac{\Delta F_A l_1}{EA_1}, \quad \Delta e_2 = \frac{\Delta F_C l_2}{EA_2}$$

As before, (see Equations 5.22a,b), we obtain

$$\Delta F_A = \frac{A_1 l_2 \Delta P}{A_1 l_2 + A_2 l_1}, \quad \Delta F_C = -\frac{A_2 l_1 \Delta P}{A_1 l_2 + A_2 l_1}$$

Upon removal of the load, we have

$$\Delta F_A = -5.5 \text{ kip}, \quad \Delta F_C = 7.3 \text{ kip}$$

$$\Delta e_1 = -\Delta e_2 = -0.0121 \text{ in}$$

The residual forces are

$$F_C^R = F_A^R = F_C^\circ - \Delta F_C = -690 \text{ lb } (3100 \text{ N})$$

Both segments remain in a state of residual compression. The shoulder B has a permanent displacement:

$$e_1 = -e_2 = -e_2^\circ + \Delta e_2 = 0.0133 - 0.0122 = 0.0011 \text{ in}$$

This simple example illustrates a common occurrence whenever a part of a member or structure suffers inelastic deformation while adjoining parts remain elastic. Here, both bars 1 and 2 remain in a state of residual compression. The state of each is illustrated by the stress-strain plot of Figure 5.10. During loading, bar l has been deformed elastically along path OA and then inelastically to point 1 at the yield stress σ_0; bar 2 has been deformed elastically from O to 2 at the yield stress $-\sigma_0$. Upon unloading, bar 1 traces path 1 to l_R, whereas bar 2 retraces path 2 to 2_R. Both remain in states of residual compression. If the clamp at A were abruptly removed, the collar B would return to its initial position; the stored energy in both bars would be released. Segment 1 has suffered a permanent extension: At the limit load, $e_1 = -e_2 = 0.0133$ in; upon removal of the load

$$\Delta e_1 = \frac{\sigma_0 l_1}{E} = -0.0107 \text{ in}$$

Hence Segment 1 has sustained a permanent elongation:

$$e = 0.0133 - 0.0107 = 0.0026 \text{ in}$$

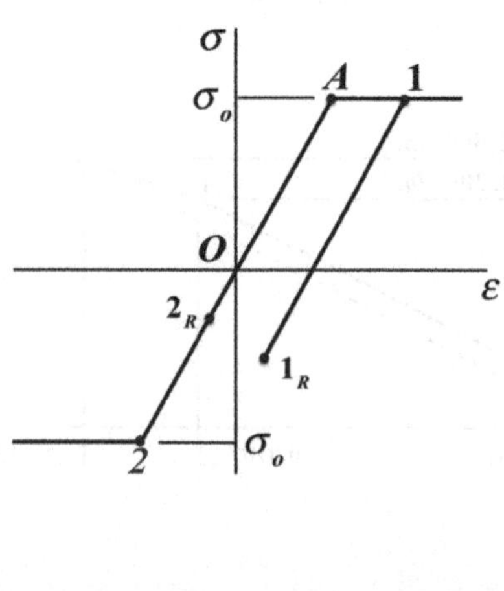

Figure 5.10

ILLUSTRATIVE EXAMPLE 5.6: AXIAL MOTION AND VIBRATION

The foregoing concepts apply as well to solid bodies in motion. We need only replace the equations of equilibrium with the equations of motion, e.g., $F \neq 0$ but $F = ma$.

A turbine blade, as depicted in Figure 5.11, has a uniform cross-section and rotates at n revolutions per minute or $\omega = \pi n/30$ rad/sec. Let us estimate the total extension e when the turbine attains that angular velocity. The equation of radial motion for the body of Figure 5.11b asserts that

$$F = \int_r^L \rho a dr$$

where ρ denotes the mass per unit length and a, the radial acceleration

$$a = r\omega^2$$

Therefore

$$F(r) = \rho\omega^2 \frac{1}{2}(L^2 - r^2)$$

Assuming the deformation as described in Section 5.3 and HI-HO steel, we have

$$\varepsilon_x(r) = \frac{F}{EA} = \frac{\rho\omega^2}{EA}\frac{1}{2}(L^2 - r^2)$$

Then the total elongation follows

$$e = \int_{R_0}^L \varepsilon_x(r)dr = \frac{\rho\omega^2}{EA}\frac{L^3}{3}\left[1 - \frac{3}{2}\frac{R_0}{L} + \frac{1}{2}\left(\frac{R_0}{L}\right)^3\right]$$

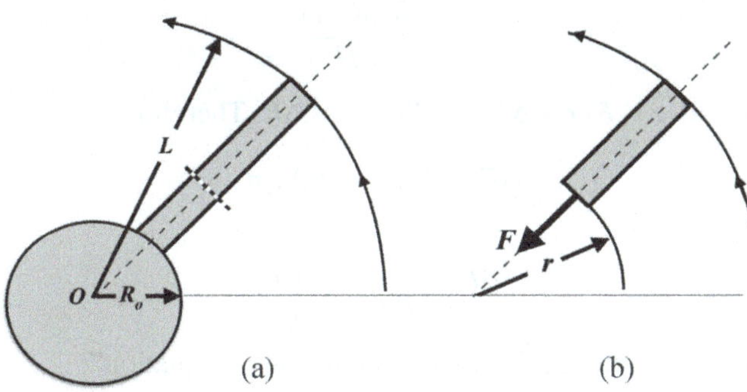

Figure 5.11

ILLUSTRATIVE EXAMPLE 5.7

Let us consider a cylindrical rod fixed at one end and free of all other external loads. The rod is composed of HI-HO material. It is set into axial motion, e.g., an impact upon the free end. We assume that the particles in a cross section remain in a plane and perpendicular to the centroidal axis, i.e., the deformation which is described in Section 5.3. This is depicted in Figure 5.12.

The axial extensional strain follows:

$$\varepsilon_x = \lim_{x \to 0}\left[\frac{u(x + \Delta x) - u(x))}{\Delta x}\right] = \frac{\partial u(x,t)}{\partial x}$$

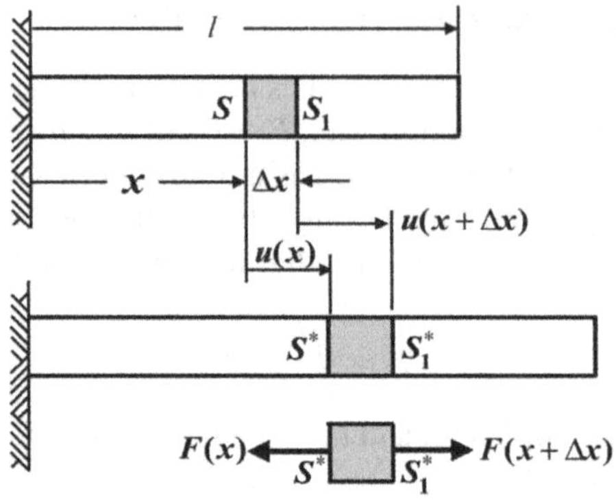

Figure 5.12

Since the material is HI-HO,

$$\sigma_x(x,t) = E\varepsilon_x = E\frac{\partial u(x,t)}{\partial x}$$

The force upon a cross-section follows:

$$F(x,t) = \iint_A \sigma_x \, dA = EA\frac{\partial u(x,t)}{\partial x} \tag{5.26}$$

where A denotes the cross-sectional area. Let us denote the mass density (per unit length) by ρ. Then the mass of the slice between sections S and S_1 is $\Delta m = \rho\Delta x$. The acceleration of the slice is

$$a = \frac{\partial^2 u(x,t)}{\partial t^2}$$

The net force upon the slice is $F(x+\Delta x,t) - F(x,t) \equiv \Delta F$. Therefore,

$$\Delta F = \Delta m a = (\rho\Delta x)a$$

In the limit, we obtain

$$\frac{\partial F}{\partial x} = \rho a = \rho\frac{\partial^2 u}{\partial t^2} \tag{5.27}$$

Substituting Equation (5.26) into (5.27), we obtain the "wave equation":

$$\frac{\partial^2 u(x,t)}{\partial x^2} - \frac{\rho}{EA}\frac{\partial^2 u(x,t)}{\partial t^2} = 0$$

One solution describes a sinusoidal wave which propagates at frequency:

$$\omega_n = p_n\sqrt{\frac{EA}{\rho}}, \quad p_n = \frac{(2n-1)}{2}$$

Here n is any integer and

$$u(x,t) = \sin p_n x \sin \omega_n t$$

This provides a relation between the sounds (frequencies) heard, the stiffness E, the density ρ of the material, and the length l, when the 'nail' is struck by the hammer.

5.6 TRUSSES

5.6.1 GENERAL DESCRIPTION

A truss is a structure composed of slender members joined together at their ends. Portions of bridges, buildings, and airframes are often structures of this type. The members of a truss are relatively slender because they are not intended to resist transverse (bending) loads. Hence, it is common practice to neglect the rotational resistance of the end connections; the effects of bending couples transmitted through these connections are neglected. Accordingly, we assume that the members are joined together by ideal pin connections. Hence, an *ideal* truss is composed of two force members; each member is acted upon by axial forces applied at the ends.

According to Saint-Venant's principle, the manner of load distribution at the end of a member has only a localized effect. Since the members of a truss are relatively slender, such effects are usually of secondary importance. The possibility of a stability failure (buckling) in the compressed members must be kept in mind. This problem is treated in Chapter 6. For the present, we assume that all loads are well below their critical values. Then the results of the previous analyses of axially loaded members (Section 5.3) are applicable to the individual members of the ideal truss.

Figure 5.13 provides a typical example of an actual truss. The photograph on the left shows a supporting tower for the transmission of electric power. Detail views are shown on the right; the lower view depicts the "gusset" plate at the site *A*. The interested reader will find many examples on the internet; search truss definition.

Figure 5.13

ILLUSTRATIVE EXAMPLE 5.8

Let us consider the simple truss depicted in Figure 5.14a. The lengths l of all members are the same; *ABC*, *BCD*, *CDE* are equilateral triangles. The forces upon each member are statically determinate. From the free body of Figure 5.14, we can determine the reactions:

$$H = O, R_A = V = P/2$$

where P denotes the external load at joint C. We can progress to the diagrams at (*c*) and (*d*) to compute the forces upon members 1, 2, 3, and 4, then to the joints C, D, and E to determine the forces upon each member.

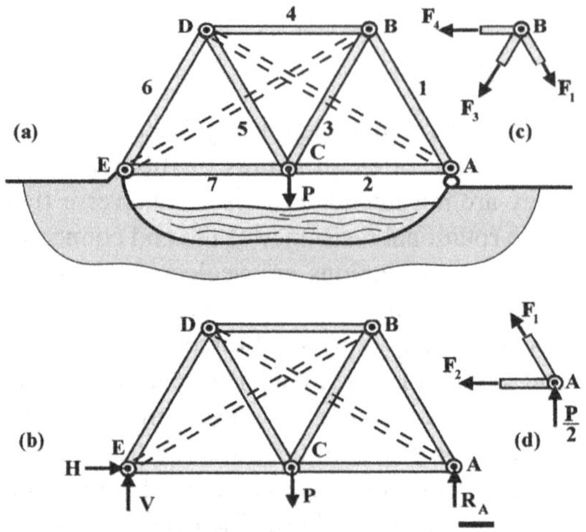

Figure 5.14

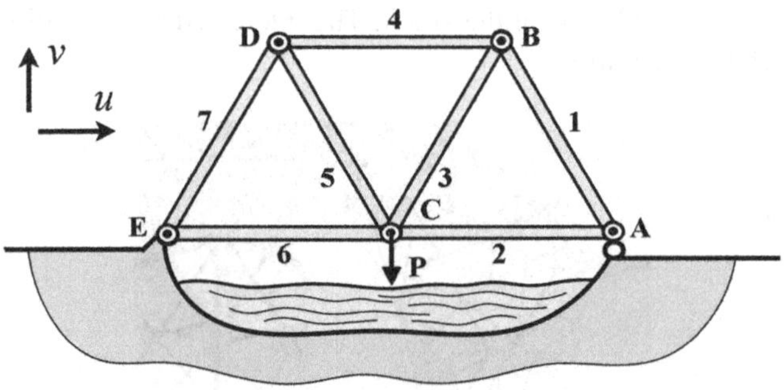

Figure 5.15

As an alternative, we can apply the Lagrangian approach, i.e., the method of virtual work. To that end, we note the 7 degrees of freedom: Seven components of displacement at the joints fully determine the configuration of this truss. Let us denote components of displacement in directions u and v as depicted in Figure 5.15

With the constraints indicated at supports A and E, we have the seven displacements: u_A, u_B, v_B, u_C, v_C, u_D, v_D. The extensions of all members can be expressed in terms of these variables, e.g.,

$$e_3 = \frac{1}{2}u_B + \frac{\sqrt{3}}{2}v_B - \frac{1}{2}u_C - \frac{\sqrt{3}}{2}v_C$$

The virtual work consists of the work expended against internal forces (δU_I) and the external load (δU_E). If the members are elastic, then the former (δU_I) is the incremental change of internal/strain energy. If the load (P) is constant (so-called "dead load") then the latter is the change of potential ($\delta U_E = P\delta v_C$).

The principle of virtual work asserts that $\delta U_I + \delta U_E = 0$. In the present case, we have

$$\delta U_I = F_1\delta e_1 + F_2\delta e_2 + F_3\delta e_3 + F_4\delta e_4 + F_5\delta e_5 + F_6\delta e_6 + F_7\delta e_7$$
$$\delta U_E = P\delta v_C$$

We now express the incremental elongations in terms of the incremental displacements, e.g.,

$$\delta e_3 = \frac{1}{2}\delta u_B + \frac{\sqrt{3}}{2}\delta v_B - \frac{1}{2}\delta u_C - \frac{\sqrt{3}}{2}\delta v_C$$

The result follows:

$$\left(\frac{1}{2}F_1 + F_2\right)\delta u_A + \left(-\frac{1}{2}F_1 + \frac{1}{2}F_3 + F_4\right)\delta u_B + \left(-F_2 - \frac{1}{2}F_3 + F_7 + \frac{1}{2}F_5\right)\delta u_C$$
$$+ \left(F_4 + \frac{1}{2}F_5 - \frac{1}{2}F_6\right)\delta u_D - \frac{\sqrt{3}}{2}(F_1 + F_3)\delta v_B + \left(-\frac{\sqrt{3}}{2}F_3 - \frac{\sqrt{3}}{2}F_5 + P\right)\delta v_c$$
$$-\frac{\sqrt{3}}{2}(F_5 + F_6)\delta v_D = 0$$

Since each displacement is independently variable (a degree of "freedom"), each parenthetical expression must vanish, e.g., the coefficient of δu_A:

$$\frac{1}{2}F_1 + F_2 = 0$$

We recognize this as the equation requiring the equilibrium of the horizontal components of the forces upon joint A. The solution of the seven equations follows:

$$F_1 = -F_3 = -\frac{P}{\sqrt{3}},\ F_2 = \frac{P}{2\sqrt{3}},\ F_4 = -\frac{P}{\sqrt{3}},\ F_5 = \frac{P}{\sqrt{3}},\ F_6 = -\frac{P}{\sqrt{3}},\ F_7 = \frac{P}{2\sqrt{3}}$$

Each of the foregoing forces and the parenthetical terms can be expressed in terms of the elongations, e.g.

$$F_1 = \frac{EAe_1}{l}$$

For simplicity, let us assume that all members are the same (moduli E and area A are the same). Then the resulting seven equations are thereby expressed in terms of the seven displacements. In the present case, our problem is simplified when we note the symmetry of the system and the deformation. Since the reaction H vanishes, the pin at E can be (theoretically) removed. Then, we can view the entire deformed configuration as symmetric about a vertical (y) line through C. The virtual work upon the members CD, CE, and DE is the same as that upon CB, CA, and BA, respectively. Thus we can express δU_I in terms of the four displacements δu_A, δu_B,δv_B, and δv_C.

Then,

$$\delta U_I = \frac{EA}{2l}\left[\left(\sqrt{3}v_B - u_B + u_A\right)\left(\sqrt{3}\delta v_B - \delta u_B + \delta u_A\right) + 4u_A\delta u_A + 8u_B\delta u_B\right.$$
$$\left. + \left(-\sqrt{3}v_C + u_B + \sqrt{3}v_B\right)\left(-\sqrt{3}\delta v_C + \delta u_B + \sqrt{3}\delta v_B\right)\right]$$
$$\delta u_E = P\delta v_C$$

Again, the principle of virtual work provides the conditions for equilibrium:

$$\delta U_I + \delta U_E = 0$$

This, in turn, requires that the coefficient of each incremental displacement (δu_A, δu_B, δv_B, δu_C) vanish. Four equations follow:

$$\frac{\sqrt{3}}{2}v_B - \frac{1}{2}u_B + \frac{5}{2}u_A = 0$$

$$5u_B - \frac{1}{2}u_A - \frac{\sqrt{3}}{2}v_C = 0$$

$$3v_B + \frac{\sqrt{3}}{2}u_A - \frac{3}{2}v_C = 0$$

$$\frac{3}{2}v_C - \frac{\sqrt{3}}{2}u_B - \frac{3}{2}v_B + \frac{PL}{EA} = 0$$

The displacements follow:

$$u_A = \frac{1}{2\sqrt{3}}\frac{PL}{EA}, \quad u_B = -\frac{1}{2\sqrt{3}}\frac{PL}{EA}$$

$$v_B = -\frac{PL}{EA}, \quad v_C = -\frac{11}{6}\frac{PL}{EA}$$

Several features of this truss are noteworthy: It is a simple truss, and every member is essential; the failure of any member would cause a collapse. Failure is most likely to result from buckling. If all members are the same, then the members 1, 4, and 7 would be vulnerable, since these bear greater compressive forces. The introduction of members *AD* and *EB* (indicated by the dotted lines) would provide a measure of safety. The forces upon the members of that modified truss would be statically indeterminate. However, the analysis via the Lagrangian method of virtual work is not significantly altered; one needs to add only the internal work/energy upon the added members. One still obtains seven simultaneous equations or 4, if one acknowledges the symmetry.

ILLUSTRATIVE EXAMPLE 5.9

In Chapter 1 (Section 1.6), we examined the elastic (HI-HO) behavior of the simple truss of Figure 5.16. We return to this example to examine the behavior under load F, which eventually causes small inelastic deformations. As before, we assume that the properties are the same for each member (modulus E, area A, and yield stress $\sigma = \pm\sigma_o$).

The forces (F_C, F_B, F_D) are statically indeterminate. Therefore, a solution requires consideration of the kinematics. Specifically, the elongations of the members are expressed in terms of the two displacements, u and v, as depicted in Figure 5.16:

$$e_C = u, \quad e_D = v, \quad e_B = \frac{1}{\sqrt{2}}(u+v)$$

Since the members have the same HI-Ho properties,

$$F_C = \frac{EA}{L}u, \quad F_D = \frac{EA}{L}v, \quad F_B = \frac{EA}{2L}(u+v) \tag{5.28}$$

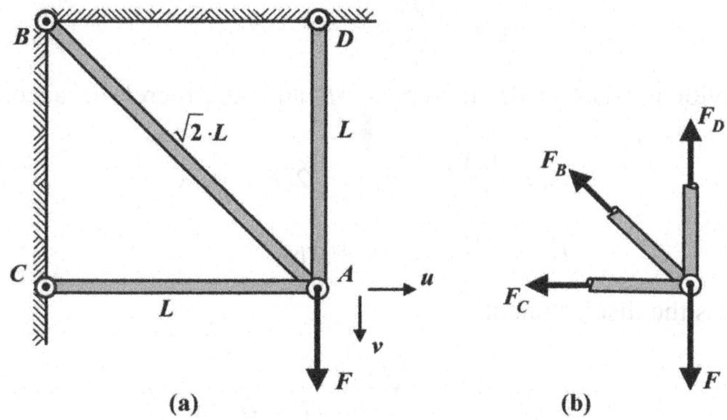

Figure 5.16

The two equilibrium equations follow:

$$\sum F_u = 0 : \left(1 + 2\sqrt{2}\right) u + v = 0$$

$$\sum F_v = 0 : u + \left(1 + 2\sqrt{2}\right) v = \frac{2\sqrt{2}FL}{EA}$$

The solution provides the displacements:

$$u = -\frac{\sqrt{2}}{2\left(2+\sqrt{2}\right)}\frac{FL}{EA} = -0.207\frac{FL}{EA} \tag{5.29a}$$

$$v = \frac{4+\sqrt{2}}{2\left(2+\sqrt{2}\right)}\frac{FL}{EA} = 0.793\frac{FL}{EA} \tag{5.29b}$$

We now accept the uniform strain and stress distributions (Section 5.3):

$$\sigma_C = \frac{E e_C}{L} = \frac{Eu}{L} = -0.207\frac{F}{A} \tag{5.30a}$$

$$\sigma_D = \frac{E e_D}{L} = \frac{Ev}{L} = 0.793\frac{F}{A} \tag{5.30b}$$

$$\sigma_B = \frac{E e_B}{\sqrt{2}L} = \frac{E}{2L}(u+v) = 0.293\frac{F}{A} \tag{5.30c}$$

These results are valid only if the members respond elastically. Since the largest stress (σ_D) occurs in member AD, yielding is initiated in that member when

$$\sigma_D = \sigma_0 = 0.793\frac{F_0}{A}, \quad F_0 = 1.26\sigma_0 A.$$

Henceforth, the material of member AD deforms inelastically.

Let us assume that the material is a structural steel, and that an ideally plastic behavior is justified; the extension of member AD progresses at constant stress (σ_0) while members AB and AC remain elastic. Equilibrium requires that

$$F_B = \sqrt{2}(F - F_D) = \sqrt{2}(F - \sigma_0 A) \tag{5.31a}$$

$$F_C = -\frac{F_B}{\sqrt{2}} = -F + \sigma_0 A \qquad (5.31b)$$

The Hookean behavior persists in the members AB and AC; therefore, according to Equations (5.28) and (5.31),

$$F_B = \frac{EA}{2L}(u+v) = \sqrt{2}(F - \sigma_0 A)$$
$$F_C = \frac{EA}{L}u = -F + \sigma_0 A$$

The solution provides the displacement,

$$v = \frac{\left(1+2\sqrt{2}\right)L}{EA}(F - \sigma_0 A) \qquad (5.32)$$

The equation traces the line AA' in Figure 5.16. The relation holds until the stress in member AB attains the yield stress (σ_0). According to Equation (5.131a), that "limit" load is

$$F_L = \frac{\left(1+\sqrt{2}\right)\sigma_0 A}{\sqrt{2}} = 1.71\sigma_0 A$$

At the limit load

$$F_D^L = A\sigma_0, \quad F_C^L = -0.707A\sigma_0, \quad F_B^L = A\sigma_0$$

At the limit load F_L, the truss can support no additional load. The plot of load F versus displacement v is shown in Figure 5.17. Of course, our analysis is founded on the assumption of small displacements ($v/L \ll 1$) and ideally plastic behavior. Eventually, the displacements invalidate our linear equation and the material exhibits some strain hardening. Both effects will result in some additional support; then the actual plot would follow an upward path, as indicated by the dotted curve $A''B$ of Figure 5.17.

Let us examine the consequences of unloading from inelastically deformed state A'' of Figure 5.17. During the unloading, the structural steel deforms elastically. Specifically, the incremental relation: between stress and strain is Hookean, such that

$$\Delta e_D = \frac{L}{EA}\Delta F_D, \quad \Delta e_C = \frac{L}{EA}\Delta F_C, \quad \Delta e_B = \frac{\sqrt{2}L}{EA}\Delta F_B \qquad (5.33)$$

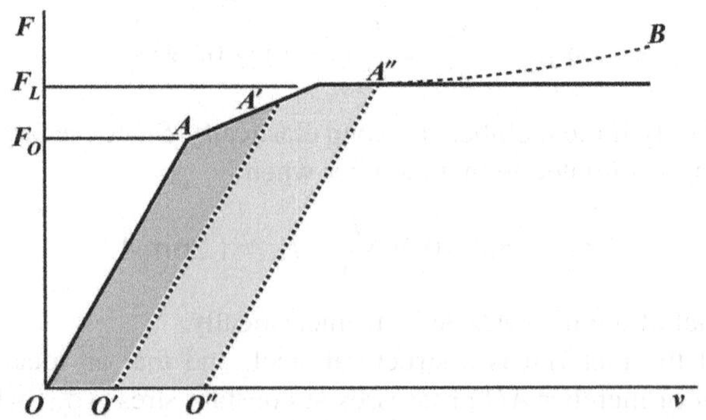

Figure 5.17

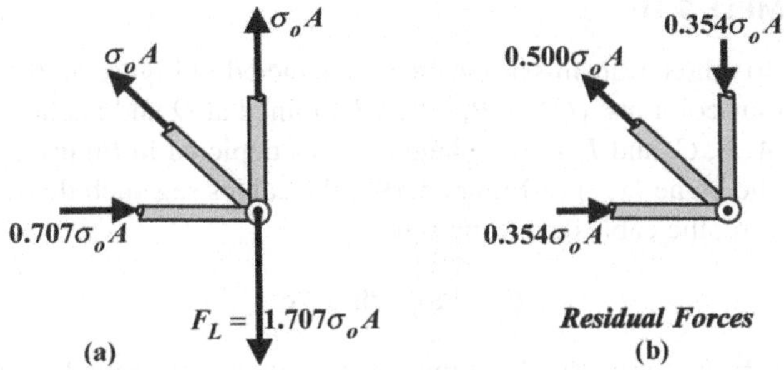

Figure 5.18

If the displacement is not excessive, then the linear kinematic relations hold:

$$\Delta e_D = \Delta v, \quad \Delta e_C = \Delta u, \quad \Delta e_B = \frac{1}{\sqrt{2}}(\Delta u + \Delta v) \tag{5.34}$$

The linear versions of the equilibrium conditions also hold:

$$\Delta F_D + \frac{1}{\sqrt{2}}\Delta F_B = \Delta F, \quad \Delta F_C + \frac{1}{\sqrt{2}}\Delta F_B = 0 \tag{5.35}$$

Substituting Equation (5.34) into (5.33) into (5.35), we obtain the two equations in the displacements (Δu, Δv) and the solution:

$$\Delta u = -0.207\,\frac{L\Delta F}{EA}, \quad \Delta v = 0.793\,\frac{L\Delta F}{EA}$$

The changes of the forces follow from Equations (5.33) and (5.34):

$$\Delta F_D = \frac{EA}{L}\Delta v, \quad \Delta F_C = \frac{EA}{L}\Delta u, \quad \Delta F_B = \frac{EA}{2L}(\Delta u + \Delta v)$$

When the load is removed, $\Delta F = -F_L$,

$$\Delta F_D = -1.354\sigma_0 A\,, \quad \Delta F_C = +0.354\sigma_0 A\,, \quad \Delta F_B = -0.500\sigma_0 A$$

The residual forces follow:

$$F_D = F_D^L + \Delta F_D,\ F_C = F_C^L + \Delta F_C,\ F_B = F_B^L + \Delta F_B$$

$$F_D^R = -0.354\sigma_0 A,\ F_C^R = -0.354\sigma_0 A,\ F_B^R = +0.500\sigma_0 A$$

Conditions at the limit (F_L) and the state of residual stresses are depicted in Figure (5.18a) and (b), respectively. Once again, we encounter a situation wherein a part of a structure has deformed inelastically, whereas another part remains elastic. The result is a state of residual stress. The unloading from states A' and A'' trace the dotted lines $A'O$ and $A''O$, respectively. The energy dissipated in the inelastic deformations is the shaded areas $OA'O' and OA''O''$, respectively.

As noted in our introductory remarks, slender members under axial compression are liable to buckle. Unless lateral deflection is inhibited by adjoining supports, such buckling usually occurs at stresses that are less than the compressive yield stress. The buckling might actually delimit loading. Buckling and its consequences are examined in Chapter 8. For the present, we presume that buckling is inhibited and that simple extensional deformations prevail.

ILLUSTRATIVE EXAMPLE 5.10

A simple tower supports three transmission cables as depicted in Figure 5.19.

It consists of the four columns, (OA, OB, OC, OD) joined at O and anchored to the horizontal concrete platform at A, B, C, and D in the plane (x,y) as depicted in Figure 5.19. The cables are directed as shown in the plane (x,y) of Figure 5.19b; the cables sag such that they slope 10° from the horizontal. Therefore, the cables exert the force

$$\mathbf{F} = P\hat{\mathbf{a}} + P\hat{\mathbf{b}} + P\hat{\mathbf{c}}$$

where unit vectors ($\hat{\mathbf{a}}$, $\hat{\mathbf{b}}$, $\hat{\mathbf{c}}$) define the directions of the tensions P applied at the apex O.

$$\begin{aligned}
\hat{\mathbf{a}} &= 0.696\hat{\mathbf{i}} - 0.174\hat{\mathbf{j}} + 0.696\hat{\mathbf{k}} \\
\hat{\mathbf{b}} &= -0.696\hat{\mathbf{i}} - 0.174\hat{\mathbf{j}} + 0.696\hat{\mathbf{k}} \\
\hat{\mathbf{c}} &= -0.174\hat{\mathbf{j}} - 0.985\hat{\mathbf{k}}
\end{aligned}$$

and $\hat{\mathbf{i}}$, $\hat{\mathbf{j}}$, and $\hat{\mathbf{k}}$ form the triad of orthonormal vectors in the directions x, y, z, respectively.

To implement our analysis, we require the orientation of each of the four legs; These are expressed by the four unit vectors $\hat{\mathbf{A}}$, $\hat{\mathbf{B}}$, $\hat{\mathbf{C}}$, and $\hat{\mathbf{D}}$ in the directions AO, BO, CO, DO, respectively.

$$\begin{aligned}
\hat{\mathbf{A}} &= -0.243\hat{\mathbf{i}} + 0.970\hat{\mathbf{j}} \\
\hat{\mathbf{B}} &= 0.243\hat{\mathbf{i}} + 0.970\hat{\mathbf{j}} \\
\hat{\mathbf{C}} &= 0.970\hat{\mathbf{j}} + 0.243\hat{\mathbf{k}} \\
\hat{\mathbf{D}} &= 0.970\hat{\mathbf{j}} - 0.243\hat{\mathbf{k}}
\end{aligned}$$

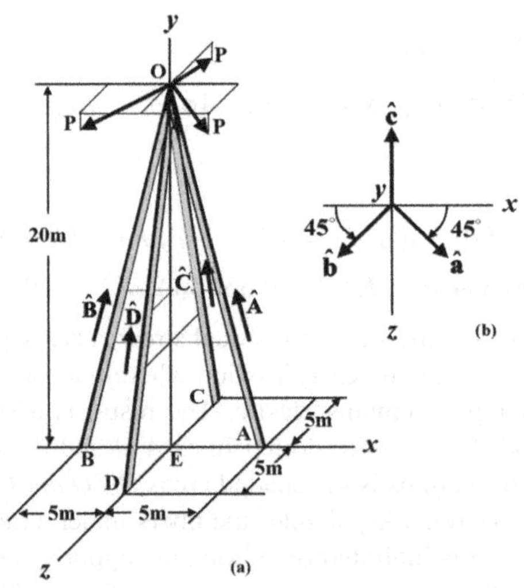

Figure 5.19

Let us denote the displacement of joint O by

$$\mathbf{v} = u\hat{\mathbf{i}} + v\hat{\mathbf{j}} + w\hat{\mathbf{k}} \tag{5.36}$$

We note that this truss has but three degrees of freedom; u, v, w determine the deformed configuration. The elongations of the three members follow:

$$e_A = \hat{\mathbf{A}} \cdot \mathbf{v} = -0.243u + 0.970v \tag{5.37a}$$

$$e_B = \hat{\mathbf{B}} \cdot \mathbf{v} = 0.243u + 0.970v \tag{5.37b}$$

$$e_C = \hat{\mathbf{C}} \cdot \mathbf{v} = 0.970v + 0.243w \tag{5.37c}$$

$$e_D = \hat{\mathbf{D}} \cdot \mathbf{v} = -0.970v - 0.243w \tag{5.37d}$$

The virtual displacement of the joint O requires the virtual work δU_I expended upon internal axial forces, F_A, F_B, F_C, and F_D, respectively.

$$\delta U_I = F_A \delta e_A + F_B \delta e_B + F_C \delta e_C + F_D \delta e_D \tag{5.38}$$

The virtual work expended upon the external forces is

$$\delta P_E = -\mathbf{F}\delta\mathbf{v} = -P\left(\hat{\mathbf{a}} + \hat{\mathbf{b}} + \hat{\mathbf{c}}\right) \cdot \delta\mathbf{v}$$

According to Equations (5.36)–(5.38), we obtain

$$\begin{aligned} \delta U_I + \delta U_E = &(-0.243F_A + 0.243F_B)\,\delta u \\ &+ (0.970F_A + 0.970F_B + 0.970F_C + 0.970F_D + 3 \times 0.174P)\,\delta v \\ &+ (0.243F_C - 0.243F_D - 0.407P)\,\delta w \end{aligned} \tag{5.39}$$

Equilibrium requires that each parenthetical coefficient vanish. The first supports our observation of symmetry with respect to the $y - z$ plane:

$$F_A = F_B$$

We are left with two equations in the three forces, $F_A = F_B$, F_C, and F_D. These forces are statically indeterminate. Therefore, we express the equations in terms of the displacements. Since the structure is symmetric with respect to the x-y plane, we would assign the same properties to members AO and BO; then $u = O$. It follows from Equations (5.37a,b) that $e_A = e_B$ and, therefore

$$F_A = F_B = \frac{EA}{l} e_A = 0.970 \frac{EA}{l} v \tag{5.40}$$

where A denotes the cross-sectional area of members OA and OB. The two remaining equations (coefficients of δv and δw vanish) follow:

$$2 \times (0.970) F_A + 0.970F_C + 0.970F_D = -3 \times 0.174P \tag{5.41a}$$

$$0.243F_C - 0.243F_D = 0.407P \tag{5.41b}$$

If we assume the same properties (E and A) for all members, then

$$F_C = \frac{EA}{l} e_C = \frac{EA}{l} 0.970v + \frac{EA}{l} 0.243w \tag{5.42a}$$

$$F_D = \frac{EA}{l}e_D = \frac{EA}{l}0.970v - \frac{EA}{l}0.243w \tag{5.42b}$$

Substituting the expressions (5.40) and (5.42) for forces, F_A, F_C, F_D into the two equilibrium Equations (5.41a,b), we obtain

$$4 \times (0.970)^2 v = -3 \times 0.174 \frac{Pl}{EA} \tag{5.43a}$$

$$2 \times (0.243)^2 w = 0.407 \frac{Pl}{EA} \tag{5.43b}$$

It follows from (5.43), (5.42) and (5.40) that

$$v = -0.139 \frac{Pl}{EA}, \quad w = 3.45 \frac{Pl}{EA}$$

$$F_C = 0.703P, \quad F_D = -.973P, \quad F_A = -0.135P$$

Not surprisingly, the two columns OA and OB support only the slight downward forces; OC is subject to a tensile force. We might suppose that the column OD is most vulnerable to buckling under the largest compressive force. Certainly, the engineer would propose a member with the larger cross-section and more resistance to buckling. Our analysis follows as before. If, for example, we set the cross-sectional area of member OD at twice that of the others, i.e., $A_D = 2A$, then Equation (5.42b) is replaced by

$$F_D = \frac{EA}{l}(2 \times 0.970)v - \frac{EA}{l}(2 \times 0.243)w$$
$$F_C = \frac{EA}{l}(0.970)v + \frac{EA}{l}(0.243)w$$

Now Equations (5.41a,b) assume the forms:

$$4.705v - 0.236w = -0.522 \frac{PL}{EA}$$

$$-0.236v + 0.177w = 0.407 \frac{PL}{EA}$$

The solution follows:

$$v = +0.00470 \frac{PL}{EA}, \quad w = 2.305 \frac{PL}{EA}$$

The forces follow according to Equations (5.40) and (4.43a,b):

$$F_A = F_B = +0.005P, \quad F_C = 0.565P, \quad F_D = -1.111P$$

Note that the increase of stiffness in members OD reduces displacement v and increases the compressive load upon member OD. However, the cross-sectional property I (see Section 6.9, Equation 6.82) of member OD would be significantly increased to inhibit buckling. Lateral members (ties) between the legs would greatly increase the stability of the tower. The engineer must also consider the very different, but possible, situation, should any of the cables be disconnected. The interested reader can pursue a similar analysis with the external loading altered in the expression for δU_E.

5.7 TWISTING OF CIRCULAR SHAFTS

The Problem: Shafts of circular cross section are commonly subjected to twisting couples. The drive shaft of an automobile, for example, is subjected to torques that tend to twist one end relative to the other. Another example is the torsion bar found in spring suspension systems. As a specific example, consider the circular cylinder shown in Figure 5.20. One end ($z = 0$) is fixed, and the free end is subjected to the twisting couple $\bar{T}$.

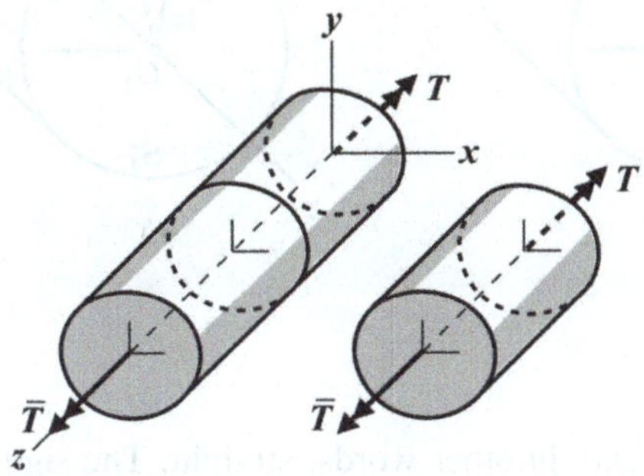

Figure 5.20

From a free-body diagram of any portion of the cylinder, Figure 5.21b, it is apparent that the net internal force system on any cross-section is simply the twisting couple $T = \bar{T}$. Our attention is presently limited to the shaft of circular cross-section, wherein all properties are symmetric about the axis. This symmetry of properties and loading implies certain symmetry in the deformation.

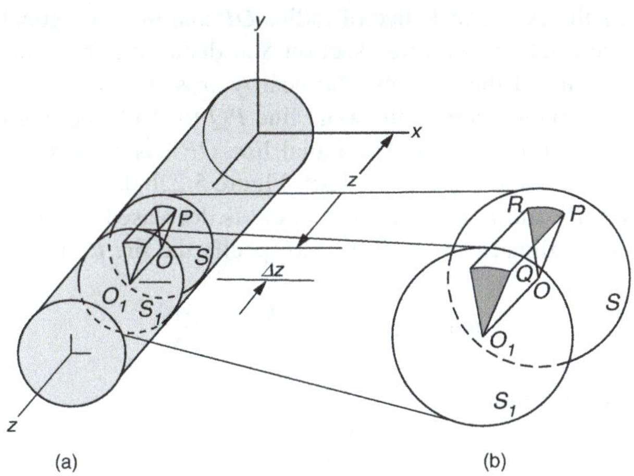

Figure 5.21

5.7.1 DESCRIPTION OF THE DEFORMATION

Our symmetric shaft appears the same when viewed from either end. This means that the midsection must remain plane; either half is a reflection of the other about that midplane. Cross sections must remain symmetric about the axis, but rotate. After deformation, a straight diameter must

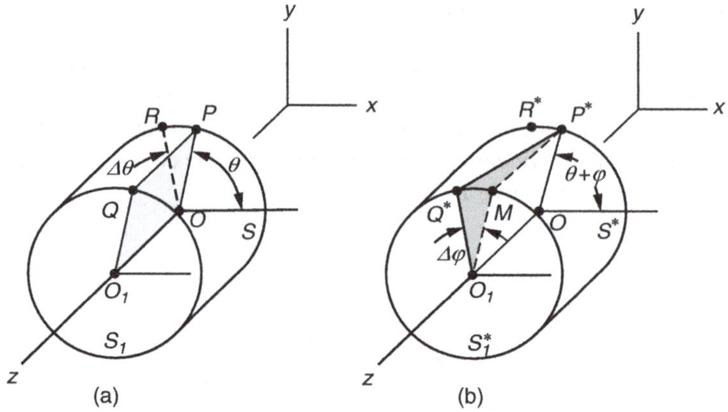

Figure 5.22

appear the same from either end, in other words, straight. The significant feature of the twisting deformation is the relative rotation of cross sections about the axis of the cylinder. We assume that, when this deformation occurs, (1) the axis of the circular cylinder remains straight and (2) all radii in a cross section remain straight and rotate through the same angle about the axis. The kinematics of this deformation were addressed in Section 2.24, Illustrative Example 2.20. We now recount the earlier description: Figure 5.21 depicts two cross-sections S and S_1, in the undeformed cylindrical shaft and an axial line QP of length Δz between these sections. Figure 5.21b depicts the slice between the sections, S and S_1, the axial line QP and the orthogonal circumferential line PR. Note that the radial position OP is arbitrary.

Figure 5.22a depicts the cylindrical slice of radius OP and the orthogonal circumferential line PR. Let $\phi(z)$ denote the rotation of a cross-section S at distance z from the fixed end, as shown in Figure 5.22b. As a result of the twisting, the nearby cross-section S_1 rotates relative to S by an amount $\Delta\phi$. The deformation carries the axial line PQ to P^*Q^*, but the line P^*R^* remains in the plane of S^*, i.e., normal to the axis. The radial line OP has rotated through the angle $\phi(z)$, but O_1Q^* has rotated through the angle $\phi(z)+\Delta\phi$. Figure 5.22b depicts the change in the relative positions of the lines P^*R^* and P^*Q^*; the right angle is decreased by the angle MP^*Q^*. Let us employ the cylindrical coordinates $(r = OP, \theta, z)$ to specify the site P; then

$$\tan\angle MP^*Q^* = \frac{MQ^*}{MP^*} = r\frac{\Delta\phi}{\Delta z}$$

By our definition of a shear strain

$$\gamma_{z\theta} = \lim_{z\to 0}\angle MP^*Q^*$$

For small strains ($\gamma_{z\theta} \ll 1$), we have the approximation:

$$\gamma_{z\theta} \doteq r\frac{d\phi}{dz} \tag{5.44}$$

Under the prescribed circumstances, the geometry and the loading, the strains of lines in a cross section are neglected; we presume that $\varepsilon_\theta = \varepsilon_r = \gamma_{r\theta} = 0$. Likewise, the components ε_x, ε_y, and γ_{xy} are neglected.

5.7.2 STRESS AND RESULTANT COUPLE

In view of the loading and strain $\gamma_{z\theta}$, we must anticipate a shear stress $\tau_{z\theta}$. Moreover, the symmetry requires that the component can depend only on the radius; i.e., $\tau_{z\theta} = \tau_{z\theta}(r)$.

The cross-section S is shown in Figure 5.23. On the element of area ΔA, the shear stress produces a force $\Delta V_\theta = \tau_{z\theta}\Delta A = \tau_{z\theta} r\, d\theta\, dr$ Since $\tau_{z\theta}$ is independent of the angle θ, the distribution is symmetric about the axis z. As shown on the cross section of Figure 5.23 for each elemental area with the force ΔV_θ, there is a diametrically opposed element with an opposite force ΔV_θ. Such distribution produces only a couple T upon the cross section; in other words, the net shear force vanishes ($V_y = V_z = 0$). The twisting couple is the integral

$$T = \lim_{\Delta A \to 0} \iint_A r\tau_{z\theta} \mathrm{d}A = \iint_A r^2 \tau_{z\theta} \mathrm{d}\theta \mathrm{d}r \tag{5.45}$$

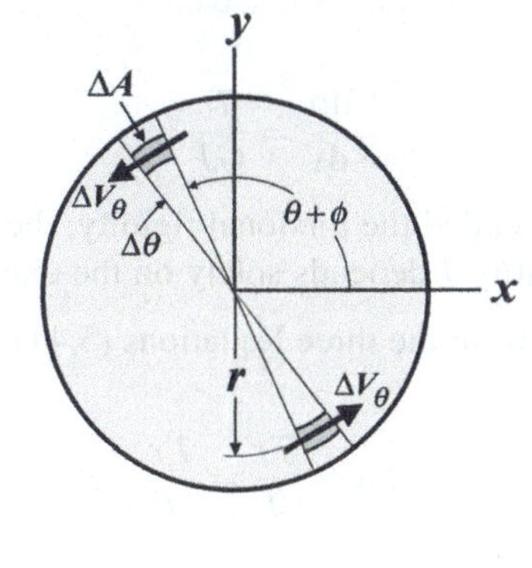

Figure 5.23

5.7.3 HOOKEAN ISOTROPIC BEHAVIOR

If the material is Hookean and isotropic, then our assumptions about the deformation - specifically $\varepsilon_x = \varepsilon_y = \varepsilon_z = \gamma_{zr} = \gamma_{r\theta} = 0$ - imply also that $\sigma_x = \sigma_y = \sigma_z = \tau_{zr} = \tau_{r\theta} = 0$. Indeed, the latter conditions are as plausible and equivalent to the previous assumptions. The non-vanishing stress is

$$\tau_{z\theta} = G\gamma_{z\theta} \tag{5.46}$$

The kinematic result (5.44), the *stress-strain* Equation (5.46), and the statical requirement (5.45) can be used to obtain the interrelationships among the stress ($\tau_{z\theta}$), applied torque (T), and rotation (ϕ). Substituting Equation (5.44) into the right side of Equation (5.46) and the latter into the right side of Equation (5.45), we obtain

$$T = \frac{\mathrm{d}\phi}{\mathrm{d}x} \iint Gr^3 \mathrm{d}\theta \mathrm{d}r \tag{5.47}$$

Our underlying assumptions presume only symmetry about the axis. Indeed, the modulus G might depend on the radius r. Also, the shaft might be tubular with a (hollow) portion of radius a and outer radius R. Therefore, Equation (5.47) takes the general form

$$T = 2\pi \frac{\mathrm{d}\phi}{\mathrm{d}x} \int_a^R G(r) r^3 \mathrm{d}r \tag{5.48}$$

If the shaft is composed of HI-HO material (Hookean isotropic and homogeneous), then G is constant, and Equation (5.48) assumes the particular form

$$T = G\frac{\pi}{2}(R^4 - a^4)\frac{\mathrm{d}\phi}{\mathrm{d}x} \tag{5.49}$$

Frequently, engineers employ the abbreviation[3]

$$\frac{\pi}{2}(R^4 - a^4) \equiv J \tag{5.50}$$

Then, the twisting of the HI-HO circular shaft is described by the equation

$$T = GJ\frac{\mathrm{d}\phi}{\mathrm{d}x} \tag{5.51a}$$

or

$$\frac{\mathrm{d}\phi}{\mathrm{d}x} = \frac{T}{GJ} \tag{5.51b}$$

The quantity GJ is sometimes called the torsional rigidity; the modulus G reflects the stiffness of the material, whereas the quantity J depends solely on the diameter(s) of the shaft. If the twist $\frac{\mathrm{d}\phi}{\mathrm{d}x}$ and strain $\gamma_{z\theta}$ are eliminated from the three Equations (5.44), (5.46), and (5.51), we obtain

$$\tau_{z\theta} = \frac{Tr}{J} = \frac{\bar{T}r}{J} \tag{5.52}$$

The right side of Equation (5.52) merely acknowledges the equilibrium condition; the couple on the cross section $T = \bar{T}$, the applied torque.

The overall deformation of the cylinder is described by the rotation of one end relative to the other, in this case, the rotation of the free end at $x = L$. From Equation (5.51b), the relative rotation of two nearby cross sections is given by

$$\Delta\phi = \frac{\mathrm{d}\phi}{\mathrm{d}x}\Delta x = \frac{T}{GJ}\Delta x$$

If the rotation of the face at $x = L$ relative to the face at $x = 0$ is denoted by ϕ_0, then

$$\phi_0 = \int_0^L \frac{T}{GJ}\mathrm{d}x \tag{5.53}$$

In the present case, $T = \bar{T}$ and G and J are constant, so that

$$\phi_0 = \frac{\bar{T}}{GJ}\int_0^L \mathrm{d}x = \frac{\bar{T}L}{GJ} \tag{5.54}$$

The stress distribution of Equation (5.52) applies to every cross section of the cylinder. Strictly speaking, the end faces at $x = 0$ and $x = L$ are also included. This would greatly restrict the utility of Equations (5.52) and (5.54) if it were not for Saint-Venant's principle. If a circular cylinder is reasonably slender and the loading applied to its ends are twisting couples, Equation (5.52) is applicable to intermediate cross sections, and Equation (5.54) gives the overall deformation.

If the end torques are actually applied according to Equation (5.52), then Equations (5.52) and (5.54) are the same as results obtained from more elaborate theories. Furthermore, it can be shown that all the other components of stress and strain vanish; that is,

$$\varepsilon_r = \varepsilon_\theta = \varepsilon_z = \gamma_{r\theta} = \gamma_{rz} = 0$$
$$\sigma_r = \sigma_\theta = \sigma_z = \tau_{r\theta} = \tau_{rz} = 0$$

For cylinders that do not have circular cross sections, the assumptions about the deformation are not valid. In particular, the noncircular cross section will not remain plane.

ILLUSTRATIVE EXAMPLE 5.11: NONHOMOGENEOUS HOOKEAN SHAFT

The foregoing analysis of torsion incorporated certain *kinematics, statics, and material* properties. The underlying description presumes symmetry about the axis and uniformity along the shaft. Accordingly, cross-sections are not distorted, and the kinematic relation (5.44) follows

$$\gamma_{z\theta} = r\frac{d\phi}{dz} \tag{5.55}$$

In view of the symmetry, the essential statical result is Equation (5.45); on every cross section, the resultant is the couple.

$$T = \iint \tau_{z\theta} r \mathrm{d}r \mathrm{d}\theta \tag{5.56}$$

These results apply to any circular shaft with properties that are symmetric about the axis. As an example, suppose that the material is Hookean and isotropic *but* the modulus is a function:

$$G = G(r)$$

One might incorporate a material (a matrix of metal or plastic) filled with small fibers of stiffer and stronger material. The latter are dispersed more densely in the outer portions to create a more efficient shaft. To illustrate, suppose that

$$G = G_0\left(1 + \frac{r}{R}\right)$$

Then, the material at the outer surface is twice as stiff as the material at the center and

$$\tau_{z\theta} = G\gamma_{z\theta} = G_0\left(1 + \frac{r}{R}\right)\gamma_{z\theta} \tag{5.57}$$

Again, the kinematics, the statics, and the material relations are all needed to effect the solution. Substituting Equation (5.55) into Equation (5.57), followed by substitution into Equation (5.56), yields

$$\begin{aligned} T &= G_0\frac{d\phi}{dz}\int_0^R\int_0^{2\pi}\left(1 + \frac{r}{R}\right) r^3 \mathrm{d}\theta \mathrm{d}r \\ &= G_0\left(\frac{9}{10}\pi R^4\right)\frac{d\phi}{dz} \end{aligned} \tag{5.58}$$

If $\frac{d\phi}{dx}$ and $\gamma_{z\theta}$ are eliminated from Equations (5.55), (5.57), and (5.58), we obtain,

$$\tau_{z\theta} = \frac{10r}{9\pi R^4}\left(1+\frac{r}{R}\right)T \tag{5.59}$$

ILLUSTRATIVE EXAMPLE 5.12: COMPOSITE SHAFT

In keeping with the general description of the symmetric circular shaft, we might form the shaft with a light cylindrical core of Hookean isotropic material and an outer cylindrical tube of harder and stiffer material. Such a shaft would be lighter, stiffer, and more wear-resistant. To illustrate, suppose that the core and outer tube are bonded together; in particular, since both are joined at the ends, both undergo the same twist. Kinematic Equation (5.44) holds; again,

$$\gamma_{z\theta} = r\frac{d\phi}{dz} \tag{5.60}$$

The torque is again given by Equation (5.45):

$$T = \iint \tau_{z\theta} r dA \tag{5.61}$$

Let us denote the radius of the core (also, the inner radius of the surrounding tube) by R_c, the modulus of the core by G_c, and the modulus of the outer shell by G_s. Then

$$\tau_{x\theta} = G_c\gamma_{z\theta}, \quad r < R_c \tag{5.62a}$$

$$\tau_{x\theta} = G_s\gamma_{z\theta}, \quad R_c < r \leq R_0 \tag{5.62b}$$

Again, utilizing the *kinematics*, the *statics*, and the *description* of the material and substituting Equation (5.60) into Equations (5.62) and (5.61), we have

$$T = \frac{d\phi}{dz}\left[G_c \iint_{A_c} r^2 dA + G_s \iint_{A_s} r^2 dA\right] \tag{5.63a}$$

The two distinct moduli for the core and shell require that we split the integral into the two parts. Here, as before, the integrals reflect the size of the sections:

$$\iint_{A_c} r^2 dA = \frac{\pi R_c^4}{2} \equiv J_c, \quad \iint_{A_s} r^2 dA = \frac{\pi}{2}\left(R_0^4 - R_c^4\right) \equiv J_s$$

Substituting these abbreviations into Equation (5.63a), we obtain

$$T = (G_c J_c + G_s J_s)\frac{d\phi}{dz} \tag{5.63b}$$

The reader can compare this result with that obtained for the homogeneous (HI-HO) shaft, Equation (5.51a). As might be anticipated, the stiffness of the composite shaft is the sum of the stiffnesses of the core and shell.

ILLUSTRATIVE EXAMPLE 5.13: TORSIONAL MOTION AND VIBRATION

The foregoing kinematics can be applied to torsional motion. Specifically, we assume that every transverse slice rotates as a rigid disc; however, the angle ϕ is dependent on time t, i.e., $\phi = \phi(z,t)$.

If the body is composed of HI-HO material, then

$$T(z,t) = GJ\frac{\partial \phi}{\partial z}$$

If the material has a homogeneous mass density ρ, then the mass moment of inertia for a slice of width Δz is

$$\Delta I = \frac{1}{2}\rho \pi R^4 \Delta z \equiv \rho J \Delta z$$

Since the slice rotates with angular acceleration, $\partial^2\phi/\partial t^2$, we have

$$T(z+\Delta z,t) - T(z,t) = \rho J\frac{\partial^2 \phi}{\partial t^2}\Delta z$$

where ρ denotes the mass per unit volume. In the limit

$$\frac{\partial T}{\partial z} = \rho J\frac{\partial^2 \phi}{\partial t^2} = GJ\frac{\partial^2 \phi}{\partial z^2}$$

Then,

$$\frac{\partial^2 \phi}{\partial z^2} - \frac{\rho}{G}\frac{\partial^2 \phi}{\partial t^2} = 0$$

This is the "wave equation," as encountered in Illustrative Example 5.7. Here, the modulus G replaces the stiffness EA in the equation governing the axial motion.

5.8 TORSION OF AN IDEALLY ELASTIC-PLASTIC SHAFT

Our description of the twisting depends only upon the axial symmetry of the shaft. In particular, the kinematic result (5.44) is applicable provided only that the properties remain symmetric—that is, dependent only on the radial coordinate r. Restated, the strain is given by the equation:

$$\gamma_{z\theta} = r\frac{\mathrm{d}\phi}{\mathrm{d}z}$$

Now, suppose that the circular shaft of Figure 5.21 is made of an ideally elastic-plastic material. The stress-strain diagram is shown in Figure 5.24.

If the shear strain does not exceed $\gamma_0 = \tau_0/G$ anywhere in the shaft, the material behavior is linearly elastic, and the results previously obtained are applicable. In particular, the shear stress varies linearly along a radius. As the twisting couple is increased, the strain $\gamma_{z\theta}$ exceeds the yield strain γ_0 in the outermost portion of the cross section, as indicated in Figure 5.25a. We denote by b the radius at which $\gamma_{z\theta} = \gamma_0$.

The inner core ($r < b$) behaves elastically because $\gamma_{z\theta} < \gamma_0$ whereas the outermost portion ($r > b$) undergoes plastic deformation because $\gamma_{z\theta} > \gamma_0$. In the elastic core the shear stress $\tau_{z\theta}$ is proportional to the shear strain $\gamma_{z\theta}$.

$$\tau_{z\theta} = G\gamma_{z\theta} = G\frac{\mathrm{d}\phi}{\mathrm{d}z}r, \quad r \le b \tag{5.64a}$$

In the plastic (outer) portion

$$\tau_{z\theta} = \tau_0, \quad r \geq b \tag{5.64b}$$

The shear stress τ_0 varies along a radius, as shown in Figure 5.25b. At the elastic-plastic interface ($r = b$), we have

$$G\frac{d\phi}{dz}b = \tau_0 \tag{5.65}$$

Using this to eliminate $G\,d\phi/dz$ from Equation (5.64a), we obtain, finally

$$\tau_{x\theta} = \begin{cases} \frac{r}{b}\tau_0, & r \leq b \\ \tau_0, & r \geq b \end{cases} \tag{5.66a,b}$$

This is equivalent to Equation (5.64); we have merely replaced the twist $d\phi/dz$ in Equation (5.64a) with the unknown radius b in Equation (5.66a). Again, we note that all preceding steps are valid as the symmetry about the axis is maintained. Our shaft may be solid or tubular. As before, we admit either by assuming an inner radius a and outer radius R. **Note: Figure 5.25 depicts only the solid shaft, i.e.,** $a - 0$

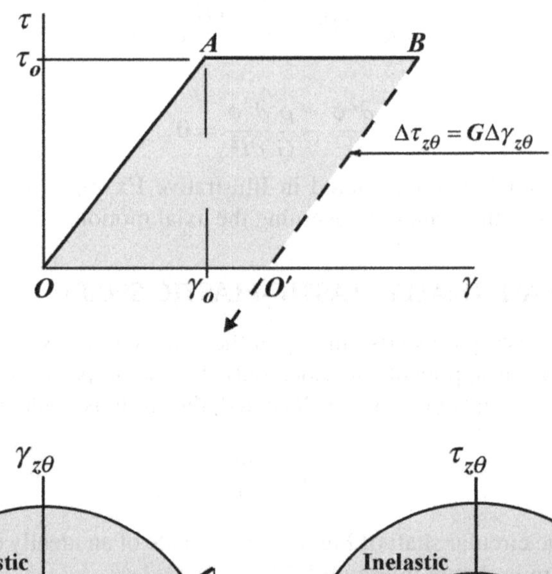

Figure 5.24

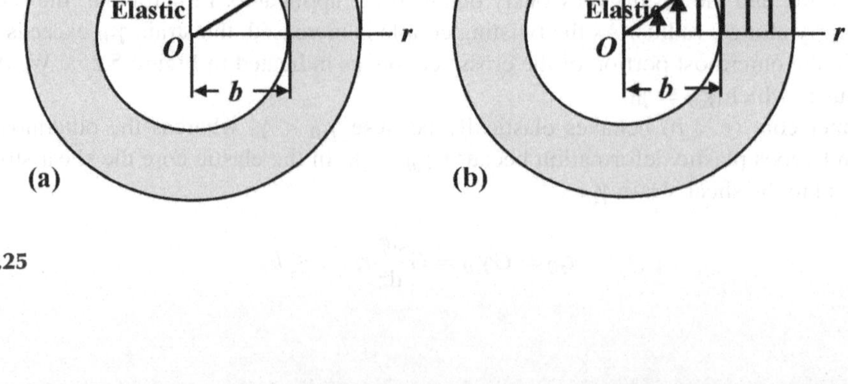

Figure 5.25

The twisting couple T on a cross section is given by

$$T = \iint_S r\tau_{z\theta}\mathrm{d}A = \int_0^{2\pi}\int_a^R \tau_{z\theta}r^2\mathrm{d}r\mathrm{d}\theta$$

and since $\tau_{z\theta}$ does not depend on θ,

$$T = 2\pi\int_a^R \tau_{z\theta}r^2\mathrm{d}r$$

In view of Equations (5.66a,b), the integral must be split into two parts: one for the interval a to b, the other for interval b to R: Then

$$\begin{aligned} T &= 2\pi\frac{\tau_o}{b}\int_a^b r^3\mathrm{d}r + 2\pi\tau_0\int_b^R r^2\mathrm{d}r \\ &= \frac{\pi\tau_0}{6}\left[4R^3 - b^3 - 3\frac{a^4}{b}\right] \end{aligned} \tag{5.67}$$

The radius of the elastic core can be determined by solving Equation (5.67) for b. For the results to be meaningful, b must lie between a and R. When $b = R$, the material at the outermost portion of the cross section is about to yield. The twisting couple at this point is

$$T_0 = \frac{\pi\tau_0}{2R}(R^4 - a^4) \tag{5.68}$$

For T less than T_0, the behavior of the material is linearly elastic, and the previously obtained results can be used.

As b decreases, the elastic core becomes smaller. In the limiting condition $b = a$, we say that the cross section is fully plastic. The twisting couple has the limiting value

$$T_L = \frac{2}{3}\pi\tau_0(R^3 - a^3) \tag{5.69}$$

Most shafts are solid; that is, $a = 0$. Then the limit torque is 33% greater than the torque that initiates yield, or

$$\frac{T_L}{T_0} = \frac{4}{3}$$

The inner portion of the ideally plastic shaft provides a measure of safety at the expense of weight and cost. Within the elastic limit, a hollow tube is more efficient than a solid one. Since the solid shaft is far more common, let us examine its behavior in further detail. After yielding is initiated, the radius b is given by Equation (5.65):

$$b = \frac{\tau_0}{G\mathrm{d}\phi/\mathrm{d}z}$$

Substituting this into Equation (5.67) and setting $a = 0$ for the solid shaft, we obtain

$$T = \frac{\pi\tau_0}{3}\left[2R^3 - \frac{\tau_0^3}{2G^3(\mathrm{d}\phi/\mathrm{d}z)^3}\right] \tag{5.70}$$

Prior to yielding, the twisting is governed by the *linear* relation (5.49), viz.

$$T = \frac{\pi}{2}R^4G\frac{\mathrm{d}\phi}{\mathrm{d}z}$$

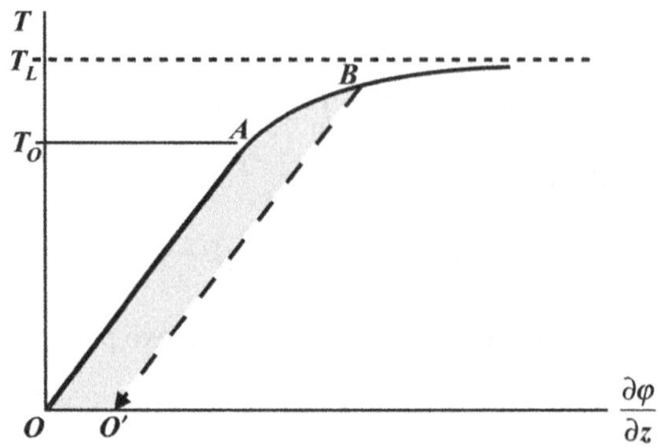

Figure 5.26

This linear equation applies provided $T < T_0$. Additional loading follows the non-linear equation (5.70). As the twist increases ($\mathrm{d}\phi/\mathrm{d}z \to \infty$), the torque approaches asymptotically the limit torque ($T \to T_L$). This progression is depicted graphically in Figure 5.26.

5.8.1 RESIDUAL STRESS

Suppose that a twisting couple is applied to the solid elastic-plastic shaft that exceeds the yield value T_0, and causes plastic deformation. The history of deformation traces the path *OAB* of Figure 5.26. Now suppose that the couple is reduced, changed in amount $\Delta\bar{T}$. This is accompanied by changes in stress, $\Delta\tau_{z\theta}$, in strain, $\Delta\gamma_{z\theta}$, and in twist, $\Delta(\mathrm{d}\phi/\mathrm{d}z)$. In keeping with the kinematic relation (5.44),

$$\Delta\gamma_{z\theta} = r\Delta\frac{\mathrm{d}\phi}{dz} \tag{5.71}$$

Also, by the *static* condition (5.45)

$$\Delta T = \Delta\bar{T} = \iint r\Delta\tau_{z\theta}\mathrm{d}A \tag{5.72}$$

To examine the material behavior, we must consider the regions $r < b$ and $b < r < R$. The former has remained Hookean; the state of stress lies on the initial linear segment *OA* in the diagram of Figure 5.24. In the outer portion ($b < r < R$), the material has yielded; the state of stress is at a point *B* of Figure 5.24. During unloading from this yielded state, as from the elastic state, the changes of stress and strain follow the linear relation.

$$\Delta\tau_{z\theta} = G\Delta\gamma_{z\theta} \tag{5.73}$$

In other words, unloading from state *B* traces the dotted curve *BO′*, parallel to *OA*. Once again, during this unloading phase, we require the kinematics, statics, and material behavior, which are embodied in Equations (5.71)–(5.73), respectively. After eliminating the stress and strain increments via Equations (5.71) and (5.73), we obtain, from (5.72), the result

$$\Delta T = G\Delta\left(\frac{\mathrm{d}\phi}{\mathrm{d}z}\right)\int_0^R\int_0^{2\pi} r^3\mathrm{d}\theta\mathrm{d}r = G\frac{\pi}{2}R^4\Delta\left(\frac{\mathrm{d}\phi}{\mathrm{d}z}\right) \tag{5.74}$$

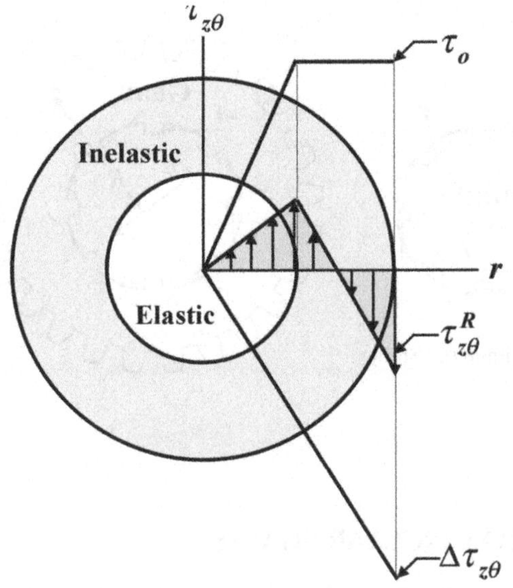

Figure 5.27

Note that this is a linear relation between the changes in torque $\Delta\bar{T}$ and twist $\Delta(d\phi/dz)$. The unloading traces the straight line BO' in Figure 5.26. If the twist and strain are eliminated from Equations (5.71), (5.73), and (5.74), we obtain

$$\Delta\tau_{z\theta} = \Delta\bar{T}\frac{2r}{\pi R^4} = \frac{\Delta\bar{T}r}{J} \tag{5.75}$$

Prior to unloading, the stress distribution was given by Equation (5.66) and depicted graphically by the plot of Figure 5.25b. The distribution $\tau_{z\theta}$ and the change $\Delta\tau_{z\theta}$ are both plotted in Figure 5.27, together with the residual stress.

$$\tau^R_{z\theta} = \tau^B_{z\theta} + \Delta\tau_{z\theta}$$

For purposes of illustration, the radius of the elastic core in Figure 5.27 is $b = R/2$. Then

$$T_B = \frac{31}{48}\pi\tau_0 R^3$$

Upon unloading,

$$\Delta T = -T_B, \quad \Delta\tau_{z\theta} = -\frac{31}{24}\tau_0\frac{r}{R}$$

Note that the residual distribution gives $T = 0$.

A few words of caution are in order: The ideally plastic-elastic material is merely a mathematical model. It provides a reasonable description of a few materials; the most notable is structural (hot-rolled) steel. It must be understood that all materials are somewhat heterogeneous. Consequently, parts of a member invariably yield at lower stresses and others yield at higher stresses, so that the elastic and plastic regions are not clearly delineated, as suggested by Figures 5.25 and 5.27. Nonetheless, the model does provide some useful guidelines and insights into the behavior of elastic-plastic members.

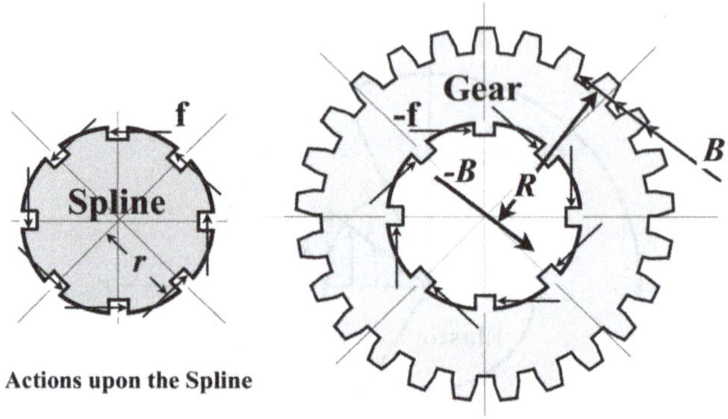

Figure 5.28

5.9 APPLICATIONS TO CIRCULAR SHAFTS

Shafts with circular cross sections are used in machinery to transmit torsional loads. The simplest situation of this kind, a circular shaft subjected to twisting couples applied to its ends, is discussed in Section 5.7. We refer to this case as simple torsion and use it as a basis for the solution of other problems involving the torsion of shafts that have circular sections. Let us first recall the essential features and limitations of the simple torsion.

For a shaft composed of a HI-HO material, the twisting couples cause a shear stress on cross sections that is given by Equation (5.52):

$$\tau = \frac{Tr}{J} \tag{5.76}$$

Also, the relative rotation of the ends is expressed by Equation (5.54):

$$\phi = \frac{TL}{GJ} \tag{5.77}$$

The simple torsion of a shaft of ideally elastic-plastic material is described in Section 5.8. In particular, we note the onset of yielding and the existence of a limit load corresponding to a fully plastic condition.

The results for simple torsion can be applied to portions of a shaft that fit our description in regard to shape, material, and loading. However, the engineer must also appreciate the limitations of the theory: Our physical model is a shaft of circular cross section. Every cross section is presumed to rotate without distortion. Consequently, the only significant strain is the shear component $\gamma_{z\theta} = r(d\theta/dz)$, which varies linearly with the radius r. In the simplest circumstance, a HI-HO material, the one significant stress is, therefore, the shear component $\tau_{z\theta} = Tr/J$, which also varies linearly. In any real application, torque must be transmitted to the shaft via some mechanical connection to gears, pulleys, or levers. Typically, a shaft and gear are joined by splines, which mesh as depicted in Figure 5.28.

Then the torque is the resultant of the numerous forces (**f**) of contact between the faces of the spline. The resultant of the eight forces **f** is a couple $\mathbf{C} = 8\mathbf{f} \times \tau = \mathbf{B} \times \mathbf{R}$, the couple upon the gear. At the very *end* of the shaft, $\tau_{z\theta} = 0$. Only at some distance from the end does the stress approach the linear distribution. This is another manifestation of Saint-Venant's principle. If the design requires an assessment of the localized stress near the connections (splines or keyways), then more refined analyses are necessary. Our theory of simple torsion is applicable under the special circumstances set forth but *also* in *portions* of actual shafts. The theory gives an adequate description of the overall twist (rotation ϕ) and, by virtue of Saint-Venant's principle, the stress and strain at points sufficiently distant from any connections and any abrupt steps. Our description is then applicable within portions that are relatively slender ($R/L \ll 1$).

ILLUSTRATIVE EXAMPLE 5.14: A STEPPED SHAFT

The shaft of Figure 5.29 consists of two cylindrical portions; part *AC* has a diameter d_1 and part *CD* has a diameter d_2. The shaft is clamped at *A*, subjected to twisting couples applied at locations *B* and *D*, and composed of HI-HO material (e.g., steel). Let us calculate the rotation of the end *D* and the shear stress on cross sections *AB*, *BC*, and *CD*.

We begin by considering the equilibrium of the three parts: *AB* is part 1, *BC* is part 2, and *CD* is part 3. The free-body diagrams of these parts are shown in Figure 5.29. A cross-section is placed just to the left of point *B* so that the couple T_1 is shown acting on part *BC*. This was an arbitrary choice; the cut may equally well be to the right of point *B*, putting couple T_1 on the right end of *AB*. The net result would be unchanged. For equilibrium,

$$T_C = T_2, \quad T_A = T_B = T_1 + T_2$$

Each of the parts in Figure 5.29 is a relatively slender circular shaft subjected to twisting couples acting at its ends; that is, each is a case of simple torsion. Equation (5.77) applies to the HI-HO material; the rotation of end *D* relative to *C* (portion 3) is

$$\phi_3 = \frac{L_3 T_C}{G J_2} = \frac{L_3 T_2}{G J_2}$$

where $J_2 = \frac{\pi}{32}(d_2)^4$. Similarly, the rotation of *C* relative to *B* and *B* relative to *A* are

$$\phi_1 = \frac{L_1 T_B}{G J_1} = \frac{L_1 (T_1 + T_2)}{G J_1}$$
$$\phi_2 = \frac{L_2 T_C}{G J_1} = \frac{L_2 T_2}{G J_1}$$

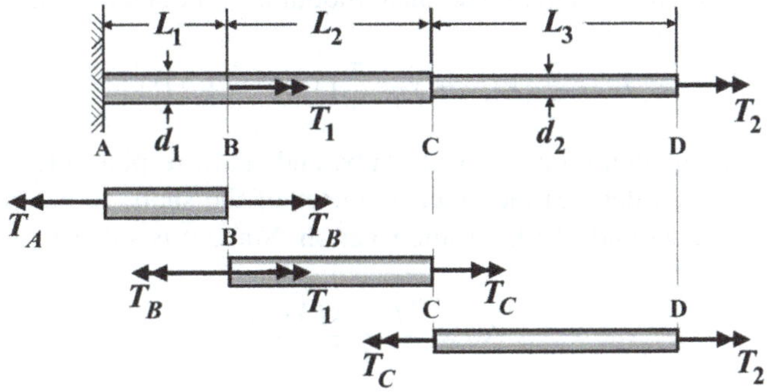

Figure 5.29

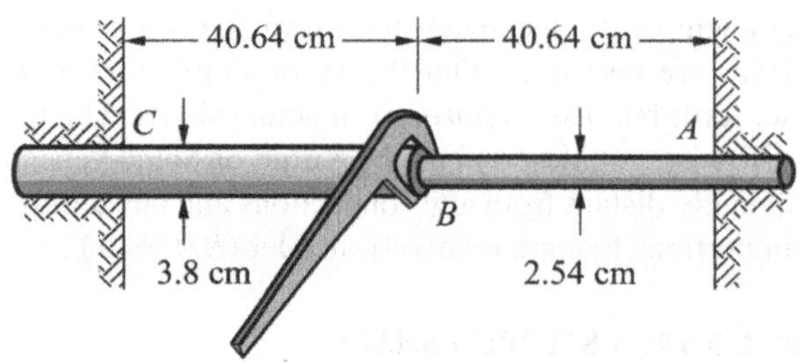

Figure 5.30

where $J_1 = \frac{\pi}{32}(d_1)^4$.

The rotation ϕ of the right end is simply the sum

$$\phi = \phi_1 + \phi_2 + \phi_3 = \frac{L_1}{GJ_1}(T_1 + T_2) + \left(\frac{L_2}{GJ_1} + \frac{L_3}{GJ_2}\right) T_2$$

Also, the shear stress on cross sections of each part can be calculated by Equation (5.76):

$$\tau_1 = \frac{T_1 + T_2}{J_1} r, \quad \tau_2 = \frac{T_2}{J_1} r, \quad \tau_3 = \frac{T_2}{J_2} r \tag{5.78}$$

Illustrative Example 5.15

A hollow steel tube is slipped onto the end of a solid steel shaft as shown in Figure 5.30. A wrench is used to apply a twisting couple $T = 11,300$ N-cm (1000 lb-in) to the end B of the hollow tube. The tube is then joined to the shaft by a set screw, and the wrench is released; the couple T is removed. The two components are made of the same (HI-HO) steel; neither experiences inelastic deformations during the initial twisting of the tube nor the subsequent removal of the couple T. We are asked to determine the maximum residual shear stress.

When the couple T is removed, the tube responds elastically to the change ΔT according to Equation (5.77):

$$\Delta\phi = \frac{\Delta T L}{GJ_T} \tag{5.79}$$

where L denotes the length (40.64 cm), G the shear modulus of the steel and J_T is the *polar moment* of Equation (5.50), viz

$$J_T = \frac{\pi}{2}\left(R^4 - a^4\right) = \frac{\pi}{2}\left(1.90^4 - 1.27^4\right)$$

Subscripts T and S denote a property of the tube and shaft, respectively. R denotes the outer diameter of the tube, and a denotes the outer diameter of the shaft. However, the solid shaft is locked to the tube and must undergo the same rotation. Since it is subject to the residual torque $(T - \Delta T)$, we have

$$\Delta\phi = \frac{(T - \Delta T)L}{GJ_S} \tag{5.80}$$

where

$$J_S = \frac{\pi}{2}a^4 = \frac{\pi}{2}(1.27)^4$$

Equating (5.79) and (5.80) we obtain

$$\Delta T = \frac{T}{(1 + J_S/J_T)} = 0.8004T$$

The residual torque follows:

$$T^R \equiv T - \Delta T = 0.1996T = 2255 \text{ N-cm}$$
$$= (200 \text{ lb-in})$$

The maximum stress in each occurs at the outer surface:

$$\tau_T(\max) = \frac{T^R R}{J_T} \quad , \quad \tau_S(\max) = \frac{T^R a}{J_S}$$

$$\frac{R}{J_T} = \frac{2R}{\pi(R^4 - a^4)} = 0.116 \text{ cm}^{-3} \quad , \quad \frac{a}{J_S} = \frac{2}{\pi a^3} = 0.311 \text{ cm}^{-3}$$

Clearly, the maximum stress occurs at the surface of the shaft, where

$$\tau_S = 2255 \times 0.311 \frac{\text{N}}{\text{cm}^2} = 701 \text{ N/cm}^2 2 = 70.1 \text{ kPa}$$

5.10 TORSION-APPLICATIONS

ILLUSTRATIVE EXAMPLE 5.16: THE COIL SPRING

Very few problems of deformable bodies can be solved *exactly*. We can formulate the equations governing continuous functions; however, mathematical, so-called closed form, solutions are rare. Often we can adapt some simple solutions to achieve useful approximations to actual mechanical or structural components. The coil spring of Figure 5.31 is such a mechanical device.

The necessary geometrical attributes are indicated:

δ denotes the pitch, the distance between successive coils.
e denotes the total elongation under force P.
N is the number of coils.
β is the angle of the deformed helix on the cylinder of radius r
$\tan\beta = (\delta + e/N)/2\pi r$
$\hat{\boldsymbol{a}}$ is the unit vector, tangent to the helix; $\hat{\boldsymbol{n}}$ is a unit vector normal to $\hat{\boldsymbol{a}}$ and directed along the radial line. Note that $\hat{\boldsymbol{n}}$ is not the principal normal to the helix. Unit vector $\hat{\boldsymbol{b}} = \hat{\boldsymbol{a}} \times \hat{\boldsymbol{n}}$ completes the triad.

The couples upon an arbitrary cross-section are depicted in Figure 5.31c. The components of forces and couples are statically equivalent to the axial force P. These are a normal force $\mathbf{F}_n$ and shear force $\mathbf{V}$:

$$\mathbf{F} = \hat{\mathbf{a}} P \sin\beta, \quad \mathbf{V} = \hat{\mathbf{b}} P \cos\beta \tag{5.81}$$

A twisting couple $\mathbf{T}$ and bending couple $\mathbf{M}$ act on the cross-section:

$$\mathbf{T} = \hat{\mathbf{a}} Pr\cos\beta, \quad \mathbf{M} = -\hat{\mathbf{b}} Pr\sin\beta \tag{5.82}$$

Note that each force and couple depends on the angle β; the angle β depends on the elongation e. This means that any relation between force P and elongation e is nonlinear. To appreciate the

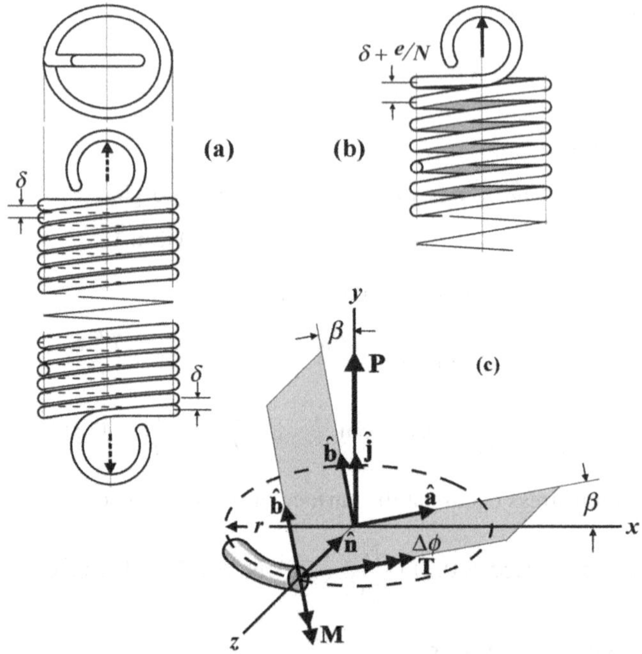

Figure 5.31

consequences, we consider the normal force **F** of Equation (5.81). As the elongation e and angle β increase, then force **F** also increases. More of the applied load P is resisted by the direct tensile force upon the cross-section; the overall stiffness of the spring increases. A spring scale for the measurement of weight can only provide accurate measurements if calibrated to accommodate the nonlinearity. The nonlinear equations can be solved,[4] however, here we limit ourselves to a linear approximation.

Firstly, we note that a transverse shear (V) on such thin elements causes little deformation; minor effects are explored in Chapter 6. If β remains small, then only the twisting couple plays a significant role. Therefore, our approximation focuses on the torsion. We also assume that the radius of the wire is much less than the radius r of the helix. Then a segment, as shown in Figure 5.31c,appears much as a straight prismatic bar. Accordingly, we adapt Equation (5.51b)

$$d\phi = \frac{T\,ds}{GJ} \doteq \frac{Pr\,ds}{GJ} \tag{5.83}$$

where ds denotes a differential of length along the helix. We are assuming that our spring is composed of steel (HI-HO). The differential rotation about the tangent $\hat{\boldsymbol{a}}$ can be treated as a vector $(d\phi)\hat{\boldsymbol{a}}$. This incremental rotation causes an incremental change in the elongation e. The displacement at the axis y is the vector

$$(d\phi)\hat{\boldsymbol{a}} \times (r\hat{\boldsymbol{n}}) = r\,d\phi\,\hat{\boldsymbol{a}} \times \hat{\boldsymbol{n}} = \frac{Pr^2 ds}{GJ}\hat{\mathbf{b}}$$

Note that the displacement is in the direction of vector $\hat{\mathbf{b}}$. As before, we presume that angle β

remains small ($\cos\beta \doteq 1$). Then, the incremental elongation follows:

$$de \doteq \frac{Pr^2}{GJ} ds$$

This must be integrated along the entire length ($2\pi rN$) to obtain

$$e = \frac{2\pi N r^3}{GJ} \tag{5.84}$$

ILLUSTRATIVE EXAMPLE 5.17: TORSION OF A BAR WITH A SQUARE CROSS-SECTION

The fundamentals of this chapter are founded on the so-called *inverse* or *semi-inverse* methods of solution: In other words, a correct assumption is made that provides the basis for the solution. Examples are the Bernoulli hypothesis for simple bending of a symmetrical beam and the St. Venant hypothesis for torsion of a cylindrical bar. These provide the exact solutions for the specific simple situations. However, with certain reservations, they can often be extended, e.g., the Bernoulli hypothesis can be applied to beams under transverse loadings (see Chapter 6). Here, we explore the application of St. Venant's hypothesis to a bar with square cross-section. According to the hypothesis, the displacement components in directions x, y, z of Figure 5.32 are, respectively,

$$u = -\phi yz, \quad v = \phi xz, \quad w = 0 \tag{5.85}$$

We assume that the square cross-section rotates without distortion as the prismatic bar is twisted under the action of the Torque T. This rotation is depicted in Figure 5.32a. If ϕ denotes the twist (rotation per unit length), then the consequent strains are

$$\gamma_{xz} = -\phi y, \quad \gamma_{yz} = \phi x$$

We assume that there is no extension axially, $\varepsilon_z = 0$, and no deformation of the cross-section:

$$\varepsilon_x = \varepsilon_y = \gamma_{xy} = 0$$

If the bar is composed of HI-HO material, then the only components of stress are

$$\tau_{xz} = -G\phi y, \quad \tau_{yz} = G\phi x \tag{5.86}$$

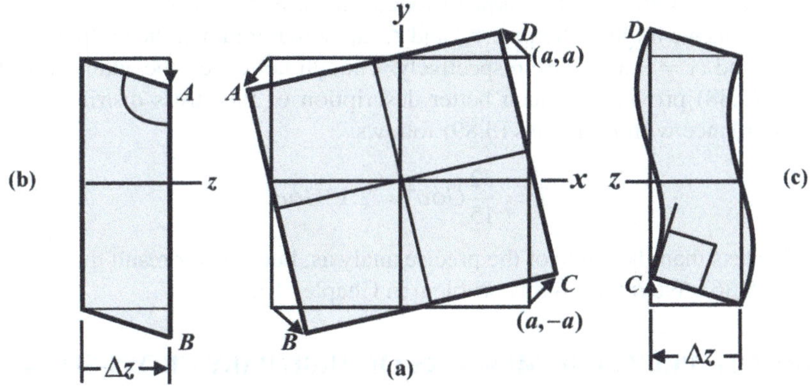

Figure 5.32

The resultant twisting couple (torque) follows:

$$
\begin{aligned}
T &= \int_{-a}^{a}\int_{-a}^{a}(-y\tau_{xz}+x\tau_{yz})\,\mathrm{d}x\mathrm{d}y \\
&= \frac{8}{3}G\phi a^4 = 2.67G\phi a^4
\end{aligned}
\tag{5.87}
$$

This result is 19% greater than a more accurate value ($2.25G\phi a^4$), as obtained by a more precise analysis.[5]

The simple approximation, Equation (5.85), asserts that the cross-section remains plane. Accordingly, the edge AB appears as the straight line on the slice in Figure 5.32b. At the corners, e.g., A and B, the shear strain and stress must be

$$
\gamma_{yz} = -\phi a \quad \text{and} \quad \tau_{yz} = -G\phi a
$$

This violates the conditions of a stress-free lateral surface, i.e., $\tau_{yz} = 0$. Various solutions of the St. Venant torsion recognize the occurrence of cross-sectional warping, an axial displacement w, as seen in Figure 5.32c. A better solution hinges on the determination, or approximation, of the warping function $w(x,y)$.

We note that symmetry requires equal values (w) on the axes (x,y) and also at the corners and that the warping eliminate, or reduce, the surface tractions of the initial approximations (5.86). To that end, we introduce another *approximation*:

$$
w = \frac{\phi}{2}xy\left(1-\frac{y^2}{a^2}\right) - \frac{\phi}{2}xy\left(1-\frac{x^2}{a^2}\right) \tag{5.88}
$$

The shear strains follow in accordance with Equations (5.85) and (5.88):

$$
\gamma_{xz} = \frac{\partial u}{\partial z} + \frac{\partial w}{\partial x} = -\phi y\left(1+\frac{y^2}{2a^2}\right) + \frac{3}{2}\phi y\frac{x^2}{a^2} \tag{5.89a}
$$

$$
\gamma_{yz} = \frac{\partial v}{\partial z} + \frac{\partial w}{\partial y} = -\phi x\left(1+\frac{x^2}{2a^2}\right) - \frac{3}{2}\phi x\frac{y^2}{a^2} \tag{5.89b}
$$

The approximation does not entirely fulfill the requisite surface conditions: The shear stresses (τ_{xz}, τ_{yz}) at surface $x = a$ are displayed graphically in Figure 5.33; the stresses $(\bar{\tau_{xz}}, \bar{\tau_{yz}})$ of the initial approximation (5.86) are also shown (dotted lines). Note that the stress components on a cross-section vary as we might anticipate. τ_{xz} and τ_{yz} are a maximum at the midpoints of the edges, $(x = 0, y = \pm a)$ and $(y = 0, x = \pm a)$, respectively. Though not exact, the addition of the warping w of Equation (5.88) provides a much better description of the stress distribution. The torque obtained in accordance with Equations (5.89) follows:

$$
T = \frac{32}{15}G\phi a^4 = 2.13G\phi a^4
$$

The latter is 5% less than the value of the precise analysis, but a better result than provided by the approximation (5.86). We return to this problem in Chapter 10.

5.11 TORSION: PRISMATIC MEMBERS OF ARBITRARY CROSS-SECTION

In Section 5.7, we described the simple torsion of a shaft with a circular cross-section[6]. The geometric and physical symmetries admit a simple description of the deformation. Indeed, we

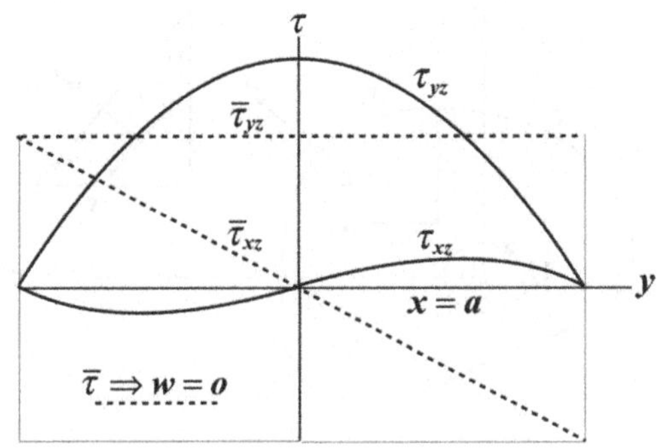

Figure 5.33

assumed the solution. The circumstances of a non-circular cross-section invalidate that simple description. Our Illustrative Example 5.17 demonstrates the warping of a non-circular section.

The solution for a HI-HO prism of arbitrary cross-section was given by B. St. Venant. His approach has been termed a *semi-inverse* method. He adopted the assumption that the section is not distorted in the plane. Therefore, if z is the axial coordinate, $\varepsilon_x = \varepsilon_y = \gamma_{xy} = 0$. The cross-section rotates, just as the circular section, *but* warps. Viz, there is a displacement $w(x,y)$ in the axial direction. As before, he assumed no axial extension, viz. $\varepsilon_z = 0$.

If the cross-section at one end ($z = 0$) is constrained against rotation, not against warping, then

$$u = -\phi yz, \quad v = \phi xz \tag{5.90}$$

As before, ϕ denotes the twist, i.e., rotation per unit length. Every cross-section is similarly deformed, i.e.

$$w = \phi \psi(x,y) \tag{5.91}$$

$\psi(x,y)$ is the unknown warping function.

For brevity, we now denote the derivatives with a comma, e.g.

$$\frac{\partial f}{\partial x} = f_{,x}, \quad \frac{\partial f}{\partial y} = f_{,y}$$

In this *shorthand* the, shear strains are expressed as

$$\gamma_{zx} = w_{,x} + u_{,z} = \phi(\psi_{,x} - y) \tag{5.92a}$$

$$\gamma_{zy} = w_{,y} + v_{,z} = \phi(\psi_{,y} + x) \tag{5.92b}$$

Since the material is HI-HO,

$$\tau_{zx} = G\phi(\psi_{,x} - y), \quad \tau_{zy} = G\phi(\psi_{,y} + x) \tag{5.93a,b}$$

Equilibrium requires (see Section 3.14)

$$\frac{\partial \tau_{zy}}{\partial y} + \frac{\partial \tau_{zx}}{\partial x} = 0, \quad \psi_{,yy} + \psi_{,xx} = 0 \tag{5.94}$$

At the boundary C of Figure 5.34, the surface is stress-free, i.e.

$$\tau_{zn} = \tau_{zx} \cos\alpha + \tau_{zy} \sin\alpha = 0 \tag{5.95}$$

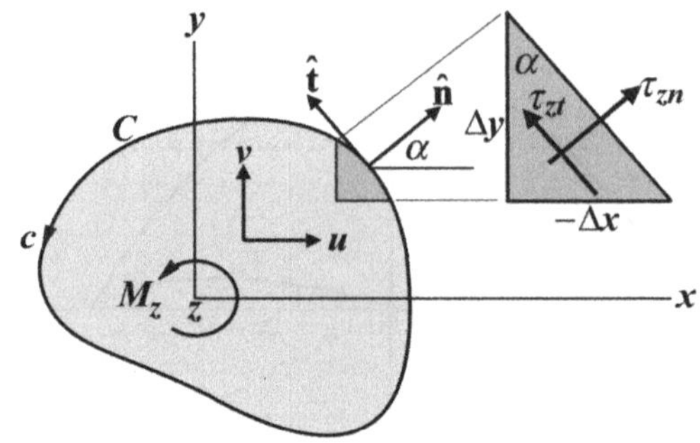

Figure 5.34

Substituting Equations (5.93a,b) into the boundary condition (5.95), we obtain

$$\psi_{,x} \cos\alpha + \psi_{,y} \sin\alpha = y\cos\alpha - x\sin\alpha \tag{5.96}$$

Together, the solution of the differential Equation (5.94), subject to the boundary condition (5.96), provides the unknown warping function $\psi(x,y)$.

An alternative formulation was given by Prandtl[7] who introduced a stress function $F(x,y)$ which satisfies equilibrium when

$$\tau_{xz} = G\phi F_{,y}, \quad \tau_{yz} = -G\phi F_{,x} \tag{5.97a,b}$$

Then

$$\gamma_{xz} = \phi F_{,y}, \quad \gamma_{yz} = -\phi F_{,x} \tag{5.98a,b}$$

However, the existence of a compatible displacement $w(x,y)$ is insured if, and only if, the conditions (2.82) are satisfied:

$$\frac{\partial^2 \gamma_{xz}}{\partial x \partial y} - \frac{\partial^2 \gamma_{yz}}{\partial x^2} = 0, \quad (F_{,yy} + F_{,xx})_{,x} = 0$$

$$\frac{\partial^2 \gamma_{xz}}{\partial y^2} - \frac{\partial^2 \gamma_{yz}}{\partial x \partial y} = 0, \quad (F_{,yy} + F_{,xx})_{,y} = 0$$

These equations are satisfied only if the parenthetical factors are constant, viz.

$$F_{,xx} + F_{,yy} = K \text{ (constant)} \tag{5.99}$$

Now we equate the St. Venant and the Prandtl expressions for the stress components, specifically, Equations (5.93a,b) and (5.97a,b), respectively:

$$\tau_{zx} = G\phi(\psi_{,x} - y) = G\phi F_{,y} \tag{5.100a}$$

$$\tau_{zy} = G\phi(\psi_{,y} + x) = -G\phi F_{,x} \tag{5.100b}$$

When the latter are substituted into the left side of (5.99), we obtain

$$F_{,xx} + F_{,yy} = -2 \tag{5.101a}$$

It remains to express the boundary condition (5.95) in terms of the stress function F. Substituting the expressions (5.97a) and (b) into Equation (5.95) provides the condition

$$F_{,y}\cos\alpha - F_{,x}\sin\alpha = 0 \tag{5.101b}$$

Let c denote the length along the boundary C of Figure 5.34. Then

$$\frac{\mathrm{d}y}{\mathrm{d}c} = \cos\alpha, \quad \frac{\mathrm{d}x}{\mathrm{d}c} = -\sin\alpha$$

Then the boundary condition (5.101b) takes the form:

$$F_{,y}\frac{\partial y}{\partial c} + F_{,x}\frac{\partial x}{\partial c} = \frac{\partial F}{\partial c} = 0$$

In words, F is constant on the boundary C: On C

$$F\,|_C = K$$

The resultant twisting couple is the integral:

$$M_z = \iint_A (\tau_{zy}x - \tau_{zx}y)\mathrm{d}A = -G\phi\iint_A (F_{,x}x + F_{,y}y)\mathrm{d}A$$

Integration by parts gives us:

$$M_z = -2G\phi KA + 2G\phi\iint_A F\mathrm{d}A = 2G\phi\iint_A (F-K)\mathrm{d}A$$

where K denotes the unspecified constant boundary value. The formulation is simplified by defining

$$\Phi = F - K$$

Then

$$M_z = 2G\phi\iint_A \Phi\mathrm{d}A \tag{5.102}$$

The stress components are expressed by (see Equation 5.97)

$$\tau_{zx} = G\phi\Phi_{,y}, \quad \tau_{zy} = -G\phi\Phi_{,x} \tag{5.103}$$

and $\Phi(x,y)$ must also satisfy the Poisson Equation (5.101a):

$$\Phi_{,xx} + \Phi_{,yy} = -2 \tag{5.104}$$

Now,

$$\Phi\,|_c = 0 \tag{5.105}$$

and the warping function is obtained according to Equations (5.100a,b), viz.

$$\psi_{,x} = \Phi_{,y} + y, \quad \psi_{,y} = -\Phi_{,x} - x \tag{5.106a}$$

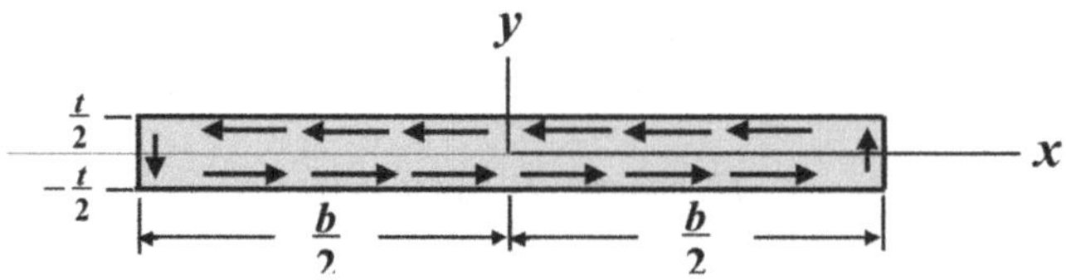

Figure 5.35

ILLUSTRATIVE EXAMPLE 5.18

Consider a bar with the narrow-rectangular cross-section of Figure 5.35 ($t \ll b$). The bar is subject to the couple M_z. Since the surface is stress-free, τ_{yz} must vanish at the upper and lower surfaces ($y = \pm t/2$). At the ends ($x = \pm b/2$), $\tau_{zx} = 0$. At any intermediate cross-section, we anticipate no significant transverse sheer (τ_{zy}), but a component $\tau_{zx} = +G\phi\Phi_{,y}$. The anticipated stress is given by

$$\Phi = \left(\frac{t}{2}\right)^2 - y^2 \tag{5.107}$$

Then

$$\tau_{zx} = -2G\phi y, \quad \tau_{zy} \doteq 0 \tag{108a,b}$$

Note that this is an approximation which is not valid near the ends ($x = \pm b/2$). If we neglect the discrepancy and employ Equation (5.102), we obtain

$$M_z = G\phi\frac{bt^3}{3} \tag{5.109}$$

On the other hand, we might neglect the stress τ_{zy} near the ends as indicated by Equation (5.108). Then we obtain

$$\iint_A (-\tau_{zx}y + \tau_{zy}x)\mathrm{d}A \doteq G\phi\frac{bt^3}{6} \tag{5.110}$$

Because the first result (5.109) violates the condition $\Phi = 0$ at ends $x = \pm b/2$, it is excessive. The second result (5.110) neglects a small contribution from stress τ_{zy} near the ends (the second term of the integral); it is too low.

ILLUSTRATIVE EXAMPLE 5.19

Our approximation of the stress function Φ of Equation (5.107) and the stress distribution of Equation (5.108) might be applied as well to other thin-walled cross-sections. Consider, for example, the thin-walled circular cross-section of Figure 5.36a. Note, however, that it is *open*; a slit extends the length of the tube. In this case

$$\Phi = \left(\frac{t}{2}\right)^2 - n^2$$

$$n \equiv r - R$$

wherein R denotes the mean radius. We need only replace b in Equation (5.109) by the circumferential length $2\pi R$ to obtain

$$M_z = \frac{2\pi G\phi R t^3}{3}$$

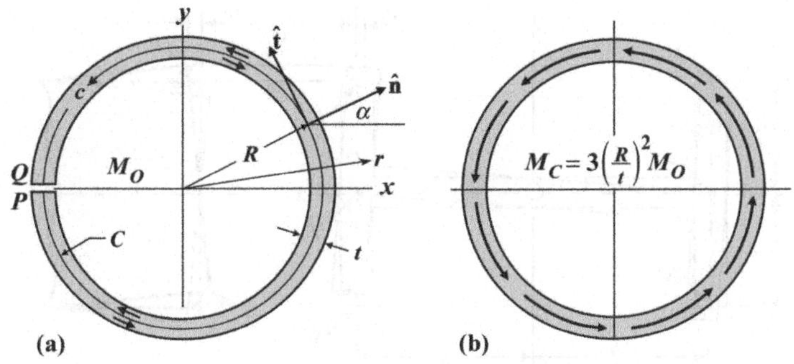

Figure 5.36

It is most revealing to examine the warping of this cross-section. *Along the median curve C*, the shear stress components τ_{zr}, τ_{nt} and τ_{zx}, τ_{zy} vanish. Therefore, by Equations (5.100a,b),

$$\psi_{,x} = y, \quad \psi_{,y} = -x$$

Here

$$x = R\cos\alpha, \quad y = R\sin\alpha, \quad c = R\alpha$$

$$\frac{dy}{dc} = \cos\alpha, \quad \frac{dx}{dc} = -\sin\alpha$$

$$\frac{d\psi}{dc} = -y\sin\alpha - x\cos\alpha = -R\left(\sin^2\alpha + \cos^2\alpha\right) = -R$$

As we traverse the path from end P to end Q, we obtain

$$\psi_Q - \psi_P = -2\pi R^2$$

The displacement increases linearly with c from P to Q. Point Q moves in the negative (z) direction; at the slit, the relative displacement is (see Equation 5.91)

$$w_Q - w_P = -2\pi\phi R^2$$

It is well worth comparing the foregoing results with those obtained for the closed tube. We need only recall the results of Section 5.7. For the closed tube

$$M_z = 2\pi G\phi R^3 t$$

The result is remarkable. The closed tube is much stiffer than the tube with the slit; the ratio follows:

$$\frac{M(\text{closed})}{M(\text{open})} = 3\left(\frac{R}{t}\right)^2 \tag{5.111}$$

Note: The factor 2, the difference between results (5.109) and (5.110), is *not the issue* in these examples. It is the ratio (5.111) that sends an important message, viz. the stiffness of the closed versus that of the open thin-walled section.

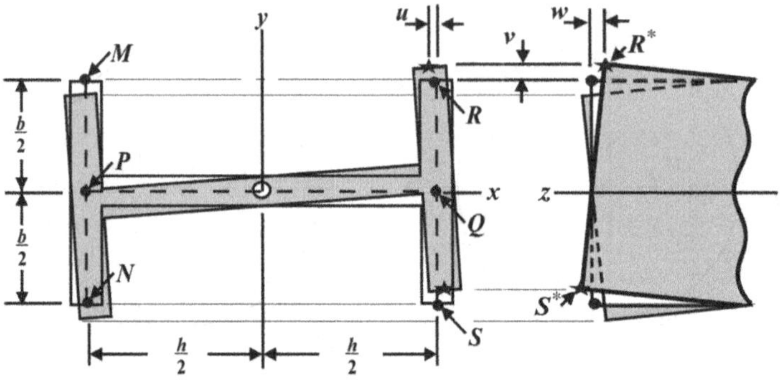

Figure 5.37

ILLUSTRATIVE EXAMPLE 5.20: TORSION OF AN I BEAM

The I cross-section of Figure 5.37a is most important in many structures. Again, the narrow cross-section supports shear stresses tangent to the surfaces. In torsion, the shear stresses vanish at the median line. As in the preceding examples (see Equations 5.100a,b), the warping function ψ has the derivatives on the median lines of each segment:

$$\text{on } PQ \quad \psi_{,x} = 0\,, \qquad \psi = 0$$

$$\text{on } MN \quad \psi_{,y} = \frac{h}{2}\,, \qquad \psi = \frac{h}{2}y$$

$$\text{on } RS \quad \psi_{,y} = -\frac{h}{2}\,, \qquad \psi = -\frac{h}{2}y$$

As in the preceding examples, we are assuming that cross sections are not constrained against warping. If, for example, a cantilevered beam were fixed at one end, that would provide resistance to warping near the constrained end. The relative rotation of the flanges RS and MN is depicted in Figure 5.37b, e.g., the rotation of RS to R^*S^*. These are exaggerated in our illustration. The actual stiffness of the beam can be approximated as in the preceding examples.

5.12 TORSION: TUBULAR MEMBERS WITH MULTIPLY CONNECTED CROSS SECTIONS

The warping function ψ and the stress function Φ of Section 5.11 apply to prismatic members of arbitrary cross section. The function Φ must be constant on a stress-free surface. Now we consider a cross section as depicted in Figure 5.38, which has two cavities bounded by curves $\bar{C}_1$ and $\bar{C}_2$ which enclose the areas A_1 and A_2. As before, we can set the function Φ to vanish at the external surface bounded by curve C. However, the two cavities are also stress-free; therefore, Φ must also be constant on each, i.e.

$$\Phi|_{C_1} = K_1, \quad \Phi|_{C_2} = K_2$$

To ensure compatibility of the strains, we require continuity of the displacement $w = \phi\,\psi(x,y)$. Specifically,

$$\oint_{\bar{C}_1} \mathrm{d}\psi = \oint_{\bar{C}_2} \mathrm{d}\psi = 0 \tag{5.112}$$

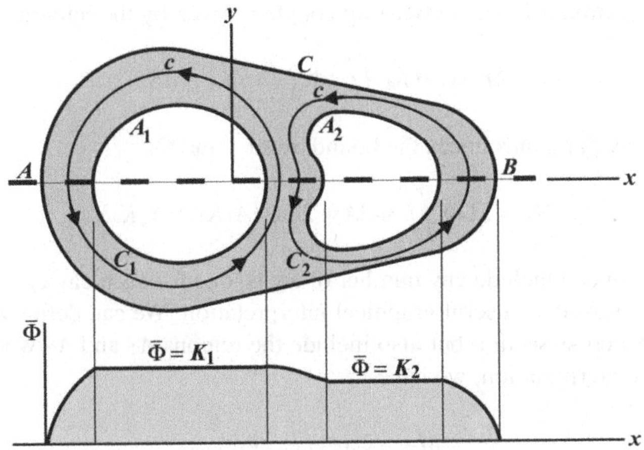

Figure 5.38

where C_1 and C_2 could be any curves that enclose the cavities. Thus, each cavity introduces a constant (K_1, K_2), but each introduces an additional equation.

Equation (5.112) can be recast as follows:

$$d\psi = \psi_{,x}\,dx + \psi_{,y}\,dy$$

According to Equations (5.100a,b),

$$d\psi = F_{,y}\,dx - F_{,x}\,dy + ydx - xdy$$

However, $F_{,y} = \Phi_{,y}$, $F_{,x} = \Phi_{,x}$; therefore,

$$d\psi = \Phi_{,y}\,dx - \Phi_{,x}\,dy + ydx - xdy$$

As we traverse a closed path (see Figure 5.34)

$$dy = \cos\alpha dc, \quad dx = -\sin\alpha dc$$

Therefore,

$$d\psi = -\Phi_{,y}\sin\alpha dc - \Phi_{,x}\cos\alpha dc + ydx - xdy$$

By Equations (5.103a,b):

$$\begin{aligned} d\psi &= -\frac{1}{G\phi}\tau_{zx}\sin\alpha dc + \frac{1}{G\phi}\tau_{zy}\cos\alpha dc + ydx - xdy \\ &= \frac{1}{G\phi}(+\tau_{zt})\,dc + ydx - xdy \end{aligned}$$

Returning to the continuity condition (5.112), we obtain

$$\oint_{C_1} \tau_{zt}\,dc = G\phi \oint_{C_1} (-ydx + xdy) = 2G\phi A_1 \tag{5.113}$$

Here C_1 can be any curve enclosing the area A_1 about a cavity.

As before (see Equation 5.103), the twisting couple is given by the integral:

$$M_z = -G\phi \iint_A (\Phi_{,x} x + \Phi_{,y} y)\,\mathrm{d}A$$

Now the integration by parts introduces the boundaries C_1 and C_2.

$$M_z = 2G\phi \iint_A \Phi \mathrm{d}A + 2G\phi\,(A_1K_1 + A_2K_2) \tag{5.114a}$$

Note that the final term can include any number of areas, one for each cavity.

The result (5.114a) admits a useful graphical interpretation. We can define a function $\bar{\Phi}$ as the function Φ within the cross section but also include the regions A_1 and A_2 wherein it has values, K_1 and K_2. With this interpretation, we have

$$M_z = 2G\phi \iint_{A_T} \bar{\Phi} \mathrm{d}A \tag{5.114b}$$

where A_T signifies the entire region enclosed by the outer boundary C. Figure 5.38b shows a plot of $\bar{\Phi}$ through the line AB of the section AB of Figure 5.38a.

ILLUSTRATIVE EXAMPLE 5.21: TORSION OF A THIN-WALLED TUBULAR MEMBER

Let us consider a member that has a thin wall as depicted in Figure 5.39a. The stress function $\bar{\Phi}$ has constant values K_1, K_2, and K_3 over the open regions with areas A_1, A_2, and A_3, respectively. On each of the walls, the shear stress is tangent to the bounding curves and given by the normal derivative

$$\tau_{zt} = -G\phi\bar{\Phi}_{,n}$$

where n is the outwardly directed normal. Since the wall is relatively thin, we assume that the gradient, $\bar{\Phi}_{,n}$, varies little across the wall; in other words, we assume that the stress is constant across the wall. Figure 5.39b shows a plot of $\bar{\Phi}$ through the section BE. By virtue of the symmetry $K_3 = K_1$

On segments ABC and DEF, we have

$$\tau_{zt} = \frac{G\phi}{t}K_1$$

On segments CA and FD, we have

$$\tau_{zt} = \frac{G\phi}{t}(K_1 - K_2)$$

On segment FA and CD, we have

$$\tau_{zt} = \frac{G\phi}{t}K_2$$

The constants K_1 and K_2 are determined by the continuity conditions (5.113): The integrals are taken along the midlines of each wall.

On the path $ABCA$,

$$\oint_{C_1} \tau_{zt}\mathrm{d}c = \oint_0^{\pi} a\frac{G\phi}{t}K_1\mathrm{d}\theta + \oint_0^{2a} \frac{G\phi}{t}(K_1 - K_2)\mathrm{d}y = G\phi a^2$$

On the path *ACDF*,

$$\oint_{C_2} \tau_{zt} dc = 2\oint_0^{2a} \frac{G\phi}{t} K_2 dx - 2\oint_0^{2a} \frac{G\phi}{t}(K_1 - K_2) = 8G\phi a^2$$

The two equations follow:

$$(2+\pi)K_1 - 2K_2 = \pi at$$
$$-K_1 + 2K_2 = 2at$$

Therefore,

$$K_1 = \frac{2+\pi}{1+\pi}at, \quad K_2 = \frac{4+3\pi}{2(1+\pi)}at$$

The stress components follow: On *ABC* and *DEF*

$$\tau_{zt} = \frac{2+\pi}{1+\pi}G\phi a = 1.24G\phi a$$

On the webs *CA* and *FD*

$$\tau_{zt} = -\frac{\pi}{2(1+\pi)} = -0.38G\phi a$$

On the segments *FA* and *CD*

$$\tau_{zt} = \frac{4+3\pi}{2(1+\pi)}G\phi a = 1.62G\phi a$$

Recall that we accepted the linear variation of $\bar{\Phi}$ across the thin walls. Consequently, in our evaluation of the couple M_z via Equation (5.114b), we can treat the function $\bar{\Phi}$ as constant to the midlines; then

$$M_z = 2G\phi(A_1K_1 + A_2K_2 + A_3K_3)$$

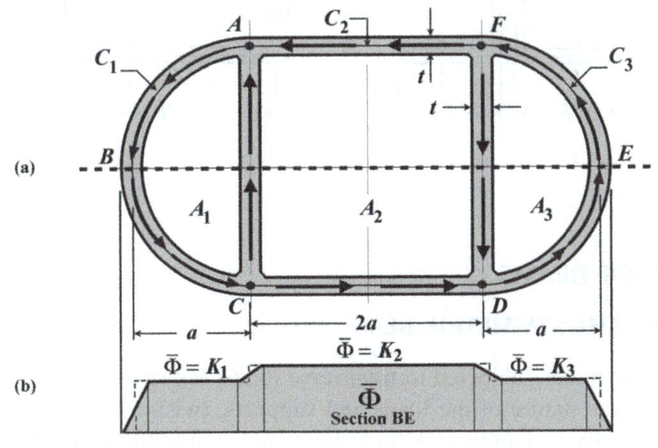

Figure 5.39

wherein the areas are those enclosed by the midlines; the plot of Figure 5.39b is essentially supplanted by the dotted trace. Hence, we obtain

$$M_z = 2G\phi\left(\pi a^2 K_1 + 4a^2 K_2\right)$$
$$= \left[\frac{16(1+\pi)+2\pi^2}{1+\pi}\right] G\phi a^3 t = 20.8 G\phi a^3 t$$

The resistance is reduced to $M_z = 9.92 G\phi a^3 t$ if the two webs are removed. A single web at the middle would have the same effect, since a central web would offer little torsional stiffness.

ILLUSTRATIVE EXAMPLE 5.22

Two beams have the cross sections depicted in Figure 5.40. Both have the same amount of area, the same amount of the same HI-HO material.

The beam with the cross section b is much more resistant to bending and buckling in the x-z plane. By the analyses and approximations of Sections 5.11 and 5.12, the resistance to twisting is much greater; for the same twist ϕ, the ratio of twisting couples is

$$M_z^b / M_Z^a = 120$$

The engineer must weigh the structural integrity versus the cost of fabrication. A large I beam would be less costly than the comparable tubular beam.

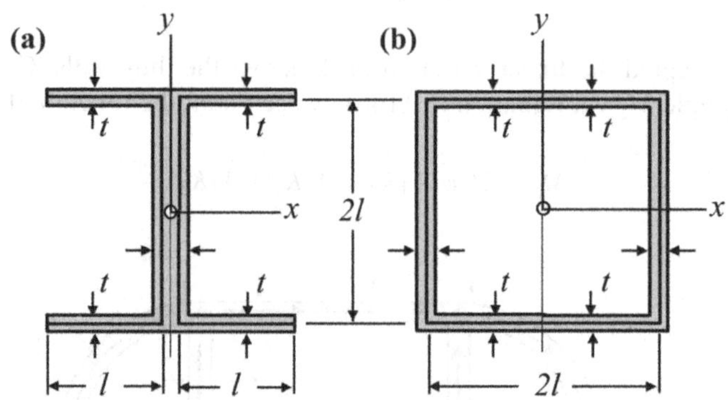

Figure 5.40

5.13 BENDING OF BEAMS

5.13.1 PROBLEM: THE SYMMETRIC BEAM

A beam is a slender member subjected to transverse loads. However, depending on the shape of the cross section and the nature of the loads and supports, twisting can accompany bending. At present, we consider only situations in which the beam is initially a straight prismatic member that has a longitudinal plane of symmetry, both geometrically and materially. We also restrict our attention to beams that are loaded and supported symmetrically with respect to this longitudinal

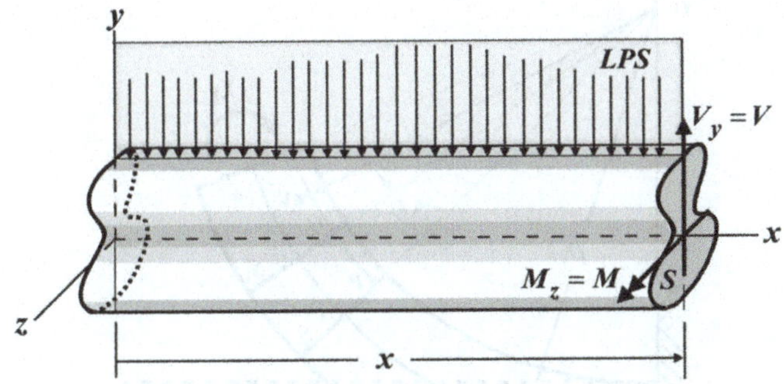

Figure 5.41

plane of symmetry (abbreviated LPS). These restrictions eliminate the possibility of the beam twisting as it bends.

If the loads (in the LPS) are perpendicular to the axis of the beam, the normal-force component on a cross section is zero. Then the net internal force system consists of a shear force and a bending couple, as indicated in Figure 5.41. The portion of the beam to one side of a cross section S is shown with a coordinate system established at one end. The cross section is located by its distance x from the end of the beam. If the xy-plane is the longitudinal plane of symmetry, the only components of the net internal force system on S that are not identically zero are V_y, and M_z. We therefore drop the subscripts and write V for V_y, and M for M_z.

Note that our description of the loads and internal force system neglects deformations. Consequently, the description applies only to small strains and small rotations. Our subsequent treatment of simple bending places no restriction upon rotations.

5.13.2 SIMPLE BENDING

Let us consider a particular situation. The beam of Figure 5.42 is held fixed at the left end and is subjected to a bending couple M applied to the right end. The couple M acts in the LPS. The coordinate system is located so that the xy-plane is coincident with the LPS and the x-axis is parallel to the axis of the member. A cross section S is located by its distance x from the left end. Equilibrium dictates that every slice is subject to the couple M.

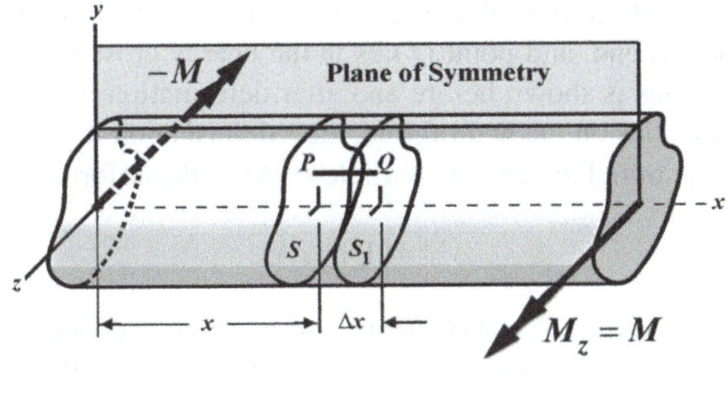

Figure 5.42

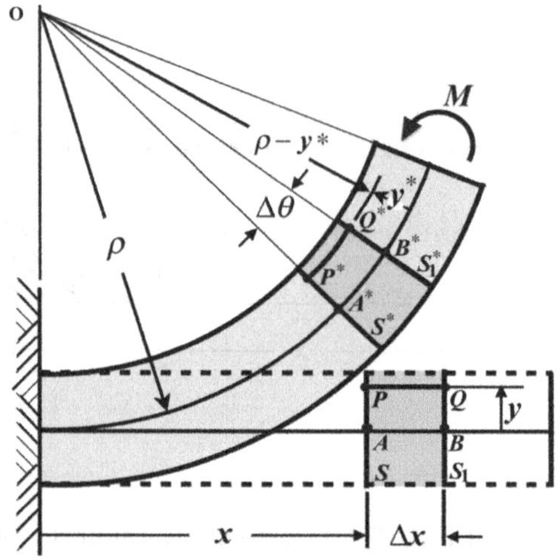

Figure 5.43

5.13.3 DESCRIPTION OF THE DEFORMATION

Because the beam and the loading are symmetric with respect to the LPS, we expect the deformed beam to be symmetric with respect to the LPS. The effect of the bending couple is to compress the upper portions and stretch the lower portions of the beam (assuming the sense of M shown in Figure (5.42). This description suggests that some axial line in the LPS experiences no extension but deforms into a plane curve in the LPS. To characterize the main features of the deformation, we assume the following:

1. There is a line of particles in the LPS parallel to the axis of the beam that experiences no extensional strain and deforms into a plane curve in the LPS. This line is called the *neutral axis* of the beam.
2. Every cross section remains plane and perpendicular to the neutral axis after deformation.[8]

The x-axis is placed along the neutral axis of the beam. The position of the x-axis in the cross section is yet to be determined.

Consider the extension of an axial line element PQ. Point P lies in a cross section S located at a distance x from the left end, and point Q lies in the nearby cross section at $x + \Delta x$. The slice between these cross sections is shown before and after deformation in the profile view of Figure 5.43. Line AB of that slice lies on the neutral line; after deformation A^*B^* has the length of AB.

The y-coordinate is the initial distance to particle P. After the deformation, this distance is

$$y^* = (1 + \bar{\varepsilon}_y)y$$

where $\bar{\varepsilon}_y$ denotes the mean value of the extensional strain. Point O denotes the center of curvature of the neutral axis at A^*, and ρ denotes the radius of curvature at A^*. Because AB is a part of the neutral axis, its length is unchanged; consequently

$$PQ = AB = A^*B^* = \Delta x = \rho\Delta\theta$$

However, the length of the deformed line element P^*Q^* is

$$P^*Q^* = (\rho - y^*)\Delta\theta$$

Then, from the definition of extensional strain, we have

$$\begin{aligned}\varepsilon_x(P) &= \lim_{Q\to P} \frac{P^*Q^* - PQ}{PQ}\\ &= \lim_{\theta\to 0} \frac{(p - y^*)\Delta\theta - \rho\Delta\theta}{\rho\Delta\theta}\\ &= -\frac{y^*}{\rho} = -(1+\bar{\varepsilon}_y)\frac{y}{\rho}\end{aligned} \tag{5.115}$$

In most practical problems $\bar{\varepsilon}_y \ll 1$, so that we can neglect the extension strain compared to unity ($\bar{\varepsilon}_y \ll 1$) and employ the approximation

$$\varepsilon_x(P) \doteq -\frac{y}{\rho} \tag{5.116}$$

Equation (5.116) is the basic result of the *kinematic* analysis. The result presumes symmetry with respect to the xy- (LPS) plane; otherwise, it is independent of the material.

We must anticipate a normal stress σ_x. In general, that stress produces a normal force ΔF on an elemental area ΔA, as depicted in Figure 5.44.

$$\Delta F = \sigma_x \Delta A$$

This contributes to the couple

$$\Delta M = -y^*\Delta F = -y^*\sigma_x\Delta A$$

Summing the contributions and passing to the limit, we obtain the normal force and couple:

$$F = \iint_S \sigma_x dA \tag{5.117}$$

$$M = -\iint_S \sigma_x y^* dA \doteq -\iint \sigma_x y dA \tag{5.118}$$

The final approximation is consistent with Equation (5.116), wherein $y^* = (1+\bar{\varepsilon}_y) \doteq y$. Equations (5.117) and (5.118) are also basic results that are independent of the material.

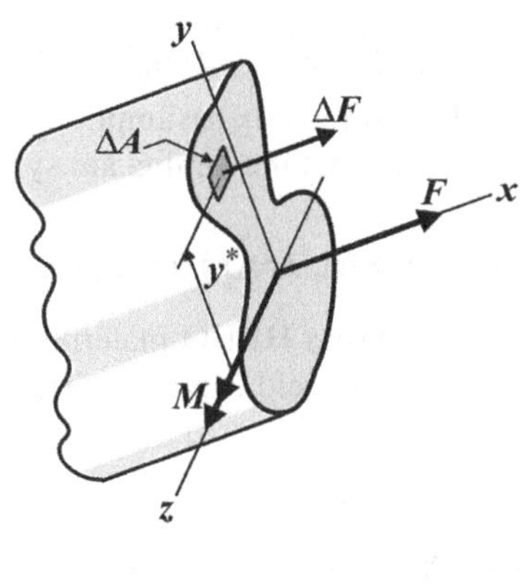

Figure 5.44

5.14 HOOKEAN ISOTROPIC BEAM

Suppose that the beam is composed of Hookean isotropic material. Then

$$\varepsilon_x = \frac{1}{E}[\sigma_x - \nu(\sigma_y + \sigma_z)] \tag{5.119a}$$

$$\varepsilon_y = \frac{1}{E}[\sigma_y - \nu(\sigma_z + \sigma_x)] \tag{5.119b}$$

$$\varepsilon_z = \frac{1}{E}[\sigma_z - \nu(\sigma_x + \sigma_y)] \tag{5.119c}$$

From the foregoing description of the kinematics, only ε_x is described explicitly, and only σ_x is anticipated. Some additional assumptions are needed. If we bend a deformable strip, transverse extensions and contractions are observed. The reader can do this with a piece of rubber, though not Hookean. The upper portion, compressed by the couple, extends in the z-direction ($\varepsilon_z > 0$), whereas the lower contracts ($\varepsilon_z < 0$). On the other hand, the absence of any stresses on the lateral surfaces suggests that the components σ_y and σ_z are negligible, if not zero. Then

$$\varepsilon_x = \frac{\sigma_x}{E} \tag{5.120}$$

$$\varepsilon_y = \varepsilon_z = -\nu\frac{\sigma_x}{E} \tag{5.121}$$

The kinematics, the statics, and the description of the material now provide the basis for a solution. If we substitute Equation (5.116) into the left side of Equation (5.120) and use that result in the integrals (5.117) and (5.118), we obtain

$$F = -\frac{1}{\rho}\iint_S Ey\mathrm{d}A \tag{5.122}$$

$$M = \frac{1}{\rho}\iint_S Ey^2\mathrm{d}A \tag{5.123}$$

5.14.1 CASE I: DOUBLY SYMMETRIC CROSS SECTION

Suppose that our beam is symmetric with respect to the xz-plane as well as the xy-plane. The I-section of Figure 5.45a is a common example. If the property E is an even function of y (and z), then the integral (5.122) vanishes if the x-axis is at the center of the section; stated otherwise, the neutral (unstretched) axis is the central line. Note that the modulus E might be variable, but $E(y)$ must be an even function. Indeed, all the foregoing assumptions and equations would apply to the laminated beam of Figure 5.45b, provided that the lamina are symmetrically positioned.

5.14.2 CASE II: HOOKEAN, ISOTROPIC, AND HOMOGENEOUS BEAM

Suppose now that the beam is composed of a HI-HO material, that is, E is constant. This is essentially the case for most metals within the yield condition. Then, according to Equation (5.122), $F = 0$ if

$$\iint_S y\mathrm{d}A = 0 \tag{5.124}$$

In other words, the coordinate y must be measured from the centroid of the cross section. The neutral axis (x) pierces the centroid of the section.

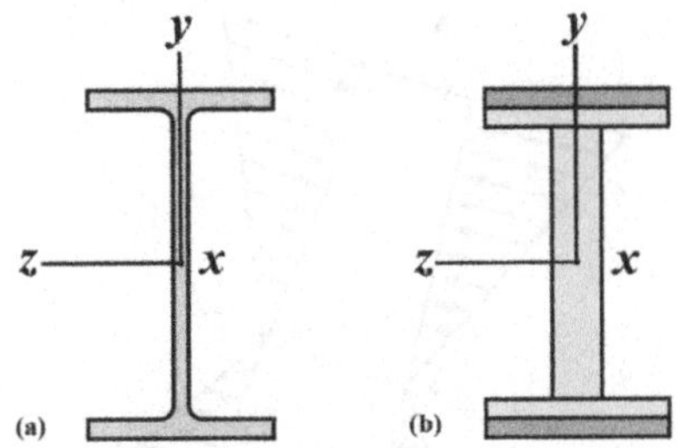

Figure 5.45

The couple M is given by Equation (5.123):

$$M = \frac{E}{\rho} \iint_S y^2 dA$$

Following usual conventions, we denote the integral by I:

$$I \equiv \iint_S y^2 dA$$

Then, the curvature and bending couple are linearly related:

$$\kappa \equiv \frac{1}{\rho} = \frac{M}{EI} \tag{5.125}$$

Equation (5.125) is a fundamental relation in elementary beam theory, sometimes called the *Bernoulli-Euler curvature formula.* It relates the bending couple on a cross section to the curvature of the neutral axis at that location if the material is Hookean, isotropic, and homogeneous (HI-HO).[9]

The extensional strain ε_x is expressed in terms of the couple by means of Equations (5.116) and (5.125)

$$\varepsilon_x = -\frac{My}{EI} \tag{5.126}$$

The stress σ_x is then given by Equation (5.120):

$$\sigma_x = -\frac{My}{I} \tag{5.127}$$

This linear distribution of normal stress is depicted in Figure 5.46.

We can acquire additional insights by examining the deformations of a rectangular cross section as depicted in Figure 5.47. According to Equations (5.116) and (5.121)

$$\varepsilon_z = \nu \frac{y}{\rho}$$

The transverse extensional strain ε_z varies with distance y, as the axial strain ε_x, but has the opposite sign. The former is associated with a curvature ν/ρ as depicted in Figure 5.47; this curvature in the plane $(y-z)$ of the cross section is called the anticlastic curvature.

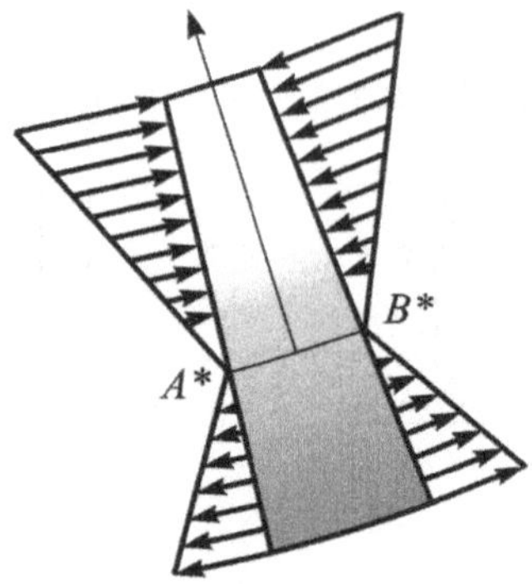

Figure 5.46

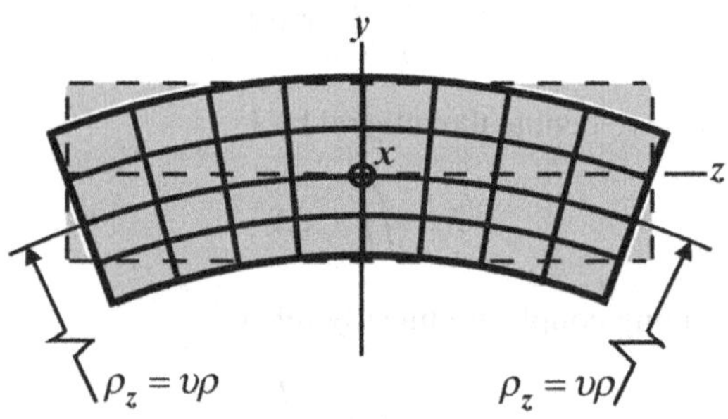

Figure 5.47

The strain and stress distributions of Equations (5.126) and (5.127) apply to all cross sections of the beam. In particular, this includes the end faces where the load is applied and where the beam is supported. Fortunately, Saint-Venant's principle enables us to use these results in many other cases if the beam is slender. However, if the end couples are applied in accordance with Equation (5.127), the results obtained here are in agreement with the solutions of more elaborate theories. In fact, it can then be shown that, for a linearly elastic material, $\sigma_y = \sigma_z = \tau_{xy} = \tau_{xz} = \tau_{yz} = 0$. Then, $\gamma_{xy} = \gamma_{yz} = \gamma_{xz} = 0$, and, by Hooke's law, $\varepsilon_y = \varepsilon_z = -\nu\varepsilon_x = \nu My/EI$.

ILLUSTRATIVE EXAMPLE 5.23

The significant features of the bending deformation of straight beams loaded in a longitudinal plane of symmetry (LPS) do not depend on the material. We still assume the existence of a neutral axis that experiences no extension strain (although it need not be the centroidal axis). Furthermore, cross sections are assumed to remain plane and perpendicular to the neutral axis. Under these conditions, Equation (5.116) holds for all materials.

Suppose that the beam has the T-shaped cross section shown in Figure 5.48, the material is HI-HO, and the beam is subjected to a simple bending couple. Then the net force vanishes and, in accordance with Equations (5.122) and (5.124),

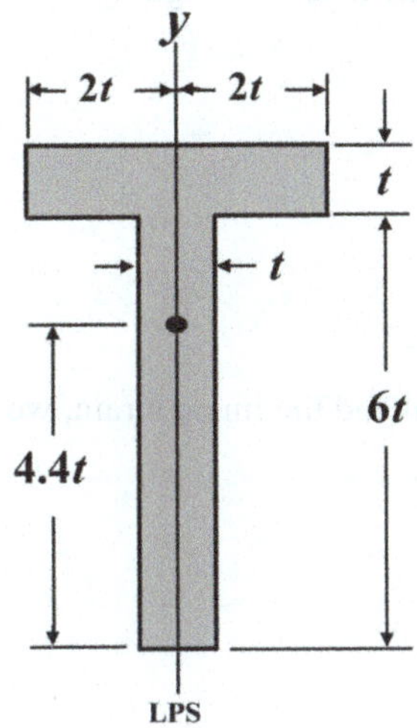

Figure 5.48

Here the stress varies linearly, in accordance with Equation (5.127):

$$\sigma_x = -\frac{M}{I}y$$

The maximum and minimum stresses occur at the bottom and top, respectively:

$$\sigma_{\max} = -\frac{M(-4.4t)}{47.7t^4} = \frac{M}{10.8t^3}$$
$$\sigma_{\min} = -\frac{M(2.6t)}{47.7t^4} = -\frac{M}{18.4t^3}$$

To assign some reasonable values, let us suppose that the member is a rolled-steel beam with $t = 0.25$ in. (0.64 cm) and σ_0 (yield stress) = 48 ksi (330 MPa). Then

$$\sigma_0 = 48{,}000\left(\frac{\text{lb}}{\text{in}^2}\right) = \frac{M}{10.8\,(0.25)^3\ \text{in.}^3}$$
$$\Longrightarrow M = 8140\ \text{in.}\cdot\text{lb} \quad (920\ \text{N}\cdot\text{M})$$

$$\iint_S y\text{d}A = 0$$

The neutral line is at the centroid, a distance $4.4t$ from the bottom. The bending couple M is related to the curvature in accordance with Equation (5.123):

$$M = \frac{E}{\rho}\iint_S y^2\text{d}A = \frac{EI}{\rho}$$

where

$$I = \iint_S y^2\text{d}A = t\int_{-4.4t}^{1.6t} y^2\text{d}y + 4t\int_{1.6t}^{2.6t} y^2\text{d}y$$
$$= 47.7t^4$$

Our steel beam is 1.75 in. deep. Two strong students could exert the couple through a 48-in. (1.2 m) crossbar (wrench) by applying forces of 170 lb (760 N).

ILLUSTRATIVE EXAMPLE 5.24: A LAMINATED BEAM

A beam is built up by laminating *very* thin layers of HI-HO materials, each slightly denser and stiffer, so that the modulus effectively varies according to the equation

$$E = E_0\left(1 + \frac{y}{h}\right)$$

Here the origin ($y = 0$) is arbitrarily placed at the bottom of the rectangular cross section, as depicted in Figure 5.49; the modulus is E_0 at the bottom ($y = 0$) and $2E_0$ at the top ($y = h$). The beam is subjected to a simple bending couple.

In accordance with the foregoing arguments, the strain varies linearly, but the neutral axis ($\varepsilon_x = 0$) is not known a *priori*. Since the material is not homogeneous, the position of the neutral line is not necessarily at the centroid. The linear function $\varepsilon_x(y)$ can be written with the origin ($y = 0$) arbitrarily placed at the bottom where $\varepsilon_0(0) = \varepsilon_0$:

$$\varepsilon_x = \varepsilon_0 - \frac{y}{\rho}$$

The force and couple are given by Equations (5.117) and (5.118):

$$F = \iint_S \sigma_x dA \text{ and } M = -\iint_S \sigma_{xy} dA$$

The former is zero in the case of simple bending. With the given material and the linear strain, we have

$$\sigma_x = E_0\left(1+\frac{y}{h}\right)\left(\varepsilon_0 - \frac{1}{\rho}y\right)$$

$$F = w\int_0^h E_0\left(1+\frac{y}{h}\right)\left(\varepsilon_0 - \frac{1}{\rho}y\right)dy = 0$$

$$M = -w\int_0^h E_0\left(1+\frac{y}{h}\right)\left(\varepsilon_0 - \frac{1}{\rho}y\right)ydy$$

Upon integrating these two equations, we obtain

$$\left(\frac{3}{2}h\right)\varepsilon_0 - \left(\frac{5}{6}h\right)\frac{h}{\rho} = 0$$

$$-\left(\frac{5}{6}h^2\right)\varepsilon_0 + \left(\frac{7}{12}h^2\right)\frac{h}{\rho} = \frac{M}{E_0 w}$$

The solution follows:

$$\varepsilon_0 = \frac{60}{13}\frac{M}{E_0 wh^2}, \quad \frac{h}{\rho} = \frac{108}{13}\frac{M}{E_0 wh^2}$$

The neutral axis is at the location (y_0) of zero strain:

$$\varepsilon_x(y_0) = 0 = \varepsilon_0 - \frac{y_0}{\rho}, \quad y_0 = \frac{5}{9}h$$

Since the material is stiffer at the top, the neutral line lies above the midline. Strain and stress distributions are depicted in Figure 5.50. Note that the stress distribution is *not* linear.

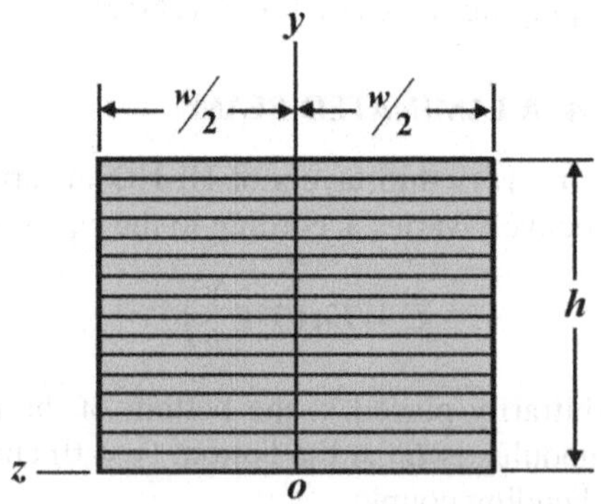

Figure 5.49

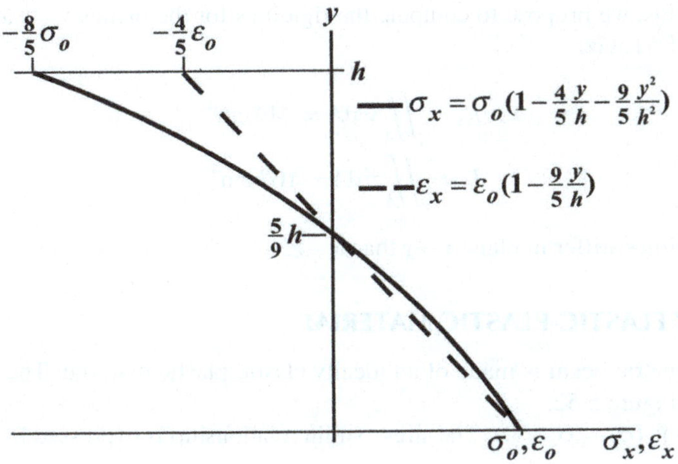

Figure 5.50

ILLUSTRATIVE EXAMPLE 5.25

The product EI is often called the flexural rigidity, since it is the constant of proportionality between the bending moment M and curvature $\kappa = 1/\rho$. An I beam is a common structural component and typically supports transverse loads acting in the direction of the web, i.e. in the plane $x-y$ of Figure 5.51.

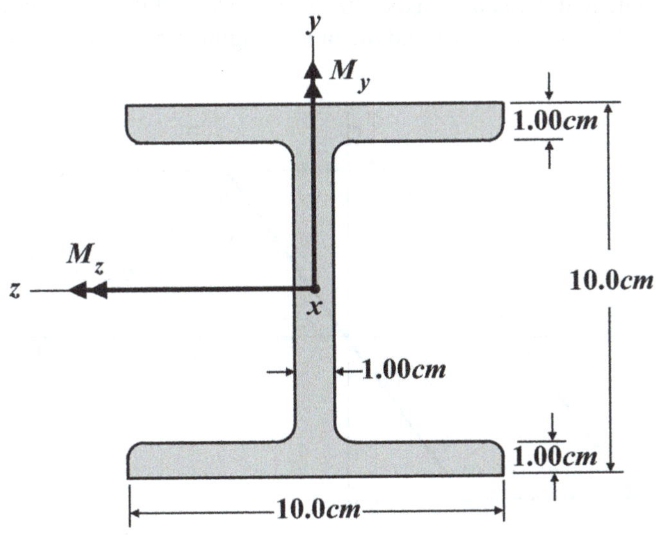

Figure 5.51

To appreciate this, we propose to compute the rigidities for the planes $x-y$ and $x-z$ for the I section of Figure 5.51, viz.

$$I_{yy} = \iint_A y^z dA = 449\ \text{cm}^4$$

$$I_{zz} = \iint_A z^z dA = 167\ \text{cm}^4$$

This beam is 2.7 times stiffer in plane $x-y$ than $x-z$.

5.15 IDEALLY ELASTIC-PLASTIC MATERIAL

Suppose the symmetric beam is made of an ideally elastic-plastic material. The stress-strain diagram is shown in Figure 5.52.

We again assume ($\sigma_y = \sigma_z = 0$). The stress-strain relationship is expressed by

$$\sigma_x = \begin{cases} -\sigma_0, & \varepsilon_x \le -\varepsilon_0 \\ E\varepsilon_x, & -\varepsilon_0 \le \varepsilon_x \le \varepsilon_0 \\ \sigma_0, & \varepsilon_x \ge \varepsilon_0 \end{cases} \tag{5.128}$$

The preceding description, specifically the *kinematic* result (5.116) the *dynamic* relations (5.117) and (5.118) and the stress-strain equations (5.128), applies to a beam of arbitrary cross section but symmetric about the xy-plane. Some further simplification results when the shape of the cross section is also symmetric about the xz-plane. To illustrate the analyses, we consider a few examples.

ILLUSTRATIVE EXAMPLE 5.26: RECTANGULAR CROSS SECTION

Suppose that the cross section is a rectangle of width b and depth h, as shown in Figure 5.53. We will tentatively assume that the neutral axis is the central axis. This seems reasonable because the axis is the horizontal line of symmetry, and the plot of Figure 5.52 shows an odd function $\sigma_x(\varepsilon_x)$.

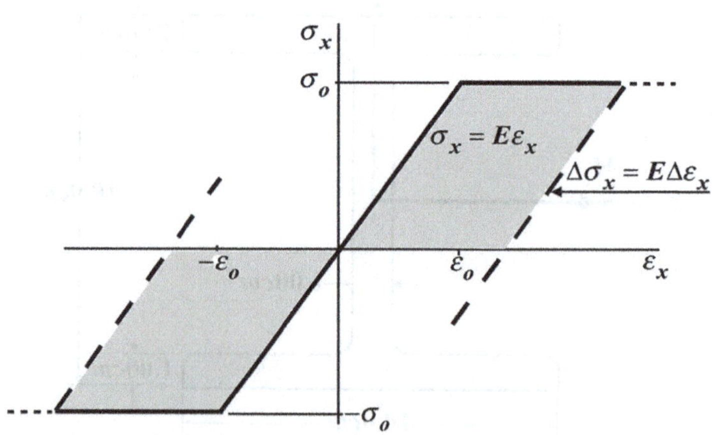

Figure 5.52

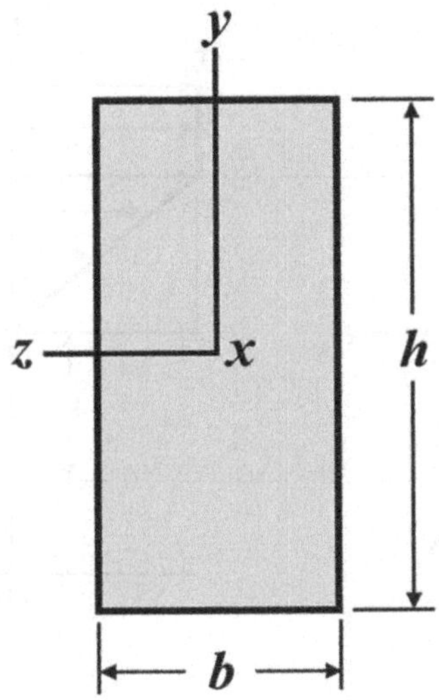

Figure 5.53

According to Equation (5.116) the strain ε_x is proportional to the distance y. Hence, the maximum and minimum extensional strains occur at the bottom and top of the beam. These are

$$\varepsilon_{x_{\max}} = -\varepsilon_{x_{\min}} = \frac{h/2}{\rho} \tag{5.129}$$

As long as the extension strain $[\varepsilon_x]$ is everywhere less than the yield strain $\varepsilon_0 = \sigma_0/E$, the behavior is linearly elastic throughout the beam, and the results of the previous analysis are applicable. In particular, the normal stress σ_x is given by Equation (5.127) and is linearly distributed on a cross section. As the bending couple is increased, the extensional strain $|\varepsilon_x|$ exceeds the yield strain ε_0 in the top and bottom portions of the beam, as indicated in Figure 5.54a. Let c denote half the depth of the elastic core. At $y = \pm c$ we have $\varepsilon_x = \mp\varepsilon_0$; hence, from Equation (5.116),

$$\varepsilon_0 = \frac{c}{\rho}, \quad \text{or} \quad \frac{1}{\rho} = \frac{\varepsilon_0}{c} = \frac{\sigma_0}{Ec} \tag{5.130}$$

The normal stress in the elastic core $|y| \leq c$ is proportional to the extensional strain:

$$\sigma_x = E\varepsilon_x = -\frac{Ey}{\rho}$$

or, with Equation (5.130),

$$\sigma_x = -\frac{y}{c}\sigma_0, \quad |y| \leq c \tag{5.131a}$$

In the plastic portions of the beam $|\varepsilon| \geq \varepsilon_0$, the material is subjected to the yield stress

$$\sigma_x = \begin{cases} -\sigma_0, & c \leq y \leq \dfrac{h}{2} \\ \sigma_0, & -\dfrac{h}{2} \leq y \leq -c \end{cases} \tag{5.131b}$$

The distribution of normal stress on a cross section is as shown in Figure 5.54b.

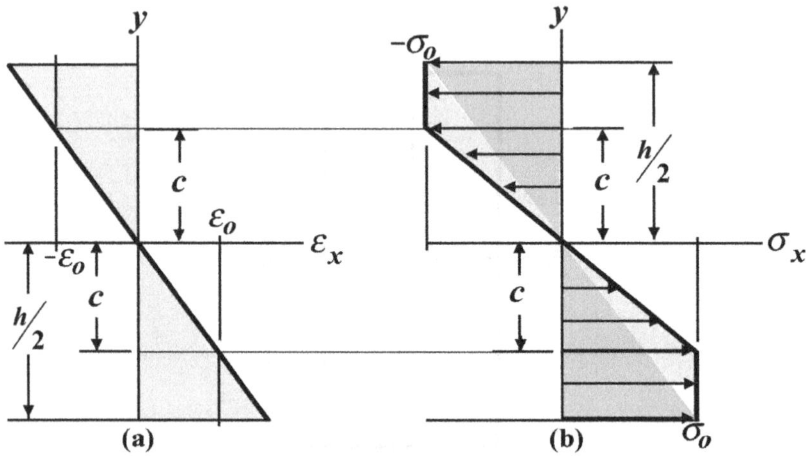

Figure 5.54

The normal force upon the cross section is given by Equation (5.117) where the stress is given by Equations (5.131a,b) and depicted in Figure 5.54b. The integral of this odd function $\sigma_x(y)$ over the rectangular region vanishes; that is, $F = 0$. Then the neutral axis is along the central axis *in this case*.

The bending couple is given by Equation (5.118). Accordingly, the integral must be evaluated in three parts, as follows:

$$\begin{aligned} M &= -b\int_{-h/2}^{-c} \sigma_0 \, y dy + b\int_{-c}^{c} \frac{\sigma_0}{c} y^2 dy + b\int_{c}^{h/2} \sigma_0 \, y dy \\ &= \sigma_0 b\left[\frac{h^2}{4} - \frac{c^2}{3}\right] \end{aligned} \tag{5.132}$$

The half-depth of the elastic core can be determined by solving Equation (5.132) for c. However, the results are meaningless unless c is between 0 and $h/2$. When $c = h/2$, the outermost portion of the cross section is about to yield. The bending couple, from Equation (5.132), is

$$M_0 = \sigma_0 \frac{bh^2}{6} \tag{5.133}$$

For M less than M_0 the behavior is linearly elastic, and the previous results apply. As c decreases, more and more of the cross section deforms plastically. In the limiting condition, $c = 0$, the cross section is said to be fully plastic. For this condition, the bending couple has the value

$$M_L = \lim_{c\to 0} = \sigma_0 b\left[\frac{h^2}{4} - \frac{c^2}{3}\right] = \sigma_0 \frac{bh^2}{4} \tag{5.134}$$

For future reference let us note the relative magnitudes of the limit couple M_L and initial-yield couple M_0:

$$\frac{M_L}{M_0} = \frac{3}{2} \tag{5.135}$$

Prior to yielding, $M < M_0$, the moment M and curvature κ follow the linear relation (5.125). After yielding, $M_0 < M < M_L$, the stress is distributed according to Equation (5.131). At the elastic-plastic interface, $y = \pm c$, the yielding is initiated, so that

$$\sigma_x = \pm\sigma_0 = E\varepsilon_x(\mp c)$$

According to Equation (5.116) and the definition $\kappa = 1/\rho$,

$$\varepsilon_x(\mp c) = \pm\frac{c}{\rho} \equiv \pm c\kappa$$

It follows that

$$c = \frac{\sigma_0}{E\kappa} \tag{5.136}$$

Substituting (5.136) into (5.132), we obtain

$$M = \sigma_0 b\left[\frac{h^2}{4} - \frac{\sigma_0{}^2}{3E^2\kappa^2}\right], \quad M_0 < M < M_L \tag{5.137}$$

This nonlinear equation governs after yielding. As the curvature increases, the second term in the bracket diminishes, and, in the limit $\kappa \to \infty$, the couple M approaches the limit value M_L.

The complete load-deformation curve is drawn in Figure 5.55. From O to A the curvature, κ and load M are linearly related by Equation (5.125). As the load M is increased beyond M_0, plastic deformation occurs in the beam. As M is increased, the curve approaches a horizontal asymptote at M_L: this is the limit load.

Let us now consider the behavior if the couple is removed after yielding has occurred-that is, from the point B of Figure 5.55. Let ΔM, $\Delta\sigma_x$, $\Delta\varepsilon_x$, and $\Delta\kappa$ denote the subsequent *changes* in moment, stress, strain, and curvature, respectively. In accordance with our previous description of this material, the changes in stress and strain follow a linear relation (see Figure 5.52) whether the material has yielded (regions $|y| > c$ in the beam) or not (region $-c < y < c$). Then, *throughout* the beam, during unloading,

$$\Delta\sigma_x = E\Delta\varepsilon_x \tag{5.138}$$

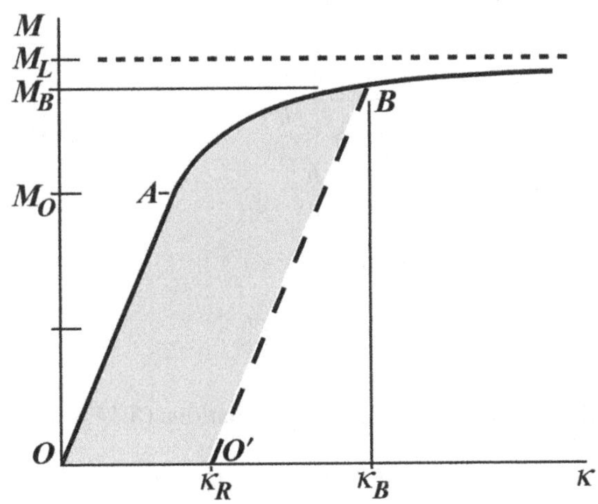

Figure 5.55

The kinematic relation (5.116) also holds, so that

$$\Delta\varepsilon_x = -y\Delta\kappa \tag{5.139}$$

The change in the couple ΔM is related to the change in stress $\Delta\sigma_x$ in the same way as M and σ_x in Equation (5.118):

$$\Delta M = -\iint \Delta\sigma_x y dA \tag{5.140}$$

Again, Equations (5.138) and (5.139) hold throughout; their substitution into Equation (5.140) gives the result

$$\Delta M = EI\Delta\kappa \tag{5.141}$$

where, in the present example, $I = bh^3/12$. In short, the unloading follows the linear relation (5.141), wherein the constant of proportionality is the same (EI) that appears in Equation (5.125), which governs the initial elastic response. The unloading traces the straight line BO', parallel to AO in Figure 5.55. If the moment M_B is removed, the curvature changes in amount

$$\Delta\kappa = \kappa_R - \kappa_B = -\frac{M_B}{EI} \tag{5.142}$$

Here κ_R is the residual curvature (see Figure 5.55)

5.15.1 RESIDUAL STRESS

During removal of the moment, the change in the stress is given by Equation (5.138) and, according to Equation (5.139), that change is linear in the coordinate y:

$$\Delta\sigma_x = -E\Delta\kappa y \tag{5.143}$$

Because the strains are presumed small, this change can be superposed upon the stress prior to unloading. To illustrate, let us suppose that yielding has progressed to one-half the depth. Prior to unloading, the stress distribution is depicted in Figure 5.56a:

$$\sigma_x^B = \begin{cases} -\sigma_0, & \dfrac{h}{4} \le y \\ -\sigma_0\dfrac{4y}{h}, & -\dfrac{h}{4} \le y \le \dfrac{h}{4}, \\ \sigma_0, & y \le -\dfrac{h}{4} \end{cases} \qquad M^B = \frac{11}{48}bh^2\sigma_0 \tag{5.144a,b}$$

At $y = \pm h/4$, $\sigma_x = \mp E\kappa_B\frac{h}{4} = \mp\sigma_0$, so that

$$\kappa_B = \frac{4}{h}\frac{\sigma_0}{E} \tag{5.145}$$

During unloading, the change of stress is given by Equations (5.143), (5.142), and (5.145):

$$\begin{aligned} \Delta\sigma_x &= -E\Delta\kappa y = -E(\kappa_R - \kappa_B)y \\ &= -E\left(\kappa_R - \frac{4}{h}\frac{\sigma_0}{E}\right)y \end{aligned} \tag{5.146}$$

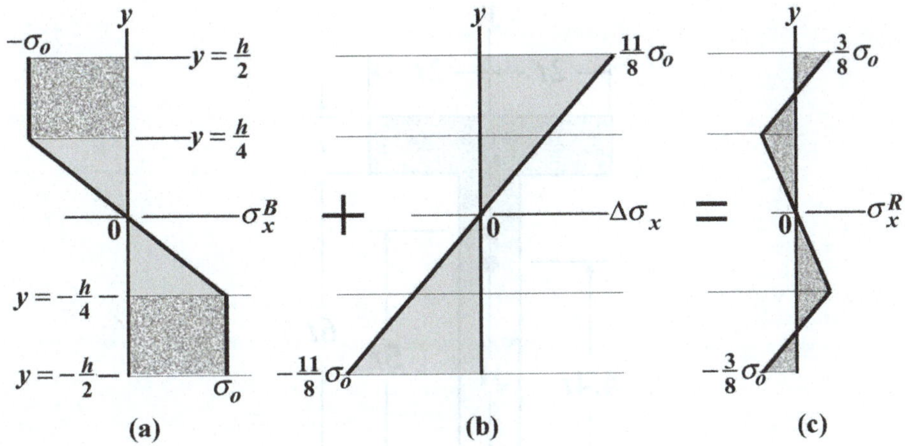

Figure 5.56

The unloaded state implies a vanishing moment:

$$M = M^B - \int_{-h/2}^{h/2} \Delta\sigma_x y dy = 0 \tag{5.147}$$

The latter Equation (5.147) is expressed in terms of one unknown, κ_R (σ_0, E, and h are known properties) in accordance with Equations (5.144) and (5.146). After integrating, we have

$$\kappa_R = \frac{5}{4}\frac{\sigma_0}{Eh}, \quad \Delta\sigma_x = \frac{11}{4}\sigma_0\frac{y}{h}$$

The residual stress distribution, $\sigma_x^R = \sigma_x^B + \Delta\sigma_x$ is depicted in Figure 5.56c; it is the sum of the distributions in (a) and (b) $\left[\sigma_x^B + \Delta\sigma_x\right]$. Again we observe a residual stress when loading causes yielding in some regions of the body while other regions remain elastic.

The unloaded beam exhibits a residual curvature κ_R. Theoretically, if one were to remove the outermost layers (those plastically deformed, $y > h/4$, $y < -h/4$), then the remaining inner layer would be straight. The process is hypothetical and the recovery of a straight elastic region is theoretical. In reality, the mechanism of yielding is heterogeneous so that no precise boundaries exist between plastic and elastic regions. Still, our analysis provides essential features of elastic-plastic interactions.

ILLUSTRATIVE EXAMPLE 5.27

Let us return to the T cross section of Figure 5.57. Previously we examined the HI-HO response in simple bending. We established the position of the centroid, the neutral line during elastic bending ($y = 4.4t$) and the maximum tensile strength at the bottom. If σ_0 denotes the yield stress, then yielding is initiated at

$$M = M_0 = 10.8t^4\sigma_0$$

Now suppose that ideally plastic deformation ensues. Specifically, the material obeys the description of Figure 5.52. At the limit state, the entire cross section has yielded. The beam is subject only to the bending couple. Since the resultant force is zero ($F = 0$), the areas under tension

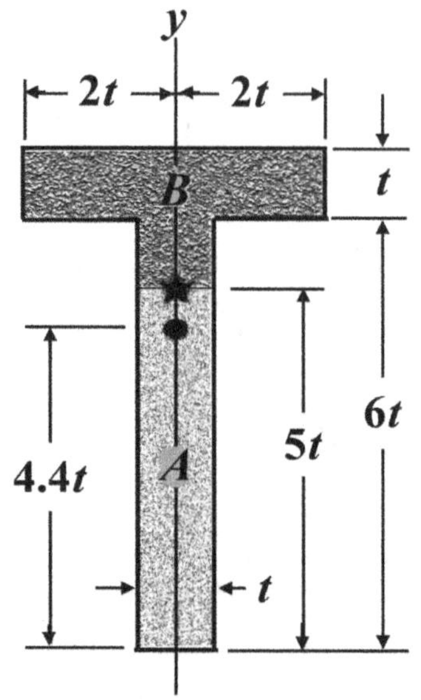

Figure 5.57

($\sigma_x = \sigma_0$) and compression ($\sigma_x = -\sigma_0$) are equal. Areas A and B in Figure 5.57 are equal. This places the neutral line at $y = 5t$. The neutral line has shifted during the progression of yielding. The limit moment is $M_L = 19t^3\sigma_0$ and

$$\frac{M_L}{M_0} = 1.76$$

Recall the corresponding ratio (1.50) for the rectangular cross section (Equation 5.135). Note: The shift of the neutral line and the ratio M_L/M_0 are both dependent on the shape of the cross section. Finally, note that these results are but a special form of the requirement $F = 0$. They hold only when the magnitudes of the tensile and compressive yield stresses are the same.

ILLUSTRATIVE EXAMPLE 5.28: DOUBLE-MODULUS MATERIAL

As another illustration, consider a material that is stiffer in tension than in compression. The stress-strain diagram for such a material is shown in Figure 5.58. The stress-strain relation follows:

$$\sigma_x = \begin{cases} E^T\varepsilon_x, & \varepsilon_x \geq 0 \\ E^C\varepsilon_x, & \varepsilon_x \leq 0 \end{cases} \quad E^T \geq E^C \tag{5.148}$$

Again let us work with a rectangular cross section of width b and depth h. From Equation (5.116) we have

$$\varepsilon_x = -\frac{y}{\rho} \tag{5.149}$$

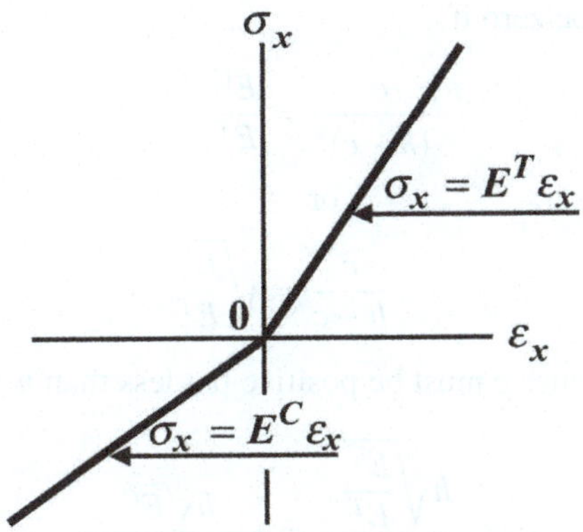

Figure 5.58

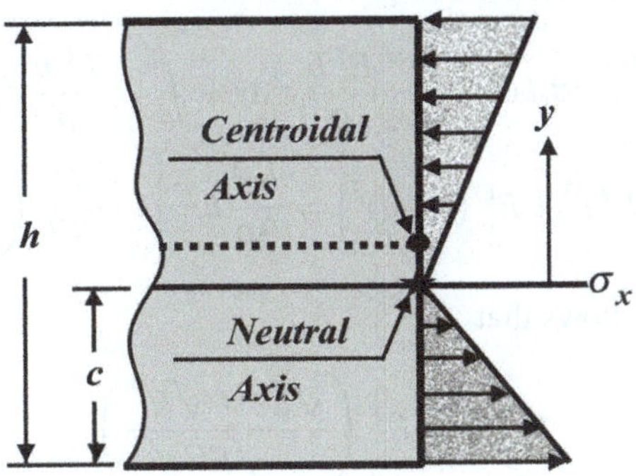

Figure 5.59

If ρ is positive at a cross section,

$$\sigma_x = \begin{cases} -\dfrac{E^T y}{\rho}, & y \leq 0 \\ -\dfrac{E^C y}{\rho}, & y \geq 0 \end{cases} \tag{5.150}$$

The neutral axis in this case will not be located at the centroid of the cross section. Suppose it is a distance c from the bottom of the beam, as shown in Figure 5.59. Then the normal force on the cross section is given by

$$\begin{aligned} F &= \int_{-c}^{h-c} \sigma_x b dy = b \int_{-c}^{0} -\frac{E^T}{\rho} y dy + b \int_{0}^{h-c} -\frac{E^c}{\rho} y dy \\ &= \frac{b}{2\rho} \left[E^T c^2 - E^C (h-c)^2 \right] \end{aligned}$$

Now the normal force F will be zero if

$$\frac{c^2}{(h-c)^2} = \frac{E^C}{E^T} \tag{5.151a}$$

or

$$\frac{c}{h-c} = \sqrt{\frac{E^C}{E^T}} \tag{5.151b}$$

(We take the positive root because c must be positive but less than h.) Then

$$c = \frac{h\sqrt{\dfrac{E^C}{E^T}}}{1+\sqrt{\dfrac{E^C}{E^T}}} = \frac{h\sqrt{E^C}}{\sqrt{E^T}+\sqrt{E^C}} \tag{5.152}$$

The bending couple on the cross section is

$$\begin{aligned} M &= -\int_{-c}^{h-c} y\sigma_x b\mathrm{d}y = \int_{-c}^{0} \frac{E^T b}{\rho} y^2 \mathrm{d}y + \int_{0}^{h-c} \frac{E^C b}{\rho} y^2 \mathrm{d}y \\ &= \frac{b}{3\rho}\left[E^T c^3 + E^C (h-c)^3\right] = \frac{bc^3}{3\rho} E^T \left[1 + \frac{E^C}{E^T}\left(\frac{h-c}{c}\right)^3\right] \end{aligned}$$

From Equation (5.151b) it follows that

$$M = \frac{bc^3}{3\rho} E^T \left(\frac{\sqrt{E^C}+\sqrt{E^T}}{\sqrt{E^C}}\right) \tag{5.153}$$

Substituting for c from Equation (5.152), we obtain

$$M = \frac{bh^3}{3\rho} \frac{E^T E^C}{\left(\sqrt{E^T}+\sqrt{E^T}\right)^2} \tag{5.154}$$

Let us *define* a reduced modulus E^R as follows:

$$E^R \equiv \frac{4E^T E^C}{\left[\sqrt{E^T}+\sqrt{E^C}\right]^2} \tag{5.155}$$

Then, upon noting that

$$\frac{bh^3}{12} = I$$

where I is the moment of inertia of this rectangular cross section about the horizontal line through the centroid, we finally obtain, from Equations (5.154) and (5.155),

$$M = \frac{E^R I}{\rho} \tag{5.156}$$

This is exactly the same as the Bernoulli-Euler curvature formula with E replaced by E^R.

Placing Equation (5.156) into Equations (5.149) and (5.150), we get

$$\varepsilon_x = -\frac{My}{E^R I}$$

and

$$\sigma_x = \begin{cases} -\dfrac{E^T}{E^R}\dfrac{My}{I}, & y \le 0 \\ -\dfrac{E^C}{E^R}\dfrac{My}{I}, & y \ge 0 \end{cases} \tag{5.157}$$

Note that these results were derived for a rectangular cross section only. For a differently shaped cross section, E^R will *generally not be defined by* Equation (5.155). Also, the location of the neutral axis will differ; the method of calculation will be the same.

ILLUSTRATIVE EXAMPLE 5.29: A REINFORCED CONCRETE BEAM

A concrete beam has the rectangular cross section shown in Figure 5.60a. It is reinforced by the steel *re-bars* which are positioned as shown. Let us treat both materials as HI-HO with the following moduli:

$$E_c = 3 \times 10^3 \text{ ksi}, \quad E_s = 30 \times 10^3 \text{ ksi}$$

Here subscripts c and s signify the concrete and steel, respectively. The concrete has a compressive *yield* strength:

$$\sigma_c = -3 \text{ ksi}$$

The tensile yield strength of the steel is

$$\sigma_s = 30 \text{ ksi}$$

We propose to choose the *re-bar* to sustain the tensile stresses under the action of a bending couple M. Because the concrete is very weak in tension, we propose to discount its resistance to tension.

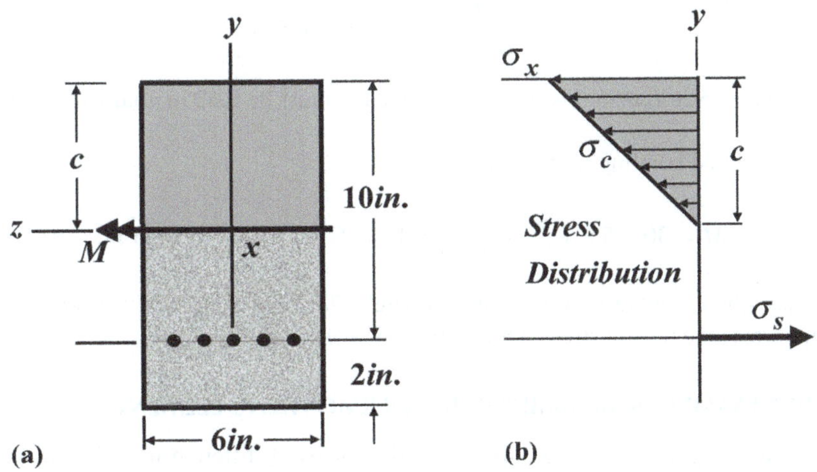

Figure 5.60

The symmetry of the cross section justifies our description of simple bending wherein the extension strain ε_x is a linear function of coordinate y. Since the properties of the materials differ, we cannot anticipate the location of the neutral axis. Accordingly, we place the origin of the unknown distance c from the top of the section (see Figure 5.60). Since the stress at the top is to attain $\sigma_x = -\sigma_c$, the stress in the concrete portions follows:

$$\begin{aligned} \sigma_x &= -E_c \kappa y, \quad \sigma_c = E_c \kappa c \\ -3 &= -3 \times 10^3 c\kappa \\ c\kappa &= 10^{-3} \end{aligned} \tag{5.158}$$

At the *re-bars*, $y = -(10 - c)$, and

$$\sigma_s = 30 \text{ ksi} = E_s(10 - c)\kappa, \quad (10 - c)\kappa = 10^{-3} \tag{5.159}$$

Solution of Equations (5.158) and (5.159) gives us

$$\kappa = 2 \times 10^{-2}, \quad c = 5 \text{ in.} \tag{5.160a}$$

These results provide the stress distribution, as depicted in Figure 5.60b. In the concrete

$$\sigma_x = -E_c \kappa y$$

In the *re-bar*

$$\sigma_x = \sigma_s$$

We anticipate little variation across the small diameter of a bar.

There is no force on the cross section; therefore,

$$F = A\sigma_s + 6\sigma_c \int_0^5 \frac{y}{c} \mathrm{d}y = 0$$

where A denotes the total cross-sectional area of the bars. The result follows.

$$A = -\frac{\sigma_c}{\sigma_s} \frac{6 \times 5^2}{5 \times 2} = \frac{15}{10} = 1.5 \text{ in.}^2$$

If we chose to place 5 *re-bar* in this position, then each must be 0.62 in. diameter (approximately 5/8 in).

Then the associated bending moment is

$$M = 30 \times 5 \times 1.5 + 6 \int_0^5 \frac{3}{5} y^2 \mathrm{d}y = 225 + 150 = 375 \text{ kip-in.}$$

In our example, the concrete sustains approximately 2/3 of the moment. Of course, in reality, concrete supports some tensile stress. Our analysis is conservative.

ILLUSTRATIVE EXAMPLE 5.30: BIMETALLIC THERMOSTATIC ELEMENT

The small beam in Figure 5.61a is composed of two strips of different materials that are bonded together. A composite cross section is depicted in Figure 5.61b. The upper strip is steel; the lower is bronze. Both behave as HI-HO materials. Their properties are given here with subscripts S and

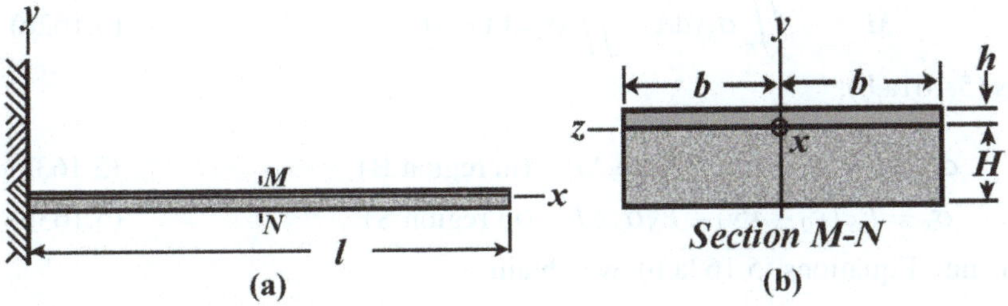

Figure 5.61

B which signify the steel and bronze properties, respectively: Coefficients of thermal expansion are denoted by α:

$$E_S = 30 \times 10^6 \text{ psi (200 GPa)}, \quad E_B = 15 \times 10^6 \text{ psi (100 GPa)}$$

$$\alpha_S = 6.5 \times 10^{-6}/°F \ (12 \times 10^{-6}/°C), \quad \alpha_B = 13 \times 10^{-6}/°F \ (23 \times 10^{-6}/°C)$$

$$h = 0.015 \text{ in. } (0.038 \text{ cm}), \quad H = 0.060 \text{ in. } (0.152 \text{ cm})$$

$$\text{Length} = l = 2.00 \text{ in. } (5.08 \text{ cm})$$

Suppose now that the strip is clamped at one end ($x = 0$) but otherwise free. It is initially straight at room temperature and is then heated so that the temperature is uniformly increased by an amount.

$$\Delta T = 400°F \ (220°C)$$

Since the coefficient α_B is greater than the coefficient α_S, the bronze tends to expand more, so that the composite strip bends when heated. Such a bimetallic strip might be used as a thermostatic device to detect temperature change or activate an electrical switch. Our task is to determine the deflection of the free end ($x = \ell$) and the stresses that accompany the rise in temperature.

We observe that the strip is a slender beam, uniform along its length, and symmetric about the middle plane (xy). Then, according to all previous arguments, each slice (Δx) behaves as every other, with the possible exception of regions near the ends ($x = 0, \ell$) and edges ($z = \pm b$). Again, according to Saint-Venant's principle, such end effects are expected to be localized. Then, throughout the beam, except near the ends, plane cross sections remain plane, and the extensional strain ε_x is a linear function of coordinate y. As a matter of convenience, we set the origin at the interface where the strain is an unknown ε_0:

$$\varepsilon_x = \varepsilon_0 - \kappa y$$

Again, we assume a simple state of stress; we neglect all but the normal component σ_x. Then, the HI-HO materials (S and B) are described by the stress-strain-temperature equations:

$$\varepsilon_x = \frac{\sigma_x}{E_B} + \alpha_B \Delta T \quad \text{(in region B)} \tag{5.161a}$$

$$\varepsilon_x = \frac{\sigma_x}{E_S} + \alpha_S \Delta T \quad \text{(in region S)} \tag{5.161b}$$

Because no external loads act on the strip, the resultant on every cross section vanishes. By Equations (5.117) and (5.118),

$$F = \iint_B \sigma_x dA + \iint_S \sigma_x dA = 0 \tag{5.162a}$$

$$M = -\iint_B \sigma_x y \mathrm{d}A - \iint_S \sigma_x y \mathrm{d}A = 0 \tag{5.162b}$$

According to Equations (5.161a,b),

$$\sigma_x = E_B(\varepsilon_0 - \kappa y) - E_B\alpha_B\Delta T \quad \text{(in region B)} \tag{5.163a}$$

$$\sigma_x = E_S(\varepsilon_0 - \kappa y) - E_S\alpha_S\Delta T \quad \text{(in region S)} \tag{5.163b}$$

Substituting these forms into Equations (5.162a,b), we obtain

$$F = 2b\int_{-H}^{0}\left[E_B(\varepsilon_0 - \kappa y) - E_B\alpha_B\Delta T\right]\mathrm{d}y + 2b\int_0^h\left[E_S(\varepsilon_0 - \kappa y) - E_S\alpha_S\Delta T\right]\mathrm{d}y \tag{5.164}$$

$$M = -2b\int_{-H}^{0}\left[E_B\left(\varepsilon_0 y - \kappa y^2\right) - E_B\alpha_B\Delta T y\right]\mathrm{d}y - 2b\int_0^h\left[E_S\left(\varepsilon_0 y - \kappa y^2\right) - E_S\alpha_S\Delta T y\right]\mathrm{d}y = 0 \tag{5.165}$$

Now we note that the properties (E_B, E_S, α_B, α_S) and temperature ΔT are constants; also, by our assumptions, ε_0 and κ are constants. The integrands are simply quadratic in y. These provide two linear equations in the two unknowns ε_0 and κ. They are put in nondimensional form by dividing F by $2bHE_B$ and M by $-bH^2E_B$. The results follow:

$$\left(1+\frac{E_S}{E_B}\frac{h}{H}\right)\varepsilon_0 + \frac{1}{2}\left(1-\frac{E_S}{E_B}\frac{h^2}{H^2}\right)\kappa H = \left(1+\frac{E_S}{E_B}\frac{h}{H}\frac{\alpha_S}{\alpha_B}\right)\alpha_B\Delta T \tag{5.166}$$

$$\left(1-\frac{E_S}{E_B}\frac{h^2}{H^2}\right)\varepsilon_0 + \frac{2}{3}\left(1+\frac{E_S}{E_B}\frac{h^3}{H^3}\right)\kappa H = \left(1-\frac{E_S}{E_B}\frac{h^2}{H^2}\frac{\alpha_S}{\alpha_B}\right)\alpha_B\Delta T \tag{5.167}$$

With the numerical values given, we obtain

$$\varepsilon_0 = 36.0\times 10^{-4}, \quad \kappa = 41.8\times 10^{-3}$$

The deflected form of the axis is a circular arc (κ = constant), as depicted in Figure 5.62a, where the deflection is exaggerated. The deflection at the end is

$$\delta = \rho - \rho\cos\theta$$

However, $\theta = \ell/\rho = \kappa\ell$:

$$\delta = \rho(1-\cos\kappa\ell)$$

As noted, the radius of curvature ρ is relatively large and the angle θ is small. In our case

$$\ell/\rho = \kappa\ell = 0.0837$$

For small θ, $\cos\theta \doteq 1 - \frac{1}{2}\theta^2$, therefore,

$$\delta \doteq \frac{1}{\kappa}\left(1 - 1 + \frac{1}{2}\kappa^2\ell^2\right) = \frac{1}{2}\kappa\ell^2 = 0.0837 \text{ in.} \quad (0.213 \text{ cm})$$

The stress distributions are linear in each region according to Equations (5.163a) and (b). These are plotted in Figure 5.62b.

The foregoing analysis is founded on several sweeping assumptions. Firstly, we assumed $\sigma_y = \sigma_z = O$; neither is true throughout. A better result is obtained if we do not assume $\sigma_z = 0$, but only that the resultant upon the edges ($z = \pm b$) vanishes. Secondly, we assumed that the strains and stresses are independent of x and z. Indeed, this is reasonably valid at a distance from the ends and the edges. Actually very large, but localized, stresses occur in narrow zones near the ends and edges. More complicated analyses are required to determine such concentrated stresses. Our prediction of the deflection is quite good.

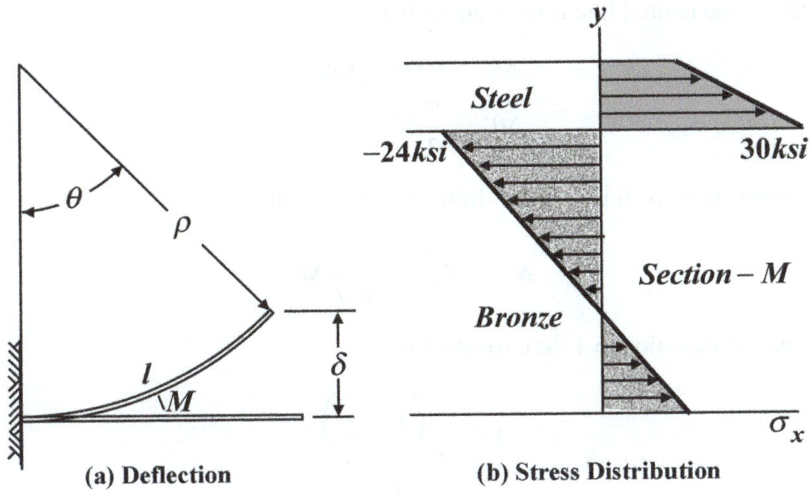

Figure 5.62

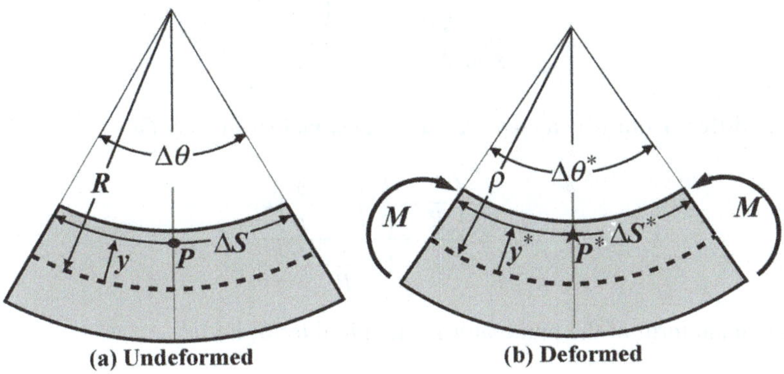

Figure 5.63

5.16 BENDING OF A CURVED BEAM

A segment of a curved beam is depicted in Figure 5.63. The beam is symmetric about the plane (*LPS*) and subjected to the bending couple M. Again, as in a straight beam, there exists an arc (a neutral line) that deforms to another arc [from (a) to (b) in Figure 5.63] without extensional strain. Here too, we can assume that a cross section remains plane and perpendicular to that neutral line. From these basic assumptions, we can express the extensional strain ε_θ in terms of the deformation of the neutral line.

Let ΔS and ΔS^* denote the undeformed and deformed lengths of the line through the particle at P and subtended by the angles $\Delta\theta$ and $\Delta\theta^*$; they are at distances y and y^* from the undeformed and deformed neutral lines. The radii of curvatures are R (initial) and ρ (deformed), as depicted in Figure 5.63. The lengths are related to the respective angles, as follows:

$$\Delta S = (R - y)\Delta\theta, \quad \Delta S^* = (\rho - y^*)\Delta\theta^* \qquad (5.168\text{a,b})$$

The length ΔS_0 of the neutral line is unchanged; hence,

$$\Delta S_0 = R\Delta\theta = \rho\Delta\theta^*$$
$$\Delta\theta^* = \frac{R}{\rho}\Delta\theta$$

The last equation serves to eliminate $\Delta\theta^*$ from (5.168b), so that

$$\Delta S^* = R\left(1 - \frac{y^*}{\rho}\right)\Delta\theta$$

With this result and Equation (5.168a), we obtain

$$\frac{\Delta S^*}{\Delta S} = \frac{\left(1 - \frac{y^*}{\rho}\right)}{\left(1 - \frac{y}{R}\right)}$$

Therefore,

$$\varepsilon_\theta \equiv \lim_{\Delta S \to 0} \frac{\Delta S^*}{\Delta S} - 1 = \frac{\left(\frac{y}{R} - \frac{y^*}{\rho}\right)}{1 - \frac{y}{R}}$$

The distance y^* differs from y by a *mean* radial extensional strain, say $\bar{\varepsilon}_R$

$$\varepsilon_\theta = \frac{\left(\frac{1}{R} - \frac{1}{\rho}\right)y - \frac{y}{\rho}\bar{\varepsilon}_R}{1 - \frac{y}{R}}$$

Clearly the dominant term of the numerator is the initial term; i.e.

$$\left(\frac{1}{R} - \frac{1}{\rho}\right)y \simeq O(\varepsilon_\theta)$$

We presume that $y/R \ll 1$, so that the term $(y/R)\bar{\varepsilon}_R$ is negligible. Otherwise stated, we neglect the stretching of the transverse line ($y \doteq y^*$) and accept the approximation

$$\varepsilon_\theta = \frac{\left(\frac{1}{R} - \frac{1}{\rho}\right)y}{1 - \frac{y}{R}} \tag{5.169}$$

Our result is consistent with the earlier results (see Equation 5.116) for the straight beam ($1/R \to 0$).

We assume that $\sigma_z = 0$ and introduce an approximation, viz $\sigma_r \doteq 0$.[10] Then the normal stress on the cross section of the HI-HO beam is

$$\sigma_\theta = E\varepsilon_\theta = E\left(\frac{1}{R} - \frac{1}{\rho}\right)\frac{y}{1 - \frac{y}{R}} \tag{5.170}$$

In the absence of a normal force,

$$F_\theta = \iint_A \sigma_\theta \mathrm{d}A = 0$$

In view of Equation (5.170), the latter condition requires

$$\iint_A \frac{y}{1-\dfrac{y}{R}} \mathrm{d}A = 0 \tag{5.171}$$

For any given cross section, the position of the neutral line is determined by the latter, Equation (5.171). The moment M is given by the integral:

$$M = -\iint_A y\sigma_\theta \mathrm{d}A = E\left(\frac{1}{\rho}-\frac{1}{R}\right)\iint_A \frac{y^2}{1-\dfrac{y}{R}} \mathrm{d}A \tag{5.172}$$

Let us define

$$I^* \equiv \iint_A \frac{y^2}{1-\dfrac{y}{R}} \mathrm{d}A \tag{5.173}$$

Then the moment, Equation (5.172) is given by

$$M = EI^*\left(\frac{1}{\rho}-\frac{1}{R}\right) \tag{5.174}$$

Once again we observe that the earlier result (5.125) is a special case. Here, the change-of-curvature $(1/\rho - 1/R)$ replaced the curvature $(1/\rho)$ and I^* replaces I. The combination of axial force (F_θ) and couple (M) is readily accommodated by superposition, as described in Chapter 7.

Let us examine the question of displacement of particles on a neutral line: A segment of the deformed and undeformed line is shown at P and P^*, respectively, in Figure 5.64. Curvatures of these segments are determined by the relations as follows:

$$\frac{1}{R} = \frac{\mathrm{d}\theta}{\mathrm{d}S}, \quad \frac{1}{\rho} = \frac{\mathrm{d}\theta^*}{\mathrm{d}S}$$

In accordance with Equation (5.174), these are related to the couple M:

$$\begin{aligned} \mathrm{d}\theta^* &= \frac{1}{\rho}\mathrm{d}S = \left(\frac{1}{R}+\frac{M}{EI^*}\right)\mathrm{d}S \\ &= \mathrm{d}\theta + \frac{MR}{EI^*}\mathrm{d}\theta \end{aligned} \tag{5.175}$$

The unit tangent vectors to the undeformed and deformed lines are

$$\hat{\mathbf{t}} = \frac{\mathrm{d}\mathbf{r}}{\mathrm{d}S}, \quad \hat{\mathbf{T}} = \frac{\mathrm{d}\mathbf{R}}{\mathrm{d}S}$$

The displacement $\boldsymbol{V}$ is determined by the differential,

$$\mathrm{d}\mathbf{V} = \mathrm{d}\mathbf{R} - \mathrm{d}\mathbf{r} = (\hat{\mathbf{T}} - \hat{\mathbf{t}})\mathrm{d}S = (\hat{\mathbf{T}} - \hat{\mathbf{t}})R\mathrm{d}\theta \tag{5.176}$$

From Figure 5.64 the tangent vectors are expressed in terms of the rotations and the constant unit vectors $(\hat{\mathbf{i}}$ and $\hat{\mathbf{j}})$:

$$\hat{\mathbf{t}} = \hat{\mathbf{i}}\cos\theta + \hat{\mathbf{j}}\sin\theta \tag{5.177a}$$

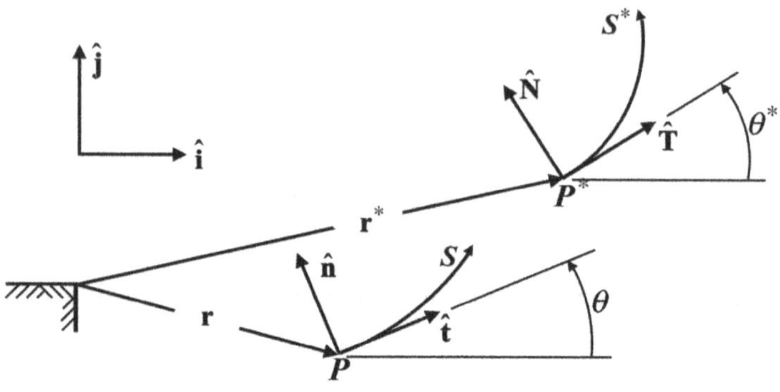

Figure 5.64

$$\hat{\mathbf{T}} = \hat{\mathbf{i}}\cos\theta^* + \hat{\mathbf{j}}\sin\theta^* \tag{5.177b}$$

In accordance with (5.175),

$$\theta^* = \theta + \Delta\theta$$

wherein $\Delta\theta$ is the relative rotation (Equation 5.175)

$$\theta^* - \theta = \Delta\theta = \int \frac{MR}{EI^*} d\theta \tag{5.178}$$

We assume $\Delta\theta \ll 1$, so that

$$\cos\theta^* \doteq \cos\theta - \Delta\theta\sin\theta \tag{5.179a}$$

$$\sin\theta^* \doteq \Delta\theta\cos\theta + \sin\theta \tag{5.179b}$$

It follows from Equations (5.176), (5.177a,b), and (5.179a,b) that

$$\boldsymbol{V} = \int \left(-\hat{\mathbf{i}}\Delta\theta\sin\theta + \hat{\mathbf{j}}\Delta\theta\cos\theta\right) R d\theta \tag{5.180}$$

In general, two integrations are needed to obtain the displacement $\boldsymbol{V}$ from the change of curvature M/EI^*, viz. Equations (5.178) and (5.180). In Chapter 6 we explore the deflection of straight beams. Then $\theta = 0$, $I^* = I$, $1/R = 0$ and $\Delta\theta = \theta^*$. Equations (5.178) and (5.180) reduce to

$$\theta^* = \int \frac{1}{\rho} dS = \int \frac{M}{EI} dS, \quad \mathbf{V} = \hat{\mathbf{j}} \int \theta^* dS$$

The first is exact; the second is an approximation which assumes $\theta^* \ll 1$.

ILLUSTRATIVE EXAMPLE 5.31

The cantilever beam of Figure 5.65 is circular, viz $R =$ constant. It is composed of a HI-HO material and subjected to a couple M at the unsupported end. Hence, every cross section experiences the same moment; the change of curvature is constant.

$$\frac{1}{\rho} = \frac{1}{R} + \frac{M}{EI^*}, \quad R = \rho\left(1 + \frac{MR}{EI^*}\right), \quad \bar{\theta} = \frac{R}{\rho}\frac{\pi}{2} = \left(1 + \frac{MR}{EI^*}\right)\frac{\pi}{2} \tag{5.181a,b,c}$$

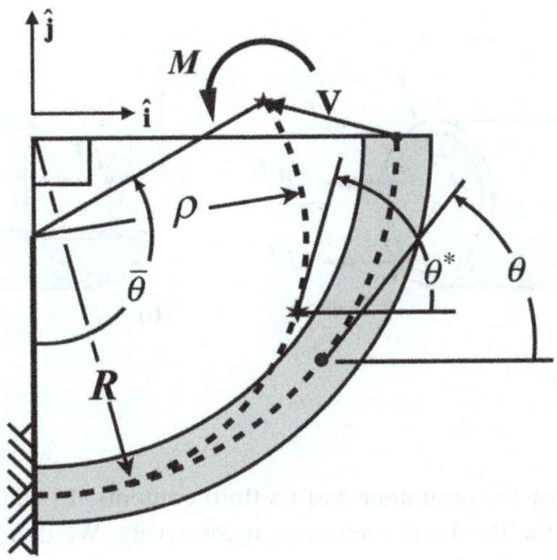

Figure 5.65

The initial circular beam is deformed to another circular beam as depicted in Figure 5.65. From the geometry, we have the deflection of the free end:

$$\mathbf{V} = \hat{\mathbf{j}}(\rho - \rho\cos\bar{\theta} - R) - \hat{\mathbf{i}}(R - \rho\sin\bar{\theta}) \tag{5.182}$$

Note that ρ and $\bar{\theta}$ are expressed in terms of the moment M and the properties of the beam, viz E and I^*. Note that Equation (5.182) incorporates no geometric approximations. If the beam were thin and flexible, Equation (5.182) would accurately determine a large deflection.

If we adopt the approximation:

$$\frac{MR}{EI^*} = \left(\frac{R}{\rho} - 1\right) \ll 1$$

then, according to Equations (5.181b) and (c)

$$\cos\bar{\theta} \doteq -\frac{MR}{EI}\frac{\pi}{2}, \quad \sin\bar{\theta} \doteq 1, \quad \rho \doteq R\left(1 - \frac{MR}{EI^*}\right)$$

We then obtain from Equation (5.182) or (5.180) the approximation

$$\mathbf{V} \doteq -\frac{MR^2}{EI^*}\left[\left(1 - \frac{\pi}{2}\right)\hat{\mathbf{j}} + \hat{\mathbf{i}}\right]$$

This result is valid for small relative rotations. We return to the problem in Chapter 9 via the energy method of Castigliano.

5.17 THIN-WALLED VESSELS

Cylindrical Vessels: Thin-walled tubes and tanks are widely used to transfer or to contain fluids. A transverse section and an axial section are depicted in Figure 5.66a and b, respectively. Each is

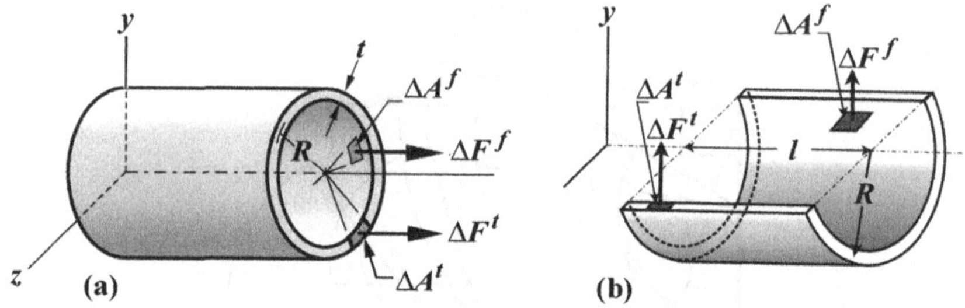

Figure 5.66

a free body of the portion of the container and its fluid contents. In each diagram, superscripts f and c denote actions upon the fluid and container, respectively. We denote the mean radius of the vessel by R, the thickness by t; $t/R \ll 1$.

5.17.1 STATICS

On each element of the fluid surface and the vessel wall, ΔA^f and ΔA^c, we have forces, ΔF^f and ΔF^c, respectively. We denote the pressure of the fluid by p, the axial stress upon the wall by σ_x, and the circumferential stress by σ_θ. For the cross section of Figure 5.66a, we have the approximations:

$$A^f = \pi\left(R - \frac{t}{2}\right)^2 \doteq \pi R^2$$

$$A^t = \pi\left(R + \frac{t}{2}\right)^2 - \pi\left(R - \frac{t}{2}\right)^2 \doteq 2\pi R t$$

For the axial section of Figure 5.66b, we have the approximations:

$$A_f = 2\ell(R-t) \approx 2\ell R, \quad A_t = 2\ell t$$

Then the equilibrium condition for the body of Figure 5.66a gives us the relation:

$$\sigma_x \approx \frac{pR}{2t} \tag{5.183a}$$

Likewise, for the equilibrium of the body of Figure 5.66b, we have

$$\sigma_\theta = \frac{pR}{t} \tag{5.183b}$$

Kinematics: A circumferential segment AB of the undeformed cross section is shown in Figure 5.67a. The deformed segment A^*B^* is depicted in Figure 5.67b. The increase in radius r is the radial displacement u. The circumferential strain follows:

$$\varepsilon_\theta = \frac{A^*B^*}{AB} - 1 = \frac{u+r}{r} - 1 = \frac{u}{r} \tag{5.184}$$

Since the wall is thin ($t/R \ll 1$), changes in thickness are very small compared to displacement. Therefore, we accept the approximation of constant ε_θ throughout. Likewise, we can assume that the axial strain ε_x is constant.

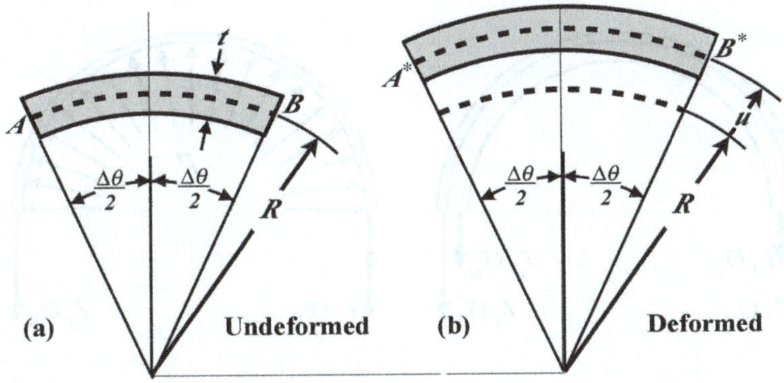

Figure 5.67

5.17.2 MATERIAL PROPERTIES

The normal stress σ_r vanishes at the exterior surface; $\sigma_r = -p$ at the interior surface. However, we can anticipate that $|p| \ll \sigma_\theta$ and $|p| \ll \sigma_x$. Since the wall is then, we assume that σ_r is negligible throughout. If the material is HI-HO, then

$$\sigma_\theta = \frac{E}{1-\nu^2}(\varepsilon_\theta + \nu\varepsilon_x)\,,\ \sigma_x = \frac{E}{1-\nu^2}(\varepsilon_x + \nu\varepsilon_\theta) \tag{5.185a,b}$$

We can now introduce the equilibrium Equations (5.183a,b) and kinematic Equations (5.184) into the stress-strain Equations (5.185a,b). The latter take the forms:

$$\nu\varepsilon_x + \frac{u}{R} = \frac{(1-\nu^2)\,pR}{Et}\,,\ \varepsilon_x + \nu\frac{u}{R} = \frac{(1-\nu^2)\,pR}{2Et}$$

Solving for ε_x and $\varepsilon_\theta = u/R$, we obtain

$$\varepsilon_x = \frac{(1-2\nu)pR}{2Et}\,,\ \varepsilon_\theta = \frac{u}{R} = \frac{(2-\nu)pR}{2Et} \tag{5.186a,b}$$

If the material yields according to the maximum shear stress criterion, then yielding is initiated at pressure

$$p_0 = \frac{t}{R}\sigma_0$$

This is a reasonable estimate for structural steel.

In some instances, a cylindrical tube/pipe might be constrained against axial movement, i.e. $\varepsilon_x = 0$. Then Equations (5.183a) and Equations (5.185a,b) are invalid. Instead, we have

$$\begin{aligned} \varepsilon_x &= 0 = \frac{1}{E}(\sigma_x - \nu\sigma_\theta) \\ \sigma_x &= \nu\sigma_\theta = \nu\frac{pR}{t} \\ \varepsilon_\theta &= \frac{u}{R} = \frac{(1-\nu^2)}{E}\frac{pR}{t} \end{aligned} \tag{5.187}$$

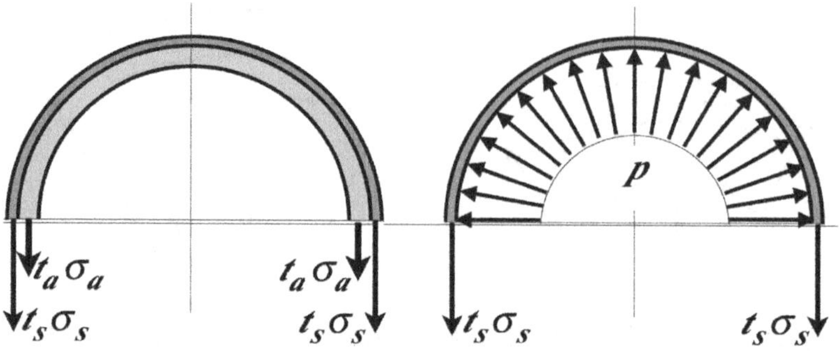

Figure 5.68

ILLUSTRATIVE EXAMPLE 5.32

An aluminum tube is initially at 70°F and has the following geometric and material properties

$$\text{I. Dia.} = 1.800 \text{ in.}, \quad E_A = 10 \times 10^3 \text{ ksi}$$
$$\text{O. Dia.} = 2.000 \text{ in.}, \quad \alpha_A = 13.0 \times 10^{-6}/^\circ F$$

A steel tube just slips over the aluminum tube when heated to 570°F. The steel tube has the following properties:

$$\text{wall thickness } t = 0.050 \text{ in. } E_S = 30 \times 10^3 \text{ ksi}, \quad \alpha_s = 6.5 \times 10^{-6}/^\circ F$$

Here, E and α denote the moduli and coefficients of thermal expansion. We require the circumferential stresses in each tube and the pressure between the tubes when the assembly returns to 70°F. Geometry requires that both reach the same expansion/contraction, i.e., the circumferential strains ε_θ must be equal:

$$\varepsilon_A = \varepsilon_S \tag{5.188}$$

Each is related to their respective stress and temperature changes:

$$\varepsilon_A = \frac{\sigma_A}{E_A}, \quad \varepsilon_S = \frac{\sigma_S}{E_S} + \alpha_S \Delta t \tag{5.189a,b}$$

Here we accept the absence of axial stress σ_x; σ_A and σ_S are the circumferential stresses (σ_θ).

A transverse slice of the assembly is depicted in Figure 5.68a. The forces on the steel and aluminum walls must equilibrate:

$$A_A\sigma_A + A_S\sigma_S = 0, \quad 0.10\sigma_A + 0.050\sigma_S = 0 \tag{5.190}$$

The substitution of Equations (5.189a,b) into (5.188) provides the relation:

$$\frac{\sigma_A}{E_A} - \frac{\sigma_S}{E_S} = \alpha_S \Delta t \tag{5.191}$$

The solution of Equations (5.190) and (5.191) gives us the results:

$$\sigma_s = -\frac{2E_S E_A}{E_S + 2E_A}\alpha_S \Delta T = 39 \text{ ksi}$$
$$\sigma_A = -19.5 \text{ ksi}$$

Figure 5.68b shows a transverse slice of the steel tube and the pressure exerted by the aluminum tube. Equilibrium dictates the result

$$p = 1950 \text{ psi}$$

5.17.3 SPERICAL SHELL

The spherical shell is a very efficient tank for compressed, often liquefied, fluids. A sphere has the most capacity (volume) for the given surface area; a spherical tank under internal pressure has the least, and uniform, *membrane* stress.

5.17.4 EQUILIBRIUM CONDITIONS

Figure 5.69a shows a central cross section of a spherical tank. Forces ΔF^f and ΔF^s act upon the elemental areas, ΔA^f and ΔA^s, of the fluid and the walls, respectively. Equilibrium requires that

$$\sigma_\theta = \frac{pR}{2t} \tag{5.192}$$

where $\Delta F^f = -p\Delta A^f$ and $\Delta F^s = \sigma_\theta \Delta A^s$, $A^f = \pi R^2$, $A^s = 2\pi Rt$. Here we assume that $t \ll R$, the mean radius of the wall.

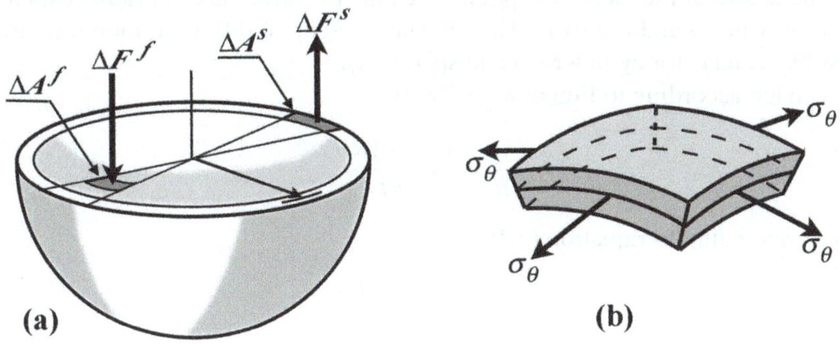

Figure 5.69

5.17.5 KINEMATICS

Each element of the sphere displaces radially; if u denotes the radial displacement, then (see Figure 5.67) every circumferential element experiences the strain:

$$\varepsilon_\theta = \frac{u}{R}$$

5.17.6 MATERIAL PROPERTIES

Every rectangular segment of the wall is subjected to the biaxial stress σ_θ as depicted in Figure 5.69b. Every circumferential line experiences the same strain ε_θ. The outer surface of the wall is stress free ($\sigma_r = \tau_{r\theta} = 0$); the inner surface is subject to an insignificant stress, viz $\sigma_r = -p$

where $p = 2\sigma_\theta t/R$. Accordingly, we neglect the radial stress σ_r. If the material is HI-HO, e.g., steel within the yield stress, then

$$\varepsilon_\theta = \frac{u}{R} = \frac{(1-\nu)}{E}\sigma_\theta$$
$$\frac{u}{R} = \frac{(1-\nu)Rp}{2Et} \tag{5.193}$$

The foregoing results presume that the shell is thin ($t/R \ll 1$). These results are valid for a spherical container without irregularities, such as pipe inlets and supports. Stress concentrations occur at any alterations and connections. If the steel of a regular tank yields according to the criterion of maximum shear stress, then yielding is initiated at pressure.

$$p_0 = \frac{2t\sigma_0}{R}$$

ILLUSTRATIVE EXAMPLE 5.33

Figure 5.70 depicts a tank of a form, which is commonly used to contain liquefied gases under pressure. The central portion $ACFD$ is cylindrical; the ends are hemispherical. Equations (5.183a,b) and (5.186a,b) hold for the cylindrical part; Equations (5.192) and (5.193) hold for the hemispherical ends. At the junctures, i.e., lines ABC and DEF, we must anticipate local stress concentration. To illustrate the nature of the local disruptions, we note the differences in radial expansions at the same internal pressure p and radius R. The thickness might be different; therefore, let us denote the thickness by t_c and t_s for cylinder (c) and sphere (s):

For the cylinder, according to Equation (5.186b),

$$\frac{u}{R} = \frac{(2-\nu)Rp}{2Et_c}$$

For the sphere, according to Equation (5.193),

$$\frac{u}{R} = \frac{(1-\nu)Rp}{2Et_s}$$

Setting $t_c = t_s$, with a typical value for ν (0.30) the expansion of the cylinder (u_c) would be much greater than the expansion of the sphere (u_s):

$$u_c = 2.4u_s$$

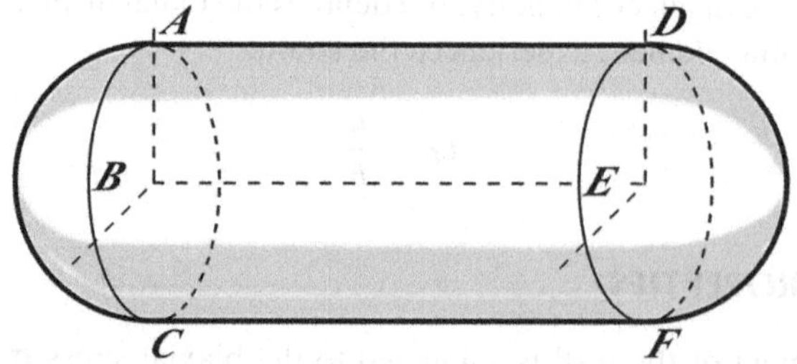

Figure 5.70

The union of the cylinder and sphere of equal thickness would force some local deformations. One might avoid the problems in part by altering the thickness, e.g., if we set $u_c = u_s$, we obtain

$$t_c = \frac{2-\nu}{1-\nu} t_s = 2.43 t_s$$

The propane tank at the author's residence has $t_c = 1.18 t_s$. However, engineers know too that the hemispherical *head* is produced by *deep-drawing* as opposed to rolling; the final products have different properties.

5.18 THICK-WALLED CYLINDERS

A thick-walled cylinder might be used to contain fluid under pressure ρ_i (e.g., a pipe) or resist external pressure ρ_o (e.g. a tube in a boiler); both loadings are included in Figure 5.71a. The cylinder is composed of HI-HO material; hence, the deformation is axisymmetric: A particle displaces radially an amount $u(r)$ as shown in Figure 5.71b. A radial segment BC of length Δr displaces an amount $u(r)$ at B and $u(r) + (\mathrm{d}u/\mathrm{d}r)\Delta r$ at C, the innermost and outermost surfaces. Therefore, the radial extensional strain follows:

$$\varepsilon_r = \lim_{\Delta r \to 0} \frac{u(r+\Delta r) - u(r)}{\Delta r} = \frac{\mathrm{d}u}{\mathrm{d}r} \tag{5.194}$$

A circumferential segment AB of length $r\Delta\theta$ is displaced radially to the extended length $(r+u)\Delta\theta$. Therefore, the circumferential extensional strain follows:

$$\varepsilon_\theta = \lim_{\Delta\theta \to 0} \frac{(r+u)\Delta\theta - r\Delta\theta}{r\Delta\theta} = \frac{u}{r} \tag{5.195}$$

The kinematic relations, (5.194) and (5.195) hold for any homogeneous medium.

Equilibrium of the small piece in Figure 5.71c requires the vanishing of radial force; terms of third-order in the increments follow:

$$\Delta\sigma_r r \Delta z \Delta\theta + \sigma_r \Delta z \Delta r \Delta\theta - \sigma_\theta \Delta z \Delta r \Delta\theta = 0$$

We divide this difference equation by $r\Delta r\Delta\theta\Delta z$ and pass to the limit:

$$\lim_{\Delta r \to 0} \left[\frac{\Delta\sigma_r}{\Delta r} + \frac{\sigma_r - \sigma_\theta}{r} \right] = \frac{\mathrm{d}\sigma_r}{\mathrm{d}r} + \frac{\sigma_r - \sigma_\theta}{r} = 0 \tag{5.196}$$

This equilibrium equation holds as well for any medium.

Let us assume that the axial stress vanishes; $\sigma_z = 0$, i.e., the vessel, or pipe, is not constrained against axial movement. Then the strain-stress relations of the HI-HO material are:

$$\varepsilon_r = \frac{1}{E}(\sigma_r - \nu\sigma_\theta)$$

$$\varepsilon_\theta = \frac{1}{E}(-\nu\sigma_r + \sigma_\theta)$$

Combining these with the kinematic relations (5.194) and (5.195), we obtain

$$\frac{\mathrm{d}u}{\mathrm{d}r} = \frac{1}{E}(\sigma_r - \nu\sigma_\theta) \tag{5.197a}$$

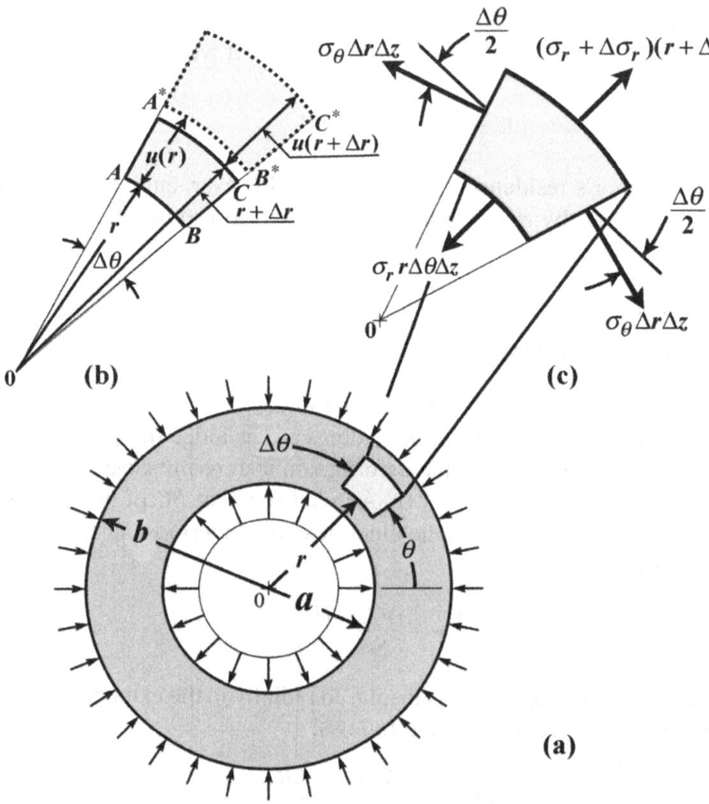

Figure 5.71

$$\frac{u}{r} = \frac{1}{E}(-v\sigma_r + \sigma_\theta) \tag{5.197b}$$

We can eliminate σ_θ by means of the equilibrium equation (5.196), viz

$$\sigma_\theta = r\frac{\mathrm{d}\sigma_r}{\mathrm{d}r} + \sigma_r \tag{5.198}$$

Then, (5.197) and (5.198) provide the following:

$$\frac{\mathrm{d}u}{\mathrm{d}r} = \frac{1}{E}\left[(1-v)\,\sigma_r - vr\frac{\mathrm{d}\sigma_r}{\mathrm{d}r}\right]$$

$$u = \frac{r}{E}\left[(1-v)\,\sigma_r + r\frac{\mathrm{d}\sigma_r}{\mathrm{d}r}\right]$$

Differentiating the last equation and substituting that into the previous, we obtain one ordinary differential equation.

$$\frac{\mathrm{d}^2\sigma_r}{\mathrm{d}r^2} + \frac{3}{r}\frac{\mathrm{d}\sigma_r}{\mathrm{d}r} = 0 \tag{5.199}$$

The solution of this equation is as follows:

$$\sigma_r(r) = A + \frac{B}{r^2}$$

Here, the unspecified constants are determined to satisfy the internal and external boundary conditions, viz

$$\sigma_r(a) = -p_i, \quad \sigma_r(b) = -p_0$$

Hence

$$\sigma_r(r) = \frac{a^2b^2}{b^2-a^2}\left[\frac{p_i}{b^2} - \frac{p_0}{a^2} + \frac{p_0 - p_i}{r^2}\right] \tag{5.200}$$

Then, from the equilibrium equation (5.198), we have

$$\sigma_\theta(r) = \frac{a^2b^2}{b^2-a^2}\left[\frac{p_i}{b^2} - \frac{p_0}{a^2} + \frac{p_i - p_0}{r^2}\right] \tag{5.201}$$

The foregoing solution holds for arbitrary wall-thickness $t = b - a$ and, in particular, for a thin wall wherein

$$\frac{t}{a} \ll 1$$

Then, in (5.201) we can substitute

$$b = a\left(1 + \frac{t}{a}\right) \approx a, \quad r \approx a$$

$$b^2 - a^2 = a^2\left(1 + 2\frac{t}{a} + \frac{t^2}{a^2}\right) - a^2 = 2at$$

Equation (5.201) provides the approximation

$$\sigma_\theta \approx \frac{a(p_i - p_0)}{t}$$

This is in accord with Equation (5.183) of the preceding section. If the cylindrical vessel is constrained against axial movement, then $\varepsilon_z = 0$; this constraint replaces the condition $\sigma_z = 0$. The solution has the same form; the modulus E is replaced by $E/(1 - \nu^2)$.

NOTES

1. Barre de Saint-Venant, "Memoire sur la Torsion des Prismes ... " Memoire des savants Etrangers, T. XIV (Paris: 1855), 297–99.
2. If a prismatic bar is subjected to uniform normal stresses on its end faces, then the preceding solution is in accord with that obtained from more elaborate theories. Furthermore, it can be shown that $\sigma_y = \sigma_z = \tau_{xy} = \tau_{xz} = \tau_{yz} = 0$, $\varepsilon_y = \varepsilon_z = -\nu\varepsilon_x$ and $\gamma_{xy} = \gamma_{xz} = \gamma_{yz} = 0$.
3. J is sometimes called the polar moment of the cross section S.
4. See a general numerical method: G. Wempner. Discrete approximations related to nonlinear theories of solids. *Int. J. Solids Struct.*, 7:1581–1599, 1971.
5. See S. Timoshenko and J. N. Goodier, *Theory of Elasticity*, second edition, p. 278. McGraw-Hill, Columbus, 1981.
6. This topic requires some familiarity with partial differential equations.
7. L. Prandtl. Zur Torsion von prismatischen Stäben. *Physik. Zeit.*, 4:758–770, 1903.
8. We can offer logical arguments to justify assumption 2, the so-called Bernoulli hypothesis. We observe that every cross section is subject to the same action, the couple M, so that every slice is expected to deform similarly. Indeed, under this uniform bending, the x-axis must be deformed to a circular arc. The beam of Figure 5.43 could also be extended to $x = -L$. Then such an extended beam would be entirely equivalent and symmetric about the yz-plane; opposing couples at the ends $x = \pm L$ would preserve that

symmetry. Then the left and right halves would be reflections about the midsection ($x = 0$). It follows that the midsection remains plane and perpendicular to the bent axis. But, because of the uniformity of properties and loading, it is plausible to assume that all cross sections remain plane and perpendicular to the bent axis, a circular arc.

9. If the material is Hookean (linearly elastic) but not isotropic and homogeneous, curvature and moment are proportional but the constant of proportionality is not that (EI) of Equation (5.125).
10. We must acknowledge an approximation not inherent in the prior solution for a straight beam. Although the radial stress σ_r vanishes at the inner and outer surfaces, it does not follow that $\sigma_r = 0$ throughout. The presence of that normal stress is essential to enforce the equilibrium of an element, as depicted in Figure 5.71. Note the component $(\sigma_\theta \Delta\theta)$ of stress σ_θ in the radial direction: this must be countered by the normal stress σ_r. The requisite *differential* equation of equilibrium is derived in Section 5.17.

6 Bending of Symmetric Beams

In Sections 5.13 and 5.14, we described the bending of symmetric beams under the action of couples only. Then every slice behaves the same. Certain results of simple bending are also applicable to nonuniform distributions of symmetric loads, wherein each segment experiences different actions and deformation. The extension of the earlier analyses and results entails a degree of approximation and also additional mathematical tools. As we extend the early analysis, we must be mindful of the limitations inherent in the approximation. Still, the resulting theory of beams remains an important and useful tool in the hands of the practicing engineer.

6.1 THE SYMMETRIC BEAM

The beam, symmetric about a plane (LPS), was described in Section 5.13 as a prelude to the special circumstance of simple bending. In practice, the engineer seldom encounters *simple bending* but often employs beams to support various transverse loads, which cause bending, strains, and stresses. We now consider such problems but limit our immediate attention to circumstances wherein the beam, supports, and loadings are symmetric about the plane.

6.2 LOADS, INTERNAL FORCES, AND EQUILIBRIUM

6.2.1 SYMMETRIC LOADS AND SUPPORTS

Loads and reactions are addressed in Section 3.1. Recall now that the reaction of a support upon any body can be represented by a resultant consisting of a force acting at a point and a couple; each can be represented by three components. Also recall that such representation provides no specific information about the distribution of the forces and therefore does not admit an analysis of the local behavior - that is, strains or stresses in the immediate neighborhood. Once again, by virtue of St. Venant's principle (Section 5.2), the overall response, as well as strains and stresses at remote locations, is little affected by the actual distribution. With an awareness of such limitations, let us consider the nature of the reactions that arise in the present circumstance. To be specific, let us denote the reactions of a support by a force **R** and couple **M**:

$$\mathbf{R} = F_x\hat{\mathbf{i}} + R_y\hat{\mathbf{j}} + R_z\hat{\mathbf{k}} \tag{6.1}$$

$$\mathbf{M} = T_x\hat{\mathbf{i}} + M_y\hat{\mathbf{j}} + M_z\hat{\mathbf{k}} \tag{6.2}$$

Since the loading is entirely symmetric about the plane (x - y *LPS*)

$$R_z = T_x = M_y = 0$$

In other words, a transverse (shear) force R_z, the twisting couple T_x, and a bending couple M_y vanish, as shown in Figure 6.la. The nonzero components are depicted in Figure 6.1b. The subscripts can now be deleted from the components of reactive force **R** and couple **M**:

$$\mathbf{R} = F\hat{\mathbf{i}} + V\hat{\mathbf{j}} \tag{6.3}$$

$$\mathbf{M} = M\hat{\mathbf{k}} \tag{6.4}$$

DOI: 10.1201/9781003665861-6

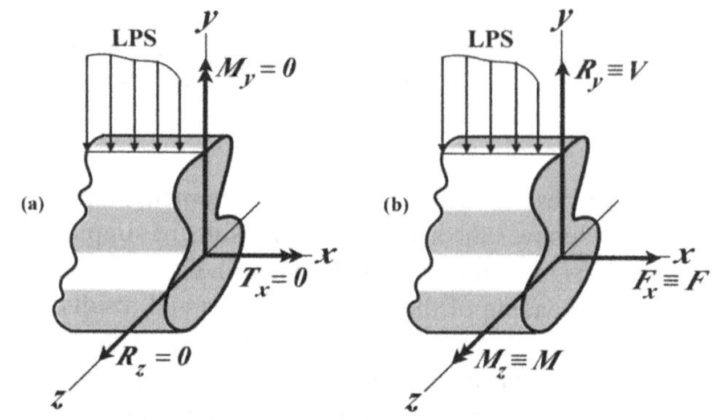

Figure 6.1

The purpose of the roller support at B in Figure 6.2a is to prevent vertical movement. It does this by applying the necessary reaction R_B. (For clarity, the load on the beam is not shown.) Such a support cannot prevent rotation; the bending couple at B is zero. Hence, the reactive force system is only a vertical force, as shown in Figure 6.2b. The pin at support A also prevents horizontal movement, since the reactive force system includes a horizontal component, as in Figure 6.2b. A beam supported as shown in Figure 6.2 is said to be *simply supported.*

The beam of Figure 6.3a has a fixed, built-in, or clamped connection at A. Such a support prevents both translation and rotation. This is accomplished by the force system shown in Figure 6.3b. The force components and the bending couple assume the necessary magnitudes to inhibit completely displacement at A. This type of end connection is frequently called a *fixed end.* At b the beam is free of support; this is called a *free* end. The name obviously implies that all the components of the force system at b are zero. A beam that is fixed at one end and free at the other is said to be cantilevered.

The simple supports of Figure 6.2 and the fixed and free ends of Figure 6.3 are the most commonplace, but other, more complicated circumstances arise when beams are employed as structural members. For example, the floor joists of a building might be supported by vertical columns that are also deformable - that is, the column resists the movement, displacement, and rotation of the joist (beam) but cannot entirely prevent movement. The situation is illustrated in Figure 6.4a, wherein the horizontal joist (beam) is affixed to the vertical column. Under transverse (vertical) loading, the flexible beam bends; the end tends to rotate. Such rotation is restrained by the column, which exerts the couple M upon the beam; the opposite action upon the column tends to bend the column. The restraint cannot entirely prevent the rotation θ. If the column is Hookean, then the

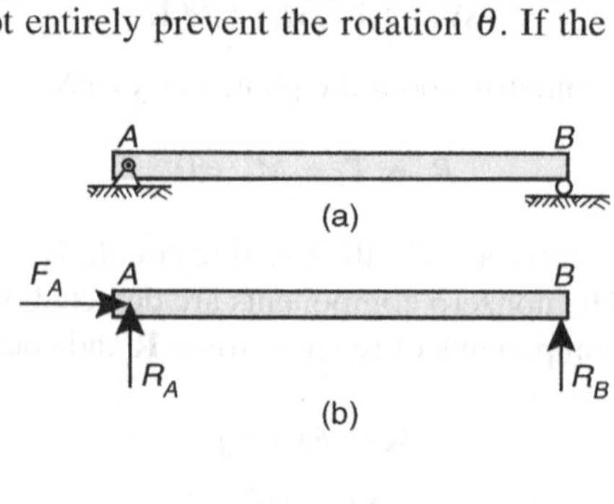

Figure 6.2

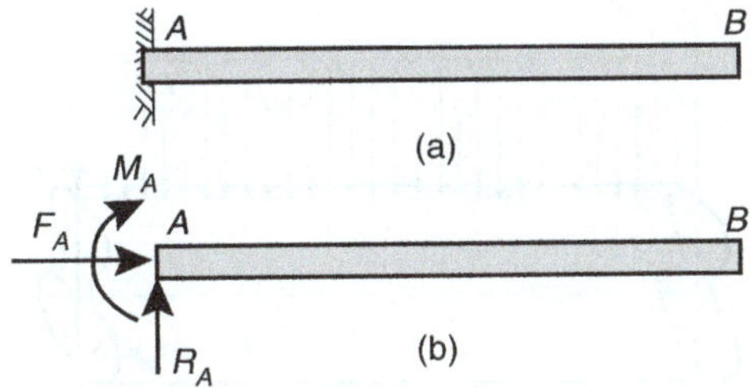

Figure 6.3

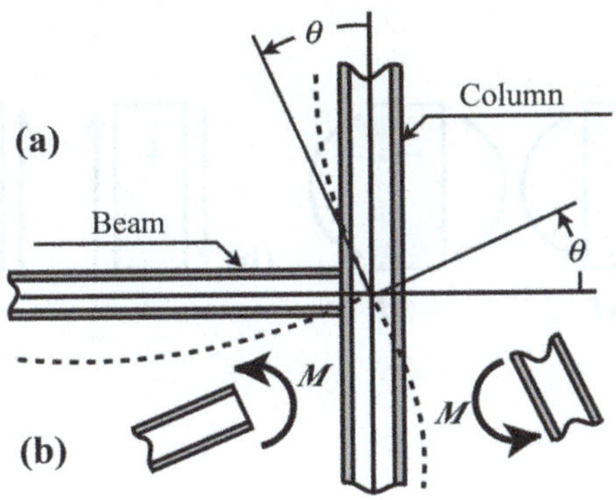

Figure 6.4

rotation is proportional to the couple, and the condition at the end of the beam can be expressed by the equation

$$M = k\theta \tag{6.5}$$

where the coefficient k depends upon the stiffness, material, and size of the column. The beam is neither fixed nor simply supported; the constraint is intermediate.

6.2.2 NET INTERNAL FORCE SYSTEM

With the loading (in the *LPS*) perpendicular to the axis of the beam, the normal force component on any cross-section is zero. Consequently, the net internal force system on any cross-section consists of, at most, a shear force and a bending couple. Figure 6.5 shows a portion of a beam to one side of a cross-section S. This cross-section is located by its distance x from the end of the beam. Since V_y and M_z are the only components of the net internal force system on S that are not identically zero, we drop the subscripts and write V for V_y and M for M_z.

At this point, it is advantageous to adopt a sign convention for the bending couple M and the shear force V. A bending couple that tends to compress the upper portions of the beam is regarded as positive. If the couple tends to stretch the upper portions, it is negative. This is illustrated in Figure 6.6a. A shear force is regarded as positive if it tends to rotate the free body on which it acts in a counterclockwise sense. If the shear force tends to rotate the free body in a clockwise sense, then it is negative. This situation is illustrated in Figure 6.6b.

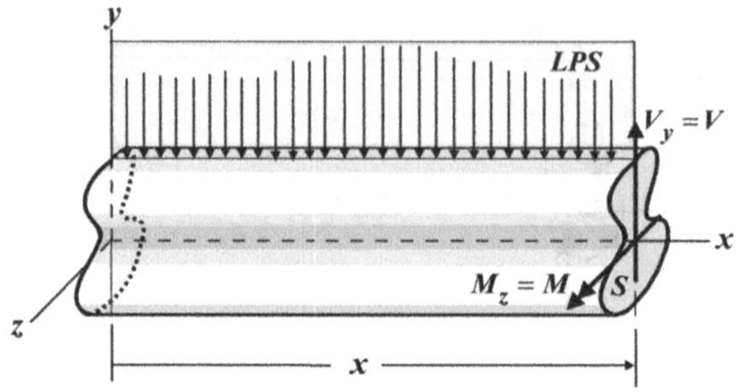

Figure 6.5

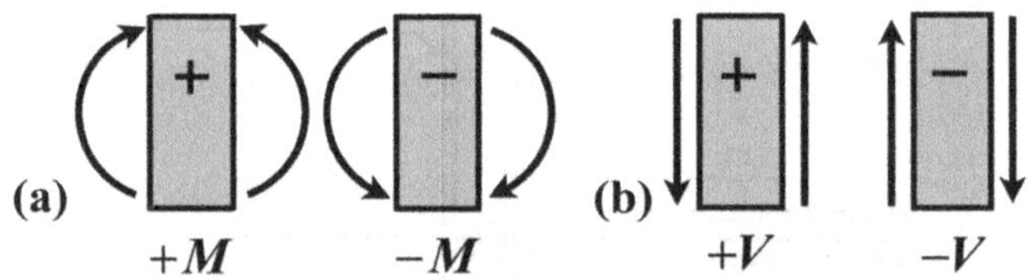

Figure 6.6

The shear force and bending couple are related to the loading. For example, consider a beam subjected to a distributed load $P(x)$ (force/length), as shown in Figure 6.7a. Additionally, one or more concentrated forces may act. For clarity, we have shown only a typical concentrated force F_i. Also, the end supports have been replaced by their reactions (R_A, M_A, R_B, M_B). The cross section S is located by the coordinate distance x. To relate the shear force and bending couple on S to the loading, we consider the free-body diagram of the portion of the beam to the left of S as shown in Figure 6.7b. Our choice of the left part is arbitrary; the portion to the right of S would serve as well. The conditions for equilibrium of the free body determine V and M as follows:

$$V(x) = -R_A + F_i + \int_0^x p(\xi)\mathrm{d}\xi \tag{6.6}$$

$$M(x) = M_A + R_A x - F_i(x - L_i) - \int_0^x (x-\xi)p(\xi)\mathrm{d}\xi \tag{6.7}$$

If there is more than one concentrated load, the F_i term in each of these equations is replaced by a sum of similar terms ($i = 1, 2, ---$).

6.2.3 DIFFERENTIAL EQUATIONS OF EQUILIBRIUM

We could eliminate the integrals in Equations (6.6) and (6.7) by differentiating with respect to x and so relate the derivatives of M and V to $P(x)$. Instead, let us obtain the equations by considering the equilibrium of a slice from the beam.

In Figure 6.8, we have isolated a slice of the beam between the cross section S at x and one nearby at $x + \Delta x$. The distributed load $p(x)$ exerts a force $\bar{p}\Delta x$, where $\bar{p}$ denotes the mean value over the segment Δx. For the slice to be in equilibrium, $\sum F_y = 0$, so that

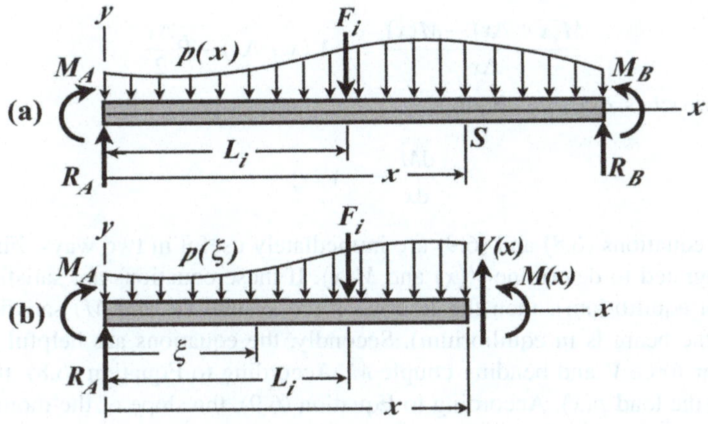

Figure 6.7

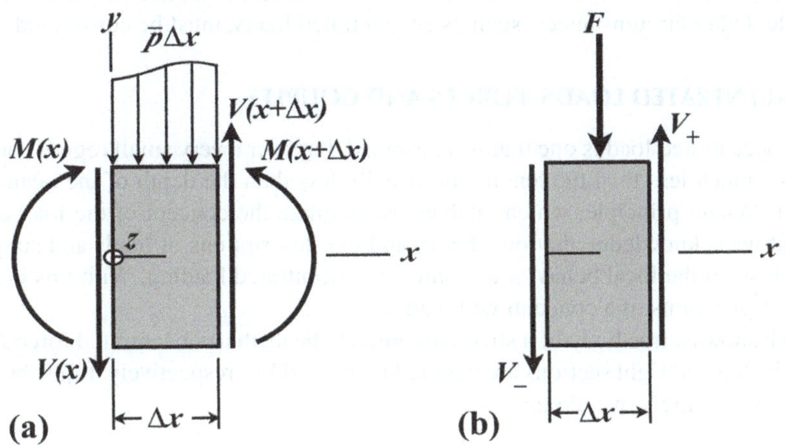

Figure 6.8

$$V(x+\Delta x) - V(x) - \bar{p}\Delta x = 0$$

or

$$\frac{V(x+\Delta x) - V(x)}{\Delta x} = \bar{p}$$

As Δx approaches zero, $\bar{p}$ approaches the actual load $p(x)$ at x, and in the limit we obtain

$$\frac{dV}{dx} = p(x) \tag{6.8}$$

Equilibrium of the slice in Figure 6.8a also requires that $\sum M_z = O$; hence,

$$M(x+\Delta x) - M(x) + V(x+\Delta x)\Delta x + \bar{p}\Delta x\frac{\Delta x}{2} = 0$$

or

$$\frac{M(x+\Delta x)-M(x)}{\Delta x}=-V(x+\Delta x)-\bar{P}\frac{\Delta x}{2}$$

Again, as Δx approaches zero, we obtain

$$\frac{\mathrm{d}M}{\mathrm{d}x}=-V(x) \tag{6.9}$$

The equilibrium equations (6.8) and (6.9) are immediately useful in two ways. Firstly, the equations can be integrated to determine $V(x)$ and $M(x)$. If these equations are satisfied everywhere (every slice is in equilibrium), then the internal force system (V and M) satisfies equilibrium (every piece of the beam is in equilibrium). Secondly, the equations are helpful in plotting the diagrams of shear force V and bending couple M. According to Equation (6.8), the slope of the shear diagram is the load $p(x)$. According to Equation (6.9), the slope of the moment diagram is the negative shear $-V(x)$. It follows, too, that the maximum and minimum values of the bending couple $M(x)$ occur where the shear V is zero. The derivations of Equations (6.8) and (6.9) and their applications presuppose that the limits exist and that the load, shear, and moment are continuous and integrable. Other circumstances, such as concentrated loads, must be considered:

6.2.4 CONCENTRATED LOADS–FORCES AND COUPLES

In reality, a concentrated load is one that is distributed, but over a very small region compared with other lengths - much less than the length and usually less than the depth of the beam. Again, we invoke the St. Venant principle, which enables us to utilize the concept of the load concentrated at a point, but we acknowledge that our theory and our descriptions of loads and supports do not admit an analysis of the local behavior at points of concentrated loading. With this in mind, let us examine the implications of a concentrated load.

Figure 6.8b shows a free body of a slice immediately beneath a concentrated force F. The shear forces upon the left and right sections are denoted by V_- and V_+, respectively. Equilibrium requires that $\sum F=0$. Accordingly, we obtain

$$V_+-V_-=F \tag{6.10}$$

In other words, the shear force exhibits a jump of amount F in the small distance of the loading. If the load were actually concentrated (it isn't) at a point, then the shear force would be discontinuous at the point of application.

A concentrated couple is also conceivable. For example, two equal but opposing forces might be applied to a bracket, as shown in Figure 6.9a. The net effect upon the beam is a concentrated couple, which acts at a point, as depicted in Figure 6.9b. The free-body diagrams of Figure 6.9a and b show the force system that must satisfy equilibrium, in particular, $\sum M=0$. The width of the slice is presumed small (compared with the other dimensions of the beam), so that the moment of the shear forces is neglected (in the limit $\varepsilon\to 0$, it vanishes). Then,

$$M_+-M_-=-C \tag{6.11}$$

The bending moment $M(x)$ exhibits a jump in the small distance ε. If the couple were concentrated, then the function $M(x)$ would be discontinuous at the point.

The reader can compare the conditions in Equations (6.10) and (6.11) with their counterparts, Equations (6.8) and (6.9), which hold where the functions $p(x)$ and $V(x)$ exist as smooth (continuous and bounded) functions.

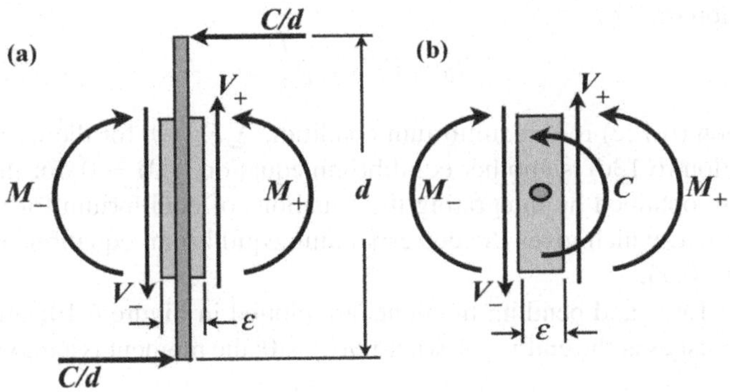

Figure 6.9

ILLUSTRATIVE EXAMPLE 6.1

The beam of Figure 6.10a is simply supported at ends $A(x=0)$ and $B(x=l)$. The load is distributed so that $p(x) = \bar{p}x/l$, where $\bar{p}$ is a constant, the magnitude at the right end ($x = l$). Substituting the load $p(x)$ into the right side of Equation (6.8) and integrating, we obtain

$$dV = p(x)dx = \bar{p}\frac{x}{l}dx \tag{6.12a}$$

$$\int_{V_0}^{V(x)} dV = \int_0^x \bar{p}\frac{x}{l}dx \tag{6.12b}$$

$$V(x) - V_0 = \bar{p}\frac{x^2}{2l} \tag{6.12c}$$

Note that the limits on the left and right sides of (6.12b) are in accord: $V_0 = V(O)$ is the value of the shear force at $x = 0$; the upper limit is indefinite, because we need the function $V(x)$ in order to proceed with Equation (6.9), namely,

$$dM = -V(x)dx = -\left(V_0 + \bar{p}\frac{x^2}{2l}\right)dx \tag{6.13a}$$

$$\int_0^{M(x)} dM = -\int_0^x \left(V_0 + \bar{p}\frac{x^2}{2l}\right)dx \tag{6.13b}$$

$$M(x) = -\left(V_0 x + \bar{p}\frac{x^3}{6l}\right) \tag{6.13c}$$

Note that the condition of simple support, $M(O) = 0$, has been introduced into the left side of (6.13b). Similarly,

$$M(l) = 0 = -\left(V_0 l + \bar{p}\frac{l^2}{6}\right) \tag{6.14a}$$

or

$$V_0 = -\bar{P}\frac{l}{6} \tag{6.14b}$$

A free body of a slice above the left support is pictured in Figure 6.10d, which shows that

$$R_A = -V_0 = \bar{p}\frac{l}{6} \tag{6.15a}$$

Likewise, by Equation (6.12c),

$$R_B = V(l) = \bar{p}\frac{l}{3} \tag{6.15b}$$

Note that Equation (6.12c) is an equilibrium condition, $\sum F = 0$, for the free-body diagram of Figure 6.10c; Equation (6.13c) is another equilibrium equation, $\sum M = 0$, for the same free-body diagram. These were obtained by integrating the equations of equilibrium for a slice (summing) from the end $x = 0$ to x, which gives the corresponding equilibrium equations for the finite body lying in the segment $(0, x)$.

Finally, the shear force and bending moments are plotted in Figure 6.10e and f. Note that the slope of the shear vanishes at the end $x = 0$, where $p(0) = 0$; the moment is a maximum at $x = l/\sqrt{3}$ where $V(l/\sqrt{3}) = 0$.

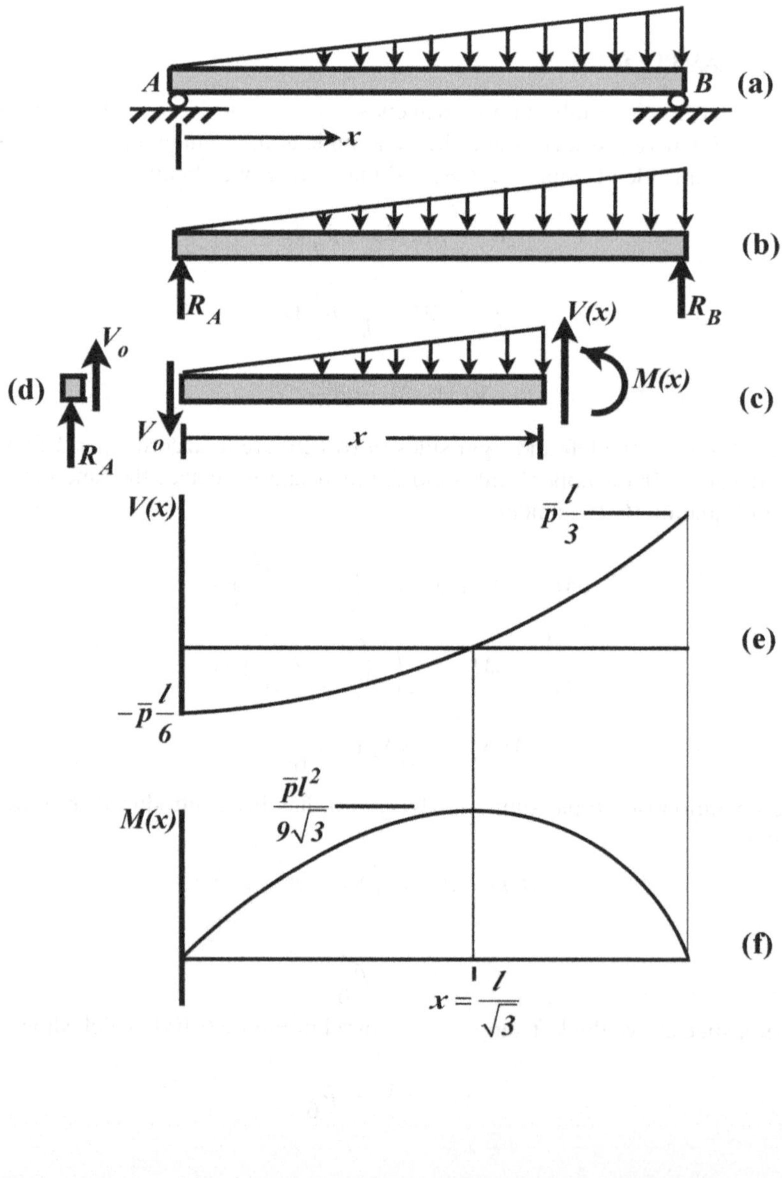

Figure 6.10

ILLUSTRATIVE EXAMPLE 6.2

The beam of Figure 6.11a serves to illustrate the relationships between load, shear force, and bending couple. The free-body diagram of the entire beam is shown in Figure 6.11b. For equilibrium, the reactions must be 333 lb and 1167 lb, as shown. To determine V and M on a cross-section between supports, we consider the free-body diagram of Figure 6.11c.

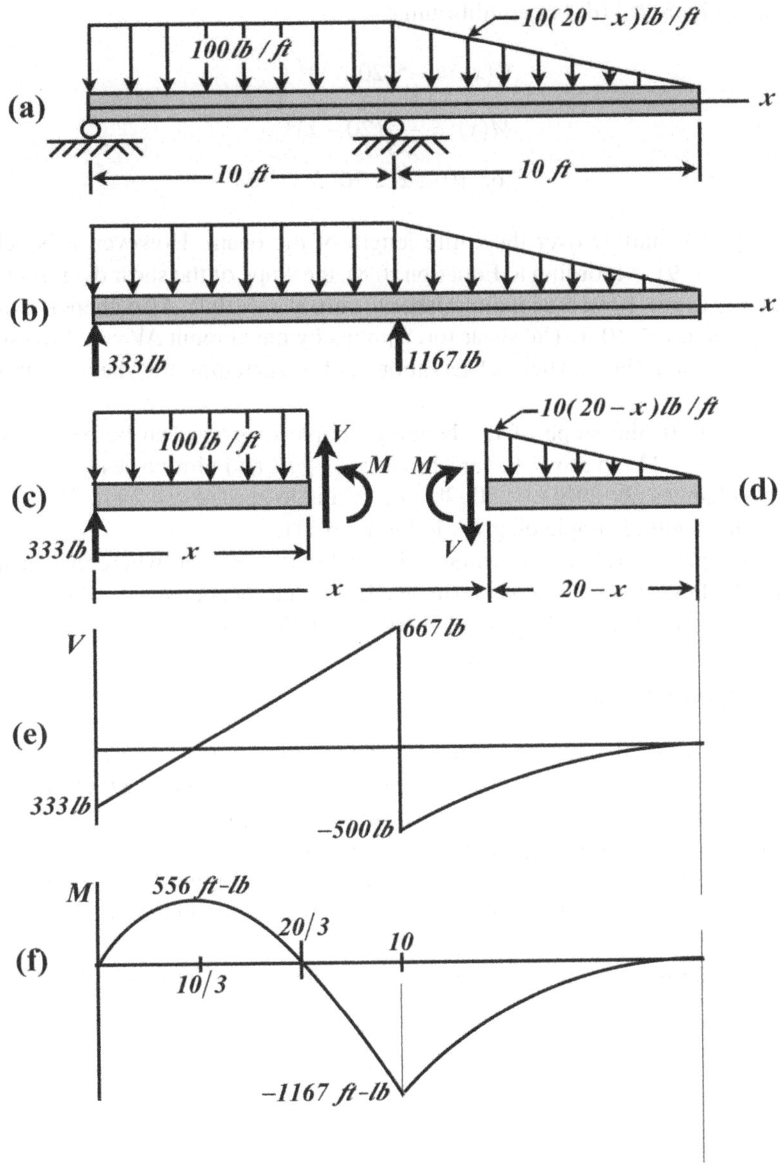

Figure 6.11

For equilibrium,

$$V(x) = -333 + 100x$$
$$M(x) = 333x - 50x^2$$
$$\text{for } 0 < x < 10 \text{ ft}$$

To obtain expressions for V and M on a cross section in the overhang, we consider the free-body diagram shown in Figure 6.11d. For equilibrium,

$$V(x) = -5(20 - x)^2$$
$$M(x) = -\frac{5}{3}(20 - x)^3$$
$$\text{for } 10 < x < 20 \text{ ft}$$

Now we can plot V and M over the entire length of the beam. However, it is helpful to use Equations (6.8) and (6.9). According to Equation (6.8), the slope of the shear diagram is constant at 100 lb/ft from $x = 0$ to $x = 10$ ft; then it diminishes to zero at $x = 20$ ft. Also, there is a discontinuity in the shear diagram at $x = 10$ ft. The shear force jumps by the amount $\Delta V = -1167$ lb, due to the concentrated force acting there. These observations aid in sketching the shear diagram shown in Figure 6.11e.

From Equation (6.9), the slope of the bending couple diagram can be read from the shear diagram. The slope is $+333$ lb at $x = 0$, vanishes at $x = \frac{10}{3}$ ft, and diminishes to -667 lb at $x = 10$ ft; here the slope changes abruptly to $+500$ lb, decreasing to zero at $x = 20$ ft. This information is used in plotting the bending couple diagram in Figure 6.11f.

The bending couple is a relative maximum (556 ft-lb) at $x = \frac{10}{3}$ ft, where the shear force vanishes. However, the largest bending couple (in absolute value) occurs at $x = 10$ ft, where the shear force has a discontinuity. The bending couple here is -1667 ft-lb.

ILLUSTRATIVE EXAMPLE 6.3

A small barge is nearly rectangular in form and carries cargo which consists of a uniform distribution of coal over portion BC and a heavy machine at D, as depicted in Figure 6.12a and b. For our purposes, we can treat the latter as a concentrated load. Neglecting the slight deformations of the barge, we treat the buoyant force as a uniform upward load.

The forces acting on the free body of the barge are shown in Figure 6.12b. From free bodies of the segments AB, AC, AD, and DE, we obtain the shear force and bending moments as plotted in Figure 6.12c and d, respectively.

Note the relationships between the variables and the derivatives (slopes in these plots), viz.

$$\frac{dV}{dx} = p(x), \quad \frac{dM}{dx} = -V(x)$$

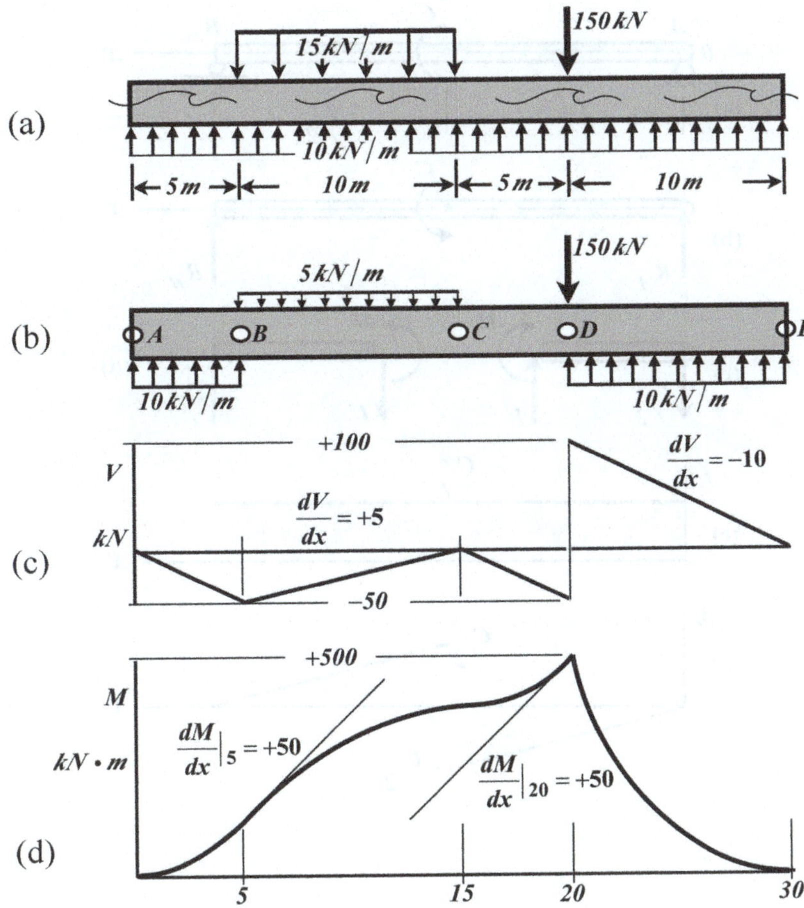

Figure 6.12

ILLUSTRATIVE EXAMPLE 6.4

Let us consider the simply supported beam of Figure 6.13, which is subjected to the concentrated couple C at the midpoint. From the free body of the entire beam in Figure 6.13b, we obtain the reactions

$$R_A = -\frac{C}{L}, \quad R_C = \frac{C}{L}$$

From the diagrams of Figure 6.13c and d we obtain

$$V = C/L, \quad 0 < x < l/2$$
$$l/2 < x < l$$
$$M = -C\frac{x}{l}, \quad 0 < x < l/2$$
$$M = C(1 - \frac{x}{l}), \quad l/2 < x < l$$

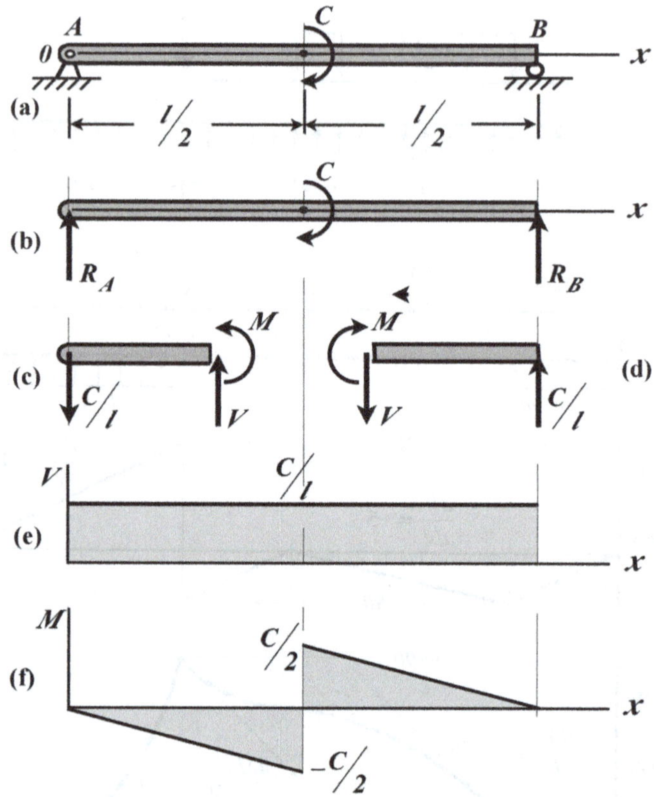

Figure 6.13

Again, we observe the relationships of Equations (6.8) and (6.9), viz.

$$\frac{\mathrm{d}V}{\mathrm{d}x} = p(x), \quad \frac{\mathrm{d}M}{\mathrm{d}x} = -V(x)$$

The latter does not hold in the neighborhood of $x = l/2$, the site of the concentrated couple.

6.3 NORMAL STRESSES IN HI-HO BEAMS

In Section 5.13 we obtained a relation (Equation 5.127) between the normal stress on a cross section and the bending couple for the case of simple bending of homogeneous linearly elastic beams:

$$\sigma_x = -\frac{My}{I} \tag{6.16}$$

This relation is based on the assumption that plane cross sections of the beam remain plane and perpendicular to the neutral axis after bending. For simple bending, the only component of the net internal force system on a cross section is the bending couple M. However, with transverse loading, a shear force V must also act to maintain equilibrium. The presence of this shear force raises a question regarding the validity of the hypothesis about the behavior of cross sections. If the cross sections remain plane and perpendicular to the neutral axis, the shear strain must be zero there. Then the shear stress must also be zero, but this is unlikely because of the presence of the shear force V. The dilemma is resolved by taking the view that the influence of the shear strain is small enough to be ignored in obtaining Equation (6.16). This is generally true for a slender beam, one whose depth is small compared to its length. In most applications, Equation (6.16) gives excellent results, although $V \neq 0$.

A word of caution is appropriate: Although transverse shear is neglected in the derivation of Equation (6.16) and, indeed, the stress component τ_{xy} is usually also negligible, shear stress on other surfaces can be very significant. If the component σ_x is the only significant stress on the coordinate surfaces (x,y,z), the state is one of simple stress, wherein the maximum shear stress is $\tau_{\max} = \sigma_x/2$ (see Section 3.10) and that maximum shear stress might initiate yielding (see Section 4.17).

ILLUSTRATIVE EXAMPLE 6.5

Suppose that the beam of Figure 6.11 has the cross section shown in Figure 6.14a and is composed of HI-HO material.

Then the normal stress σ_x is given by Equation (6.16). The maximum and minimum values can occur at $x = \frac{10}{3}$ ft or at $x = 10$ ft, and depend on the values of the variable y, that is, the distance from the neutral line. We must first locate the centroidal line: To that end, we divide the cross section into two 1-in. × 2-in. rectangles, as shown in Figure 6.14b. Then, if $\bar{y}$ locates the centroid, we have

$$\bar{y} = \frac{A_1\bar{y}_1 + A_2\bar{y}_2}{A_1 + A_2} = \frac{2\left(\frac{1}{2}\right) + (2)(2)}{2+2} = \frac{5}{4}\ \text{in.}$$

With the origin at the centroid, $-\frac{5}{4}$ in. $\leq y \leq \frac{7}{4}$ in. The moment of inertia of the cross section (about the line through the centroid) is also needed. Using the parallel-axis theorem[1] for each of the rectangular areas, we get

$$I_1 = \tfrac{1}{12}(2)(1)^3 + (2)\left(\tfrac{3}{4}\right)^2 = \tfrac{31}{24}\ \text{in.}^4$$
$$I_2 = \tfrac{1}{12}(1)(2)^3 + (2)\left(\tfrac{3}{4}\right)^2 = \tfrac{43}{24}\ \text{in.}^4$$

Hence,

$$I = I_1 + I_2 = \tfrac{37}{12}\ \text{in.}^4$$

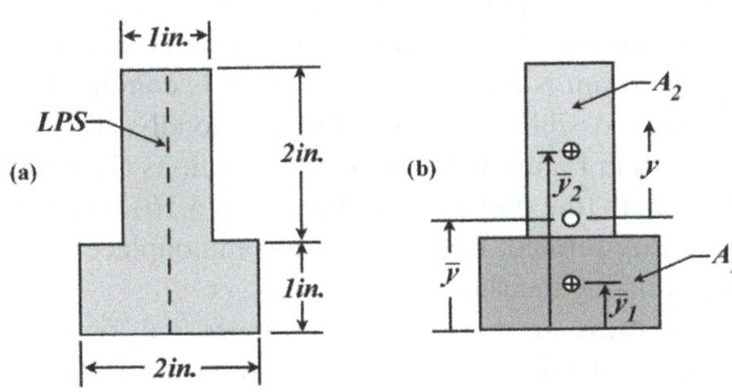

Figure 6.14

At $x = \frac{10}{3}$ ft, $M = 556$ ft-lb, so that the maximum and minimum normal stresses *at that location are*

$$\sigma_x = -\frac{(556 \times 12)(-\frac{5}{4})}{\frac{37}{12}} = +2700 \text{ lb/in.}^2$$

$$\sigma_x = -\frac{(556 \times 12)(\frac{7}{4})}{\frac{37}{12}} = -3790 \text{ lb/in.}^2$$

At $x = \frac{10}{3}$ ft the maximum (tensile) stress occurs at the bottom ($y = -\frac{5}{4}$ in.); the minimum (compressive) stress occurs at the top ($y = +\frac{7}{4}$ in.).

At $x = 10$ ft, $M = -1167$ ft-lb, so that the maximum and minimum normal stresses *at that location are*

$$\sigma_x = -\frac{(-1167 \times 12)(\frac{7}{4})}{\frac{37}{12}} = +7950 \text{ lb/in.}^2$$

$$\sigma_x = -\frac{(-1167 \times 12)(-\frac{5}{4})}{\frac{37}{12}} = -5680 \text{ lb/in.}^2$$

From the computations, we must conclude that the beam is likely to fail near the simple support ($x = 10$ ft). The maximum tensile stress is 7950 lb/in.^{2} (55 MPa) and the maximum shear stress is (7950 lb/in.^{2})/2 = 3980 lb/in.^{2} (27 MPa). Of course, these values are based upon the effects of bending. They do not account for the local conditions - for example, 'stress concentrations' - at the supports.

Additional applications of Equation (6.16) are presented in Section 5.13. Stresses caused by bending of non-Hookean beams are discussed in Section 5.14.

6.4 SHEAR STRESSES IN HI-HO BEAMS

As previously noted, the shear strain γ_{xy} is neglected in the derivation of the equation for the normal stress σ_x. Still, a shear stress τ_{xy} must exist whenever the force V is present, since

$$V = \iint_A \tau_{xy} dA$$

In some circumstances, the shear component can be important. Specific situations are noteworthy: Firstly, the beam might be composed of an anisotropic material that has little resistance to that stress. The common wooden beam is an example; the grain is commonly parallel to the axis and shearing along the grain is a possible mode of failure. Secondly, the beam might have a cross section like the I-beam shown in Figure 6.15. Such cross sections are desirable because they provide much stiffness and strength [the property I in Equation (6.16) is large] with less material and weight, and they are also easily formed. I-beams are so commonplace that we use the I-section to develop a procedure for estimating the shear stress.

Note first that the shear stress must vanish at the upper and lower surfaces of the flanges, because these are free surfaces; that is, $\tau_{yx} = \tau_{xy} = 0$. Then it is unlikely that the shear stress τ_{xy} is appreciable in the flanges. It follows that most of the shear force, and the significant shear stress, act upon the web. This is typical of "thin-walled" members; shear stresses upon the thin-walled

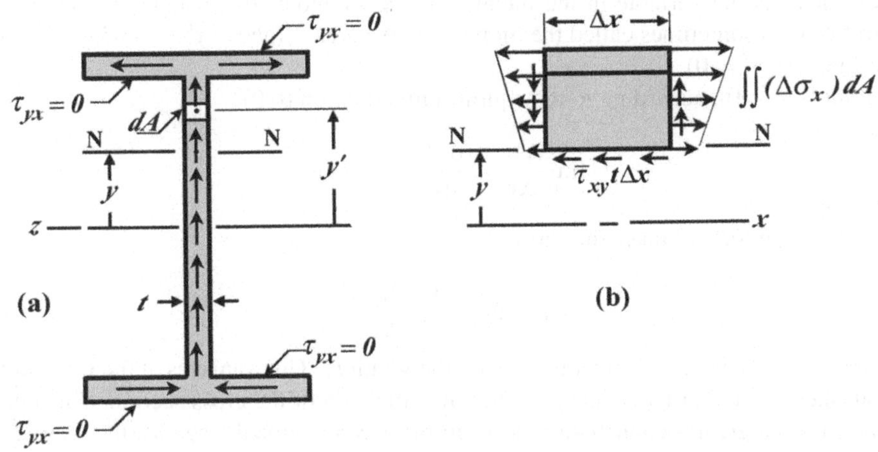

Figure 6.15

sections act effectively tangent to the walls, as τ_{xy} acts in the $y-$ direction upon the web. This means that most of the shear force is carried by the web.

To estimate the shear stress τ_{xy} at the distance y, we accept the normal stress σ_x given by Equation (6.16) and appeal to the conditions for equilibrium of a piece of the beam. In particular, we isolate a piece of the beam as follows. Firstly, we consider a slice lying between sections at x and at $x+\Delta x$; secondly, we cut the slice by the plane $N-N$ at distance y above the neutral line in Figure 6.15a. The free body isolated by these cuts is shown in Figure 6.15b: On the free body, we show only the forces acting in the direction of the axis x: On the shaded portion A of the cross section, the *net* force is

$$\iint \Delta\sigma_x \mathrm{d}A$$

On the surface cut by plane $N-N$ the force is

$$\bar{\tau}_{yx} t \Delta x$$

where $\bar{\tau}_{yx} = \bar{\tau}_{xy}$ denotes the mean value of the shear stress and t denotes the thickness of the web. The sum of the components must vanish for equilibrium:

$$\sum F_x = \iint_A \Delta\sigma_x \mathrm{d}A - \bar{\tau}_{yx} t \Delta x = 0 \tag{6.17a}$$

The normal stress σ_x is given by Equation (6.16), and the change $\Delta\sigma_x$ is due to the change in the moment ΔM, so that Equation (6.17a) takes the form

$$\bar{\tau}_{yx} = -\frac{\Delta M}{\Delta x} \frac{\iint_A y' \mathrm{d}A}{It} \tag{6.17b}$$

For brevity, we denote the integral in Equation (6.17b) by

$$Q(y) = \iint_A y' \mathrm{d}A \tag{6.18}$$

Note that y' denotes the variable in the integral, since y denotes the distance to the cut $N-N$. The integral $Q(y)$ is sometimes called the moment of the area A (above $y'=y$) with respect to the centroidal line ($y=y'=0$).

In the limit ($\Delta x \to 0$), according to the equilibrium equation (6.9),

$$\lim_{x \to 0} \frac{\Delta M}{\Delta x} = \frac{dM}{dx} = -V(x) \tag{6.19}$$

Therefore, Equation (6.17b) takes the form

$$\bar{\tau}_{yx} = \bar{\tau}_{xy} = \frac{V(x)Q(y)}{It} \tag{6.20}$$

The value $\bar{\tau}_{yx} = \bar{\tau}_{xy}$ is the mean value (across the width t). This vanishes at the top and bottom as $Q(y)$ vanishes. Note that $Q(y)$ changes little over the web of the cross section. The area of the flanges provides the greater contribution; also, in most conventional I-beams, the web is thinner than the flanges. An approximation is obtained by dividing the force V by the area of the flange.

ILLUSTRATIVE EXAMPLE 6.6

Shear Stress in a Standard I-Beam

An American Standard wide-flange I-beam has the cross section shown in Figure 6.16; this is an 18-WF-50 (18-in. depth-wide-flange beam-50 lb/ft).[2] The beam is 30.0 ft long, is simply supported at both ends, and carries a uniform load of 240 lb/ft. The maximum (absolute) value of the component τ_{xy} occurs near end supports, where the shear force is

$$|V_0| = \frac{240 \times 30.0}{2} = 3600 \text{ lb}$$

The property I in Equation (6.20) can be calculated from the given dimensions ($I = 793$ in.^{4}); however, a table gives the value:

$$I = 800 \text{ in.}^4$$

Note that the actual cross section of the hot-rolled beam does not have the precise shape illustrated; for example, corners have fillets, and flanges exhibit slight taper. We use the tabulated value, since it is based on the actual cross-sectional shape.

To compute the maximum value of τ_{xy}, we require the integral $Q(0)$ for the darker shaded region (A) of Figure 6.16. From the given dimensions,

$$Q(0) = 0.358 \int_0^{8.43} y\mathrm{d}y + 7.50 \int_{8.43}^{9.00} y\mathrm{d}y = 50.0 \text{ in.}^3$$

The shear $\bar{\tau}_{xy}$ can now be computed by means of Equation (6.20):

$$\bar{\tau}_{xy}(0) = \frac{3600 \times 50.0}{800 \times 0.358} = 628 \text{ lb/in}^2 \quad (4.33 \text{ MPa})$$

Let us also calculate the value at the point B - that is, at the top of the web. There

$$Q(8.43) = 37.3 \text{ in.}^3$$

so that

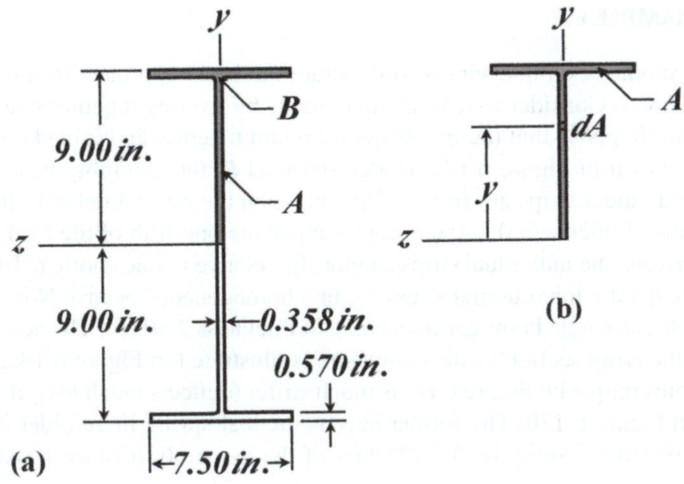

Figure 6.16

$$\bar{\tau}_{xy} = \frac{3600 \times 37.3}{800 \times 0.358} = 469 \text{ lb/in}^2 \quad (3.23 \text{ MPa})$$

We note again that the component $\bar{\tau}_{xy}$ does not vary greatly over the web; it is 25% less at B. Also, we note that the web supports most of the shear force. An approximation is obtained by dividing the shear force V by the area of the web:

$$\bar{\tau}_{xy}(\text{approx.}) = \frac{3600}{(16.86 \times 0.358)} = 596 \text{ lb/in}^2 \quad (4.11 \text{ MPa})$$

This is only 5% less than the maximum value at the centroid.

Note that this value of $\bar{\tau}_{xy}(0)$ is *not* the maximum shear stress in this beam. At the center, the bending moment is

$$M_{\max} = 27{,}000 \text{ ft-lb.} \quad (36{,}000 \text{ N} \cdot \text{M})$$

Then the maximum value of the component σ_x occurs at the bottom ($y = -9.00$) and at the middle where, according to Equation (6.16),

$$\sigma_x = \frac{27{,}000 \times 12 \times (-9.00)}{800}$$
$$= 3645 \text{ lb/in}^2 \quad (25.1 \text{ MPa})$$

At this location, the maximum shear stress occurs on planes inclined 45° to the axis (see Section 3.10). That maximum value is $3645/2 = 1820$ lb/in.^{2} (12.5 MPa), approximately three times greater than the maximum value of the component $\bar{\tau}_{xy}$. Yielding would probably occur at the location of the maximum shear stress (see Section 4.17).

ILLUSTRATIVE EXAMPLE 6.7

Built-Up Beam: Another example serves to illustrate the basic approach to the computation of longitudinal shear stress. Consider a small member made by riveting together five strips 0.500 cm × 2.50 cm × 30 cm. Suppose that the five strips were laid together and joined only by the clamp at the left end, as shown in Figure 6.17a. Under end load P, the assembly bends as depicted in Figure 6.17b; the individual strips are free to slip, one upon the other. In effect, the assembly acts merely as five beams of thickness 0.500 cm, each supporting one-fifth of the load P. When joined by equally spaced rivets, the individual strips cannot slip relative to one another. The rivets provide the shear resistance (as the longitudinal stress $\bar{\tau}_{xy}$ in a homogeneous beam). Now the assembly is unified; it acts much as a single homogeneous beam of thickness 2.50 cm. The action is depicted in Figure 6.17c, and the cross section of the composite is illustrated in Figure 6.18a. The reader can well imagine that the composite Figure 6.17c is much stiffer (deflects much less) than the assembly of disjoint strips in Figure 6.17b. The former acts as the leaf spring in an older automobile. The deflection of the riveted assembly (c) is 1/25 that of the assembly (c) (see Illustrative Example 6.11).

Let us now pose the problem: The rivets are spaced at 2.50-cm intervals along the axis. The rivets are 0.625 cm in diameter and are made of steel. Specifications require that the mean transverse shear on the rivets cannot exceed $\bar{\tau} = 70$ MPa : What is the maximum admissible load P?

We approach this problem as we approached the homogeneous beam, but this beam is not quite homogeneous. We isolate not an *arbitrary* slice but a typical slice of width $\Delta x = 2.50$ cm, as shown in Figure 6.18. From our previous analysis, we can anticipate that the maximum shear occurs at the interfaces nearest the centroidal axis. Accordingly, the cut (counterpart of $N - N$ in Figure 6.15) is made at that interface in Figure 6.18b. On this free-body diagram, we show the forces acting in the axial (x) direction: the normal force $\iint \Delta\sigma_x dA$ and the shear force $\bar{\tau}_{xy} A_R$ upon the rivet; A_R denotes the cross-sectional area of the rivet.

Equilibrium requires that

$$\bar{\tau}_{xy} A_R = +\iint \Delta\sigma_x \mathrm{d}A \tag{6.21}$$

As before, the increment $\Delta\sigma_x$ is caused by the increment ΔM, in accordance with Equation (6.16):

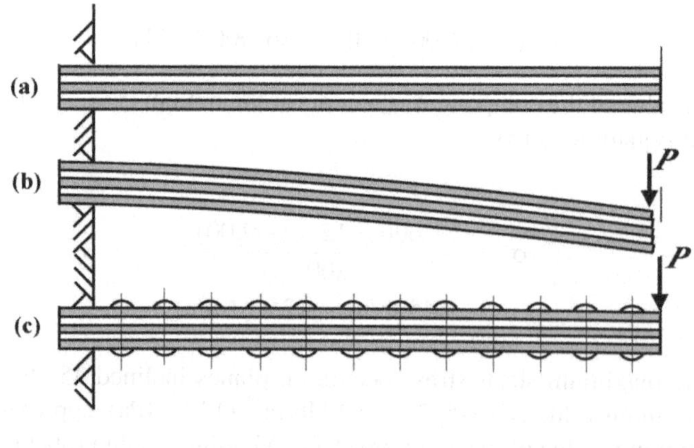

Figure 6.17

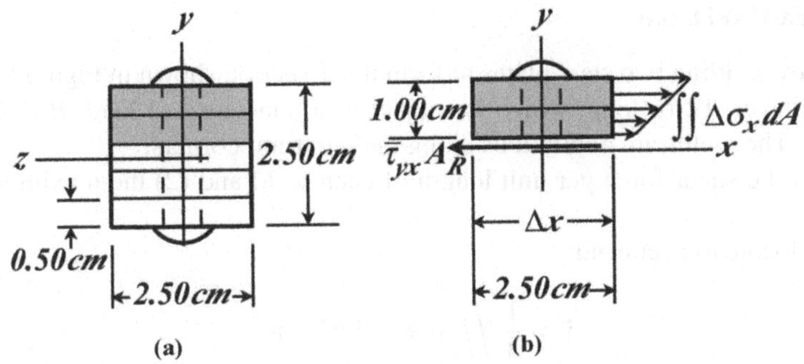

Figure 6.18

$$\Delta\sigma_x = -\frac{\Delta My}{I} \tag{6.22}$$

Now the increment ΔM is related to the shear V by equilibrium. The condition differs from Equation (6.9) in that we cannot pass to the limit. Instead, we have the condition for our typical slice:

$$\Delta M = -V\Delta x \tag{6.23}$$

By substituting Equation (6.23) into (6.22) and that result into Equation (6.21), we obtain

$$\bar{\tau}_{xy} = +\frac{V\Delta x}{IA_R}\iint_A ydA \tag{6.24}$$

As before, the integral in Equation (6.24) extends over the cross-sectional areas of the two strips in Figure 6.18a. We need only calculate that integral and I and substitute the numerical values into Equation (6.24):

$$Q = \iint_A ydA = 2.5\int_{0.25}^{1.25} ydy = 1.875 \text{ cm}^3$$

$$I = \frac{2.5 \times (2.5)^3}{12} = 3.255 \text{ cm}^4$$

$$A_R = \frac{\pi}{4}(0.625)^2 = 0.3068 \text{ cm}^2$$

$$\bar{\tau}_{xy} = 70 \times 10^2 \text{ N/cm}^2$$

$$V = \frac{IA_R\bar{\tau}_{xy}}{\Delta xQ} = \frac{3.255 \times 0.3068 \times 70 \times 10^2}{2.5 \times 1.875} \text{ N} = 1491 \text{ N}$$

Since $V = P$, this is the maximum admissible force P. Note that the foregoing analysis entails approximations and provides only a mean value, $\bar{\tau}_{yx}$. Also, the use of Equation (6.22) presumes HI-HO behavior of the steel.

ILLUSTRATIVE EXAMPLE 6.8

A beam is made by welding two steel plates to form the T section shown in Figure 6.19. The beam is used as a cantilever, 12 ft. long, which is subject to a concentrated load, $P = 2000$ lb. at the unsupported end. The welds are continuous along the reentrant corners.

We require (1) the shear force per unit length of each weld and (2) the maximum shear stress in the plate.

First we must locate the centroid:

$$\bar{y} = \frac{1}{A}\iint_A y dy = 4.625 \text{ in.}$$

where $\bar{y}$ denotes the centroid as shown in Figure 6.19a and A denotes the entire area of the section.

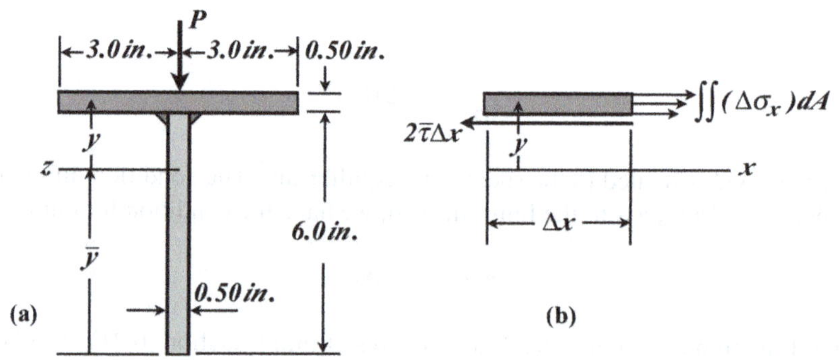

Figure 6.19

The essential free body is depicted in Figure 6.19b. The requisite equation expresses the equilibrium of the shear forces ($2\bar{\tau}\Delta x$) and the normal forces upon the shaded area ($\Delta MQ/I$):

$$2\bar{\tau}\Delta x = \Delta MQ/I$$

$$Q = \iint_{A_1} y dA$$

Here A_1 denotes the cross-sectional area of the flange and $\bar{\tau}$ denotes the force per unit length upon each weld. Again, we pass to the limit to obtain

$$\bar{\tau} = \frac{VQ}{2I}$$

Here, the respective integrals yield

$$Q = 4.875 \text{ in.}^3, \quad I = 24.90 \text{ in.}^4$$

The result follows:

$$\bar{\tau} = \frac{2000 \text{ lb.} \times 4.875 \text{ in.}^3}{2 \times 24.90 \text{ in.}^4} = 196 \text{ lb/in.}$$

The maximum absolute value of the normal stress occurs at the bottom where

$$|\sigma_x| = \frac{2000 \text{ lb.} \times 144 \text{ in.} \times 4.625 \text{ in.}}{24.90 \text{ in.}^4}$$
$$= 53,500 \text{ psi}$$

Therefore (see Section 3.10), the maximum shear stress is

$$\tau_{max} \approx \frac{53,490}{2} = 26,750 \text{ psi}$$

6.5 DEFLECTION OF SYMMETRIC HOOKEAN BEAMS

6.5.1 THE DEFLECTION EQUATION

The Bernoulli-Euler curvature formula was obtained (Section 5.13, Equation 5.125) for the case of simple bending of a linearly elastic beam:

$$\kappa = \frac{1}{\rho} = \frac{M}{EI} \tag{6.25}$$

As in the discussion of bending stresses just concluded, we extend the use of Equation (6.25) to cases where the shear stress is not zero on every cross section. Again, this leads to very useful results for slender beams, since the effects due to shear are generally insignificant. We simply treat κ and M as functions of the coordinate x that locates a cross section. Note, too, that the curvature κ and moment M are proportional for any Hookean beam; only the coefficient (the flexural rigidity EI) is different for anisotropic or nonhomogeneous beams.

The deformation of the beam is determined by the deformed shape of the neutral axis. Initially, the neutral axis is along the x-axis, as in Figure 6.20. After the beam deforms, the neutral axis becomes a curve in the xy-plane, as shown in Figure 6.20. The equation of this curve determines the transverse displacement of particles on the neutral axis. This equation is therefore called the deflection function, or, more simply, the deflection. In Figure 6.20, Δs is an element of arc length along the deformed neutral axis and ρ is the radius of curvature of the neutral axis at x. From the figure, we see that $\Delta s = \rho\Delta\theta$, and in the limit, as Δs approaches zero,

$$\kappa = \frac{1}{\rho} = \frac{d\theta}{ds} \tag{6.26}$$

The angle θ is the rotation of the tangent to the neutral axis. If we substitute Equation (6.25) into Equation (6.26), the result is

$$\kappa = \frac{d\theta}{ds} = \frac{M}{EI} \tag{6.27}$$

The relation (6.27) is applicable to symmetric HI-HO beams wherein the strains are small. By definition, the neutral axis is unextended, i.e.,

$$\Delta s = \Delta x \tag{6.28}$$

Therefore,

$$\kappa = \frac{d\theta}{dx} == \frac{M}{EI} \tag{6.29}$$

It remains to relate the rotation θ to the displacement w, shown in Figure 6.20. We note that, as $\Delta x \to 0$, $\Delta\theta \to 0$, and

$$\frac{dw}{ds} = \frac{dw}{dx} = \sin\theta \tag{6.30a}$$

Let us now collect the differential equations that govern the four variables: dynamic variables V and M and kinematic variables θ and w. The dynamic variables are governed by the equilibrium equations (6.8) and (6.9). The kinematic variables are governed by (6.29) and (6.30a). The first three equations are linear differential equations; the fourth is not. If the rotations are small, e.g., 10° or 0.20 rad. then,

$$\sin\theta \doteq \theta \tag{6.30b}$$

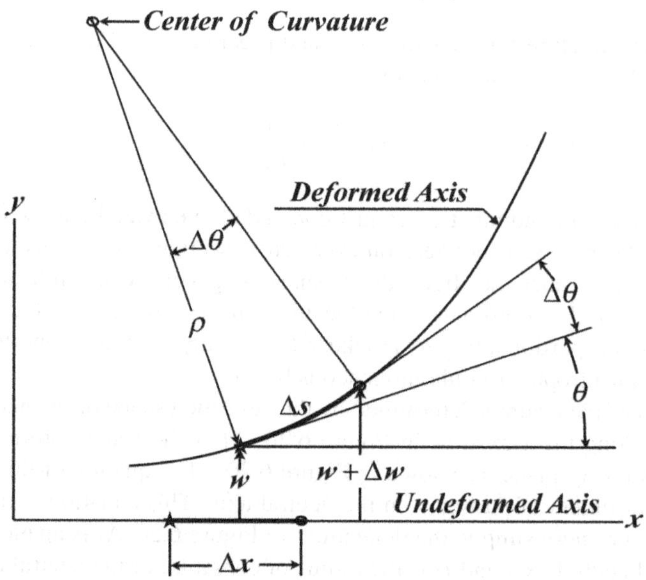

Figure 6.20

Then the four *linear* equations follow:

$$\frac{dV}{dx} = p(x) \tag{6.31}$$

$$\frac{dM}{dx} = -V(x) \tag{6.32}$$

$$\frac{d\theta}{dx} = \frac{M(x)}{EI} \tag{6.33}$$

$$\frac{dw}{dx} = \theta(x) \tag{6.34}$$

To obtain the functions $V(x)$, $M(x)$, $\theta(x)$, and $w(x)$, we must successively integrate the four first-order linear equations. This introduces some arbitrariness in the form of constants of integration. The evaluation of these constants is accomplished by requiring the force V, couple M, rotation θ, and/or deflection w to satisfy the conditions imposed by the supports. Let us therefore examine the support conditions at the more common types of supports.

6.5.2 SUPPORT CONDITIONS

At a clamped or fixed support, the axis cannot rotate, nor can it displace. For example, the beam of Figure 6.21 is clamped at A; therefore

$$w_A = \theta_A = 0$$

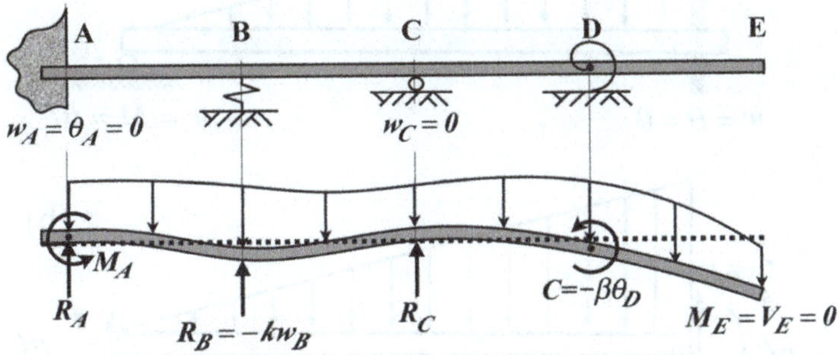

Figure 6.21

At the free end E, the force V and couple M must vanish:

$$V_E = M_E = 0$$

A pin or roller prevents transverse displacement but does not inhibit rotation; a reactive force prevents displacement. Thus at site C of Figure 6.21, there is a reaction R_C; $w_C = 0$

Displacement w and/or rotation θ can be resisted by an adjoining structure. These circumstances are depicted schematically in Figure 6.21 at the sites B and D, respectively. If the supporting structures are linearly elastic, then the reactions are proportional to the respective motions, i.e.,

$$R_B = -kw_B, \quad C = -\beta\theta_D$$

Here the constants, β and k, depend upon the properties of the adjoining structure. The following examples illustrate the application of the differential equations, (6.31), (6.32), (6.33), and (6.34), and the support conditions in order to determine the variables, V, M, θ, and w.

ILLUSTRATIVE EXAMPLE 6.9

The beam in Figure 6.22a is fixed at the left end ($x = 0$) and simply supported at the right end ($x = L$). The conditions at the left end are

$$\theta(0) = 0, \quad w(0) = 0 \tag{6.35}$$

At the right end, the simple support can exert no couple but prevents any displacement:

$$M(L) = 0, \quad w(L) = 0 \tag{6.36a,b}$$

The beam is subjected to the linearly varying load

$$p(x) = \bar{p}\left(1 - \frac{x}{L}\right) \tag{6.37}$$

where $\bar{p}$ is a constant (240 lb/ft), the magnitude of the load at the left end ($x = 0$).

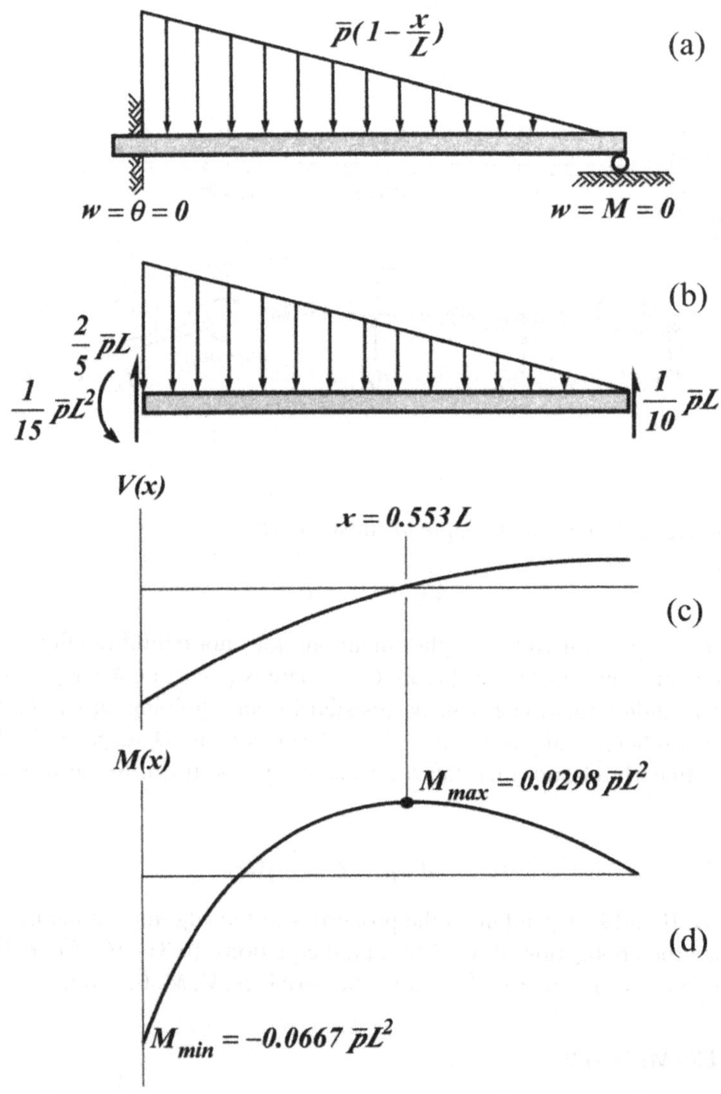

Figure 6.22

We begin the solution by substituting the load $p(x)$, Equation (6.37), into the first differential, Equation (6.31), and integrating between the corresponding limits $(0,x)$:

$$dV = p(x)dx = \bar{p}\left(1 - \frac{x}{L}\right)dx \tag{6.38a}$$

$$\int_{V_0}^{V(x)} dV = \bar{p}\int_0^x \left(1 - \frac{x}{L}\right)dx \tag{6.38b}$$

$$V = V_0 + \bar{p}\left(x - \frac{x^2}{2L}\right) \tag{6.38c}$$

Note that the limits correspond: $V(O) = V_0$, the unknown force exerted on the left end, $x = 0$. The resulting Equation (6.38c) is now substituted into the second differential equation, Equation (6.32), which is integrated between the corresponding limits $(0, x)$:

$$dM = -V dx = -\left[V_0 + \bar{p}\left(x - \frac{x^2}{2L}\right)\right]dx \tag{6.39a}$$

$$\int_{M_0}^{M} dM = -\int_0^x \left[V_0 + \bar{p}\left(x - \frac{x^2}{2L}\right)\right]dx \tag{6.39b}$$

$$M = M_0 - V_0 x - \bar{p}\left(\frac{x^2}{2} - \frac{x^3}{6L}\right) \tag{6.39c}$$

Here $M(O) = M_0$, the unknown couple exerted upon the left end, $x = 0$. This last result, Equation (6.39c), is now substituted into the third differential equation, Equation (6.33), which is integrated between the corresponding limits $(0, x)$:

$$EI d\theta = M dx = -\left[M_0 - V_0 x - \bar{p}\left(\frac{x^2}{2} - \frac{x^3}{6L}\right)\right]dx \tag{6.40a}$$

$$EI\int_0^{\theta(x)} d\theta = \int_0^x \left[M_0 - V_0 x - \bar{p}\left(\frac{x^2}{2} - \frac{x^3}{6L}\right)\right]dx \tag{6.40b}$$

$$EI\theta(x) = M_0 x - V_0\frac{x^2}{2} - \bar{p}\left(\frac{x^3}{6} - \frac{x^4}{24L}\right) \tag{6.40c}$$

Note that the end condition, $\theta(0) = 0$, has been inserted in the lower limit of Equation (6.40b, c). Finally, the result, Equation (6.40c), is substituted into the fourth differential equation, Equation (6.34), which is integrated between the corresponding limits $(0, x)$:

$$dw = \theta(x)dx \tag{6.41a}$$

$$EI\int_0^{w(x)} d\theta = \int_0^x \left[M_0 x - V_0\frac{x^2}{2} - \bar{p}\left(\frac{x^3}{6} - \frac{x^4}{24L}\right)\right]dx \tag{6.41b}$$

$$EIw(x) = M_0\frac{x^2}{2} - V_0\frac{x^3}{6} - \bar{p}\left(\frac{x^4}{24} - \frac{x^5}{120L}\right) \tag{6.41c}$$

Here, too, an end condition, $w(0) = 0$, has been enforced.

The functions $V(x)$, $M(x)$, $\theta(x)$, and $w(x)$ are now determined, except for the two unknown constants, M_0 and V_0. We presume that the dimensions of the beam (L and I), the material (E), and the load (p) are given; also, we have assumed that the properties (E and I) are constant. It remains only to determine the two constants by means of the remaining conditions at the right end, Equations (6.36a,b); these are the only requirements not yet satisfied. They provide two linear algebraic equations in the two unknowns:

$$M(L) = M_0 - V_0 L - \bar{p}\frac{L^2}{3} = 0$$

$$EIw(L) = M_0\frac{L^2}{2} - V_0\frac{L^3}{6} - \bar{p}\frac{L^4}{30} = 0$$

The second is simplified when multiplied by $2/L^2$; then the two equations take the forms:

$$M_0 - V_0 L = \bar{P}\frac{L^2}{3}$$

$$M_0 - V_0\frac{L}{3} = \bar{P}\frac{L^2}{15}$$

The solution of these provides the two actions, V_0 and M_0, at the left end:

$$V_0 = -\frac{2}{5}\bar{p}L, \quad M_0 = \bar{p}\frac{L^2}{15}$$

Also, by means of Equation (6.38c),

$$V(L) = \tfrac{1}{10}\bar{p}L$$

The free body of the entire beam is shown in Figure 6.22b, and the plots of force $V(x)$ and couple $M(x)$ are shown in Figure 6.22c and d.

So that we can appreciate the orders of magnitude, let us compute the maximum normal stress σ_x at $x = 0$ and the deflection $w(L/2)$ at midspan. For this purpose, we use an American Standard wide-flange (12-WF-40) beam:

$$I = 310 \text{ in.}^4, \quad \text{depth} = 12.0 \text{ in.}, \quad L = 20.0 \text{ ft}$$

Then, by Equation (6.16), the maximum normal stress occurs at the left end and at the top ($y = 6.00$ in.):

$$\sigma_{\max} = -\frac{M_{\min}(6.00)}{310}$$

$$= -\frac{(-76,800)(6.00)}{310} = 1490 \text{ lb/in.}^2 \quad (10.3 \text{ MPa})$$

The deflection at midspan is given by Equation (6.41c) with $x = 10$ ft. The modulus for steel is approximately $E = 30 \times 10^6$ lb/in.^{2} Then,

$$w\left(\frac{L}{2}\right) = \frac{1}{EI}\left[\frac{M_0}{2} - \frac{V_0 L}{12} - \bar{p}\left(\frac{L^2}{96} - \frac{L^2}{960}\right)\right]\frac{L^2}{4}$$

$$= \frac{10^{-6}}{30 \times 310}\left[-\frac{1}{30} + \frac{1}{30} - \frac{9}{960}\right]\frac{(240)^4 \times 20}{4}$$

$$= -0.0167 \text{ in.} \quad (-0.0424 \text{ cm})$$

Now suppose that we remove the support at the right end of the beam in Figure 6.22. Equations (6.38c) and (6.39c) still hold, since the load $p(x)$ is unchanged. We need only enforce the conditions at the right end, viz.

$$V(L) = 0, \quad M(L) = 0$$

Substituting Equations (6.38c) and (6.39c) into these conditions, we obtain

$$V_0 = -\frac{\bar{p}L}{2}, \quad M_0 = -\frac{\bar{p}L^2}{6}$$

Note that the force $V(x)$ and couple $M(x)$ are fully determined by these equilibrium conditions. The problem is said to be statically determinate; the internal force system can be determined without recourse to kinematic conditions. However, the problem is not completely solved. We still need the kinematic equations to determine the rotation $\theta(x)$ and the deflection $w(x)$. The two additional differential equations and their solutions are Equations (6.40c) and (6.41c). With the present values of the constants V_0 and M_0, Equation (6.41c) takes the form

$$w(x) = -\frac{\bar{p}L^4}{12EI}\left(\frac{x^2}{L^2} - \frac{x^3}{L^3} + \frac{x^4}{2L^4} - \frac{x^5}{10L^5}\right)$$

With the previous dimensions and properties, we compute the maximum normal stress σ_x and deflection $w(L)$. The former occurs at the fixed end, where

$$M_0 = -\frac{240 \times 20 \times 240}{6} = -192{,}000 \text{ in.} \cdot \text{lb} \quad (21.7 \times 10^3 \text{ N} \cdot \text{m})$$
$$\sigma_x = \frac{192{,}000 \times 6.00}{310} = 3720 \text{ lb/in.}^2 \quad (25.6 \text{ MPa})$$

The deflection at the free end is

$$w(L) = -\frac{\bar{p}L^4}{30EI} = 0.238 \text{ in.} \quad (0.605 \text{ cm})$$

This deflection is 14 times as great as the midspan deflection of the beam with the support at the right end.

ILLUSTRATIVE EXAMPLE 6.10

It is worth noting that the moment-curvature relations (6.26) and (6.29) are *not* restricted to small rotations or displacements. That relation requires only the linearity of the strain-stress relations, i.e., Hookean behavior. To illustrate, let us consider a thin strip of high-carbon steel: length, $L =$ 20 in., thickness, $t = 0.020$in., width, $d = 0.400$in. Imagine a common 'hacksaw' blade.

Suppose that we apply couples to the two ends, bend the strip into a circle such that opposing ends overlap 0.250 in., where they are bonded together, as depicted in Figure 6.23.

The deformed strip is (approximately) a circle of radius

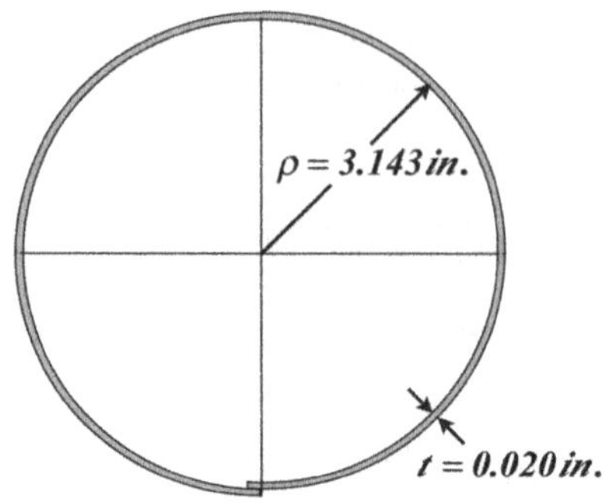

Figure 6.23

$$\rho = \frac{19.75}{2\pi} = 3.143 \text{ in.}$$

The modulus E and moment-of-inertia I are, respectively,

$$E = 30 \times 10^6 \text{ psi}, \quad I = \frac{dt^3}{12} = 2.667 \times 10^{-7} \text{ in.}^4$$

According to Equation (6.25)

$$M = \frac{EI}{\rho} = \frac{3 \times 2.667}{3.143} = 2.546 \text{ in-lb}$$

Equation (6.16) provides the maximum normal stress:

$$\sigma_x = \frac{My}{I} = \frac{2.546 \times 1.00 \times 10^{-2}}{2.667 \times 10^{-7}} = 95,500 \text{ psi}$$

This stress is well within the tensile yield stress of the high-carbon steel.

Note that the foregoing example does not entail the geometric approximation (6.30b). Indeed, if we apply the nonlinear Equation (6.30a), we have

$$\mathrm{d}w = \sin\theta \mathrm{d}s = \rho \sin\theta \mathrm{d}\theta$$

Integrating from end to end

$$w = \rho \int_0^{2\pi} \sin\theta \mathrm{d}\theta = 0$$

This is the correct result as we affixed the ends.

ILLUSTRATIVE EXAMPLE 6.11

Recall the example of the beam in Section 6.4 (Figure 6.17). The beam is a cantilever, fixed at the left end ($x = 0$) and subjected to the concentrated load at the right ($x = L$). The conditions at the left end are

$$\theta(0) = 0, \quad w(0) = 0$$

The conditions at the right end can be stated as follows:

$$V(L) = -P, \quad M(L) = 0$$

As before, four successive integrations lead to the expressions $[p(x) = 0]$:

$$\begin{aligned}
V(x) &= V_0 = -P \\
M(x) &= M_0 - V_0 x = -PL\left(1 - \frac{x}{L}\right) \\
\theta(x) &= -\frac{PL^2}{EI}\left(\frac{x}{L} - \frac{x^2}{2L}\right) \\
w(x) &= -\frac{PL^3}{2EI}\left(\frac{x^2}{L^2} - \frac{x^3}{3L^3}\right) \\
w(L) &= -\frac{PL^3}{3EI}
\end{aligned} \tag{6.42}$$

Let us examine the difference in flexibility between the assembly of Figure 6.17b and the unified composite of Figure 6.17c. In the former, each strip acts independently. If t denotes the thickness and b is the width of the individual strip, then the property I of the rectangular cross-section is $I = bt^3/12$

Since each strip of the assembly supports one-fifth of the load P, the deflection is obtained by substituting the property I and $P/5$ into Equation (6.42):

$$w(L) = -\frac{PL^3 12}{15Ebt^3} = -\frac{4}{5}\frac{PL^3}{Ebt^3} \tag{6.43}$$

The property I of the composite beam is

$$I = \frac{b(5t)^3}{12}$$

The deflection of the composite beam is obtained by substituting this expression and the total load P into Equation (6.42):

$$w(L) = -\frac{PL^3}{3Eb}\frac{12}{(125)t^3} = -\frac{4}{125}\frac{PL^3}{Ebt^3} \tag{6.44}$$

Comparison of the deflections, Equations (6.43) and (6.44), indicates that the assembly is 25 times as flexible as the composite.

ILLUSTRATIVE EXAMPLE 6.12

The torque wrench shown in Figure 6.24 is designed so that the pointer moves 1.00 cm on the scale to indicate a 4500 N-cm couple acting upon a bolt head A. The wrench is made of steel ($E = 200$ MPa $= 20 \times 10^6$ N/cm^2). We must determine the width h.

- The force P acts upon a beam at a distance L to produce the torque: $PL = 4500\ \text{N} \cdot \text{cm}$
- The moment on this cantilever is $M = P(L - x)$

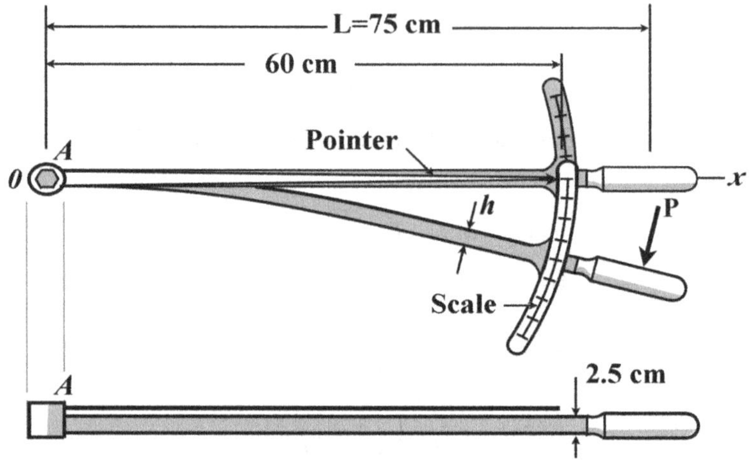

Figure 6.24

We turn to Equation (6.33):

$$EId\theta = Mdx = P(L-x)dx, \quad EI\theta = P\left(Lx - \frac{x^2}{2}\right)$$

Here, we assume that there is no rotation of the bolt, i.e., $\theta_A = 0$. Then, according to Equation (6.34)

$$EIdw = EI\theta dx, \quad EIw = P\left(L\frac{x^2}{2} - \frac{x^3}{6}\right), \quad EIw(l) = PLl^2\left(\frac{1}{2} + \frac{l}{6L}\right)$$

Note that Equation (6.34) serves as well as (6.30a), since $w(x)$ is to trace the arc at x as opposed to the displacement perpendicular to the original axis x. The required values follow:

$$I = \frac{2.5h^3 \text{ cm}}{12}, \quad EI = 4.167h^3 \frac{\text{N}}{\text{cm}} \times 10^6$$

$$PLl^2\left(\frac{1}{2} - \frac{l}{6L}\right) = 45 \times 36 \times (0.3667) \times 10^4 = 594 \times 10^4 \text{ N} \cdot \text{cm}^3$$

When we enforce the conditions, i.e., $P \times L = 4500$ N-cm, $w(l) = 1$ cm, we obtain

$$h^3 = \frac{5.94 \text{ cm}^3}{4.1677} = 1.426 \text{ cm}^3, \quad h = 1.125 \text{ cm} \quad (0.44 \text{ in.})$$

ILLUSTRATIVE EXAMPLE 6.13

A Simple Frame: consists of two beams, joined as shown in Figure 6.25. The beam BD is clamped at D and rigidly joined to the column AC at B; i.e., both members must undergo the same rotation (θ_B) at B when the uniform load $\bar{p}$ is imposed on the beam BD. The column AC supports the beam BD at B and resists the rotation at the end B. The load causes an insignificant extension of either

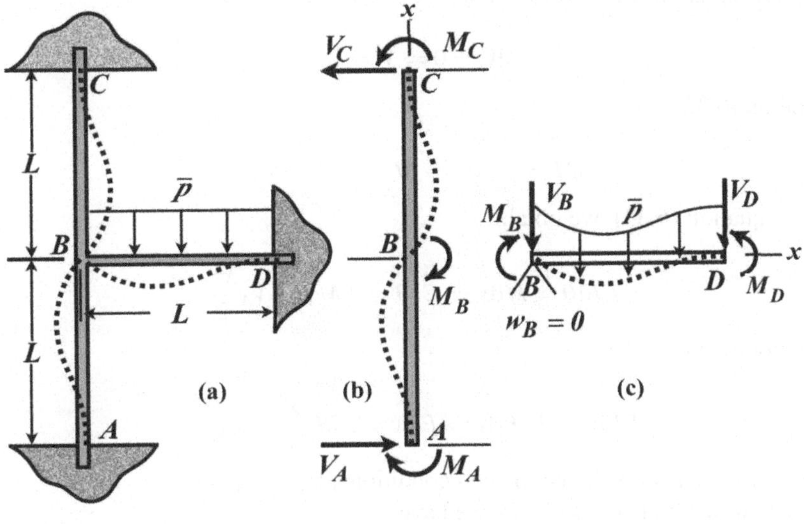

Figure 6.25 A simple Frame: A Union of Two Beams

member; therefore, we can justly neglect the displacement at B. Both members have the same properties, i.e., the same rigidity EI.

The column AC is clamped at both ends. It deflects as a beam subjected to the central couple $\bar{M}$ exerted by the beam. The free body is shown in Figure 6.25b. The free body of the beam BD is shown at (c).

We are presented with two problems: (a) The beam AC subject to the central couple $\bar{M}$ and the similar end conditions at A and C:

$$\theta_A = w_A = \theta_C = w_C = 0 \tag{6.45}$$

The beam BD which is subjected to the load $\bar{p}$ and the end conditions:

$$w_B = w_D = \theta_D = 0 \tag{6.46}$$

The solutions are joined by the conditions:

$$\theta_B \text{ (beam)} = \theta_B \text{ (column)}$$
$$M_B \text{ (column)} = M_B \text{ (beam)}$$

Our solution of the first (the column) is simplified when we acknowledge the antisymmetry of the loading and the deflection, viz.

$$M_C = -M_A, \quad w_B = 0, \quad V_A = V_C$$

For equilibrium of the entire column, we have

$$V_A + V_C = 0, \quad M_B + M_A - M_C - LV_C - LV_A = 0$$

or

$$2M_A - 2LV_A = -M_B \tag{6.47}$$

We proceed with Equation (6.31) for the interval $0 \leq x < L$ (A to B):

$$\mathrm{d}V = 0 \Rightarrow V = V_A$$

Then by Equation (6.32)

$$\mathrm{d}M = -V\mathrm{d}x \Rightarrow M = +M_A - V_A x$$

Proceeding to Equation (6.33), we obtain

$$EI\mathrm{d}\theta = M\mathrm{d}x \Rightarrow EI\theta = +M_A x - V_A \frac{x^2}{2}$$

Then by Equation (6.34)

$$EI\mathrm{d}w = EI\theta\mathrm{d}x \Rightarrow EIw = +M_A \frac{x^2}{2} - V_A \frac{x^3}{6}$$

In these conditions, we acknowledge the end conditions, $\theta_A = w_A = 0$.

From the conditions (6.45) and (6.47), we have

$$w(L) = w_B = 0 \Rightarrow +M_A - \frac{1}{3} V_A L = 0$$

$$+M_A - V_A L = -\frac{1}{2} M_B$$

or

$$V_A = \frac{3}{4} \frac{M_B}{L}, \quad M_A = \frac{1}{4} M_B$$

Finally

$$EI\theta_B = -\frac{1}{8} M_B L \tag{6.48}$$

Note: By our sign convention, θ_B and M_B are both clockwise.

Let us turn to the deflection of the beam of Figure 6.25c where by Equation (6.31)

$$\mathrm{d}V = \bar{p}\mathrm{d}x \Rightarrow V = V_B + \bar{p}x$$

Then according to Equations (6.32)–(6.34) we obtain

$$\begin{aligned} M &= M_B - V_B x - \bar{p}\frac{x^2}{2} \\ EI\theta &= EI\theta_B + M_B x - V_B \frac{x^2}{2} - \bar{p}\frac{x^3}{6} \\ EIw &= EI\theta_B x + M_B \frac{x^2}{2} - V_B \frac{x^3}{6} - \bar{p}\frac{x^4}{24} \end{aligned} \tag{6.49}$$

The clamped conditions at D provide two equations:

$$\theta_D = 0 \Rightarrow EI\theta_B + M_B L - V_B \frac{L^2}{2} = \bar{p}\frac{L^3}{6} \tag{6.50a}$$

$$w_D = 0 \Rightarrow EI\theta_B + M_B\frac{L}{2} - V_B\frac{L^2}{6} = \bar{p}\frac{L^3}{24} \tag{6.50b}$$

The simultaneous solution of the Equations (6.48) and (6.50a,b) follows:

$$M_B = -\frac{1}{18}\bar{p}L^2 \tag{6.51a}$$

$$V_B = -\frac{11}{24}\bar{p}L \tag{6.51b}$$

$$EI\theta_B = -\frac{1}{144}\bar{p}L^3 \tag{6.51c}$$

If the union at B did not inhibit a relative rotation, that is, acted as a hinge, then $M_B = 0$ and

$$V_B = -\frac{3}{8}\bar{p}L, \quad EI\theta_B = -\frac{1}{48}\bar{p}L^3$$

Finally, the deflection of the beam BD is given by Equation (6.49) with the constants of Equations (6.51a–c):

$$EIw(x) = -\frac{\bar{p}L^3}{144}(x + 4x^2 - 11x^3 + 6x^4)$$

ILLUSTRATIVE EXAMPLE 6.14

To conserve material and weight, a cantilever (HI-HO) beam has a rectangular cross-section with constant depth h and variable width $b = h(1 + x/L)$ as shown in Figure 6.26.

We want to compare the weights, maximum normal stress, and deflection with those of a uniform beam with constant width $2h$. Let subscripts t and u signify the tapered and uniform beams, respectively.

If ρ denotes the density, then the weights (w) follow:

$$\begin{aligned} w_t &= \rho L h\left(\tfrac{3}{2}\right) \\ w_u &= \rho L(2h) \end{aligned} \qquad \frac{w_t}{w_u} = \frac{3}{4}$$

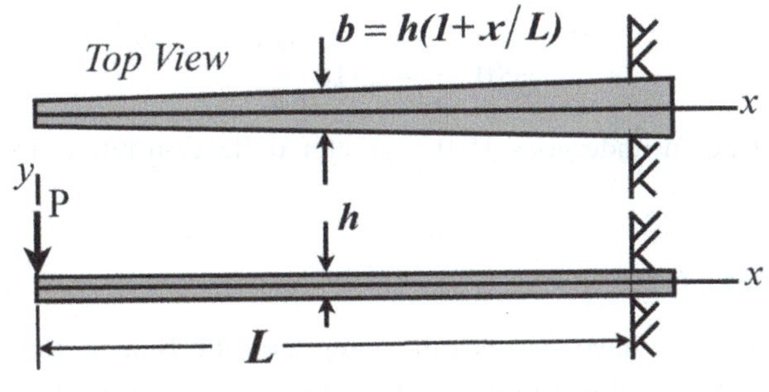

Figure 6.26 Tapered Bem

Both beams are subject to the same maximum moment at the clamped end and have the same cross-sections at that end. Therefore, both are subject to the same normal stress:

$$\sigma(\max) = \frac{PL(h/2)}{\frac{1}{12}(2h)h^3} = 3\frac{PL}{h^3}$$

The maximum deflection of the uniform beam (at the loaded end) is

$$w_u = -\frac{PL^3}{3EI} = -\frac{2PL^3}{Eh^4}$$

To determine the deflection of the tapered beam, we require the variable I:

$$I = \frac{1}{12}bh^3 = \frac{h^4}{12}\left(1 + \frac{x}{L}\right)$$

Next, we turn to Equation (6.33), viz

$$\mathrm{d}\theta = \frac{M}{EI}\mathrm{d}x = -\frac{12P}{Eh^4}\left(\frac{x}{1+x/L}\right)\mathrm{d}x$$

Integrating, we obtain

$$\theta - \theta_0 = -\frac{12P}{Eh^4}L^2\left[\frac{x}{L} - \ln\left(1 + \frac{x}{L}\right)\right]$$

Next, we turn to Equation (6.34),

$$\mathrm{d}w = \mathrm{d}\theta\mathrm{d}x$$

Therefore, we have

$$w - w_0 = \theta_0 x - \frac{12PL^2}{Eh^4}\left[\frac{x^2}{2L} - L\left(1 + \frac{x}{L}\right)\ln\left(1 + \frac{x}{L}\right) + x\right]$$

Enforcing the end conditions, $\theta(L) = w(L) = 0$, we have

$$\theta_0 = \frac{12PL^2}{Eh^4}(1 - \ln 2), \; w_0 = \frac{12PL^3}{Eh^4}(\frac{1}{2} - \ln 2)$$

The maximum deflection is at $x = 0$

$$w(0) = -2.318\frac{PL^3}{Eh^4}$$

Therefore, the tapered beam undergoes 15.9% greater deflection but weighs 25% less than the uniform beam.

ILLUSTRATIVE EXAMPLE 6.15

In the previous examples, we have successfully integrated the four first-order equations (6.31)–(6.34). In the case of a beam subject only to constraints at either end, we must impose four end conditions. Alternatively, one can successively differentiate those four equations to obtain one fourth-order equation:

$$\frac{\mathrm{d}^4 w}{\mathrm{d}x^4} = \frac{\mathrm{d}^3\theta}{\mathrm{d}x^3} = \frac{1}{EI}\frac{\mathrm{d}^2 M}{\mathrm{d}x} = -\frac{1}{EI}\frac{\mathrm{d}V}{\mathrm{d}x} = -\frac{1}{EI}p(x) \tag{6.52a}$$

This form is needed whenever the imposed loading depends upon the displacement $w(x)$. For example, a beam might be supported by an elastic foundation that exerts a force $-kw(x)$ resisting the displacement. Then, the governing equation follows:

$$\frac{d^4w}{dx^4} = \frac{k}{EI}w = -\frac{1}{EI}p(x) \tag{6.52b}$$

Depending on the end constraints, four end conditions must be imposed upon the rotation and/or moment and/or shear force. These are expressed in terms of the function $w(x)$:

$$\theta = \frac{dw}{dx}, \quad M = EI\frac{d^2w}{dx^2}, \quad V = -EI\frac{d^3w}{dx^3} \tag{6.53}$$

We return to a fourth-order equation to examine the buckling of columns in Section 6.9. For the interested reader, we record the homogeneous solution of Equation (6.52):

$$w_H = A_1 \sin\lambda x + A_2 \cos\lambda x + A_3 \cosh\lambda x + A_4 \sinh\lambda x$$

where $\lambda^4 = k/4EI$. The function w_H must be augmented with the particular solution to accommodate the load $p(x)$. Constants (A_1, A_2, A_3, A_4) are determined by the conditions at the ends.

ILLUSTRATIVE EXAMPLE 6.16

The determination of the deflection is not as straightforward if concentrated loads act at *intermediate* locations on a beam. For example, consider the simply supported beam subjected to the concentrated force as shown in Figure 6.27. the load $p(x)$ is zero everywhere but at the location $x = a$. The differential equations, Equations (6.31)–(6.34), apply in the interval $0 < x < a$ and in the interval $a < x < L$, but a discontinuity in shear occurs at point $x = a$. The expressions for each of the variables (V, M, θ, w) are different in the two segments. Accordingly, we identify the expressions in the first $(0 < x < a)$ by the subscript 1 and in the second $(a < x < L)$ by the subscript 2. Integrating successfully the four equations and using the conditions $w(0) = M(0)$ in the first, we obtain in $0 < x < a$:

$$V_1(x) = V_0 \tag{6.54a}$$

$$M_1(x) = -V_0 x \tag{6.54b}$$

$$EI\theta_1(x) = EI\theta_0 - V_0\frac{x^2}{2} \tag{6.54c}$$

$$EIw_1(x) = EI\theta_0 x - V_0\frac{x^3}{6} \tag{6.54d}$$

In the interval $(a < x < L)$ our lower limit is $x = a$, where we denote the values by $V_2(a)$, $M_2(a)$, $\theta_2(a)$, and $w_2(a)$. Then the successive integrations lead to the expressions in $a < x < L$:

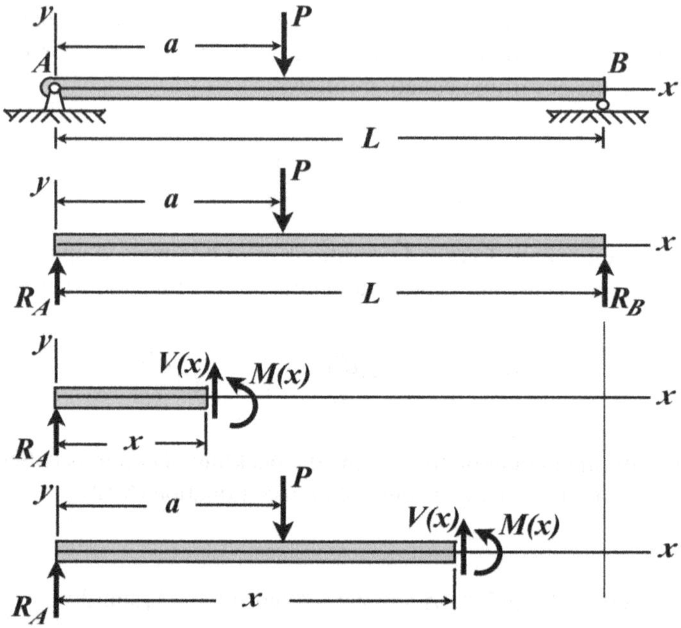

Figure 6.27 Effects of a Concentrated Load

$$V_2(x) = V_2(a) \tag{6.55a}$$

$$M_2(x) = M_2(a) - V_2(a)(x-a) \tag{6.55b}$$

$$EI\theta_2(x) = EI\theta_2(a) + M_2(a)(x-a) - V_2(a)\frac{(x-a)^2}{2} \tag{6.55c}$$

$$EIw_2(x) = EIw_2(a) + EI\theta_2(a)(x-a) + M_2(a)\frac{(x-a)^2}{2} - V_2(a)\frac{(x-a)^3}{6} \tag{6.55d}$$

The complete deflection is given by the two Equations (6.54d) and (6.55d), which contain six unknown constants, V_0, θ_0, $V_2(a)$, $M_2(a)$, $\theta_2(a)$, and $w_2(a)$. We presume that the load (P) and properties (E, I) are known.

As before, the two conditions at the right end provide two linear algebraic equations in the unknown constants:

$$M_2(L) = 0, \; w_2(L) = 0 \tag{6.56a,b}$$

Four additional equations are needed. These are obtained by acknowledging the conditions of continuity or discontinuity, which must hold at the location of the concentrated load ($x = a$).

Firstly, let us consider the dynamic requirements at concentrated loads, as set forth in Section 6.2. By Equation (6.10), the shear force must exhibit a jump at the concentrated load; therefore, we have

$$V_2(a) - V_1(a) = P \tag{6.56c}$$

Since there is no concentrated couple, the moment is everywhere continuous; therefore,

$$M_2(a) = M_1(a) \tag{6.56d}$$

Secondly, let us consider the kinematic requirements. Underlying all our work has been the assumption that the materials are continuous and cohesive media. The distinguishing characteristic of a cohesive medium is that lines and surfaces of particles in the body cannot be torn apart or made to pass through one another. Then the deflection of the neutral axis to either side of the concentrated load must be the same; otherwise, we will have a break in the neutral axis. Furthermore, the rotation of the neutral axis to either side of the concentrated load must be the same. This follows from our assumption that cross sections remain perpendicular to the neutral axis. If there were a corner in the deformed shape of the neutral axis, then two cross sections near this corner would have to pass through each other, as illustrated in Figure 6.28.

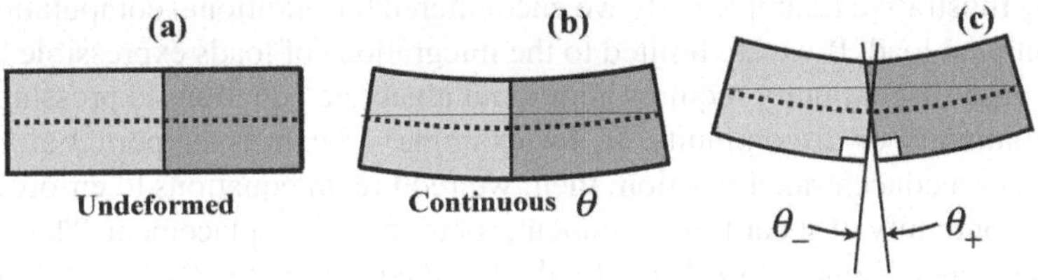

Figure 6.28

The required continuity of rotation θ and deflection w provides the remaining equations:

$$\theta_2(a) = \theta_1(a), \; w_2(a) = w_1(a) \tag{6.56e,f}$$

With the expressions in Equations (6.54) and (6.55) for the variables in the respective segments, the six Equations (6.56) assume the forms:

$$M_2(L) = M_2(a) - V_2(a)(L-a) = 0$$

$$EIw_2(L) = EIw_2(a) + EI\theta_2(a)(L-a) + M_2(a)\frac{(L-a)^2}{2} - V_2(a)\frac{(L-a)^3}{6} = 0$$

$$V_2(a) = P + V_1(a) = P + V_0$$

$$M_2(a) = M_1(a) = -V_0 a$$

$$\theta_2(a) = \theta_1(a) = \theta_0 - \frac{V_0 a^2}{2EI}$$

$$w_2(a) = w_1(a) = \theta_0 a - \frac{V_0 a^3}{6EI}$$

The solution follows:

$$V_1 = V_0 = -\frac{P(L-a)}{L}, \quad V_2 = \frac{Pa}{L}$$

$$M_1(a) = M_2(a) = +\frac{P(L-a)a}{L}$$

$$\theta_0 = -\frac{Pa(L-a)(2L-a)}{6EIL}, \quad \theta_1(a) = \theta_2(a) = -\frac{Pa(L-a)(L-2a)}{3LEI}$$

$$w_2(a) = w_1(a) = -\frac{Pa^2(L-a)^2}{3LEI}$$

The greatest deflection occurs where the rotation is zero. This is generally *not* under the load. In fact, $\theta(a)$ vanishes only if the load acts at the middle $(x = L/2)$.

This example illustrates an apparent complication when concentrated loads act at intermediate locations. By the procedure described, each load requires an additional expression for each of the four variables and four additional constants, which are determined by the four conditions of continuity and/or discontinuity at the site of each concentrated load. Operational procedures are available to simplify the treatment of such problems. These mathematical and procedural tools are presented in Section 6.6.

6.6 CONCENTRATED AND DISCONTINUOUS LOADS

In the preceding Illustrative Example 6.16, we encountered the additional computations associated with the concentrated load. If we are limited to the integrations of loads expressible by continuous functions, then each discontinuity requires additional algebraic equations expressing the requisite conditions of continuity or discontinuity. If, for example, a beam is supported at n intermediate sites, each imposes a concentrated reaction; then, we require $4n$ equations to enforce the requisite continuity or discontinuity of shear force, moment, rotation, and displacement. The solution would require the simultaneous solution of $4(n+1)$ algebraic equations to satisfy the 4 end conditions plus $4n$ conditions of continuity/discontinuity at each intermediate support.

The Figure 6.29 depicts four typical discontinuous loadings: (a) concentrated couple, (b) concentrated force, (c) abruptly applied uniform loading, and (d) abruptly applied, but variable loading.

In any treatment of "concentrated loads," forces, or couples, we must acknowledge an imprecise description and the need for a refined analysis to describe deformations and stresses in the immediate neighborhood (see Section 4.19 - Stress Concentration). With that in mind, let us consider suitable representations and their integrals; these maintain the requisite continuities and simplify the solutions:

From a practical point of view, the concentrated load is actually distributed about a point $(x = a)$ over some small region; let us say that the width is 2ε. Indeed, the concept of the concentrated load is often employed because the actual distribution is unknown and/or because the behavior near the load is not required. What is relevant is the total load and the fact that the region is small $(2\varepsilon \ll L)$. Mathematically, if the resultant load is just a force of *unit* magnitude, then

$$\int_{x<a-\varepsilon}^{x>a+\varepsilon} p(x)\mathrm{d}x = 1$$

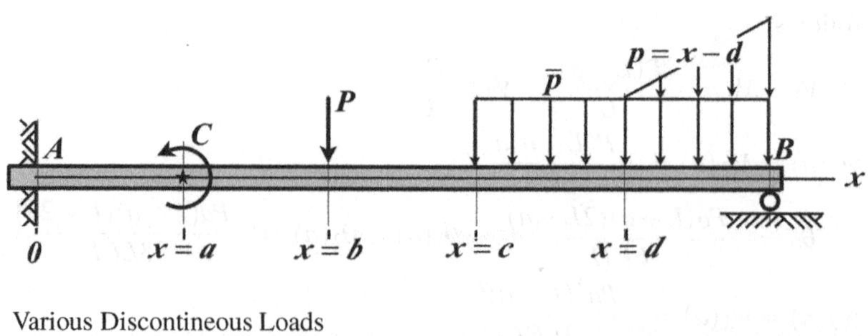

Figure 6.29 Various Discontineous Loads

Likewise, if the resultant is only a (counterclockwise) couple of *unit* magnitude, then

$$\int_{x<a-\varepsilon}^{x>a+\varepsilon} p(x)\mathrm{d}x = 0$$

and

$$\int_{x<a-\varepsilon}^{x>a+\varepsilon} (a-x)p(x)\mathrm{d}x = 1$$

Any distributions that satisfy these requirements are therefore acceptable. Distributions that satisfy the requirements for the concentrated couple and force, respectively, are plotted in Figure 6.30a and b.

The constant force $(+1/\varepsilon^2)$ over the region $a-\varepsilon < x < a$ and the constant force $(-1/\varepsilon^2)$ over the region $a < x < a+\varepsilon$ in Figure 6.30a have a resultant that is a simple couple (one-unit magnitude). The net force is zero.

The pyramid-like distribution of Figure 6.30b has a resultant that is a simple force (one-unit magnitude). The net moment is zero. Clearly, these distributions satisfy our requirements for a concentrated couple and force, respectively. Moreover, each is represented by a simple, albeit discontinuous function. We can integrate them. To simplify the matter, bear in mind that the integral of a function can be interpreted graphically as the area under the plotted function.

Let us begin with the function plotted in Figure 6.30a, which represents a concentrated couple of unit magnitude. Let's call it $C\langle x-a\rangle$. The bracketed argument $\langle x-a\rangle$ is simply a signal that the action is at $x=a$, where the argument $(x-a)$ vanishes; the letter C helps us remember that it is a couple. For all $x < a-\varepsilon$, the integral is zero. From $x=a-\varepsilon$ to $x=a$, the integral (area under

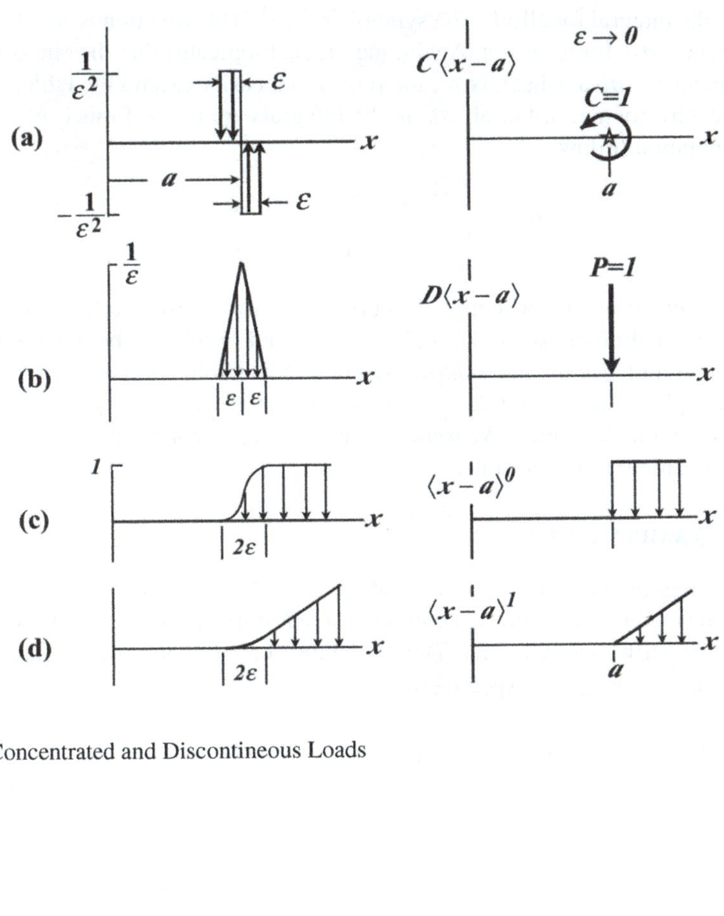

Figure 6.30 Concentrated and Discontineous Loads

the curve) grows linearly, *exactly* as the function plotted in Figure 6.30b. Moreover, at $x = a$ the integral reaches the value $1/\varepsilon$, *exactly* the value plotted in Figure 6.30b. From $x = a$ to $x = a + \varepsilon$, the integral of $C\langle x-a\rangle$ decreases linearly to zero at $x = a + \varepsilon$, *exactly* as the function of Figure 6.30b. Subsequently, for all $x > a + \varepsilon$, the integral of $C\langle x-a\rangle$ remains zero; the area under the entire plot is zero. Indeed, we have just observed that the integral of $C\langle x-a\rangle$ is the pyramidlike function of Figure 6.30b. Let's call it $D\langle x-a\rangle$ (D for Δ).

The "delta" function $D\langle x-a\rangle$ of Figure 6.30b represents our concentrated force acting about $x = a$. Again, the bracketed argument $\langle x-a\rangle$ is merely a signal to indicate that the action is at $x = a$. Now, the integral of the function $D\langle x-a\rangle$ is zero for all $x < a - \varepsilon$. On integrating to any $x > a + \varepsilon$, the value of the integral climbs to unity, and the area under the delta is the triangular area $\frac{1}{2} \times (2\varepsilon) \times (1/\varepsilon) = 1$. For all $x > a + \varepsilon$, the value remains unity. The plot of that function is the step shown in Figure 6.30c. Let's call this function $\langle x-a\rangle^0$. Again, the bracketed argument signals the location of the step. For $x > a$ the function is $(x-a)^0 = 1$.

The functions $C\langle x-a\rangle$, $D\langle x-a\rangle$, and $\langle x-a\rangle^0$ are listed in Table 6.1 along with their plots. This is a table of integrals. We have shown that $\langle x-a\rangle^0$ is the integral of $D\langle x-a\rangle$, which is the integral of $C\langle x-a\rangle$. Now, let's extend that table. But first note that the magnitude of ε is irrelevant for our purposes, so that in our further examination we can neglect ε compared with distances such as a. For example, the step of Figure 6.30c can be treated as a sharp jump at $x = a$ rather than an ascent from $x = a - \varepsilon$ to $x = a + \varepsilon$. The function $\langle x-a\rangle^0$ is zero for all $x < a$, and its integral is also zero. For all $x > a$,

$$\int_0^{x>a} \langle x-a\rangle^0 \mathrm{d}x = \int_a^{x>a} 1\mathrm{d}x = (x-a)$$

Let us denote the integral for all x by the symbol $\langle x-a\rangle^1$. This function is zero for all $x < a$, and it is the function $(x-a)^1$ for all $x > a$. Again, the bracket indicates that the change occurs where $x = a$ (the argument $x - a$ vanishes). Now, the reader can easily extend the table. All subsequent integrals are also zero for $x < a$. For all $x > a$, the integrals are of the form $1/n!(x-a)^n$. Indeed, the appropriate notation follows:

$$\frac{1}{n!}\langle x-a\rangle = \begin{cases} 0, & x < a \\ \dfrac{1}{n!}(x-a)^n, & x > a \end{cases}$$

With the representation of concentrated couples and forces by the functions $C\langle x-a\rangle$ and $D\langle x-a\rangle$, respectively, Equations (6.31), (6.32), (6.33), and (6.44) can be integrated. The correct continuity or discontinuity of the integrals or their derivatives follows. The determination of shear force, bending couple, rotation, and deflection requires no more effort than the analyses for the more familiar continuous functions. We need only use the representations, notations, and integrals of Table 6.1. An example illustrates this.

ILLUSTRATIVE EXAMPLE 6.17

To illustrate the foregoing representations and integrations, let us consider a beam loaded as shown in Figure 6.31. The load includes the concentrated force R imposed by the simple support at $x = 3a$ and the step in the distribution at $x = 2a$: These are represented by the "delta" function $D\langle x-3a\rangle$ and the "step" function $\langle x-a\rangle^0$; respectively:

$$p(x) = 2p - p\langle x-2a\rangle^0 - R\cdot D\langle x-3a\rangle$$

A free body diagram of the entire beam is shown to identify the reactions, V_0 and M_0 at A and R at C. Successive integrations of Equations (6.31)–(6.34) contain the three reactions:

$$V = V_0 + 2px - p\langle x-2a\rangle^1 - R\langle x-3a\rangle^0$$

$$M = M_0 - V_0 x - px^2 + \frac{1}{2}p\langle x-2a\rangle^2 + R\langle x-3a\rangle^1$$

$$EI\theta = M_0 x - \frac{1}{2}V_0 x^2 - \frac{1}{3}px^3 - \frac{1}{6}p\langle x-2a\rangle^3 - \frac{1}{2}R\langle x-3a\rangle^2$$

$$EIw = \frac{1}{2}M_0 x^2 - \frac{1}{6}V_0 x^3 - \frac{1}{12}px^4 - \frac{1}{24}p\langle x-2a\rangle^4 - \frac{1}{6}R\langle x-3a\rangle^3$$

The solution requires the determination of constants M_0, V_0, and R by enforcing the conditions at the end ($x = 4a$) and the support ($x = 3a$). Note that the conditions at the fixed end ($x = 0$,

Table 6.1
Function-based representation of force systems

Label	Plot	Integral
$C\langle x-a\rangle$	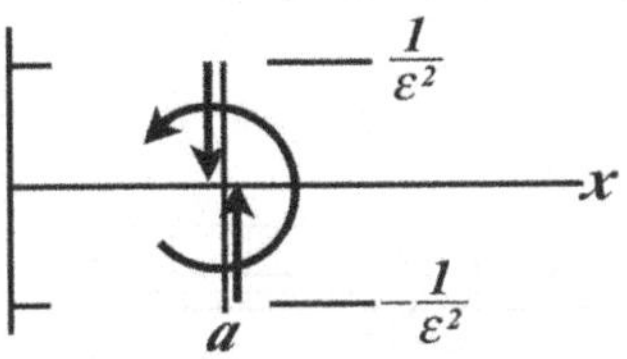	
$D\langle x-a\rangle$	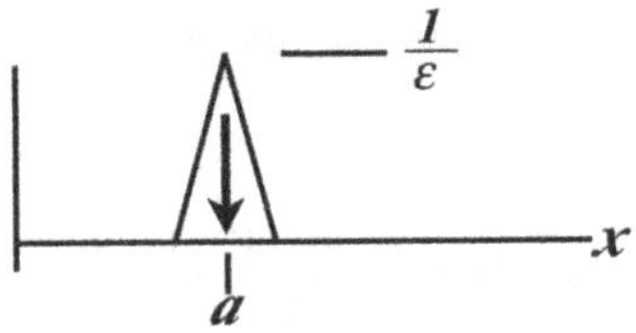	$D\langle x-a\rangle = \int_0^x C\langle x-a\rangle dx$
$\langle x-a\rangle^0$	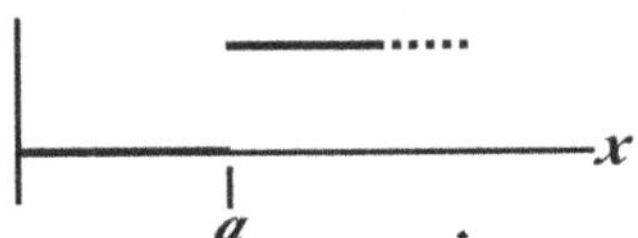	$\langle x-a\rangle^0 = \int_0^x D\langle x-a\rangle dx$
$\langle x-a\rangle^1$	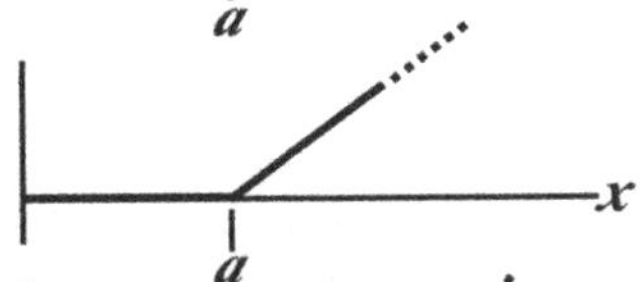	$\langle x-a\rangle^1 = \int_0^x \langle x-a\rangle^0 dx$
$\frac{1}{2}\langle x-a\rangle^2$	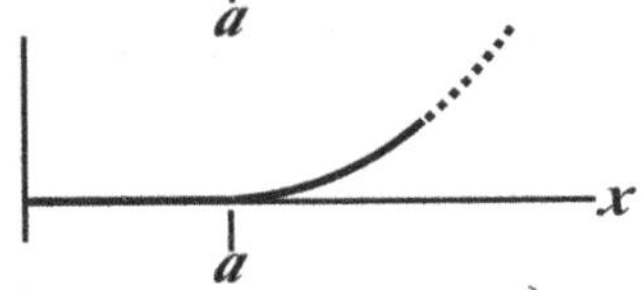	$\frac{1}{2}\langle x-a\rangle^2 = \int_0^x \langle x-a\rangle^1 dx$
$\frac{1}{6}\langle x-a\rangle^3$	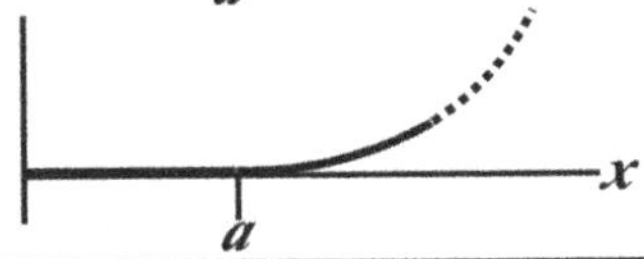	$\frac{1}{6}\langle x-a\rangle^3 = \int_0^x \frac{1}{2}\langle x-a\rangle^2 dx$

$\theta = w = 0$) were introduced at the lower limits of our integrals. The three remaining conditions follow:

$$M(4a) = 0 = M_0 - 4aV_0 + aR - 14a^2 p$$
$$V(4a) = 0 = V_0 - R + 6ap$$
$$w(3a) = 0 = \frac{9}{2}a^2 M_0 - \frac{9}{2}a^3 V_0 - \frac{1}{24}(162 - 1)a^4 p$$

Note that the second equation $[V(4a) = 0]$ is the condition for equilibrium of the entire beam (Figure 6.31b). It results from integrating the equilibrium condition of a slice (width dx).

The simultaneous solution provides the result:

$$V_0 = -\frac{351.5}{108}ap = -3.26ap$$
$$M_0 = -\frac{190.5}{108}a^2 p = -1.76a^2 p$$
$$R = \frac{296.5}{108}ap = 2.75ap$$

Had we not employed the functions and integrals of Table 6.1, we would be confronted with the additional conditions and continuity at B and at C and the discontinuity of $V(x)$ at C.

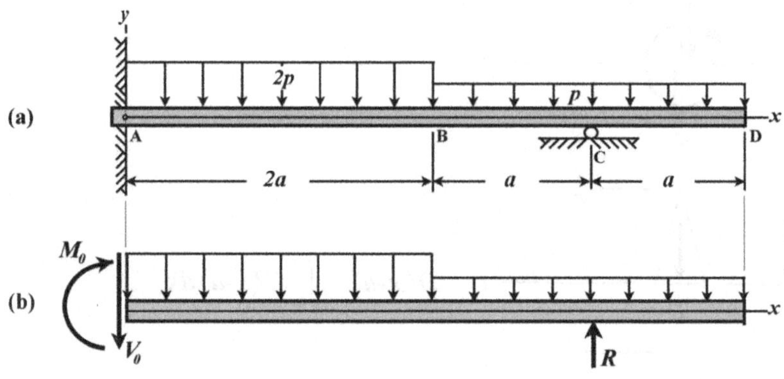

Figure 6.31 Beam Subject to Discontinuous Loading

6.7 DEFLECTION DUE TO SHEAR

The Bernoulli-Euler theory of Hookean beams ($\kappa = M/EI$) neglects the transverse shear strain (γ_{xy}). Actual beams under transverse loads are subjected to shear force V and stress τ_{xy}; the stress in a HI-HO beam is estimated, according to Equation (6.20):

$$\tau = \frac{VQ(y)}{It}$$

The corresponding strain in the HI-HO beam is

$$\bar{\gamma}_{xy} = \frac{VQ(y)}{GIt}$$

The shear stress and the strain vanish at the top and bottom and are maximum at the centroidal axis:

$$\bar{\gamma}_{xy}(0) \equiv \bar{\gamma}_0 = \frac{VQ_0}{GIt} \tag{6.57}$$

where $Q_0 = Q(O)$.

From Figure 6.32, it is evident that the shear strain (which is assumed to be small) causes additional deflection of the neutral axis.

$$\Delta w_s = \bar{\gamma}_0 \Delta x$$

or in the limit

$$\frac{dw_s}{dx} = \bar{\gamma}_0$$

In view of Equation (6.57), the *additional* rotation due to shear is (approximately)

$$\frac{dw_s}{dx} \doteq \frac{V(x)Q_0}{GIt} \tag{6.58}$$

The additional deflection can now be computed by integrating Equation (6.58). This additional deflection is generally quite small compared to that caused by bending and is usually neglected. The shear deflection is significant when the beam is relatively deep compared to its length. Even then the accuracy of such an estimate is dubious, since the basis neglected the shear strain. Nevertheless, in the absence of a more precise analysis, such a calculation provides a correction to the deflection equation when the shear is not insignificant.

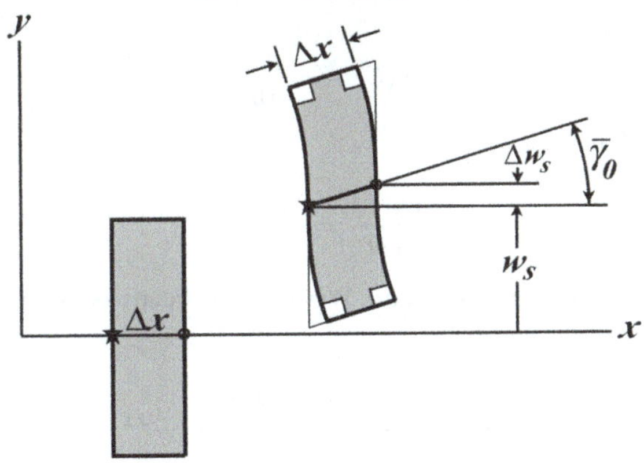

Figure 6.32 Transverse Shear Deformation

ILLUSTRATIVE EXAMPLE 6.18

As an illustration, consider the cantilever shown in Figure 6.33. The same beam and loading used in Illustrative Example 6.9 are used here, but without the support at the right end.

$$V = -\frac{\bar{p}L}{2}\left(1 - \frac{x}{L}\right)^2$$

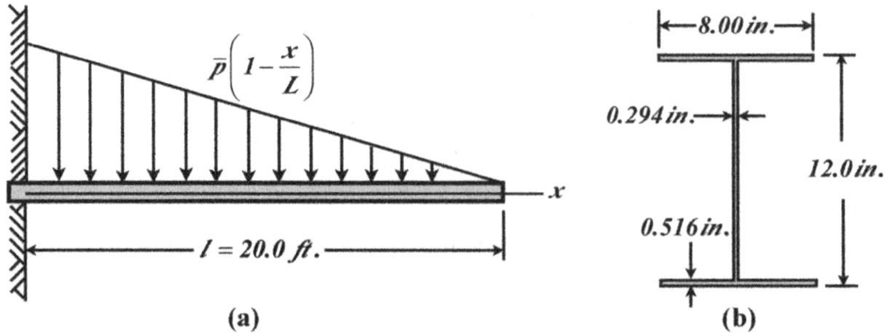

Figure 6.33

Then, by Equation (6.58)

$$\frac{dw_s}{dx} = -\frac{Q_0}{GIt}\frac{\bar{p}L}{2}\left(1-\frac{x}{L}\right)^2$$

Integrating, we obtain

$$w_s = +\frac{Q_0\bar{p}L^2}{GIt6}\left[\left(1-\frac{x}{L}\right)^3 - 1\right]$$

At the end $x = L$,

$$w_s(L) = -\frac{Q_0\bar{p}L^2}{6GIt}$$

The deflection caused by bending, according to the Bernoulli-Euler theory, is

$$w_B(L) = -\frac{\bar{p}L^4}{30EI}$$

The ratio at the end $x = L$ is

$$\frac{w_s(L)}{w_B(L)} = \frac{5EQ_0}{GtL^2}$$

The cross section of the American Standard WF (12-40) beam is shown in Figure 6.33b. The integral Q_0 follows:

$$Q_0 = (8 \times 0.516 \times 5.742) + \frac{(5.484)^2}{2} \times (.294) = 28.1 \text{ in.}^3$$

For this steel beam,

$$I = 310 \text{ in.}^4, \quad t = 0.294 \text{ in.}, \quad \frac{E}{G} \doteq 2.50, \quad L = 20.0 \text{ ft}$$

The ratio follows:

$$\frac{w_s(L)}{w_B(L)} = 0.0207$$

We see that the deflection due to the shear is approximately 2%.

6.8 LIMIT LOADS FOR IDEALLY PLASTIC BEAMS

In Section 5.14.1, we considered the behavior of a beam composed of an ideally plastic material when subjected to simple bending. On the assumption that plane cross sections remain plane and perpendicular to the neutral axis, we obtained a relationship between the bending couple and the curvature. A distinctive feature of this relation is the existence of a limiting condition, which occurs when the bending couple attains a value M_L. The beam cannot support a greater moment, since the action of this couple causes unrestricted rotations of nearby cross sections.

We now apply this concept to ideally plastic beams subjected to transverse loads, ignoring the effects of shear. When the bending couple at any cross section of the beam reaches the limiting value, a plastic "hinge" forms, and unrestricted rotation can occur. Here, we make no attempt to calculate the deflection after yielding but concern ourselves with the load-carrying capacity of the beam as it is limited by the occurrence of plastic hinges and subsequent plastic collapse. The method of calculating the limit load for an ideally plastic beam is illustrated by the following examples.

ILLUSTRATIVE EXAMPLE 6.19

The cantilever beam of Figure 6.34a is made of structural steel, which can be regarded as ideally plastic with $\sigma_0 = 40,000$ lb/in.^{2} (the same magnitude in tension and in compression). Its cross section is shown in Figure 6.35b). A single concentrated load P is applied to the end of the beam so that the bending couple on any cross section is given by

$$M(x) = -P(L-x)$$

For a given load, the greatest bending couple occurs at $x = 0$. We therefore expect a plastic hinge to form there when $M(O) = -M_L$.

To calculate M_L, we suppose that the cross section is in the fully plastic state; then the stress is distributed as illustrated in Figure 6.34c. The entire cross section has yielded; the portion above the neutral axis is subjected to the normal stress $\sigma = +\sigma_0$, whereas the remainder is acted upon by the stress $\sigma = -\sigma_0$.

The position of the neutral axis is determined by the condition that the normal force on the cross-section is zero.

$$F = \iint_a \sigma_x dA = 0$$

If p locates the neutral axis from the bottom edge of the cross section as in Figure 6.34b, then

$$F = \sigma_0 A_2 - \sigma_0 A_1 = +\sigma_0 [4 + (5-p)] - \sigma_0 p = 0$$

so that $p = 4.5$ in. The fully plastic bending couple is then

$$M_L = -\iint_A \sigma_x y dA = +\sigma_0 \iint_{A_1} y dA - \sigma_0 \iint_{A_2} y dA$$

where A_2 and A_1 are the areas above and below the neutral axis, respectively. When these integrals are evaluated, the result is $M_L = -570,000$ in.-lb. The load, which causes this bending couple at $x = 0$, is the limit load,

$$P_L = -\frac{M_L}{L} = 4750 \text{ lb} \quad (2.1. \times 10^3 \text{ N})$$

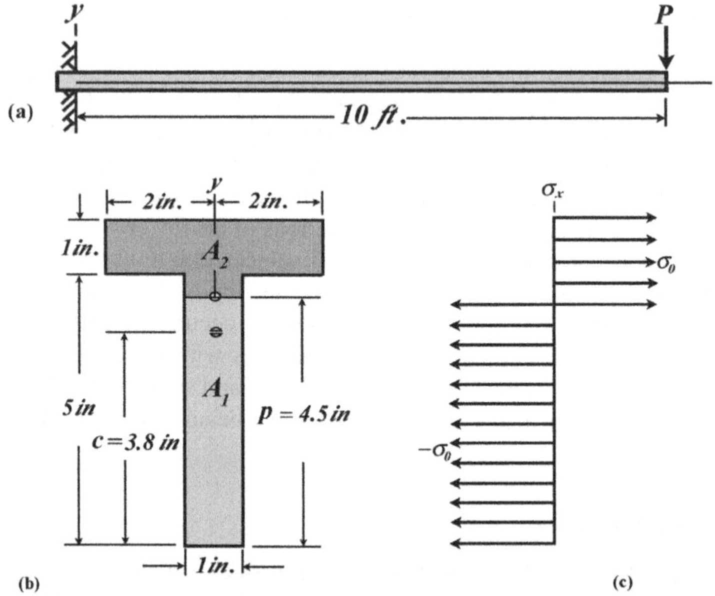

Figure 6.34

As a matter of interest, let us compute the yield couple M_0. This couple is the bending couple that initiates yielding. For HI-HO behavior, the neutral axis is at the centroid of the cross section; in Figure 6.34b, it is located at distance $c = 3.833$ in. Accordingly, the largest (absolute value) stress occurs at the bottom, where, for incipient yielding, $\sigma_0 = -M_0 c/I$. In this case $I = 30.75$ in.^{4}, so yielding is initiated when

$$M_0 = \frac{\sigma_0 I}{c} = -321{,}000 \text{ in.-lb}$$

The behavior of the beam is linearly elastic as long as the absolute value of the bending couple is everywhere less than M_0 . Where the value exceeds M_0 , plastic deformation ensues.

Since the first cross section to yield is at $x = 0$, the load that initiates yielding in the beam is

$$P_0 = -\frac{M_o}{L} = 2675 \text{ lb} \quad (11.9 \times 10^3 \text{ N})$$

As the load increases beyond P_0, the bending couple at $x = 0$ increases. When $P = P_L$, the region adjacent to the wall behaves as a plastic hinge. The beam is then, effectively, a mechanism that can rotate about the plastic hinge, as indicated in Figure 6.35. This constitutes plastic collapse. The load P_L is, therefore, the limit load.

Note that yielding is not restricted to the immediate vicinity of the hinge. In Figure 6.35, some yielding occurs throughout the region $O < x < 52.4$ in., since $M > M_0$ in that part of the beam when $P = P_L$.

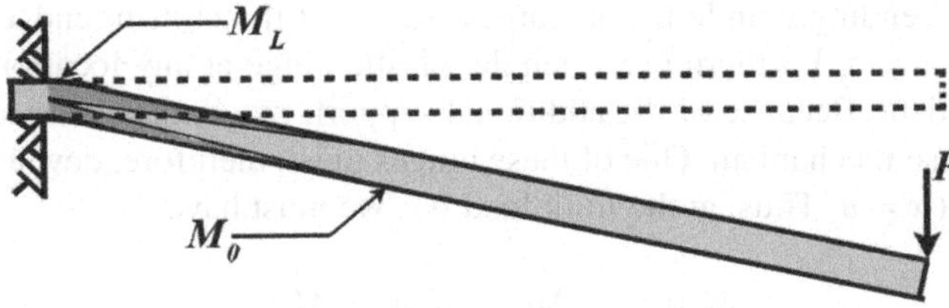

Figure 6.35

ILLUSTRATIVE EXAMPLE 6.20

Suppose that the cantilevered beam of the preceding example is provided with an additional support at the right end, as shown in Figure 6.36, and is then subjected to a uniform load W. For equilibrium of the free body in Figure 6.36b, we obtain

$$M(x) = -\frac{W}{2}(L-x)^2 + R_B(L-x) \tag{6.59}$$

where R_B is the reaction of the support at B. The bending couple diagram of Figure 6.36c indicates, in a general way, the variation of the bending couple.

At the outset, we cannot assign values because R_B remains unknown. However, it is clear from

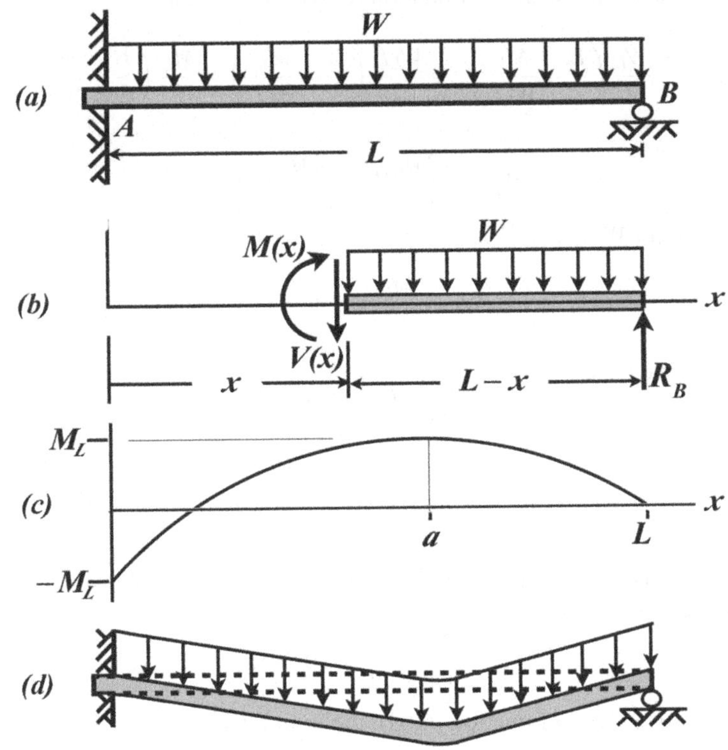

Figure 6.36 Plastic Collapse

Figure 6.36c that the bending couple has its largest values at the built-in end $x = 0$ and at some intermediate location $x = a$. Furthermore, a single, plastic hinge at any location will *not* convert the beam to a mechanism. Because of the additional support, two hinges are needed to transform the beam into a collapse mechanism. One of these hinges must, therefore, develop at the fixed end $x = 0$ and the other, at $x = a$. Thus, at the limit load W_L, we must have

$$M(0) = -M_L, \quad M(a) = M_L \tag{6.60}$$

Putting the first of these conditions in Equation (6.59) yields

$$-M_L = -\frac{W_L}{2}L^2 + R_B L$$

Solving this equation for R_B and eliminating it from Equation (6.59), we obtain (at the limit load)

$$M(x) = -\frac{W_L}{2}(L-x)^2 + \frac{W_L L}{2}(L-x) - \frac{M_L}{L}(L-x) \tag{6.61a}$$

The second plastic hinge is at the intermediate location $x = a$ where M is a maximum - that is, where $\mathrm{d}M/\mathrm{d}x = 0$. Differentiating Equation (6.61a) we have

$$\left.\frac{\mathrm{d}M}{\mathrm{d}x}\right|_{x=a} = W_L(L-a) - \frac{W_L L}{2} + \frac{M_L}{L} = 0 \Rightarrow (L-a) = \left(\frac{L}{2} - \frac{M_L}{W_L L}\right) \tag{6.61b}$$

Returning to the second of Equations (6.60), when $x = a$, M takes on the value M_L. Substituting this in Equation (6.61a), with the aid of Equation (6.61b), we obtain

$$M_L = -\frac{W_L}{2}\left(\frac{L}{2} - \frac{M_L}{W_L L}\right)^2 + \frac{W_L L}{2}\left(\frac{L}{2} - \frac{M_L}{W_L L}\right) - \frac{M_L}{L}\left(\frac{L}{2} - \frac{M_L}{W_L L}\right) \tag{6.62}$$

which reduces to

$$W_L^2 - \frac{12M_L}{L^2}W_L + \frac{4M_L^2}{L^4} = 0$$

Solving this equation for W_L gives

$$W_L = \frac{M_L}{L^2}(6 \pm 4\sqrt{2}) \tag{6.63}$$

By Equation (6.61b) we have

$$L - a = \frac{L}{2} - \frac{L}{6 \pm 4\sqrt{2}}$$

or

$$a = \frac{2 \pm \sqrt{2}}{3 \pm 2\sqrt{2}}L$$

The negative sign yields a value $a > L$, which has no physical meaning in the present problem. Therefore, the hinge is at

$$a = \frac{2+\sqrt{2}}{3+2\sqrt{2}}L = 0.585L$$

Then, taking the positive sign in Equation (6.63), the limit load is

$$W_L = (6 + 4\sqrt{2})\frac{M_L}{L^2}$$

Using the value of M_L calculated in the preceding example, we arrive finally at

$$W_L = 461 \text{ lb/in.} \quad (807 \text{ N/cm})$$

This example suggests that limit load analyses of indeterminate beams and frames can be very laborious. This is generally true because we cannot usually foretell the correct collapse mechanism - that is, the location of the plastic hinges.

6.9 BUCKLING OF A SYMMETRIC HOOKEAN COLUMN

Under compressive loading, thin members are susceptible to buckling, that is, to assuming a quite different bent configuration. When buckling occurs, the member can sustain little, if any, additional load. The thin column is a simple, but striking, example because the initial configuration is straight, and the buckled form is clearly bent.

The Hookean column of Figure 6.37 is a particular case, wherein the lower end ($x = 0$) is fixed; the upper end ($x = L$) is free but subject to the axial load P, which remains constant and vertical. The cross section is symmetric about the xy-plane, which is also the plane of least rigidity (EI) and, therefore, the plane of buckling.

To determine the critical load, which precipitates the buckling, and also the buckled form, we must examine the equilibrium conditions of such *deformed* states. The key word is *deformed*. In our earlier considerations of equilibrium (see Section 6.2, Equations 6.8 and 6.9), deformation was neglected (a slice of the beam, the free body of Figure 6.6, was undeformed). Now, we examine a deformed slice of the column, as depicted in Figure 6.38. The notations are those in our previous analyses of symmetric Hookean beams (see Section 6.2, Figures 6.8 and 6.9). Only the notations n and t are added to signify the directions, normal and tangent, to the *bent* axis x.

Equilibrium of the free body in Figure 6.38 requires that

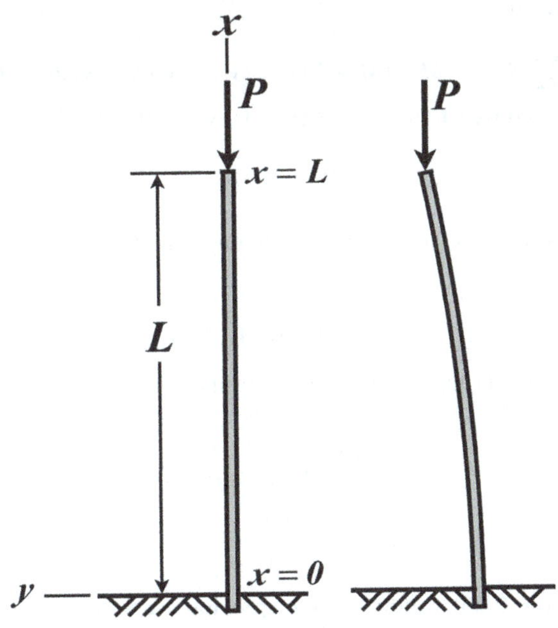

Figure 6.37

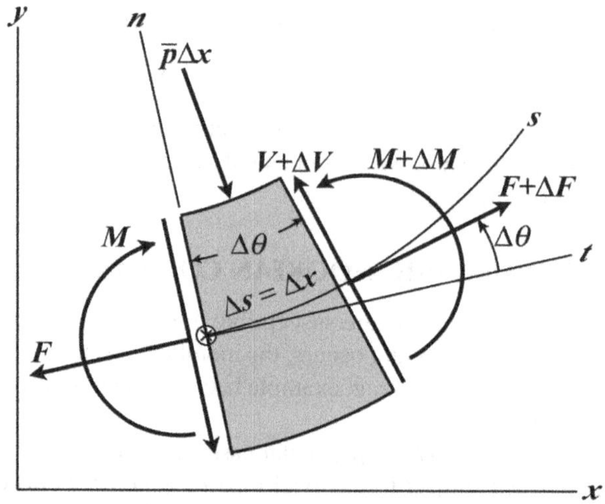

Figure 6.38 Actions on the Deformed Element

$$\sum F_n = -V + (V+\Delta V)\cos\Delta\theta + (F+\Delta F)\sin\Delta\theta - \bar{p}\cos\varepsilon\Delta\theta = 0$$

Here, the orientation of the lateral (normal) load $\bar{p}$ is intermediate, $0 < \varepsilon < 1$. As before, we divide by Δs, pass to the limit ($\Delta s \to 0$), and obtain

$$\frac{\mathrm{d}V}{\mathrm{d}s} + F\frac{\mathrm{d}\theta}{\mathrm{d}s} = p \tag{6.64}$$

Equilibrium also requires that

$$\sum M_0 \doteq -M + (M+\Delta M) + (V+\Delta V)\Delta s\cos\Delta\theta = 0$$

Again, we divide by Δs and pass to the limit ($\Delta s \to 0, \Delta\theta \to 0$) and obtain the result:

$$\frac{\mathrm{d}M}{\mathrm{d}s} = -V \tag{6.65}$$

Now, the strain of the axis (ε_0) is small compared to unity, so that we can again justify the approximation.

$$\mathrm{d}s = (1+\varepsilon_0)\mathrm{d}x \doteq \mathrm{d}x$$

Then the two Equations (6.64) and (6.65) take the forms:

$$\frac{\mathrm{d}V}{\mathrm{d}x} + F\frac{\mathrm{d}\theta}{\mathrm{d}x} = p \tag{6.66}$$

$$\frac{\mathrm{d}M}{\mathrm{d}x} = -V \tag{6.67}$$

Only the former, Equation (6.66), differs from the previous version (see Equations 6.31 and 6.32).

The column is again a symmetric Hookean beam, but the loading differs; now $p = 0$, $F \neq 0$. The moment-curvature relation, Equation (6.33), holds, namely,

$$\frac{d\theta}{ds} \doteq \frac{d\theta}{dx} = \frac{M}{EI} \tag{6.68}$$

Again, the axis must be at the centroid of the cross section. Since we intend to examine only the onset of buckling, the rotation θ remains small and the kinematic relation, Equation (6.34), prevails:

$$\frac{dw}{dx} = \theta \tag{6.69}$$

A slightly bent form of the column must satisfy four equations: the two equilibrium equations (6.66) and (6.67), the moment-curvature relation (6.68), and the kinematic equation (6.69). Only the first, Equation (6.66), is strictly nonlinear; it contains the product $F(d\theta/dx)$. However, the force F is constant ($= -P$) at the onset of buckling. Moreover, we intend to explore only equilibrium states adjacent (slightly bent) to the prebuckled (straight) state.

Accordingly, the force can be assumed constant, $F = -P$. Then the buckled state is governed by four first-order linear equations in the four variables w, θ, M, V. These must be augmented by four end conditions.

Note that our present problem differs from the previous problem of the beam (see Section 6.5) in two respects: Now, $p = 0$ and one equation, Equation (6.66), contains an additional, crucial term, $-P d\theta/dx$.

Before indulging in the mathematics of our problem, let us complete our formulation with the end conditions for the column of Figure 6.37. At the lower end, the conditions are the kinematic constraints, which prevent rotation and displacement:

$$\theta = 0, \; w = 0 \quad \text{at } x = 0 \tag{6.70a,b}$$

To understand the mechanics at the upper end, consider the displaced end shown in Figure 6.39. The only action upon the end is the force P; therefore,

$$M = 0$$
$$V = P \sin\theta$$
$$F = -P \cos\theta$$

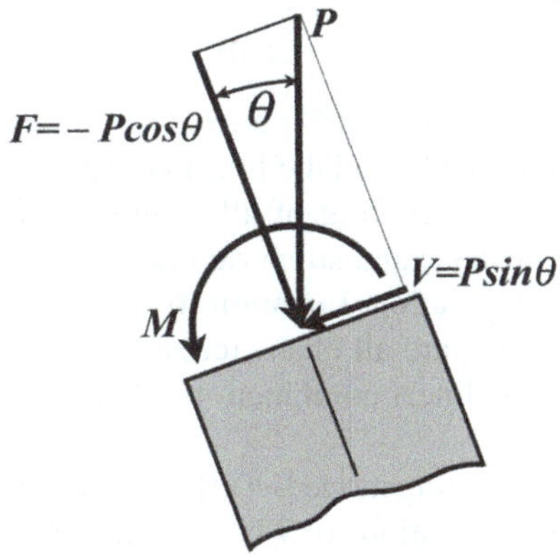

Figure 6.39 Actions on the End

Again, we note that our interests are limited to *small* rotation, so that first-order approximations are justified ($\sin\theta \doteq \theta, \cos\theta \doteq 1$). Also note that we already anticipated the final equation, $F = -P$. The remaining end conditions follow:

$$M = 0 \quad \text{at } x = L \tag{6.70c}$$

$$V = P\theta \quad \text{at } x = L \tag{6.70d}$$

Mathematically, our present problem differs from that of a beam under lateral load p. In that case, one could successively integrate the four equations and determine the four constants directly. Now, the presence of the additional term ($F\mathrm{d}\theta/\mathrm{d}x$) in Equation (6.66) couples these equations. To effect the solution, we rephrase the problem in terms of displacement w as follows: Firstly, we eliminate θ by substituting Equation (6.69) into Equation (6.68) and obtain

$$M = EI\frac{\mathrm{d}^2w}{\mathrm{d}x^2} \tag{6.71}$$

Secondly, we substitute Equation (6.71) into Equation (6.67) and obtain

$$V = -EI\frac{\mathrm{d}^3w}{\mathrm{d}x^3} \tag{6.72}$$

Note that Equations (6.69), (6.71), and (6.72) express the variables θ, M, and V in terms of displacement w. Finally, we substitute Equations (6.72), (6.69), $F = -P$, and $p = 0$, into the remaining differential equation (6.66) and obtain

$$-EI\frac{\mathrm{d}^4w}{\mathrm{d}x^4} - P\frac{\mathrm{d}^2w}{\mathrm{d}x^2} = 0, \qquad \frac{\mathrm{d}^4w}{\mathrm{d}x^4} + \left(\frac{P}{EI}\right)\frac{\mathrm{d}^2w}{\mathrm{d}x^2} = 0 \tag{6.73}$$

In like manner, we express the four end conditions, Equations (6.70a,b,c,d) in terms of the displacement w and derivatives:

$$\frac{\mathrm{d}w}{\mathrm{d}x} = 0, \quad w = 0 \qquad \text{at } x = 0 \tag{6.74a,b}$$

$$\frac{\mathrm{d}^2w}{\mathrm{d}x^2} = 0, \quad \frac{\mathrm{d}^3w}{\mathrm{d}x^3} + \frac{P}{EI}\frac{\mathrm{d}w}{\mathrm{d}x} = 0 \qquad \text{at } x = L \tag{6.74c,d}$$

Equilibrium of the slightly buckled column of Figure 6.37 is governed by the one fourth-order linear differential equation, Equation (6.73), and the four linear end conditions, Equations (6.74a–d). Certain features are common to the buckling of all symmetric Hookean columns that are uniform in their properties (EI is constant) and supported only at their ends. These problems are all governed by the same differential equation, Equation (6.73); it is homogeneous; that is, the right side is zero ($p = 0$). These problems are all characterized by four homogeneous end conditions, such as Equations (6.74a–d); certain linear combinations of displacement, w, $\mathrm{d}w/\mathrm{d}x$, $\mathrm{d}^2w/\mathrm{d}x^2$, and $\mathrm{d}^3w/\mathrm{d}x^3$, vanish at the ends.

To obtain the critical load and buckling mode [$w(x)$], we require the solution of Equation (6.73), which satisfies the end conditions, Equations (6.74a–d). The general solution is

$$w = A\sin\lambda x + B\cos\lambda x + Cx + D \tag{6.75}$$

Derivatives $(\mathrm{d}w/\mathrm{d}x, \mathrm{d}^2w/\mathrm{d}x^2, \mathrm{d}^3w/\mathrm{d}x^3)$ are required in the end conditions. The fourth derivative $(\mathrm{d}^4w/\mathrm{d}x^4)$ appears in the differential equation. These results follow:

$$\frac{\mathrm{d}w}{\mathrm{d}x} = \lambda A\cos\lambda x - \lambda B\sin\lambda x + C \tag{6.76a}$$

$$\frac{\mathrm{d}^2w}{\mathrm{d}x^2} = -\lambda^2 A\sin\lambda x - \lambda^2 B\cos\lambda x \tag{6.76b}$$

$$\frac{\mathrm{d}^3w}{\mathrm{d}x^3} = -\lambda^3 A\cos\lambda x + \lambda^3 B\sin\lambda x \tag{6.76c}$$

$$\frac{\mathrm{d}^4w}{\mathrm{d}x^4} = \lambda^4 A\sin\lambda x + \lambda^4 B\cos\lambda x \tag{6.76d}$$

When Equations (6.76b,d) are substituted into the differential equation, we find that the equation is satisfied everywhere if and only if

$$\lambda^2 = \frac{P}{EI}, \quad \lambda = \sqrt{\frac{P}{EI}} \tag{6.77}$$

In the particular case at hand, Figure 6.37, the end conditions, Equations (6.74a–d) take the form of four linear homogeneous algebraic equations:

$$\begin{array}{lllll} (\lambda)A & & +C & & = 0 \\ & B & & +D & = 0 \\ -(\lambda^2\sin\lambda L)A - & (\lambda^2\cos\lambda L)B & & & = 0 \\ & & \lambda^2 C & & = 0 \end{array} \tag{6.78a,b,c,d}$$

Every problem of this kind (buckling of a uniform Hookean column with end supports) reduces to a linear homogeneous algebraic system, such as Equations (6.78a–d), in terms of the four constants A, B, C, and D and the parameter $\lambda(=\sqrt{P/EI})$. The homogeneous equations have a nontrivial solution ($A = B = C = D = 0$ is trivial, the straight configuration) if and only if the determinant of the coefficients vanishes; that is,

$$\begin{vmatrix} \lambda & 0 & 1 & 0 \\ 0 & 1 & 0 & 1 \\ (-\lambda^2\sin\lambda L) & (-\lambda^2\cos\lambda L) & 0 & 0 \\ 0 & 0 & \lambda^2 & 0 \end{vmatrix} = 0 \tag{6.79}$$

Of course $\lambda = 0$, $(P = 0)$ is also trivial; only positive values are relevant. We can, therefore, factor -λ^2 from the third row and λ^2 from the fourth. We follow the rules for expanding a determinant to obtain an equation in the parameter $\lambda(\sqrt{P/EI})$.

Expanding this determinant, we obtain

$$\cos\lambda L = 0 \tag{6.80}$$

The nontrivial roots are

$$\lambda L = \frac{\pi}{2}, \quad \frac{3\pi}{2}, \ldots, \frac{(2n-1)\pi}{2} \tag{6.81}$$

The only important root is the first, which gives the least load that admits a buckled configuration; that root gives the critical load

$$P_{CR} = \frac{\pi^2}{4}\frac{EI}{L^2} \tag{6.82}$$

Other roots characterize unstable states of equilibrium, which cannot be realized.

The buckled mode is determined by the four constants that must satisfy Equations (6.78a,b,c,d); in this case

$$A = C = 0, \quad B = -D$$

Hence the form (mode) of the buckled column is

$$w(x) = D(1 - \cos\lambda x) \tag{6.83}$$

Note that the amplitude of the mode (D) remains unknown. This is always true in the solution of the linearized formulation.

It is never enough to achieve a mathematical result in engineering. We must examine the implications and limitations. Figure 6.40a depicts the column and a dial gage arranged to measure lateral deflection $w(L)$. Our theory suggests that the straight column sustains load P in the straight configuration OA $[w(L) = O]$ in Figure 6.40b and then encounters the bifurcation A at the critical load and assumes the buckled form with unrestricted amplitude.

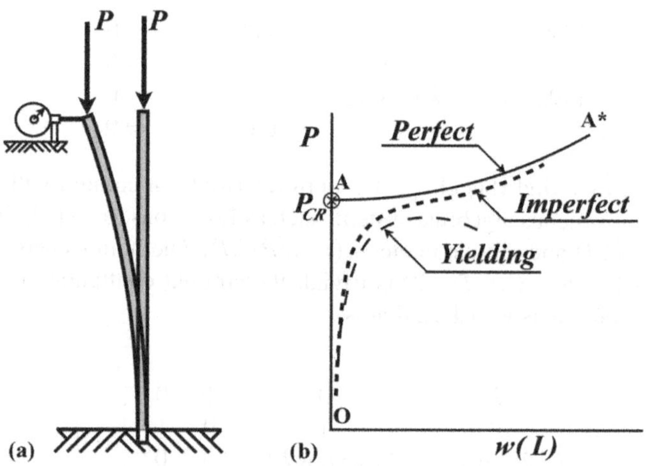

Figure 6.40 Post-critical Response

In reality, as the deflection increases, nonlinearities must be considered ($\sin\theta \neq \theta$) and *slight* additional load is required to cause further deflection. The trace for a perfect column is AA^*. Of course, no column is perfect; all possess some initial crookedness. Also, it is impossible to align the load perfectly. Therefore, the actual imperfect column experiences some bending at the onset of loading. That deflection grows and becomes excessively large at the critical load; this behavior is indicated by the dotted curve of Figure 6.40b. From a practical viewpoint, the critical load signals the useful limit, since deflection becomes excessive with very little additional loading; P remains practically constant at the critical load P_{CR}. In many instances, buckling is immediately followed by yielding of the material (e.g., structural steel) and collapse ensues.

6.9.1 EFFECT OF IMPERFECTIONS

To gain some insight into the consequences of initial imperfection, let us suppose that the column of Figure 6.37 is initially slightly bent. For simplicity, let us suppose an initial deflection $w_o(x)$ in the form of the buckling mode.

$$w_0(x) = \bar{w}\left(1 - \cos\frac{\pi x}{2L}\right) \tag{6.84}$$

Let $w(x)$ denote any additional deflection, so that the actual bent form has the net deflection.

$$w_1(x) = w_0(x) + w(x) \tag{6.85}$$

The deflection w_1 is related to the net rotation θ_1 according to Equation (6 69):

$$\frac{dw_1}{dx} = \theta_1 = \frac{dw_0}{dx} + \frac{dw}{dx} \tag{6.86}$$

Neither couple M nor force V act initially. The former is proportional to the *change* of curvature associated with the additional deflection. Therefore, relations of couple and force are like Equations (6.71) and (6.72):

$$M = EI\frac{d^2w}{dx^2}, \quad V = -EI\frac{d^3w}{dx^3} \tag{6.87a}$$

As before, we substitute Equations (6.86) and (6.87b) into the equilibrium equation (6.66) and set $p = 0$ and $F = -P$ to obtain

$$-EI\frac{d^4w}{dx^4} - P\frac{d^2w}{dx^2} = P\frac{d^2w_0}{dx^2} \tag{6.88a}$$

Now, our initial deflection w_0 has the form of Equation (6.84); therefore, Equation (6.88a) takes the following form:

$$\frac{d^4w}{dx^4} + \frac{P}{EI}\frac{d^2w}{dx^2} = -\frac{\pi^2}{4L^2}\frac{P}{EI}\bar{w}\cos\frac{\pi x}{2L} \tag{6.88b}$$

This equation differs from Equation (6.73) by virtue of the "loading" on the right side. We require the appropriate particular solution to augment the homogeneous solution, Equation (6.75). The particular solution, in this special case, also satisfies the end conditions and, therefore, constitutes the complete solution:

$$w(x) = w_p(x) = \frac{\bar{w}}{\left(\frac{\pi^2 EI}{4L^2 P} - 1\right)}\left(1 - \cos\frac{\pi x}{2L}\right) \tag{6.89}$$

Note that the deflection becomes unbounded as $P \to P_{CR} = \pi^2 EI/4L^2$. Again, as the deflection grows, the error of our linear theory increases ($\sin\theta \neq \theta$); then the nonlinearities must be accommodated, and the refined theory shows that some additional load can be sustained *but* only as the deflection becomes excessive. Such severe deflection is usually accompanied by yielding or other failure of the material.

6.9.2 SIGNIFICANCE OF HIGHER MODES

Our solution to the problem of Figure 6.37 indicates an infinity of equilibrium states $[w = D(l - \cos \lambda x)]$ under the loads $\lambda = (2n - l)\pi/2L$. As noted, only the first mode at the least load ($P_{CR} = EI\lambda^2, n = 1$) can be realized; the higher modes are unstable. To examine the question, let us consider yet another example, the column of Figure 6.41, which is simply supported at both ends:

$$w = \frac{\mathrm{d}^2 w}{\mathrm{d}x^2} = 0 \quad \text{at } x = 0, L$$

The solutions in this case are

$$w_n = A \sin \lambda_n x, \quad \lambda_n = n\pi/L, \quad n = 1, 2, \ldots \tag{6.90}$$

The critical load is again the smallest. The buckling mode follows ($n = 1$):

$$P_{CR} = P_1 = \lambda_1^2 EI = \pi^2 EI/L^2 \tag{6.91}$$

The buckling mode follows:

$$w(x) = A \sin \pi x/L \tag{6.92}$$

However, all solutions satisfy equilibrium conditions. The second mode ($n = 2$) is shown dotted in Figure 6.41b. Indeed, with a simple support at the midpoint, that second mode is the buckling mode, which occurs at the load

$$P_2 = \lambda_2^2 EI = 4\pi^2 EI/L^2 \tag{6.93}$$

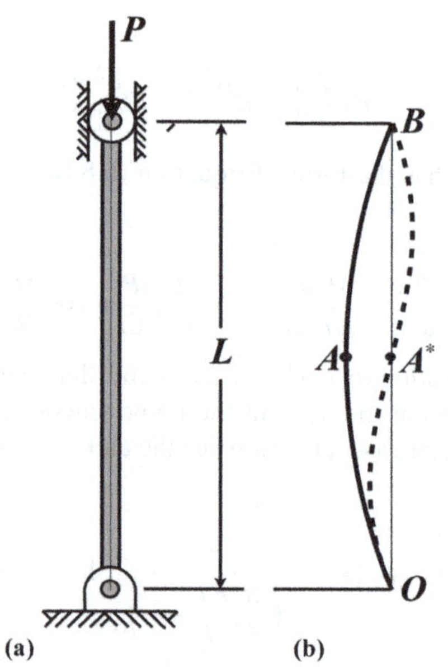

Figure 6.41

The midpoint is a point of inflection ($\mathrm{d}^2 w/\mathrm{d}x^2 = M/EI = 0$), and each half of our original column behaves as a simply supported column of length $L/2$.

Let us now consider the implications of the column with an intermediate support. Theoretically, that support needs to exert no lateral force; a very weak support ought to suffice. We might imagine that we manually hold the midpoint and, therefore, attain the higher load $P_2 = 4P_1$. What would happen if we suddenly removed our support or, in a practical circumstance, a weak support failed? We then have our original column (without the benefit of intermediate support) subjected to $P_2 = 4P_1 = 4P_{CR}$. In this state, a perfect column *is* in equilibrium - *but unstable* equilibrium. It is in a state such as a ball sitting precisely at the top of a hill. Like the ball, the column tends toward a stable state. Because the load is far in excess of the sustainable load, abrupt collapse ensues. There is an important practical lesson in the example. A weak intermediate support, such as an adjoining beam that supports a column, can greatly increase the critical load, but can also create a state of instability and collapse.

6.9.3 EULER'S FORMULA AND END CONDITIONS

The analysis of thin Hookean columns, the determination of the critical load, bifurcation, and also postbuckling behavior were first investigated by Leonhard Euler.[3] Euler's formula for the critical load has the general form

$$P_{CR} = C\frac{\pi^2 EI}{L^2} \tag{6.94}$$

The constant C is determined by the conditions of support at the ends. It has been called the *end-fixity* factor. Four cases and the critical loads are shown in Figure 6.42.

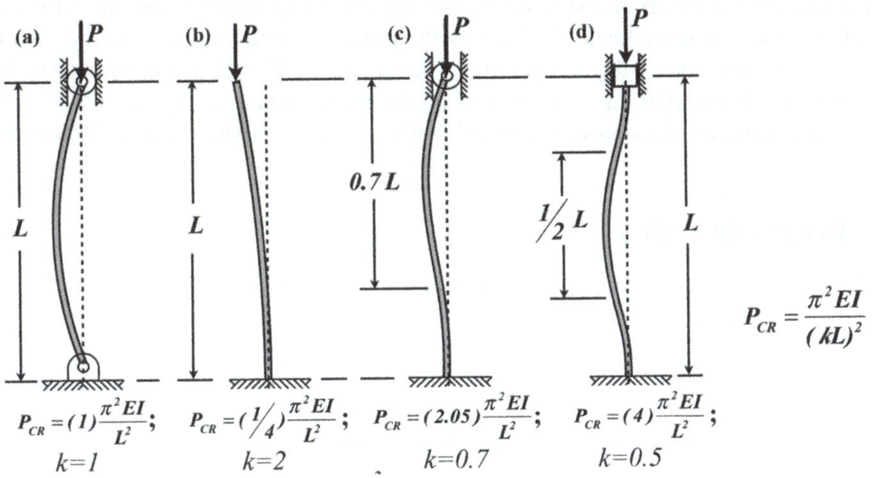

Figure 6.42 Columns with Various End Supports

The situations depicted in Figure 6.42 are ideal; the ends are either totally fixed, completely free, or ideally pinned. In reality, columns are connected to adjoining members; for example, a vertical column might be joined to a horizontal beam that resists lateral deflection and rotation as it bends and twists, but does not completely prevent the movement. The adjoining member acts as a spring; the lateral and torsional moduli depend upon the properties of the member. Such a circumstance is illustrated in Figure 6.43, where a torsional spring resists rotation at the lower end. Here, the conditions at the lower end ($x = 0$) are

$$w(0) = 0$$

$$M(0) = EI\frac{d^2w}{dx^2}\bigg|_0 = \beta\frac{dw}{dx}\bigg|_0$$

The solution is obtained according to the procedure of Section 6.9. The critical load is determined by the smallest root of the equation

$$\tan\lambda L = \frac{\alpha\lambda L}{\alpha + \lambda^2 L^2} \tag{6.95}$$

where

$$\lambda^2 L^2 = \frac{PL^2}{EI}, \quad \alpha = \frac{\beta L}{EI}$$

If $\beta = \alpha = 0$, the lower end is simply supported and $\lambda L = \pi$ (see Figure 6.42a). If $\beta \to \infty, \alpha \to \infty$, the lower end is fixed and $\lambda L = \sqrt{2.05}\pi$ (see Figure 6.42c). Sometimes Equation (6.94) is written as

$$P_{CR} = \frac{\pi^2 EI}{(kL)^2} \tag{6.96}$$

where $k = 1/\sqrt{C}$. In this form, kL is called the *reduced length* of the column. The significance of this terminology can be seen in Figure 6.42 For the pin-connected column, $k = 1$. In this case, the column buckles into one arch of a sine wave. The equivalent length of a column is the portion of length that assumes the same buckled shape as the pin-connected column. For the clamped-free column, we would need twice the actual length; hence, $k = 2$. For the clamped-clamped column, half its actual length buckled into one arch of a sine wave; hence, in this case, $k = \frac{1}{2}$ - For the clamped-pinned column, the equivalent length turns out to be approximately 0.7 of the actual length.

6.9.4 SLENDERNESS RATIO

One further modification is usually made in Equation (6.96) by dividing through by the cross-sectional area of the column. Then

$$\sigma_{CR} = \frac{P_{CR}}{A} \tag{6.97}$$

is called the *critical stress*. We can also define a parameter r:

$$r^2 \equiv \frac{I}{A} \tag{6.98}$$

This is the so-called radius of gyration of the cross-section. With this notation, Equation (6.97) can be written

$$\sigma_{CR} = \frac{\pi^2 E}{\left(\frac{kL}{r}\right)^2} \tag{6.99}$$

The quantity kL/r is called the *slenderness ratio* of the column. (Some authors call L/r the slenderness ratio.) A plot of critical stress versus slenderness ratio as determined by Equation

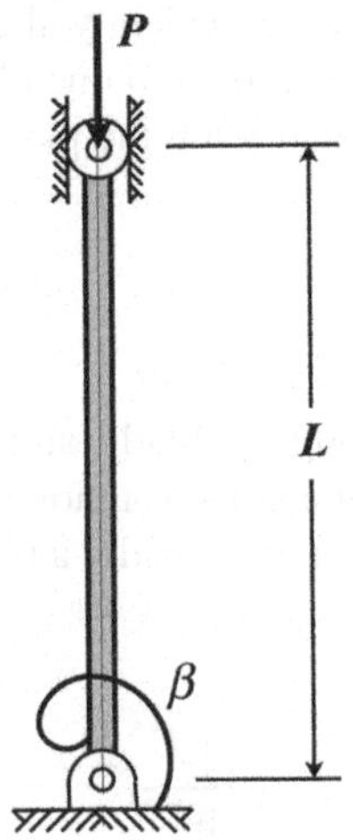

$$P_{CR} = \lambda^2 EI$$
$$\lambda \rightarrow \pi/L \quad as\,\beta \rightarrow 0$$
$$\lambda \rightarrow \sqrt{2.05}\,\pi/L \quad as\,\beta \rightarrow \infty$$

Figure 6.43 Torsional end Constraint

(6.99) is shown in Figure 6.44 (The scales are not indicated, since they depend on the value of E for any particular column.)

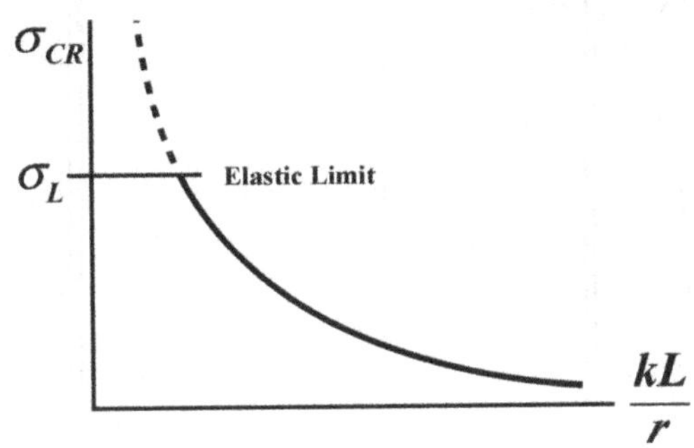

Figure 6.44

As the slenderness ratio decreases, the critical stress grows quite large. When it exceeds the proportional limit for the material, the Euler formula is no longer valid, since it is based on linearly elastic behavior. The portion of the curve that is not valid is shown in dashed. For very small values of slenderness ratio, the column becomes a squat compression block. In this case, buckling is no longer a problem, since failure is the result of compression (fracture for some very brittle materials or excessive deformation for ductile materials). However, there is an intermediate range of slenderness for which the Euler formula gives values above the proportional limit, but failure can still occur as the result of buckling. This situation is considered in Section 6.10.

One final point must be emphasized concerning the elastic buckling of columns. The critical load (or stress) is the *smallest* nonzero value for which a buckled equilibrium configuration is possible. Extreme care must, therefore, be exercised in choosing the values of k and r that apply in a particular case. These quantities, in general, can vary widely if the plane of bending is changed. For example, a thin rectangular cross-section will bend much more easily in one direction than another because of the difference in moments of inertia (I) or, equivalently, the difference in radii of gyration (r). Also, an end connection can act as a pin with respect to one plane of bending but as a clamp with respect to another. To give the smallest critical load, the plane of buckling must be chosen so that k/r is a maximum.

ILLUSTRATIVE EXAMPLE 6.21

To illustrate the final remarks, let us consider a possible, though unlikely, situation. An I beam is employed as a column. The end supports provide no restraint against rotation in the x-y plane, but total constraint in the x-z plane, as depicted in Figure 6.45. Let us consider a typical I beam which has $I_y = 8I_z$, ($I_y \equiv \iint y^2 dA$, $I_z \equiv \iint z^2 dA$).

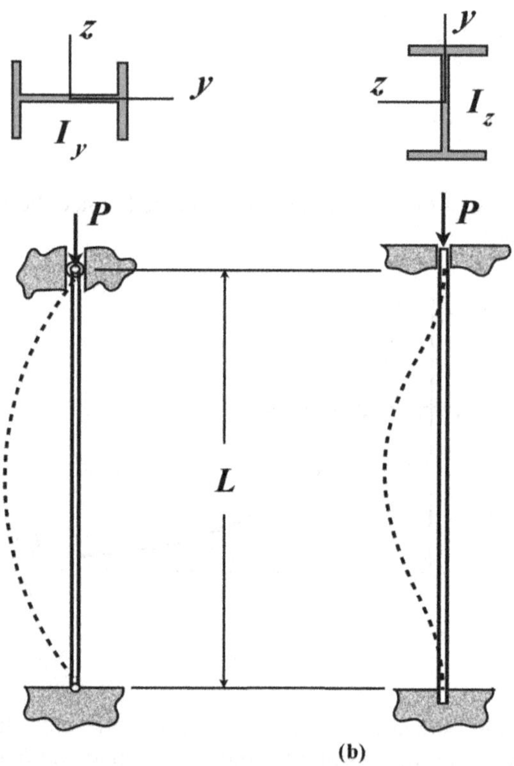

Figure 6.45

Applying the equations of Figure 6.42a and d, we obtain the critical loads ${P_{CR}}^a$ and ${P_{CR}}^b$ for buckling in planes x-y and x-z:

$$ {P_{CR}}^a = \frac{\pi^2 E I_y}{L^2}, \quad {P_{CR}}^b = \frac{4\pi^2 I_z}{L^2}, \quad {P_{CR}}^a / {P_{CR}}^b = 2 $$

Our example serves to emphasize the role of the cross-sectional form, viz., the factor I.

ILLUSTRATIVE EXAMPLE 6.22

In Section 5.6, we considered trusses, a most prevalent example of axially loaded members. The members are typically viewed as pin-connected, hence subject to buckling according to Equation (6.91). In each example, we can identify a member subject to the greatest compressive force, apply Equation (6.91), and thereby predict a likely mode of failure. Such a prediction would be conservative since the connections do provide some constraint against relative rotations ($M \neq 0$).

Members 1, 4, and 6 in the truss of Figure 5.15 are subject to the compressive force $P/\sqrt{3}$. Therefore, the structure is likely to fail at the load:

$$P = \sqrt{3}\pi^2 EI/l^2$$

Member CA in the truss of Figure 5.16 is subject to the compressive force $0.207F$. It is subject to failure at the load:

$$F = \frac{\pi^2 EI}{0.207l^2}$$

The member OD in the Tower of Figure 5.19 is subject to the compressive force of $0.973P$. Then failure might be anticipated at

$$P = \frac{\pi^2 EI}{(0.973)l^2}$$

In each case, the failure depends upon the flexural rigidity, EI; the form of the cross-section (I), not the area (A), is most significant.

6.9.5 COLUMNS WITH INTERMEDIATE SUPPORTS

If a column has an intermediate support (for example, a column is connected to a beam that acts as an elastic support), then the deflection to either side is given by functions $w_{-(x)}$ and $w_{+(x)}$, which satisfy Equation (6.73). Each function has the form of Equation (6.75); each has four unspecified constants. As in the case of a beam with an intermediate support (see Section 6.5, "Support Conditions") or a concentrated load (see Illustrative Example 6.16), each support entails four additional conditions of continuity and/or discontinuity in the variables (deflection w, rotation w', moment EIW'', and/or shear $-EIw'''$).[4] Consequently, each support introduces four additional constants and also four additional equations. Therefore, the solution for the critical load (and buckled mode) entails a determinant, which is $4(1+n) \times 4(1+n)$, where n is the number of such additional supports. In general, the critical load for most *linearly* elastic systems is characterized by a bifurcation of equilibrium states (point A of Figure 6.40) and a system of *homogeneous* linear equations. The critical load is the least value that provides a nontrivial solution (buckled mode), specifically, the vanishing of the determinant $[4(1+n) \times 4(1+n)]$.

ILLUSTRATIVE EXAMPLE 6.23

Consider the column shown in Figure 6.46a, which has an intermediate simple support: Simple supports at both ends impose the conditions of vanishing displacement ($w = 0$) and moment ($w'' = 0$); these four conditions are augmented by the four conditions at the intermediate support. If we designate the displacement of the upper segment w_1, and the lower w_2, then we have the continuity conditions $w_1' = w_2'$ and $w_1'' = w_2''$ as well as the two conditions $w_1 = w_2 = 0$.

If we express both displacements in terms of a coordinate x with origin at one end, then we have eight conditions, ostensibly leading to an 8×8 determinant. Our task is simplified if we express each displacement in terms of the coordinate measured from the simply supported ends, as depicted in Figure 6.46b and c. Then the four conditions at these ends are satisfied by omitting the constants B and D in the general solution, Equation (6.75):

$$w_1 = A_1 \sin\lambda x_1 + C_1 x_1$$
$$w_2 = A_2 \sin\lambda x_2 + C_2 x_2$$

We impose the conditions for the union of $w_1(1-\alpha)L$ and $w_2(\alpha L)$ at the site of the support:

$$w_1[(1-\alpha)L] = w_2(\alpha L) = 0:$$
$$w_1[(1-\alpha)L] = A_1 \sin(1-\alpha)L + C_1(1-\alpha)L = 0$$
$$w_2(\alpha L) = A_2 \sin\alpha L + C_2 \alpha L = 0$$
$$w_1'[(1-\alpha)L] = -w_2'\alpha L:$$
$$\lambda A_1 \cos\lambda(1-\alpha)L + C_1 + \lambda A_2 \cos\lambda\alpha L + C_2 = 0$$
$$w_1''[(1-\alpha)L] - w_2''(\alpha L) = 0:$$
$$-\lambda^2 A_1 \sin(1-\alpha)L + \lambda^2 A_2 \sin\alpha L = 0$$

The determinant of the coefficients follows:

$$\begin{vmatrix} \sin\lambda(1-\alpha)L & (1-\alpha)L & 0 & 0 \\ 0 & 0 & \sin\lambda\alpha L & \alpha L \\ \lambda\cos\lambda(1-\alpha)L & 1 & \lambda\cos\lambda\alpha L & 1 \\ -\sin\lambda(1-\alpha)L & 0 & \sin\lambda\alpha L & 0 \end{vmatrix} = 0$$

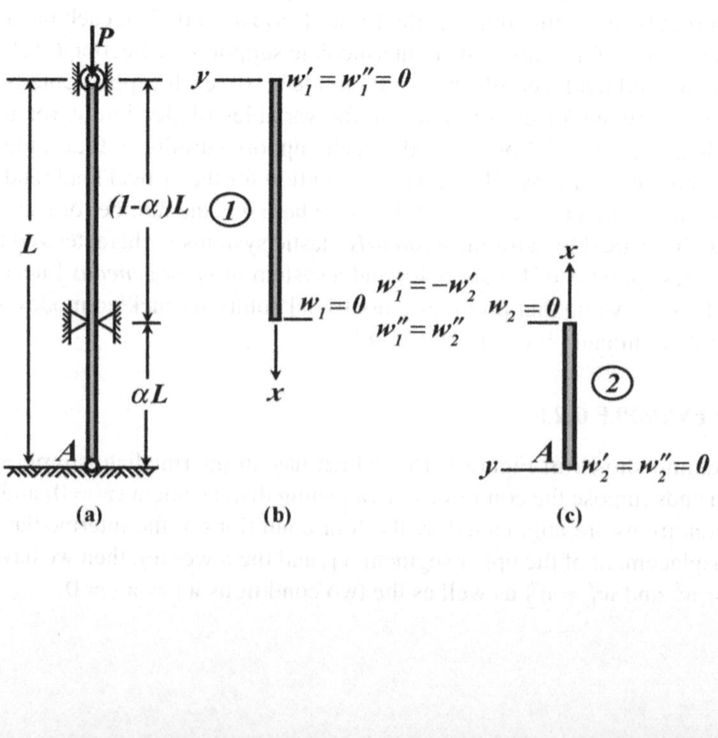

Figure 6.46

When the determinant is expanded and simplified via the trigonometric relations, viz.

$$\cos\lambda(1-\alpha)L = \cos\lambda L\cos\lambda\alpha L + \sin\lambda L\sin\lambda\alpha L$$
$$\sin\lambda(1-\alpha)L = -\cos\lambda L\sin\lambda\alpha L + \sin\lambda L\cos\lambda\alpha L,$$

we obtain the condition for $\lambda = \sqrt{P_{CR}/EI}$:

$$\sin\lambda\alpha L(\sin\lambda L\cos\lambda\alpha L - \cos\lambda L\sin\lambda\alpha L) = \lambda L\alpha(1-\alpha)\sin\lambda L \tag{6.100}$$

If the support is placed at the midpoint, $\alpha = 1/2$ we obtain

$$\sin^2\frac{\lambda L}{2} = \frac{\lambda L}{4}\sin\lambda L$$

The smallest positive root is $\lambda L = 2\pi$:

$$\lambda^2 = \frac{4\pi^2}{L^2} = \frac{P_{CR}}{EI}, \quad P_{CR} = \frac{4\pi^2}{EIL^2}$$

This is the result given by Equation (6.93) and depicted by the dotted line of Figure 6.41b.

If the support is placed close to the end A, the practical effect is a constraint against rotation; then the buckling mode and critical load should be essentially that depicted in Figure 6.42c. Indeed, if we set $\lambda L = \sqrt{2.05}\pi$ ($P_{CR} = 2.05\pi^2 EI/L^2$) in Equation (6.100), we obtain the solution $\alpha \doteq 0.02$.

6.9.6 EFFECTS OF AN INTERMEDIATE SUPPORT

Columns seldom appear as the sole support for an axial load, but as a component in a structural framework. Adjoining members serve to inhibit buckling, e.g., lateral supports. The engineer must anticipate the effects and the consequences if the support should fail.

ILLUSTRATIVE EXAMPLE 6.24

The simply supported column in Figure 6.47a is supported at the middle by a HI-HO cross-beam, which acts as an elastic spring with modulus k.

If the support is minimal, then the buckled form is symmetrical about the middle, similar to the dotted curve in Figure 6.47c. Then it suffices to consider the segment ($x \leq 0 < L/2$), as shown in Figure 6.47b. In that interval, the deflection must satisfy Equation (6.73); the general solution follows:

$$w(x) = A\sin\lambda x + B\cos\lambda x + Cx + D \tag{6.101}$$

where $\lambda^2 = P/EI$. The conditions at ends, $x = 0$ and $x = L/2$ are

$$w(0) = 0, \quad M(0)/EI = w''(0) = 0 \qquad ((6.102a,b))$$
$$w'(L/2) = 0 \qquad (6.102c)$$
$$w'''(L/2) - (k/EI)w(L/2) = 0, (V = -kw) \qquad (6.102d)$$

The conditions (6.102a,b) require that

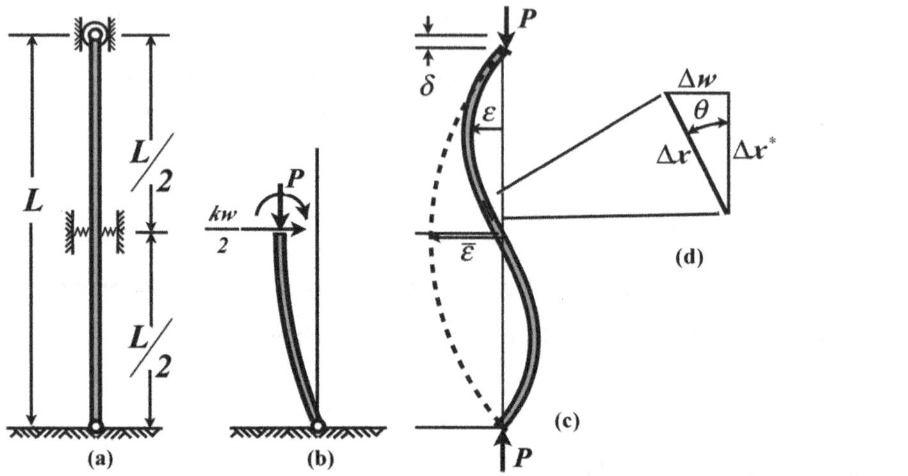

Figure 6.47 The Intermediate Support

$$B = D = 0$$

Conditions (6.102c,d) require that

$$A\lambda \cos \lambda L/2 + C = 0 \tag{6.102a}$$

$$A\left[\lambda^3 \cos \lambda L/2 + (k/EI)\sin \lambda L/2\right] + (k/EI)(L/2)C = 0 \tag{6.102b}$$

Again, the determinant at the coefficients (A and C) must vanish:

$$\tan \lambda L/2 = \frac{\lambda L}{2} - \frac{(\lambda L)^3 EI}{kL^3} \tag{6.103}$$

For any given modulus k, Equation (6.103) determines (λL) and the critical load:

$$(\lambda L)^2 = \frac{P_{CR}L^2}{EI}$$

According to Equation (6.102c),

$$C = -(\lambda \cos \lambda L/2)A$$

$$w(x) = A(\sin \lambda x - x\lambda \cos \lambda L/2)$$

Table 6.2 gives representative values of moduli and the corresponding critical loads. At $\lambda L = \pi$ the buckled form is that of the dotted curve ($\sin \pi x/L$) of Figure 6.47c. At $\lambda L = 2\pi$ the buckled form is that of the solid curve ($\sin 2\pi x/L$). At $\lambda L = 1.8\pi$ the form departs significantly from the dotted curve, exhibiting the effect of the support. In all cases, the elastic support results in significant increases in the critical load.

Note that the foregoing analysis assumes that the buckled form (mode) remains symmetric with respect to the midpoint ($x = L/2$). At $P = 4\pi^2 EI/L^2$ the column is most likely to buckle in the mode depicted by the solid lines of Figure 6.47c.

Table 6.2
Effect of Support k on Critical Load P_{CR}

$\frac{kL^3}{EI}$	0	25	57	79
λL	π	$\sqrt{2}\,\pi$	$1.8\,\pi$	$2.0\,\pi$
$\frac{P_{CR}L^2}{\pi^2 EI}$	1.0	2.0	3.24	4.0

6.9.7 FAILURE OF A SUPPORT

Our previous Illustrative Example (6.24) indicates that an intermediate support can provide much structural stability. However, failure of supporting members can precipitate collapse.

ILLUSTRATIVE EXAMPLE 6.25

Let us return to the situation depicted in Figure 6.47a. Without the support, the column would buckle at, or near, $P = \pi^2 EI/L^2$. With adequate support, the column would buckle at, or near, $P = 4\pi^2 EI/L^2$.

Now, let us suppose that the support does inhibit the displacement at the midpoint. The critical load ($4\pi^2 EI/L^2$) causes the column to assume the slightly deformed configuration:

$$w_1 = \varepsilon \sin 2\pi x/L \tag{6.104}$$

The buckling causes a slight axial displacement δ of the load and a commensurate reduction in the potential:

$$\Delta V_1 = -P\delta \tag{6.105}$$

In Figure 6.47d, we depict a segment of the buckled column; Δx^* denotes the axial length:

$$\Delta x^* = \Delta x \cos\theta$$

Since the rotation is small, we employ the approximations:

$$\theta \doteq \frac{dw}{dx} = w', \quad \cos\theta \doteq 1 - \frac{1}{2}\theta^2$$

Then we have the approximation:

$$\Delta x^* \doteq \Delta x \left[1 - \frac{1}{2}(w')^2\right]$$

The overall decrease in length (δ) follows:

$$\delta = L - \int_0^L \left[1 - \frac{1}{2}(w')^2\right] dx = \frac{1}{2}\int_0^l (w')^2 dx \tag{6.106}$$

When the column assumes the slightly deformed state, it experiences an increase in internal (strain) energy. The strain energy under simple extension/compression is $u = \frac{1}{2}E\varepsilon_x{}^2$ (see Section 4.11). According to the Bernoulli theory of bending (see Section 5.13) $\varepsilon_x = yd\theta/ds$. Our approximation of the internal energy follows:

$$\Delta U_1 = \frac{1}{2}EI\int_0^L (w'')^2 \mathrm{d}x \tag{6.107}$$

According to Equations (6.104)–(6.106), the reduction in the potential of the load follows:

$$\Delta V_1 = -P\delta = -\frac{2\pi^2}{L^2}P\varepsilon^2\int_0^L \cos^2\frac{2\pi x}{L}\mathrm{d}x \tag{6.108}$$

According to Equations (6.104) and (6.107), the increase in internal (strain) energy follows:

$$\Delta U_1 = \frac{8\pi^4}{L^4}EI\varepsilon^2\int_0^L \sin^2\frac{2\pi x}{L}\mathrm{d}x \tag{6.109}$$

Our system is conservative; therefore,

$$\Delta V_1 + \Delta U_1 = 0$$

We note that the integrals of Equations (6.108) and (6.109) are equal. Our result follows:

$$P_{CR} = \frac{4\pi^2 EI}{L^2} \tag{6.110}$$

We have arrived at our previous result via an energetic analysis, but also have an approximation of the internal energy of the slightly buckled column, viz. Equation (6.109)

Now suppose that the support fails. The column abruptly assumes the form of the dotted curve in Figure 6.47c. At the failure, the potential of the load is given by Equation (6.108); the internal energy is given by Equation (6.109): At this same position δ, the dotted curve has the form

$$w_2 = \bar{\varepsilon}\sin\frac{\pi x}{L}$$

where

$$\delta = \frac{1}{2}\int_0^L (w_2{}')^2\mathrm{d}x = \frac{\bar{\varepsilon}^2\pi^2}{2L^2}\int_0^L \left(\cos\frac{\pi x}{L}\right)^2\mathrm{d}x \tag{6.111}$$

Also,

$$\delta = \frac{1}{2}\int_0^L (w_1{}')^2\mathrm{d}x = \frac{2\varepsilon^2\pi^2}{L^2}\int_0^L \left(\cos\frac{2\pi x}{L}\right)^2\mathrm{d}x \tag{6.112}$$

Note that the integrals in Equations (6.111) and (6.112) have the same value.
Therefore, equating Equations (6.111) and (6.112), we obtain

$$\bar{\varepsilon}^2 = 4\varepsilon^2$$

The column is in the configuration of the dotted curve after the support has failed. The internal energy is now

$$\Delta U_2 = 2EI\frac{\varepsilon^2\pi^4}{L^4}\int_0^L \left(\sin\frac{\pi x}{L}\right)^2\mathrm{d}x$$

Let us compare the internal energies ΔU_2 and ΔU_1:

$$\frac{\Delta U_2}{\Delta U_1} = \frac{1}{4}\frac{\int_0^L \left(\sin\frac{\pi x}{L}\right)^2 \mathrm{d}x}{\int_0^L \left(\sin\frac{2\pi x}{L}\right)^2 \mathrm{d}x} = \frac{1}{4}$$

At this instant, the internal energy is 1/4 that of the previous state. It must deform more to sustain the load. Yet, the load continues to descend and perform additional work upon the column, which further decreases the potential. We are describing an imminent collapse. The conservative (elastic) system is no longer in a state of minimum potential, i.e., stable equilibrium.

ILLUSTRATIVE EXAMPLE 6.26

A symmetric HI-HO column is simply supported at both ends ($x = 0, L$). The column is embedded in an elastic medium which resists lateral deflection $w(x)$: specifically, the deflection causes a proportional resistive load $p(x) = \alpha w(x)$, where α is the effective modulus of the medium.

The resisting load appears on the right side in Equations (6.66) and (6.73): Therefore, the deflection is governed by the linear differential equation,

$$\frac{\mathrm{d}^4 w}{\mathrm{d}x^4} + \frac{P}{EI}\frac{\mathrm{d}^2 w}{\mathrm{d}x^2} + \frac{\alpha}{EI} w(x) = 0$$

With the prescribed end conditions, $w(0) = w(L) = \mathrm{d}w/\mathrm{d}x(0) = \mathrm{d}w/\mathrm{d}x(L) = 0$, the solution is

$$w = A \sin n\,\pi x/L$$

$$P = \frac{\pi^2 EI}{L^2}\left(n^2 + \frac{\gamma}{n^2}\right)$$

where γ is a non-dimensional modulus:

$$\gamma \equiv \frac{\alpha L^4}{\pi^4 EI}$$

For small value γ ($\gamma < 4$) the critical load occurs at $n = 1$.

$$P_{CR} = \frac{\pi^2 EI}{L^2}(1 + \gamma)$$

For $\gamma > 4$ the critical load occurs at $n = 2$.

$$P_{CR} = \frac{\pi^2 EI}{L^2}\left(4 + \frac{\gamma}{4}\right)$$

6.10 BUCKLING OF COLUMNS WITH NONLINEAR (INELASTIC) BEHAVIOR

There are circumstances in which the Euler formula predicts a critical stress beyond the proportional limit of the material. In this range, the Euler formula is not valid; it gives values that are much higher than those attainable. Figure 6.48 indicates the deviations from the Euler equation for some typical engineering materials. The ordinate is the critical stress divided by the material's modulus of elasticity, so that all materials follow the same curve in the linear range of their stress-strain diagram.

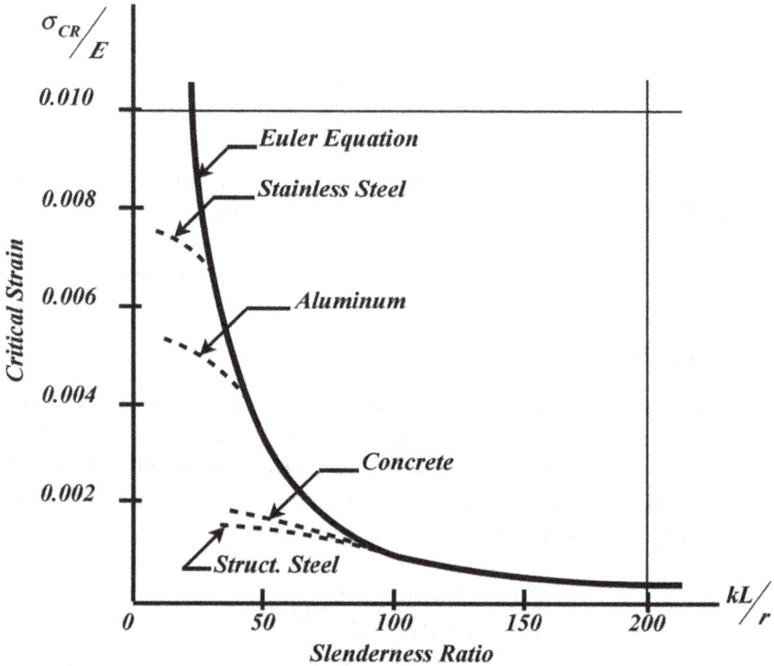

Figure 6.48

6.10.1 EMPIRICAL RELATIONS

Many empirical formulas have been devised for the design of columns in the intermediate range. Widely used by engineers, these are often found in building codes, where their use is mandatory. Such empirical formulas are simple algebraic relations between critical stress and slenderness ratio, which are fitted to experimental data. One of the earliest formulas used was of the form

$$\sigma_{CR} = \frac{a}{1 + b\left(\frac{kL}{r}\right)^2}$$

where a and b are constants determined from test data. The simplest approximation is obtained by expressing the critical stress as a linear function of the slenderness ratio. Very widely used, it has the form

$$\sigma_{CR} = a - b\left(\frac{kL}{r}\right)$$

A parabolic approximation is occasionally used:

$$\sigma_{CR} = a - b\left(\frac{kL}{r}\right)^2$$

The constants in all these empirical formulas must be determined from experimental data, since their values are different for different materials. It is customary to adjust these values to ensure against the uncertainties of actual service conditions. For example, if a and b are experimentally determined values for the coefficients in one of the formulas, then the a and b used in the formula are those that give a curve somewhat lower than the experimental one.

These empirical formulas are restricted to a range of slenderness ratios that depends both on the particular formula and the material. By a suitable choice of constants, some of the empirical formulas can be extended well into the range of high slenderness, obviating the need for Euler's formula in these cases.

6.10.2 ENGESSER'S TANGENT MODULUS THEORY

Another approach to the intermediate range of buckling is to modify the calculation of critical stresses to obtain reliable results for all engineering materials in the nonlinear stress-strain range. One of the simplest and most successful methods was proposed by F. Engesser in 1889.

Let Figure 6.49 represent the stress-strain diagram for the material as obtained from a compression test. Suppose that σ_{CR} is the actual critical stress for the column we are considering, and ε_{CR} is the corresponding strain. The slope of the stress-strain diagram at this point is E_T. Then, for stresses and strains very close to σ_{CR} and ε_{CR}, we can approximate the stress-strain diagram by the straight line

$$\sigma - \sigma_{CR} = E_T\,(\varepsilon - \varepsilon_{CR})$$

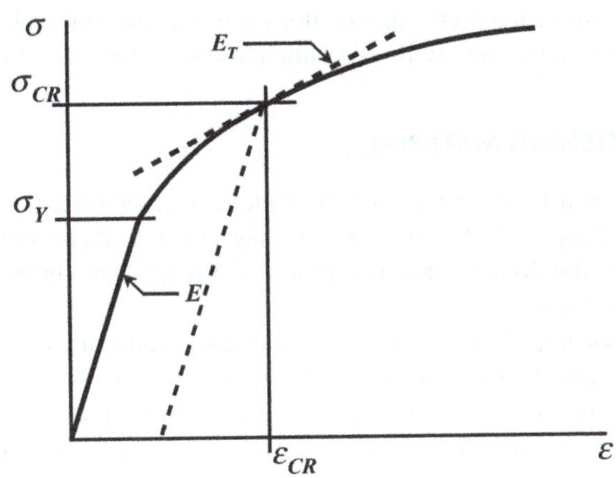

Figure 6.49 Tangent Modulus

At the critical load, the column is axially compressed, and the extensional strain is constant everywhere on a cross section; its value is ε_{CR}. If cross sections remain plane when the column bends, the extensional strain due to bending varies linearly over the cross section. The changes of strain due to bending represent infinitesimal deformations. Therefore, the resulting state of strain can be obtained by the superposition of the strains for the two states.

If the centroidal axis of the column (the neutral axis with respect to the bending deformation alone) remains at the critical stress during buckling, the strain there will be ε_{CR}; then, measuring y from this axis, the extensional strain anywhere in the cross section is

$$\varepsilon = \varepsilon_{CR} - \frac{y}{\rho}$$

The bending couple due to these stresses is

$$M = -\iint_A y\sigma \mathrm{d}A = -\sigma_{CR}\iint_A y\mathrm{d}A + \frac{E_T}{\rho}\iint_A y^2\mathrm{d}A$$

The first integral is zero, since y is measured from the centroid of the cross section. The second is the moment of inertia, I. Thus

$$M = \frac{E_T I}{\rho}, \qquad \frac{1}{\rho} = \frac{M}{E_T I}$$

This is exactly the same formula used in the analysis of the Hookean column, except that E_T replaces E. By following the same procedure, we obtain the Euler-Engesser formula.

$$\sigma_{CR} = \frac{\pi^2 E_T}{\left(\frac{kL}{r}\right)^2} \tag{6.113}$$

Equation (6.113) is analogous to Equation (6.99). If σ_{CR} is in the linearly elastic range, then E_T is the slope in this range, namely, E, and Equation (6.113) reduces to Equation (6.99). The quantity E_T is called the *tangent modulus*, and Equation (6.113) is sometimes referred to as the *tangent modulus formula*. Although Engesser's theory does not account for inelastic unloading during buckling, it provides good agreement with experimental results for many materials.

6.10.3 STRAIN HARDENING MATERIAL

The following example illustrates the use of the Euler-Engesser formula. Suppose a column is made of an aluminum alloy with the stress-strain diagram in compression, as shown in Figure 6.50. In the same figure, the dotted curve is a plot of the tangent modulus versus stress with the tangent-modulus scale at the top.

Note that, to the proportional limit, the tangent modulus is constant and equal to the modulus of elasticity. Then, as the slope of the stress-strain diagram decreases, so does the tangent modulus. If the column is designed to have a given critical stress, we enter the curve at that value of stress and find the corresponding tangent modulus. The slenderness ratio is then determined from

$$\frac{kL}{r} = \sqrt{\frac{\pi^2 E_T}{\sigma_{CR}}}$$

If, on the other hand, the slenderness ratio is known and we want the critical stress, a trial-and-error procedure can be used. Guess at the critical stress (try the Euler value, for instance) and find the corresponding value of E_T from the tangent modulus curve. If the guess is correct, these values will satisfy.

$$\frac{\sigma_{CR}}{E_T} = \frac{\pi^2}{\left(\frac{kL}{r}\right)^2}$$

If not, adjust σ_{CR} up or down as indicated and repeat the calculation with the new guess until the desired accuracy is achieved.

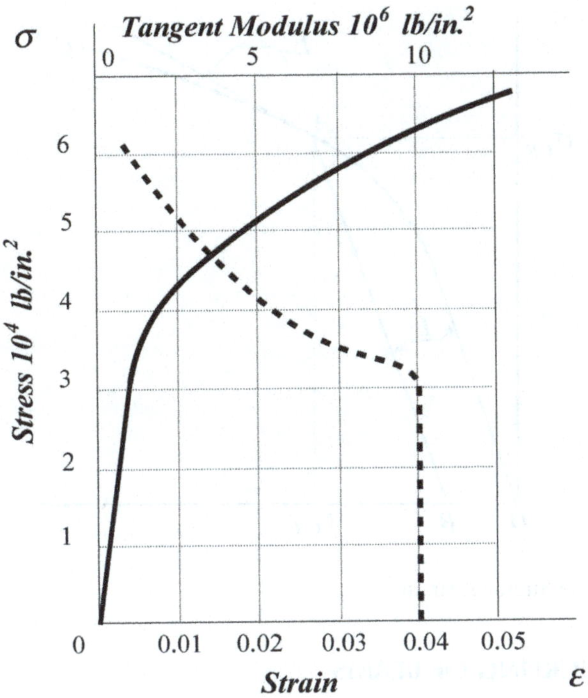

Figure 6.50

ILLUSTRATIVE EXAMPLE 6.27

In Chapter 4, we observed that a typical metal exhibits a nonlinear and inelastic behavior, as exemplified by the plot of Figure 6.51. At a stress σ_{CR}, which exceeds the yield stress, unloading follows the linear path AB with slope E, whereas loading follows the path AC with slope E_T.

If a column buckles at the critical stress σ_{CR} (point A), then the consequent moment (M)-curvature ($d\theta/dx$) relation is that of a material with double modulus (E_R) (see Illustrative Example 5.28)

$$M = +\frac{d\theta}{dx}\left(E_T \iint_{A_+} y^2 dA + E \iint_{A_-} y^2 dA\right)$$

Here, A_+ and A_- denote the areas above and below the neutral axis. If we now neglect the shift of the axis [$c = h$ in Figure 5.59] and assume a doubly symmetric cross-section, e.g., an I-beam, then

$$\iint_{A_+} y^2 dA \doteq \iint_{A_-} y^2 dA = \frac{1}{2} I$$

It follows that

$$M = \frac{d\theta}{dx}\left(\frac{E_T + E}{2}\right) I \equiv \frac{d\theta}{dx} E_A I$$

Then the appropriate modulus in the Euler formula (6.94) would be E_A.

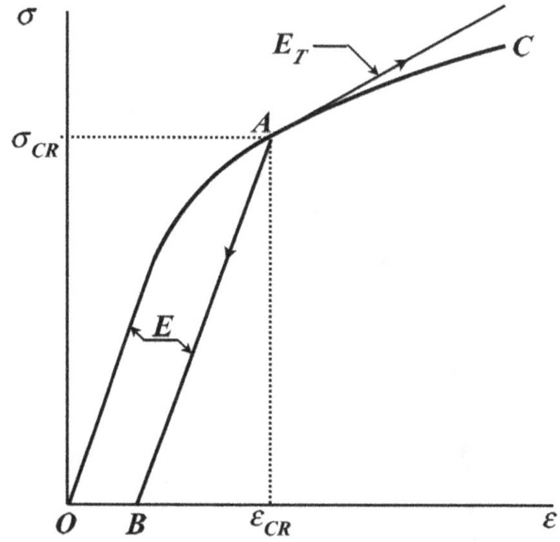

Figure 6.51 Compressive Stress vs. strain

6.11 LATERAL BUCKLING OF BEAMS

Beams with narrow cross-sections are subject to lateral buckling, as the cantilever beam depicted in Figure 6.52. To predict the buckling, we must examine equilibrium in the slightly deformed, i.e., buckled state. To that end, let us examine the geometry of the beam of Figure 6.53; a cross-section has been displaced and rotated through the small angles (θ_x, θ_y, and θ_z). Unit vectors $(\hat{\mathbf{i}}, \hat{\mathbf{j}}, \hat{\mathbf{k}})$ are rotated to unit vectors $(\hat{\mathbf{I}}, \hat{\mathbf{J}}, \hat{\mathbf{K}})$. Since the rotation is *small*, it can be treated as a vector (an approximation):

$$\hat{\mathbf{I}} \approx \hat{\mathbf{i}} + \boldsymbol{\theta} \times \hat{\mathbf{i}} \approx \hat{\mathbf{i}} - \theta_y \hat{\mathbf{k}} + \theta_z \hat{\mathbf{j}}$$
$$\hat{\mathbf{J}} \approx \hat{\mathbf{j}} + \boldsymbol{\theta} \times \hat{\mathbf{j}} \approx \hat{\mathbf{j}} + \theta_x \hat{\mathbf{k}} - \theta_z \hat{\mathbf{i}}$$
$$\hat{\mathbf{K}} \approx \hat{\mathbf{k}} + \boldsymbol{\theta} \times \hat{\mathbf{k}} \approx \hat{\mathbf{k}} - \theta_x \hat{\mathbf{j}} + \theta_y \hat{\mathbf{i}}$$

Conversely,

$$\hat{\mathbf{i}} \approx \hat{\mathbf{I}} + \theta_y \hat{\mathbf{K}} - \theta_z \hat{\mathbf{J}} \tag{6.114a}$$
$$\hat{\mathbf{j}} \approx \hat{\mathbf{J}} - \theta_x \hat{\mathbf{K}} + \theta_z \hat{\mathbf{I}} \tag{6.114b}$$
$$\hat{\mathbf{k}} \approx \hat{\mathbf{K}} + \theta_x \hat{\mathbf{J}} - \theta_y \hat{\mathbf{I}} \tag{6.114c}$$

The initial system $(\hat{\mathbf{i}}, \hat{\mathbf{j}}, \hat{\mathbf{k}})$, the rotated system $(\hat{\mathbf{I}}, \hat{\mathbf{J}}, \hat{\mathbf{K}})$, and the rotations $(\theta_x, \theta_y, \theta_z)$ are depicted in Figure 6.53. If $\mathbf{R}_0$ denotes the position of the deformed neutral axis, then

$$\hat{\mathbf{I}} = \frac{d\mathbf{R_0}}{dx} \equiv \mathbf{R_{0,x}}$$

Note: The definitions of the curvatures and twist do not hinge on the small rotations, but involve only gradients of the rotations. The curvatures (κ_x, κ_y) and twist $(\kappa_{xy} = \tau)$ are expressed by

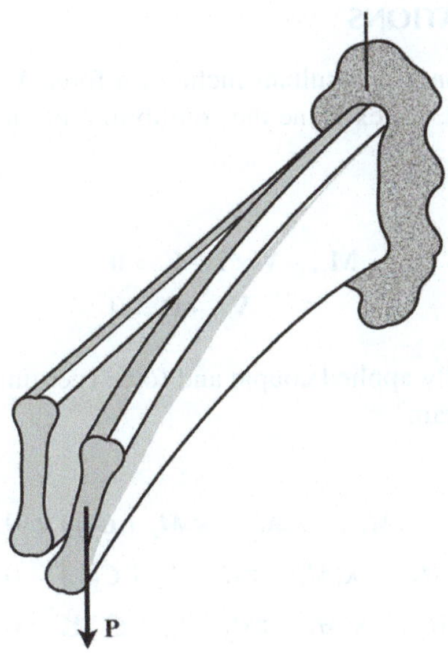

Figure 6.52 Lateral Buckling

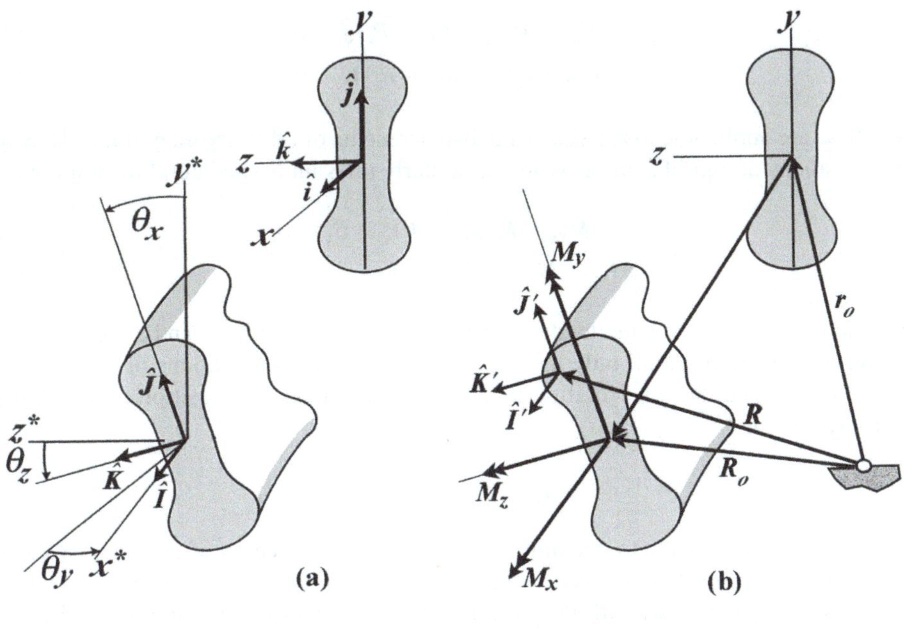

Figure 6.53 Displaced Cross-section

$$\kappa_z \equiv \hat{\mathbf{I}}_{,x} \cdot \hat{\mathbf{K}} \approx -\theta_{y,x} = -\hat{\mathbf{K}}_{,x} \cdot \hat{\mathbf{I}} \tag{6.115a}$$

$$\kappa_y \equiv \hat{\mathbf{I}}_{,x} \cdot \hat{\mathbf{J}} \approx +\theta_{z,x} = -\hat{\mathbf{J}}_{,x} \cdot \hat{\mathbf{I}} \tag{6.115b}$$

$$\tau \equiv \kappa_{xy} = \hat{\mathbf{J}}_{,x} \cdot \hat{\mathbf{K}} = -\hat{\mathbf{K}}_{,x} \cdot \hat{\mathbf{J}} \approx \theta_{x,x} \tag{6.115c}$$

6.11.1 EQUILIBRIUM EQUATIONS

On the axis of the cross section, the resultant includes a force **V** and couple **M** as depicted in Figure 6.53b. As in Section 6.2, we examine the equilibrium of an element, a slice of length Δx, and obtain the results:

$$\mathbf{M}_{,x} - \mathbf{V} \times \hat{\mathbf{i}} + \mathbf{C} = \mathbf{0}$$
$$\mathbf{V}_{,x} + \mathbf{P} = 0$$

Where **C** and **P** denote externally applied couple and force (per unit length). Taking into account the curvatures and twist, we obtain[5]

$$(M_{x,x} - \kappa_y M_y - \kappa_z M_z + C_x)\hat{\mathbf{I}} = 0 \quad (6.116a)$$
$$(M_{y,x} + \kappa_y M_x - \tau M_z - V_z + C_y)\hat{\mathbf{J}} = 0 \quad (6.116b)$$
$$(M_{z,x} - \kappa_z M_x + \tau M_y + V_y + C_z)\hat{\mathbf{K}} = 0 \quad (6.116c)$$

$$(V_{x,x} - \kappa_y V_y - \kappa_z V_z + P_x)\hat{\mathbf{I}} = 0 \quad (6.117a)$$
$$(V_{y,x} - \kappa_y V_x - \tau V_z + P_y)\hat{\mathbf{J}} = 0 \quad (6.117b)$$
$$(V_{z,x} + \kappa_z V_x + \tau V_y - P_z)\hat{\mathbf{K}} = 0 \quad (6.117c)$$

Note: These are applicable to rotations and displacements of arbitrary magnitude. They provide six scalar equations of equilibrium; two are particularly relevant to the lateral buckling, viz.

$$M_{x,x} - M_y\kappa_y - M_z\kappa_z + C_x = 0 \quad (6.118a)$$
$$M_{y,x} + M_x\kappa_y - M_z\tau - V_z + C_y = 0 \quad (6.118b)$$

Note that the last equation is an extended version of Equation (6.67) (y and z are interchanged). The equilibrium equations entail only the geometry of the axis and rotations of the cross-section, i.e., the position $\mathbf{R}_0$ and the rigid rotation $\boldsymbol{\theta}$. These motions place the position of an arbitrary site in the cross-section at

$$\mathbf{R} = \mathbf{R}_0 + y\hat{\mathbf{J}} + z\hat{K}$$

In Chapter 5, Section 5.11, to accommodate the torsion of beams of arbitrary cross-section, we introduced the warping function (ψ), which caused an axial displacement $\phi\psi(y,z)$ where $\phi(x)$ denotes the twist. However, we did not consider the circumstances in which the twist was variable, i.e., $\phi = \phi(x)$. Let us now consider such situations. Then the displaced position requires the additional term (underlined):

$$\mathbf{R} = \mathbf{R}_0 + y\hat{\mathbf{J}} + z\hat{\mathbf{K}} + \phi(x)\psi(y,z)\hat{\mathbf{I}}$$

To ascertain the effect, let us examine the additional virtual work; specifically, we evaluate the work of the stress components and the corresponding internal energy. To that end, we require the tangent vectors at an arbitrary site in the cross-section (see Figure 6.53b):

$$
\begin{aligned}
\hat{\mathbf{I}}' &= \mathbf{R}_{o,x} + y\frac{d\hat{\mathbf{J}}}{dx} + z\frac{d\hat{\mathbf{K}}}{dx} + \phi_{,x}\psi\hat{\mathbf{I}} \\
&= (1 - y\kappa_y - z\kappa_z)\hat{\mathbf{I}} - z\tau\hat{\mathbf{J}} + y\tau\hat{\mathbf{K}} + \phi_{,x}\psi\hat{\mathbf{I}} \\
\hat{\mathbf{J}}' &= \hat{\mathbf{J}} + \phi\psi_{,y}\hat{\mathbf{I}} \\
\hat{\mathbf{K}}' &= \hat{\mathbf{K}} + \phi\psi_{,z}\hat{\mathbf{I}}
\end{aligned}
$$

We recall the definitions of Equations (6.115), viz $\theta_{y,x} \equiv -\kappa_z$, $\theta_{z,x} \equiv \kappa_y$, and $\theta_{x,x} \equiv \tau$. The relevant strains (caused by twist and flexure) follow:

$$\varepsilon_x = \hat{\mathbf{I}}' \cdot \hat{\mathbf{I}} - 1 \approx -z\kappa_z - y\kappa_y + \phi_{,x}\psi \tag{6.119a}$$

$$\gamma_{xz} = \hat{\mathbf{I}}' \cdot \hat{\mathbf{K}} + \hat{\mathbf{K}}' \cdot \hat{\mathbf{I}} \approx y\tau + \phi\psi_{,z} \tag{6.119b}$$

$$\gamma_{xy} = \hat{\mathbf{I}}' \cdot \hat{\mathbf{J}} + \hat{\mathbf{J}}' \cdot \hat{\mathbf{I}} \approx -z\tau + \phi\psi_{,y} \tag{6.119c}$$

For brevity, we now denote the derivative $(\partial F/\partial y)$ with $(F_{,y})$. Here, κ_z, κ_y, and τ are defined by Equations (6.115a,b,c). The reader will recognize the lead terms in ε_x as the strains caused by flexure in direction x-z and x-y (κ_z, κ_y); the lead terms of γ_{xz} and γ_{yz} are the shear strains caused by the twist (τ). Additional internal energy results from the final terms containing the warping function ψ.

The independent displacements are the three rotations $(\theta_x, \theta_y, \theta_z)$. We understand the role of the rotations and the associated strains $(\kappa_x, \kappa_y, \tau)$. We now address the role of the warping $\phi(x)\psi(y,z)$. Our attention is restricted to cross-sections with double symmetry; therefore

$$\iint_A y\mathrm{d}A = \iint z\mathrm{d}A = 0$$

The virtual work of the relevant stress upon the associated strains in Equations (6.19a–c) follows:

$$\delta W_E = \iint \left[\sigma_x\psi\delta\phi_{,x} + (\tau_{xz}\psi_{,z} + \tau_{xy}\psi_{,y})\delta\phi\right]\mathrm{d}A$$

Let us define

$$F \equiv \iint_A (\sigma_x)\psi\mathrm{d}A \tag{6.120a}$$

$$H \equiv \iint_A (\tau_{xz}\psi_{,z} + \tau_{xy}\psi_{,y})\mathrm{d}A \tag{6.120b}$$

The virtual work of these two forces (F and H) between any two sites (Q and P) is

$$\delta W = \int_Q^P (F\delta\phi_{,x} + H\delta\phi)\mathrm{d}x$$

The initial term can be integrated by parts, as follows:

$$\delta W_E = F\delta\phi\,|_Q^P + \int_Q^P (-F_{,x} + H)\delta\phi\mathrm{d}x$$

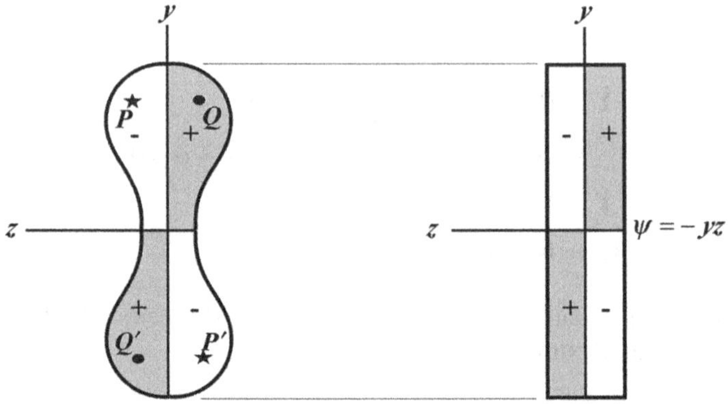

Figure 6.54 Doubly Symmetric Cross-Section

The variation $\delta\phi$ is an arbitrary continuous function, yet the virtual work must vanish in an equilibrium state. This requires that actions F and H satisfy the differential equation:

$$F_{,x} - H = 0 \tag{6.120c}$$

Our general problem of the deformed beam entails 3 displacements (w_0, u_0, v_0), 3 rotations $(\theta_x, \theta_y, \theta_z)$ *and* the 1 warping (ϕ): Accordingly, we have the 7 equations of equilibrium, viz Equations (6.116), (6.117), and (6.120c).

Our problem now acknowledges 5 forces (V_x, V_y, V_z, $\mathscr{F}$, and $\mathscr{H}$) and 5 associated "strains" ($\bar{\varepsilon}_x$, $\bar{\gamma}_{xy}$, $\bar{\gamma}_{xz}$, $\phi_{,x}$, and ϕ). Here the overbar denotes the mean values of those strains. Additionally, we have 3 moments (M_x, M_y, M_z) and 3 associated "strains" $(\kappa_x, \kappa_y, \tau)$. The strain energy (per unit length) follows (See Equations 6.119a–c):

$$\begin{aligned} U =& \frac{E}{2}\left(\bar{\varepsilon}_x^2 + I_y\kappa_y{}^2 + I_z\kappa_z{}^2\right) + \frac{E}{2}\phi_{,x}{}^2 \iint_A \psi^2 \mathrm{d}A + E\left(\kappa_z\phi_{,x}\iint z\psi \mathrm{d}A + \kappa_y\phi_{,x}\iint y\psi \mathrm{d}A\right) \\ &+ \frac{G}{2}\left[2\tau\phi\iint (y\psi_{,z} - z\psi_{,y})\,\mathrm{d}A + \tau^2\iint (y^2+z^2)\mathrm{d}A + \phi^2\iint \left(\psi_{,y}{}^2 + \psi_{,z}{}^2\right)\mathrm{d}A + \bar{\gamma}_{xy}^2 + \bar{\gamma}_{xz}^2\right] \end{aligned}$$

Several integrals are solely dependent on the cross-section. A warping function (ψ) for a doubly-symmetric cross-section has these symmetries:

$$\psi(y,z) = -\psi(-y,z) = \psi(-y,-z) = -\psi(y,-z)$$

This may be appreciated by viewing Figure 6.54; there $\psi(P) = \psi(P')$, $\psi(Q) = \psi(Q')$.

Consequently,

$$\iint y\psi \mathrm{d}A = \iint z\psi \mathrm{d}A = \iint \psi \mathrm{d}A = 0$$

Let us define

$$\Gamma \equiv \iint_A \psi^2 \mathrm{d}A$$
$$I_y \equiv \iint_A y^2 \mathrm{d}A, \quad I_z \equiv \iint_A z^2 \mathrm{d}A$$
$$I \equiv I_y + I_z$$
$$J \equiv \iint_A [\psi_{,z} y - \psi_{,y} z]\, \mathrm{d}A + I$$

also[6]

$$\iint [\psi_{,y}{}^2 + \psi_{,z}{}^2]\, \mathrm{d}A = I - J$$

With these properties of the cross-section, the internal energy assumes the abbreviated form:

$$U = \frac{E}{2}\left[\bar{\varepsilon}_x{}^2 + I_y \kappa_y{}^2 + I_z \kappa_z{}^2 + \Gamma \phi_{,x}{}^2\right] + \frac{G}{2}\left[\bar{\gamma}_{xy}^2 + \bar{\gamma}_{xz}^2 + I\tau^2 + 2(J-I)\tau\phi - (J-I)\phi^2\right]$$

By the arguments of Section 4.11 (see Equation 4.39), we obtain

$$M_x = \frac{\partial U}{\partial \tau} = GI\tau + G(J-I)\phi \tag{6.121a}$$

$$M_z = \frac{\partial U}{\partial \kappa_y} = EI_y \kappa_y, \quad M_y = -\frac{\partial U}{\partial \kappa_z} = -EI_z \kappa_z \tag{6.121b,c}$$

$$\mathscr{F} = \frac{\partial U}{\partial \phi_{,x}} = E\Gamma \phi_{,x}, \quad \mathscr{H} = \frac{\partial U}{\partial \phi} = G(J-I)(\tau - \phi) \tag{6.121d,e}$$

$$V_x = E\bar{\varepsilon}_x, \; V_y = G\bar{\gamma}_{xy}, \; V_z = G\bar{\gamma}_{xz} \tag{6.121f,g,h}$$

Recall that $\kappa_y \approx \theta_{z,x}$, $\kappa_z \approx -\theta_{y,x}$, $\tau = \theta_{x,x}$

ILLUSTRATIVE EXAMPLE 6.28

Let us consider the case of a narrow beam subject to the couples $\mathbf{M} = +M\hat{\mathbf{k}}$ as depicted in Figure 6.55a. Then, by means of Equation (1.114c), we have

$$M_x \approx -M\theta_y, \quad M_y \approx M\theta_x, \quad M_z = M \tag{6.122a,b,c}$$

These approximations are fully justified since we are examining the onset of buckling, i.e., a state of equilibrium adjacent to the initial state ($\theta = 0$). Placing the moments (6.122a,b) into the constitutive Equations (6.121a and c) we obtain

$$\theta_{x,x} + \frac{J-I}{I}\phi + \frac{M}{GI}\theta_y = 0 \tag{6.123a}$$

$$\theta_{y,x} - \frac{M}{EI_z}\theta_x = 0 \tag{6.123b}$$

Inserting the constitutive Equations (6.121d,e) into the equilibrium Equation (6.120c) provides a third differential equation in the variables $(\theta_x, \theta_y, \phi)$:

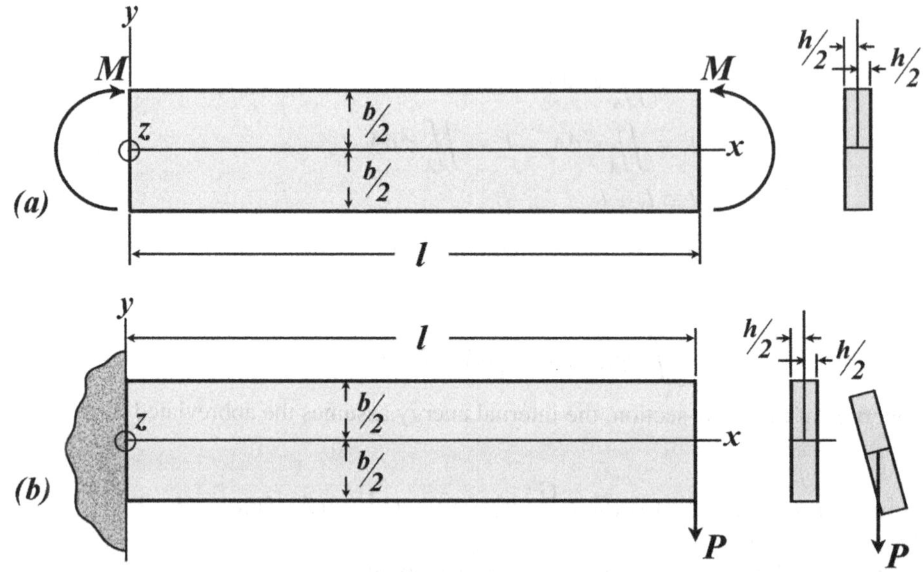

Figure 6.55

$$\phi_{,xx} + \frac{G(I-J)}{E\Gamma}(\theta_{x,x} - \phi) = 0 \tag{6.124}$$

Eliminating the variables θ_y and ϕ, we have the one differential equation:

$$\theta_{x,xxxx} + \left(\frac{M^2}{EI_z GI} - \frac{C}{C_1}\right)\theta_{x,xx} - \frac{M^2}{C_1 EI_z}\theta_x = 0 \tag{6.125}$$

where

$$C \equiv GJ, \quad C_1 \equiv \frac{EI\Gamma}{I-J}$$

To assess the roles of the coefficients, let us introduce a dimensionless variable, $\eta = x/l$. Then Equation (6.125) takes the form:

$$\theta_{x,nnnn} + (B^2M^2 - 2K)\theta_{x,nn} - Q^2M^2\theta_x = 0 \tag{6.126}$$

where

$$B^2 \equiv \frac{l^2}{EI_z GI}, \quad 2K \equiv \frac{l^2 GJ(I-J)}{EI\Gamma}, \quad Q^2 \equiv \frac{l^4(I-J)}{E^2 I\Gamma I_z}$$

Then

$$\frac{B^2}{Q^2} = \frac{E\Gamma}{Gl^2(I-J)}$$

To appreciate the magnitude, let us consider a narrow rectangular cross-section of width h and depth b, e.g., $h/b = 1/6$.

$$J \approx \frac{bh^3}{3}, \quad I = \frac{hb^3}{12}\left[1 + \left(\frac{h}{b}\right)^2\right]$$

$$I - J \approx \frac{hb^3}{12}\left[1 - 3\left(\frac{h}{b}\right)^2\right]$$

$$\Gamma = \frac{h^3 b^3}{144}$$

In this case

$$\frac{B}{Q} = (0.49)\left(\frac{h}{l}\right) \ll 1$$

Therefore, we can justifiably neglect the coefficient B^2M^2 in Equation (6.126) and focus on the following:

$$\theta_{x,nnnn} - 2K\theta_{x,nn} - Q^2M^2\theta_x = 0 \tag{6.127}$$

We require a solution of this equation that satisfies the end conditions: they impose no constraints against flexure κ_z nor warping ϕ:

$$M_y = EI_{zz}\theta_{y,x} = 0, \quad \iint_A \sigma_x \psi dA = E\Gamma\phi_{,x} = 0 \quad \text{at} \quad \eta = 0, 1$$

These follow from Equations (6.121c,d).

By Equation (6.123a,b), these are fulfilled if

$$\theta_x(0) = \theta_{x,xx}(0) = \theta_x(1) = \theta_{x,xx}(1) = 0$$

These conditions are met by the solution:

$$\theta_x = a \sin(m\eta)$$

if

$$m^4 + 2Km^2 - Q^2M^2 = 0$$
$$m = \sqrt{-K + \sqrt{K^2 + Q^2M^2}} \tag{6.128}$$

The least value, i.e., the critical couple M is provided by $m = \pi$. The solution of (6.128) provides the result:

$$M_{CR} = \frac{\pi}{l}\sqrt{GJEI_z\left(1 + \underline{\frac{C_1}{C}\frac{\pi^2}{l^2}}\right)} \tag{6.129}$$

For a narrow rectangular cross-section

$$\frac{C_1}{C}\frac{\pi^2}{l^2} \approx 0.6\left(\frac{b}{l}\right)^2 \ll 1$$

Indeed, for most narrow beams, the underlined term of Equation (6.129) can be neglected.

6.11.2 LATERAL BUCKLING OF A BEAM WITH NARROW RECTANGULAR CROSS-SECTION

The constitutive Equation (6.121a) relates the twisting couple M_x to the twist $\tau \equiv \theta_{x,x}$ and the warping ϕ. That result can be traced to the expressions for the shear strains $(\gamma_{xy}, \gamma_{xz})$ in Equation (6.119b,c) and the corresponding stress components (τ_{xy}, τ_{xz}). However, on the narrow rectangular cross section, τ_{xz} is negligible (see Section 5.11). A more accurate description is given in Illustrative Example 5.18, viz

$$M_x = GJ\theta_{x,x} \tag{6.130}$$

where

$$J = \frac{bh^3}{3}$$

Note too that warping $\phi = -zy$ is incorporated in Equation (6.130). Replacing Equation (6.121a) by Equation (6.130) and deleting the warping ϕ, we obtain the simpler approximation of Equation (6.123a):

$$\theta_{x,x} + \frac{M}{GJ}\theta_y = 0 \tag{6.131}$$

Together with Equation (6.123b), we have the governing equation:

$$\theta_{x,xx} + \frac{M^2}{GJEI_z}\theta_x = 0 \tag{6.132}$$

This simple description is attributed to Prandtl,[7] who also presented the following application to the cantilever beam.

ILLUSTRATIVE EXAMPLE 6.29

The cantilever beam is subject to the end load P as shown in Figure 6.55b. Then

$$M = P(l - x)$$

Equation (6.132) assumes the form

$$\theta_{x,xx} + Q^2(l-x)^2\theta_x = 0 \tag{6.133}$$

where

$$Q^2 \equiv \frac{P^2}{GJEI_z} \tag{6.134}$$

The solution entails a change of the independent variable, viz $s = l - x$. Then Equation (6.133) assumes the form:

$$\theta_{x,ss} + Q^2 s^2 \theta_x = 0 \tag{6.135}$$

Equation (6.135) has a general solution in the form of Bessel functions.[8] The least value of Q^2 which satisfies the end conditions, viz $\theta_x(s = l) = 0$, $M_y(s = 0)$ follows[9]:

$$Q^2 \approx \frac{16}{l^2} \Rightarrow P \approx \frac{4}{l^2}\sqrt{GJEI_z}$$

NOTES

1. The parallel-axis theorem is reviewed in Appendix B.
2. The weight of the beam is neglected in our examples but can be included if circumstances warrant.
3. Euler obtained his formula for the critical load in a memoir of 1757. The works of Euler are chronicled by I. Todhunter and K. Pearson, *Elasticity and Strength of Materials*, vol. 1, pp. 32–57. Cambridge University Press, New York, 1886.
4. Here, for brevity we signify a derivative with the apostrophe (').
5. These equations have been attributed to Kirchhoff.
6. The final equation follows from the definition of J and I when one observes that ψ is a "harmonic" function, viz. $\psi,_{yy} + \psi,_{zz} = 0$.
7. L. Prandtl. *Kippersheinungen: Ein Fall von instabilem elastischem Gleichgewicht.*, Dissertation, Technische Hochschule München, Munich, 1900.
8. E. Jahnke and F. Emde, *Table of Functions*. Dover, New York, 1945.
9. Further details are given by S. P. Timoshenko and James M. Gere, Theory of elastic stability (2nd ed.), Dover, New York: McGraw-Hill, 1961.

7 Combined Loads; Superposition

With few exceptions, the preceding problems involved a single type of loading, such as axial force, twisting couple, *or* bending couple; each entailed a simple deformation, such as extension, twist, *or* flexure. Here, we recognize that most actual members are subjected to a combination of loadings. For example, a drive shaft is intended to transmit torque but is also subjected to bending couples imposed by attachments and supporting bearings. In reality, most of the geometric changes are *relatively* small; specifically, strains and rotations are small. Consequently, the geometric relations are *linear*. If the description of the material is also linear, that is, the material is Hookean, then all relevant equations are linear. The important consequence is the superposition of solutions. Stated otherwise, we can add the effects of two (or more) loads to obtain the effects of the combined loading. The purpose of this chapter is to extend the utility of the foregoing results greatly and to provide some examples of superposing.

7.1 COMBINED LOADS

In many instances, structural members and machine parts are subjected to a variety of loads acting simultaneously. These loads may be of the same type; for example, a bar may be acted upon by several axial loads at the same time. On the other hand, the same member may support axial, twisting, and bending loads simultaneously. Loading of this kind is referred to as *combined loading*.

We have already discussed methods for determining the response of a member to axial loads, twisting couples, and transverse loads. Now let us turn our attention to problems involving combined loadings by superposing the effects of each individual load. The following examples serve to illustrate the concept, the procedure, and the limitations of *superposition*.

The linearly elastic cylinder of Figure 7.1a is stretched by an axial force of magnitude $P_1 + P_2$. If the strains are small, we have

$$\sigma = \frac{P_1 + P_2}{A}, \qquad e = \frac{(P_1 + P_2)L}{EA}$$

where L is the length of the cylinder, A is the cross-sectional area, and e is the total extension. Now consider the same bar subjected to the load P_1 alone, as in Figure 7.1b, and P_2 alone, as in Figure 7.1c. For P_1 alone

$$\sigma_1 = \frac{P_1}{A}, \; e_2 = \frac{P_1 L}{EA}$$

whereas for P_2 alone,

$$\sigma_2 = \frac{P_2}{A}, \qquad e_2 = \frac{P_2 L}{EA}$$

From the form of these relations, we see that

$$\sigma = \sigma_1 + \sigma_2 \quad \text{and } e = e_1 + e_2$$

Hence the response of the bar loaded by the axial force $P_1 + P_2$ can be obtained by adding, or *superposing*, the effects of the loads P_1 and P_2 acting individually. If the force P_1 were applied first and if the initial area A were significantly changed (contracted), then the true stress would not be

DOI: 10.1201/9781003665861-7

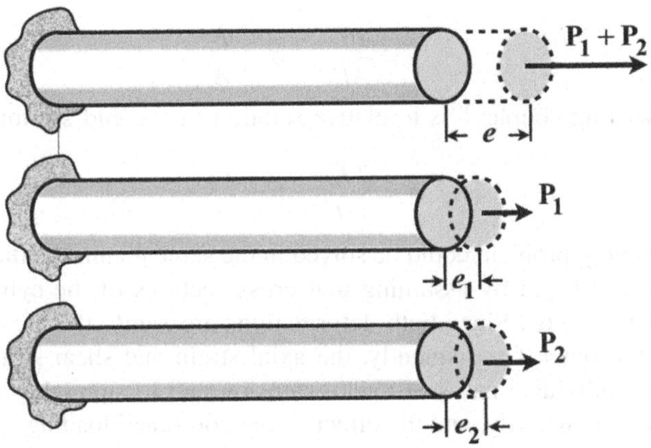

Figure 7.1

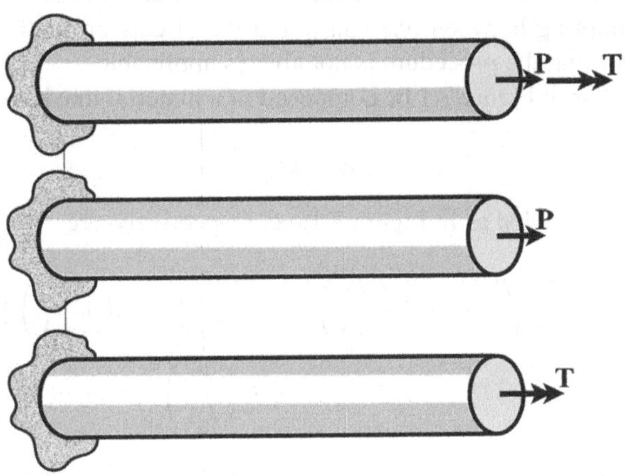

Figure 7.2

simply $\sigma_1 = P_1/A$. If the actual deformed area were A^*, then the true stress would be $\sigma_1 = P_1/A^*$. Moreover, the second force P_2 would also cause further change. If the final area were A^{**}, then the final stress would be

$$\sigma = \frac{P_1 + P_2}{A^{**}} \neq \sigma_1 + \sigma_2$$

The resulting stress is not precisely the sum of the stresses associated with each force acting alone. The difference is caused by the geometric change; the change of area invalidates the simple *linear* relation. Fortunately, such geometric changes are usually so small that they are negligible. The linear equations are adequate, and the stresses can be added.

As another example, consider the problem illustrated in Figure 7.2. The cylindrical shaft is composed of a linearly elastic material. It is simultaneously subjected to an axial force P and a twisting couple T. We already know the effects caused by each force system when it is separately applied. The force P acting alone causes an axial extension and a normal stress of amounts.

$$e = \frac{PL}{AE}, \quad \sigma = \frac{P}{A}$$

The effect of the twisting couple T is a relative rotation of the end sections and a shear stress. These are

$$\phi = \frac{TL}{GJ}, \quad \tau = \frac{Tr}{J}$$

The combined-loading problem could be solved in the same manner as the individual problems in Chapter 5. We could begin by assuming that cross sections of the cylinder displace axially and also rotate about the axis. Since both deformations are *small*, the presence of one does not significantly affect the other. Consequently, the axial strain and shear strain are essentially the same as those in the individual cases of axial extension and torsion. However, both are present when P and T act simultaneously, and the effect of the combined loading is therefore the sum of the effects that result when each load acts separately.

These examples illustrate the superposition of effects. An inspection of these two examples reveals that the addition of effects has its origin in the fact that one system of loads does not cause any change that will modify the effects of a second system of loads, and vice versa. In other words, there is no coupling between one cause and the effects of another cause. The following counterexamples show that the procedure is not always applicable.

Let the cylindrical bar of Figure 7.1 be composed of a material that has a nonlinear stress-strain relation of the form

$$\sigma = k\varepsilon^{1/2}$$

The elongations of a bar loaded as in Figure 7.1a–c, respectively, are

$$e = L\left(\frac{P_1 + P_2}{kA}\right)^2 = L\left(\frac{P_1}{kA}\right)^2 + L\left(\frac{P_2}{ka}\right)^2 + 2L\left(\frac{P_1}{kA}\right)\left(\frac{P_2}{kA}\right)$$

$$e_1 = L\left(\frac{P_1}{kA}\right)^2, \quad e_2 = L\left(\frac{P_2}{kA}\right)^2$$

It is evident that in general $e \neq e_l + e_2$, and so the principle cannot be employed to obtain extensions in this case. This can be seen graphically in Figure 7.3. The curve OA in each plot, Figure 7.3a–c, depicts the effect of applying the load $P = P_1 + P_2$. Curves OB and OC in Figure 7.3a and b, depict the effects of applying loads P_1 and P_2, respectively. The *addition* of OC to OB leads to incorrect result at the asterisk in Figure 7.3a. The converse $(OC + OB)$ is depicted in Figure 7.3b. (Addition is commutative.). Finally, Figure 7.3c shows that the sum $e_1 + e_2 \neq e$, the correct extension. The nonlinear behavior of the material is responsible for the difference in effects.

If a body undergoes extensive changes in size or shape, superposition is not generally applicable. To illustrate this, we consider the structure of Figure 7.4a, which consists of two identical linearly elastic wires OA and OB. A vertical load P applied at O causes the structure to deform as shown in Figure 7.4b. If δ denotes the vertical displacement of the point O, the elongation of AO or BO is

$$e = \sqrt{L^2 + (L+\delta)^2} - \sqrt{2}L$$

Denoting the axial force in each wire by F, we find from equilibrium that

$$F = P\frac{\sqrt{L^2 + (L+\delta)^2}}{2(L+\delta)}$$

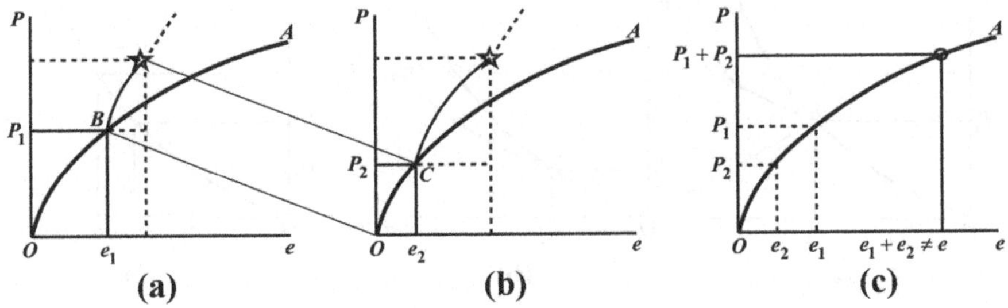

Figure 7.3

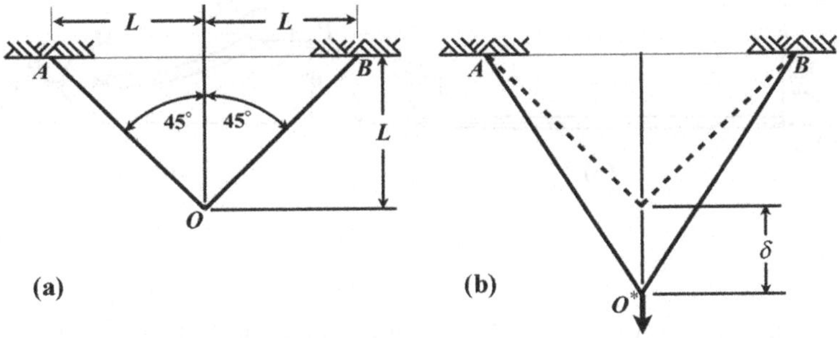

Figure 7.4

The elongation e is related to the force F by

$$e = \frac{F\sqrt{2}L}{AE}$$

where A is the cross-sectional area of the wire. Eliminating e and F between these equations, we find that

$$P = \frac{2(L+\delta)}{\sqrt{L^2+(L+\delta)^2}} \frac{AE}{\sqrt{2}L}\left[\sqrt{L^2+(L+\delta)^2}-\sqrt{2}L\right] \tag{7.1}$$

Although the material is linearly elastic, the load-displacement relation embodied in Equation (7.1) is nonlinear, as illustrated in Figure 7.5. The effect of a combined load $P = P_1 + P_2$ is the displacement δ in Figure 7.5c. This is not the sum of the displacements δ_1 and δ_2 caused by P_1 and P_2 acting separately. Again, the addition of the responses to each loading, when acting alone, would yield the erroneous results, as shown in Figure 7.5a and b. When the loads are applied consecutively, the first load changes the geometry of the structure significantly (if δ_1 is large) and so alters the response to the second load. *Note:* If $\delta/L << 1$, then Equation (7.1) has the linear approximation, $P \doteq EA\delta/\sqrt{2}L$.

Another counterexample is illustrated in Figure 7.6a. A cantilevered beam is stretched by an axial load P and is bent by a couple M_0. According to the theory of Section 6.5,

$$\kappa = \frac{d^2w}{dx^2} = \frac{M}{EI} = \frac{M_0}{EI} + \frac{P}{EI}(w_0 - w) \tag{7.2}$$

In this case, the presence of the axial load P causes an addition to the bending couple in the amount $P(w_0 - w)$. However, this additional bending couple exists only because the beam bends and gives

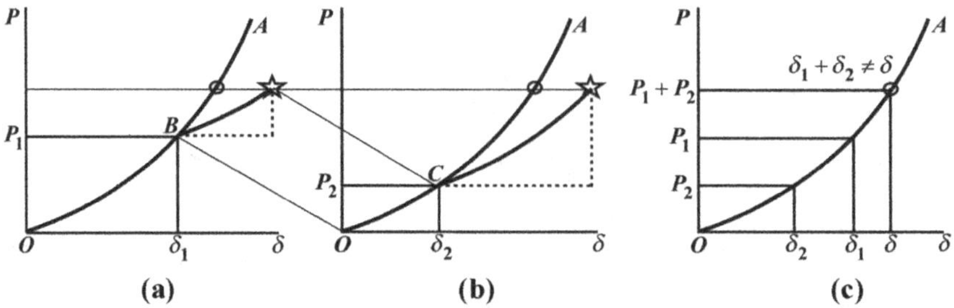

Figure 7.5

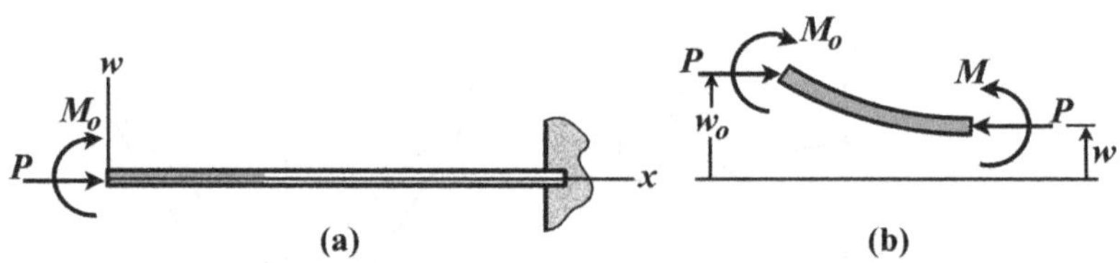

Figure 7.6

a transverse deflection so that $w_0 - w \neq 0$. On the other hand, the axial load P acting alone causes no transverse deflection and therefore cannot contribute a bending couple. The difficulty stems from coupling between the deformation due to the bending load M_0 and the effects of the axial load P. In other words, because the bending couple M_0 causes deflection w, the axial load P causes a bending couple $+P(w_0 - w)$, which, through Equation (7.2), also affects the transverse deflection.

Each of the preceding examples illustrates a type of situation in which the superposition procedure cannot be applied. Each example involves a nonlinear relationship: The example of Figure 7.3 involves a *nonlinear* material; the stress-strain relation is nonlinear. The example of Figure 7.4 involves a relatively large displacement; δ/L is *not* small compared to unity. This is a geometric *nonlinearity*. The example of Figure 7.6 involves a nonlinearity in the equilibrium equation, namely, a *product* of the applied force and the displacement $P(w_0 - w)$. This is essentially a geometric *nonlinearity*. Superpositions (additivity of effects) are invalidated by nonlinearities that might be attributed to the behavior of the material or to relatively large changes of *geometry*. Material nonlinearities arise in the stress-strain relations. Geometric nonlinearities manifest themselves in kinematic relations (e.g., strain-displacement equations) and also in dynamic relations (e.g., equilibrium equations). These cases are excluded in the following applications of superposition.

7.2 METHOD OF SUPERPOSITION

If a body is composed of a linearly elastic material, the stresses can be expressed as linear functions of the strains. If the body is subjected independently to each of two different equilibrated loads, then the sum of the two effects is essentially the same as the effect caused by the simultaneous application of the two loads, provided that the following conditions hold:

1. Neither force system creates strains or relative rotations that significantly alter the geometric configuration of the body.

2. Neither force system contributes an effect dependent on the presence and the effects of the other force system.

Mathematically speaking, superposition requires linear equations in the dependent variables; internal forces, stresses, strains, and displacements must all be linearly related in the equations governing the system (the material, equilibrium, and kinematics).

7.3 SUPERPOSITION OF STRESSES

ILLUSTRATIVE EXAMPLE 7.1

Combined Axial Loads: The steel truss in Figure 7.7a has a turnbuckle on the rod *OA*. Before applying the load *P*, the truss is prestressed via the turnbuckle, which has 20 threads per inch. We require the number (n) of turns, in order to equalize the axial stress in the three rods at the load $P = 8{,}000$ lb

To fully exemplify superposition, let us separate the problem into two parts: First, we determine the prestresses caused by $'n'$ turns. Second, we determine the stresses caused by the load *P*. We then superpose the two results and compute the number n which satisfies the condition of equal stresses.

Under axial force F a HI-HO rod experiences an elongation $e = FL/EA$. The rod *OA* is also contracted via the turnbuckle. These elongations are also expressed in terms of the *small* horizontal and vertical displacements of joint O, u, and v, respectively. Denoting the elongations of the rods *OA*, *OB*, and *OC* by subscripts *A*, *B*, and *C*, respectively, we have (in inches)

$$e_A = -\frac{n}{20} + \frac{120F_A}{EA_A} = \frac{4}{5}u + \frac{3}{5}v \tag{7.3a}$$

$$e_B = \frac{72F_B}{EA_B} = v, \quad e_C = \frac{120F_C}{EA_C} = -\frac{4}{5}u + \frac{3}{5}v \tag{7.3b,c}$$

From the symmetry of Figure 7.7b, we observe the equilibrium requirement: $F_C = F_A$. Also, we note that $A_C = A_A$. Then the addition of Equations (7.3a) and (7.3c) yields the following:

$$-\frac{n}{20} + \frac{240F_A}{EA_A} = \frac{6}{5}v$$

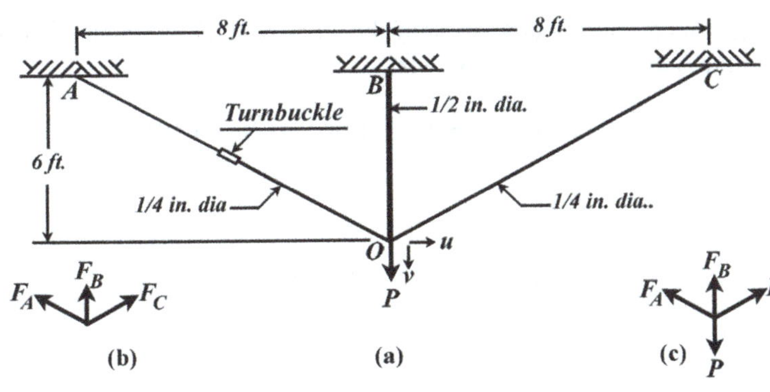

Figure 7.7

Displacement $'v'$ is eliminated by means of Equation (7.3b); then

$$\frac{240F_A}{EA_A} - \frac{6 \times 72F_B}{5EA_B} = \frac{n}{20} \tag{7.4}$$

Equilibrium of the forces requires that $F_B = -6F_A/5$. Therefore, we have

$$\left(240 + \frac{6 \times 72 \times 6A_A}{5 \times 5A_B}\right) F_A = \frac{EA_A}{20} n$$

Since $A_A/A_B = 1/4$, we obtain these results:

$$F_A = \frac{EA_A}{5.32} n \times 10^{-3}, \quad F_B = -\frac{EA_A}{4.43} n \times 10^{-3}$$

For steel $E = 3 \times 10^6$ psi; therefore

$$\sigma_A = \frac{F_A}{A_A} = 5.64 \times 10^3 n \text{ psi} \tag{7.5a}$$

$$\sigma_B = \frac{F_B}{A_B} = -1.69 \times 10^3 n \text{ psi} \tag{7.5b}$$

Next we turn to the truss under load P, as depicted in Figure 7.7c. Equation (7.4) holds with $n = 0$. However, the equilibrium condition requires

$$F_B = P - \frac{6}{5} F_A$$

Therefore,

$$F_A = \frac{45}{554} P \Rightarrow \sigma_A = \frac{45}{554} \frac{P}{A_A} v = 1.65P \tag{7.6a}$$

$$F_B = \frac{500}{554} P \Rightarrow \sigma_B = \frac{500}{554} \frac{P}{A_B} v = 4.60P \tag{7.6b}$$

To meet the requirement, the superposition of the stresses of (7.5a,b) and (7.6a,b) must result in equal stresses in the rods OA, OB, and OC, viz

$$5.64 \times 10^3 n + 1.65P = -1.69 \times 10^3 n + 4.60P$$

At $P = 8,000$ lb, we obtain $n = 3.2$.
With the prestress,

$$\sigma_A = \sigma_B = \sigma_C = 31,300 \text{ psi}$$

Without the prestress,

$$\sigma_A = 13,200 \text{ psi}, \quad \sigma_B = 36,800 \text{ psi}$$

Prestressing reduces the stress in member OB by 15%.

ILLUSTRATIVE EXAMPLE 7.2

Combined Axial Force and Bending Couple: The bracket depicted in Figure 7.8a is composed of steel, HI-HO to the yield stresses 280 MPa (tensile) and -350 MPa (compressive). It must support a force F, as shown, but must sustain no permanent deformation. It has the T-shaped cross section shown at (b). We require the force F that would initiate yielding.

From the free body at (c) we find that the action upon a cross section (A-A) consists of a tensile force F at the centroid and a couple $M = FD$, where D denotes the distance from the centroid to the line of action.

From the given dimensions of Figure 7.7b, we determine the area A, distance d to the centroid, and then the integral I with respect to the centroidal axis:

$$A = 2.5\ \text{cm}^2, \quad d = \frac{1.25 \times 1.75 + 1.25 \times 0.25}{2.5} = 1.00\ \text{cm}$$

$$I = \iint y^2 dA = 2.5 \int_{-1}^{-0.5} y^2 dy + 0.5 \int_{-0.5}^{2.0} y^2 dy = 2.08333\ldots\ \text{cm}^4$$

The stress at any point of the cross section can be obtained by superposing that attributed to the axial force $F(\sigma_x = F/A)$ and the couple $M(\sigma_x = -My/I)$:

$$\sigma_x = \frac{F}{A} - \frac{3.5Fy}{I}$$

The maximum and minimum values occur at $y = -1$ cm and $y = 2$ cm, respectively.

$$\sigma_x(\max) = \frac{F}{2.5} + \frac{3.5F}{2.083} = 2.08F$$

$$\sigma_x(\min) = \frac{F}{2.5} - \frac{3.5 \times 2F}{2.083} = -2.96F$$

The limiting values (yielding in tension and compression) follow:

$$F = \frac{280 \times 10^3}{2.08} N = 13.5\ \text{kN}$$

$$F = \frac{350 \times 10^3}{2.96} N = 11.8\ \text{kN}$$

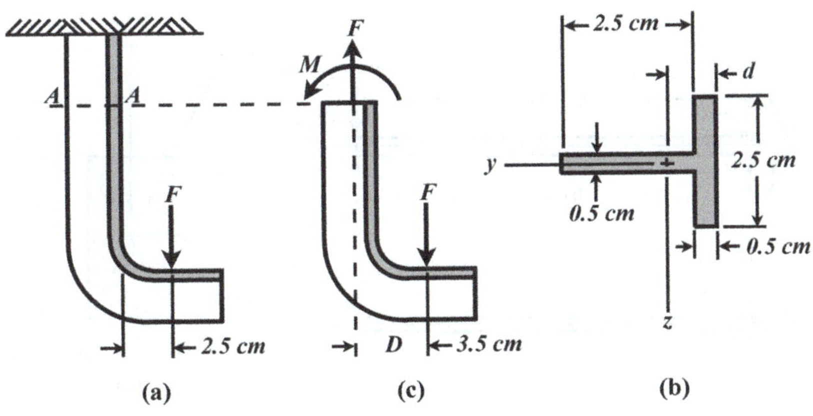

Figure 7.8

To avoid permanent deformation, the load must be less than 11.8 kN.

ILLUSTRATIVE EXAMPLE 7.3

Combined Bending in Two Planes of Symmetry: The cantilever beam of Figure 7.9 is linearly elastic and supports an end load P, which acts perpendicular to the axis of the beam and along a diagonal of the rectangular cross-section.

The line of action is not in a longitudinal plane of symmetry but passes through the central axis. To determine the effects of the load, we superpose the effects of the components, as shown in Figure 7.9b and c; each acts in a plane of symmetry:

$$P_y = \left(\frac{2}{\sqrt{5}}\right) 100 \text{ lb}, \quad P_x = \left(\frac{1}{\sqrt{5}}\right) 100 \text{ lb}$$

The maximum stress σ_x'; caused by P_y, alone, is at the end A and at the top in Figure 7.9a, where $y = 2$ in. and $M_x = -P_y L$:

$$\sigma_x' = -\frac{M_x y}{I_{yy}} = \frac{P_y L(2)}{I_{yy}}$$

The maximum normal stress σ_x'', produced by P_x, alone, is at end A and at the top of Figure 7.9c, where $z = 1$ in. and $M_y = P_x L$:

$$\sigma_x'' = \frac{M_y z}{I_{zz}} = \frac{P_x L(1)}{I_{zz}}$$

For the rectangular cross section,

$$I_{yy} \equiv \iint y^2 dA = \frac{32}{3} \text{ in.}^4, \quad I_{zz} \equiv \iint z^2 dA = \frac{8}{3} \text{ in.}^4$$

Both contributions, σ_x', and σ_x'', are a maximum at point Q of Figure 7.9. By superposition, the maximum normal stress is at Q, where

$$\sigma_x = \sigma_x' + \sigma_x'' = \frac{P_y L(2)}{I_{yy}} + \frac{P_x L(1)}{I_{zz}} = 1207 + 1207 = 2415 \text{ psi}$$

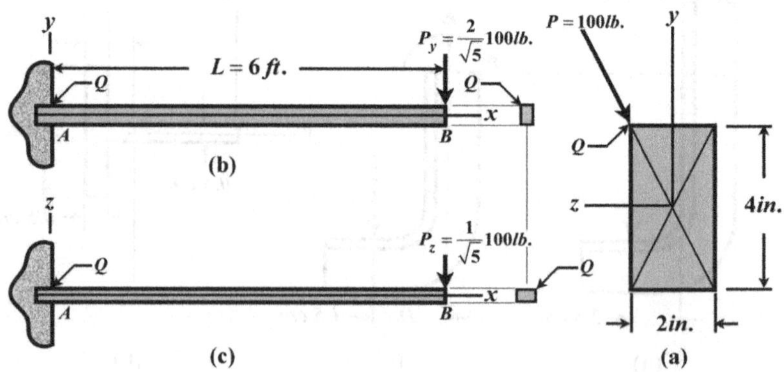

Figure 7.9

ILLUSTRATIVE EXAMPLE 7.4

Stresses in Bending and Torsion: The member *OBC* in Figure 7.10a is formed of standard $\frac{3}{4}$-in. steel pipe. Segment *OB* is joined to *BC* by a right elbow at *B* (assume a rigid joint) and clamped to a fixed support at *O*.

A concentrated force *Q* is applied perpendicular to the plane (*xz*) at end *C*. We can assume that the steel is HI-HO to the yield condition. Tension and torsion tests provide the yield stresses and moduli:

$$\sigma_0 = 48 \times 10^3 \text{ lb/in.}^2, \quad \tau_0 = 24 \times 10^3 \text{ lb/in.}^2, \quad E = 30 \times 10^6 \text{ lb/in.}^2$$
$$G = 12 \times 10^6 \text{ lb/in.}^2$$

Inner and outer diameters of the pipe are

$$\text{I.D.} = 0.824 \text{ in.}, \quad \text{O.D.} = 1.050 \text{ in.}$$

The cross-sectional integrals and areal moments of inertia follow:

$$I = \iint_A y^2 dA = 0.0370 \text{ in.}^4$$
$$J = \iint (y^2 + z^2)\, dA = 2I = 0.0741 \text{ in.}^4$$

Since the steel appears to yield in shear — that is, according to the Tresca criterion (see Section 4.17, Equations 4.48) — we want to compute the maximum shear stress.

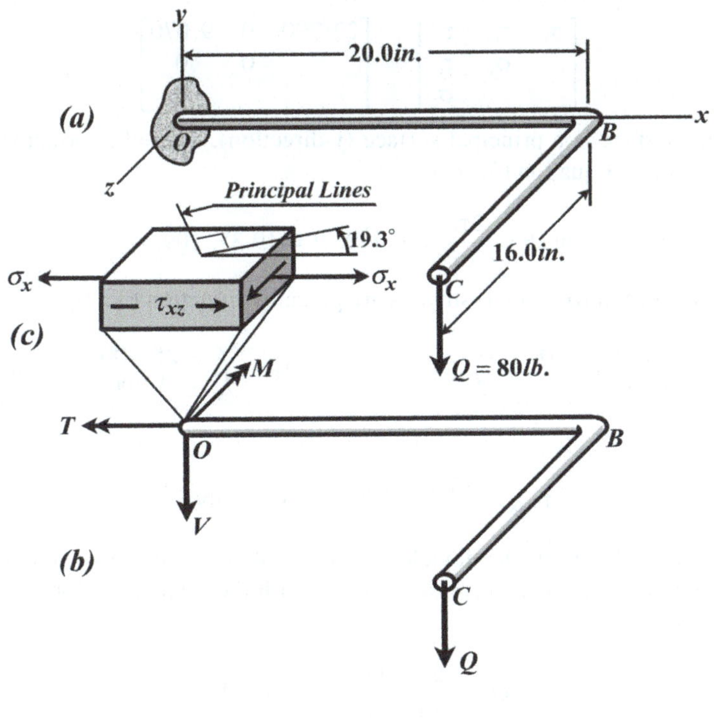

Figure 7.10

We can determine the internal force system on any cross section by considering the equilibrium of a free body such as that depicted in Figure 7.10b. Only the nonzero components are shown on the cross section through O, the clamped end. At O these assume their greatest values. From the equilibrium equations, we have

$$\sum M_x = 0: \quad T = 80 \times 16 = 1280 \text{ in.-lb}$$
$$\sum M_z = 0: \quad M = -80 \times 20 = -1600 \text{ in.-lb}$$
$$\sum F_y = 0: \quad V = -80 \text{ lb}$$

At this cross section, the maximum value of the normal stress σ_x occurs at the top (0, 0.525, 0). By Equation (5.127)

$$\sigma_X = -\frac{My}{I} = -\frac{(-1600)(0.525)}{0.0370} = 22,700 \text{ lb/in.}^2$$

At this cross section the maximum value of the shear stress (τ_{xz}) also occurs at the same location. That stress is caused by the twisting couple T and is given by Equation (5.52):

$$\tau_{xz} = \frac{Ty}{J} = \frac{(1280)(0.525)}{0.0741} = 9,070 \text{ lb/in.}^2$$

We note that the transverse shear force V causes a small shear stress τ_{xy} at the central line ($y = 0$). However, at that location, the dominant stress component σ_x is zero. Therefore, the maximum shear stress occurs at the site (0, 0.525, 0), where the state of stress has the two nonzero values just computed, namely,

$$\begin{bmatrix} \sigma_x & \tau_{xy} & \tau_{xz} \\ & \sigma_y & \tau_{yz} \\ & & \sigma_z \end{bmatrix} = \begin{bmatrix} 22,700 & 0 & 9,070 \\ & 0 & 0 \\ & & 0 \end{bmatrix}$$

The free upper surface is a principal surface (y-direction). The other principal directions are determined according to Equation (3.37b):

$$\tan 2\theta = \frac{2\tau_{xz}}{\sigma_x} = 0.799, \ \theta = 19.3°, \ 109.3°$$

The corresponding normal (principal) stresses are given by Equation (3.34):

$$\sigma \begin{pmatrix} \max \\ \min \end{pmatrix} = \frac{\sigma_x}{2} + \frac{\sigma_x}{2} \cos 2\theta + \tau_{xy} \sin 2\theta = \begin{pmatrix} +25,900 \\ -3,200 \end{pmatrix} \text{ lb/in.}^2$$

Recall that

$$\tau_{\max} = \frac{\sigma_{\max} - \sigma_{\min}}{2} = 14,550 \text{ lb/in}^2$$

The data for this material indicate that it yields at $\tau_{\max} = 24,000 \text{ lb/in}^2$. Since the behavior is linear to the onset of yielding, the stress increases linearly with the load. Therefore, we can anticipate yielding at the load.

$$Q_0 = \frac{24,000}{14,550} \times 80 = 132 \text{ lb}$$

7.4 DEFLECTIONS DUE TO COMBINED LOADS

ILLUSTRATIVE EXAMPLE 7.5

Deflections from Bending and Twisting: Let us now compute the deflection of the end C of the member in Figure 7.10. We do this by superposing (adding) the deflections caused by the bending and the twisting. We first divide the member at B and consider separately the segments OB and BC, as in Figure 7.11a and b, respectively. The segment OB is a cantilever subjected to an end load $Q = 80$ lb and a twisting couple $T = 1600$ in.-lb. The former causes a deflection of the end B, according to the flexure theory of Chapter 6:

$$\delta_1 = \frac{QL^3}{3EI} = \frac{80 \times (20)^3}{3 \times 30 \times 10^6 \times 0.0370} = 0.192 \text{ in.}$$

The twisting couple T causes a rotation of the cross section B, according to the torsion theory of Chapter 5:

$$\phi = \frac{TL}{GJ} = \frac{1280 \times 20}{12 \times 10^6 \times 0.0741} = 0.0288$$

The segment BC is also a cantilever but subject only to the bending caused by the end load Q. Disregarding (temporarily) any movement of the end B, the deflection of end C relative to end B is again given by the flexure theory:

$$\delta_3 = \frac{QL^3}{3EI} = \frac{80 \times 16^3}{3 \times 30 \times 10^6 \times 0.0370} = 0.0984 \text{ in.}$$

Next, we add the displacements caused by the bending of segment OB, the twisting of segment OB, and the bending of segment BC. The bending of part OB alone causes B and C to displace the amount 0.192 in. If no other deformation occurred, the portion BC would translate in the y-direction, carrying BC to $B'C'$, as shown in Figure 7.11c. However, the end B rotates the additional amount $\phi = 0.0288$, which would displace C' to position C''; that displacement is given by the first-order approximation ($\phi \ll 1$)

$$\delta_2 = \phi L = 0.0288 \times (16) = 0.461 \text{ in.}$$

Finally, the bending of the segment BC causes the additional deflection $\delta_3 = 0.0984$, so that the total deflection of end C is

$$\delta = \delta_1 + \delta_2 + \delta_3 = 0.192 + 0.461 + 0.098 = 0.751 \text{ in.}$$

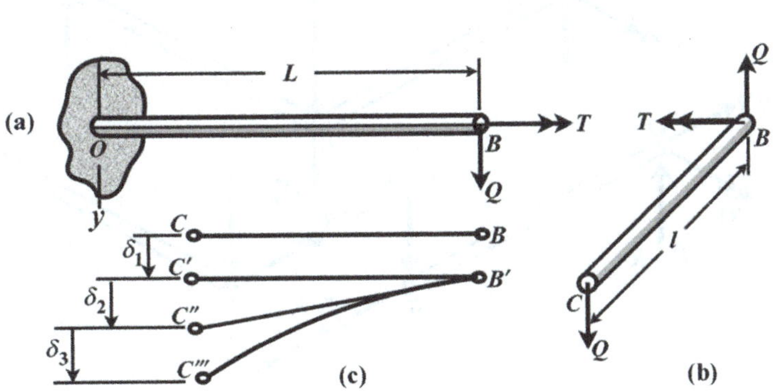

Figure 7.11

The successive movements, corresponding to the three terms, are depicted in Figure 7.11c.

ILLUSTRATIVE EXAMPLE 7.6

The steel bar in Figure 7.12a forms a right angle at point B. It is clamped at O and simply supported at C. A force $F = 5$ kN acts normal to the plane (x-z) at the corner B. The bar has a circular cross-section; in the notations of Sections 5.7 and 5.13, the properties follow:

$$I = 40.0\ \text{cm}^4, \quad J = 80.0\ \text{cm}^4, \quad E = 200 \times 10^3\ \text{MPa} = 200 \times 10^6\ \text{N/cm}^2, \quad \nu = 0.25$$

We require the deflection of corner B.

From the equilibrium conditions for the free body of Figure 7.12b, we have

$$Q = F - R, \quad M = 75(F - R), \quad T = 75R$$

The rotation at end B, i.e. twist, is

$$\phi_B = \frac{TL}{GJ} = \frac{2(1+\nu)TL}{EJ} = \frac{2.5 \times 75^2 R}{80 \times 200 \times 10^5} = 0.8789 \times 10^{-5}$$

The deflection at B is caused by bending.

$$w_B = -\frac{(F-R)L^3}{3EI} = -\frac{(F-R)75^3}{3(200 \times 10^5)40} = -17.58 \times 10^{-5}(F - R)$$

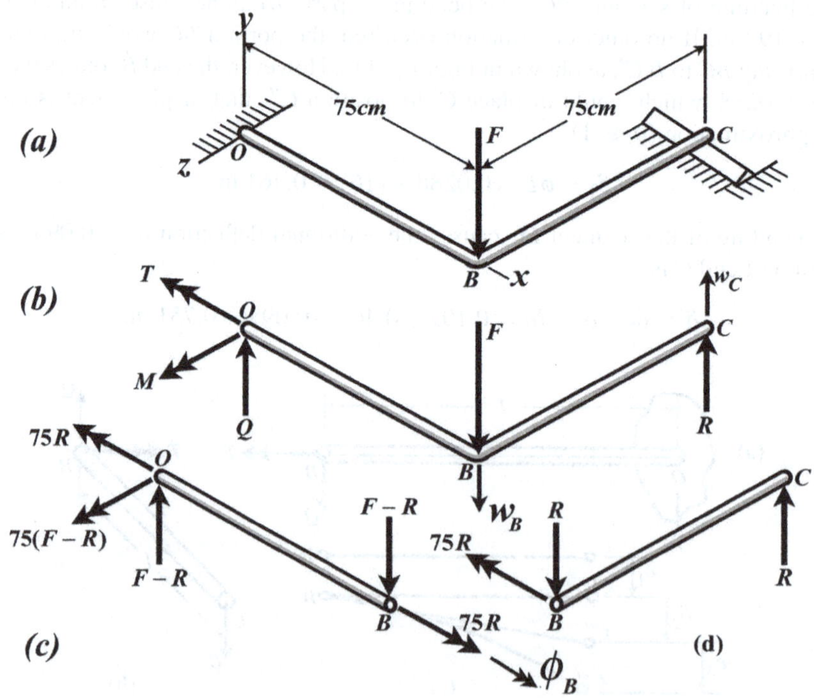

Figure 7.12

The deflection at C *relative* to B is caused by the rotation θ_B, but also by the bending of CB:

$$w_{CB}^{B} = \frac{RL^3}{3EI} = \frac{R\,75^3}{3(200 \times 10^5)40} = 17.58 \times 10^{-}5\,R$$

The deflection (w_c at C must vanish. This is obtained by superposing the deflection at B and the deflection at C relative to B. The latter has two parts: Due to the twist at B, we have

$$w_{CB}^{T} = \phi_B 75$$

Therefore, the total displacement at C is

$$w_C = -w_B + \phi_B 75 + w_{CB}^{B}$$
$$O = [-17.58(F - R) + 7.5 \times 8.789R + 17.58R]\,10^{-5}$$

With $F = 5 \times 10^3$ N, we obtain

$$R = 870\text{ N} = (196\text{ lb})$$
$$w_B = 17.58(F - R) \times 10^{-5} = 0.726\text{ cm } (0.286\text{ in})$$

7.5 COMBINED LOADS ON SYMMETRIC BEAMS

ILLUSTRATIVE EXAMPLE 7.7

Deflection of a Beam: A simply supported beam with $EI = 10^8$ lb-in.^{2} is loaded as shown in Figure 7.13a.

To find the deflection at point A by the method of superposition, we consider the beams of Figure 7.13b and c. Note that we have two beams in these figures with different loads but the *same end conditions* as the original beam. From Illustrative Example 6.16 the deflection at x of a simply supported beam carrying a concentrated load P at a distance a from the left end of the beam is

$$w = \begin{cases} -\dfrac{P}{6EIL}(L-a)\left[(2L-a)\,ax - x^3\right], & x \le a \\ -\dfrac{P}{6EIL}(L-x)\left[(2L-x)\,ax - a^3\right], & x \ge a \end{cases}$$

Therefore, the deflection of the beam in Figure 7.12b at $x = 6$ ft is

$$w_{1|A} = 0.18\text{ in.}$$

and for the beam in Figure 7.7c,

$$w_{2|A} = 0.57\text{ in.}$$

On adding these two, we obtain the deflection at $x = 6\,ft$ for the beam in Figure 7.12a

$$w_{|A} = w_{1|A} + w_{2|A} = 0.75\text{ in.}$$

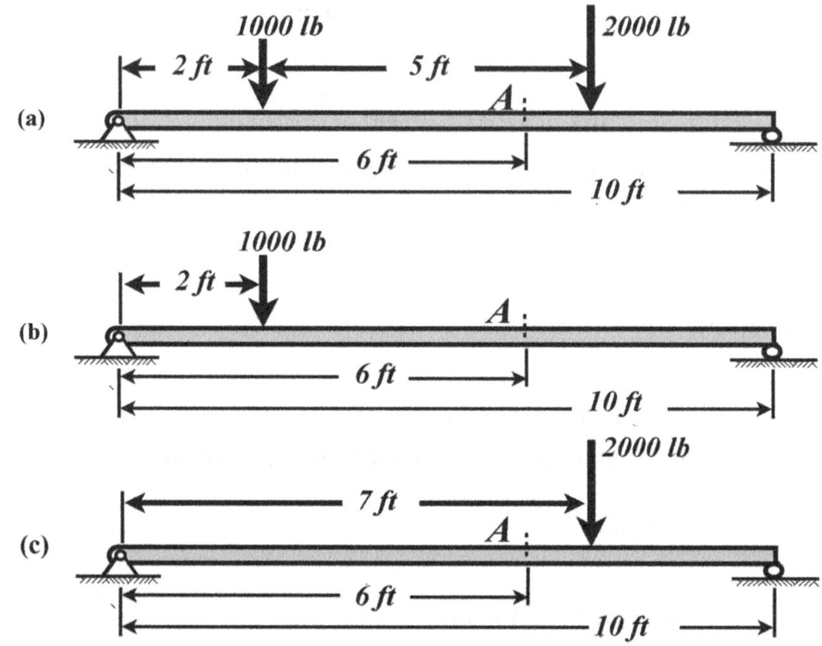

Figure 7.13

ILLUSTRATIVE EXAMPLE 7.8

A Statically Indeterminate Problem: The reactions on the beam of Figure 7.14 are statically indeterminate. We can employ the superposition principle to obtain the reaction at *B*. To do this, we consider the two statically determinate beams of Figure 7.14b and c.

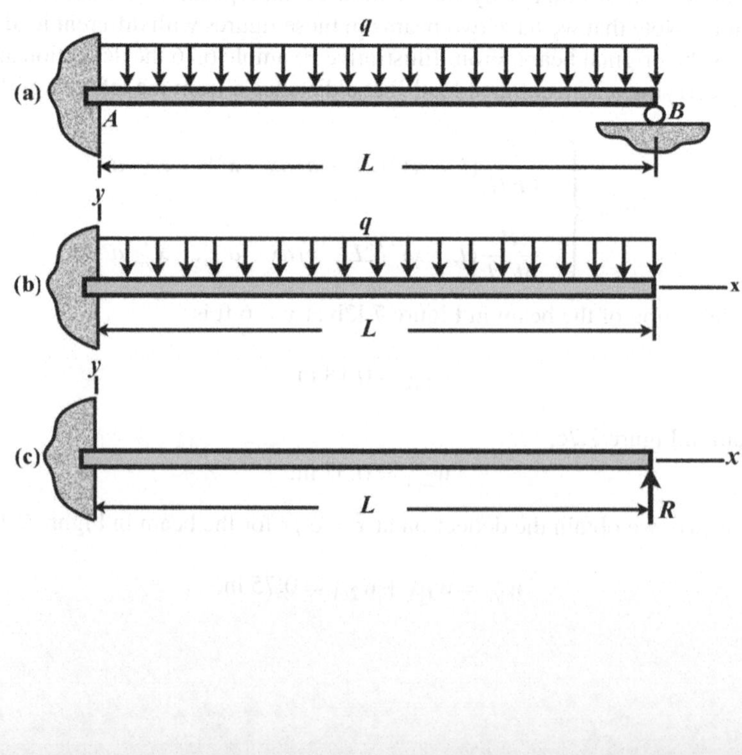

Figure 7.14

The first is a cantilevered beam loaded by the uniform load q. The second cantilevered beam is subjected only to an unknown force R acting at B. (This force is created by the roller in the original problem.) For the first beam,

$$w_1(L) = -\frac{qL^4}{8EI}$$

For the second beam,

$$w_2(L) = \frac{RL^3}{3EI}$$

Since $w_1(L) + w_2(L) = 0$ (due to the roller at B), we have

$$-\frac{qL^4}{8EI} + \frac{RL^3}{3EI} = 0$$

or

$$R = \frac{3}{8}qL$$

ILLUSTRATIVE EXAMPLE 7.9

Combined Bending in Two Planes of Symmetry: Let us revisit the cantilevered beam of Figure 7.9. The end load P acts perpendicular to the axis of the beam and along a diagonal of the rectangular cross section. This load is not in a longitudinal plane of symmetry; however, it does pass through the central axis. Suppose that the material of the beam (wood) is linearly elastic with $E = 1.5 \times 10^6$ lb/in^2.

To determine the effect of the load P, we superimpose the effects of its components $P_y = (2/\sqrt{5})100$ lb and $P_z = (1/\sqrt{5})100$ lb shown in Figure 7.9b and c. Notice that each component does act in a longitudinal plane of symmetry. The load P, acting alone would cause bending in the xy-plane. The deflection of end B that would be caused by the load P_y, is

$$w_y(B) = -\frac{P_y L^3}{3EI_{yy}}$$

Similarly, the deflection that would be caused by the load P_z acting alone is

$$w_z(B) = -\frac{P_z L^3}{3EI_{zz}}$$

in the xz-plane. Here, I_{zz} and I_{yy} denote the area moments of inertia with respect to coordinates z and y:

$$I_{zz} = \frac{8}{3}\text{ in.}^4, \quad I_{yy} = \frac{32}{3}\text{ in.}^4$$

Substituting numerical values, we have

$$w_y(B) = -0.696\text{ in.}$$
$$w_z(B) = -1.39\text{ in.}$$

7.6 THERMAL STRESSES

Previously (Section 4.9) *small* strains from changes of temperature were superposed upon other small strains of Hookean solids. Those linear strain-stress-temperature relations were subsequently employed (Illustrative Examples 5.3 and 5.30) to illustrate procedures of analysis. We now consider some additional problems in which the stresses and strains originate with changes of temperature.

ILLUSTRATIVE EXAMPLE 7.10

A steel pipe carries hot water. The pipe passes through two fixed walls 20 ft apart, as shown in Figure 7.15a. The pipe is rigidly fastened to the walls at A and B. When the pipe was installed (stress-free), the temperature was 60°F. Because of poor insulation the temperature of the water entering at A drops from 210° at A to 160° at B in a linear manner. Let us determine the normal stress on a cross section of the pipe. The temperature at x (in inches) is

$$Q(x) = 160 + 50\left(1 - \frac{x}{240}\right)$$

The temperature change from the initial value (60°) is

$$\Delta Q(x) = 50\left(3 - \frac{x}{240}\right)$$

Since the pipe tends to expand, the walls will exert an axial force P, as shown in Figure 7.15b.

This must be accompanied by a stress σ_x. By the assumptions previously used in axial extension, the stress, σ_x, is constant on a cross section. Again, we assume that $\sigma_y = \sigma_z = 0$. Then, the stress-strain temperature relation (see Section 4.9) yields

$$\varepsilon_x = \frac{\sigma_x}{E} + \alpha \Delta Q$$

For this steel the modulus E and coefficient α are

$$E = 30 \times 10^6 \text{ lb/in.}^2, \quad \alpha = 6.5 \times 10^{-6}/\text{oF}$$

Then

$$\varepsilon_x = \frac{1}{3} \times 10^{-7}\sigma_x + 3.25 \times 10^{-4}\left(3 - \frac{x}{240}\right)$$

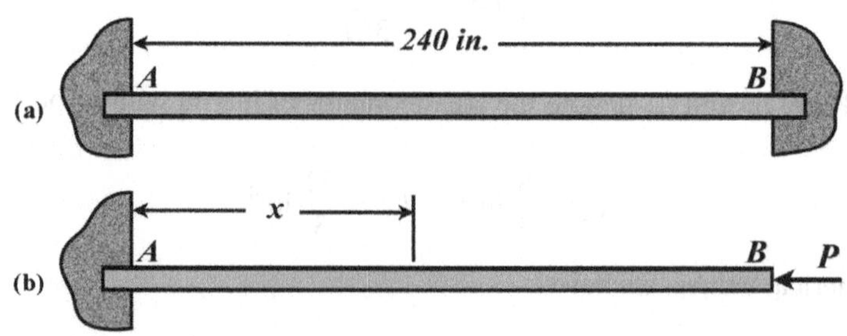

Figure 7.15

The length of pipe between A and B cannot change; therefore,

$$e = \int_0^{240} \varepsilon_x \mathrm{d}x = 0$$

or

$$\int_0^{240} 3.25 \times 10^{-4} \left(3 - \frac{x}{240}\right) \mathrm{d}x + 8 \times 10^{-6} \sigma_x = 0.195 + 8 \times 10^{-6} \sigma_x = 0$$

Hence, the normal stress in the pipe is

$$\sigma_x = -24{,}400 \text{ lb/in.}^2$$

Note that σ_x does not depend on the cross-sectional area of the pipe.

ILLUSTRATIVE EXAMPLE 7.11

A cylindrical aluminum tube has an internal diameter of 2 in. at ambient temperature and a thickness of 0.10 in. It just slips over a steel tube, also 0.10 inches thick when heated 200°F, We can assume that there is insufficient friction to inhibit relative axial movement. The ends are free so that neither is subject to axial stress. We require the circumferential stress in each when cooled to the ambient temperature.

Let D_S denote the common diameters of the interface when assembled at the elevated temperature. If α_A denotes the coefficient of thermal expansion of the aluminum, then

$$(2.00)(1 + \alpha_A T) = D_S$$

where T denotes the temperature increase (200°F). Let D denote the common diameter when both reach the ambient temperature. Then,

$$\sigma_A = E_A \varepsilon_A = E_A (D - 2.00)/2.00$$
$$\sigma_S = E_S \varepsilon_S = E_S (D - D_S)/2.00$$

Here σ_A and σ_S denote the stresses at the interface. Eliminating D, we obtain

$$\sigma_S - \frac{E_S}{E_A} \sigma_A = -E_S \alpha_A T \tag{A}$$

Equilibrium dictates that the force on the composite vanish (see Figure 7.16)

$$\bar{\sigma_A} + \bar{\sigma_B} = 0$$

where the bar signifies the mean value on each cylinder. A kinematic analysis reveals that the *strain* variation on a cross is approximately 5%. If we accept a comparable variation in the stresses, then

$$\bar{\sigma_A} = 0.95\sigma_A, \quad \bar{\sigma_S} = 1.05\sigma_S$$

and

$$\sigma_A = -1.11\sigma_S \tag{B}$$

Simultaneous solution of the two Equations (A) and (B) provides our result:

$$\left(1 + 1.11 \frac{E_S}{E_A}\right) \sigma_S = -E_S \alpha_A T$$

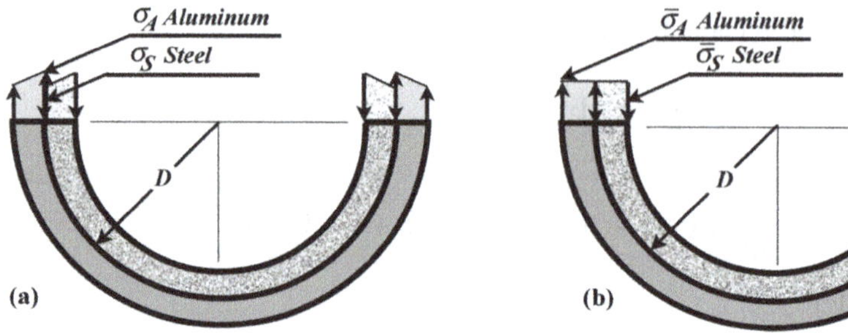

Figure 7.16

Typical properties are $E_S = 30 \times 10^6$ psi, $E_A = 10 \times 10^6$ psi, $\alpha_A = 13 \times 10^6/^\circ$F. Therefore,

$$\sigma_S = -18{,}000 \text{ psi}, \quad \sigma_A = 20{,}000 \text{ psi}$$
$$\bar{\sigma}_S = -19{,}000 \text{ psi} = -\bar{\sigma}_A$$

Equilibrium of either component provides the compressive radial stress at the interface (−190 psi)

ILLUSTRATIVE EXAMPLE 7.12

A bimetallic strip has the initial circular form as shown in Figure 7.17a.

The inner metal is a copper alloy designated B; the outer is steel S. Properties are as follows:

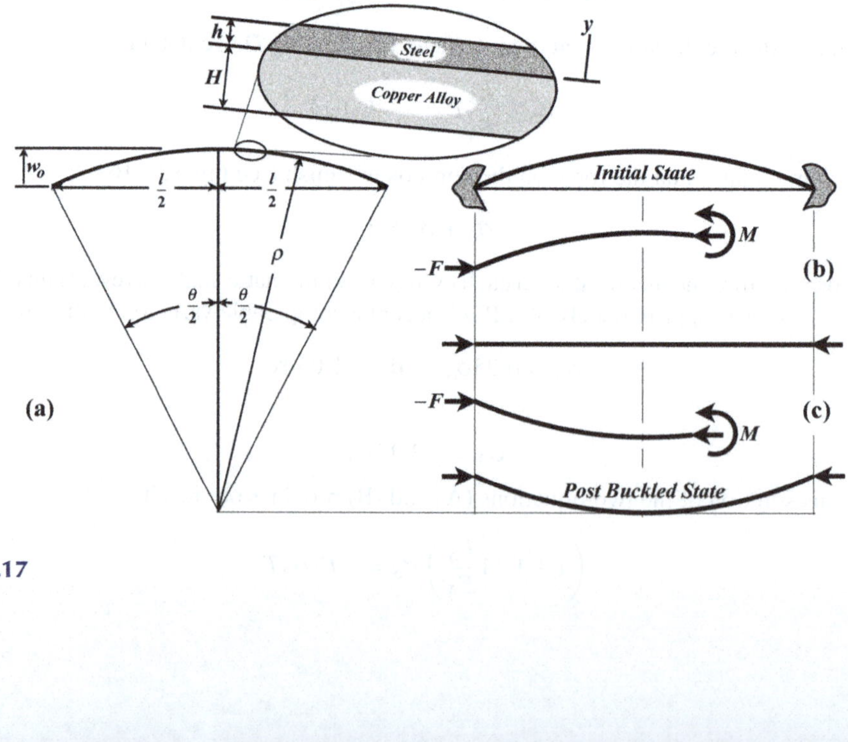

Figure 7.17

$$H = 2\times 10^{-1}, \qquad h = \sqrt{2}\times 10^{-1}\ \text{cm}$$
$$E_B = 100\times 10^5\ \text{N/cm}^2, \qquad E_S = 200\times 10^5\ \text{N/cm}^2$$
$$\alpha_B = 24\times 10^{-6}/^\circ C, \qquad \alpha_S = 12\times 10^{-6}/^\circ C$$

When unconstrained, uniform heating causes deformation, extensional strain and curvature. With the reference surface at the interface,

$$\varepsilon = \varepsilon_0 - y\kappa$$

The notations and sign conventions are those of the previous Chapters 4 and 6. κ is the *change* of curvature; the initial curvature κ_0 is negative, the initial displacement w_0 is positive.

The strip is initially stress-free when placed between the simple supports. When heated, the strip exerts a force F upon the support, which then imposes the axial force $-F$ on the strip. Our intent is to select an initial curvature such that the straight configuration coincides with the critical load at a specific temperature increase ΔT. This assures a decisive negative displacement at the chosen temperature.

From Illustrative Example 5.29, we recall the relations for a HI-HO bimetallic strip:

$$\sigma_x = E_B(\varepsilon_0 - \kappa y) - E_B\alpha_S\Delta T, \quad -H < y < 0 \tag{7.7a}$$
$$\sigma_x = E_S(\varepsilon_0 - \kappa y) - E_S\alpha_S\Delta T, \quad 0 < y < h \tag{7.7b}$$

We define the following dimensionless variables, f (force), m (moment), ε_0 (extensional strain at $y = 0$), $\bar{\kappa}$ (curvature change), ΔT (temperature increase) and constants, $(A, B, C, D, \bar{I}, \text{and} K)$:

$$f \equiv \frac{F}{E_B H}, \qquad m \equiv \frac{M}{E_B H^2} \qquad \bar{\kappa} \equiv H\kappa, \qquad \bar{T} \equiv \alpha_B \Delta T$$

$$A \equiv 1 + \frac{E_S h}{E_B H} = 2.4142, \quad B \equiv \frac{1}{2}\left(1 + \frac{E_S h^2}{E_B H^2}\right) = 1, \qquad C \equiv \left(1 + \frac{E_S \alpha_s h}{E_B \alpha_B H}\right) = 1.707$$
$$D \equiv \frac{1}{2}\left(1 - \frac{E_S h^2}{E_B H^2}\right) = 0, \quad J \equiv \frac{1}{3}\left(1 + \frac{E_S h^3}{E_B H^3}\right) = 0.56904, \quad K \equiv \frac{1}{2}\left(1 - \frac{E_S \alpha_S h^2}{E_B \alpha_B H^2}\right) = 0.25000$$

The force F and the couple M (per unit width N/cm) are the integrals:

$$F = \int_{-H}^{h} \sigma_x\, dy, \quad M = -\int_{-H}^{h} \sigma_x y\, dy$$

where the stress σ_x is given by Equations (7.7a,b); in the dimensionless quantities:

$$f = A\varepsilon_0 + B\bar{\kappa} - C\bar{T} \tag{7.8a}$$
$$m = D\varepsilon_0 + J\bar{\kappa} - K\bar{T} \tag{7.8b}$$

Here $\bar{\kappa}$ and $\bar{T}$ are the dimensionless versions of the *change of curvature and change of temperature.*

Our solution requires a geometrical analysis of the circular arc. In Figure 7.17, ρ denotes the radius of curvature, w the central displacement, L the arc length and l the chord, The relations follow:

$$L = \rho\theta, \quad l = 2\rho \sin\frac{\theta}{2}, \quad w = \rho(1 - \cos\frac{\theta}{2}), \quad \kappa \equiv \frac{1}{\rho} = \frac{2\sin\frac{\theta}{2}}{l}$$

$$\frac{w}{l} = \frac{1 - \cos\frac{\theta}{2}}{2\sin\frac{\theta}{2}} \approx \frac{\theta}{8} \approx \frac{l\kappa}{8}$$

For pinned ends, the buckling load follows:

$$P_{CR} = \frac{\pi^2}{l^2}\overline{EI} = \frac{\pi^2 E_B H^3 J}{l^2} = 7{,}020 \text{ N/cm} \Rightarrow \frac{P_{CR}}{E_B H} \equiv p_{CR} = 3.510 \times 10^{-3}$$

Here we adopt the approximation:

$$\overline{EI} \approx \int_{-H}^{0} E_B y^2 dy + \int_{0}^{h} E_s y^2 dy$$

In the straight configuration, with or without the end supports, $M = m = 0$. Then (see Equation 7.8b),

$$\bar{T} = \frac{J}{K}\bar{\kappa} \tag{7.9}$$

and (see Equation 7.8a)

$$f = A\varepsilon_0 + \left(B - \frac{CJ}{K}\right)\bar{\kappa} \tag{7.10}$$

At a temperature $\Delta T = 110°$ ($\bar{T} = 2.64 \times 10^{-3}$) Equation (7.9) gives us $\kappa = 1.160 \times 10^{-3}$. If the strip expands freely ($f = 0$) from the initial curvature $\bar{\kappa}_0 = 1.160 \times 10^{-3}$ to the straight configuration, we obtain from Equation (7.10)

$$A\varepsilon_0 = 3.347 \times 10^{-3}$$

If constrained by the simple supports, we must superpose a force

$$f = -3.347 \times 10^{-3}, \qquad F = -6{,}694 \text{ N/cm}$$

If the temperature change is $\Delta T = 120°\text{C}$, we obtain

$$\kappa = 6.3264 \times 10^{-3}, \qquad F = -7{,}020 \text{ N/cm}$$

This is slightly above the buckling force (7,020 N/cm) and suggests a deflection beyond the straight position; if the initial arc had curvature $\kappa_0 = -6.3264 \times 10^{-3}$

$$w_0 \approx -\frac{l^2\kappa_0}{8} = 0.056 \text{ cm } (0.020 \text{ in})$$

Note that the progression of deformation from the initial configuration ($\kappa = \varepsilon_0 = 0$) to the straight position ($\kappa = \kappa_0$) requires a far more complicated analysis which must accommodate the nonlinear equilibrium equation (see Figure 7.17b and c), viz.

$$M - Fw = 0$$

This appears to be the innocent linear equation which is employed in our analysis of buckling (see Section 6.9). There the force is the constant $F = -P_{CR}$ and $M = EIw''$. Here F and w are variables; initially $(\Delta T = 0), F = 0, w = w_0$. In the straight configuration $F \approx -P_{CR}$ and $w = 0$. At any intermediate state, these variables must satisfy the non-linear equation.

The bimetallic strip might serve as a thermostat to activate a switch at $\Delta T \approx 120°$ and $w \leq 0.020$ in. A far better element is the shallow bimetallic shell depicted in Figure 7.18a. The inner metallic layer has a higher coefficient of thermal expansion. As the element is heated, the shell becomes increasingly shallow. The dimensions (a, b, and h) can be adjusted such that the shell abruptly snaps to a nearly inverted form; this will occur near the flat position. The effect is a very pronounced movement and a decisive switching mechanism. A characteristic plot of temperature versus displacement is depicted in Figure 7.19. The snap-buckling is exhibited by the line AB at temperature T_1 and the subsequent return by line CD at temperature T_2. This is a non-linear problem founded on the *linear* stress-strain-temperature relations but requires also *nonlinear* mathematical equations. These can be solved via the computational procedure originally devised by the author and dubbed the "generalized arc length" method.[1]

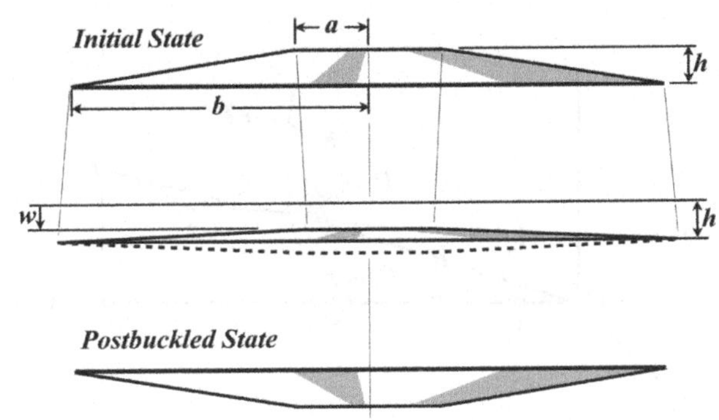

Figure 7.18

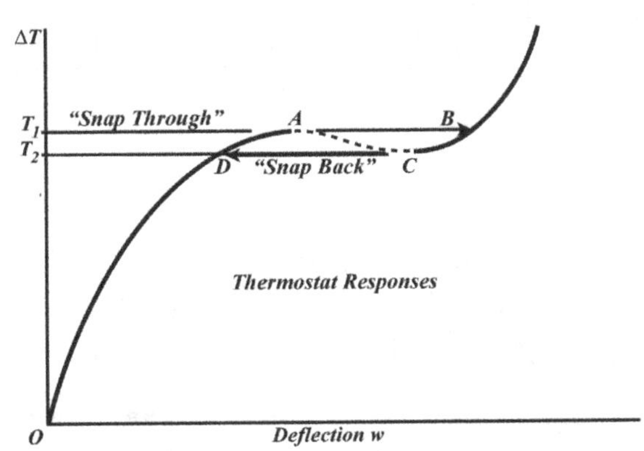

Figure 7.19

7.7 BENDING OF NONSYMMETRIC BEAMS

In some circumstances, the engineer employs members that do not have symmetric cross sections. Often these are required to support loads that cause bending. Then the axis is not generally deformed to a plane curve but to a three-dimensional spatial curve. Still, the curvature and strains are small enough to admit superposition.

Suppose that an axial line lies initially along the axis of Figure 7.20a. Under loads, that line is deformed to the curve A^*B^*.

Let us place the line AB on the neutral axis, i.e. at the centroid of the cross-section. The position to a point P^* on the deformed line A^*B^* is given by the vector $\mathbf{R}$. Since the neutral line is not stretched ($\varepsilon_0 = 0$), the unit tangent vector $\hat{\mathbf{T}}$ is

$$\hat{\mathbf{T}} = \frac{\mathrm{d}\mathbf{R}}{\mathrm{d}x}$$

The curvature κ is given by the second derivative:

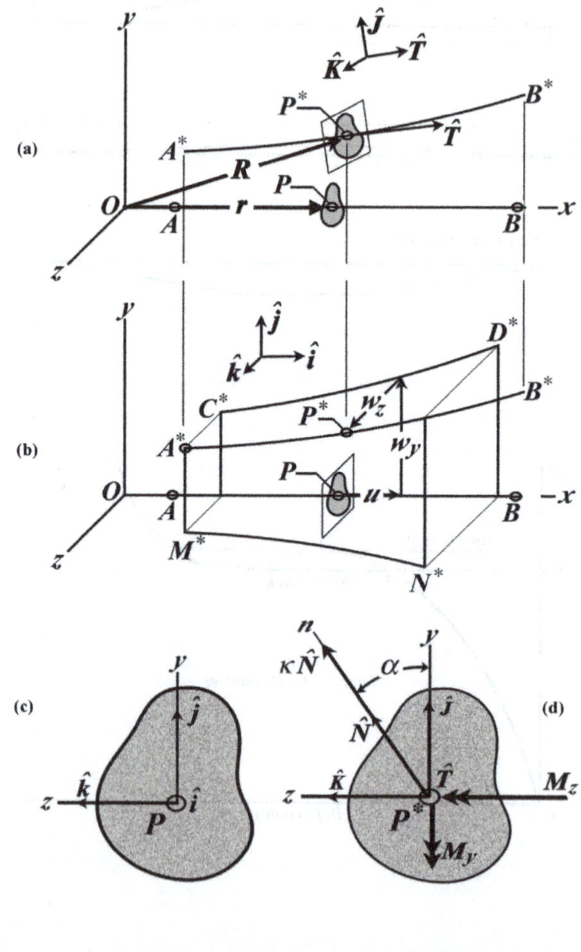

Figure 7.20

$$\frac{d^2\mathbf{R}}{dx^2} = \kappa\hat{\mathbf{N}}$$

where the unit vector $\hat{\mathbf{N}}$ is normal to vector $\hat{\mathbf{T}}$. In accordance with the Bernoulli hypothesis (the normal cross-section at P remains normal to the deformed curve at P^*):

$$\varepsilon_x = -\kappa n$$

where n is the distance along the normal $\hat{\mathbf{N}}$ as depicted in Figure 7.20d. Note that the vector $K\hat{\mathbf{N}}$ can be decomposed as follows:

$$\kappa\hat{\mathbf{N}} = \kappa_y\hat{\mathbf{J}} + \kappa_z\hat{\mathbf{K}}$$

Presumably, the vectors $\hat{\mathbf{J}}$ and $\hat{\mathbf{K}}$ are parallel to the initial x-y and x-z planes, respectively, as we anticipate no twisting. κ_y and κ_z are curvatures as perceived in the x-y and x-z planes, i.e., these are the curvatures of the projections, C^*D^* and M^*N^*, of the curve A^*B^* in Figure 7.20b. From Figure 7.20d,

$$\kappa_y = \kappa\cos\alpha, \quad \kappa_z = \kappa\sin\alpha \tag{7.11a,b}$$

If our beam is HI-HO and we assume a simple state of stress ($\sigma_y = \sigma_z = 0$), then

$$\sigma_x = E\varepsilon_x = -E\kappa n$$

The components of moments M_z and M_y (see Figure 7.20d) follow:

$$M_z = -\iint_A \sigma_x y\mathrm{d}A = E\kappa\iint_A ny\mathrm{d}A \tag{7.12a}$$

$$M_y = -\iint_A \sigma_x z\mathrm{d}A = E\kappa\iint_A nz\mathrm{d}A \tag{7.12b}$$

The coordinate n is expressed in terms of y and z as follows:

$$n = y\cos\alpha + z\sin\alpha \tag{7.13}$$

When Equations (7.11a,b) and (7.13) are substituted into Equations (7.12a,b), we obtain

$$M_z = EI_{yy}\kappa_y + EI_{yz}\kappa_z \tag{7.14a}$$

$$M_y = EI_{yz}\kappa_y + EI_{zz}\kappa_z \tag{7.14b}$$

where

$$I_{yy} = \iint_A y^2\mathrm{d}A, \quad I_{yz} = \iint_A yz\mathrm{d}A, \quad I_{zz} = \iint_A z^2\mathrm{d}A$$

The integral I_{yz} couples the moments, M_z and M_y, to both components of curvature. However, we can show (see Appendix B) that the three integrals (I_{yy}, I_{yz}, I_{zz}) transform as components of a Cartesian tensor, i.e. just as the stress components (σ_x, τ_{xy}, σ_y) (see Section 3.12). This means that one can always identify two principal directions; these provide orthogonal coordinates for which $I_{yz} = 0$. In our subsequent analyses of bending, we employ these principal directions. Then,

$$M_z = EI_{yy}\kappa_y, \quad M_y = EI_{zz}\kappa_z \tag{7.15a,b}$$

Note that these relations incorporate no restrictions upon the magnitudes of rotations or displacements.

The relations between the curvatures and the *small* rotations and between the rotations and the *small* deflections are similar to the earlier results (see Section 6.5, Equations 6.26 and 6.34). We then have the curvatures (κ_y and κ_z), rotations (θ_y and θ_z), and displacements (w_y and w_z),

$$\frac{\mathrm{d}\theta_y}{\mathrm{d}x} \approx \kappa_y, \quad \frac{\mathrm{d}\theta_z}{\mathrm{d}x} \approx \kappa_z \tag{7.16a,b}$$

$$\frac{\mathrm{d}w_y}{\mathrm{d}x} \approx \theta_y, \quad \frac{\mathrm{d}w_z}{\mathrm{d}x} \approx \theta_z \tag{7.17a,b}$$

Let us now turn to the equations of equilibrium. The transverse shear forces, V_y and V_z, and the bending couples, M_y and M_z, are shown in Figure 7.21.

Now, as before, let us presume small rotations, so that a slice of the beam appears as shown in Figure 7.22a and b.

The notations and sign conventions are those indicated and are consistent with the previous development (see Figure 6.8). The free-body diagram of a slice and the equilibrium conditions provide four differential equations, two like Equation (6.8) and two like Equation (6.9):

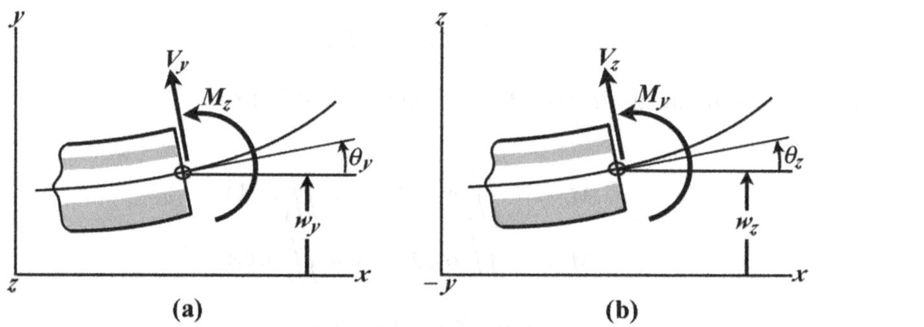

Figure 7.20

Figure 7.21

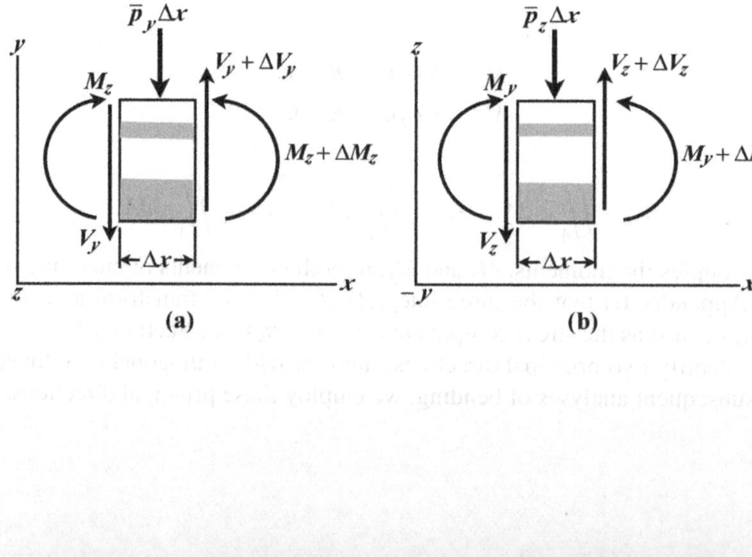

Figure 7.22

$$\frac{dV_y}{dx} = p_y(x), \quad \frac{dV_z}{dx} = p_z(x) \tag{7.18a,b}$$

$$\frac{dM_z}{dx} = -V_y(x), \quad \frac{dM_y}{dx} = -V_z(x) \tag{7.19a,b}$$

These equations of equilibrium are linear and uncoupled; the variables V_y and M_z are related only to the load p_y, whereas Vz and M_y are related only to p_z. We must now turn to the relations between the kinematic variables (κ_z, κ_y) and the internal actions (M_z,M_y), which depend upon the behavior of the material. Let us assume the simplest behavior, a HI-HO material. As before, we assume that the transverse normal stresses, σ_y and σ_z, are negligible. Then the Equations (7.15a,b) apply. This provides a system of ten linear equations, viz. (7.18a,b), (7.19a,b), (7.15a,b), (7.16a,b), and (7.17a,b), which relate the ten variables, viz. V_y, V_Z, M_y, M_z, κ_y, κ_z, θ_y, θ_z, w_y, w_z. Five equations and five variables relate to the bending in plane x-y and five to bending in plane x-z. In order,

$$\frac{dV_y}{dx} = p_y(x), \quad \frac{dV_z}{dx} = p_z(x) \tag{7.20a,b}$$

$$\frac{dM_z}{dx} = -V_y(x), \quad \frac{dM_y}{dx} = -V_z(x) \tag{7.21a,b}$$

$$M_z = EI_{yy}\kappa_y, \quad M_y = EI_{zz}\kappa_z \tag{7.22a,b}$$

$$\frac{d\theta_y}{dx} = \kappa_y, \quad \frac{d\theta_z}{dx} = \kappa_z \tag{7.23a,b}$$

$$\frac{dw_y}{dx} = \theta_y, \quad \frac{dw_z}{dx} = \theta_z \tag{7.24a,b}$$

If we successively substitute the last in the preceding equation into the preceding, etc., then we obtain

$$\frac{d^4w_y}{dx^4} = -\frac{p_y(x)}{EI_{yy}}, \quad \frac{d^4w_z}{dx^4} = -\frac{p_z(x)}{EI_{zz}} \tag{7.25a,b}$$

In the process, we relate the actions (M_z, V_y, M_y, V_z) to the displacements (w_y, w_z):

$$M_z = EI_{yy}\frac{d^2w_y}{dx^2}, \quad M_y = EI_{zz}\frac{d^2w_z}{dx^2} \tag{7.26a,b}$$

$$V_y = -EI_{yy}\frac{d^3w_y}{dx^3}, \quad V_z = -EI_{zz}\frac{d^3w_z}{dx^3} \tag{7.27a,b}$$

In addition to the equations governing the flexure, we can superpose the equation governing extension, viz.

$$F = EA\varepsilon_0 \tag{7.28}$$

where ε_0 is the strain of the centroidal axis. Superposing the stresses caused by extension and bending, we obtain[2]

$$\sigma_x = \frac{F}{A} - \frac{M_z y}{I_{yy}} - \frac{M_y z}{I_{zz}} \tag{7.29}$$

Here the origin is at the centroid, and the coordinates (y, z) are in principal directions. The three terms on the right represent the contributions from extension and bending in each of the principal planes. Limitations of the foregoing theory must be noted. Equations (7.22a,b), (7.25a,b), (7.26a,b), (7.27a,b), (7.28), and (7.29) apply only to HI-HO materials. The linear equations of

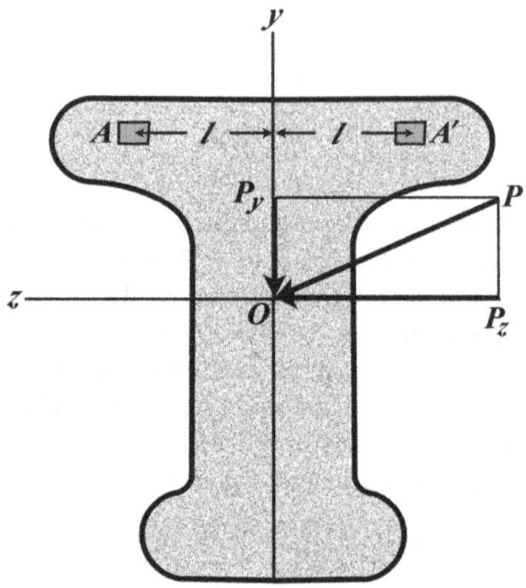

Figure 7.23

equilibrium, (7.20a,b) and (7.21a,b), neglect the rotation and curvature of the element. Also, the linear kinematic equations (7.24a,b) are valid only for small rotations.

One further note of caution is necessary: Any transverse load, whether distributed $p(x)$ or concentrated P, can be decomposed into components in the principal directions so that bending in these directions is uncoupled; then superposition of the bending effects is possible. However, if the principal lines are not also lines of symmetry, the loads could cause twisting. For example, the transverse load P in Figure (7.23) can be decomposed into components P_y, and P_z. The component P_y causes bending in the xy-plane (LPS), *but* the xz-plane is not a longitudinal plane of symmetry. Without additional theory and analysis, we do not know whether loading P_z causes twisting about the x-axis. There exists a point O such that loads acting through the point cause no twist. That point is termed the *shear center*, in general, the point is *not* the centroid. A limited consideration of the question is contained in the following section.

By virtue of employing the principal axes ($I_{yz} = 0$), the bending in the planes, x-y and x-z, are uncoupled. The Equations (7.20a)–(7.27a) are independent of the analogous equations (7.20b)–(7.27b). Solution of each system follows the procedures of Chapter 6 (see Equations 6.31–6.34). Note that the algebraic Equations (7.22a,b) enable the elimination of curvatures, κ_y and κ_z; the result is two sets of four first-order linear differential equations.

In many practical problems, a cross section does have one line of symmetry; then, that line and the perpendicular line are the principal lines. The proof is as follows: If the y line is a line of symmetry, as in Figure 7.23, then each elemental area, such as A at $z = l$, has a counterpart, such as A' at $z = -l$; their contributions to the integral I_{yz} cancel.

ILLUSTRATIVE EXAMPLE 7.13

A cantilever beam is made of steel. The cross-section is the thin-walled ($t/a \ll 1$) of Figure 7.24. A transverse load $\tilde{F}$ acts at the end at corner O, so that little twisting occurs; O is the so-called "shear center."

We require the direction of the force **F** such that there is no deflection in the y direction.

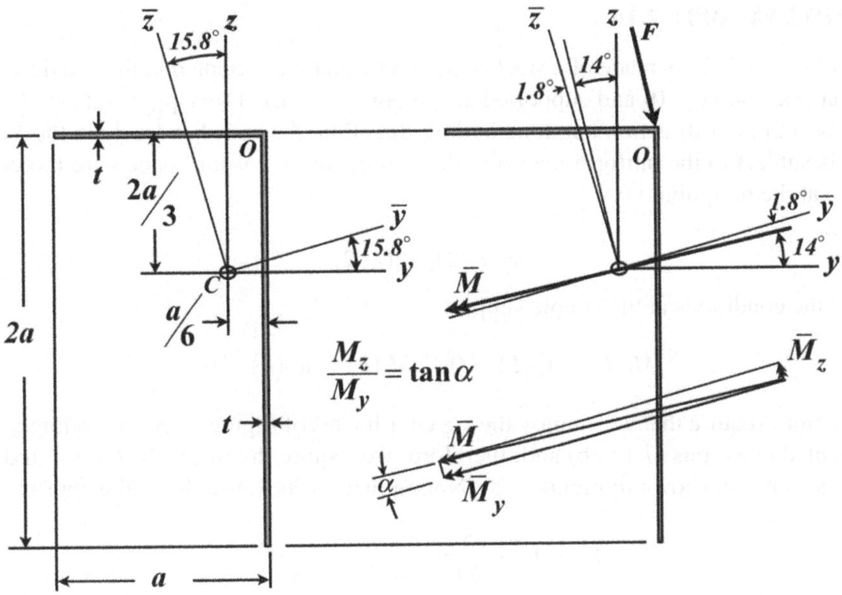

Figure 7.24

With respect to the centroid C, we have the integrals,

$$I_{yy} = \frac{ta^3}{4}, \quad I_{zz} = \frac{4ta^3}{3}, \quad I_{yz} = -\frac{ta^3}{3}$$

To determine the principal directions $(\bar{y}, \bar{z})$ we turn to the transformation of Appendix B:

$$\tan 2\theta = \frac{2I_{yz}}{I_{yy} - I_{zz}} = \frac{8}{13}, \quad 2\theta = 31.6^\circ \Rightarrow \theta = 15.8^\circ$$

$$\bar{I}_{yy} = I_{yy}\cos^2\theta + 2I_{yz}\cos\theta\sin\theta + I_{zz}\sin^2\theta = 0.15565ta^3$$

$$\bar{I}_{zz} = I_{yy}\sin^2\theta - 2I_{yz}\cos\theta\sin\theta + I_{zz}\cos^2\theta = 1.42768ta^3$$

We require no bending in the x-y plane, i.e. $\kappa_y = 0$. Then (see Equations 7.15a,b)

$$\kappa_y = 0 = -\bar{\kappa}_z\sin\theta + \bar{\kappa}_y\cos\theta = -\frac{\bar{M}_y}{E\bar{I}_{zz}}\sin\theta + \frac{\bar{M}_z}{E\bar{I}_{yy}}\cos\theta$$

$$\frac{\bar{M}_z}{\bar{M}_y} = +\frac{\bar{I}_{yy}}{\bar{I}_{zz}}\tan\theta = \tan\alpha = +0.030858 \Rightarrow \alpha = 1.77^\circ$$

Note: As shown in Figure 7.20d, vector $\mathbf{M}_y$ has the direction of the negative y axis. The bending moment is oriented relative to the $\bar{y}$ axis at angle 1.78° and 14° relative to the y axis as shown in Figure 7.24. Force F is inclined 14° to the negative z-axis as depicted.

ILLUSTRATIVE EXAMPLE 7.14

The beam of Figure 7.25 is made of a steel angle with equal legs oriented in the y and z directions. It is fixed at one end ($x = 0$) and supported at the other ($x = L$). The support (at $x = L$) provides simple support in the y direction but none in the z direction; the end slips freely in the z direction. The beam is subject to the uniform load p in the y direction. We want to compare the deflections (w_y and w_z) at the midpoint, i.e.

$$w_y(L/2)/w_z(L/2)$$

We note the conditions at the simple support,

$$M_y(L) = V_z(L) = 0, \quad M_z(L) = w_y(L) = 0$$

These conditions require that we employ the y-z coordinates of Figure 7.25. Accordingly, we must use the coupled equations (7.14a,b) and, therefore, we require the integrals, I_{yy}, I_{zz}, and I_{yz}, with respect to the centroid shown in Figure 7.26. Noting that $t/l \ll 1$, we obtain the approximations:

$$I_{yy} = I_{zz} = \frac{5}{24}tl^3, \quad I_{yz} = -\frac{1}{8}tl^3$$

By Equations (7.20a,b) we obtain

$$V_y = px + \bar{V}_y, \quad V_z = 0$$

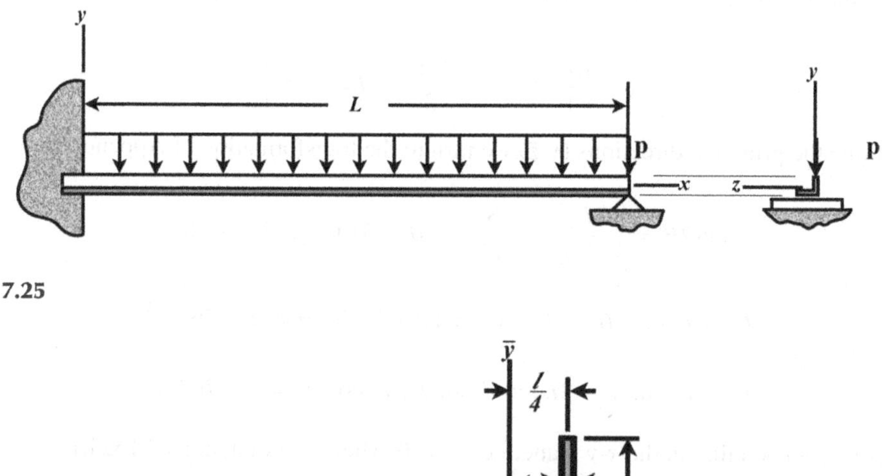

Figure 7.25

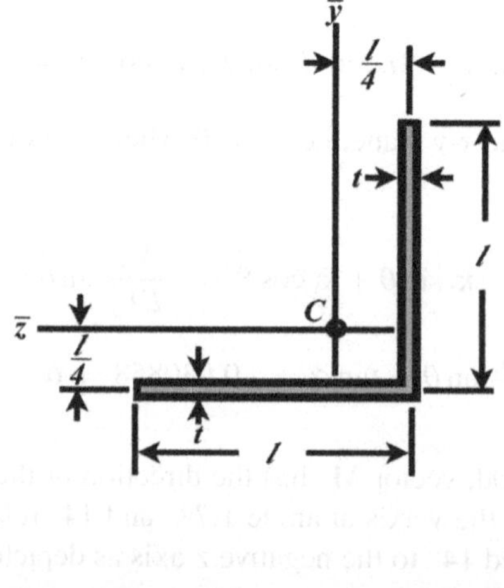

Figure 7.26

The overline (¯) signifies the value at $x = 0$. Then, Equations (7.21a,b) provide

$$M_z = -\frac{px^2}{2} - \bar{V}_y x + \bar{M}_z, \quad M_y = \bar{M}_y$$

Our next step, to Equations (7.23a,b), requires the curvatures, κ_y and κ_z, in terms of the moments, M_z and M_y. Equations (7.22a,b) are not applicable, as they apply only to principal lines. The centroidal lines ($\bar{y}$ and $\bar{z}$) are shown in Figure 7.26.

Inversion of Equations (7.14a,b) provides the required relations:

$$\kappa_y = \frac{M_z}{E\bar{I}_{yy}} - \frac{M_y}{E\bar{I}_{yz}} \tag{7.30a}$$

$$\kappa_z = \frac{M_y}{E\bar{I}_{zz}} - \frac{M_z}{E\bar{I}_{yz}} \tag{7.30b}$$

where

$$\bar{I}_{yy} \equiv \frac{1}{I_{zz}}(I_{yy}I_{zz} - I_{yz}{}^2) = \frac{2}{15}tl^3$$

$$\bar{I}_{zz} \equiv \frac{1}{I_{yy}}(I_{yy}I_{zz} - I_{yz}{}^2) = \frac{2}{15}tl^3$$

$$\bar{I}_{yz} \equiv \frac{1}{I_{yz}}(I_{yy}I_{zz} - I_{yz}{}^2) = -\frac{2}{9}tl^3$$

Equations (7.23a,b) take the form

$$Etl^3\frac{d\theta_y}{dx} = \frac{15}{2}M_z + \frac{9}{2}M_y$$

$$Etl^3\frac{d\theta_z}{dx} = \frac{9}{2}M_z + \frac{15}{2}M_y$$

$$Etl^3\frac{d\theta_y}{dx} = -\frac{15}{4}px^2 - \frac{15}{2}\bar{V}_y x + \frac{15}{2}\bar{M}_z + \frac{9}{2}\bar{M}_y$$

$$Etl^3\frac{d\theta_z}{dx} = -\frac{9}{4}px^2 - \frac{9}{2}\bar{V}_y x + \frac{9}{2}\bar{M}_z + \frac{15}{2}\bar{M}_y$$

Integrating, we obtain

$$Etl^3\theta_y = -\frac{5}{4}px^3 - \frac{15}{4}\bar{V}_y x^2 + \frac{15}{2}\bar{M}_z x + \frac{9}{2}\bar{M}_y x$$

$$Etl^3\theta_z = -\frac{3}{4}px^3 - \frac{9}{4}\bar{V}_y x^2 + \frac{9}{2}\bar{M}_z x + \frac{15}{2}\bar{M}_y x$$

Substituting these results into Equations (7.24a,b) and integrating, we obtain

$$Etl^3 w_y = -\frac{5}{16}px^4 - \frac{5}{4}\bar{V}_y x^3 + \frac{15}{4}\bar{M}_z x^2 + \frac{9}{4}\bar{M}_y x^2$$

$$Etl^3 w_z = -\frac{3}{16}px^4 - \frac{3}{4}\bar{V}_y x^3 + \frac{9}{4}\bar{M}_z x^2 + \frac{15}{4}\bar{M}_y x^2$$

We have 3 unknowns: Moments $\bar{M}_y$, $\bar{M}_z$, $\bar{V}_y$. The end condition $\bar{V}_z(L) = 0$ is satisfied. The remaining conditions at $x = L$ follow:

$$w_y(L) = 0 = -\frac{5}{16}pL^4 - \frac{5}{4}\bar{V}_y L^3 + \frac{15}{4}\bar{M}_z L^2 + \frac{9}{4}\bar{M}_y L^2$$

$$M_z(L) = 0 = -p\frac{L^2}{2} - \bar{V}_y L + \bar{M}_z = 0$$

$$M_y(L) = 0 \Longrightarrow \bar{M}_y = 0,\ \bar{V}_y = -\frac{5pL}{8},\ \bar{M}_z = -\frac{pL^2}{8}$$

The deflections at the midpoint follow

$$Etl^3 w_y = -\frac{10}{256}pL^4, \quad Etl^3 w_z = -\frac{6}{256}pL^4, \quad \frac{w_z(L/2)}{w_y(L/2)} = \frac{3}{5}$$

It should be noted that *no* external forces or couples act in the x-z plane.

7.8 THE SHEAR CENTER

Transverse loads upon beams cause bending, but can also cause twisting if not aligned appropriately. In the case of symmetric cross-sections, the loads must act in a plane of symmetry; then bending occurs in that plane. In the case of an unsymmetrical cross-section, loads must act through the so-called "shear center". Definition: That point through which the resultant shear force acts without an accompanying twisting couple.

The location of the shear center depends upon our ability to describe the distribution of shear stresses upon the cross-section, which comprises the twisting couple. In all cases, some assumptions and approximations are required. The following example serves to illustrate the procedure.

ILLUSTRATIVE EXAMPLE 7.15

A beam has the triangular cross-section of Figure 7.27a. Transverse loads act in the direction y; these cause shear stresses τ_{xy}. It is likely, certainly possible, that some stresses τ_{xz} occur; we know only that the mean value of the latter must vanish, i.e. $V_z = 0$. Our first approximation is the neglect of those components, i.e. assume $\tau_{xz} = 0$ throughout. Our second approximation is the Bernoulli hypothesis, i.e. a plane cross-section remains plane and normal to the neutral axes x. Then, it follows that bending in a y direction causes the normal stress

$$\sigma_x = -y\frac{M_z}{I_{yy}} \tag{7.31}$$

Additionally, we have the equilibrium equation

$$\frac{dM_z}{dx} = -V_y \tag{7.32}$$

To access the stress distribution (τ_{xy}), we isolate the piece delineated in Figure 7.27b. The free-body of that piece is depicted in Figure 7.27c as viewed from the $-z$ direction. Note that any stress components τ_{xz} are neglected. Equilibrium provides the result:

$$\bar{\tau}_{xy}\Delta x\Delta z = \Delta z \int_{\bar{y}}^{y=\frac{6}{8}z} \Delta\sigma_x dy$$

$$\tau_{xy}\Delta z = -\frac{\Delta z}{I_{yy}}\frac{\Delta M_z}{\Delta x}\int_{\bar{y}}^{y=\frac{6}{8}z} ydy$$

Note: The integral is comparable to that obtained in Section 6.4; here the result applies only to the slice of width Δz. Introducing Equation (7.32), we obtain

$$\tau_{xy}\Delta z = \frac{V_y}{I_{yy}}\left(\frac{9}{32}z^2 - \frac{\bar{y}^2}{2}\right)\Delta z$$

Note that the distribution is symmetric with respect to the line $\bar{y} = 0$.

Next we evaluate the shear force (per unit width) ΔV_y:

$$\Delta V_y = 2\int_0^{\bar{y}=\frac{3}{4}z} (\tau_{xy}\Delta z)d\bar{y}$$

$$\Delta V_y = \frac{9}{32}\frac{V_y}{I_{yy}}z^3\Delta z \tag{7.33}$$

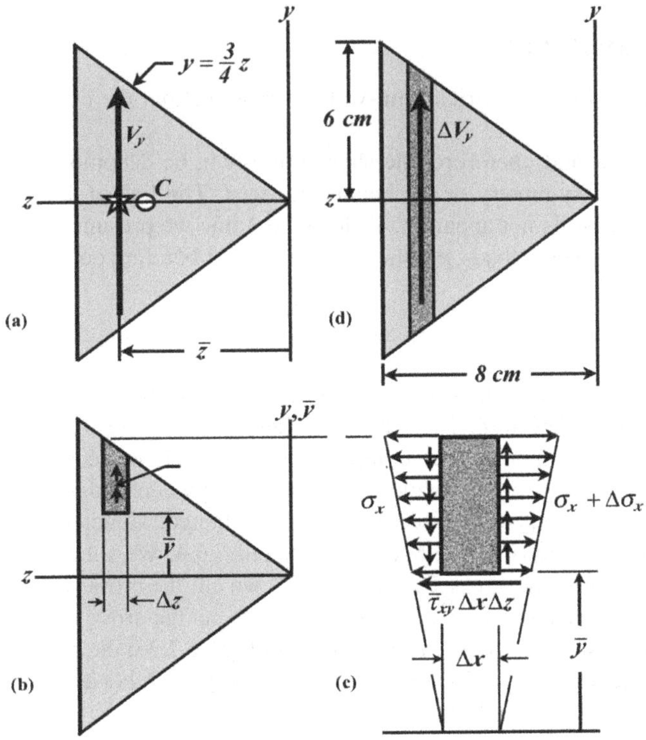

Figure 7.27

For this cross-section

$$I_{yy} = 288\ (\text{cm})^4 \quad (7.34)$$

We can now determine the position of the shear center, the distance $\bar{z}$ as shown in Figure 7.27a.

$$\bar{z}\Delta V_y = \int_0^8 \Delta V_y z \mathrm{d}z = \frac{9}{32}\frac{V_y}{I_{yy}}\frac{z^5}{5}$$

Introducing the numerical values, we obtain $\bar{z} = 6.40$ cm. Note that the centroid is at $z = 5.33$ cm. To avoid twisting, the loads should act through the plane at $\bar{z}$ in Figure 7.27a.

7.9 SHEAR STRESSES IN THIN-WALLED BEAMS

As in the earlier discussion (Section 6.4) of shear stress in symmetric beams, we accept the normal stress given by the Bernoulli theory for HI-HO beams and appeal to equilibrium. With respect to principal lines of the cross section, the normal stress from bending only ($F = 0$) is given by the final two terms of Equation (7.29):

$$\sigma_x = -\frac{M_z y}{I_{yy}} + \frac{M_y z}{I_{zz}} \quad (7.35)$$

Also, we recall the equilibrium equations (7.19a,b)

$$\frac{\mathrm{d}M_z}{\mathrm{d}x} = -V_y, \quad \frac{\mathrm{d}M_y}{\mathrm{d}x} = -V_z \quad (7.36)$$

ILLUSTRATIVE EXAMPLE 7.16

Let us now consider a nonsymmetrical thin-walled cross section, the channel depicted in Figure 7.28.

The transverse loads and, therefore, the shear forces can be decomposed. The component V_z must act on the axes of symmetry or the beam will twist. The line of action for the component V_y, which causes no twist, is not apparent. To locate the line we presume bending in the xy-plane under the action of transverse loads p_y, shear force V_y, and bending couple M_z; according to the Bernoulli theory:

$$\sigma_x = -\frac{M_z y}{I_{yy}} \quad (7.37)$$

For simplicity, let us assume that the web and flange have the same thickness $t \ll d, t \ll b$. On the lateral surfaces of the web, $\tau_{zx} = \tau_{xz} = 0$; therefore, the stress τ_{xz} is assumed negligible throughout the *thin* web. Likewise, the component $\tau_{yx} = \tau_{xy}$ is neglected in the *thin* flange. In general, the shear stress on the thin-walled section is everywhere tangent to the wall; here, the significant components are τ_{xy} in the web and τ_{xz} in the flange. To determine their distributions, we appeal to the equilibrium of a free body, as in Section 6.4. We slice the beam by sections at x and $x + \Delta x$. To determine the stress $\bar{\tau}_{xy}$, on the web, we cut the cross section by a plane N-N at coordinate y. To determine the stress τ_{xz} on the flange, we cut the cross section by a plane at M-M at coordinate z. The cuts (N-N and M-M) are shown in Figure 7.28; the respective free bodies and the net force components in the x-direction are shown in Figure 7.29a and b.

For equilibrium of the body in Figure 7.29b, we require

$$\sum F_x = \bar{\tau}_{xz} t \Delta x + \iint_A \Delta\sigma_x \mathrm{d}A = 0 \quad (7.38)$$

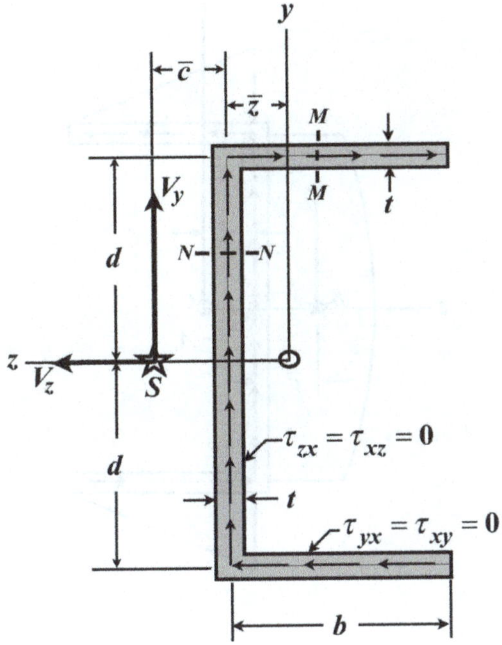

Figure 7.28

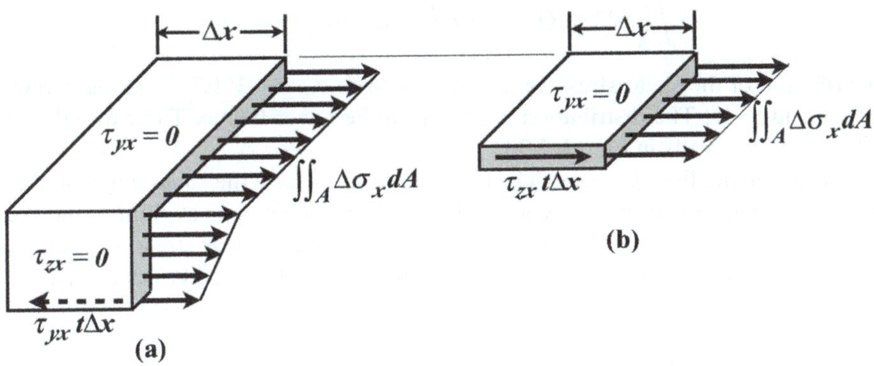

Figure 7.29

The integral extends over the shaded area A on the portion of the section shown on the body. The change $\Delta\sigma_x$ is caused by the change of moment ΔM_z in accordance with Equations (7.37) and (7.36).

$$\bar{\tau}_{xz} = \lim_{\Delta x \to 0} \frac{\Delta M_z}{\Delta x} \frac{\iint_A y dA}{I_{yy} t} = -\frac{V_y}{I_{yy} t} \iint_A y dA \qquad (7.39a,b)$$

Let us evaluate the stress for the flange of Figure 7.29b. Here, we recognize that $t \ll y$ and take

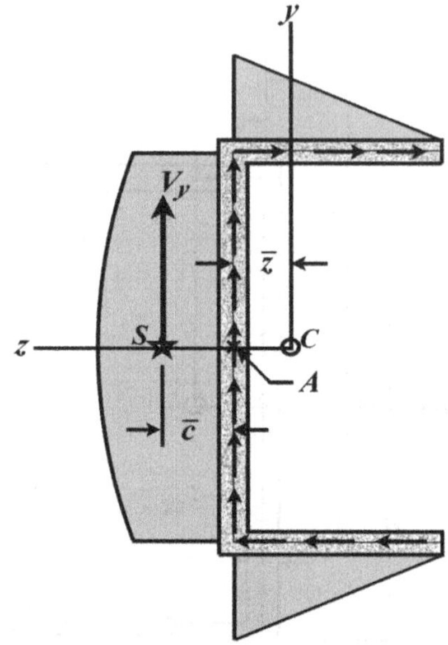

Figure 7.30

$y = d$ throughout the upper flange. Then

$$\iint_A y\mathrm{d}A \equiv Q(z) = td\int_{\bar{z}-b}^{z} \mathrm{d}z = td(z-\bar{z}+b)$$

The distribution of the shear stress varies linearly from $\tau_{xz} = -V_y bd/I_{yy}$ at the corner $z = \bar{z}$ to $\tau_{xz} = 0$ at the end $\bar{z} - b$. The distribution is similar in the lower flange. The entire distribution is plotted adjacent to the walls in Figure 7.30.

Let us now locate the line of action. The resultant V_y must have the same moment as the distribution. Since any axis will suffice, we choose the most convenient axis at point A; the shear stress acting along the web acts through point A. The moments of the shear stresses on the upper and lower flanges are the same; the total is equal to the moment of the force V_y (clockwise):

$$-2\int_{\bar{z}-b}^{\bar{z}} \tau_{xz} y\mathrm{d}A = V_y\bar{c} - 2V_y\frac{d^2t}{I_{yy}}\int_{\bar{z}-b}^{\bar{z}}(z-\bar{z}+b)\mathrm{d}z$$

$$\bar{c} = \frac{tb^2d^2}{I_{yy}} = \frac{b^2}{2(b+d/3)}$$

Note: In the limit $(b \to 0)$, $\bar{c} \to 0$; also, if $b \geq d$, $O.500b > \bar{c} \geq 0.375b$ For a typical channel, American Standard 12×30, $b = 0.528d$, $\bar{c} = 0.3065b$. The point S at $z = \bar{z} + \bar{c}$ is the shear center. We anticipate little or no twisting if the shear force acts through this point.

ILLUSTRATIVE EXAMPLE 7.17

The channel depicted in Figure 7.31 has the cross-section of the traditional curtain rod; these rods have very thin walls pressed from thin sheets of ductile steel. We can determine the shear center by the procedures of the preceding example:

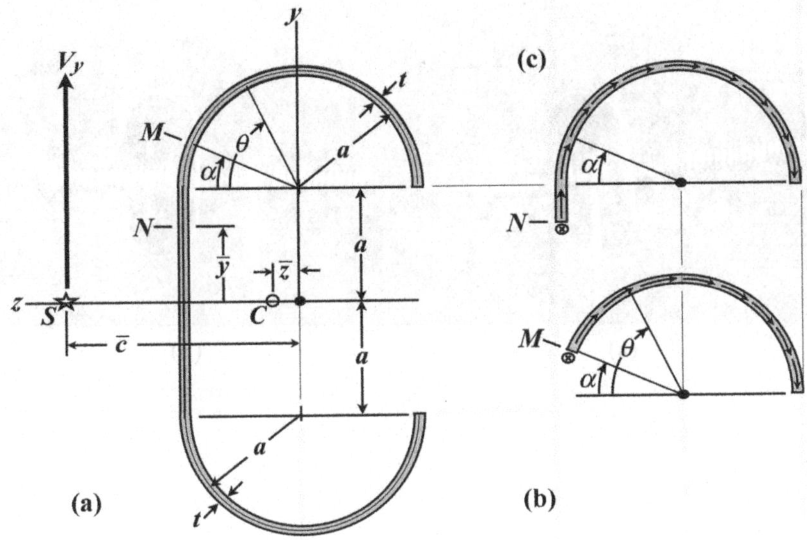

Figure 7.31

Cuts at sites M and N isolate the free-body segments shown in Figure 7.31b and c, respectively. The former provides the shear "flow" t (force per unit length) in the circular segment ($0 \leq \alpha \leq \pi$):

$$t(\alpha) = \frac{V_y}{I_{yy}} a^2 \int_{\alpha}^{\pi} (1 + \sin\theta) d\theta = \frac{V_y}{I_{yy}} a^2 [(\pi - \alpha) + (1 + \cos\alpha)] \tag{7.40}$$

The free-body of Figure 7.31c provides the expression for the straight segment ($-a \leq y \leq a$):

$$t(\bar{y}) = \frac{V_y}{I_{yy}} \left[\int_{\bar{y}}^{a} y dy + a^2 \int_0^{\pi} (1 + \sin\theta) d\theta \right] = \frac{V_y}{I_{yy}} \left[\left(\frac{a^2}{2} - \frac{\bar{y}^2}{2} \right) + a^2(\pi + 2) \right] \tag{7.41}$$

We require the moment of this shear "flow" with respect to any convenient point. Let us use the origin of the coordinates (y, z). Note that the moment upon the lower part ($y \leq 0$) and upper ($y \geq 0$) are equal. Then,

$$M_0 = \frac{V_y 2a^3}{I_{yy}} \left[\int_0^{\pi} t(\theta)(1 + \sin\theta) a d\theta + a \int_0^{a} t(y) dy \right] = 21.7 \frac{V_y a^4}{I_{yy}} \tag{7.42}$$

The shear force produces this moment if the line of action passes through the "shear center" at a distance $\bar{c}$ of Figure 7.31.

$$\bar{c} V_y = M_0 \Rightarrow \bar{c} = 21.7 \frac{a^4}{I_{yy}}$$

Here $I_{yy} = 13.81a^3$. Therefore,

$$\bar{c} = 1.57a \tag{7.43}$$

We conclude this example with the images in Figure 7.32. A simple curtain rod is supported as a cantilever and subjected to transverse loads applied (a) near the centroid and (b) near the shear center.

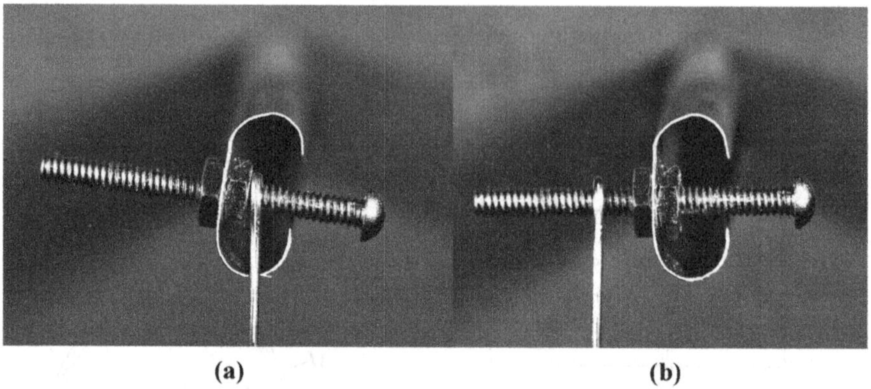

Figure 7.32

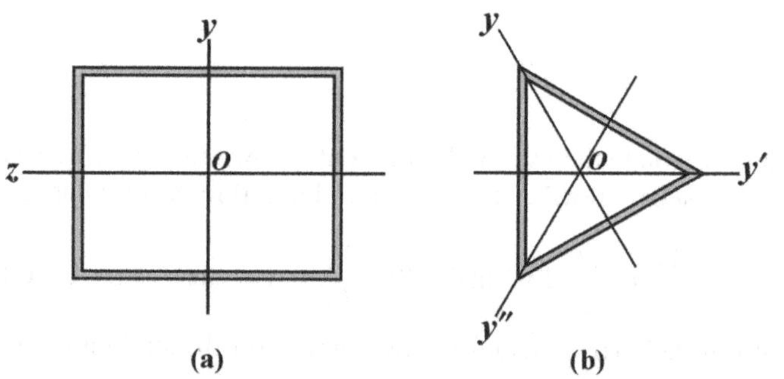

Figure 7.33

The shear centers for sections such as those of Figure 7.33a and b can be identified from symmetry. A transverse shear force upon the section of Figure 7.33a can be decomposed into components in directions y and z, which are axes of symmetry. A shear force upon the section of Figure 7.33b can be decomposed into components in two of the directions y, y', or y'', which are axes of symmetry. In each case, the actions upon the cross section can be represented by a force at the shear center and a couple. Then, we can superpose the shear stresses caused by the transverse shear force and the twisting couple.

If a thin-walled cross-section is multiply connected, i.e., tubular, but does not possess symmetry with respect to two lines, then the shear center is not evident. The determination of the shear center requires additional considerations; specifically, we must provide for continuity of the resulting deformation, i.e., warping of the cross-section. The criteria can be exemplified with the cross-section of Figure 7.34.

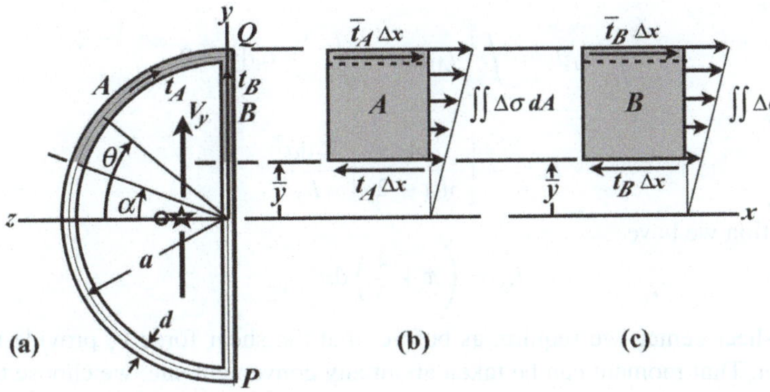

Figure 7.34

ILLUSTRATIVE EXAMPLE 7.18

Unlike the cross-sections in the previous examples, there is *no* open-end: We can only proceed as before, if we suppose that the section were "cut" at some site, say the corner Q. Let us denote the shear "flow" acting upon segment A at the site Q by $\bar{t}_A$ and that acting upon segment B by $\bar{t}_B$. Equilibrium dictates that $\bar{t}_B = -\bar{t}_A$. We have introduced one unknown quantity as a result of the "cut."

Proceeding as before, we obtain the shear "flows" by enforcing the equilibrium of the free bodies depicted in Figure 7.34b and c. In each case, the shear forces must equilibrate the normal forces, viz.

$$(t_A - \bar{t}_A)\Delta x = \iint_A \Delta\sigma_x dA, \quad (t_B - \bar{t}_B)\Delta x = \iint_A \Delta\sigma_x dA$$

In each case, the areas (A) are the faces of the shaded segments A and B in Figure 7.34a.

As before, we observe the origin of the stress $\Delta\sigma_x$, i.e. Equations (7.37), (7.38), and (7.39). Here, the "flows" replace the shear stresses: $t_A = \tau_{x\alpha} d$, $t_B = \tau_{xy} d$. Acknowledging that $\bar{t}_B = -\bar{t}_A$, we arrive at the relations between the "flows" and the resultant force V_y:

$$t_A = \frac{V_y d}{I_{yy}} \int_\alpha^{\pi/2} a^2 \sin\alpha d\alpha + \bar{t}_A, \qquad t_B = \frac{V_y d}{I_{yy}} \int_{\bar{y}}^{a} y dy - \bar{t}_A$$

After integrating we obtain

$$t_A = \frac{V_y d a^2}{I_{yy}} \cos\alpha + \bar{t}_A, \quad t_B = \frac{V_y d}{I_{yy}} \left(\frac{a^2}{2} - \frac{\bar{y}^2}{2} \right) - \bar{t}_A \tag{7.44a,b}$$

To evaluate the unknown, the interaction $\bar{t}_A$, we must enforce the continuity at site Q. One might say, we must "reattach" the walls at Q. To that end, we recall the continuity condition of Section 5.12, viz. Equation (5.113):

$$\int_O \tau_{xt} dc = 0 \tag{7.45a}$$

Although our assessment of the shear stresses derives from the Bernoulli hypothesis (plane cross-sections remain plane), we do now acknowledge a warping caused by the shear and impose the condition of continuity around the cell $(PAQP)$ where $\tau_{xt} d \equiv t$. Specifically,

$$d\int_O \tau_{xt} dc = \int_{-\frac{\pi}{2}}^{\frac{\pi}{2}} t_A a d\alpha + \int_a^{-a} -t_B d\bar{y} = 0 \tag{7.45b}$$

Our result follows:

$$\bar{t}_A = -\left[\frac{4}{3(\pi+2)}\right]\frac{V_y da^2}{I_{yy}}$$

For this cross-section we have

$$I_{yy} = \left(\pi + \frac{2}{3}\right) da^3$$

To locate the shear center, we require, as before, that the shear force V_y provide the moment of this distribution. That moment can be taken about any convenient site; we choose the origin of coordinates (y, z):

$$M = 2\int_0^{\pi/2} a t_A(\alpha) d\alpha = \left[\frac{2(\pi+6)}{3(\pi+2)}\right]\frac{V_y da^4}{I_{yy}}$$

If $\bar{z}$ denotes the location of the line of action, we require

$$\bar{z}V_y = M = \frac{2(\pi+6)a}{(\pi+2)(3\pi+2)}V_y$$

$$\bar{z} = 0.311a$$

In the present context, we can provide an alternative interpretation of the integrals (7.45a,b). Recall that we introduced a cut at Q and an unknown by reattaching the walls at Q. Consider the integral (see Section 5.12):

$$\Delta\bar{w} = \frac{1}{G}\int_P^Q \tau_{xt} dc$$

In the absence of twist (ϕ), this is a measure of the axial displacement of point Q relative to P. The cross-section of Figure 7.34 was cut at Q; continuity is not in question elsewhere. Then our requirement can be expressed by the condition:

$$\int_{PAQ} t_A dc = \int_{-\frac{\pi}{2}}^{\frac{\pi}{2}} t(\alpha) a d\alpha = \int_{PBQ} t_A dy \tag{7.46}$$

It follows that (746) and (7.45b) are the same criterion.*Note*: The direction of the path is clockwise, whereas dy is counterclockwise.

If a cross-section has *no* symmetry then we must first locate the centroid; this serves to uncouple bending and axial extension. Second, we must determine the principal lines at the centroid; the bending actions are only uncoupled in the planes of the principal lines (x-$\bar{y}$ and x-$\bar{z}$).

In the preceding example, we introduced one cut in a tubular member with one cell, i.e. one cavity. The same procedure can be employed when the cross-section has multiple cells; however, each additional cell requires an additional cut. In the following example, we apply the foregoing methodology to an unsymmetrical, tubular cross-section with two cells:

ILLUSTRATIVE EXAMPLE 7.19

The unsymmetrical cross-section of Figure 7.35 has the centroid at point P:

$$\bar{y} = \frac{2a}{\pi+2}, \quad \bar{z} = \frac{(\pi+6)}{2(\pi+2)}a$$

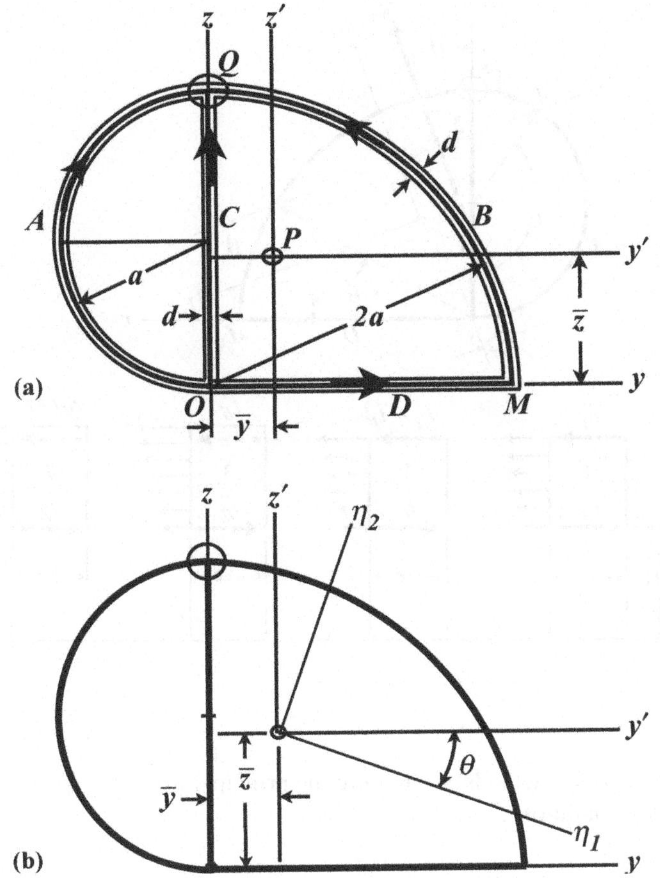

Figure 7.35

With respect to axes y and z, originating at point O, we obtain the integrals:

$$I_{yy} \equiv \iint_A y^2 dA = \left(\frac{5\pi}{2} + \frac{8}{3}\right) da^3$$

$$I_{yz} \equiv \iint_A yz dA = 2da^3$$

$$I_{zz} \equiv \iint_A z^2 dA = \left(\frac{7\pi}{2} + \frac{8}{3}\right) da^3$$

We next obtain the corresponding integrals for axes y' and z' which originate at the centroid P; these are given by the transfer of axes (see Appendix B):

$$I'_{yy} = I_{yy} - \bar{y}^2 A = 8.9647 da^3$$

$$I'_{zz} = I_{zz} - \bar{z}^2 A = 5.5356 da^3$$

$$I'_{yz} = I_{yz} - \bar{y}\bar{z} A = -1.5559 da^3$$

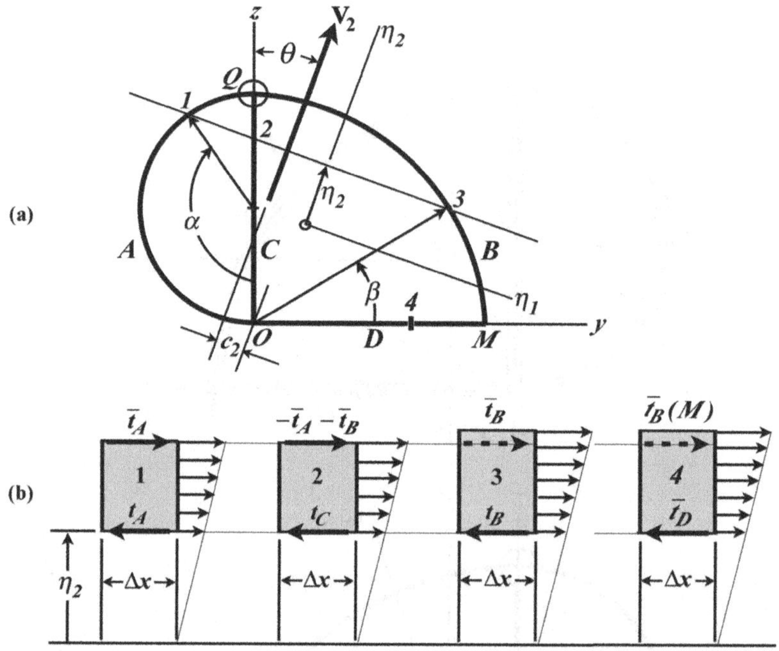

Figure 7.36

To implement the foregoing methods, we require the principal axes (η_1, η_2) [Appendix *B*]. These are determined by the condition[3]:

$$\tan 2\theta = \frac{-2I'_{yz}}{I'_{yy} - I'_{zz}} = 0.90745,\ \theta = 21.11^\circ \tag{7.47}$$

Finally, we compute the integrals $(\bar{I}_{11}, \bar{I}_{22})$ for the principal axes (η_1, η_2), shown in Figure 7.35b:

$$\bar{I}_{11} = 9.5654da^3, \quad \bar{I}_{zz} = 4.9348da^3 \tag{7.48a,b}$$

This enables us to uncouple the bending actions and to ascertain independently the two coordinates of the shear center. Let us determine the line of action for the component V_2 of the shear force V_z which acts through the shear center (see Figure 7.36a). i.e., when acting on that line, V_2 is to cause no twist.

If we remove the juncture at *Q*; this requires the "cuts." At the upper end of the wall *A* we introduce the shear "flow" $\bar{t}_A$ at the cut and at the upper end of the wall *B* the "flow" $\bar{t}_B$. Then by virtue of equilibrium the "flow" at the top of the web *C* is $\bar{t}_C = -\bar{t}_A - \bar{t}_B$.

Proceeding as before, we consider the free bodies of segments of walls *A*, *B*, *D*, and web *C*; these are the segments above cuts at sites 1, 2, 3, and 4. The sites are identified in Figure 7.36a; these segments, 1, 2,3, and 4 are shown in Figure 7.36b. Again, we require that the shear forces equilibrate the differential of the bending stresses upon the respective faces of each body, e.g.

$$t_A \Delta x = \iint_A \Delta\sigma_x dA + \bar{t}_A \Delta x$$

In each case, we are led to similar results, e.g.

$$t_A = \frac{V_2 d}{I_{22}} \iint_A \eta_2 dA + \bar{t}_A$$

To facilitate the integrations, we express η_2 in terms of the coordinates y and z.

$$\eta_2 = y \sin\theta + z\cos\theta - L$$

where

$$L \equiv \bar{y}\sin\theta + \bar{z}\cos\theta \tag{7.49}$$

After executing the integration, much algebra, and some arithmetic, we arrive at the results:
On wall A at site 1,

$$t_A(\alpha) = \frac{V_2 a^2 d}{I_{22}}\left[\sin\theta(1+\cos\alpha) + (\pi-\alpha)\cos\theta + \cos\theta\sin\alpha - \frac{L}{a}(\pi-\alpha)\right] + \bar{t}_A \tag{7.50}$$

On web C at site 2

$$t_C(z) = \frac{V_2 a^2 d}{I_{22}}\left[\frac{1}{2}(4-\frac{z^2}{a^2})\cos\theta - \frac{L}{a}(2-\frac{z}{a})\right] - \bar{t}_A - \bar{t}_B \tag{7.51}$$

On segment OM of wall B at site 4,

$$t_B(y) = \frac{V_2 a^2 d}{I_{22}}\left[\frac{1}{2}(4-\frac{y^2}{a^2})\sin\theta + 4\sin\theta + 4\cos\theta - \frac{L}{a}\left(\pi+2-\frac{y}{a}\right)\right] + \bar{t}_B \tag{7.52}$$

On segment MQ of wall B at site 3

$$t_D(y) = \frac{V_2 a^2 d}{I_{22}}\left[4\sin\theta(1-\sin\beta) + 4\cos\theta\cos\beta - \frac{L}{a}(\pi-2\beta)\right] + \bar{t}_B \tag{7.53}$$

The next task is the determination of constants $\bar{t}_A$ and $\bar{t}_B$. Again. we turn to the condition for continuity of axial displacement (the reattachment at Q), viz. the integrals around each cell:

$$\int_O t dc = 0$$

Each integral provides one of two equations in terms of the unknowns t_A and $\bar{t}_B$. This requires some integrations, more algebra, and arithmetic, which provide the results:

$$\bar{t}_A = -0.4298\frac{V_2 a^2}{I_{zz}}, \quad \bar{t}_B = -1.1075\frac{V_2 a^2}{I_{zz}} \tag{7.54a,b}$$

With the constants $\bar{t}_A$ and $\bar{t}_B$ the functions $t_A(\alpha)$, $t_B(\beta)$, and $t_C(z)$ are fully determined. We now determine the moment about the origin O:

$$M_O = a^2\int_{-\frac{\pi}{2}}^{\frac{\pi}{2}} t_A(\alpha)(1-\cos\alpha)d\alpha - 4a^2\int_0^{\pi/2} t_B(\beta)d\beta = 2.477\frac{V_2 a^2 d}{I_{22}} = c_2 V_2$$

The force V_2 exerts this moment if the distance normal to the line of action is at $c_2 = 0.502a$. Distance c_2 is shown in Figure 7.36a.

It is unlikely that loading of this beam would be in the plane of η_2. Therefore, implementation would require that we determine also the appropriate line of action for a component V_1 in direction η_1. The procedure would be the same as the foregoing, which provided the distance c_2. The intersection of the two lines of action is the "shear center."

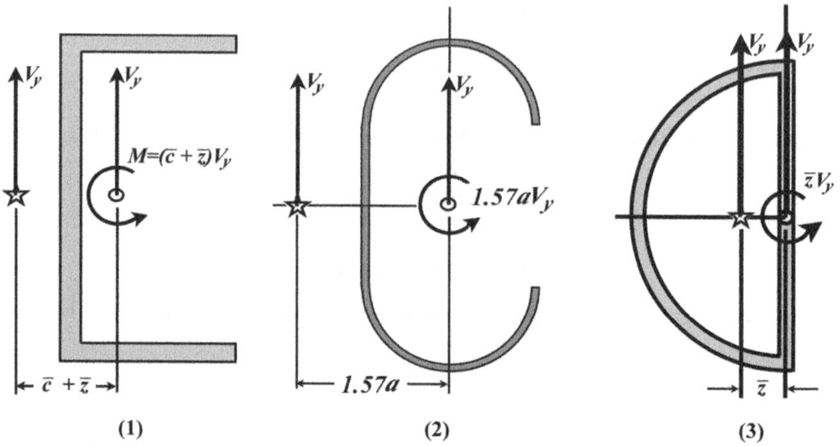

Figure 7.37

7.10 BENDING AND TWISTING OF BEAMS SUBJECT TO TRANSVERSE LOADS

The preceding Section 7.9 focused on the position and direction of a transverse shear, which would impose no twisting couple. However, in practice, the transverse loads are seldom in those specific planes. Consider beams with the cross-sections of the Illustrative Examples 7.16–7.18. These are redrawn in Figure 7.37 with the transverse shear V_y acting through the respective shear centers. In those positions, there is no twisting couple.

In Figure 7.37(1), we would anticipate a load distributed upon the flange. If it were to pass through the centroid, the shear force V_y would be accompanied by a twisting couple, $M = (\bar{c}+\bar{z})V_y$. The channel of Figure 7.37(2) would most likely be loaded upon the upper or lowermost surfaces; the shear V_y would then be accompanied by the twisting couple $M = 1.57aV_y$. A beam with the cross section of Figure 7.37(3) might be subject to loads acting in the plane of the straight web; that would impose a twisting couple $M = \bar{z}V_y$.

In each of the situations depicted in Figure 7.37, the probable loadings would cause the bending in the x-y plane which accompany the shear force V_y *and* also the twisting caused by the couple M. The former deformations are determined by the methods of Section 6.5, i.e., as the deflections of a symmetric (HI-HO) beam.

NOTES

1. Presented at the ASCE annual meeting, Chicago, 1969 and published in the International Journal of Solids and Structures, 7, 1971.
2. Note: Here M_y is contrary to the "right hand" rule for couples, but consistent with Figures 7.22 and 7.24b.
3. Here angle θ is taken in the clockwise sense as depicted in Figure 7.35.

8 Extension and Flexure of Thin Plates

8.1 INTRODUCTION

The most fundamental applications (Chapter 5) entail deformations which are uniform with respect to one coordinate x: These are simple extension (Section 5.3), simple torsion (Section 5.7), and simple bending (Section 5.13). The subsequent applications to beams (Chapters 6 and 7) describe the extension, torsion, and bending, which vary with respect to the one coordinate. The one dimensionality of those applications can be attributed to the form of the bars, shafts, and beams which are relatively thin bodies in the lateral dimensions y and z. By contrast, a thin plate is relatively small in but one dimension: If h denotes the thickness (in direction z) and l denotes a typical width (in direction x or y), then $h/l \ll 1$.

When the thin plate is subjected to loadings, extension, torsion, and bending are variable with respect to the two directions, x and y. The underlying mechanics, concepts, and approximations are analogous to those that form the basis for the theory of slender beams. Only the mathematics is necessarily more complicated: Each variable (extensional strains, curvatures, forces, and couples) is dependent upon two variables x and y. The mathematics now entails the partial derivatives ($\partial/\partial x$ and $\partial/\partial y$) and the solution of partial differential equations.

Our intent here is to lay the mechanical and mathematical foundations. Solutions are provided for a few simple situations. In a subsequent chapter (Chapter 10), we utilize those foundations to demonstrate the method of finite elements as a means of approximation.

8.2 EXTENSIONAL DEFORMATION

We limit our treatment to a thin plate of thickness h and properties which are symmetrical about a middle (x-y) plane, i.e., properties are even functions of the coordinate z. Moreover, we direct our immediate attention to so-called "in-plane" loading (symmetrical about the plane), such that the deformations are symmetric with respect to that mid-plane. The deformation of a typical element is shown in Figure 8.1.

As mentioned in Chapter 2, the components of the displacement $\mathbf{V}$ in x and y are denoted by u and v, respectively. For small strains and rotations, the strain components are given by the linear Equations (2.78):

$$\varepsilon_x = \frac{\partial u}{\partial x}, \quad \varepsilon_y = \frac{\partial v}{\partial x}, \quad \gamma_{xy} = \frac{\partial u}{\partial x} + \frac{\partial v}{\partial x} \qquad (8.1\text{a,b,c})$$

8.3 HI-HO BEHAVIOR OF THE THIN PLATE

Since our plate is thin, stresses at the surface are negligible or vanish; such applied tractions are much less than interior stresses. Accordingly, we assume a state of plane stress: $\sigma_z = \tau_{zx} = \tau_{zy} = 0$.[1] The Hookean-Isotropic behavior is expressed by Equations (4.26a–c)

DOI: 10.1201/9781003665861-8

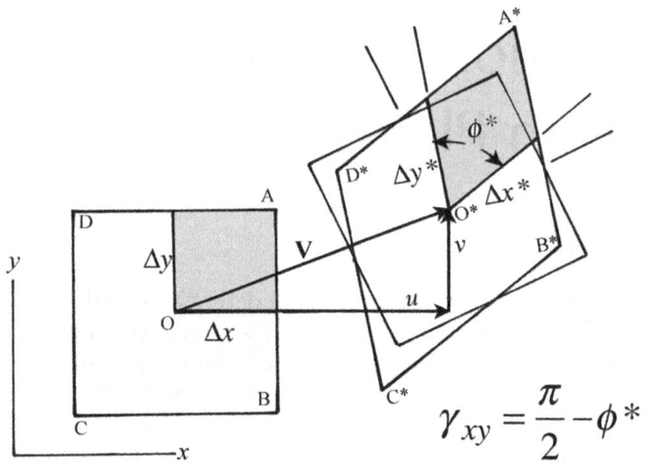

Figure 8.1 Displacement and Strain in a Plane

$$\sigma_x = \frac{E}{1-\nu^2}(\varepsilon_x + \nu\varepsilon_y) \tag{8.2a}$$

$$\sigma_y = \frac{E}{1-\nu^2}(\varepsilon_y + \nu\varepsilon_x) \tag{8.2b}$$

$$\tau_{xy} = G\gamma_{xy} = \frac{E}{2(1+\nu)}\gamma_{xy} \tag{8.2c}$$

These components of stress are expressed in terms of the two components (u,v) of displacement. By substituting (8.1) into (8.2), we obtain

$$\sigma_x = \frac{E}{1-\nu^2}\left(\frac{\partial u}{\partial x} + \nu\frac{\partial v}{\partial y}\right) \tag{8.3a}$$

$$\sigma_y = \frac{E}{1-\nu^2}\left(\frac{\partial v}{\partial y} + \nu\frac{\partial u}{\partial x}\right) \tag{8.3b}$$

$$\tau_{xy} = \frac{E}{2(1+\nu)}\left(\frac{\partial u}{\partial y} + \frac{\partial v}{\partial x}\right) \tag{8.3c}$$

8.4 EQUILIBRIUM OF THE PLANE STATE

Under the state of "in-plane" loading, symmetric with respect to the plane, the only non-vanishing equations of equilibrium follow (See Section 3.14.1)

$$\frac{\partial \sigma_x}{\partial x} + \frac{\partial \tau_{xy}}{\partial y} = 0, \qquad \frac{\partial \tau_{xy}}{\partial x} + \frac{\partial \sigma_y}{\partial y} = 0 \tag{8.4a,b}$$

Again, note that these presume small strains and rotations. We recall that loadings and properties, hence deformations and stresses, are symmetrical about the (x-y) plane. Then, each of the equations, (8.1)–(8.4), can be integrated through the thickness ($-h/2 \le z \le h/2$) to obtain similar equations in terms of the average values denoted by the overbar ($\bar{}$):

$$\bar{\varepsilon}_x \equiv \frac{1}{h}\int_{-h/2}^{h/2} \varepsilon_x dz, \quad \bar{\varepsilon}_y \equiv \frac{1}{h}\int_{-h/2}^{h/2} \varepsilon_y dz, \quad \bar{\gamma}_{xy} \equiv \frac{1}{h}\int_{-h/2}^{h/2} \varepsilon_{xy} dz \tag{8.5a,b,c)}$$

$$\bar{u} \equiv \frac{1}{t}\int_{-h/2}^{h/2} u dz, \quad \bar{v} \equiv \frac{1}{t}\int_{-h/2}^{h/2} v dz \tag{8.5d,e)}$$

$$\bar{\sigma}_x = \frac{1}{h}\int_{-h/2}^{h/2} \sigma_x dz \equiv \frac{N_x}{h} \tag{8.6a}$$

and similarly

$$\bar{\sigma}_y \equiv \frac{N_y}{h}, \quad \bar{\tau}_{xy} \equiv \frac{N_{xy}}{h} \tag{8.6a,b}$$

Here, N_x, N_y, and N_{xy} are the normal forces, and shear force per unit length, respectively.

Our complete system of equations consists of the kinematic equation (8.1), stress-strain/displacement relations (8.2/8.3), and the equilibrium conditions (8.4); we need only place the bar ($\bar{}$) over each variable to signify the average value.

Substituting (8.3) into (8.4) expresses the two equations in terms of two displacements, $\bar{u}(x,y)$ and $\bar{v}(x,y)$:

$$2\frac{\partial^2 \bar{u}}{\partial x^2} + (1+\nu)\frac{\partial^2 \bar{v}}{\partial x \partial y} + (1-\nu)\frac{\partial^2 \bar{u}}{\partial y^2} = 0 \tag{8.7a}$$

$$(1-\nu)\frac{\partial^2 \bar{v}}{\partial x^2} + (1+\nu)\frac{\partial^2 \bar{u}}{\partial x \partial y} + 2\frac{\partial^2 \bar{v}}{\partial y^2} = 0 \tag{8.7b}$$

We note that our equilibrium equations (8.4), hence (8.7) also contain no loadings; hence, these do not accommodate body forces nor surface tractions. Loading is presumably applied only at the edges. For simplicity, we suppress the overbar ($\bar{}$) in subsequent developments.

8.5 EDGE CONDITIONS OF THE PLANE STATE

An effective device for assessing alternative edge conditions is the concept of virtual work. Here, we have two possible movements at an edge, normal and tangential components of displacement: Figure 8.2 depicts an element of the edge, the virtual displacements (δu_n, δu_t) and the associated actions (N_n, N_{nt}) in the normal and tangential directions, respectively. The virtual work of actions (per unit length) follows:

$$\delta W = N_n \delta u_n + N_{nt} \delta u_t$$

Possible conditions are the following:

Clamped: $u_n = u_t = 0$. In general, forces N_n and N_{nt} are needed to enforce the constraint; they emerge from the solution.

Free: $N_n = N_{nt} = 0$. The displacement of the edge, u_n and u_t, follows from the solution.

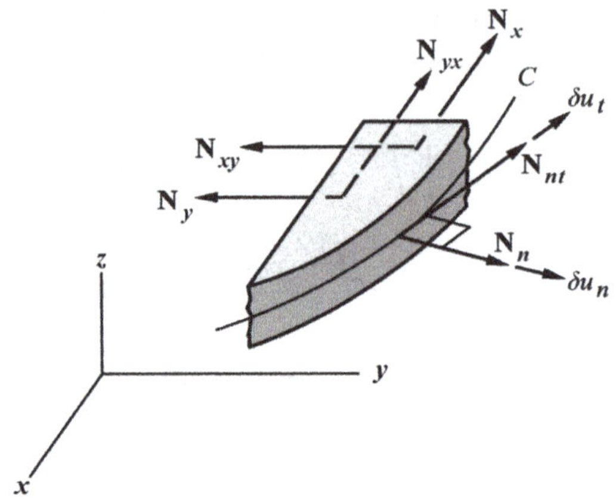

Figure 8.2 Edge Conditions

Combinations are conceivable; for example, normal movement might be prevented ($\delta u_n = 0$) while tangential movement (sliding) might be unconstrained ($N_{nt} = 0$).

Implementation of conditions at an edge requires that the edge displacements, or forces, be expressed in terms of the variables in the governing equations (e.g., the rectangular components u, v, N_x, N_y, N_{xy}). The transformation of displacement components follows familiar vectorial equations. The transformation of the stresses is given by Equations (3.32) and (3.33), or (3.34) and (3.35).

8.6 INTERNAL (STRAIN) ENERGY OF THE PLANE STATE

With a practical view toward the methods of approximation (e.g. finite elements) we record the expression of internal energy (per unit area) of the HI material in a state of "plane stress" ($\sigma_z = \tau_{zx} = \tau_{zy} = 0$): The requisite expression follows from Equations (4.19c) and (4.20b,c) wherein plane stress ($\sigma_z = \tau_{zx} = \tau_{zy}$) implies that

$$\varepsilon_z = -\frac{v}{1-v}(\varepsilon_x + \varepsilon_y), \quad \gamma_{zx} = \gamma_{zy} = 0$$

The internal energy (density per unit volume) follows:

$$\phi = \frac{E}{2(1-v^2)}(\varepsilon_x^2 + 2v\varepsilon_x\varepsilon_y + \varepsilon_y^2) + \frac{E}{4(1+v)}\gamma_{xy}^2 \tag{8.8}$$

8.7 AIRY STRESS FUNCTION

A solution to the linear equations governing plane stress must provide displacements that satisfy the differential equations (8.7) and the prescribed edge conditions. An alternative approach is provided by the Airy[2] function F. Stress components are expressed in terms of the Airy function, as follows:

$$\sigma_x = \frac{\partial^2 F}{\partial y^2}, \quad \sigma_y = \frac{\partial^2 F}{\partial x^2}, \quad \tau_{xy} = \frac{\partial^2 F}{\partial x \partial y} \tag{8.9a,b,c}$$

Provided that the function F and second derivatives are continuous, these components satisfy the equilibrium equations (8.4). Of course, it is not sufficient that the stress components satisfy equilibrium. They must also be related to strain and displacement components as given by the linear algebraic Equations (8.2) and (8.3). The existence of a compatible displacement is ensured if, and only if, the strains satisfy the compatibility Equations (2.82). In the present case $[\gamma_{zx} = \gamma_{zy} = 0,\ F = F(x,y)]$, only the first is needed, viz.

$$\frac{\partial^2 \varepsilon_x}{\partial y^2} + \frac{\partial^2 \varepsilon_y}{\partial x^2} - \frac{\partial^2 \gamma_{xy}}{\partial x \partial y} = 0 \tag{8.10}$$

When Equations (8.9) are substituted into the strain-stress relations and then into the compatibility condition (8.10), the latter provides the equations governing the Airy function:

$$\frac{\partial^4 F}{\partial x^4} + 2\frac{\partial^4 F}{\partial x^2 \partial y^2} + \frac{\partial^4 F}{\partial x^4} = 0 \tag{8.11}$$

Again, we must note that our derivation does not provide for body forces or tractions upon the lateral surfaces.

Equation (8.11) is the so-called biharmonic equation, which has been extensively treated; a variety of solutions are known. In the following, we offer but a few simple solutions that provide insights into a few significant structural problems.

8.7.1 EXAMPLE: CANTILEVER BEAM

To illustrate the foregoing theory of plane stress, we examine the solution for the rectangular region depicted in Figure 8.3. The engineer views this as a deep cantilever beam with a rectangular cross-section of width h, depth $2c$, and length l.

The stress components can be derived via (8.9) from the Airy function:

$$F = -\frac{Pxy^3}{6I} + \frac{Pc^2xy}{2I}$$

where

$$I \equiv h\int_{-c}^{+c} y^2 dy = \frac{2hc^3}{3}$$

According to (8.9), we obtain

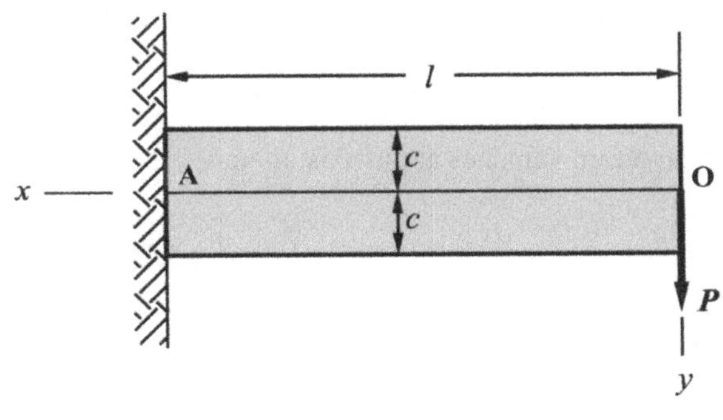

Figure 8.3 Cantilever Beam

$$\sigma_x = -\frac{Pxy}{I}, \quad \sigma_y = 0, \quad \tau_{xy} = \frac{P}{2I}(c^2 - y^2)$$

The displacements can be obtained by the integration of Equation (8.3):

$$u = -\frac{Py}{2EI}\left(x^2 + v\frac{y^2}{3}\right) - \frac{Py}{2GI}\left(\frac{y^2}{3} - c^2\right) - dy$$
$$v = \frac{Px}{2EI}\left(\frac{x^2}{3} + vy^2\right) + d(x - l) - \frac{Pl^3}{6EI}$$

The unspecified constant 'd' depends on the manner of support at A. The solution does not admit clamping along the entire edge $x = l$, but the manner of fixation at the point A ($x = l,\ y = 0$). Since the shear strain does not vanish at point A ($\partial u/\partial y + \partial v/\partial x \neq 0$),it is not possible to fix both the horizontal line ($\partial v/\partial x = 0$) and the vertical line ($\partial u/\partial x = 0$). In keeping with the elementary theory of a beam, let us set

$$\left.\frac{\partial v}{\partial x}\right|_{x=l,y=0} = 0 \Rightarrow d = -\frac{Pl^2}{2EI}$$

Then the deflection of the midline is

$$v(x,0) = \frac{Px^3}{6EI} - \frac{Pl^2x}{2EI} + \frac{Pl^3}{3EI}$$

The displacement at the loaded end ($x = O$) is $v(0,0) = Pl^3/3EI$, in agreement with the Bernoulli theory of beams. Note, however, that the cross-sections do not remain plane, as a consequence of the transverse shear strain γ_{xy}.

8.8 APPROXIMATION OF THE CANTILEVER BEAM

One can always obtain an approximation of the equilibrium configuration via the principle of stationary potential. A good approximation calls for a good "guesstimate" of the displacement; our choice must satisfy any imposed constraints. Such constraints are $u = v = \partial v/\partial x = 0$ at the point A of Figure 8.3; these are the conditions imposed in the preceding example. A simple description of the deflection $v(x,0)$ is a cubic:

$$v(x,0) = A(l-x)^3 + B(l-x)^2$$

If we suppose that the shear strain vanishes at the axis, then

$$\left.\frac{\partial u}{\partial y}\right|_{y=0} = -\left.\frac{\partial v}{\partial x}\right|_{y=0} = +3A(l-x)^2 + 2B(l-x)$$

This suggests the approximation

$$u(x,y) = 3A(l-x)^2y + 2B(l-x)y$$
$$v(x,y) = A(l-x)^3 + B(l-x)^2$$

These give the strain components

$$\varepsilon_x = \frac{\partial u}{\partial x} = -6A(l-x)y - 2By,\ \varepsilon_y = \gamma_{xy} = 0$$

The total strain energy follows from Equation (8.8):

$$U = \frac{Eh}{2(1-\nu^2)}\int_0^l \int_{-c}^{c} \varepsilon_x^2 dydx = \frac{2EIl}{1-\nu^2}(3A^2l^2 + 3ABl + B^2)$$

The total potential W must also include the potential Ω of Load P:

$$\Omega = -Pv(0,0) = -(Al^3 + Bl^2)P$$

A stationary of the potential requires that

$$\frac{\partial}{\partial A}(U+\Omega) = 0,\ \frac{\partial}{\partial B}(U+\Omega) = 0$$

These provide two linear equations in the unknown parameters, A and B:

$$2Al + B = \frac{1-\nu^2 PL}{6EI}$$

$$3Al + 2B = \frac{1-\nu^2 PL}{2EI}$$

It follows that

$$A = -\frac{(1-\nu^2)P}{6EI},\ B = \frac{(1-\nu^2)Pl}{2EI}$$

The deflection at the end follows:

$$v(0,0) = \frac{(1-\nu^2)Pl^3}{3EI}$$

This result differs from that given in the preceding example by the factor $(1-\nu^2)$. Had we invoked the assumption $\sigma_y = 0$, then $u = E\varepsilon_x^2/2$ and the result is in full accord with the earlier result.

8.9 STRESS CONCENTRATION AT A HOLE

An important consideration in the design of any structural or mechanical component is the so-called "stress-concentration" which occurs at an abrupt reentrant corner or hole. The reality of such phenomena is illustrated by a solution for the problem of a HI-HO plate with a circular hole, as depicted in Figure 8.4.

We recall from Section 3.14.2 the components of stress expressed in terms of the polar coordinates (r,θ). These satisfy the equations of equilibrium (3.50a,b) when derived from the Airy function F:

$$\sigma_r = \frac{1}{r}\frac{\partial F}{\partial r} + \frac{1}{r^2}\frac{\partial^2 F}{\partial \theta^2},\quad \sigma_\theta = \frac{\partial^2 F}{\partial r^2},\quad \sigma_{r\theta} = -\frac{\partial}{\partial r}\left(\frac{1}{r}\frac{\partial F}{\partial \theta}\right)$$

The Airy function must satisfy the compatibility condition (8.11), but in polar coordinates. The latter can be obtained via the "chain—rule":

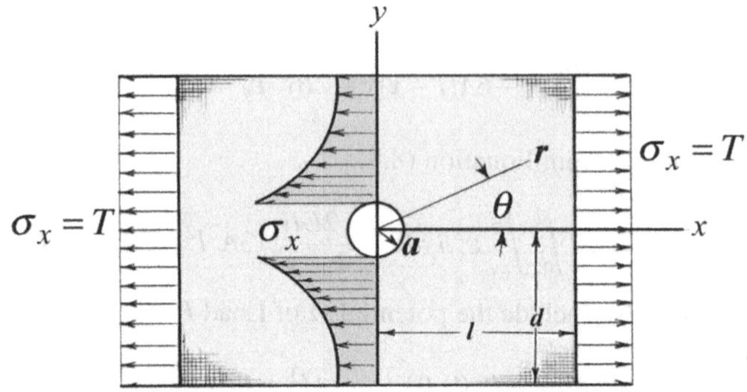

Figure 8.4 Stress Concentration

$$\frac{\partial}{\partial x} = \frac{\partial r}{\partial x}\frac{\partial}{\partial r} + \frac{\partial \theta}{\partial x}\frac{\partial}{\partial \theta} = \cos\theta\frac{\partial}{\partial r} - \frac{\sin\theta}{r}\frac{\partial}{\partial \theta}$$
$$\frac{\partial}{\partial y} = \frac{\partial r}{\partial y}\frac{\partial}{\partial r} + \frac{\partial \theta}{\partial y}\frac{\partial}{\partial \theta} = \sin\theta\frac{\partial}{\partial r} + \frac{\cos\theta}{r}\frac{\partial}{\partial \theta}$$

The result is the biharmonic equation expressed in polar coordinates:

$$\left(\frac{\partial^2}{\partial r^2} + \frac{1}{r}\frac{\partial}{\partial r} + \frac{1}{r^2}\frac{\partial^2}{\partial \theta}\right)\left(\frac{\partial^2 F}{\partial r^2} + \frac{1}{r}\frac{\partial F}{\partial r} + \frac{1}{r^2}\frac{\partial^2 F}{\partial \theta^2}\right) = 0$$

The requisite solution is given by the Airy function:

$$F = Mr^2 + N\log\frac{r}{a} + (Ar^2 + Br^4 + Cr^{-2} + D)\cos 2\theta$$

The six constants (M, N, A, B, C, D) are determined by the boundary conditions. We assume that the hole is very small compared with length $2l$ and width $2d$, such that, in limit $\frac{a}{r} \to 0$:

$$\sigma_r \to T\cos^2\theta, \quad \sigma_\theta \to T\sin^2\theta, \quad \tau_{r\theta} \to -T\cos\theta\sin\theta$$

This is the state of simple tension at the extremities of the plate. The constant B must be excluded so that the stresses are bounded $r \to \infty$. At the edges of the hole $\sigma_r = \tau_{r\theta} = 0$. Details can be found elsewhere[3]; we cite here the important feature, the stress on the line $x = 0$; at $y = a$:

$$\sigma_x(0, a) = 3T$$

This stress concentration is depicted by the plot of $\sigma_x(0, y)$

8.10 FLEXURAL DEFORMATION

In the preceding Sections (8.1–8.9), we addressed the "in-plane" loadings upon a thin plate and the consequent extensional deformations. We turn now to the flexure and torsion that occur when thin plates are subjected to transverse loadings. If a plate is thin, the transverse displacements vary little across the thickness. Usually it suffices to describe the displacement of a midplane; typically, but not necessarily, that is the middle surface. Here, we set the Cartesian coordinates (x,y) at the reference plane and denote the displacement in the z direction by $w(x,y)$. Let us address the flexural deformation of the reference surface.

First we must acknowledge that few deformations are possible without also stretching the surface. We can roll a plane surface into a cylinder or a cone without stretching the surface. These are exceptions: Cylinders and cones are characterized by straight lines, so-called generators. Stated otherwise, they have no curvature in these directions. Extensional deformations (stretching) are small in most practical circumstances. However, extensional deformations do occur and must be acknowledged when curvatures become large.

Suppose that the rectangular elements of Figures 8.5 and 8.6 are displaced and bent without significant extensional deformation. An initial triad $(\hat{\mathbf{i}},\hat{\mathbf{j}},\hat{\mathbf{k}})$ is displaced to the triad $(\hat{\mathbf{I}},\hat{\mathbf{J}},\hat{\mathbf{N}})$; the former are the unit vectors in the initial orthogonal directions (x,y,z). The latter are in the directions of the displaced lines. Vector $\bar{\mathbf{N}}$ is normal to the deformed (bent) surface. Figure 8.5 depicts bending in the x direction; the x line has turned through the angle θ_x and exhibits the curvature κ_x:

$$\kappa_x \equiv \frac{1}{r_x} = \lim_{\Delta x \to 0} \frac{\Delta\theta_x}{\Delta x} \equiv \frac{\partial\theta_x}{\partial x} \tag{8.12a}$$

Figure 8.6 exhibits a similar flexure of a y line:

$$\kappa_y = \frac{1}{r_y} \equiv \frac{\partial\theta_y}{\partial y} \tag{8.12b}$$

If $\mathbf{R}_0$ denotes the position of a particle on the deformed but unstretched surface, then

$$\hat{\mathbf{I}} = \frac{\partial \mathbf{R}_0}{\partial x}, \quad \hat{\mathbf{J}} = \frac{\partial \mathbf{R}_0}{\partial y} \tag{8.13a,b}$$

Also, from Figures 8.5 and 8.6, we have

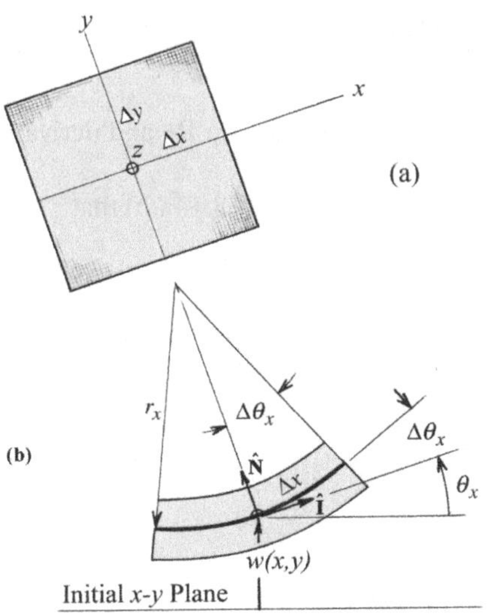

Figure 8.5

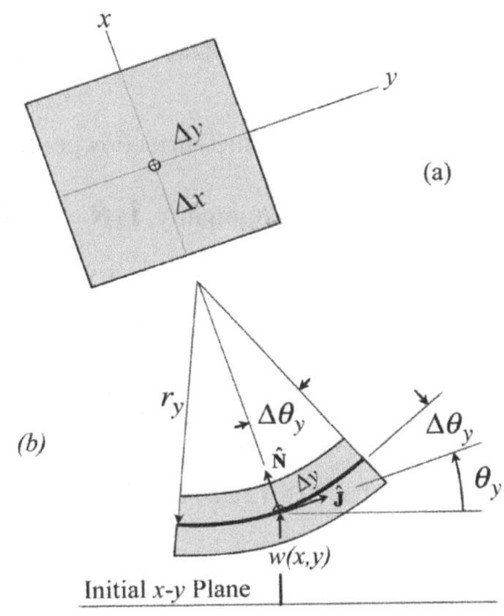

Figure 8.6

$$\frac{\partial \hat{\mathbf{I}}}{\partial x} = \frac{\partial \theta_x}{\partial x}\hat{\mathbf{N}} = \frac{\partial^2 \mathbf{R}_0}{\partial x^2}, \quad \kappa_x = \frac{\partial^2 \mathbf{R}_0}{\partial x^2}\cdot \hat{\mathbf{N}}$$
$$\frac{\partial \hat{\mathbf{J}}}{\partial y} = \frac{\partial \theta_y}{\partial y}\hat{\mathbf{N}} = \frac{\partial^2 \mathbf{R}_0}{\partial y^2}, \quad \kappa_y = \frac{\partial^2 \mathbf{R}_0}{\partial y^2}\cdot \hat{\mathbf{N}}$$

Since $\hat{\mathbf{I}}\cdot\hat{\mathbf{N}} = \hat{\mathbf{J}}\cdot\hat{\mathbf{N}} = 0$, we obtain

$$\mathbf{N}\cdot\frac{\partial \hat{\mathbf{I}}}{\partial x} = -\hat{\mathbf{I}}\cdot\frac{\partial \hat{\mathbf{N}}}{\partial x} = \frac{\partial \theta_x}{\partial x} = \kappa_x \tag{8.14a}$$
$$\mathbf{N}\cdot\frac{\partial \hat{\mathbf{J}}}{\partial y} = -\hat{\mathbf{J}}\cdot\frac{\partial \hat{\mathbf{N}}}{\partial y} = \frac{\partial \theta_y}{\partial y} = \kappa_y \tag{8.14b}$$

Figure 8.7 depicts a rectangular element which is twisted. Specifically, in Figure 8.7a we see the tangent vector $\hat{\mathbf{I}}(x,y)$ turning with Δy:

$$\Delta\hat{\mathbf{I}} \doteq \Delta\theta_x\hat{\mathbf{N}}, \quad \frac{\Delta\hat{\mathbf{I}}}{\Delta y} \doteq \frac{\Delta\theta_x}{\Delta y}\hat{\mathbf{N}}$$

In the limit ($\Delta y \to 0$), we have

$$\frac{\partial \hat{\mathbf{I}}}{\partial y} = \frac{\partial \theta_x}{\partial y}\hat{\mathbf{N}} = \frac{\partial}{\partial y}\left(\frac{\partial \mathbf{R_0}}{\partial x}\right) = \frac{\partial^2 \mathbf{R}_0}{\partial y \partial x} \tag{8.15a}$$

In like manner (see Figure 8.7b), we obtain

$$\frac{\partial \hat{\mathbf{J}}}{\partial x} = \frac{\partial \theta_y}{\partial x}\hat{\mathbf{N}} = \frac{\partial}{\partial x}\left(\frac{\partial \mathbf{R_0}}{\partial y}\right) = \frac{\partial^2 \mathbf{R}_0}{\partial x \partial y} \tag{8.15b}$$

Let us define the "torsion":

$$\kappa_{xy} = \hat{\mathbf{N}}\cdot\frac{\partial^2 \mathbf{R}_0}{\partial x \partial y} = \hat{\mathbf{N}}\cdot\frac{\partial^2 \mathbf{R}_0}{\partial y \partial x} \tag{8.16}$$

Here we are assuming the surface is continuous and smooth, i.e., the function $\mathbf{R}_0$ and derivatives are continuous.

In view of the orthogonality, $\hat{\mathbf{I}}\cdot\hat{\mathbf{N}} = \hat{\mathbf{J}}\cdot\hat{\mathbf{N}} = 0$. It follows from Equations (8.14a,b) that

$$\hat{\mathbf{I}}\cdot\frac{\partial \hat{\mathbf{N}}}{\partial x} = -\hat{\mathbf{N}}\cdot\frac{\partial \hat{\mathbf{I}}}{\partial x} = -\kappa_x = -\frac{\partial \theta_x}{\partial x} \tag{8.17a}$$
$$\hat{\mathbf{J}}\cdot\frac{\partial \hat{\mathbf{N}}}{\partial y} = -\hat{\mathbf{N}}\cdot\frac{\partial \hat{\mathbf{J}}}{\partial y} = -\kappa_y = -\frac{\partial \theta_y}{\partial x} \tag{8.17b}$$

It follows from (8.15a,b) and (8.16) that

$$\hat{\mathbf{I}}\cdot\frac{\partial \hat{\mathbf{N}}}{\partial y} = -\hat{\mathbf{N}}\cdot\frac{\partial \hat{\mathbf{I}}}{\partial y} = -\kappa_{xy} = -\frac{\partial \theta_x}{\partial y} \tag{8.17c}$$
$$\hat{\mathbf{J}}\cdot\frac{\partial \hat{\mathbf{N}}}{\partial x} = -\hat{\mathbf{N}}\cdot\frac{\partial \hat{\mathbf{J}}}{\partial x} = -\kappa_{yx} = -\frac{\partial \theta_y}{\partial x} \tag{8.17d}$$

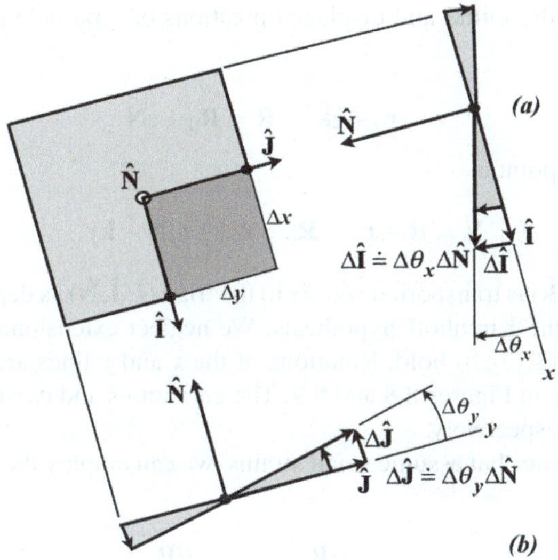

Figure 8.7

If w denotes the displacement in the initial direction z, then (see Figures 8.5 and 8.6)

$$\tan\theta_x = \frac{\partial w}{\partial x}, \quad \tan\theta_y = \frac{\partial w}{\partial y}$$

If the *rotations* remain *small, i.e.* $\theta_x \ll 1$, $\theta_y \ll 1$, then we can accept the approximations;

$$\theta_x \doteq \frac{\partial w}{\partial x}, \quad \theta_y \doteq \frac{\partial w}{\partial y} \tag{8.18a,b}$$

According to Equations (8.18a,b) and (8.17a–d) we obtain the approximations:

$$\kappa_x \doteq \frac{\partial^2 w}{\partial x^2}, \quad \kappa_y \doteq \frac{\partial^2 w}{\partial y^2} \tag{8.19a,b}$$

$$\kappa_{xy} = \kappa_{yx} \doteq \frac{\partial^2 w}{\partial x \partial y} \doteq \frac{\partial^2 w}{\partial y \partial x} \tag{8.19c}$$

The Equations (8.16) and (8.19c) assume a continuous and smooth surface, i.e., continuity of functions $\mathbf{R}_0$, w and the derivatives.

8.11 KIRCHHOFF THEORY OF PLATES

The Kirchhoff[4] theory of plates is founded upon the assumption that a line (z), initially normal to the reference plane (x-y), remains normal to the deformed reference surface. The consequent deformations are depicted in Figures 8.8 and 8.9, which illustrate the flexures and twist, respectively. The assumption is analogous to the Bernoulli assumption for the flexure of beams. Additionally, we assume that extension of the normal is negligible[5]

Let $\mathbf{r}_0$ and $\mathbf{R}_0$ denote the initial and displaced positions of a point on the reference surface, $\hat{\mathbf{k}}$ and $\hat{\mathbf{N}}$ the unit normals, $\hat{\mathbf{i}}, \hat{\mathbf{j}}$ and $\hat{\mathbf{I}}, \hat{\mathbf{J}}$ the unit vectors tangent to the initial and deformed x and y

lines, respectively. Then the initial and displaced positions of a particle at z on the normal are $\mathbf{r}$ and $\mathbf{R}$:

$$\mathbf{r} = \mathbf{r}_0 + z\hat{\mathbf{k}}, \quad \mathbf{R} = \mathbf{R}_0 + z\hat{\mathbf{N}} \tag{8.20a,b}$$

The displacement at that point is

$$\mathbf{V} = \mathbf{R} - \mathbf{r} = (\mathbf{R}_0 - \mathbf{r}_0) + z(\mathbf{N} - \mathbf{k})$$

The orthogonal triad $(\hat{\mathbf{i}}, \hat{\mathbf{j}}, \hat{\mathbf{k}})$ is transported *rigidly* to the triad $(\hat{\mathbf{I}}, \hat{\mathbf{J}}, \hat{\mathbf{N}})$ as depicted in Figures 8.8 and 8.9, in accordance with the Kirchhoff hypothesis. We neglect extensional strains of the surface; therefore, the Equations (8.13a,b) hold. Rotations of the x and y lines are θ_x and θ_y as shown in Figures 8.5, 8.6 and again in Figures 8.8 and 8.9. The curvatures and twist are given by Equations (8.17a,b) and (8.17c,d), respectively.

If we admit large rotations but assume small strains, we can employ the development of Section 2.21, where

$$\mathbf{G}_x = \frac{\partial \mathbf{R}}{\partial x}, \quad \mathbf{G}_y = \frac{\partial \mathbf{R}}{\partial y}$$

Then, according to Equations (2.65a,b,d)

$$\varepsilon_x = \hat{\mathbf{I}} \cdot \frac{\partial \mathbf{R}}{\partial x} - 1, \quad \varepsilon_y = \hat{\mathbf{J}} \cdot \frac{\partial \mathbf{R}}{\partial y} - 1$$

$$\gamma_{xy} = 2\bar{\gamma}_{xy} = \hat{\mathbf{I}} \cdot \frac{\partial \mathbf{R}}{\partial y} + \hat{\mathbf{J}} \cdot \frac{\partial \mathbf{R}}{\partial x}$$

Employing Equations (8.20) and (8.17a,b,c,d), we obtain

$$\varepsilon_x = -z\kappa_x, \quad \varepsilon_y = -z\kappa_y \tag{8.21a,b}$$

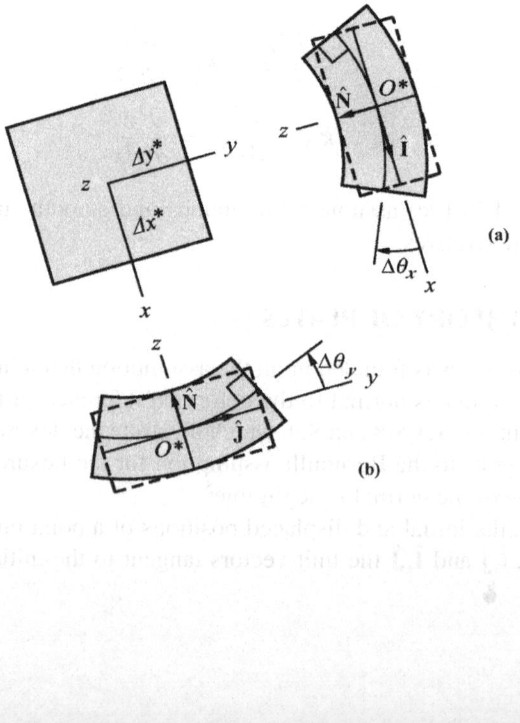

Figure 8.8

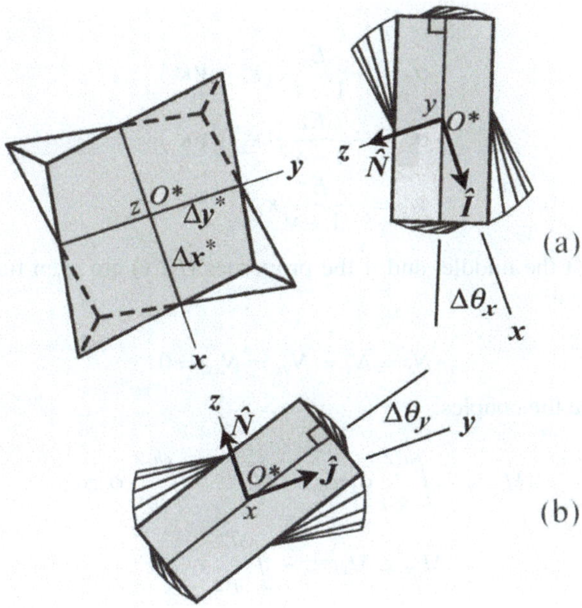

Figure 8.9

$$\gamma_{xy} = -2z\kappa_{xy} \tag{8.21c}$$

Figures 8.8 and 8.9 indicate the negative strains at the upper surfaces ($z > 0$), corresponding to the positive curvatures and twist ($\kappa_x, \kappa_y, \kappa_{xy}$).

Finally, we note that these results (8.21a–c) are not limited to small rotations. Indeed, one could roll a flat sheet into a cylinder providing only that the thickness is sufficiently small $z/r \ll 1$.

If the rotations are sufficiently small ($\theta_x \ll 1, \theta_y \ll 1$), then the approximations (8.18) and (8.19) are applicable. Then we have the approximations:

$$\varepsilon_x \doteq -z\frac{\partial^2 w}{\partial x^2}, \quad \varepsilon_y \doteq -z\frac{\partial^2 w}{\partial y^2} \tag{8.22a}$$

$$\gamma_{xy} \doteq -2z\frac{\partial^2 w}{\partial x \partial y} \tag{8.22b}$$

8.12 FLEXURAL STRESSES IN THE HI-HO KIRCHHOFF PLATE

Since our plate is thin, we neglect the components of stress upon the z surfaces. With $\sigma_z = \tau_{zx} = \tau_{zy} = 0$, we again employ the stress-strain relations (8.2) for an isotropic Hookean plate. Substituting (8.21) into (8.2), we obtain

$$\sigma_x = -\frac{Ez}{1-\nu^2}(\kappa_x + \nu\kappa_y) \tag{8.23a}$$

$$\sigma_y = -\frac{Ez}{1-\nu^2}(\kappa_y + \nu\kappa_x) \tag{8.23b}$$

$$\tau_{xy} = -\frac{Ez}{1+\nu}\kappa_{xy} \tag{8.23c}$$

If the x-y plane is at the middle, and if the properties (E, ν) are even functions of z, then the integrals (8.6) vanish, i.e.

$$N_x = N_y = N_{yx} = N_{xy} = 0$$

It remains to determine the couples:

$$M_x \equiv -\int_{-h/2}^{h/2} \sigma_x z dz, \quad M_y \equiv -\int_{-h/2}^{h/2} \sigma_y z dz \tag{8.24a,b}$$

$$M_{xy} \equiv M_{yx} \equiv -\int_{-h/2}^{h/2} \tau_{xy} z dz \tag{8.24c}$$

According to these definitions, the components (M_x, M_{yx}, M_y) transform just as the components of stress, as expressed by Equations (3.34) and (3.35) (see Section 3.12):

$$M_n = \frac{1}{2}(M_x + M_y) + \frac{1}{2}(M_x - M_y)\cos 2\theta + M_{xy}\sin 2\theta \tag{8.25a}$$

$$M_{nt} = -\frac{1}{2}(M_x - M_y)\sin 2\theta + M_{xy}\cos 2\theta \tag{8.25b}$$

If the material is homogeneous, HI-HO[6] (E and ν do not vary with z), then

$$M_x = \frac{Eh^3}{12(1-\nu^2)}(\kappa_x + \nu\kappa_y) \tag{8.26a}$$

$$M_y = \frac{Eh^3}{12(1-\nu^2)}(\kappa_y + \nu\kappa_x) \tag{8.26b}$$

$$M_{xy} = M_{yx} = \frac{Eh^3}{12(1+\nu)}\kappa_{xy} \tag{8.26c}$$

These actions upon a rectangular element are shown in Figure 8.10. Couples are represented as vectors in accordance with the right-hand (screw) rule.

8.13 EQUILIBRIUM CONDITIONS

In addition to the couples of Figure 8.10, an element is subject to transverse shear forces V_x and V_y upon the x and y edges, respectively. All actions appear as vectors in Figure 8.11; ⊙ and ⊗ are the heads and tails of forces. All actions are force, or couple, per unit length. P denotes the transverse external force per unit area; $\bar{P}$ denotes the mean value.

Summing forces in the z direction, we obtain

$$\Delta x \Delta V_x + \Delta y \Delta V_y = \bar{P}\Delta x \Delta y$$

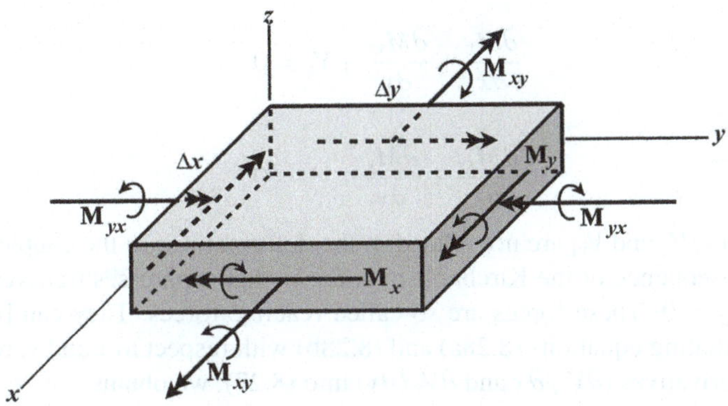

Figure 8.10

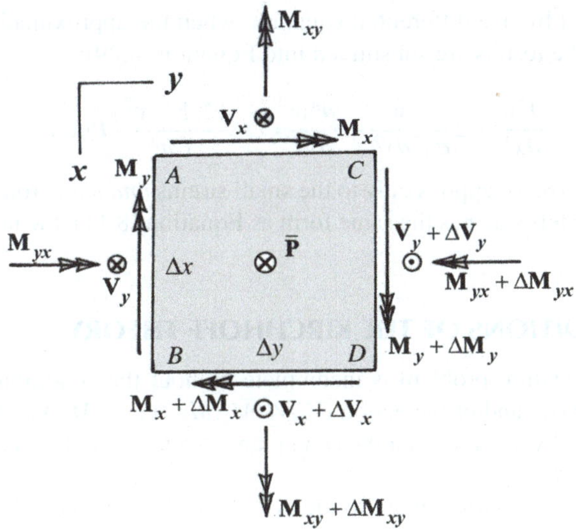

Figure 8.11

Dividing the equilibrium equation by $\Delta x\Delta y$ and passing to the limit ($\Delta x \to 0$, $\Delta y \to 0$), we obtain the one differential equation:

$$\frac{\partial V_x}{\partial x} + \frac{\partial V_y}{\partial y} = P(x,y) \tag{8.27}$$

It must be noted that no deformation is taken into account. This means that the linear equation (8.27) applies to only small strains and small curvatures. Summing moments about the edge AC, we obtain

$$\Delta y\,\Delta M_x + \Delta x\,\Delta M_{yx} + \Delta x\,\Delta y\,V_x = 0$$

Again we anticipate the limit and omit higher order terms, e.g $\Delta x\Delta y\Delta V_x$. In the limit, we obtain the differential equation:

$$\frac{\partial M_x}{\partial x} + \frac{\partial M_{yx}}{\partial y} + V_x = 0 \tag{8.28a}$$

Likewise, we obtain

$$\frac{\partial M_{xy}}{\partial x} + \frac{\partial M_y}{\partial y} + V_y = 0 \tag{8.28b}$$

Note that shear forces, V_x and V_y, are not related to the deformations as the couples via Equations (8.26). This is a consequence of the Kirchhoff hypothesis that suppresses transverse shear deformation, i.e. $\gamma_{yz} = \gamma_{xz} = 0$. These forces are so-called reactive forces. They can be eliminated as follows: By differentiating equations (8.28a) and (8.28b) with respect to x and y, respectively, and substituting those derivatives ($\partial V_x/\partial x$ and $\partial V_y/\partial y$) into (8.27), we obtain

$$\frac{\partial^2 M_x}{\partial x^2} + 2\frac{\partial^2 M_{xy}}{\partial x \partial y} + \frac{\partial^2 M_y}{\partial y^2} = -P(x,y) \tag{8.29}$$

Our theory is expressed by one differential equation, when the approximations (8.19) are substituted into (8.26), and the results are substituted into Equation (8.29):

$$\frac{\partial^4 w}{\partial x^4} + 2\frac{\partial^4 w}{\partial x^2 \partial y^2} + \frac{\partial^4 w}{\partial y^4} = -\frac{12(1-\nu^2)}{Eh^3}P(x,y) \tag{8.30}$$

Note that this Equation (8.30) applies only to the small strains *and* small rotations of a thin HI-HO plate. Note too that the left side has the same form as Equation (8.11); the latter is the well-known biharmonic equation.

8.14 EDGE CONDITIONS OF THE KIRCHHOFF THEORY

Our statement of the flexural problem is incomplete without the conditions which are imposed upon the displacement (w) and/or the action ($V_x, V_y, M_x, M_y, M_{xy} = M_{yx}$) at the edges. Figure 8.12 depicts an edge defined by a curve C on the (x-y) plane; n and c denote distance along the normal and tangent, respectively.

Once again, we appeal to virtual work to identify the necessary and sufficient conditions. If w denotes the transverse displacement, then rotation about t in the direction of M_n is $\partial w/\partial n$, about n in the direction of M_{nt} is $\partial w/\partial c$. An incremental (virtual) displacement δw produces the virtual work:

$$\delta W = \int_c \left(V_n \delta w + M_n \frac{\partial \delta w}{\partial n} + M_{nt} \frac{\partial \delta w}{\partial c} \right) \mathrm{d}c$$

Note that the final term can be integrated by parts: If A and B denote the ends of the boundary C (e.g. corners of the plate), then

$$\delta W = (M_{nt}\delta w)_B - (M_{nt}\delta w)_A + \int_c \left[\left(V_n - \frac{\partial M_{nt}}{\partial c} \right) \delta w + M_n \frac{\partial \delta w}{\partial n} \right] \mathrm{d}c \tag{8.31}$$

Quite simply, if δw is varied along the curve, $\partial \delta w/\partial c$ is not also variable. The union of shear force and the torsional couple M_{nt} is a consequence of the Kirchhoff constraint; the normal line (a-b in Figure 8.12) can not turn independently of the tangent line (t). From Equation (8.31), we can identify the possible edge conditions: If the edge is clamped, then $w = \partial w/\partial n = O$; the actions ($V_n - \partial M_{nt}/\partial c$) and M_n are determined by the solution, as required to constrain the movement. If

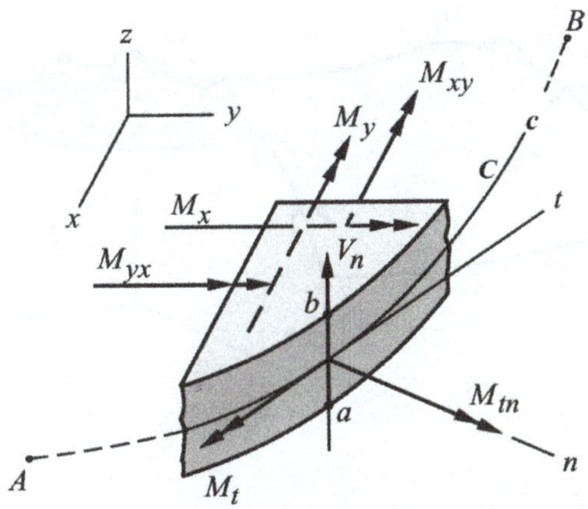

Figure 8.12

the edge is entirely free, then the actions must vanish: $(V_n - \partial M_{nt}/\partial c) = M_n = O$. Combinations are possible. For example, a simple support inhibits displacement, $w = 0$, but admits rotation $(\partial w/\partial n)$, hence $M_n = 0$. The effective shear force upon the edge of the Kirchhoff plate is

$$V_n^* \equiv V_n - \frac{\partial M_{nt}}{\partial c} \tag{8.32}$$

In accordance with (8.28)

$$V_n^* = -\frac{\partial M_n}{\partial n} - 2\frac{\partial M_{nt}}{\partial c} \tag{8.33}$$

The latter can be expressed in terms of the displacement via Equations (8.25), (8.26), and (8.19).

8.15 SIMPLE BENDING

We recall the moment-curvature relations (8.26) for a HI-HO plate, but without the approximations of small rotation:

$$M_x = D(\kappa_x + \nu\kappa_y)$$
$$M_y = D(\kappa_y + \nu\kappa_x)$$
$$M_{xy} = M_{yx} = (1-\nu)D\kappa_{xy}$$

wherein

$$D \equiv Eh^3/12(1-\nu^2)$$

A state of bending under the action of the couple M_y alone ($M_x = M_{xy} = 0$), provides the deformation:

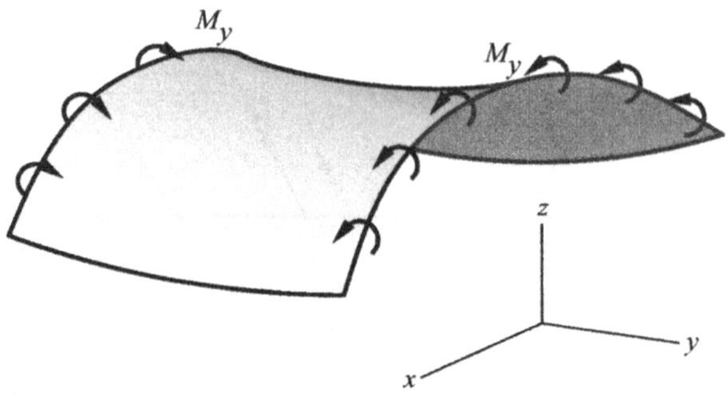

Figure 8.13

$$\kappa_y = \frac{M_y}{(1-\nu^2)D}, \quad \kappa_x = -\nu\kappa_y, \quad \kappa_{xy} = 0$$

The deformed plate has the form of a saddle, as depicted in Figure 8.13. The curvature in the x-z plane, $\kappa_x = -\nu\kappa_y$, is termed the "anticlastic curvature"; compression of y fibers in the top and extension in the lower causes the opposite deformations of the x fibers (so-called Poisson effect).

This state is one which satisfies all equations governing the small strains of a HI-HO body. Stated otherwise, one need not invoke the approximations of the plate theory, viz. the Kirchhoff hypothesis and the assumptions upon transverse stresses ($\sigma_z = \tau_{zx} = \tau_{zy} = 0$). It is an "exact" solution of the linear equations of three dimensions.[7]

8.16 SIMPLE TORSION

A simple torsion also fulfills all conditions for a state of small strain of a HI-HO body. This is given by

$$\kappa_{xy} = \kappa_{yx} = \frac{M_{yx}}{(1-\nu)D}$$

The small displacements associated with the twist are

$$w = \kappa_{xy}xy = \frac{M_{yx}}{(1-\nu)D}xy$$

The twisted form is a hyperbolic paraboloid, as depicted in Figure 8.14.

The edge couples upon the body of Figure 8.14 are the resultants of the shear stresses, ($\tau_{xy} = \tau_{yx}$), in accordance with (8.24c). If we interpret the torsion in the context of the Kirchhoff theory, then a normal (line z) remains perpendicular to the deformed plane, i.e., normal to the deformed x and y lines. A z line is depicted on the edge of the plate in Figure 8.15. Because the edge lines (z and x) rotate rigidly, the couples $M_{yx}\Delta x$ on the adjoining segments A and B of Figure 8.15 can be replaced by the couples $V\Delta x^2 = M_{yx}\Delta x$. As we progress along the edge, the forces V and $-V$ cancel; only at the corners C and D the forces $V\Delta x = M_{yx} = M_{xy}$ are not canceled. Similarly, the statical alternatives on edge EC produce the like force M_{yx} at C; altogether, we obtain the forces $\pm 2M_{yx} = \pm 2M_{xy}$ at each corner as shown. Note that the argument hinges on the equivalence of the couples, which hinges upon the constraint, normality of the z line.

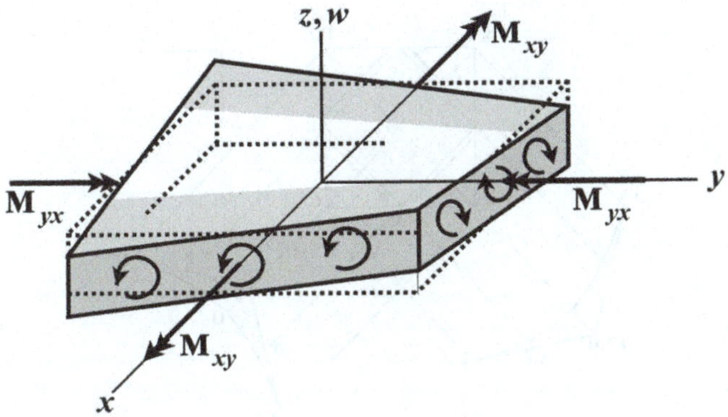

Figure 8.14

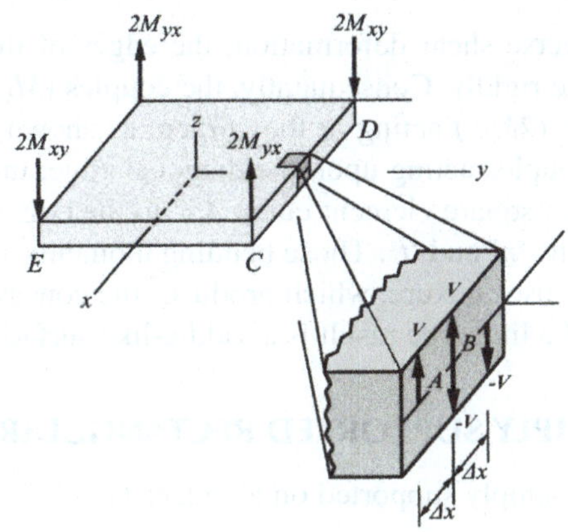

Figure 8.15

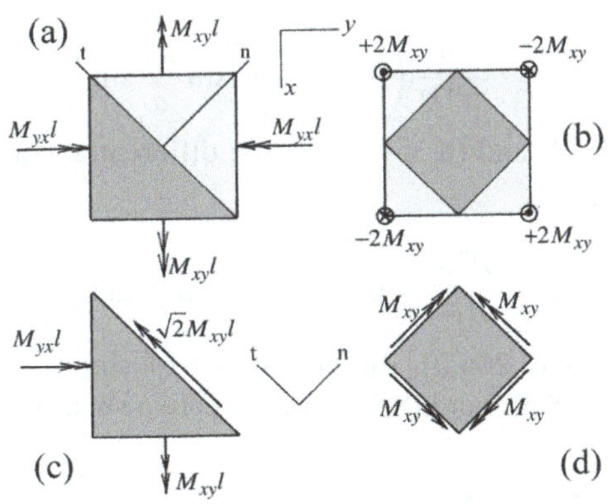

Figure 8.16

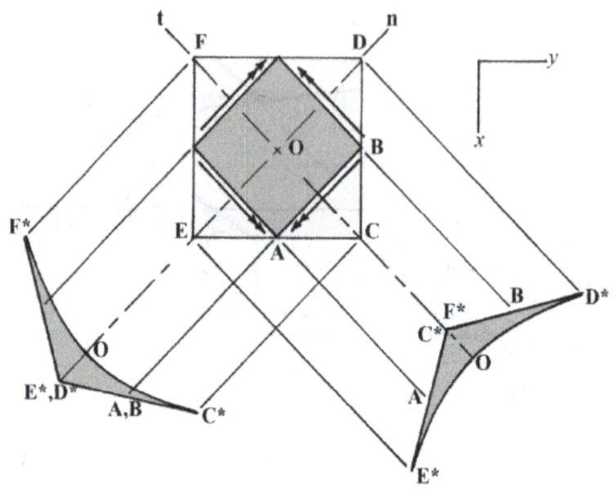

Figure 8.17

Because there is no transverse shear deformation, the edges of the rectangular plate remain straight, i.e., these edges rotate rigidly. Consequently, the couples $(M_{xy}l)$ of Figure 8.16a are statically equivalent to the forces $(2M_{xy})$ acting at the corners as shown in Figure 8.16b. In Figure 8.16c, we note the bending couples acting upon the diagonal edge; these equilibrate the twisting couples. Figure 8.16d depicts a square element cut at 45° to the edges; this element is subject to simple bending in the directions 'n' and 't'. Those bending moments act in the opposite direction. The moment on the 't' edge causes flexure, which produces the concavity of a 't' line; that on the 'n' edge causes a convexity of a line. The result is a saddle-like surface as seen in Figure 8.17.

8.17 FLEXURE OF A SIMPLY SUPPORTED RECTANGULAR PLATE

A rectangular plate is HI-HO, simply supported on all edges ($x = \pm a$, $y = \pm b$) and subjected to a transverse load:

$$P = -p\sin\frac{\pi x}{a}\sin\frac{\pi y}{b} \tag{8.34}$$

The solution of Equation (8.30) governs the Kirchhoff plate:

$$w = -\frac{p}{\pi^4 D}\left(\frac{a^2b^2}{a^2+b^2}\right)^2 \sin\frac{\pi x}{a}\sin\frac{\pi y}{b} \tag{8.35}$$

The reader can verify that (8.34) and (8.35) satisfy the differential equation (8.30). Also, we note that

$$w = \frac{\partial^2 w}{\partial x^2} = \frac{\partial^2 w}{\partial y^2} = 0, \quad \text{at } x = 0, a \quad \text{and} \quad y = 0, b$$

Then, according to Equations (8.26a,b), the conditions of simple support are fulfilled on the edge.According to Equations (8.33) and (8.26c) the effective shear force upon the edges $x = O, a$ are

$$\begin{aligned} V_x^*(0,y) &= \left(-\frac{\partial M_x}{\partial x} - 2\frac{\partial M_{yx}}{\partial y}\right)_{x=0,a} = -D\left(\frac{\partial^3 w}{\partial x^3} + 2(1-\nu)\frac{\partial^3 w}{\partial x \partial y^2}\right)_{x=0,a} \\ &= \mp\frac{p}{a\pi}\left(\frac{a^2b^2}{a^2+b^2}\right)^2\left(\frac{1}{a^2} + \frac{2(1-\nu)}{b^2}\right)\sin\frac{\pi y}{a} \end{aligned}$$

The twisting couple on the edge ($x = 0, a$) follows from (8.26c) and (8.35):

$$M_{xy}|_{x=0,a} = \mp \frac{(1-\nu)p}{ab\pi^2}\left(\frac{a^2b^2}{a^2+b^2}\right)\cos\frac{\pi y}{b} \tag{8.36}$$

We recall the arguments of Section 8.16 (see Figure 8.15) wherein the twisting couple along an edge causes a concentrated force upon the corner; here the couples (8.36) upon the edges ($x = 0, a$) produce the force at each corner:

$$R = \mp(1-\nu)\frac{p}{\pi^2}\frac{a^3b^3}{\pi^2(a^2+b^2)^2}$$

These act in the directions as shown in Figure 8.18. The similar couples upon the edges, $y = 0, b$, contribute the same forces. Altogether, the edges and corners are subjected to the distributed shear and concentrated forces, as depicted in Figure 8.18.

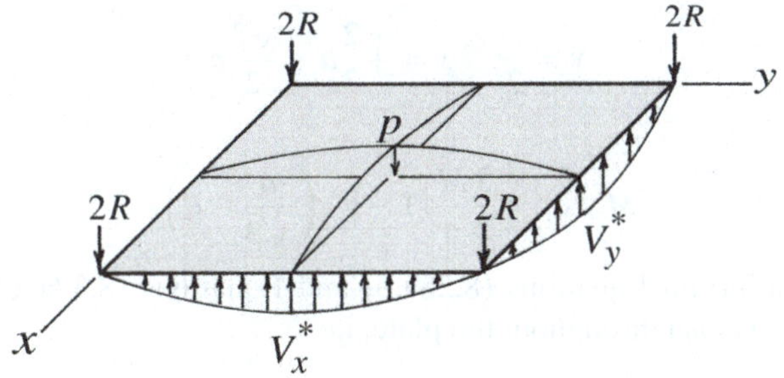

Figure 8.18

As a simple experiment,the reader can lay a thin board upon a rectangular frame and press upon the central part; he/she should observe the tendency of the corners to lift.

8.18 BENDING OF A TRIANGULAR PLATE

The plate of Figure 8.19 has the form of an equilateral triangle which is simply supported on the edges. Each edge is subjected to a uniform bending couple M. We chose this example because it has the simple solution.

$$w = -\frac{M}{4aD}\left[x^3 - 3y^2x - a(x^2+y^2) + \frac{4}{27}a^2\right]$$

The bending and twisting couples (M_x, M_y, M_z) follow, in accordance with Equations (8.26):

$$M_x = \frac{M}{2a}[-3(1-\nu)x + (1+\nu)a] \tag{8.37a}$$

$$M_y = \frac{M}{2a}[3(1-\nu)x + (1+\nu)a] \tag{8.37b}$$

$$M_{xy} = \frac{3M}{2a}(1-\nu)y \tag{8.37c}$$

By the transformation of Equations (8.25), we obtain the components in the directions n and t of Figure 8.19.

$$M_n = \frac{M}{4a}\left[2(1+\nu)a + 3(1-\nu)x + 3\sqrt{3}(1-\nu)y\right]$$
$$M_{nt} = \frac{3M}{4a}(1-\nu)\left(\sqrt{3}x - y\right) \tag{8.38}$$

Here, we note that the equation of the line BC is

$$y = \frac{1}{\sqrt{3}}\left(\frac{2}{3}a - x\right)$$

It follows that on line BC

$$M_n = M$$

If c denotes distance from C along edge CB, then

$$y = \frac{c}{2}, \quad x = +\frac{2}{3}a - \frac{\sqrt{3}}{2}c$$

Then, on CB

$$M_{nt}(c) = \frac{3}{2}\frac{M}{a}(1-\nu)\left(\frac{a}{\sqrt{3}} - c\right)$$

According to the equilibrium Equations (8.28a,b) and Equations (8.37a–c), it follows that the transverse shear forces vanish throughout the plate, i.e.

$$V_x = V_y = 0$$

Equilibrium also dictates that the shear force vanish at an edge, e.g. on edge BC

$$V_n = 0$$

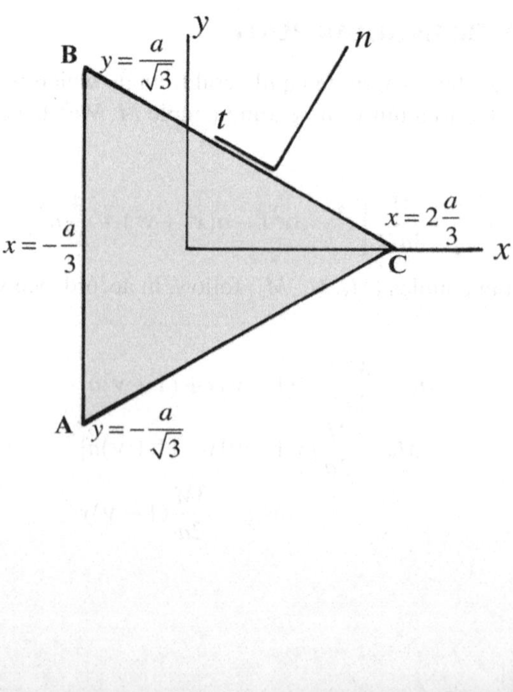

Figure 8.19

Since $V_n = 0$, the effective shear on BC follows from Equation (8.32)

$$V_n^* = -\frac{\partial M_{nt}}{\partial c} = \frac{3}{2}\frac{M}{a}(1-\nu)$$

Also according to the arguments of Section 8.16 (see Figure 8.15), the distribution of twisting couple M_{nt} on CB produces concentrated forces at ends C and B.

$$R(c=0) = \frac{\sqrt{3}M}{2}(1-\nu)$$

$$R(c=\frac{2a}{\sqrt{3}}) = -\frac{\sqrt{3}M}{2}(1-\nu)$$

The signs indicate that both act in downward ($-z$) directions. By symmetry, the same forces result from the twisting couples acting upon edges AB and AC; i.e. each corner is subjected to the concentrated force $R = \sqrt{3}M(1-\nu)$. These forces equilibrate the constant shear forces V_n^*, upon each edge.

8.19 EFFECT OF TRANSVERSE SHEAR DEFORMATION

Numerous authors[8] have introduced a generalization of the Kirchhoff hypothesis to accommodate a transverse shear deformation. A relaxed assumption asserts only that the normal line (z) remains straight but turns relative to the reference surface. In Figures 8.20 and 8.21 the orthonormal triad $(\hat{\mathbf{i}},\hat{\mathbf{j}},\hat{\mathbf{k}})$ is rigidly transported to the triad $(\hat{\mathbf{I}},\hat{\mathbf{J}},\hat{\mathbf{N}})$ where $\hat{\mathbf{I}}$ and $\hat{\mathbf{J}}$ remain tangent to the deformed x and y lines and $\hat{\mathbf{N}}$ remains normal to the deformed surface. Normal $\hat{\mathbf{N}}$ turns through the angles θ_x and θ_y, but the line z turns through the angles β_x and β_y; the differences are the transverse shear strains:

$$\gamma_{xz} = \theta_x - \beta_x, \quad \gamma_{yz} = \theta_y - \beta_y$$

An arbitrary particle has the initial position:

$$\mathbf{r} = \mathbf{r}_0 + \hat{\mathbf{k}}z$$

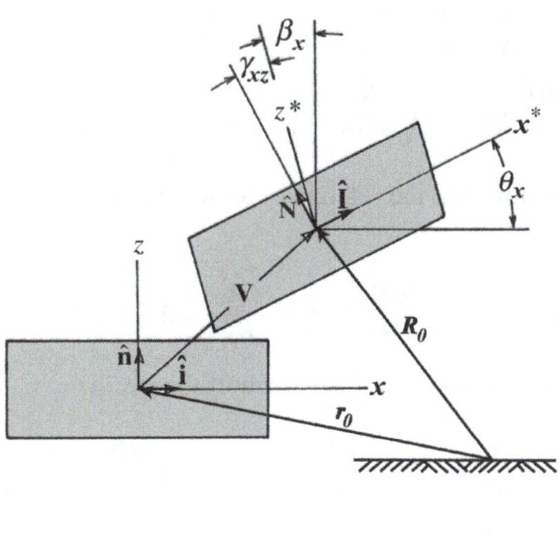

Figure 8.20

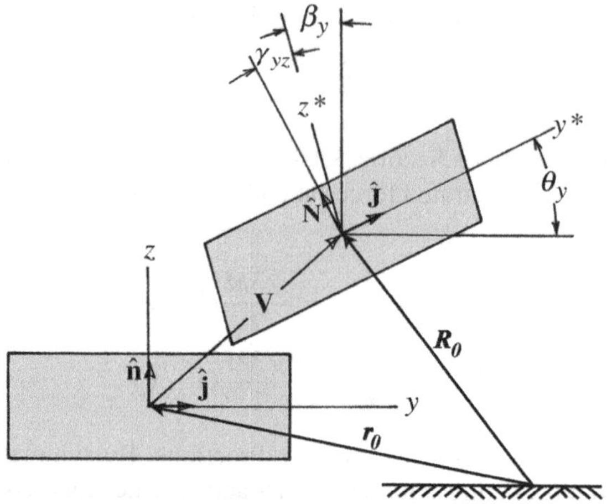

Figure 8.21

where subscript '0' signifies the position at the reference surface. In the case of small strain ($\gamma_{xz} \ll 1, \gamma_{yz} \ll 1$), the position in the deformed plate is

$$\mathbf{R} = \mathbf{R}_0 + (\hat{\mathbf{N}} + \gamma_{xz}\hat{\mathbf{I}} + \gamma_{yz}\hat{\mathbf{J}})\, z$$

Note that

$$\frac{\partial \hat{\mathbf{N}}}{\partial x} = -\frac{\partial \theta_x}{\partial x}\hat{\mathbf{I}} - \frac{\partial \theta_y}{\partial x}\hat{\mathbf{J}}, \quad \frac{\partial \hat{\mathbf{N}}}{\partial y} = -\frac{\partial \theta_x}{\partial y}\hat{\mathbf{I}} - \frac{\partial \theta_y}{\partial y}\hat{\mathbf{J}}$$

The strain components follow:

$$\varepsilon_x = \frac{\partial \mathbf{R}}{\partial x} \cdot \hat{\mathbf{I}} - 1 \doteq \left(\frac{\partial \hat{\mathbf{N}}}{\partial x} \cdot \hat{\mathbf{I}} + \frac{\partial \gamma_{xz}}{\partial x} \right) z = \left(-\frac{\partial \theta_x}{\partial x} + \frac{\partial \gamma_{xz}}{\partial x} \right) z = -\frac{\partial \beta_x}{\partial x} z \tag{8.39a}$$

$$\varepsilon_y = \frac{\partial \mathbf{R}}{\partial y} \cdot \hat{\mathbf{J}} - 1 \doteq \left(-\frac{\partial \theta_y}{\partial y} + \frac{\partial \gamma_{yz}}{\partial y} \right) z = -\frac{\partial \beta_y}{\partial x} z \tag{8.39b}$$

$$\gamma_{xy} = \frac{\partial \mathbf{R}}{\partial y} \cdot \hat{\mathbf{I}} + \frac{\partial \mathbf{R}}{\partial x} \cdot \hat{\mathbf{J}} \doteq -\left(\frac{\partial \beta_x}{\partial y} + \frac{\partial \beta_y}{\partial x} \right) z \tag{8.39c}$$

In the foregoing, we assume that the reference surface undergoes no extensional strain; therefore

$$\frac{\partial \mathbf{R}_0}{\partial x} = \hat{\mathbf{I}}, \quad \frac{\partial \mathbf{R}_0}{\partial y} = \hat{\mathbf{J}}$$

Note that the curvatures and torsion of the Kirchhoff theory, viz

$$\kappa_x = \frac{\partial \theta_x}{\partial x}, \quad \kappa_y = \frac{\partial \theta_y}{\partial y}, \quad \kappa_{xy} = \frac{1}{2}\left(\frac{\partial \theta_x}{\partial x} + \frac{\partial \theta_y}{\partial y} \right)$$

are now supplanted by

$$\bar{\kappa}_x = \frac{\partial \beta_x}{\partial x} = \kappa_x - \frac{\partial \gamma_{xz}}{\partial x}, \quad \bar{\kappa}_y = \frac{\partial \beta_y}{\partial y} = \kappa_y - \frac{\partial \gamma_{yz}}{\partial y} \tag{8.40a,b}$$

$$\bar{\kappa}_{xz} \equiv \frac{1}{2}\left(\frac{\partial \beta_x}{\partial y} + \frac{\partial \beta_y}{\partial x}\right) = \kappa_{xy} - \frac{1}{2}\left(\frac{\partial \gamma_{xz}}{\partial y} + \frac{\partial \gamma_{yz}}{\partial x}\right) \tag{8.40c}$$

The extensional strains of Equations (8.39a–c) (see Equations 8.21a–c) take the forms:

$$\varepsilon_x = -\bar{\kappa}_{xz}, \quad \varepsilon_y = -\bar{\kappa}_{yz}, \quad \gamma_{xz} = -2\bar{\kappa}_{xyz} \tag{8.41a,b,c}$$

For the HI-HO material, the moments in Equations (8.26a–c) take the forms:

$$M_x = D\bar{\kappa}_x, \quad M_y = D\bar{\kappa}_y, \quad M_{xy} = (1-\nu)D\bar{\kappa}_{xy} \tag{8.42a,b,c}$$

where

$$D \equiv \frac{Eh^3}{12(1-\nu^2)}$$

It remains to express the actions V_x, V_y in terms of the additional variables, β_x and β_y.

The transverse shear forces per unit length (see Sections 8.13 and 8.14) are the integrals:

$$V_x = \int_{-t/2}^{h/2} \tau_{xz} dz, \quad V_y = \int_{-t/2}^{h/2} \tau_{yz} dz \tag{8.43a,b}$$

For the HI-HO material

$$\tau_{xz} = \frac{E}{2(1+\nu)}\gamma_{xz}, \quad \tau_{yz} = \frac{E}{2(1+\nu)}\gamma_{yz}$$

Our underlying assumption implies that the strains, γ_{xz} and γ_{yz}, do not vary through the thickness. That implies that the stress components, τ_{xz} and τ_{yz}, are also independent of coordinate z. This presents a contradiction since the shear stresses (τ_{xz}, τ_{yz}) usually vanish at the surfaces. A better approximation of these stresses is a parabolic distribution:

$$\tau_{xz} = \frac{3}{2}\frac{V_x}{h}\left[1 - \left(\frac{2z}{h}\right)^2\right], \quad \tau_{yz} = \frac{3}{2}\frac{V_y}{h}\left[1 - \left(\frac{2z}{h}\right)^2\right] \tag{8.44a,b}$$

Since we are concerned with the HI-HO material, we might adopt the corresponding distribution for the strains $(\gamma_{xz}, \gamma_{yz})$, viz.

$$\gamma_{xz} \doteq \bar{\gamma}_{xz}\left[1 - \left(\frac{2z}{h}\right)^2\right], \quad \gamma_{yz} \doteq \bar{\gamma}_{yz}\left[1 - \left(\frac{2z}{h}\right)^2\right] \tag{8.45a,b}$$

Then Equations (8.43a,b) provide the results:

$$V_x = \frac{Eh}{3(1+\nu)}\bar{\gamma}_{xz}, \; V_y = \frac{Eh}{3(1+\nu)}\bar{\gamma}_{yz} \tag{8.46a,b}$$

Here, $\bar{\gamma}_{xz}$ and $\bar{\gamma}_{yz}$ are the midplane values, i.e. $\bar{\gamma}_{xz} = \theta_x - \beta_x$, $\bar{\gamma}_{yz} = \theta_y - \beta_y$.

Again, we must acknowledge a contradiction, since the results (8.41a–c) presume a constant distribution, whereas (8.45a,b) reflects the parabolic distribution. A refinement requires some compromise which appeals to the concepts of work expended by the couples (M_x, M_y, M_{xy}), shear forces

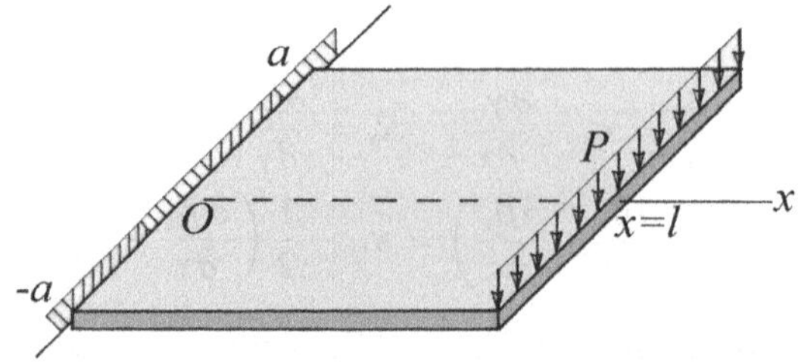

Figure 8.22

(V_x, V_y) and the associated stress distributions $(\sigma_x, \sigma_y \tau_{xy}, \tau_{xz}, \text{ and } \tau_{yz})$. Here, we intend only to identify the effects and provide the bases for the approximations of the elements in Chapter 10. The interested reader is directed to the earlier works.[9]

We recall (see Section 8.14) that the edge conditions of the Kirchhoff theory introduce an "effective" shear force (Equations 8.32 and 8.33). That occurs because the shear forces (V_x, V_y) are reactive; they do not work in the absence of an associated strain $(\bar{\gamma}_{xz}, \bar{\gamma}_{yz})$. In the shear deformable theory, these forces (or the associated displacements) must be prescribed. At the edge, as seen in Figure (8.12), one prescribes the three actions $(M_t, M_{tn}, \text{and } V_n)$ or the kinematic counterparts (β_n, β_t, w).

Finally, we note the approximation of small strains and small rotations $(\theta_x \ll 1,\ \theta_y \ll 1)$. Then

$$\theta_x \doteq \frac{\partial w}{\partial x}, \quad \theta_y \doteq \frac{\partial w}{\partial y}$$

$$\kappa_x = \frac{\partial w^2}{\partial x^2}, \quad \kappa_y = \frac{\partial w^2}{\partial y^2}, \quad \kappa_{xy} = \frac{\partial^2 w}{\partial x \partial y}$$

8.20 EFFECT OF TRANSVERSE SHEAR DEFORMATIONS

It is well known that the inclusion of the transverse shear strains $(\gamma_{xy}, \gamma_{yz})$ has little effect upon the deflection of a thin HI-HO plate. In other words, the Kirchhoff theory is usually adequate. Very thick plates are exceptional. Nonhomogenous plates, particularly those with soft inner cores, may exhibit significant shear deformations, e.g., sandwich plates. Here, we cite an example given by E. Reissner[10] The cantilever plate of Figure 8.22 is an extreme example, wherein the transverse shear force sustains the load P distributed uniformly on the end.

$$\int_{-a}^{a} V_x \mathrm{d}x = P$$

Here, we give only the result, an approximation, since it does not fully satisfy the edge conditions (see E.Reissner [??] and G.Wempner,[11] pp. 304–307). The-1pc transverse deflection is given by that solution in the form:

$$w = \frac{3Pl^3}{aEh^3}\left[\left(\frac{x^2}{l}\right) - \frac{1}{3}\left(\frac{x}{l}\right)^3 - \nu\left(\frac{y}{l}\right)^2\left(1-\frac{x}{l}\right) + \nu\left(\frac{xy^2}{l^3}\right)\right]$$
$$+\frac{3Pl^3}{aEh^3}\left[\frac{1}{3}\left(\frac{h}{l}\right)^2\left(1+\nu\frac{\sqrt{12}(a/h)}{\sinh\left(\sqrt{12}(a/h)\right)}\right)\frac{x}{l}\right] \tag{8.47a}$$

$$w(l,0) = \frac{2Pl^3}{aEh^3}\left[1+\frac{h^2}{2l^2}\left(1+\nu\frac{\sqrt{12}(a/h)}{\sin h\left(\sqrt{12}(a/h)\right)}\right)\right] \tag{8.47b}$$

The initial terms in the bracket of equation (8.47a) are obtained via the Bernoulli-Euler theory of beams. The final term represents the contribution of the transverse shear. To gain further insight, let us consider a plate of proportions $a/h = l/h = 10$. Then, we obtain

$$w(l,0) \approx \frac{2Pl^3}{aEh^3}(1+\underline{0.005})$$

Though the plate in our numerical example is relatively thick, the contribution from shear deformation (underlined) would be inconsequential in most cases.

8.21 STRAIN ENERGY OF THE HI-HO PLATE

The strain energy per unit volume of the HI-HO material in the absence of the transverse normal stress ($\sigma_z = 0$) is the quadratic form (See Section 4.11):

$$\bar{u} = \frac{E}{2(1-\nu^2)}\left[\varepsilon_x^2+\varepsilon_y^2+2\nu\varepsilon_x\varepsilon_y+\frac{1}{2}(1-\nu)\gamma_{xy}^2\right]+\frac{E}{4(1+\nu)}(\gamma_{xz}^2+\gamma_{yz}^2)$$

Substituting the expressions of Section 8.19 for the five components of strain and integrating through the thickness, we obtain the strain energy per unit area:

$$\begin{aligned} U = \int_{-\frac{h}{2}}^{\frac{h}{2}} \bar{u}dz =& \frac{Eh}{2(1-\nu^2)}\left(\bar{\varepsilon}_x^2+\bar{\varepsilon}_y^2+2\nu\bar{\varepsilon}_x^2\bar{\varepsilon}_y^2+\frac{1}{2}(1-\nu)\bar{\gamma}_{xy}^2\right) \\ &+\frac{Eh^3}{24(1-\nu^2)}\left[\bar{\kappa}_x^2+\bar{\kappa}_y^2+2\nu\bar{\kappa}_x^2\bar{\kappa}_y^2+2(1-\nu)\bar{\kappa}_{xy}^2\right] \\ &+K\frac{Eh}{(1+\nu)}(\bar{\gamma}_{xz}^2+\bar{\gamma}_{yz}^2) \end{aligned} \tag{8.48}$$

The factor K depends on the approximation of $\gamma_{xz}(z)$ and $\gamma_{yz}(z)$;

$$\frac{1}{6} < K < \frac{1}{4}$$

A variation δU represents the incremental change, which is also the work performed by the internal actions;

$$\begin{aligned} \delta U =& N_x\delta\bar{\varepsilon}_x+N_y\delta\bar{\varepsilon}_y+N_{xy}\delta\bar{\gamma}_{xy} \\ &+M_x\delta\bar{\kappa}_x+M_y\delta\bar{\kappa}_y+(M_{xy}+M_{yx})\bar{\kappa}_{xy} \\ &+V_x\bar{\gamma}_{xy}+V_y\bar{\gamma}_{yz} \end{aligned} \tag{8.49}$$

It follows that

$$N_x = \frac{\partial U}{\partial\bar{\varepsilon}_x},\ N_y = \frac{\partial U}{\partial\bar{\varepsilon}_y}, N_{xy} = \frac{\partial U}{\partial\bar{\gamma}_{xy}}$$

$$M_x = \frac{\partial U}{\partial \bar{\kappa}_x},\ M_y = \frac{\partial U}{\partial \bar{\kappa}_y}, M_{xy} = \frac{\partial U}{\partial \bar{\kappa}_{xy}}$$

$$V_x = \frac{\partial U}{\partial \bar{\gamma}_{xz}},\ V_y = \frac{\partial U}{\partial \bar{\gamma}_{yz}}$$

The reader can confirm that the results are given by Equation (8.46) with $K = \frac{1}{6}$.

8.22 THE BUCKLING OF PLATES

The behavior of a thin plate is not unlike that of a beam. A plate is also prone to buckle under compressive forces acting in the plane. Like the column, a perfect plate is in equilibrium under extreme compressive force; however, no plate is perfect. At a critical load, the plane configuration is unstable. At that critical load, there exists an adjacent, slightly bent configuration. It is a so-called bifurcation point at which the plane state is no longer stable; the system bifurcates along a path toward a stable state. Here we seek the means to determine that critical load. That requires that we examine the conditions for an adjacent (bent) equilibrium state. In most cases, the adjoining state is stable; it is a state of minimum potential. In some instances, the adjoining states are unstable, whereupon the system abruptly collapses. These situations are explored in Chapter 9.

To establish the governing equations of the slightly bent plate, we view the deformed element of Figure 8.23. The membrane forces (N_x, N_{xy}, N_y) act upon the curved element. Consequently, components act in the transverse (z) directions and augment the shear forces (V_x, V_y) of Equation (8.27). The resultants are the differentials upon opposing faces; these must equilibrate the external force $\mathbf{P}$ (force/area):

$$\frac{\mathbf{dN}_x}{\partial x} + \frac{\mathbf{dN}_y}{\partial y} = -\mathbf{P} \tag{8.50}$$

where

$$\mathbf{N}_x = N_x\hat{\mathbf{I}} + N_{xy}\hat{\mathbf{J}} + V_x\hat{\mathbf{K}},\ \ \mathbf{N}_y = N_{yx}\hat{\mathbf{I}} + N_y\hat{\mathbf{J}} + V_y\hat{\mathbf{K}}$$

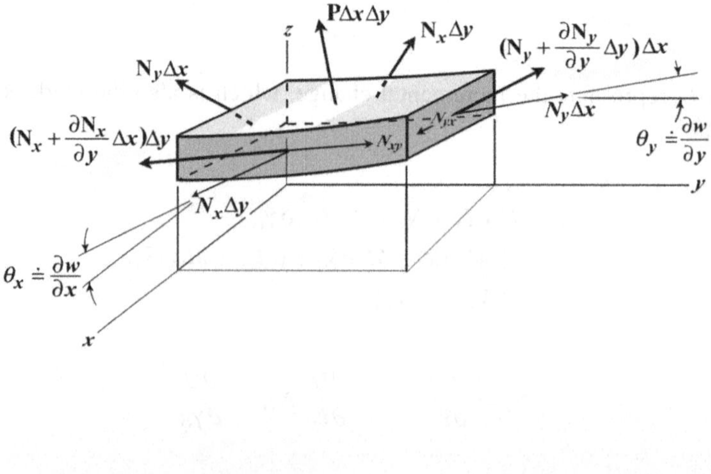

Figure 8.23

$\hat{\mathbf{I}}$, $\hat{\mathbf{J}}$, $\hat{\mathbf{K}}$ are unit vectors in the directions of the rotated edges. Differentiating, we obtain

$$\frac{\partial \mathbf{N}_x}{\partial x} = N_{x,x}\hat{\mathbf{I}} + N_{xy,x}\hat{\mathbf{J}} + V_{x,x}\hat{\mathbf{K}} + N_x\frac{\partial \hat{\mathbf{I}}}{\partial x} + N_{xy}\frac{\partial \hat{\mathbf{J}}}{\partial x} + V_x\frac{\partial \hat{\mathbf{K}}}{\partial x}$$

$$\frac{\partial \mathbf{N}_y}{\partial y} = N_{yx}\hat{\mathbf{I}} + N_{y,y}\hat{\mathbf{J}} + V_{y,y}\hat{\mathbf{K}} + N_{yx}\frac{\partial \hat{\mathbf{I}}}{\partial y} + N_y\frac{\partial \hat{\mathbf{J}}}{\partial x} + V_y\frac{\partial \hat{\mathbf{K}}}{\partial y}$$

When these are substituted into Equations (8.50) we can identify components in the directions of Vectors $\hat{\mathbf{I}}$, $\hat{\mathbf{J}}$, and $\hat{\mathbf{K}}$. We are interested in the direction, $\hat{\mathbf{K}}$ viz.

$$\hat{\mathbf{K}}\cdot\frac{\partial \mathbf{N}_x}{\partial x} = V_{x,x} + N_x\hat{\mathbf{K}}\cdot\frac{\partial \hat{\mathbf{I}}}{\partial x} + N_{xy}\hat{\mathbf{K}}\cdot\frac{\partial \hat{\mathbf{J}}}{\partial x}$$

$$\hat{\mathbf{K}}\cdot\frac{\partial \mathbf{N}_y}{\partial y} = V_{y,y} + N_{yx}\hat{\mathbf{K}}\cdot\frac{\partial \hat{\mathbf{I}}}{\partial y} + N_y\hat{\mathbf{K}}\cdot\frac{\partial \hat{\mathbf{J}}}{\partial y}$$

The vector products are the curvatures and twist of the deformed surface, viz

$$\hat{\mathbf{K}}\cdot\frac{\partial \hat{\mathbf{I}}}{\partial x} = -\frac{\partial \hat{\mathbf{K}}}{\partial x}\cdot\hat{I} \equiv \kappa_x$$

$$\hat{\mathbf{K}}\cdot\frac{\partial \hat{\mathbf{J}}}{\partial Y} = -\frac{\partial \hat{\mathbf{K}}}{\partial y}\cdot\hat{J} \equiv \kappa_y$$

$$\hat{\mathbf{K}}\cdot\frac{\partial \hat{\mathbf{J}}}{\partial x} = \hat{\mathbf{K}}\cdot\frac{\partial \hat{\mathbf{I}}}{\partial x} \equiv \kappa_{xy}$$

The components of Equation (8.50) in the transverse direction of $\hat{\mathbf{K}}$ follow:

$$V_{x,x} + V_{y,y} + N_x\kappa_x + N_y\kappa_y + (N_{xy} + N_{yx})\kappa_{xy} = -P_z \tag{8.51}$$

Equation (8.51) replaces (8.27). The *additional* terms augment the *left* side of Equation (8.29):

$$\frac{\partial^2 M_x}{\partial x^2} + 2\frac{\partial^2 M_{xy}}{\partial x \partial y} + \frac{\partial^2 M_y}{\partial y^2} - (N_x\kappa_x + 2N_{xy}\kappa_{xy} + N_y\kappa_y) = -P(x,y)$$

Note that $P(x,y)$ is directed in the direction $-z$.

If we now acknowledge the small rotations which carry the plate to an adjacent state, then (see Equations 8.19a–c and 8.26a–c)

$$\frac{\partial^4 w}{\partial x^4} + 2\frac{\partial^4 w}{\partial x^2 \partial y^2} + \frac{\partial^4 w}{\partial y^4} - \frac{1}{D}\left[N_x\frac{\partial^2 w}{\partial x^2} + 2N_{xy}\frac{\partial^2 w}{\partial x \partial y} + N_y\frac{\partial^2 w}{\partial y^2}\right] = -\frac{1}{D}P(x,y) \tag{8.52}$$

where

$$D \equiv \frac{Eh^3}{12(1-\nu^2)}$$

The reader might compare this equation with Equation (8.30).

Equation (8.52), together with various edge conditions, has been studied for more than 100 years. Numerous solutions and references are given by Timoshenko.[12] The determination of the

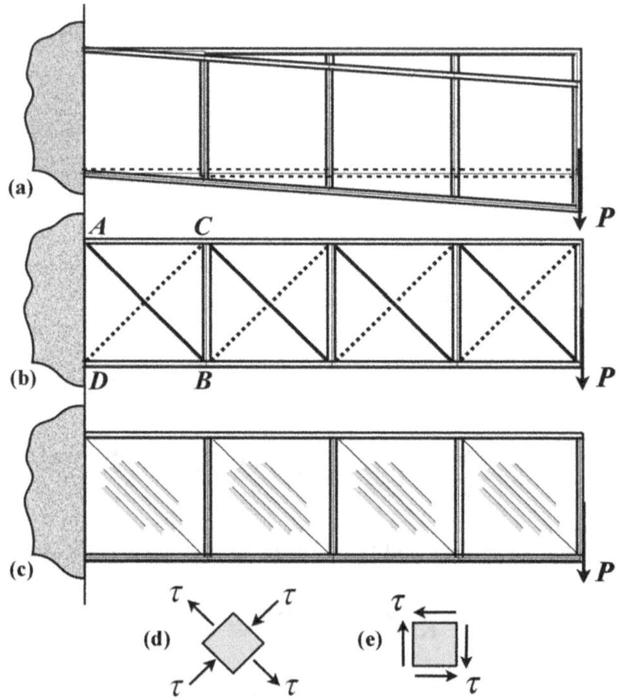

Figure 8.24

critical loads requires solutions of the homogeneous equation ($P = 0$) under the given loads (N_x, N_{xy}, N_y).

The buckling of plates and shells has been a special interest in the aeronautical industry where efforts to minimize weight invariably lead to thin-walled structural components. A simple example is the girder depicted in Figure 8.24. A frame of relatively thin horizontal and vertical members would act much as linkage and deforms as depicted in Figure 8.24a. Most components of an air frame require panels which are usually quite thin. The panels provide the resistance against the shear, i.e. the panels must play the role of the diagonal braces. The analogy is interesting: The brace AB in Figure 8.24b might be a relatively thin rod or even a flexible cable; under the load P it is subject to a tensile force. A thin rod in the position CD would be prone to buckle under the compressive force. The panels of Figure 8.24c are subject to the shear stress τ of Figure 8.24e which transforms to the tensile and compressive stresses ($\sigma_A = -\sigma_C = \tau$) of Figure 8.24d. The thin panel buckles as suggested in the illustration. An extensive treatment of this problem is presented by Timoshenko and Gere??.

ILLUSTRATIVE EXAMPLE 8.1

To provide further insights into the mechanics of buckling, let us revisit[13] the problem of lateral buckling of narrow plates as depicted in Figure 6.55 and also in Figure 8.3. Our present approach appeals to the Kirchhoff theory for the plate. Again, we view the actions upon a cross-section, as shown in Figure 8.25. On the edge of the displaced plate, we see the actions: Bending couples, m_x and m_y, twisting couple m_{xy} and a shear force V_x (all per unit length). Lower case letter $'m'$ is used

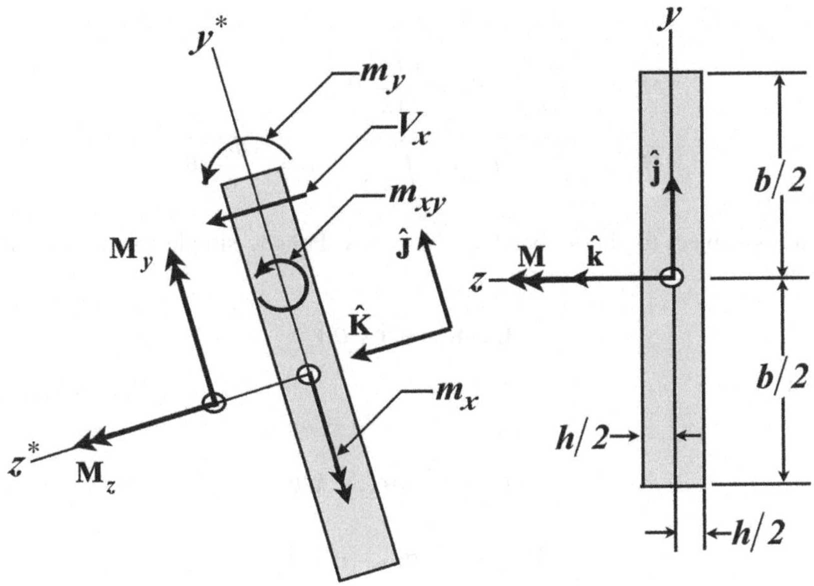

Figure 8.25

to distinguish these couples from the resultants M_x and M_y. We recall the equilibrium equations:

$$V_x = -\frac{\partial m_x}{\partial x} - \frac{\partial m_{xy}}{\partial y}$$

Also, we recall the constitutive equations:

$$m_x = D(\kappa_x + \nu\kappa_y), \quad m_y = D(\kappa_y + \nu\kappa_x), \quad m_{xy} = \frac{Gh^3}{6}\kappa_{xy}$$

$$D \equiv \frac{Eh^3}{12(1-\nu^2)}$$

If θ_x and θ_y denote rotations about axes x and y, respectively, then

$$\kappa_x = -\theta_{y,x}, \quad \kappa_{xy} = \theta_{x,x}, \quad \kappa_y = -\theta_{x,y}$$

We anticipate a width $'b'$ which is much less than length l (see Figure 8.3 where $b = 2c$). If we neglect m_y through-out, then $\kappa_y = -\nu\kappa_x$ and $\theta_{x,y} = \nu\theta_{y,x}$.

$$m_x = \frac{Eh^3}{12}\kappa_x = -\frac{Eh^3}{12}\theta_{y,x}, \qquad m_{xy} = \frac{Gh^3}{6}\theta_{x,x}, \qquad V_x = \frac{\nu Eh^3}{12(1+\nu)}\theta_{y,xx}$$

Note that V_x is an even function of y; hence it does not contribute to the couple M_x in Equation (8.53b). The resultants upon the edge follow:

$$M_y = -\int_{-b/2}^{b/2} m_x \mathrm{d}y = \frac{Ebh^3}{12}\theta_{y,x} \tag{8.53a}$$

$$M_x = +\int_{-b/2}^{b/2} m_{xy}\mathrm{d}x + \int_{-b/2}^{b/2} yV_x \mathrm{d}x = \frac{Gbh^3}{6}\theta_{x,x} \tag{8.53b}$$

If the plate is subject to simple bending, $\mathbf{M} = M\hat{\mathbf{k}}$. For the simple rotations to the adjacent (buckled) state

$$\hat{\mathbf{k}} = \hat{\mathbf{K}} + \theta_x \hat{\mathbf{J}} - \theta_y \hat{\mathbf{I}}$$

It follows that

$$M_y = \frac{Ebh^3}{12}\theta_{y,x} = M\theta_x$$

$$M_x = \frac{Gbh^3}{6}\theta_{x,x} = -M\theta_y$$

Eliminating θ_y, we obtain

$$\theta_{x,xx} + \frac{M^2}{GJEI_z}\theta_x = 0 \tag{8.35}$$

where (see Section 5.11, Equation 5.110)

$$J \approx \frac{bh^3}{6}, \quad I_z \equiv \frac{bh^3}{12}$$

This is Equation (6.132). Our approach via the plate theory is in full accord with the previous results. In particular, the present derivation supports the previous results for the loads depicted in Figure 6.55.

NOTES

1. HI-HO: Homogeneous Isotropic-Hookean.
2. *On the Strains in the Interior of Beams*, Brit. Assoc. Advanc. Sci., Report 32, 1862; 82; 1863; Phil. Trans. R. Soc. London 153, 1863.
3. see: S. P. Timoshenko. *Theory of Elasticity*, p. 78. McGraw-Hill, Columbus, 1951; G.A. Wempner, *Mechanics of Solids*, p. 78. McGraw Hill, Columbus, 1973; Mechanics of Solids, with Applications to Thin Bodies, Sijthoff and Noordhoff, The Netherlands, Rockville Maryland, 1981.
4. G. R. Kirchhoff. *Vorlesungen ü. mathematische Physik, Mechanik*, third edition. Leipzig: B. G. Teubner, 1833.
5. Extensional strain ε_z may have the magnitude of ε_x and ε_y; however,the normal stress σ_z is usually negligible. Hence the work upon extensional strain ε_z is insignificant.
6. Hookean Isotropic Homogeneous.
7. G. A. Wempner, *Mechanics of Solids*, p. 275. McGraw-Hill, Columbus, 1973.
8. W. Flügge, *Stresses in Shells*, Springer, Berlin and Heidelberg, 1962., 12(2): A69–A77, 1945.
9. Reissner, E. (1945). The effect of transverse shear deformation on the bending of elastic plates. Journal of Applied Mechanics, 12(2), A69–A77; Green, A. E. (1949). On Reissner's theory of bending of elastic

plates. Quarterly of Applied Mathematics, 7(2), 223–228; Mindlin, R. D. (1951). Influence of rotatory inertia and shear on flexural motions of isotropic, elastic plates. Journal of Applied Mechanics, 18(1), 31–38.

10. E. Reissner. On the bending of elastic plates. *Quart. Appl. Math.*, 5:55–68, 1947.
11. G. A. Wempner, *Mechanics of Solids*, p. 304. McGraw-Hill, New York, 1973; Sijthhoff & Noordhoff, Alphen aan den Rijm, The Netherlands, 1981.
12. S. P. Timoshenko and James M. Gere. *Theory of Elastic Stability*, second edition. Dover, New York, 1961.
13. In Chapter 9 we again revisit this problem via the energy criterion of E. Trefftz. Zur Theorie der Stabilität des elastishen, Gleigewichts, I.A.M.M., 13, 1933.

9 Work, Energy, and Stability

In our introductory Chapter 1, we noted two alternative approaches to mechanics, viz. the methods of Newton[1] and Lagrange.[2] Both provide means to establish a state of equilibrium. We have applied both to simple discrete systems, e.g. the trusses of Chapters 1 and 5; those applications employed the most rudimentary tool of work and energy, viz. virtual work. Equilibrium is established via criteria placed upon the work or energy expended during an incremental movement of the mechanical system. The alternative approaches entail significant practical differences: The Lagrangean approach, i.e., virtual work, requires no free-body diagram. Internal forces that expend no energy are not involved. This has distinct advantages in complicated mechanisms with many components (see Example 1.1). In the Lagrangean approach, it is the work expended and energy transferred that play the key roles; kinematics is the essential tool.

Considerations of work and energy provide not only a different avenue to equilibrium, but a means to assess the stability of equilibrium. A body may be in a state of equilibrium but unstable in the sense that a minor disturbance would cause severe disfiguration, as in the case of a ball resting at the topmost point of a hill (see Figure 9.21).

This chapter aims to set the basic concepts, which provide different and powerful methods not employed in the preceding chapters. Only enough examples and exercises are provided to reinforce the concepts and to establish a firm foundation for further study.

9.1 PRINCIPLE OF VIRTUAL WORK

The notion of *virtual work* provides a very effective viewpoint and method to assure a state of equilibrium. The concept of virtual work relies on a consideration of small changes in the configuration of the mechanical system and the work to effect such changes.

A configuration is defined by the collective positions of all particles that comprise the system. Any change of configuration involves kinematics, that is, a change in the geometry. Of course, kinematics always plays a role in the study of deformable bodies. In the principle and method of virtual work, kinematics plays the major role, whereas the free body is not essential.

9.1.1 VIRTUAL DISPLACEMENT

A virtual displacement is any infinitesimally small change (first order in the geometric variables) in the configuration of a mechanical system, *but* a change that is consistent with physical constraints.

For example, the teeter-totter (or seesaw) illustrated in Figure 9.1 is a simple mechanical system consisting of a board (assumed rigid) that can rotate only about a pivot (assumed frictionless). The configuration of this system is fully determined by specifying the angle θ. Here, the constraint is only the pivot. A virtual displacement that moves that point is *not* admissible.

Another mechanical system is the block and tackle of Figure 9.2a and b. A block and tackle is a system consisting of pulleys (assumed frictionless) and ropes (assumed inextensible).

In both cases (for simplicity), we assume that the pulleys are small and the intermediate distances are large; then the slight inclinations of the ropes can be neglected. Each is a system with one degree of freedom; if the displacement δ_A of the knot A is prescribed, then the displacement δ_O of axle O follows. The relationship is an exercise in kinematics. In case (a) $\delta_A = 2\delta_O$; in case (b) $\delta_A = 3\delta_O$.

DOI: 10.1201/9781003665861-9

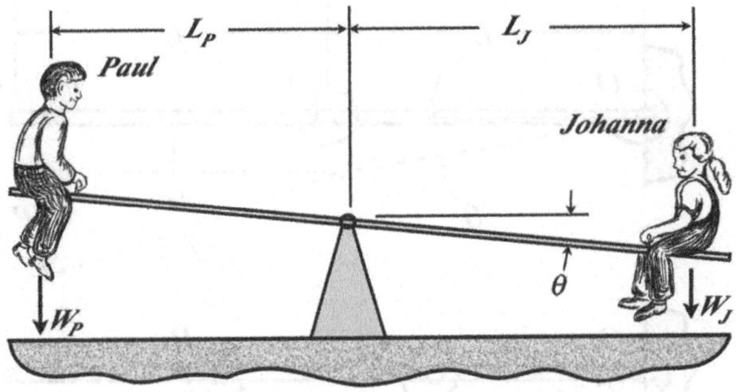

Figure 9.1 Tetter-Totter

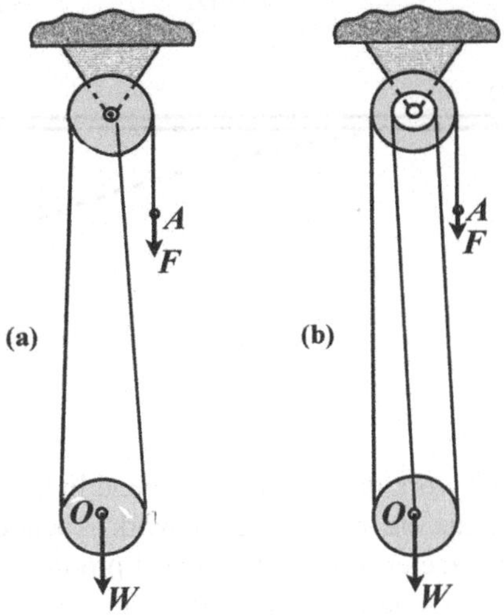

Figure 9.2

The system depicted in Figure 9.3a consists of two rigid links joined by pins at A and at O. Torsional springs resist but do not prevent relative rotations at these joints. This system has two degrees of freedom, since two angles, θ_1 and θ_2, define an admissible configuration. Similarly, the system in Figure 9.3b consists of three rigid links, pinned at O, A, and B; again, torsional springs resist relative rotations. This system has three degrees of freedom, since angles, θ_1, θ_2, and θ_3 are needed to define a configuration. We can elaborate on the linkages of Figure 9.3a and b and create systems with any number of degrees of freedom. The limit of the process is a continuous flexible beam, as shown in Figure 9.3c. At each point, the rotation can be different, but continuous; we could say that the beam has infinitely many degrees of freedom. A virtual displacement of the two-link system (Figure 9.3a) consists of any *two* independent, but infinitesimally small, changes in the angles θ_1 and θ_2. Here, *independent* is a key word; the virtual displacement can be *any* change consistent with the constraints. In other words, any movement of these systems (Figure 9.3) is an admissible virtual displacement if it does not break a physical constraint, a pivot in Figure 9.3a or b or a kink in the beam of Figure 9.3c.

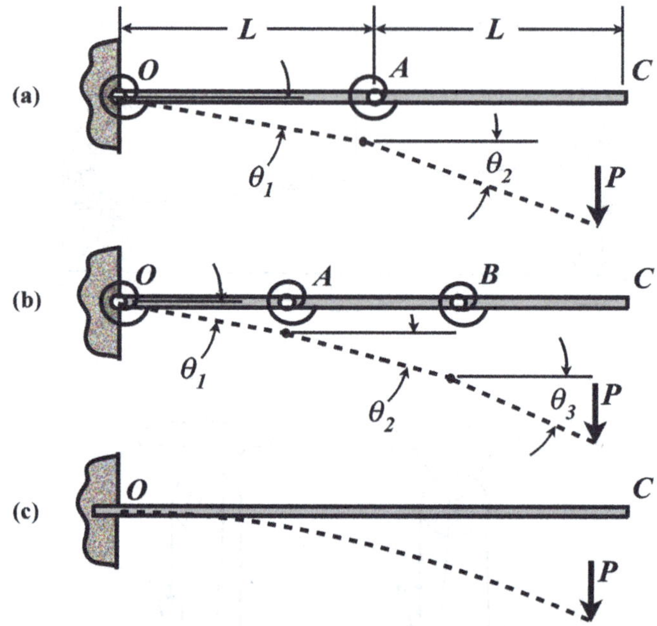

Figure 9.3

9.1.2 VIRTUAL WORK

We define the virtual work as the work that we would expend in the virtual displacement.[3] This is hypothetical, since we only imagine the motion and the work expended. For example, a small (positive) change in the angle θ in the teeter-totter in Figure 9.1 produces a small upward displacement δ_P of Paul's center of gravity and a small downward displacement δ_J of Johanna's.

$$\delta_P = L_P \delta\theta, \quad \delta_J = L_J \delta\theta$$

These are the first-order approximations, consistent with the concepts of a virtual displacement. The virtual work (the work we would do) vanishes in a state of equilibrium.

$$W_P L_P - W_J L_J = 0 \Rightarrow \frac{L_J}{L_P} = \frac{W_P}{W_J} \tag{9.1}$$

If Paul and Johanna are to teeter comfortably, this determines the appropriate position on the pivot.

The relationships between the motions of knot A and axle O in Figure 9.2a and b were noted previously. If a small downward displacement $\Delta\delta_O$ is given to the axle O and downward forces W and F act at axle O and knot A, then the virtual work in case (a) is

$$\Delta W = (2F - W)\Delta\delta_O = 0 \Rightarrow \frac{F}{W} = \frac{1}{2} \tag{9.2}$$

In case (b),

$$\Delta W = (3F - W)\Delta\delta_O = 0 \Rightarrow \frac{F}{W} = \frac{1}{3} \tag{9.3}$$

The alternative approach, viz. Newtonian, requires a free-body diagram of each pulley with the forces exerted upon the ropes.

Now, consider the virtual work for the two-degree-of-freedom system of Figure 9.3a. Under the action of the end load P, the system is deformed to a configuration, as indicated by the dotted lines. An admissible virtual displacement corresponds to any infinitesimally small changes in the *two* angles θ_1 and θ_2. Such (positive) changes, $\Delta\theta_1$, and $\Delta\theta_2$, produce a downward displacement of the loaded end:

$$\delta = L[\sin(\theta_1 + \Delta\theta_1) + \sin(\theta_2 + \Delta\theta_2) - \sin\theta_1 - \sin\theta_2] = (L\cos\theta_1)\Delta\theta_1 + (L\cos\theta_2)\Delta\theta_2$$

The virtual work (we do) upon this *external* force P is

$$\delta w_E = -LP[(\cos\theta_1)\Delta\theta_1 + (\cos\theta_2)\Delta\theta_2] \tag{9.4}$$

The number of *independent* variables is always equal to the number of degrees of freedom; for example, the system of three links (Figure 9.3b) has three independent changes ($\Delta\theta_1$, $\Delta\theta_2$, and $\Delta\theta_3$).

When the system of Figure 9.3a is displaced, one or both links must rotate. Because our principle requires the work for *any* admissible virtual displacement, θ_1 and θ_2 must be arbitrary. Then relative rotations are admissible, indeed required, at both joints. When this occurs, we expend work against the springs. Virtual work upon *internal* forces is inherent in a displacement of any deformable body. That work depends upon the physical attributes of the springs (the properties of the material). If these are linear (Hookean) springs, then the resisting couples are proportional to the relative rotations. For simplicity, assume the same torsional stiffness for both. The virtual work upon these internal couples is

$$\delta w_1 = \beta[\theta_1\Delta\theta_1 + (\theta_2 - \theta_1)(\Delta\theta_2 - \Delta\theta_1)] \tag{9.5}$$

9.1.3 PRINCIPLE OF VIRTUAL WORK

With the foregoing definitions of virtual displacement and virtual work, we now state, quite simply, the principle of virtual work. *A mechanical system is in equilibrium if, and only if, the virtual work vanishes for arbitrary admissible virtual displacements.*

The two-degree-of-freedom system in Figure 9.3a requires that the virtual work vanish for two arbitrary admissible virtual displacements. That virtual work consists of the work against external and internal forces, that is, the sum of the work given by Equations (9.4) and (9.5).

$$\delta w_E + \delta w_I = [-LP\cos\theta_1 + \beta(2\theta_1 - \theta_2)]\Delta\theta_1 + [-LP\cos\theta_2] + \beta(\theta_2 - \theta_1)]\Delta\theta_2 = 0$$

Because the virtual work must vanish for arbitrary virtual displacements ($\Delta\theta_1$ and $\Delta\theta_2$ are independent), both bracketed terms must vanish for equilibrium. Accordingly, we obtain two equations in the variables θ_1 and θ_2:

$$2\beta\theta_1 - \beta\theta_2 = LP\cos\theta_1$$
$$-\beta\theta_1 + \beta\theta_2 = LP\cos\theta_2$$

These are nonlinear equations; however, viewing this as a very simplified model of a cantilever beam, we might assume that the angles θ_1 and θ_2 are small. Then a first-order approximation ($\cos\theta_1 \approx \cos\theta_2 = 1$) provides the linear equations and their solution:

$$2\beta\theta_1 - \beta\theta_2 = LP$$
$$-\beta\theta_1 + \beta\theta_2 = LP$$
$$\theta_1 = \frac{2LP}{\beta}, \quad \theta_2 = \frac{3LP}{\beta}$$

Note that the method requires no free body. Moreover, no interactions (forces and/or couples) at interconnections (pivots, axles, or joints) are involved in the solutions of the foregoing examples. In many problems, such interactions are not required. The avoidance of interactions via the principle of virtual work is then a simplification and is, therefore, an advantage. The following example serves to highlight the differences.

ILLUSTRATIVE EXAMPLE 9.1

The small "vise-grip" in Figure 9.4 exerts a force $\boldsymbol{F}$ on the object (dotted) at point A by virtue of the force $\boldsymbol{P}$ acting along line $'n'$ through point E.Each of the four links is assumed rigid.

The essential dimensions follow:

$$OA = 2.40 \text{ in.}, \quad OB = 1.00 \text{ in.}$$
$$BC = 0.80 \text{ in.}, \quad BE = 3.40 \text{ in.}$$
$$OA \text{ is parallel to } BC$$

We seek the ratio F/P.

Because a rigid motion of the device is irrelevant, we consider the upper link (OD) fixed. The essential kinematics, i.e., the small virtual motions are needed.

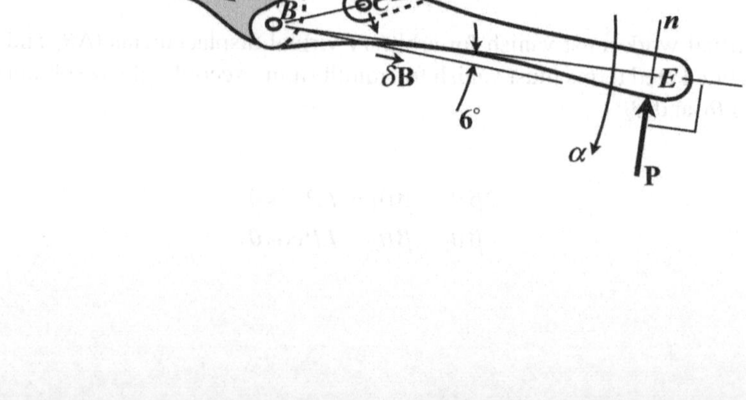

Figure 9.4

The jaw AOB can only rotate about the pin O ($\delta\theta$). This, in turn, causes the virtual displacement of A (δA) perpendicular to OA (in the direction of $\boldsymbol{F}$) and of $B(\delta B)$ in a direction perpendicular to OB. Link CD can only rotate about the fixed pin D; pin C displaces (δC) in a direction perpendicular to CD.

$$\delta B = (OB)\delta\theta = (1.00)\delta\theta$$

The link BCE can translate and rotate; the virtual rotation is denoted by $\delta\alpha$. The virtual displacement $\boldsymbol{\delta C}$ can be expressed as the sum of the displacement $\boldsymbol{\delta B}$ and the relative displacement, caused by the rotation $\delta\alpha$:

$$\boldsymbol{\delta C} = \boldsymbol{\delta B} + \boldsymbol{\delta\alpha} \times \boldsymbol{BC}$$

Since link CD is assumed rigid and fixed, at D, the displacement $\boldsymbol{\delta C}$ in direction $'m'$ vanishes:

$$\begin{aligned} \delta C_m &= \delta B(\cos 42^\circ) - \delta\alpha(BC)\sin 10^\circ \\ &= (1.00)(\cos 42^\circ \delta\theta) - \delta\alpha(0.80)\sin 10^\circ \\ \delta\alpha &= 5.349\delta\theta \end{aligned}$$

The displacement of point E is

$$\boldsymbol{\delta}\mathbf{E} = \boldsymbol{\delta}\mathbf{B} + \boldsymbol{\delta\alpha} \times \mathbf{BE}$$

The component in the direction (n) of force $\boldsymbol{F}$ is

$$\begin{aligned} \boldsymbol{\delta E_n} &= \delta\mathrm{B}(\sin 6^\circ) - \mathrm{BE}\delta\alpha \\ &= -(1.00)(\sin 6^\circ) - 3.4\delta\alpha \\ &= -18.29\delta\theta \end{aligned}$$

The virtual displacement of A (in the direction of force $\boldsymbol{F}$ is

$$\delta A = 2.4\delta\theta$$

The virtual work is

$$\delta W = -2.4F\delta\theta + 18.29P\delta\theta = 0$$

Our result follows

$$\frac{F}{P} = 7.6$$

Note that no internal forces enter the foregoing analysis. The alternative, i.e. the Newtonian approach, requires a consideration of the forces at B and C and the equilibrium of the forces upon the two free bodies, the jaw AOB and the handle BCE.

A word of caution is in order. The preceding description of virtual displacement presumes that the paths of particles are smooth. Likewise, the description of virtual work presumes that the forces vary continuously. Special circumstances do arise. For example, the roller of Figure 9.5 is at rest, in equilibrium, in the V-shaped groove. The only admissible displacement of the center is parallel to one or the other plane. In either case, the virtual work cannot vanish, because the normal force exerted by the other surface vanishes abruptly. Any admissible displacement, however small, abruptly violates equilibrium.

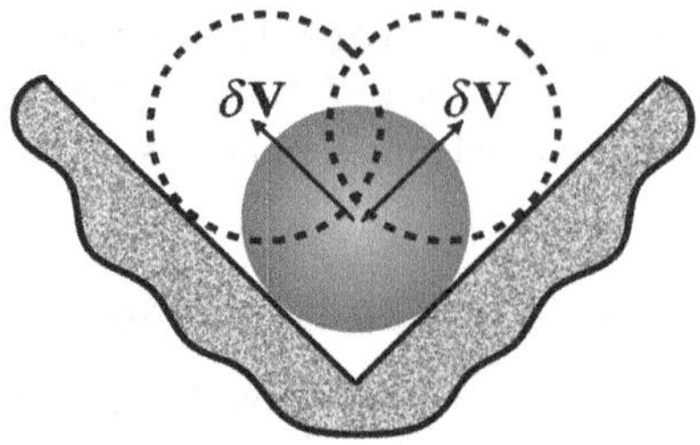

Figure 9.5

ILLUSTRATIVE EXAMPLE 9.2

In Chapter 5, we applied the principle of virtual work to numerous trusses. In Chapter 6, symmetric beams were treated via the Newtonian approach. We now reexamine the latter via the principle of virtual work.

Equilibrium of an Element

Let us first apply the criterion to a small deformed segment, as illustrated in Figure 9.6. The initial length is Δx, and the deformed length is $(1+\varepsilon_0)\Delta x$. Figure 9,6 shows the element of the beam with the forces and couples that act on it. If we discount a virtual *deformation*, the element has three degrees of freedom; it can displace in each of two directions and rotate in the plane of symmetry. Let us assume the virtual displacements δv and δu are normal and tangent to the middle line, as depicted in Figure 9.6b. Additionally, we take a virtual rotation $\delta\beta$ in the sense of the angle θ. The virtual work (we do) in the arbitrary displacement (δv, δu) and rotation $\delta\beta$) is

$$\begin{aligned}\delta w = &\left[-\Delta F\cos\frac{\Delta\theta}{2}+(2V+\Delta V)\sin\frac{\Delta\theta}{2}\right]\delta u\\ &-\left[(2F+\Delta F)\sin\frac{\Delta\theta}{2}+\Delta V\cos\frac{\Delta\theta}{2}-\bar{p}\Delta x\right]\delta v\\ &-\left[\Delta M+2\rho\sin\frac{\Delta\theta}{2}\left(V+\frac{\Delta V}{2}\cos\frac{\Delta\theta}{2}\right)\right]\delta\beta\end{aligned} \tag{9.6}$$

The final term incorporates the distance OA between the ends: $OA = 2\rho\sin\Delta\theta/2$.

Again, the virtual work must vanish for arbitrary virtual displacements. Therefore, each bracketed term must vanish independently. We suppose a practical application wherein the length Δx is relatively small, $\varepsilon_0 \ll 1$, and forces F and V and couple M are continuous. Then $\Delta\theta \approx \Delta x/\rho = \kappa\Delta x \ll 1$, $\Delta F \ll F$, and $\Delta V \ll V$,

$$\cos\frac{\Delta\theta}{2}\approx 1, \quad \text{and} \quad \sin\frac{\Delta\theta}{2}\approx\frac{\Delta\theta}{2}$$

With these approximations, the three equations of equilibrium (bracketed terms vanish) have the approximate forms:

$$\frac{\Delta F}{\Delta x}-\kappa V\approx 0, \quad \frac{\Delta V}{\Delta x}+\kappa F\approx\bar{p}, \quad \frac{\Delta M}{\Delta x}+V\approx 0 \tag{9.7a,b,c}$$

These three equations are valid for a small, but *finite*, element; that is, Δx is relatively small compared to the length of the beam. In other words, these might be used in an approximation of the

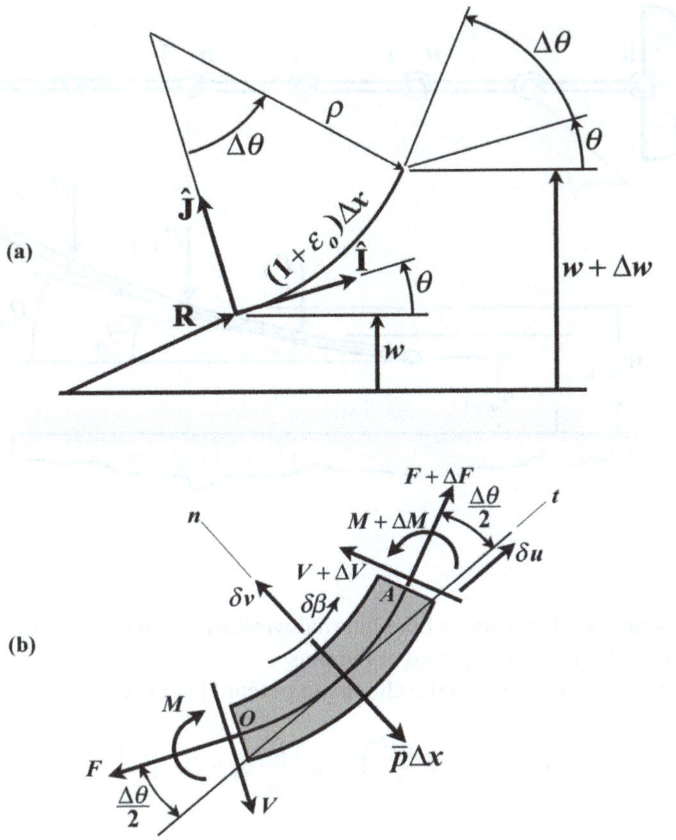

Figure 9.6 Beam Element

beam by a finite number of small discrete elements. However, to describe the continuous beam we pass to the limit, $\Delta x \to O$, and obtain the differential equations (see Equations 6.8 and 6.9)

$$\frac{\mathrm{d}F}{\mathrm{d}x} = \kappa V, \quad \frac{\mathrm{d}V}{\mathrm{d}x} + \kappa F = \bar{p}, \quad \frac{\mathrm{d}M}{\mathrm{d}x} = -V \tag{9.8a,b,c}$$

Note that Equation (9.8b) is a refinement of the previous approximation (Equation 6.8), which did not account for the curvature. These equations apply to the symmetric beam, subject to the Bernoulli assumption, viz. transverse shear strain is neglected. Therefore, the shear force V is inactive; it can be eliminated from Equations (9.8a,b) via Equation (9.8c) to obtain

$$\frac{\mathrm{d}F}{\mathrm{d}x} = -\kappa \frac{\mathrm{d}M}{\mathrm{d}x}, \quad \frac{\mathrm{d}^2 M}{\mathrm{d}x^2} - \kappa F = -\bar{p}$$

The latter equation is consistent with Equations (6.66) and (6.67) wherein $\kappa = \mathrm{d}\theta/\mathrm{d}x$.

ILLUSTRATIVE EXAMPLE 9.3

The beam depicted in Figure 9.7 is composed of $'m'$ rigid links, which are joined by frictionless pins. Relative rotations are resisted by similar torsional springs; these are linear with constant

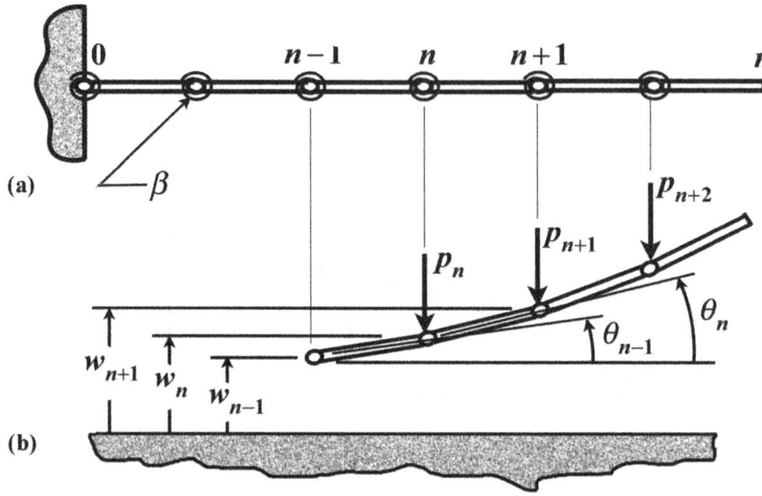

Figure 9.7

modulus β. The virtual work consists of the internal work δI against the torsional springs and the external work against the loads p_n applied at the pins.
The virtual work against one spring is the change in potential energy:

$$\delta I_n = \delta\left(\beta\frac{\Delta\theta_n^2}{2}\right) = \delta\left[\frac{\beta}{2}(\theta_n - \theta_{n-1})^2\right] \tag{9.9}$$

From the geometry

$$\theta_n \approx (w_{n+1} - w_n)/\Delta x$$

It follows that

$$\delta I_n = \frac{\beta}{\Delta x^2}(w_{n+1} - 2w_n + w_{n-1})(\delta w_{n+1} - 2\delta w_n - \delta w_{n-1}) \tag{9.10}$$

The virtual work upon the beam includes the work upon the external loads; upon the one load p_n:

$$\delta E_n = p_n \delta w_n$$

The total virtual work is the sum for all links:

$$\delta W = \sum_1^{m-1} \frac{\beta}{\Delta x^2}(w_{n+1} - 2w_n + w_{n-1})(\delta w_{n+1} - 2\delta w_n - \delta w_{n-1}) + \sum_1^m p_n \delta w_n \tag{9.11}$$

The virtual work must vanish for arbitrary virtual displacement; specifically for any one δw_n. Note that a displacement δI_n also produces work upon the adjoining spring, i.e. at $'n-1'$ and at $'n+1'$; that is reflected in the terms (δw_{n+1} and δw_{n-1}). Therefore, a virtual displacement δw_n causes the work:

$$\delta W = \frac{\beta}{\Delta x^2}\left[(w_{n+2} - 2w_{n+1} + w_n) - 2(w_{n+1} - 2w_n + w_{n+1}) + (w_n - 2w_{n-1} + w_{n-2}) + p_n\right]\delta w_n$$

The virtual work vanishes if, and only if, the bracketed term vanishes:

$$\frac{\beta}{\Delta x^2}(w_{n+2} - 4w_{n+1} + 6w_n - 4w_{n-1} + w_{n-2}) + p_n = 0 \tag{9.12}$$

If p_n represents the load on a segment of length Δx, then $p_n = \bar{p}\Delta x$; where $\bar{p}$ is the mean value of the distributed load p. Also, if we set $\beta = EI/\Delta x$, then Equation (9.12) is the difference equation that converges to the differential equation:

$$EI\frac{\mathrm{d}^4 w}{\mathrm{d}x^4} = -p(x) \tag{9.13}$$

This is Equation (6.52a).

Let us now consider the work upon the nodes $m-2$, $m-1$:

$$\begin{aligned}\sum_{m-2}^{m} \frac{\beta}{\Delta x^2}[(w_{m-1} - 2w_{m-2} + w_{m-3})(\delta w_{m-1} - 2\delta w_{m-2} + \delta w_{m-3})\\ +(w_m - 2w_{m-1} + w_{m-2})(\delta w_m - 2\delta w_{m-1} + \delta w_{m-2})\\ +p_{m-2}\delta w_{m-2} + p_{m-1}\delta w_{m-1}\end{aligned}$$

The virtual work upon the displacement δw_m must vanish:

$$\frac{\beta}{\Delta x^2}(w_m - 2w_{m-1} + w_{m-2}) = 0 \tag{9.14}$$

The virtual work upon the displacement δw_{m-1} must vanish:

$$\frac{\beta}{\Delta x^2}(w_{m-3} - 4w_{m-2} + 5w_{m-1} - 2w_m) = -p_{m-1} \tag{9.15}$$

Adding Equations (9.14) and (9.15), we obtain

$$\frac{\beta}{\Delta x^2}(w_{m-3} - 3w_{m-2} + 3w_{m-1} - w_m) = -p_{m-1} \tag{9.16}$$

Again, we set $\beta = EI/\Delta x$, $p_{m-1} = \bar{p}\Delta x$; then in the limit $\Delta x \to 0$, we obtain from Equation (9.14)

$$EI\frac{\mathrm{d}^2 w}{\mathrm{d}x^2} = 0 \quad \text{at } x = l \tag{9.17}$$

and from Equation (9.16)

$$EI\frac{\mathrm{d}^3 w}{\mathrm{d}x^3} = 0 \quad \text{at } x = l \tag{9.18}$$

The former asserts the vanishing of bending moment; the latter asserts the vanishing of the shear force (see Equations 6.53b,c).

The linkage of Figure 9.7 is an approximation of a continuous beam via finite elements. One virtue of such models, by the principle of virtual work, is a consistent version of the end conditions. That is evident in Equations (9.14) and (9.16), which converge to the conditions $M = V = 0$ at $x = l$.

ILLUSTRATIVE EXAMPLE 9.4

Let us apply the principle of virtual work to the continuous symmetric beam of Figure 9.8. All notations follow our previous conventions.[4]

The only active external force is the load $p(x)$, so that the virtual work (we do) upon the external forces is

$$\delta W_E = +\int_0^L p(x)\delta\bar{w}\mathrm{d}x \tag{9.19}$$

Here $\delta\bar{w}$ (a function of position x) is an arbitrary virtual displacement, normal to the *initial* position of the axis.

The virtual work expended against the internal forces, according to the Bernoulli hypothesis, consists of the work of the normal force F in the virtual extension $\delta\varepsilon_0$ and the bending couple M in the virtual curvature $\delta\kappa$.

$$\delta W_I = \int_0^L (F\delta\varepsilon_0 + M\delta\kappa)\mathrm{d}x \tag{9.20}$$

The major task now is to express the strain and curvature increments, $\delta\varepsilon_0$ and $\delta\kappa$, in terms of the virtual displacements. We refer to Figure 9.6 and employ the components tangent and normal to the bent axis (δu and δv). Then the virtual displacement (vector) of a particle on the axis is

$$\delta \boldsymbol{R}_0 = \delta u\,\hat{\mathbf{I}} + \delta v\,\hat{\mathbf{J}} \tag{9.21}$$

Also,

$$\frac{\mathrm{d}\boldsymbol{R}_0}{\mathrm{d}x} = (1+\varepsilon_0)\,\hat{\mathbf{I}} \tag{9.22}$$

$$\frac{\mathrm{d}}{\mathrm{d}x}\delta\boldsymbol{R}_0 = \delta\varepsilon_0\,\hat{\mathbf{I}} + (1+\varepsilon_0)\delta\theta\,\hat{\mathbf{J}} \tag{9.23a}$$

$$= \delta\varepsilon_0\,\hat{\mathbf{I}} + \delta\theta\,\hat{\mathbf{J}} \tag{9.23b}$$

The virtual change of the unit vector $\hat{\boldsymbol{I}}$ is $\delta\theta\hat{\boldsymbol{J}}$, caused by rotation $\delta\theta$. Approximation (9.23b) is based on the assumption of small strain ($\varepsilon_0 \ll 1$). Now, we observe that

$$\frac{\mathrm{d}\hat{\mathbf{I}}}{\mathrm{d}x}\cdot\hat{\mathbf{J}} \equiv \kappa = -\frac{\mathrm{d}\hat{\mathbf{J}}}{\mathrm{d}x}\cdot\hat{\mathbf{I}} \tag{9.24a,b}$$

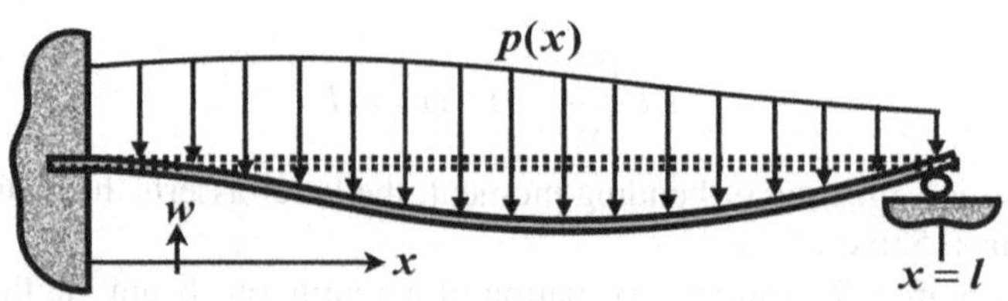

Figure 9.8

When we differentiate Equation (9.21), use Equations (9.24a,b), and then equate the result to (9.23b), we obtain

$$\frac{\mathrm{d}}{\mathrm{d}x}(\delta \boldsymbol{R}_0) = \frac{\mathrm{d}\delta u}{\mathrm{d}x}\hat{\mathbf{I}} + \kappa\delta u\hat{\mathbf{J}} + \frac{\mathrm{d}\delta v}{\mathrm{d}x}\hat{\mathbf{J}} - \kappa\delta v\hat{\mathbf{I}} \approx (\delta\varepsilon_0)\hat{\mathbf{I}} + \delta\theta\hat{\mathbf{J}}$$

It follows that

$$\delta\varepsilon_0 = \frac{\mathrm{d}\delta u}{\mathrm{d}x} - \kappa\delta v$$
$$\delta\theta = \frac{\mathrm{d}\delta v}{\mathrm{d}x} + \kappa\delta u \tag{9.25}$$

$$\delta\kappa = \frac{\mathrm{d}\delta\theta}{\mathrm{d}x} = \frac{\mathrm{d}^2\delta v}{\mathrm{d}x^2} + \frac{\mathrm{d}}{\mathrm{d}x}(\kappa\delta u) \tag{9.26}$$

Substituting Equations (9.25) and (9.26) into Equation (9.20) and adding that result to Equation (9.19), we obtain

$$\delta W_I + \delta W_E = \int_0^L \left[F\frac{\mathrm{d}}{\mathrm{d}x}\delta u - F\kappa\delta v + M\frac{\mathrm{d}^2}{\mathrm{d}x^2}\delta v + M\frac{\mathrm{d}}{\mathrm{d}x}(\kappa\delta u) + p\delta w\right]\mathrm{d}x \tag{9.27a}$$

The first, third, and fourth terms of the integrand can be integrated by parts, the third twice, to obtain

$$\begin{aligned}\delta W_I + \delta W_E =& F\delta u\Big|_0^L + M\left(\frac{\mathrm{d}}{\mathrm{d}x}\delta v + \kappa\delta u\right)\Big|_0^L - \frac{\mathrm{d}M}{\mathrm{d}x}\delta v\Big|_0^L \\ &+ \int_0^L \left[-\left(\frac{\mathrm{d}F}{\mathrm{d}x} + \kappa\frac{\mathrm{d}M}{\mathrm{d}x}\right)\delta u + \left(\frac{\mathrm{d}^2M}{\mathrm{d}x^2} - \kappa F\right)\delta v + p\delta w\right]\mathrm{d}x\end{aligned} \tag{9.27b}$$

We are now in a position to enforce the following principle: The virtual work, Equation (9.27b), must vanish for *arbitrary* virtual displacements, the independent variations of the components $\delta u(x)$ and $\delta v(x)$ $[\delta w = \delta v\cos\theta + \delta u\sin\theta]$. This means that either may be zero over all but some small region. Also, the values at the ends can assume any values consistent with the constraints. In the specific example depicted in Figure 9.8, δu and δv must be zero at the fixed end ($x = O$). δu can have any value at the right end ($x = L$); slip along the roller is admissible. Also, the rotation $\delta\theta$ at the right end ($x = L$) can assume any value.

If we are to have no virtual work for the arbitrary virtual displacement, then the coefficients of the components $\delta u(x)$ and $\delta v(x)$ in the integrand must vanish everywhere, throughout ($0 < x < L$).

$$\frac{\mathrm{d}F}{\mathrm{d}x} + \kappa\frac{\mathrm{d}M}{\mathrm{d}x} - p\sin\theta = 0 \tag{9.28}$$

$$\frac{\mathrm{d}^2M}{\mathrm{d}x^2} - \kappa F + p\cos\theta = 0 \tag{9.29}$$

These are the differential equations of equilibrium. Also, because the tangential displacement and rotation are independently variable at the right end, their coefficients must vanish:

$$F(L) = 0, \quad M(L) = 0$$

The additional end conditions are the prescribed kinematic constraints imposed by the fixture at the left ($x = O$) and the simple support at the right ($x = L$), namely,

$$v(0) = \frac{\mathrm{d}v}{\mathrm{d}x}(0) = 0, \quad v(L) = 0$$

Note that the shear force V does not appear explicitly in the foregoing equations. This happens because the Bernoulli hypothesis is a constraint against transverse shear strain; hence, the force V is inactive. We recall Equation (9.8c)

$$V = -\frac{\mathrm{d}M}{\mathrm{d}x}$$

With this equation, the shear can be eliminated from Equations (9.8a) and (9.8b) to obtain the equivalents of Equations (9.28) and (9.29). The differences are only in the loading: The previous load $\bar{p}$ is *normal* to the deformed axis, so that $\bar{p} = p\cos\theta$; the tangential component, $p\sin\theta$, is absent from Equation (9.8a)

Note that the foregoing derivation does not entail the material properties. The following example also serves to demonstrate that the principle of virtual work applies as well to nonconservative systems.

ILLUSTRATIVE EXAMPLE 9.5

The symmetric beam of Figure 9.9 is composed of structural steel, which can be treated as ideally plastic at the yield point. So-called "plastic collapse" is possible when the bending couples attain the value $M = \pm M_L$ (see Section 6.8). In this case, the limiting condition ("plastic collapse") can only occur when three so-called "hinges" occur. "Hinges" will occur at the ends ($x = 0, L$) and at an intermediate site. From the free-body of Figure 9.9b, we obtain the shear forces at the ends. Then from the body at Figure 9.9c, we obtain

$$M(x) = -M_L + \frac{\bar{p}l}{6}x - \bar{p}\frac{x^3}{6l}$$

The maximum value occurs at

$$\frac{\mathrm{d}M}{\mathrm{d}x} = 0, \quad x = l/\sqrt{3}, \quad M(\frac{l}{\sqrt{3}}) = M_L$$

The beam with "hinges" at $x = 0, l/\sqrt{3}, l$ is depicted in Figure 9.10a. At this stage, the elastic-plastic deformations between "hinges" are negligible. We now view the system as two rigid links, $(0 < x \le l/\sqrt{3})$ and $(l/\sqrt{3} \le x \le l)$. This linkage has but one degree of freedom; the configuration is described by the one variable θ, as depicted schematically in Figure 9.10b. For brevity, let $l/\sqrt{3} = a$; then

$$\delta\alpha = \frac{a}{l-a}\delta\theta, \; \delta w = -x\delta\theta, \; x \le a; \qquad \delta w = -\frac{l-x}{l-a}a\delta\theta, \; x \ge a$$

The virtual work of the moments (M_L) at the three hinges follows:

$$\delta W_I = M_L\delta\theta + M_L\left(1 + \frac{a}{l-a}\right)\delta\theta + M_L\left(\frac{a}{l-a}\right)\delta\theta = \frac{2lM_L}{l-a}\delta\theta$$

The external work follows:

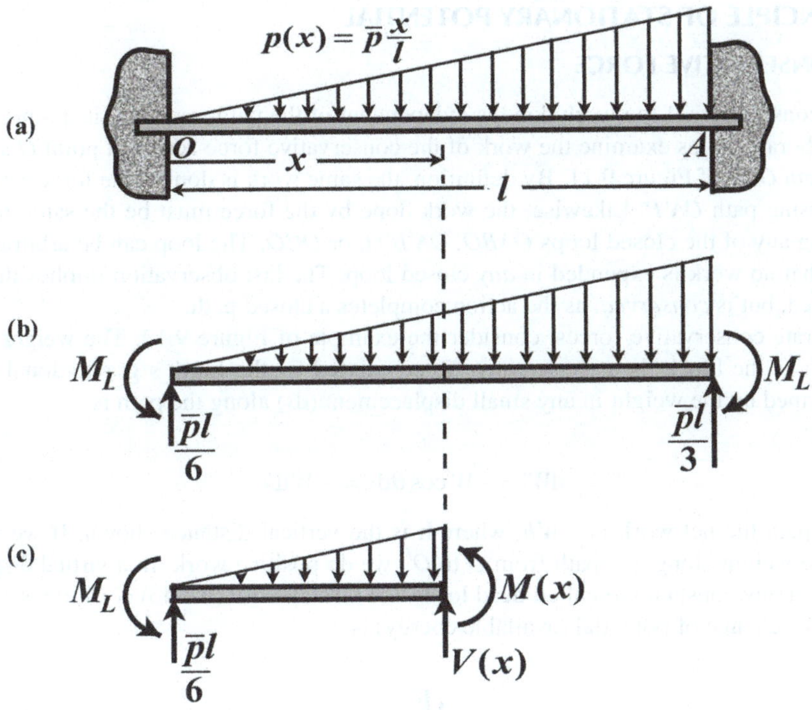

Figure 9.9

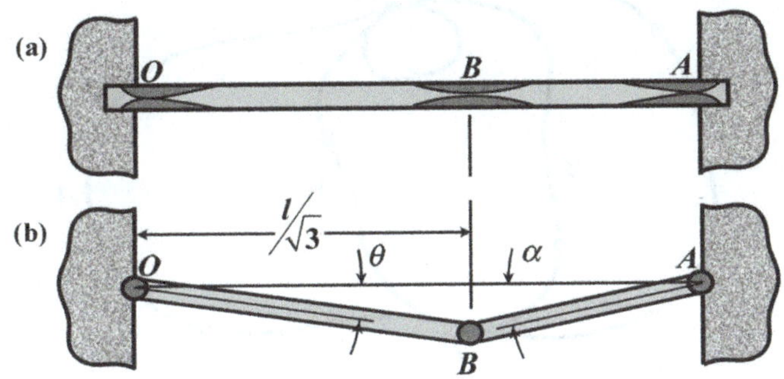

Figure 9.10 Plastic Hinges

$$
\begin{aligned}
\delta W_E &= \int_0^l p(x)\delta w \mathrm{d}x = -\bar{p}\frac{a(l^3 - la - a^3)}{6l(l-a)}\delta\theta \\
\delta W &= \delta W_I + \delta W_E = 0 \\
\bar{p} = p_L &= \frac{12l^2 M_L}{a(l^3 - la^2 - a^3)} = \frac{43.8 M_L}{l^2}
\end{aligned}
$$

9.2 PRINCIPLE OF STATIONARY POTENTIAL

9.2.1 CONSERVATIVE FORCE

A force is conservative if the work done is independent of the path traced by its point of application. To elaborate, let us examine the work of the conservative force acting at point O as it moves along the path OAP of Figure 9.11. By definition, the same work is done if the force acts upon the point traversing path $OA'P$. Likewise, the work done by the force must be the same if the point travels along any of the closed loops $OABO$, $OA'B'O$, or OCO. The loop can be arbitrarily small. It follows that no work is expended in *any* closed loop. The last observation implies that work is not dissipated, but is *conserved*, as the action completes a closed path.

To illustrate conservative forces, consider the example of Figure 9.12. The weight W of any object, such as the block, is a conservative force exerted by the Earth's gravitational field. The work performed by the weight in any small displacement (ds) along the path is

$$dW = -W\cos\theta ds = -Wdh$$

Along any path the net work is $-Wh$, where h is the vertical distance shown. If we physically transport the weight along any path from O to O', we do positive work. In a virtual displacement of a weight or any constant (so-called dead load), the virtual work (we do) constitutes a change in potential. The change of potential (available energy) is

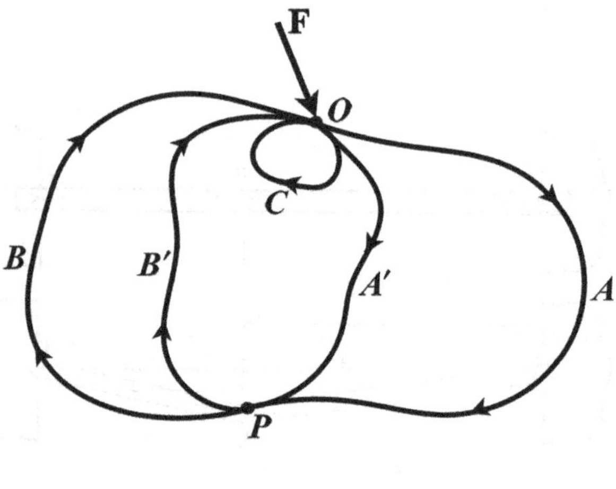

Figure 9.11

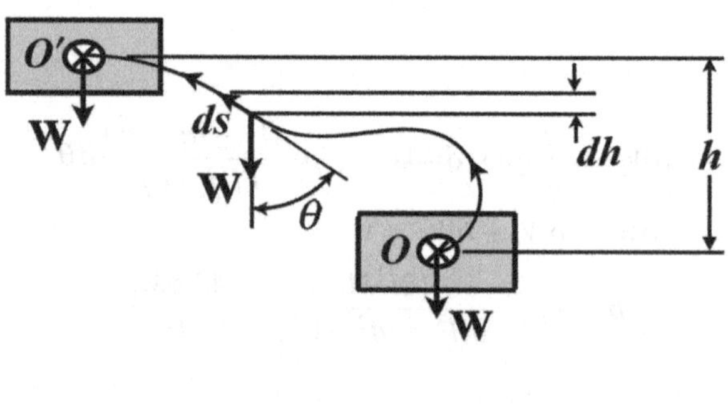

Figure 9.12

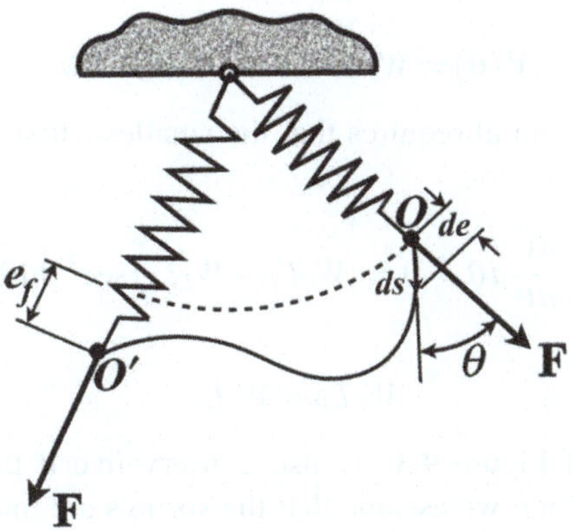

Figure 9.13

$$\Delta w = Wh$$

The elastic spring of Figure 9.13 is also a conservative mechanical system. The work we do in moving end O a small amount ds is

$$\mathrm{d}w = F\cos\theta \mathrm{d}s = F\mathrm{d}e$$

where de is the small extension of the spring. Since the spring is elastic, the force F depends only on the extension; that is, $F = F(e)$. Therefore,the work expended upon the spring in traversing *any* path from O to O' is

$$w = \int_0^{e_f} F(e)\mathrm{d}e$$

Again, the work we do constitutes a change of potential; the potential is energy stored in the spring. No work is dissipated.

9.2.2 PRINCIPLE OF STATIONARY POTENTIAL

A mechanical system is conservative if all the forces, internal and external, are conservative. Constant loads, such as weights, are the most common form of conservative external forces. The internal forces (stresses) in elastic bodies are conservative internal forces. If a system is conservative, then the virtual work (as defined in Section 9.1) is the change in potential. It is a small change associated with an arbitrary infinitesimal (first-order in the kinematic variables) change of the configuration. If a function (or functional) is unchanged by such small variations of its variables, it is said to be stationary. Accordingly, the principle of virtual work, when applied to a conservative system, is called the principle of stationary potential: *A conservative mechanical system is in equilibrium if and only if the total potential is stationary,*i.e. unchanged by an infinitesimal change in configuration.

The teeter-totter of Figure 9.1 is a conservative mechanical system. The potential of the system is the gravitational potential of the weights, W_P and W_J. Because we are concerned only with changes of potential, we can refer the potential to any state. Let us regard the potential as zero when the board is horizontal. Then the total potential is

$$V(\theta) = W_P L_P \tan\theta - W_J L_J \tan\theta$$

The criterion of stationary potential requires that the smallest (first-order) change dV vanish for arbitrary change dθ; that is,

$$dV = \frac{dV}{d\theta} d\theta = 0 = (W_P L_P - W_J L_J)\sec^2\theta \mathrm{d}\theta = 0$$

Again,

$$W_P L_P = W_J L_J$$

The mechanical system of Figure 9.3a is also conservative if the force P is a constant and the springs are elastic. As before, we assume that the springs are undeformed when the links are horizontal; also, we assume that these springs are linear (Hookean). If the horizontal position of the links is taken to be the state of zero potential of the external force P, then the total potential is

$$V(\theta_1, \theta_2) = V_E + V_I = -[PL(\sin\theta_1 + \sin\theta_2)] + \frac{\beta}{2}\left[\theta_1^2 + (\theta_2 - \theta_1)^2\right]$$

The stationary condition requires that

$$\mathrm{d}V = \frac{\partial V}{\partial \theta_1}\mathrm{d}\theta_1 + \frac{\partial V}{\partial \theta_2}\mathrm{d}\theta_2 = 0$$

This must vanish for *arbitrary* variations; dθ_1 and dθ_2 are independent. Therefore, we obtain the two equations of equilibrium:

$$\frac{\partial V}{\partial \theta_1} = -PL\cos\theta_1 + \beta\theta_1 - \beta(\theta_2 - \theta_1) = 0$$
$$\frac{\partial V}{\partial \theta_2} = -PL\cos\theta_2 + \beta(\theta_2 - \theta_1) = 0$$

These are precisely the equations obtained previously (Section 9.1)

ILLUSTRATIVE EXAMPLE 9.6

Hookean Truss: Each member of the truss in Figure 9.14 is composed of steel; the behavior is isotropic and Hookean to the yield point. All have the same cross-sectional area A. We require the displacement of joints O and B and the maximum *mean* normal stress (We do not have the information to determine local stresses at the joints). The truss is subject to the constant load P.

This is a conservative system; the configuration is determined by the displacements at joints O and B, viz u_O, v_O, u_B, and v_B. The elongations of the members follow:

$$e_1 = u_B, \quad e_2 = \frac{1}{\sqrt{2}}(u_B - v_B), \quad e_3 = u_O \tag{9.30a,b,c}$$

$$e_4 = \frac{1}{\sqrt{2}}(u_O + v_O) - \frac{1}{\sqrt{2}}(u_B + v_B) \tag{9.30d}$$

If e_i $(i = 1,2,3,4)$ denotes any of the elongations and l_i the length of that member, then the total potential follows:

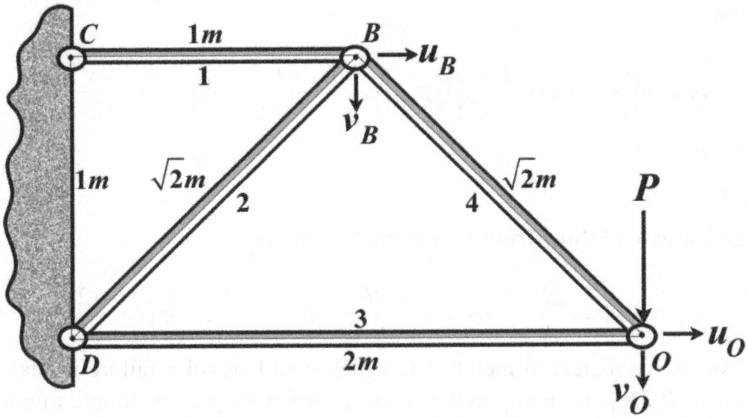

Figure 9.14

$$V = \sum_{i=1}^{4} \frac{EA}{2}\frac{e_i^2}{l_i} - \mathrm{P}v_O$$

$$V = \frac{EA}{2m}\left[u_B^2 + \frac{1}{2\sqrt{2}}(u_B - v_B)^2 + \frac{1}{2}u_O^2 + \frac{1}{2\sqrt{2}}(u_O + v_O)^2 - \frac{1}{\sqrt{2}}(u_O + v_O)(u_B + v_B) + \frac{1}{2\sqrt{2}}(u_B + v_B)^2\right] - \mathrm{P}v_O$$

The potential must be stationary with respect to arbitrary variations of the independent variations $(\delta u_B, \delta v_B, \delta u_O, \delta v_O)$

$$\frac{\partial V}{\partial u_B} = \frac{EA}{2\sqrt{2}m}\left[2(1+\sqrt{2})u_B - u_O - v_O\right] = 0$$

$$\frac{\partial V}{\partial v_B} = \frac{EA}{2\sqrt{2}m}\left[2v_B - u_O - v_O\right] = 0$$

$$\frac{\partial V}{\partial u_O} = \frac{EA}{2\sqrt{2}m}\left[-u_B - v_B + (1+\sqrt{2})u_O + v_O\right] = 0$$

$$\frac{\partial V}{\partial v_O} = \frac{EA}{2\sqrt{2}m}\left[-u_B - v_B + u_O + v_O\right] - \mathrm{P} = 0$$

In matrix form:

$$\begin{bmatrix} 2(1+\sqrt{2}) & 0 & -1 & -1 \\ 0 & 2 & -1 & -1 \\ -1 & -1 & (1+\sqrt{2}) & +1 \\ -1 & -1 & +1 & +1 \end{bmatrix} \begin{bmatrix} u_B \\ v_B \\ u_O \\ v_O \end{bmatrix} = \begin{bmatrix} 0 \\ 0 \\ 0 \\ \dfrac{2\sqrt{2}mP}{EA} \end{bmatrix}$$

The results follow.

$$u_B = \frac{p}{\sqrt{2}}, \quad v_B = \frac{1+\sqrt{2}}{\sqrt{2}} p, \quad u_O = -\frac{p}{\sqrt{2}}, \quad v_O = \frac{3+2\sqrt{2}}{\sqrt{2}} p$$

where $p \equiv \dfrac{2\sqrt{2}mP}{EA}$

The strains and stresses follow from Equations (9.30a–d):

$$\sigma_1 = E\varepsilon_1 = \frac{2P}{A}, \quad \sigma_2 = -\frac{\sqrt{2}P}{A}, \quad \sigma_3 = -\frac{P}{A}, \quad \sigma_4 = \frac{\sqrt{2}P}{A}$$

The "mean" stress is greatest in member 1, which could signal a failure in that member or at the connections C or B. Depending upon the cross-sectional properties, failure might occur via the buckling of member 3.

Several features of this formulation are noteworthy, since every problem of a discrete Hookean body with small displacements shares these attributes: The governing equations are linear in the kinematic variables (u, v); the internal potential is quadratic in those variables. The matrix of coefficients is symmetric. The external potential of the constant load(s) is linear. In general, if the system (e.g., a truss) has n degrees of freedom (e.g., displacements v_i of joints, $i = l, \ldots, n$), then

$$V_i = \sum_i^n \sum_j^n \frac{1}{2} a_{ij} v_i v_j \tag{9.31}$$

$$V_E = -\sum_i^n P_i v_i \tag{9.32}$$

Alternatively, we might employ customary matrix notations: $\boldsymbol{a}$ denotes the square matrix with symmetrical elements a_{ij}, $\mathbf{v}$ denotes the column (matrix) with elements v_i, and $\mathbf{P}$ denotes the column (matrix) with elements P_i. Of course, some elements might be zero (e.g., some joints may not be loaded). Then

$$V_I = \frac{1}{2} \mathbf{v}^{\mathbf{T}} \mathbf{a} \mathbf{v} \tag{9.33}$$

$$V_E = -\mathbf{P}^{\mathbf{T}} \mathbf{v} \tag{9.34}$$

The linear equations governing such systems have the alternative forms

$$\frac{\partial V_I}{\partial v_i} = \sum_j a_{ij} v_j = -\frac{\partial V_E}{\partial v_i} = P_i \tag{9.35a}$$

or

$$\mathbf{a}\mathbf{v} = \mathbf{P} \tag{9.35b}$$

9.3 BERNOULLI-EULER THEORY OF SYMMETRIC BEAMS

If a symmetric beam — for example, the beam of Figure 9.8 - is a homogeneous Hookean beam, then the internal potential U_I (per unit length) is obtained in accordance with the theory of Bernoulli-Euler:

$$\varepsilon_x = \varepsilon_0 - \kappa y$$

where ε_0 and κ are the extensional strain and curvature of the centroidal line, respectively. Then, the external energy (per unit length) takes the form (see Equation 4.35):

$$U_I = \iint_A \frac{E}{2}(\varepsilon_0 - \kappa y)^2 \mathrm{d}A$$

With the axis at the centroid,

$$U_I = \frac{EA}{2}\varepsilon_0^2 + \frac{EI}{2}\kappa^2 = \frac{EA}{2}\varepsilon_0^2 + \frac{EI}{2}\left(\frac{\mathrm{d}\theta}{\mathrm{d}x}\right)^2$$

The total internal potential is obtained by integrating throughout the beam:

$$V_t = \frac{E}{2}\int_O^L \left[A\varepsilon_0^2 + I\left(\frac{\mathrm{d}\theta}{\mathrm{d}x}\right)^2\right] \tag{9.36}$$

If u and v denote the displacements tangent and normal to the deformed axis (see Figure 9.6), then by Equations (9.25) and (9.26),

$$\delta\varepsilon_0 = \frac{\mathrm{d}}{\mathrm{d}x}(\delta u) - \kappa\delta v \tag{9.37}$$

$$\delta\kappa = \frac{\mathrm{d}^2}{\mathrm{d}x^2}(\delta v) + \frac{\mathrm{d}}{\mathrm{d}x}(\kappa\delta u) \tag{9.38}$$

The potential of the constant load $p(x)$ is

$$V_E = +\int_O^L p(x)\bar{v}\mathrm{d}x \tag{9.39}$$

The variation ("virtual" change of potential) is

$$\delta V = \delta V_I + \delta V_E = \int_0^L [EA\varepsilon_0\delta\varepsilon_0 + EI\kappa\delta\kappa + p\delta\bar{v}]\mathrm{d}x \tag{9.40a}$$

$$\delta V = \int_O^L \left\{EA\varepsilon_0\left[\frac{\mathrm{d}}{\mathrm{d}x}(\delta u) - \kappa\delta v\right] + EI\kappa\left[\frac{\mathrm{d}^2}{\mathrm{d}x^2}(\delta v) + \frac{\mathrm{d}}{\mathrm{d}x}(\kappa\delta u)\right] + p[\delta v\cos\theta + \delta u\sin\theta]\right\} \tag{9.40b}$$

This variation δV differs from Equation (9.27a) only because the force F and couple M are conservative actions. They are now determined by the deformation; specifically,

$$F = EA\varepsilon_0, \quad M = EI\kappa$$

Let us now simplify our analysis by restricting it to small rotations and deleting nonlinear terms.Then, the component v ($\approx w$) is normal to the initial axis ($\cos\theta \approx 1$). Instead of Equations (9.37) and (9.38) we have the approximations:

$$\varepsilon_0 = \frac{\mathrm{d}u}{\mathrm{d}x}, \quad \kappa = \frac{\mathrm{d}\theta}{\mathrm{d}x} \approx \frac{\mathrm{d}^2 w}{\mathrm{d}x^2}$$

The potential follows from Equations (9.36) and (9.39):

$$V = \int_O^L \left[\frac{EA}{2} \left(\frac{\mathrm{d}u}{\mathrm{d}x} \right)^2 + \frac{EI}{2} \left(\frac{\mathrm{d}^2 w}{\mathrm{d}x^2} \right)^2 + pw \right] \mathrm{d}x \tag{9.41}$$

This integral is a *functional*; the potential depends on two *functions*. Note that, as in Equation (9.31), the internal energy is a quadratic in the functions $u(x)$ and $w(x)$, and, as in Equation (9.32), the external potential is linear in the function $w(x)$. The enforcement of the principle requires that this functional V be stationary ($\delta V = O$) for arbitrary variations ($\delta u, \delta w$) of the functions $u(x)$ and $w(x)$. Specifically,

$$\delta V = \int_0^L \left[EA \left(\frac{du}{\mathrm{d}x} \right) \frac{\mathrm{d}}{\mathrm{d}x}(\delta u) + EI \left(\frac{\mathrm{d}^2 w}{\mathrm{d}x^2} \right) \frac{\mathrm{d}^2}{\mathrm{d}x^2} \delta w + p \delta w \right] \mathrm{d}x = 0$$

As mentioned before, the first term and second term must be integrated by parts to obtain

$$\begin{aligned} \delta V = EA \frac{du}{\mathrm{d}x} \delta u \Big|_0^L &+ EI \left(\frac{\mathrm{d}^2 w}{\mathrm{d}x^2} \right) \frac{\mathrm{d}}{\mathrm{d}x} (\delta w \Big|_0^L \\ - EI \frac{\mathrm{d}}{\mathrm{d}x} \left(\frac{\mathrm{d}^2 w}{\mathrm{d}x^2} \right) \delta w \Big|_0^L & \\ + \int_0^L \left[- \left(EA \frac{\mathrm{d}^2 u}{\mathrm{d}x^2} \right) \delta u + \left(EI \frac{\mathrm{d}^4 w}{\mathrm{d}x^4} + p \right) \delta w \right] \mathrm{d}x = 0 & \end{aligned}$$

For the beam illustrated in Figure 9.8, we have geometric constraints at the ends.

$$\delta u|_0 = \delta w|_0^L = \frac{\mathrm{d}}{\mathrm{d}x}(\delta w) \Big|_0 = 0$$

No constraint is imposed upon the displacement $u(L)$ or upon the rotation $\mathrm{d}w/\mathrm{d}x(L)$. Since these are independent, it follows that

$$EA \frac{du}{\mathrm{d}x} \Big|_L = EI \left(\frac{\mathrm{d}^2 w}{\mathrm{d}x^2} \right) \Big|_L = 0$$

These are the conditions that the axial force F and couple M vanish at $x = L$. Throughout the beam, the displacements $\delta u(x)$ and $\delta w(x)$ are arbitrary continuous functions. Therefore, the stationary criterion is satisfied if and only if each parenthetical term of the integrand vanishes:

$$EA \frac{\mathrm{d}^2 u}{\mathrm{d}x^2} = \left(\frac{\mathrm{d}F}{\mathrm{d}x} \right) = 0 \tag{9.42}$$

$$EI \left(\frac{\mathrm{d}^4 w}{\mathrm{d}x^4} \right) = \left(- \frac{\mathrm{d}V}{\mathrm{d}x} \right) = -p(x) \tag{9.43}$$

These two linear differential equations govern the stretching (u) and small transverse deflection (w) of the uniform Hookean beam.Two end conditions on u and/or $\mathrm{d}u/\mathrm{d}x$ accompany Equation (9.42). Four end conditions (two at each end) on w, $\mathrm{d}w/\mathrm{d}x$, $\mathrm{d}^2w/\mathrm{d}x^2$, or $\mathrm{d}^3w/\mathrm{d}x^2$ (deflection, rotation, bending couple, or shear force) accompany Equation (9.43).

Whether the mechanical system is *discrete* (the configuration is defined by a number of discrete variables as the truss of the preceding section) or *continuous* (the configuration is defined by continuous functions as the beam), the potential is expressed as a function (or functional) of the kinematic variables (e.g., displacements), which define the configuration. In any case, the internal forces do not enter explicitly. Even the behavior of the material is expressed by an internal energy (e.g., a Hookean member of a truss has energy $U_I = EAe^2/2L$).

9.4 RAYLEIGH'S METHOD

The principle of stationary potential provides a means to formulate the mathematical description of a continuous body; the potential of the beam, Equation (9.41), leads to Equations (9.42) and (9.43). In practice, the principle provides the basis for approximating a *continuous* body by a *discrete* model. Such is the method of finite elements, wherein the continuous body is subdivided into small but finite elements that are joined at interelement nodes. The displacements within each element are approximated, usually by simple polynomials, which, in turn, are prescribed by the nodal displacements. In this way, the continuous body is replaced by a discrete mechanical system; the *functional*, which gives the potential (such as Equation (9.41)), is replaced by a function of the discrete variables. Long before the advent of the finite element (circa 1956), Lord Rayleigh (1877) introduced the notion of approximating the displacements of a continuous body by functions that reduced the potential to a function of discrete variables. The variables in any such scheme are obtained by the stationary condition for the approximated potential. Such approximations are tantamount to physical constraints; they serve to inhibit the continuous body from assuming its most natural configuration of least potential. Accordingly, the discrete models (approximations) tend to be excessively stiff. We illustrate Rayleigh's method with a Bernoulli-Euler beam.

ILLUSTRATIVE EXAMPLE 9.7

A Cantilever Beam

The cantilever beam in Figure 9.15 is clamped at the left end ($x = O$) and simply supported at the right ($x = L$). It is subjected to the uniform load $p(x) = \bar{p}$ and the concentrated load P at the midpoint ($x = L/2$).

At best, we choose an approximation of $w(x)$ that satisfies the geometric constraints, $w(0) = w'(0) = 0$ and $w(L) = 0$ ($w' \equiv \mathrm{d}w/\mathrm{d}x$), and the natural conditions at the end, $M(L) = EIw''(L) = 0$ ($w'' \equiv \mathrm{d}^2w/\mathrm{d}x^2$). At least, we must adopt a function $w(x)$ that satisfies the geometric constraints. To illustrate, let

$$w(x) = Ax^2 + Bx^3 + Cx^4$$

To satisfy the geometric requirements,

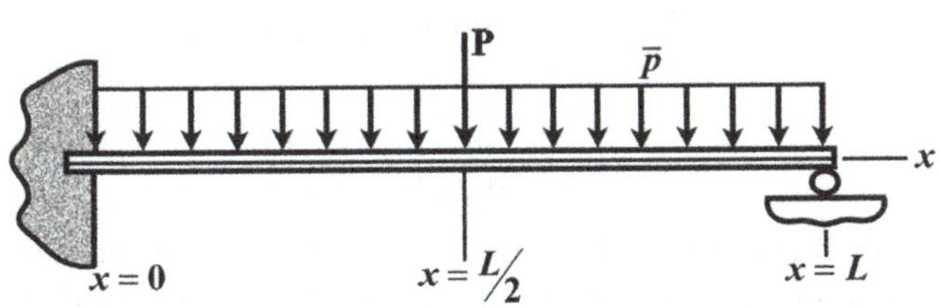

Figure 9.15

$$w(L) = 0, \quad C = -\frac{A}{L^2} - \frac{B}{L}$$

$$w(x) = A\left(1 - \frac{x^2}{L^2}\right)x^2 + B\left(1 - \frac{x}{L}\right)x^3$$

Omitting the extensional energy of Equation (9.41), we obtain our approximation to the potential of this cantilever:

$$\begin{aligned} V &= \int_0^L \left[\frac{EI}{2}(w'')^2 + \bar{p}w\right] \mathrm{d}x + Pw(L/2) \\ &= \frac{EI}{5}(42LA^2 + 44L^2AB + 12L^3B^2) + \frac{\bar{p}}{60}(8L^3A + 3L^4B) + \frac{P}{16}(3L^2A + L^3B) \end{aligned}$$

The stationary criteria provide the two equations that determine constants A and B:

$$\frac{\partial V}{\partial A} = \frac{EI}{5}(84LA + 44L^2B) + \frac{4\bar{p}L^3}{30} + \frac{3PL^2}{16} = 0$$

$$\frac{\partial V}{\partial B} = \frac{EI}{5}(44L^2A + 24L^3B) + \frac{\bar{p}L^4}{20} + \frac{PL^3}{16} = 0$$

or

$$84A + 44LB = -\frac{2\bar{p}L^2}{3EI} - \frac{15PL}{16EI}$$

$$44A + 24LB = -\frac{\bar{p}L^2}{4EI} - \frac{5PL}{16EI}$$

The solution follows:

$$A = -\frac{\bar{p}L^3}{16EI} - \frac{7}{64}\frac{PL}{EI}$$

$$B = \frac{5\bar{p}L^2}{48EI} + \frac{3P}{16EI}$$

To compare with the exact solution of the continuum, let us examine the values at the midpoint ($x = L/2$):

$$w(L/2) = -\frac{\bar{P}L^4}{192EI} - \frac{9}{1024}\frac{PL^3}{EI}$$

The exact value is

$$w(L/2) = -\frac{\bar{p}L^4}{192EI} - \frac{7}{764}\frac{PL^3}{EI}$$

Note that the contribution from the distributed load $\bar{p}$ is the same, because the quartic chosen embraces the exact solution. The contribution from the concentrated load P is 4% less; the approximate system is too stiff. The exact solution of the latter is a function with discontinuity in the third derivative ($V = -EIw'''$) at $x = L/2$ (see Sections 6.5 and 6.6).

9.5 COMPLEMENTARY ENERGY DENSITY

Recall that the potential energy density of an elastic body (see Section 2.7) can be expressed in terms of alternative strain components, the components of Cauchy-Green (Section 2.13) or the engineering components (Section 2.21). In either case, one must also employ the complementary stress components, the Trefftz components or Juamann components, respectively (see Section 3.15). In our present treatment of the complementary energy, we choose the latter. Then the energy density is a function of the engineering components:

$$u(\bar{\varepsilon}_x, \bar{\varepsilon}_y, \bar{\varepsilon}_z, \bar{\gamma}_{xy}, \bar{\gamma}_{xz}, \bar{\gamma}_{yz}) \tag{9.44}$$

and the Juamann stresses follow:

$$\begin{aligned} t_x &= \frac{\partial u}{\partial \bar{\varepsilon}_x}, \quad & t_{xy} &= \frac{\partial u}{\partial \bar{\gamma}_{xy}}, \quad & t_{xz} &= \frac{\partial u}{\partial \bar{\gamma}_{xz}} \\ t_y &= \frac{\partial u}{\partial \bar{\varepsilon}_y}, \quad & t_{yz} &= \frac{\partial u}{\partial \bar{\gamma}_{yz}}, \quad & t_z &= \frac{\partial u}{\partial \bar{\varepsilon}_z} \end{aligned} \tag{9.45}$$

We define the complementary energy density as follows:

$$\begin{aligned} u_c \equiv\, & t_x\bar{\varepsilon}_x + t_{xy}\bar{\gamma}_{xy} + t_{xz}\bar{\gamma}_{xz} \\ & + t_{yx}\bar{\gamma}_{yx} + t_y\bar{\varepsilon}_y + t_{yz}\bar{\gamma}_{yz} \\ & + t_{zx}\bar{\gamma}_{zx} + t_{zy}\bar{\gamma}_{zy} + t_z\bar{\varepsilon}_z - u \end{aligned} \tag{9.46}$$

Assuming that the stress/strain equations can be inverted to express the strains in terms of the stresses, we have

$$u_c(t_x, t_{xy}, t_{xz}, t_{yx}, t_y, t_{yz}, t_{zx}, t_{zy}, t_z) \tag{9.47}$$

The differential of Equation (9.47) can be equated to the differential of Equation (9.46), as follows

$$\begin{aligned} \frac{\partial u_c}{\partial \tau_x}\delta t_x + \frac{\partial u_c}{\partial \tau_{xy}}\delta t_{xy} \cdots \frac{\partial u_c}{\partial t_{yz}}\delta t_z &= \delta t_x\bar{\varepsilon}_x + \delta t_{xy}\bar{\varepsilon}_{xy} \cdots \delta t_z\bar{\varepsilon}_z \\ &+ \left[t_x\delta\bar{\varepsilon}_x + t_{xy}\delta\bar{\gamma}_{xy} \cdots t_z\delta\bar{\varepsilon}_z - \frac{\partial u}{\partial \bar{\varepsilon}_x}\delta\bar{\varepsilon}_x - \frac{\partial u}{\partial \bar{\varepsilon}_{xy}}\delta\bar{\varepsilon}_{xy} \cdots \frac{\partial u}{\partial \bar{\varepsilon}_z}\delta\bar{\varepsilon}_z \right] \end{aligned} \tag{9.48}$$

In view of Equations (9.45), the bracketed term of Equation (9.48) vanishes. Since the six components of stress are independent variables in Equations (9.47) and (9.48), it follows that

$$\begin{aligned} \bar{\varepsilon}_x &= \frac{\partial u_c}{\partial t_x}, \quad & \bar{\gamma}_{xy} &= \frac{\partial u_c}{\partial t_{xy}}, \quad & \bar{\gamma}_{xz} &= \frac{\partial u_c}{\partial t_{xz}} \\ \bar{\gamma}_{yx} &= \frac{\partial u_c}{\partial t_{yx}}, \quad & \bar{\varepsilon}_y &= \frac{\partial u_c}{\partial t_y}, \quad & \bar{\gamma}_{yz} &= \frac{\partial u_c}{\partial t_{yz}} \\ \bar{\gamma}_{zx} &= \frac{\partial u_c}{\partial t_{zx}}, \quad & \bar{\gamma}_{zy} &= \frac{\partial u_c}{\partial t_{zy}}, \quad & \bar{\varepsilon}_z &= \frac{\partial u_c}{\partial t_z} \end{aligned} \tag{9.49}$$

The forms of Equations (9.45) and (9.49) reveal the complementary character of the potentials, u and u_c.

9.6 CASTIGLIANO'S THEOREM

Let us define an internal complementary potential[5] for an elastic body:

$$V_c \equiv \iiint_v u_c \mathrm{d}v$$

We consider a variation of the stresses, which does not violate the equilibrium.

$$\begin{aligned}\delta V_c = \iiint_v \Big(& \frac{\partial u_c}{\partial t_x}\delta t_x + \frac{\partial u_c}{\partial t_{xy}}\delta t_{xy} + \frac{\partial u_c}{\partial t_{xz}}\delta t_{xz} \\ & + \frac{\partial u_c}{\partial t_{yx}}\delta t_{yx} + \frac{\partial u_c}{\partial t_y}\delta t_y + \frac{\partial u_c}{\partial t_{yz}}\delta t_{yz} \\ & + \frac{\partial u_c}{\partial t_{zx}}\delta t_{zx} + \frac{\partial u_c}{\partial t_{zy}}\delta t_{zy} + \frac{\partial u_c}{\partial t_z}\delta t_z \Big) \mathrm{d}v\end{aligned} \tag{9.50}$$

According to Equation (9.49), this differential has the alternative form:

$$\begin{aligned}\delta V_c = \iiint_v (& \bar{\varepsilon}_x \delta t_x + \bar{\gamma}_{xy}\delta t_{xy} + \bar{\gamma}_{xz}\delta t_{xz} \\ & + \bar{\gamma}_{yx}\delta t_{yx} + \bar{\varepsilon}_y \delta t_y + \bar{\gamma}_{yz}\delta t_{yz} \\ & + \bar{\gamma}_{zx}\delta t_{zx} + \bar{\gamma}_{zy}\delta t_{zy} + \bar{\varepsilon}_z \delta t_z) \mathrm{d}v\end{aligned} \tag{9.51}$$

We now call upon the expressions for the strains in terms of the displacements, viz. Equations (2.65):

We restrict our development to small deformations: specifically, we adopt the *linear approximation of the strain-displacement* Equations (2.65), viz.

$$\begin{aligned}& \bar{\varepsilon}_x = \hat{\boldsymbol{i}} \cdot (\boldsymbol{G}_x - \hat{\boldsymbol{i}}) = \hat{\boldsymbol{i}} \cdot (\hat{\mathbf{R}}_{,x} - \hat{\mathbf{r}}_{,x}) \\ & \bar{\varepsilon}_y = \hat{\boldsymbol{j}} \cdot (\boldsymbol{R}_{,y} - \hat{\boldsymbol{r}}_{,y}), \quad \bar{\varepsilon}_z = \hat{\boldsymbol{k}} \cdot (\boldsymbol{R}_{,z} - \hat{\boldsymbol{r}}_{,z}) \\ & \bar{\gamma}_{xy} = \frac{1}{2}(\hat{\boldsymbol{i}} \cdot \boldsymbol{G}_y + \hat{\boldsymbol{j}} \cdot \boldsymbol{G}_x) = \frac{1}{2}(\hat{\boldsymbol{i}} \cdot \boldsymbol{R}_{,y} + \hat{\boldsymbol{j}} \cdot \boldsymbol{R}_{,x}) \\ & \bar{\gamma}_{xz}\frac{1}{2}(\hat{\boldsymbol{i}} \cdot \boldsymbol{R}_{,z} + \hat{\boldsymbol{k}} \cdot \boldsymbol{R}_{,x}), \quad \bar{\gamma}_{yz} = (\hat{\boldsymbol{j}} \cdot \boldsymbol{R}_{,z} + \hat{\boldsymbol{k}} \cdot \boldsymbol{R}_{,y})\end{aligned}$$

We note the symmetry of these components, i.e. $\bar{\gamma}_{xy} = \bar{\gamma}_{yx}$, etc. Then, Equation (9.51) assumes the form

$$\begin{aligned}\delta V_C = \iiint_v \Big[& \delta t_x \hat{\boldsymbol{i}} \cdot (\boldsymbol{R} - \boldsymbol{r})_{,x} + \delta t_{xy} \hat{\boldsymbol{j}} \cdot (\boldsymbol{R} - \boldsymbol{r})_{,x} + \delta t_{xz} \hat{\boldsymbol{k}} \cdot (\boldsymbol{R} - \boldsymbol{r})_{,x} \\ & + \delta t_{yx} \hat{\boldsymbol{i}} \cdot (\boldsymbol{R} - \boldsymbol{r})_{,y} + \delta t_y \hat{\boldsymbol{j}} \cdot (\boldsymbol{R} - \boldsymbol{r})_{,y} + \delta t_{yz} \hat{\boldsymbol{k}} \cdot (\boldsymbol{R} - \boldsymbol{r})_{,y} \\ & + \delta t_{zx} \hat{\boldsymbol{i}} \cdot (\boldsymbol{R} - \boldsymbol{r})_{,z} + \delta t_{zy} \hat{\boldsymbol{j}} \cdot (\boldsymbol{R} - \boldsymbol{r})_{,z} + \delta t_z \hat{\boldsymbol{k}} \cdot (\boldsymbol{R} - \boldsymbol{r})_{,z} \Big] \mathrm{d}v \\ = \iiint & [\delta \boldsymbol{\sigma}_x \cdot (\boldsymbol{R} - \boldsymbol{r})_{,x} + \delta \boldsymbol{\sigma}_y \cdot (\boldsymbol{R} - \boldsymbol{r})_{,y} + \delta \boldsymbol{\sigma}_z \cdot (\boldsymbol{R} - \boldsymbol{r})_{,z}] \mathrm{d}v\end{aligned}$$

By the divergence theorem (integration by parts):

$$\begin{aligned}\delta V_c = & -\iiint [(\delta \boldsymbol{\sigma}_{x,x} + \delta \boldsymbol{\sigma}_{y,y} + \delta \boldsymbol{\sigma}_{z,z}) - (\boldsymbol{R} - \boldsymbol{r})] \mathrm{d}v \\ & - \iint_s [(\delta \boldsymbol{\sigma}_x l_n + \delta \boldsymbol{\sigma}_y m_n + \delta \boldsymbol{\sigma}_z n_n)] \cdot (\boldsymbol{R} - \boldsymbol{r}) \mathrm{d}s\end{aligned} \tag{9.52}$$

Here, (l_n, m_n, n_n) denote the direction cosines of the normal to the surface. If the stress variations $(\delta\boldsymbol{\sigma}_x, \delta\boldsymbol{\sigma}_y, \delta\boldsymbol{\sigma}_z)$ satisfy equilibrium, then the bracketed term of the first integral vanishes. Then Equation (6.52) expresses the variation δV in terms of a variation of the boundary stresses, which do not violate equilibrium.

$$\delta V_c = \iint_S \delta\boldsymbol{\sigma} \cdot (\boldsymbol{R} - \boldsymbol{r}) \mathrm{d}s \tag{9.53}$$

To obtain a version of Castigliano's theorem, we follow the logic of H.L. Langhaar.[6] We suppose that the body is supported on some portion S_v of the surface, so that rigid-body displacement of the entire body is not possible; there $\boldsymbol{R} = \boldsymbol{r}$, the initial position. The body may be subjected to certain concentrated tractions on the surface $S_t = S - S_v$. In the accepted practice, we regard such concentrated force as distributed on a *small* spot. We know only the resultant, not the distribution, *but* understand that local stress and deformation are beyond the scope of our theory. Let S_0 denote the small surface which is subject to the resultant $\boldsymbol{P}_0$; let $\boldsymbol{v}_0 \approx \boldsymbol{R}_0 - \boldsymbol{r}_0$ denote the mean displacement of the *spot*. Then Equation (9.53) takes the form:

$$\delta V_c = \boldsymbol{v}_0 \cdot \delta\boldsymbol{P}_0 \tag{9.54}$$

The variations of the stress $(\delta\boldsymbol{\sigma}_x, \delta\boldsymbol{\sigma}_y, \delta\boldsymbol{\sigma}_z)$ within v are a consequence of the variation of load $(\delta\boldsymbol{P}_c)$. Since $\delta\boldsymbol{P}_0$ is arbitrary, let it be collinear with an applied force P. Then the displacement v_P in the direction of that force is given in accordance with Equation (9.54):

$$v_P = \frac{\partial V_c}{\partial P} \tag{9.55}$$

Note that the elastic properties may be quite arbitrary; however, the result (9.55) is limited to *small* displacements.

9.6.1 DISCRETE FORMS OF CASTIGLIANO'S PRINCIPLE

Let us consider an elastic structure wherein the configuration can be defined by $'n'$ discrete variables, v_i e.g. the joints of a truss. We presume that the structure is subject to conservative loads $P_i\ (v_i)$. By the principle of stationary potential

$$P_i = \frac{\partial V_I}{\partial v_i}, \quad i = 1 \cdots n \tag{9.56}$$

Note that v_i and P_i are generalized variables, e.g. v_i might be a rotation where upon P_i would be a moment.

In the manner of Equation (9.46), we define a complimentary energy.

$$V_c = \sum_{n}^{i=1} P_i v_i - V_I \tag{9.57}$$

As before, we assume that Equation (9.56) can be inverted to express the kinematic variables v_i in terms of the kinetic variables P_i. Then, $V_c = V_c(P_i)$ and

$$\mathrm{d}V_c = \sum_{i=1}^{n} \frac{\partial V_c}{\partial P_i} \mathrm{d}P_i = \sum_{i=1}^{n} v_i \mathrm{d}P_i - \sum_{i=1}^{n} \left[P_i - \frac{\partial V_I}{\partial v_i} \right] \mathrm{d}v_i$$

In view of Equation (9.56), the bracketed term vanishes. Since P_i are independent variables, we obtain

$$v_i = \frac{\partial V_c}{\partial P_i} \tag{9.58}$$

The latter is the counterpart of Equation (9.55) for the discrete system.

In general, it is difficult to obtain an explicit expression of the complementary energy $V_c(P_i)$. However, in the case of a Hookean body, this is often achievable. Then the internal energy V_I is a quadratic form in the generalized displacements v_i (see Equations 9.31 and 9.33):

$$V_I = \frac{1}{2}\sum_{i=1}^{n}\sum_{j=1}^{n} a_{ij}v_iv_j \quad (\text{or } V_I = \boldsymbol{v}^T\boldsymbol{a}\boldsymbol{v}) \tag{9.59}$$

Then the forces are given by the linear equations

$$P_I = \frac{1}{2}\sum_{j=1}^{n} a_{ij}v_j \quad (\text{or } \boldsymbol{P} = \boldsymbol{a}\boldsymbol{v})$$

The inverse is

$$v_i = \sum_{j=1}^{n} a_{ij}^{-1}P_j \quad (\text{or } \boldsymbol{v} = \boldsymbol{a}^{-1}\boldsymbol{P}) \tag{9.60}$$

In accordance with Equation (9.57), the complementary energy is

$$V_c = \sum_{i=1}^{n} P_iv_i - V_I = \sum_{i=1}^{n}\sum_{j=1}^{n} a_{ij}v_iv_j - V_I = V_I \tag{9.61}$$

According to Equation (9.60) and (9.61), we have

$$V_c = V_I = \frac{1}{2}\sum_{i=1}^{n}\sum_{j=1}^{n} a_{ij}^{-1}P_iP_j \quad (\text{or } \frac{1}{2}\boldsymbol{P}^T\boldsymbol{a}^{-1}\boldsymbol{P}) \tag{9.62}$$

In other words, the internal energy V_I and its complement V_c are equal in the case of linear behavior. This is true only if the material is Hookean and the displacements, and especially the rotations are small. Then the relations between forces (P_I) and displacements (v_i) are linear, and, according to Equation (9.62), the complementary energy is the internal energy but expressed in terms of the forces. Then, in keeping with Equations (9.58) and (9.62),

$$v_i = \frac{\partial V_I}{\partial P_i} = \sum_{j=1}^{n} a_{ij}^{-1}P_j \quad (\text{or } \boldsymbol{v} = \boldsymbol{a}^{-1}\boldsymbol{P}) \tag{9.63}$$

In this form, the result (Equation 9.63) is known as Castigliano's theorem. Let us state the theorem: If V_I $(P_1, \ldots, P_n)$ is the internal energy (potential) expressed in terms of the discrete external "forces" (P_i) then the *conjugate* "displacement" (v_i) is given by Equation (9.63). Again, a force and a displacement are conjugate in the sense that the increment of work done by said force is expressed by $P_i\mathrm{d}v_i$.

In general, it is difficult to express the internal energy in terms of the applied forces. The internal energy is variable throughout a body, depending on the state of strain. Although we can usually invert the strain-stress relations and express the internal energy in terms of stresses, the stresses are readily expressed in terms of external forces only in certain simple statically determinate problems. To illustrate the theorem, we consider some simple problems of this type.

ILLUSTRATIVE EXAMPLE 9.8

Let us return to the truss of Figure 5.16, reproduced here as Figure 9.16a. To determine the deflection v by means of Castigliano's theorem, we require the internal energy expressed in terms of the load W.

The axial forces, F_B, F_C, and F_D, in members AB, AC, and AD are statically indeterminate. We temporarily remove member AD as shown in Figure 9.16b. Equilibrium of the modified system provides the results

$$F_B = \sqrt{2}(W - F_D), \quad F_C = -(W - F_D)$$

Let us presume Hookean behavior. Then the internal energy V_I ($= V_c$) of the modified system follows:

$$V_I = \frac{L}{2EA}\left(F_c^2 + \sqrt{2}F_B^2\right) = \frac{L}{2EA}(W - F_D)^2(2\sqrt{2}+1)$$

By Castigliano's theorem, the displacement associated with force F_D is

$$-v = \frac{\partial V_I}{\partial F_D} = -\frac{L}{EA}(W - F_D)(2\sqrt{2}+1)$$

In the present case, $v = e_D$, the elongation of member AD:

$$e_D = \frac{F_D L}{EA} = v = \frac{(W - F_D)L}{EA}(2\sqrt{2}+1)$$

Solving for F_D, we obtain

$$F_D = \frac{(4+\sqrt{2})}{2(2+\sqrt{2})}W$$

The internal energy can now be expressed in terms of the load W:

$$V_I = V_B + V_C + V_D = \frac{(4+\sqrt{2})}{2(4+2\sqrt{2})}\frac{W^2L}{EA}$$

The displacement v follows in accordance with (9.63):

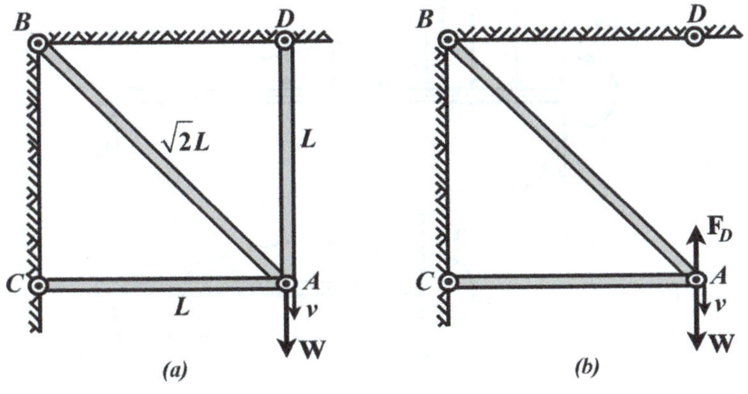

Figure 9.16

$$v = \frac{\partial V_I}{\partial W} = \frac{(4+\sqrt{2})}{(4+2\sqrt{2})}\frac{WL}{EA}$$

This is in agreement with Equation (5.29b).

ILLUSTRATIVE EXAMPLE 9.9

Let us return to the symmetric Hookean beam of Figure 6.22, reproduced here as Figure 9.17a. Suppose that we require the displacement at an arbitrary site, $x = a$, which is caused by the distributed load, $p = \bar{p}x/L$. To achieve this via Castigliano's theorem, we employ a "dummy" load Q at $x = a$, use Equation (9.55), and subsequently set $Q = 0$.

To obtain the reaction upon the right end ($x = 0$), we consider a segment, as in Figure 9.17c. Equilibrium provides the moment:

$$M(x) = R_B x - \frac{\bar{p}x^3}{6L}, \qquad x \le a$$
$$M(x) = R_B x + Q(x-a) - \frac{\bar{p}x^3}{6L}, \quad x \ge a$$

According to the Bernoulli theory, the internal energy is attributed solely to the bending. The energy per unit length is

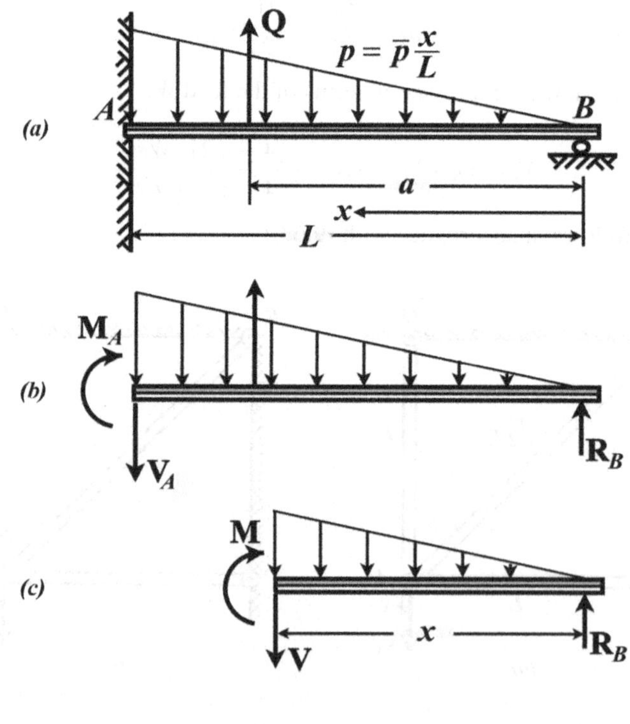

Figure 9.17

$$u_I = \iint_A \frac{E\varepsilon^2}{2}\mathrm{d}A = \iint_A \frac{E}{2}(-\kappa y)^2\mathrm{d}A = \frac{EI}{2}\kappa^2$$

Since $\kappa = M/EI$

$$u_I = \frac{M^2}{2EI}$$

The internal energy for the entire beam is the integral:

$$\begin{aligned} V_I &= \int_0^L \frac{M^2}{2EI}\mathrm{d}x = V_c \\ V_I &= \frac{1}{2EI}\left[\frac{R_B^2 a^3}{3} - \frac{R_B \bar{p} a^5}{15L} + \frac{\bar{p}^2 a^7}{252L^2}\right. \\ &\quad + \frac{R_B^2(L^3 - a^3)}{3} + 2R_B Q\left(\frac{L^3}{3} + \frac{a^3}{6} - \frac{al^2}{2}\right) \\ &\quad - \frac{R_B\bar{p}}{15L}(L^5 - a^5) + \frac{Q^2}{3}(L-a)^3 \\ &\quad \left. - \frac{Q\bar{p}}{3L}\left(\frac{L^5}{5} + \frac{a^5}{20} - \frac{aL^4}{4}\right) + \frac{\bar{p}^2}{252L^2}(L^7 - a^7)\right] \end{aligned} \tag{9.64}$$

By means of Equation (9.55), we obtain

$$v_B = 0 = \frac{\partial V_I}{\partial R_B}, \quad R_B = \frac{\bar{p}L}{10} - \frac{Q}{L^3}(L^3 - \frac{3}{2}aL^2 + \frac{1}{2}a^3) \tag{9.65}$$

Inserting this result into Equation (9.64), we have the energy V_I expressed in terms of our "dummy" load Q. To obtain v_a via Equation (9.55), we need only the terms $V_I = V_C$, which are linear in Q; the quadratic terms will vanish when we subsequently set $Q = 0$. This modified version follows:

$$\begin{aligned} \bar{V}_I(Q) &= \frac{Q}{2EI}\left[-\frac{2a^3\bar{p}}{30L^2}\left(L^3 - \frac{3}{2}aL^2 + \frac{1}{2}a^3\right)\right. \\ &\quad + \frac{\bar{p}a^5}{15L^4}\left(L^3 - \frac{3}{2}al^2 + \frac{1}{2}a^3\right) \\ &\quad - \frac{\bar{p}}{15L^2}\frac{(L^3 - a^3)}{L^2}\left(L^3 - \frac{3}{2}aL^2 + \frac{1}{2}a^3\right) \\ &\quad + \frac{\bar{p}L}{5}\left(\frac{L^3}{3} + \frac{a^3}{6} + \frac{aL^2}{2}\right) \\ &\quad + \frac{\bar{p}(L^5 - a^5)}{15L^4}\left(L^3 - \frac{3}{2}aL^2 + \frac{1}{2}a^3\right) \\ &\quad \left. - \frac{\bar{p}}{3L}\left(\frac{L^5}{5} + \frac{a^5}{2a} + \frac{aL^4}{4}\right)\right] \end{aligned} \tag{9.66}$$

The displacement at $x = a$ in the direction Q is obtained via Equation (9.55), viz

$$v(a) = \frac{\partial \bar{V}_I}{\partial Q} = \frac{1}{2EI}F(L,a)$$

where $F(L,a)$ is the bracketed term of Equation (9.66). Our interest is solely in the demonstration of Castigliano's theorem and verifying the result, as obtained in Example 6.51. Accordingly, we give the result with $a = L/2$:

$$v\left(\frac{L}{2}\right) = -\frac{9}{3840}\frac{\bar{p}L^4}{EI}$$

Values used in the earlier example are

$$I = 310\text{ in.}^4, \quad L = 240\text{ in.}$$
$$E = 30 \times 10^6\text{ lb/in.}^2, \quad \bar{p} = 20\text{ lb/in.}$$

These provide the previous result:

$$v\,(120\text{ in.}) = -0.0167\text{ in.}$$

ILLUSTRATIVE EXAMPLE 9.10

The semicircular member of Figure 9.18a is made of steel with a circular cross-section ($r =$ 1.00 cm., $R = 50$ cm, $E = 20 \times 10^4$ MPa). It is clamped stress free at A and B when the transverse load $P = 500$ N is applied at the central point O, perpendicular to the plane AOB. We require the consequential deflection at O.

Since $r \ll R$, a slice behaves much as a straight beam. We can superpose the strain ε_θ caused by the bending couple M and the strain $\gamma_{r\theta}$ caused by the twisting couple T (see Figure 9.18b). Then the strain energy $\bar{u}$ (per unit length) can be expressed in terms of those actions:

$$\bar{u} = \iint_A \left(\frac{E}{2}\varepsilon_\theta^2 + \frac{G}{2}\tau_{r\theta}^2\right) dA$$

where A denotes the cross-sectional area. In terms of couples M and T,

$$\bar{u} = \frac{1}{2EI}M^2 + \frac{1}{2GJ}T^2 \tag{9.67}$$

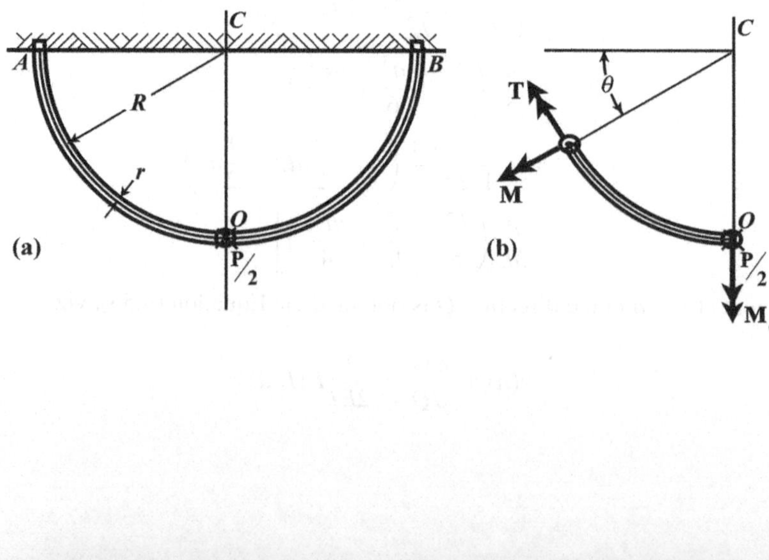

Figure 9.18

In view of the symmetry with respect to line OC, it suffices to consider the quadrant of Figure 9.18b. Symmetry dictates that the shear load is $P/2$ and the twisting couple $T = 0$ at 0; moreover, the rotation about line OC vanishes.

From the free body in Figure 9.18b, we have

$$M = \frac{PR}{2}\cos\theta - M_O \sin\theta \tag{9.68a}$$

$$T = M_O \cos\theta - \frac{PR}{2}(1 - \sin\theta) \tag{9.68b}$$

Substituting Equations (9.68a,b) into Equation (9.67) and integrating, we obtain

$$\begin{aligned} V_I &= \int_0^{\pi/2} \left(\frac{1}{2EI}M^2 + \frac{1}{2GI}T^2\right) Rd\theta \\ &= \frac{R}{2EI}\left[P^2R^2\left(\frac{\pi}{16}\right) - PRM_0\left(\frac{1}{2}\right) + M_0^2\frac{\pi}{4}\right] + \frac{R}{2GJ}\left[M_0^2\left(\frac{\pi}{4}\right) - M_0PR\frac{1}{2} + P^2R^2\left(\frac{3\pi}{16} - \frac{1}{2}\right)\right] \end{aligned} \tag{9.69}$$

The couple M_O remains unknown; it is determined by the condition of vanishing rotation at the line of symmetry, viz

$$\begin{aligned} \frac{\partial V_I}{\partial M_O} &= 0 \\ &= -PR^2\left(\frac{1}{4EI} + \frac{1}{4GJ}\right) + M_O R\frac{\pi}{4}\left(\frac{1}{4EI} + \frac{1}{4GJ}\right) = 0 \end{aligned}$$

Setting $E = 2.5G$ and $J = 2I$, we obtain

$$M_O = \frac{PR}{\pi}$$

Equation (9.69) is now expressed in terms of P alone and the known properties (E, G, R, r).

The transverse displacement at O, i.e. at the load P follows:

$$\frac{\partial V_I}{\partial P} = \frac{PR^3}{GJ}\left(\frac{19\pi^2 - 40\pi - 36}{80\pi}\right) = 0.51 \text{ cm } (0.20 \text{ in.})$$

ILLUSTRATIVE EXAMPLE 9.11

Let us return to the coil spring of Illustrative Example 5.16. The essential features are revealed in Figure 9.19. Acting on every element of the coil are transverse shear and extensional forces.

Neither contributes significantly to the extension (v). Typically, the wire diameter (d) is much smaller than the coil diameter ($2r$); then the pitch angle β is also small. Moreover, any small segment bends and twists under the actions of the couples, M and T, respectively. Since $d \ll 2r$, we can treat the deformation and energy as a straight beam (see Equation 9.67). The total internal energy in $'N'$ coils follows:

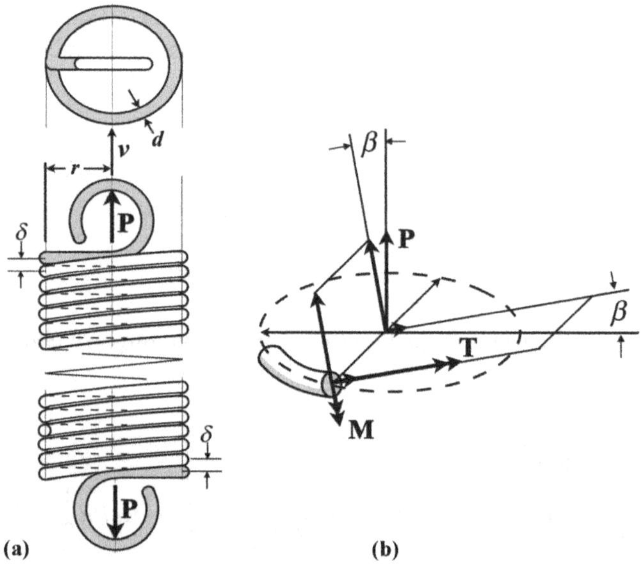

Figure 9.19

$$V_I = N\int_0^{2\pi}\left[\frac{1}{2EI}M^2 + \frac{1}{2GJ}T^2\right] r\mathrm{d}\theta \tag{9.70}$$

From the geometry of Figure 9.19b, we obtain

$$M = Pr\sin\beta, \quad T = Pr\cos\beta \tag{9.71}$$

Substituting these expressions into Equation (9.70) and integrating, we obtain

$$V_I = N\pi P^2 r^3\left[\frac{\sin^2\beta}{EI} + \frac{\cos^2\beta}{GJ}\right]$$

The elongation v follows via Castigliano's theorem:

$$v = \frac{\partial V_I}{\partial P} = 2\pi NPr^3\left[\frac{\sin^2\beta}{EI} + \frac{\cos^2\beta}{GJ}\right]$$

If, as in Illustrative Example 5.16, we assume the angle β to be small ($\sin\beta \ll 1$, $\cos\beta \approx 1$), we obtain the earlier result, viz.

$$v \approx \frac{2\pi NPr^3}{GJ}$$

9.7 STABILITY

The term *instability* implies that a situation is precarious and subject to abrupt, perhaps unanticipated, change. Such situations can arise in mechanical systems, notably structures. Unstable states are always undesirable; they usually signal failure and sometimes signal catastrophic collapse.

Several simple examples serve to illustrate such instability: A long, straight, and slender rod, such as a yardstick, can sustain axial thrust; at a certain load, the rod bends and, more importantly, continues to bend with little additional resistance. That certain load is the *critical* load. A toggle switch in an electrical circuit requires a force to move it from the on to the off position; again, at a certain force, the switch snaps *abruptly* to the off(or on) position. Our last example suggests a more critical situation, because now the critical force does not initiate as a slight movement but causes a *large* movement — that is, from completely on to completely off. The thin rod typifies the most common form of instability; the bending, which accompanies the critical load, is a form of buckling. The very abrupt and large movement of the toggle switch is a form of "snap buckling." Both can occur in various mechanical systems and, especially, structures. The purpose of the present chapter is (1) to introduce and explore the phenomena and (2) to set forth the most common means to ascertain critical loads, as exemplified by the simple column of traditional structures.

Our focus is on the stability of mechanical systems and structures, which are in a state of static equilibrium. However, the term applies as well to the systems in motion. For example, the aeronautical engineer must be concerned with the dynamic stability of an aircraft; aerodynamic forces can cause an unacceptable "flutter," i.e. uncontrolled oscillatory motions of an airfoil. Civil engineers were reminded of similar instabilities when the bridge at the Tacoma Narrows[7] was destroyed by the uncontrolled oscillatory motions caused by crosswinds.

A system is in a state of stable equilibrium if positive work is required to cause an unacceptable alteration. Of course, there are degrees of instability: A system would be *very* unstable if only a slight disturbance caused failure, more stable if more work is required to produce an unacceptable state.

ILLUSTRATIVE EXAMPLE 9.12

Consider the homely example of Figure 9.20. A person is perched on one of the higher rungs of a ladder. His/her safety is dependent upon the position (h), the angle (α), and the frictional coefficients, μ_1 at the base and μ_2 at the wall.

Let us suppose that he/she has ascended to the highest recommended position, e.g. $h = 3H/4$. Applying the equilibrium conditions, we arrive at the result.

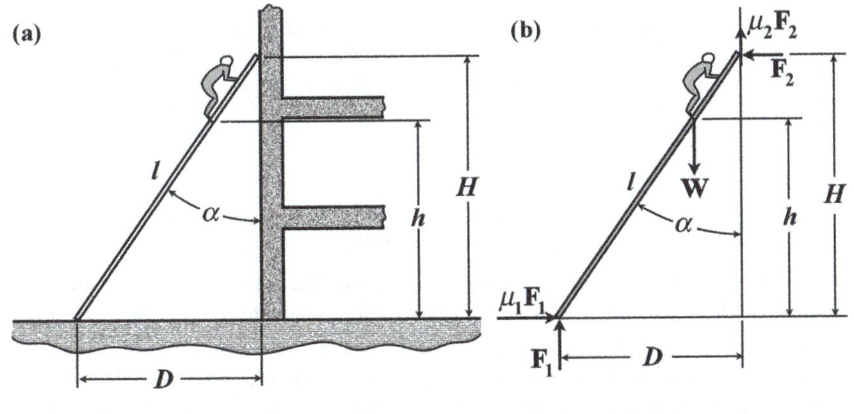

Figure 9.20

$$\tan\alpha = \frac{4\mu_1}{3-\mu_1\mu_2}$$

Typically, the friction at the base is greater than at the wall, e.g. $u_1 = 0.30$ and $u_2 = 0.15$. Then

$$\tan\alpha \approx 0.406 \quad \alpha \approx 22^\circ$$

Of course, the circumstance depicts a state of impending motion. Moreover, μ_1 and μ_2 are coefficients of static friction. When in motion, the frictional forces are typically less. This means that a slight disturbance or movement would doubtlessly cause the ladder to slide and fail. Clearly, our climber is at risk unless the angle α is decreased.

The system of Figure 9.20 is non-conservative; some of the work to cause a displacement would be dissipated. Let us apply the criterion of positive work, i.e., the system is stable if positive work is required to effect a displacement: Let $\delta\alpha$ denote the small change of angle α. The increment of work by all forces follows:

$$\delta w = W\delta h - \mu_2 F_2 \delta H + \mu_1 F_1 \delta D$$

In the equilibrium state, we have

$$\mu_2 F_2 = \frac{\mu_1 W}{1+\mu_1\mu_2}, \quad \mu_1 F_1 = \frac{W}{1+\mu_1\mu_2}$$

From the kinematics, we have

$$\frac{4h}{3l} = \frac{H}{l} = \cos\alpha, \quad \frac{D}{l} = \sin\alpha$$

$$\delta h = -\frac{3}{4} l\delta\alpha\sin\alpha, \quad \delta H = -l\delta\alpha\sin\alpha$$

$$\delta D = l\delta\alpha\cos\alpha$$

Therefore

$$\delta w = Wl\delta\alpha\left[-\frac{3}{4}\sin\alpha + \frac{\mu_1\mu_2}{1+\mu_1\mu_2}\sin\alpha + \frac{\cos\alpha}{1+\mu_1\mu_2}\right]$$

$$\delta\bar{w} \equiv \frac{(1+\mu_1\mu_2)\delta w}{\mu_1 wl\cos\alpha} = \left[-\frac{(3-\mu_1\mu_2)}{4\mu_1}\tan\alpha + 1\right]\delta\alpha$$

At

$$\tan\alpha = \frac{4\mu_1}{3-\mu_1\mu_2}$$

the bracketed term vanishes. If $\tan\alpha$ exceeds that value, the term is negative and the work δw becomes negative. Then for positive $\delta\alpha$ (negative δh) the system fails; the climber falls. The foregoing analyses employ coefficients μ_1 and μ_2 for the static state. Any disturbance that might initiate movement, i.e., the climber shifts his weight, would invoke the lesser frictional resistance. In short, safety would dictate a lesser angle. Alternatively, a coworker might stand upon the lowest rung, increasing force F_1 and the frictional force $u_1 F_1$.

9.8 STABILITY OF EQUILIBRIUM: MINIMUM POTENTIAL

Let us now focus on the equilibrium, and stability, of conservative mechanical systems, i.e., internal and external forces derived from potentials. For example, the internal forces (stresses) within an elastic structure are conservative; the weight of any body is a conservative force derived from gravitational potential. *A conservative mechanical system is in a stable state if, and only if, the total potential is a relative minimum.*

To illustrate the instability of equilibrium, consider the rollers resting upon the curved surface of Figure 9.21a. Each is resting upon a smooth surface at a location with a horizontal tangent; each has the free-body diagram of Figure 9.21b; each is in equilibrium; $N = W$ (the weight). However, we know that the roller at A cannot stay in this position (the rollers and surface are smooth.) If, indeed, one could place the roller at A with the aid of a little friction, as light disturbance would cause it to roll down the hill. On the other hand, the roller in the valley C is stable; if it were displaced slightly to C', it would roll back. What distinguishes the unstable from the stable states?

The stable state C requires that we do *positive* work to move the roller along the surface to the adjacent position C'. The unstable state A requires no work to dislodge the roller; indeed, if we imagine that we move the roller from point A along the surface to the adjacent point A', then we do *negative* work. The state B on the plateau is distinguished by the fact that zero work is performed in transporting the roller to the adjacent position B'.

The simple system of Figure 9.21 is a conservative mechanical system. The roller is acted upon by the gravitational force W and the workless normal force N. Any work that we perform upon the roller (against the weight) as we transport it along the surface constitutes a change in potential. If we must increase the potential (location C), then the system is stable. If we decrease the potential (location A), then the system is unstable. If we do not change the potential (location B), then the system is in a state of neutral equilibrium. It is important to note that our test compares the potentials in the immediate neighborhood of the state. In reality, a mechanical system may be strictly stable, as is the roller at C, but a sufficiently large disturbance could move the system to another state of lower potential with a finite change of the configuration (i.e., kick the roller from location C to D).

Based upon our experiences, such as the example of Figure 9.21, we reassert the criterion for stability. *A conservative mechanical system is in a state of equilibrium if and only if the potential is a relative minimum.*

The principle of minimum potential is a cornerstone in the theory and practice of structural mechanics. The principle and certain related phenomena are illustrated by the following mechanical systems with one degree of freedom:

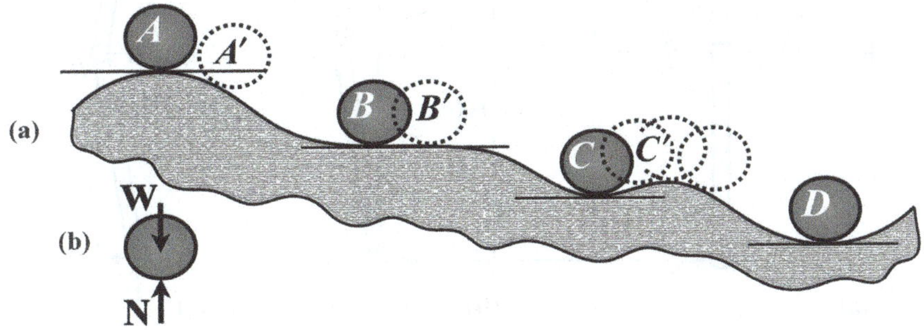

Figure 9.21 Stability at a Minimum Potential

ILLUSTRATIVE EXAMPLE 9.13

The bifurcation of equilibrated states, the buckling and postbuckled states of a column-like structure are exhibited by the mechanical system of Figure 9.22:

The two rigid links, *AB* and *BC*, are joined by a frictionless pin at *B* and constrained by the extensional spring and torsional spring; moduli are k and β, respectively. The system is symmetric about the vertical line, i.e., the lateral spring resists extension and compression, i.e. $F = kw$.

The torsional spring resists the relative rotation by a couple $C = \beta(2\theta)$. Pin *A* is constrained to move vertically while the frictionless pin *C* is fixed. Let us consider the equilibrium under the constant axial force *P*.

The system has one degree of freedom; the configuration is determined by the one variable θ. The internal energy V_I is contained in the springs:

$$V_I = \frac{kl^2}{2}\sin^2\theta + \frac{\beta}{2}(2\theta)^2$$

The potential of the load *P* is

$$V_E = 2Pl\cos\theta$$

The total potential of the system follows:

$$V = V_I + V_E = \frac{kl^2}{2}\sin^2\theta + \frac{\beta}{2}(2\theta)^2 + 2Pl\cos\theta \tag{9.72}$$

By the principle of stationary potential, the system is in equilibrium if

$$\delta V = \frac{\mathrm{d}V}{\mathrm{d}\theta}\delta\theta = (kl^2\sin\theta\cos\theta + 4\beta\theta - 2Pl\sin\theta)\delta\theta = 0 \tag{9.73}$$

i.e.

$$kl^2\sin\theta\cos\theta + 4\beta\theta - 2Pl\sin\theta = 0 \tag{9.74}$$

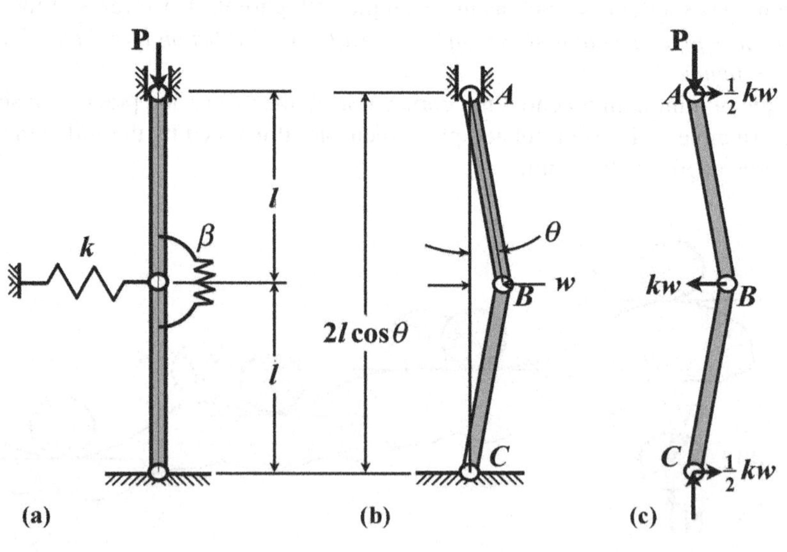

Figure 9.22 Elastic Mechanism

Evidently, $\theta = 0$ is an equilibrium state *for all P*. However, the state is not stable for all P.

Let us explore states for which $\theta \neq 0$: According to Equation (9.73), equilibrium states exist for $\theta \neq 0$, if

$$\frac{kl}{2}\cos\theta + \frac{2\beta}{l}\frac{\theta}{\sin\theta} = P \tag{9.75}$$

In the various plots of Figure 9.23, the ordinate is the dimensionless "load,"

$$p \equiv \frac{P}{P_{cr}}, \quad P_{cr} \equiv \frac{kl}{2} + \frac{2\beta}{l}$$

The equilibrium state $\theta = 0$ intersects the various branches $\theta \neq 0$, at the "bifurcation" point O ($p = 1$). The states $\theta = 0$, $p < 1$ are stable states. The stability or instability of the various branches can be ascertained via the principle of *minimum potential*.

Note that the trace for $kl^2 = 8\beta$ indicates 3 equilibrium states at points A ($\theta = 0$) and at points B and C at $p = 0.8$; the curve for $\beta = 0$ exhibits 2 equilibrium states at loads $p < 1$. Note too that these curves both exhibit negative slopes for $\theta > 0$, whereas the slope of curve OS ($k = 0$) is everywhere positive. All exhibit the zero slope ($\mathrm{d}p/\mathrm{d}\theta = 0$) at the bifurcation point. The differences hinge upon the second derivative, viz.

$$\begin{aligned}\left.\frac{\mathrm{d}^2 p}{\mathrm{d}\theta^2}\right|_{\theta=0} &= \left\{-\frac{kl}{2}\cos\theta + \frac{2\beta}{l}\left[\frac{\theta}{\sin\theta} + \frac{2(\theta\cos^2\theta - \cos\theta\sin\theta)}{\sin^3\theta}\right]\right\}_{\theta=0} \\ &= -\frac{kl}{2} + \frac{2\beta}{3l}\end{aligned} \tag{9.76}$$

We can only speculate that the states for which $kl/2 > 2\beta/3l$ are unstable at the critical load $P = P_{cr}$ ($p = 1$), i.e. at the bifurcation point. Moreover, we can also presume that all systems are unstable at $p > 1$ and $\theta = 0$. To confirm these presumptions, we consider the potentials.

The criteria of minimum potential requires that a small variation, i.e. $\delta\theta$ causes an increase in V, viz.

$$\delta V > 0 \Rightarrow \text{stability}$$

From Equation (9.72)

$$\begin{aligned}\Delta V &= \left.\frac{\mathrm{d}V}{\mathrm{d}\theta}\right|_0 \delta\theta + \left.\frac{\mathrm{d}^2 V}{\mathrm{d}\theta}\right|_0 \delta\theta^2 \\ &= \left(\frac{kl^2}{2} + 2\beta - Pl\right)\delta\theta^2 + \frac{1}{6}\left(\frac{Pl}{2} - kl^2\right)\delta\theta^4\end{aligned} \tag{9.77}$$

For small values $\delta\theta$, the initial term is dominant. The necessary condition for stability at $\theta = 0$ is $\mathrm{d}^2V/\mathrm{d}\theta^2 > 0$[8]:

$$P < \frac{kl}{2} + 2\frac{\beta}{l} = P_{cr}$$

All systems are stable at $\theta = 0$ for $P < P_{cr}$. At $P = P_{cr}$ the leading term of Equation (9.77) vanishes; the stability then hinges on the term $O(\delta\theta^4)$.[9] With $P = P_{cr}$ the system is stable if

$$\left(\frac{P_{cr}l}{2} - kl^2\right) = \beta - \frac{3}{4}kl^2 > 0 \tag{9.78}$$

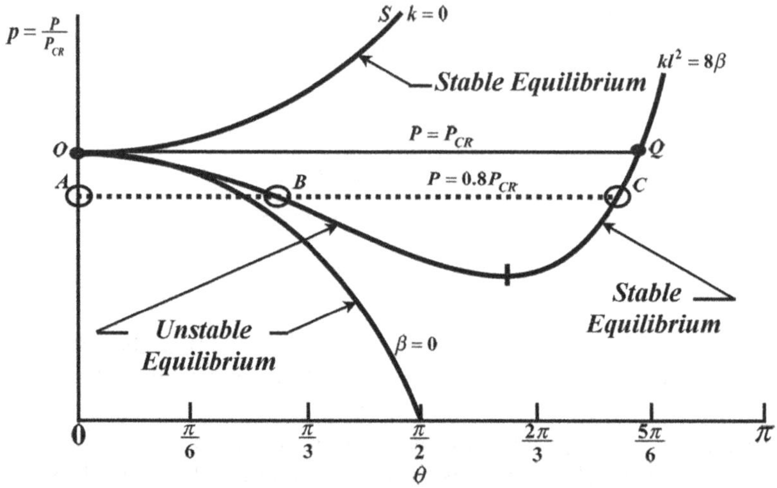

Figure 9.23

Note that this condition for stability at the critical load is also the criterion for positive slopes (see Equation 9.76) of these various branches of the equilibrium states of Figure 9.23.

The plot of load versus deflection for the case $kl^2 = 8\beta$ is redrawn in Figure 9.24 together with the potentials at $P = P_{cr}$ and at $P = 0.8P_{cr}$. At $P = 0.8P_{cr}$ the state A ($\theta = 0$) and the state C are stable, as indicated by the minimum potential at A' and C'. Of course, all points on the plot $OBCQ$ represent equilibrated states, but only those on the ascending position, e.g. C, are stable states. At the critical load P_{cr}, the state is unstable; a slight disturbance would cause the system to snap to the stable state at Q. This is indicated by the plot of the potential $O'Q'$.

Consider now the equilibrated state at A ($\theta = 0$, $P = 0.8P_{cr}$). Note that the "ball" resting in the shallow valley at A'. A sufficient disturbance would cause the "ball" to jump over the "hill" at B' into the valley at C'. The system is said to "snap-through" to the severely deformed state at Q. This form of instability, i.e. collapse, is in marked contrast to the buckling of systems wherein (see Equation 9.78)

$$\frac{\beta}{l} > \frac{3}{4}kl^2$$

The latter buckle at the critical load, e.g. branch as in Figure 9.23, as an elastic column, but can support slight additional load. Snap buckling, often total collapse, can occur in thin shells (see Section 9.9).

We must note another important distinction between structural systems, viz. those which are stable at the critical load, e.g. $\beta/L > 3/4\, kl^2$, and those which are unstable at the critical load, e.g. $\beta/L < 3/4\, kl^2$. An important practical difference is the effects of imperfections; these may be structural defects, such as deformities, e.g. the misalignment of the links in Figure 9.22. An irregularity in loading can also cause premature deformations, such as a slight eccentricity of the axial load upon a column. To appreciate these effects, consider the load/deflection plots of our example, as redrawn in Figure 9.25. Ideally the loading upon the perfect system traces the line OA to the critical load; postbuckled curves trace the solid lines. An imperfect system exhibits some deformations from the onset of load. These deformations grow dramatically as the load approaches the critical value; the realistic plots are depicted by the dotted lines. The system, which is unstable

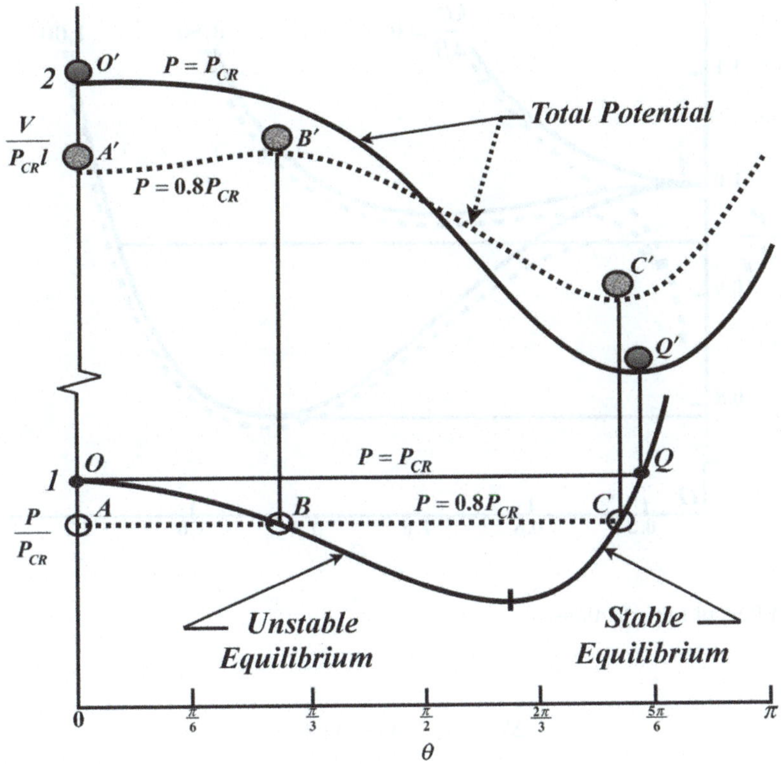

Figure 9.24

at the critical load, never attains the "critical" load; indeed, snap buckling (collapse) often occurs at loads much less than the critical value.

9.9 ENERGY CRITERIA OF TREFFTZ AND KOITER

The Illustrative Example 9.13 serves as a simple prototype that exhibits the most common features of structural instabilities. Like most structures, the ideal system undergoes imperceptible deformations prior to buckling. The pre-buckled states, e.g. the straight column, are states of minimum potential. In our example, the potential change has the form of Equation (9.77):

$$\Delta V = A_1(\delta\theta) + A_2(\delta\theta^2) + A_4(\delta\theta^4) + \cdots$$

Note: that the expression contains no cubic term. This is often the case with symmetric structures. Subsequently, we consider the consequences of a cubic term.

A typical structure possesses many degrees of freedom. Then an expansion of the energy change assumes the form:

$$\Delta V = \sum_{0}^{n} A_i W_i + \sum_{i}^{n}\sum_{j}^{n} A_{ij} W_i W_j + \sum_{i}^{n}\sum_{j}^{n}\sum_{k}^{n}\sum_{l}^{n} A_{ijkl} W_i W_j W_k W_l$$

where W_n denotes any of n generalized coordinates. In general,

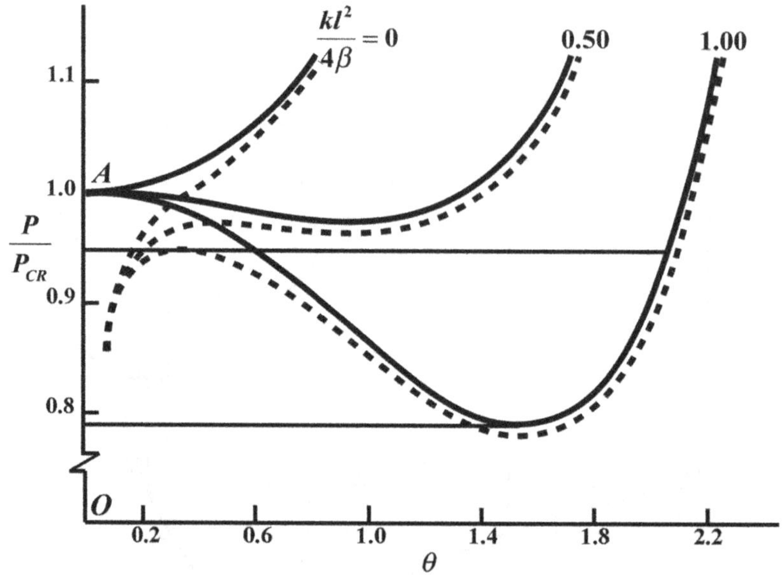

Figure 9.25 Effect of Imperfections

$$\Delta V = A_1 + A_2 + A_4 + \cdots$$

Vanishing of the lead term, $A_1 = 0$ is the stationary criterion which signals an equilibrium state. Then the dominant term is the term A_2, i.e. the term of the second order. That state is stable if $A_2 > 0$, unstable if $A_2 < 0$. The critical state is $A_2 = 0$; this is the Trefftz criterion for instability of the conservative system. More than 70 years ago, many recognized the marked difference between the buckling of different structures. Near a predicable critical load, a column buckles. Strictly speaking, the elastic column is in a stable post-buckled state ($A_4 > 0$). In most cases, this does constitute the failure, since the column can sustain little additional load. That is the type of buckling exhibited by our prototype of Figures 9.22 and 9.23 with $k = 0$.

A very different behavior is apparent in thin shells under compressive loads. W.T. Koiter (1945) focused on the cylindrical shell under axial compression and the energy conditions at the critical state ($A_2 = 0$). Of course, his theses[10] addressed the cylinder as a continuous body: Equilibrium is governed by different equations, and the energy is embedded in the integrals of continuous functions. The mathematics differs, but underlying concepts remain. He noted the role of the higher order terms. Most importantly, he noted that the negative changes ($A_3 = 0$ or $A_4 < 0$) signaled an unstable post-buckled state and, in most cases, abrupt collapse. That is the condition in our prototype of Figure 9.22 when $3\kappa l^2 > 4\beta$.

ILLUSTRATIVE EXAMPLE 9.14

A thin cylindrical shell under axial compression exhibits the response depicted in Figure 9.26. The ideal shell, i.e. the perfect cylinder, is remarkably stiff. The ideal Hookean shell traces the linear load-deflection curve *OB*. To the critical load (P_{cr}) the equilibrium states are stable. Above that load, the states are unstable ($A_2 < 0$). At the critical load, the equilibrium is unstable ($A_4 < 0$). The postbuckling states of that ideal shell would trace the descending dotted path $A'CD$; these

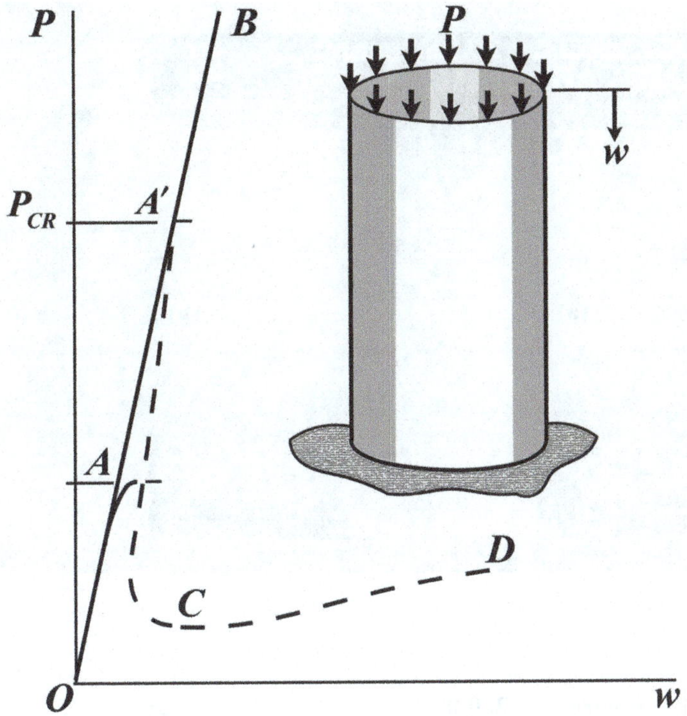

Figure 9.26

are unstable to the minimum at C. Of course, the real imperfect shell will buckle at the much lower load (A). Subsequent stable states (CD) are not realized; a metal shell will suffer inelastic deformations.[11]

To provide additional insights, we cite the circumstances depicted in Figure 9.27. Here, the very thin cylinder of Figure 9.27a is rolled from a sheet of drafting paper. That thin cylinder supports the tome by Brush and Almroth. The Figure 9.27b contributes to our appreciation of the very different pre- and post-buckling responses: This figure depicts a Yoshimura][12] model constructed from the same drafting paper. The model shows us that the thin shell can be deformed without extensional strains via a pattern of folds. The bending resistance of the thin wall is very minimal; hence, we must expect much less load-bearing ability when the shell buckles. Note, however, that the actual buckled configurations may differ from the Yoshimura model.

Finally, we cite results of a simple experiment in which we imposed axial loading upon typical thin-walled beverage cans. Our only control over the specimens was a visual inspection to eliminate obvious irregularities, dents, or other damages. Figure 9.28 depicts the load-deflection traces for six typical cans. Tests were performed on 17 cans: the average buckling load was 250 lb; the average post-buckled load was 113 lb.

Since there were no tight controls on the specimen, nor end conditions, we give no "theoretical" values for the "critical load." This experiment does suggest a cautionary approach to predicting the buckling of thin shells under external loadings.

Note too that the actual buckling load depends on any irregularities in form and/or loading. Near the buckling load, a slight disturbance can precipitate collapse.

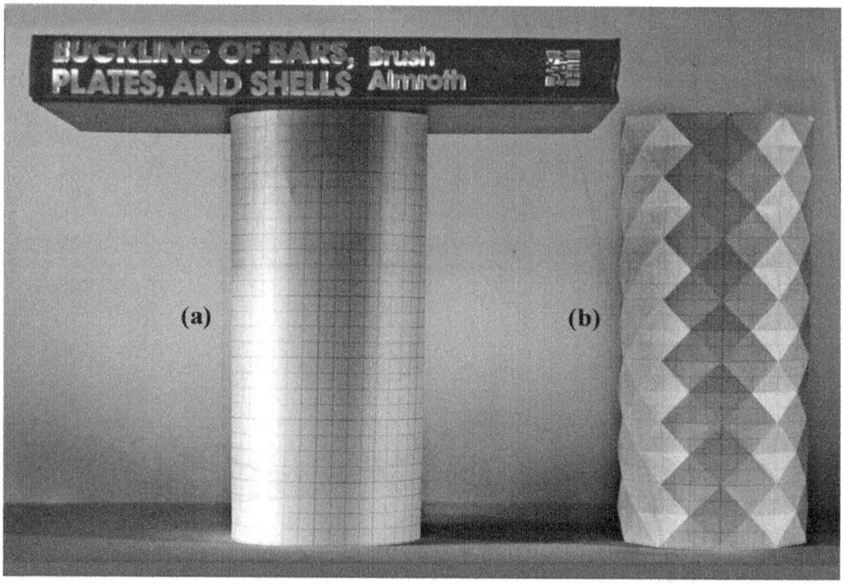

Figure 9.27

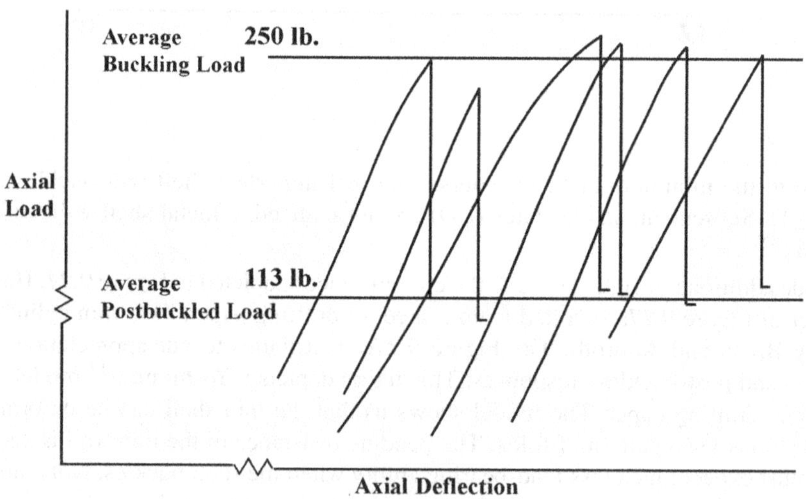

Figure 9.28

The two cans in Figure 9.29 are typical buckled specimens; note that the inelastic deformations are immediate.

ILLUSTRATIVE EXAMPLE 9.15

The mechanical system in Figure 9.30 consists of three rigid links, *OA*, *ABC*, and *CQ*. These are pinned at *A* and *C*; and to fixed pins at *O* and *Q*. Torsional springs (constant *k*) resist the relative

Figure 9.29

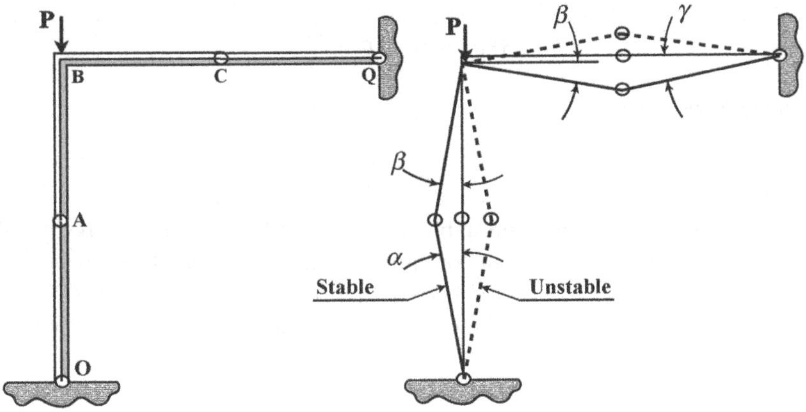

Figure 9.30

rotations at A and C. Segments OA, AB, BC, and CQ have the same lengths l.

This is a mechanism with one degree of freedom: Any one of the angles, α, β, or γ, determines the other two. Two nonlinear constraints apply:

1. The horizontal distance between O and Q is fixed ($2l$), viz.

$$-\sin\alpha + \sin\beta + \cos\beta + \cos\gamma = 2 \tag{9.79}$$

2. The vertical distance between O and Q is fixed ($2l$), viz.

$$\cos\alpha + \cos\beta - \sin\beta + \sin\gamma = 2 \tag{9.80}$$

Let us examine the stability of the system under the constant vertical load P acting at joint B: The potential of load P is

$$U_p = lP(\cos\alpha + \cos\beta) \\ \approx lP(2 - \frac{1}{2}\alpha^2 - \frac{1}{2}\beta^2 + \cdots) \tag{9.81}$$

The potential of the springs is

$$U_S = \frac{1}{2}k(\alpha+\beta)^2 + \frac{1}{2}k(\beta+\gamma)^2 \tag{9.82}$$

To ascertain the critical load, we need only examine a first-order departure from the initial (unbuckled) state ($\alpha = \beta = \gamma = 0$). Then Equations (9.79) and (9.80) provide the approximations:

$$\alpha \approx \beta \approx \gamma$$

The second-order approximation of the total potential follows:

$$\delta^2 V_P = \delta^2 U_P + \delta^2 U_S \\ = -lP\alpha^2 + 4k\alpha^2$$

The Trefftz criterion for stability of the initial state requires

$$\delta^2 V_P > 0, \quad P < \frac{4k}{l}$$

Hence the critical value,

$$P_{CR} = \frac{4k}{l} \tag{9.83}$$

The next important question concerns the stability of the post-buckled states. Specifically, does the system behave as a column, i.e. buckle but continue to support the load? Does the system collapse, as the system of Figure 9.22 with $\beta = 0$ or as the cylindrical shell of Figure 9.26?

Any post-buckled state must satisfy the equilibrium equations. These derive via the principle of stationary potential, viz.

$$\delta(U_P + U_S) = 0$$

We take the view that β and γ are dependent on α in accordance with the constraints (9.79) and (9.80). Then the stationary condition follows.

$$-lP\left(\sin\alpha + \frac{d\beta}{d\alpha}\sin\beta\right) + k(\alpha+\beta)\left(1 + \frac{d\beta}{d\alpha}\right) + k(\beta+\gamma)\left(\frac{d\beta}{d\alpha} + \frac{d\gamma}{d\alpha}\right) = 0 \tag{9.84}$$

From the constraints (9.79) and (9.80), we have

$$-\cos\alpha + \frac{d\beta}{d\alpha}(\cos\beta - \sin\beta) - \frac{d\gamma}{d\alpha}\sin\gamma = 0 \\ -\sin\alpha - \frac{d\beta}{d\alpha}(\sin\beta - \cos\beta) + \frac{d\gamma}{d\alpha}\cos\gamma = 0$$

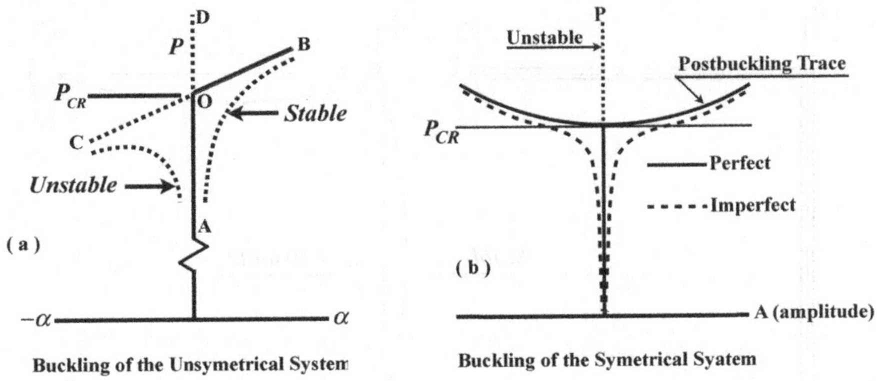

Figure 9.31

At the *small* initial step $\alpha \approx \beta \approx \gamma$, we have the approximations

$$\cos\alpha \approx \cos\beta \approx \cos\gamma \approx 1, \quad \sin\alpha \approx \sin\beta \approx \sin\gamma \approx \alpha$$

$$(1-\alpha)\frac{d\beta}{d\alpha} - \alpha\frac{d\gamma}{d\alpha} = 1 \tag{9.85a}$$

$$-(1+\alpha)\frac{d\beta}{d\alpha} + \frac{d\gamma}{d\alpha} = \alpha \tag{9.85}$$

The solution of the latter follows:

$$\frac{d\beta}{d\alpha} = \frac{1+\alpha^2}{1-2\alpha-\alpha^2}, \quad \frac{d\gamma}{d\alpha} = \frac{1+2\alpha-\alpha^2}{1-2\alpha-\alpha^3}$$

Substituting these into the equilibrium Equation (9.84), and noting the small rotations, i.e. $\beta \approx \gamma \approx \alpha$, we obtain

$$P = \frac{4k}{l(1-\alpha)} \tag{9.86}$$

If $\alpha < 0$, $P < P_{CR}$; the post-buckled state is unstable. If $\alpha > 0$, $P > P_{CR}$, the post-buckled state is stable.

Equation (9.86) provides the initial slope of the post-buckled path, i.e. line *COB* in Figure 9.31a. If the mechanism buckles in direction $-\alpha$, the states are unstable; in direction $+\alpha$, the states are stable. The perfect system is stable on path *OA* ($\alpha = 0$). The post-buckled states are stable in direction *OB*, unstable in direction *OC*. Of course, the real imperfect system exhibits the deflection from the onset of loading. Depending on the irregularities, misalignments of eccentricities, the system can buckle in either direction. In the one case ($-\alpha$) the buckling signals immediate collapse; in the other case ($+\alpha$), the system incurs large deflections but sustains the load ($P > P_{CR}$). The unstable states trace the dotted paths of Figure 9.31b; the stable states follow the solid lines. Figure 9.31b depicts the corresponding paths for the symmetric system; for example, a column might buckle in either direction depending on initial imperfections.

Our simple mechanism exhibits a behavior much like the frame of Figure 9.32. This was initially studied experimentally and reported by Roorda-5pc.[13] Subsequently, Koiter[14] provided the

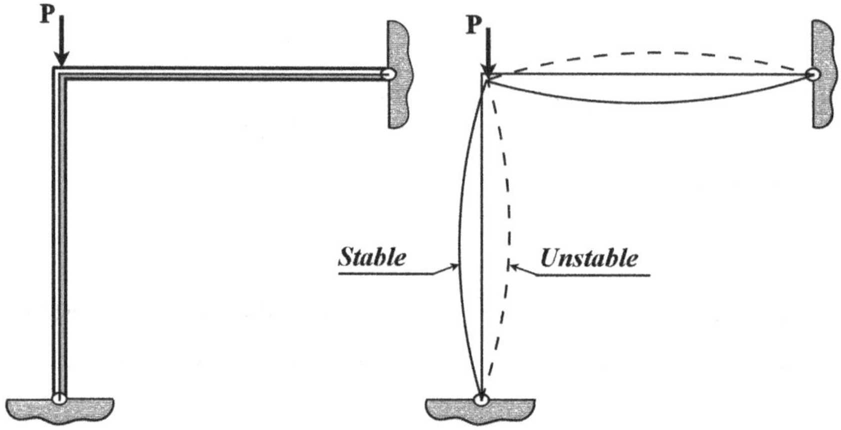

Figure 9.32

theoretical analysis. As reported by Brush and Almroth,[15], the incremental change of potential at the onset of buckling contains a cubic term in one parameter (β). Hence, the negative value ($A_3\beta^3 < 0$) signals the unstable state, whereas a positive value ($A_3\beta^3 > 0$) signals the stable state.

ILLUSTRATIVE EXAMPLE 9.16

To fully illustrate the energetic approach to stability, let us consider the HI-HO column depicted in Figure 9.33a It is simply supported at both ends and by a linear elastic spring (modulus K) at midspan. The spring could be a transverse Hookean beam. If that support were absent, i.e. $K = 0$, the column would buckle in the mode of Figure 9.33b, i.e. $w = A\sin\pi x/l$. If the spring were stiff enough, the column would buckle in the higher mode of Figure 9.33c. One can pose the question: Under what circumstances does one observe the transition from the mode at (a) to that at (b) and the critical load from $P_{CR} = \pi^2 EI/l^2$ to $P_{CR} = 4\pi^2 EI/l^2$? Also, by the criteria of Trefftz and Koiter, is the supported column stable at the critical load, i.e. does it exhibit snap-through (collapse) at the critical load?

Prior to buckling, the column is presumed straight under the axial load P and extensional strain $\varepsilon_0 = P/EA$. To apply the energy criterion, we require the potential change when the column experiences small lateral deflection $w(x)$, i.e. buckles.

Without the lateral spring, the column would buckle in the mode (b) at $P = P_{cr} = EI\pi^2/l^2$. For sufficiently small value of K, the column will buckle in a similar mode 1 of Figure 9.33a; the configuration (w) is symmetric about the midpoint A. Segment AB is a reflection of AO. It suffices to examine the potential of the OA as seen in Figure 9.33c subject to the geometric constraints:

$$\left(\frac{dw}{dx}\right)_{l/2} = 0, \quad \text{at } A; \quad w|_0 = 0 \quad \text{at } O \tag{9.87a,b}$$

A small segment Δx of the buckled column undergoes the small rotation $\theta \approx dw/dx$. The axial displacement of that segment is

$$\Delta u = (1 - \cos\theta)\Delta X \approx \frac{1}{2}\left(\frac{dw}{dx}\right)^2 dx$$

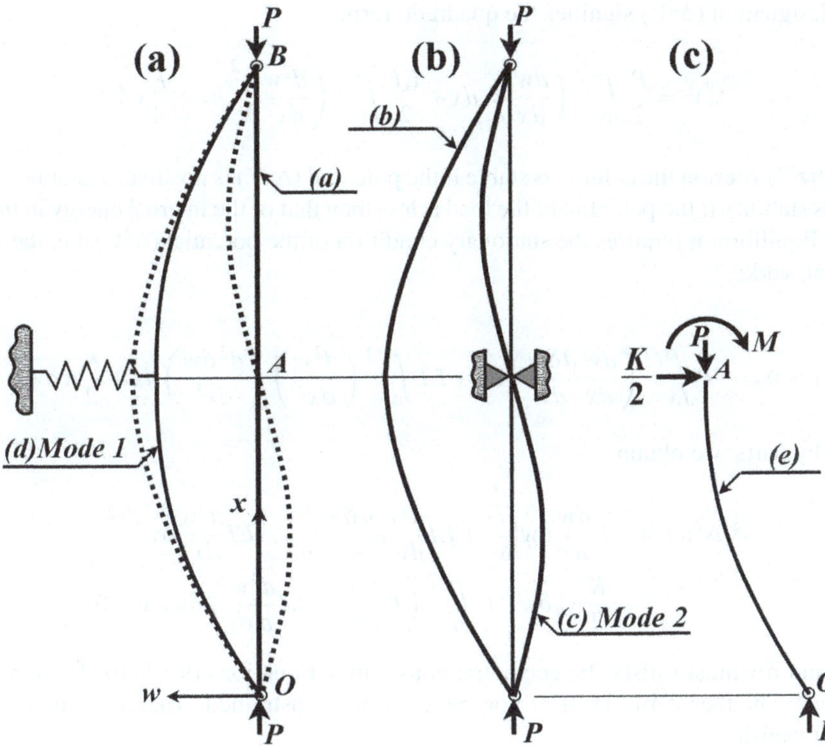

Figure 9.33 Column with Midspan Support

The axial displacement at O relative to A is the integral:

$$u_0 = \int_0^{l/2} \frac{1}{2} \left(\frac{dw}{dx} \right)^2 dx \tag{9.88}$$

The corresponding decrease in the potential of the force P at O follows:

$$\Delta V_E = -\frac{P}{2} \int_0^{l/2} \left(\frac{dw}{dx} \right)^2 dx \tag{9.89}$$

the lateral deflection w cases flexure ($\kappa \approx d^2w/dx^2$) and the attendant internal energy:

$$\Delta V_I = \frac{EI}{2} \int_0^{l/2} \left(\frac{dw}{dx} \right)^2 dx \tag{9.90}$$

Also, the lateral spring experiences the internal energy:

$$\Delta V_s = \frac{K}{4} w_A^2 \tag{9.91}$$

Altogether the potential change is the sum:

$$\Delta V = \Delta V_E + \Delta V_I + \Delta V_s \equiv \Delta^2 V \tag{9.92}$$

The latter designation ($\Delta^2 V$) signifies the quadratic form:

$$\Delta^2 V = \frac{P}{2}\int_0^{l/2}\left(\frac{dw}{dx}\right)^2 dx + \frac{EI}{2}\int_0^{l/2}\left(\frac{d^2w}{dx^2}\right)^2 dx + \frac{K}{4}w_A^2 \tag{9.93}$$

By the Trefftz[16] criterion the column is stable if the potential ($\Delta^2 V$) is positive, unstable is negative. This implies stability if the potential of the load is less than that of the internal energy in the column and spring. Equilibrium requires the stationary condition of the potential ($\Delta^2 V$), i.e. the vanishing of the virtual work:

$$\delta(\Delta^2 V) = 0 = -P\int_0^{l/2}\left(\frac{dw}{dx}\frac{d\delta w}{dx}\right)dx + EI\int_0^{l/2}\left(\frac{d^2w}{dx^2}\right)\left(\frac{d^2\delta w}{dx^2}\right)dx + \frac{K}{2}w\delta w_A \tag{9.94}$$

Integrating by parts, we obtain

$$\begin{aligned}\delta(\Delta^2 w) = &-P\frac{dw}{dx}\delta w\Big|_0^{l/2} + EI\frac{d^2w}{dw^2}\frac{d\delta w}{dx}\Big|_0^{l/2} - EI\frac{d^3w}{dx^3}\delta w\Big|_0^{l/2}\\ &+\frac{K}{2}w_A\delta w_A + \int_0^{l/2}\left(P\frac{d^2w}{dx^2} + EI\frac{d^4w}{dx^4}\right)\delta w\, dx = 0\end{aligned} \tag{9.95}$$

Both w and δw must satisfy the geometric constraints. Equations (9.87a,b). There remain two end conditions: neither $d\delta w/dx$ at O nor δw at A are constrained. Therefore, their respective factors must vanish:

$$EI\frac{d^2w}{dx^2} = 0, \text{ at } O, \quad -EI\frac{d^3w}{dx^3} + \frac{K}{2}w_A = 0, \text{ at } A \tag{9.96a,b}$$

Equation (9.96b) acknowledges the shear force imposed by the spring at end A in Figure 9.33c. Equation (9.96a) asserts the absence of a moment at the simple support O.

Since δw is arbitrary throughout $(0 < x, l/2)$, the integral in Equation (9.95) vanishes if and only if,

$$\frac{d^4w}{dx^4} + \frac{P}{EI}\frac{d^2w}{dx^2} \tag{9.97}$$

We recognize Equation (9.97) as Equation (6.73). In the present case, we require the solution subject to the geometric constraints (Equations 9.96a,b). The solution follows:

$$w(x) = A\left(\sin\frac{\lambda x}{l} - \frac{\lambda x}{l}\cos\frac{\lambda}{2}\right) \tag{9.98}$$

Equation (9.96b) requires that

$$\frac{Kl^3}{EI} = \frac{4\lambda^3}{\lambda - 2\tan(\lambda/2)} \tag{9.99}$$

where

$$\left(\frac{\lambda}{l}\right)^2 = \frac{P_{cr}}{EI} \tag{9.100}$$

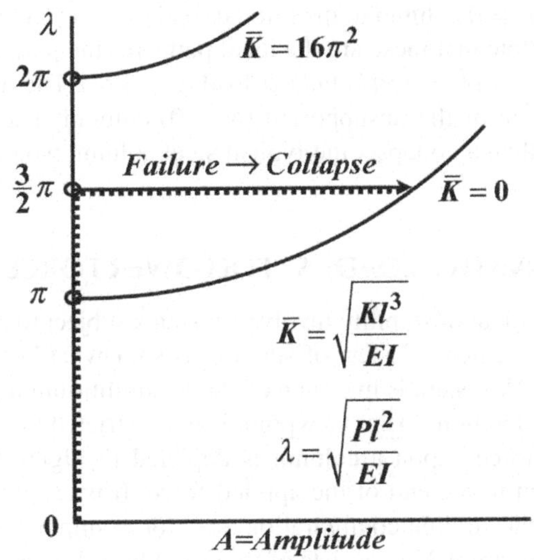

Figure 9.34 Buckling of a Symmetrical Column

Substitution of the solution, Equation (9.98) and (9.99) into (9.93) serves to affirm the Trefftz criteria; viz.

$$\Delta^2 V < 0,\ \ P < P_{cr} \Rightarrow \text{stability}$$
$$\Delta^2 V = 0,\ \ P = P_{cr}$$
$$\Delta^2 V > 0,\ \ P > P_{cr} \Rightarrow \text{instability}$$

At the critical load, the second terms ($\Delta^2 V$) vanish, but higher order terms are positive. Therefore, the postbuckled configurations are stable in accordance with the Koiter Criterion.

The Equation (9.99) is a nonlinear equation in (λ). At $\lambda = 2\pi$, we obtain

$$\bar{K} \equiv \frac{Kl^3}{EI} = 16\pi^2$$

This corresponds to mode 2 of Figure 9.33c. Larger values of the factor $\bar{K}$ are irrelevant, since mode 2 has a node at the site of the spring ($w_A = 0$). Values of $(\lambda/\pi)^2$ are given for various K by Brush and Almroth.[17]

As the stiffness of the spring increases, the potential of the buckled mode 1 approaches that of the buckled mode 2. Near the load $P_{CR} = 4\pi^2 EI/l^2$, as $\bar{K}$ approaches $16\pi^2$, the *real* column could buckle in either mode depending on imperfections, i.e. initial crookedness.

At the critical load, the terms $\delta^1 V$ and $\delta^2 V$ vanish. The remaining term of Equation (9.91c), the integral, is clearly positive; therefore, the postbuckled configurations are stable in accordance with the Koiter criterion.

Let us consider the consequences if the midspan support should fail. We suppose that the column is prone to buckle in mode 2. We can only speculate because theoretically mode 2 is possible with little midspan support. Let us assume the stiffness of $\bar{K} = 16\pi^2$, $\lambda = 2\pi$. The path of loading to the theoretical load ($\lambda = 2\pi$)) and postbuckled states is shown in Figure 9.34. Also shown is the path for the unsupported ($\bar{K} = 0$ column to the critical load ($\lambda = \pi$) and the subsequent postbuckled states of equilibrium. Note that these are the ideal paths for the perfect columns and loadings. Now suppose that the support ($\bar{K} = 16\pi^2$) fails at load $\lambda = 3\pi/2$.The unsupported column must snap to the equilibrium state of the unsupported ($\bar{K} = 0$) column. The amplitude A (deflection w) would be nothing less than a collapse; the typical steel column would undergo permanent, i.e plastic, deformation.

9.10 A NONCONSERVATIVE LOAD; A "FOLLOWER FORCE"

Most of the preceding treatment of stability involved systems subject to conservative loads, forces derived from a potential. Our general view of stability is somewhat broader: In our introduction (Section 9.7), we said that "A system is in a state of stable equilibrium if positive work is required to cause an unacceptable alteration." That viewpoint is not restricted to conservative systems.

A so-called "follower force" upon a column is depicted in Figure 9.35a where the applied force P always acts tangent to the end of the applied force. It is easily established that the only equilibrium state is the initial straight configuration (see, for example, Figure 9.35b).

The simple system of Figure 9.35c consists of two rigid bars, hinged at the joint and the base. Both hinges are supported by linear springs (modulus k) which resist relative rotations. This latter system is reappraised in Figure 9.36.

Both systems have been treated from a dynamic viewpoint. In brief, a mass distribution is assigned to the components, and the oscillatory motions are then analyzed. The critical load is that which causes the amplitude to become unbounded. The case of the continuous column (a) was investigated by Beck[18]; the two-bar linkage (c) by Ziegler.[19]

ILLUSTRATIVE EXAMPLE 9.17

The two-bar linkage of Figure 9.35c is redrawn in Figure 9.36. Here, we show a deformed state wherein the straight linkage OC assumes the position $OD'A'$ at (a) and at (c), respectively. Our sole

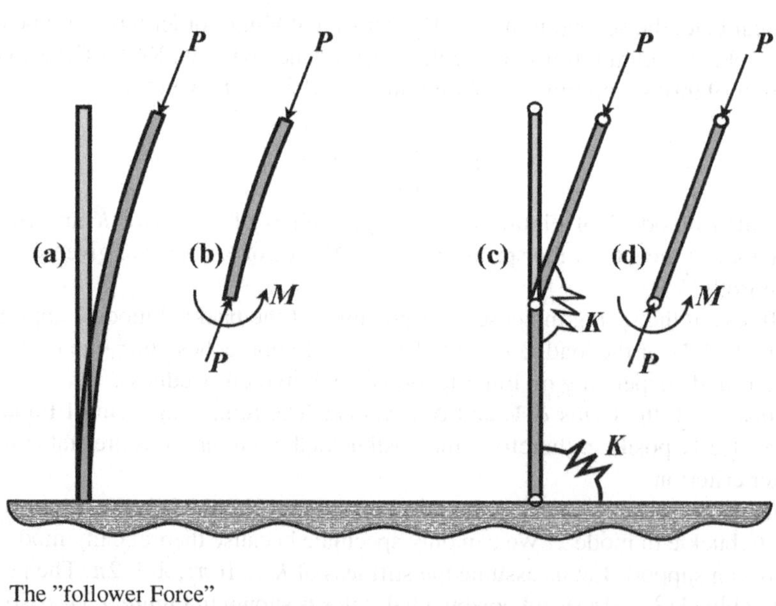

Figure 9.35 The "follower Force"

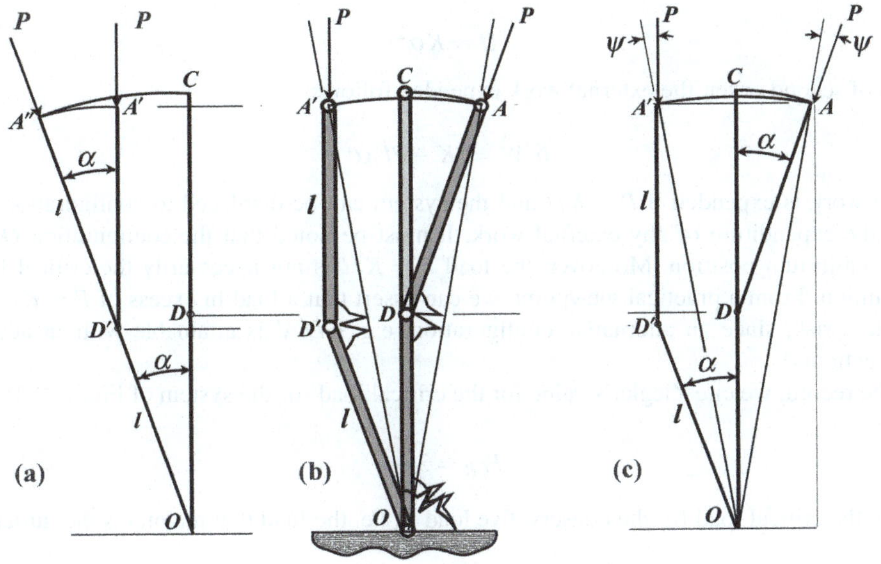

Figure 9.36

object is to demonstrate that at a load $P > Kl$, we can arrive at the state $OD'A'$ without expending any work.

If we were to rigidly rotate OA to the position OA'' no work is performed by the load P. Since the line of action would remain along the common axis OA of both bars. The force P would act perpendicular to the arc CA'' of Figure 9.36a. In a subsequent rotation of $D'A''$ to the position $D'A'$ force P again performs no work. To achieve the final position $OD'A'$ external work must be expended against the resistance of both torsional springs.

Now suppose that the position $OD'A'$ is achieved via an alternate path. In Figure 9.36c we perceive an initial rotation of DC to the position DA; no work is done by force P as it acts normal to the arc CA. Subsequently, we perceive a rigid rotation of the assembly ODA about O; in this motion, the point of application traverses the arc AA'. Force P acts at the angle ψ with the normal and, therefore, does positive work. After the necessary geometric and algebraic exercise, we obtain the work done by the "follower force" P:

$$OA = l\sqrt{1+\cos\alpha}, \quad \psi = \alpha - \arctan\left(\frac{\sin\alpha}{1+\cos\alpha}\right)$$

Note: For convenience, we have taken equal rotations α. If we consider only a small perturbation, i.e. $\alpha \ll 1$, then

$$OA \approx 2l\left(1-\frac{\alpha^2}{8}\right), \quad \psi \approx \frac{\alpha}{2}$$

The work performed by the force P follows:

$$W_P = OA \times \alpha P \sin\psi, \quad \delta^2 W = Pl\alpha^2$$

The internal work against the two springs is

$$U = K\alpha^2$$

In terms of second order, the external work expended follows:

$$\delta^2 W = (K - Pl)\alpha^2$$

Negative work is expended if $P > K/l$ and the system can be displaced to configuration $OD'A'$ without the expenditure of any external work. It must be noted that the configuration $OD'A'$ is *not* an equilibrium position. Moreover, the load $P = K/l$ is not necessarily the critical load by any definition. From a practical viewpoint, we can assert that a load in excess of $P = K/l$ would constitute a risk, since an alternative configuration, e.g. $OD'A'$ is attainable without additional external action(s).

For the record, we cite Ziegler's value for the critical load for the system of Figure 9.35c:

$$P_{CR} = 3\frac{K}{l}$$

Note that the critical load for the conservative load P, i.e. the load that maintains the initial direction is

$$P_{CR} = 0.382\frac{K}{l}$$

Also, a critical load for the column of Figure 9.35a is given by Timoshenko and Gere[20]

$$P_{CR} = \frac{2.0\pi^2 EI}{l^2}$$

whereas the value for the conservative, i.e. "dead" load is

$$P_{CR} = 0.25\frac{\pi^2 EI}{l^2}$$

In both cases, these results indicate that the systems are less likely to become unstable under the 'follower' force.

9.11 STABILITY; SOURCES FOR ADDITIONAL INFORMATION

Our treatment of instability is necessarily limited. The interested reader can find much additional information in the following sources:

B. Budiansky. *Buckling of Structures: IUTAM Symposium, Cambridge, USA*. Springer, Berlin Heidelberg, 1974.

J. Arbocz and T. Weller. *Buckling Experiments: Buckling of Thin-Walled Structures, Singer.* Wiley, Hoboken, vol. 1 and vol. 2, 2007 and 2008.

NOTES

1. I. Newton. *Philosophiae naturalis principia mathematica*, Londini, 1687.
2. J. L. Lagrange. *Méchanique analytique*. Paris: Veuve Desaint, 1788. For additional historic remarks, see C. Lanczos. *The Variational Principles of Mechanics*. University of Toronto Press, Toronto, 1949; H. L. Langhaar. *Energy Methods in Applied Mechanics*. John Wily, New York, 1962; G. A. E. Oravas and L.

McLean. Historical development of energetical principles in elastomechanics. *Appl. Mech. Rev.*, 19(8,II): 919–933, 1966.
3. Others might give a definition as the work done by the imposed forces. Our definition facilitates the subsequent treatment of conservative systems and changes in potential. More specifically, our definition is such that the virtual work on a conservative mechanical system *is* the variation of its potential.
4. This example is instructive but not essential to subsequent developments.
5. More general versions of the complementary potential are applicable to finite deformations. See B. Fraeijs de Veubeke, A new variational principle for finite elastic displacements. *Int. J. Eng. Sci.*, 10(9): 745–763, 1972; G. A. Wempner. The complementary potentials of elasticity, extremal properties, and associated functionals. *J. Appl. Mech.*, 59(3): 568–571, 1992.
6. H. L. Langhaar. *Energy Methods of Applied Mechanics*. John Wiley, Hoboken, 1962.
7. See Section 1.2
8. E. Trefftz. Zur Theorie der Stabilität des Elastichen Gleichgewichts. *Zeit. Angew. Math. Mech.*, 1933.
9. W. T. Koiter. *"Over der Stabiliteit" van Elastisch Evenwicht*, Thesis, University of Delft, H. J. Paris, 1945.
10. W. T. Koiter. *On the Stability of Elastic Equilibrium*, Thesis (in Dutch), Delft Inst. Tech., 1945.
11. P. O. Brush and O. Almroth Bo. *Buckling of Bars, Plates and Shells*, p. 332. McGraw-Hill, Columbus, 1975.
12. Y. Yoshimmura, On the Mechanism of Buckling of a Circular Cylindrical Shell under axial Compression, NACA-TM1390, 1955.
13. J. Roorda. Stability of structures with small imperfections. *Journal of the Engineering Mechanics Division*, 91(1): 87–106, 1965.
14. Koiter, W. T. Postbuckling analysis of a simple two-bar frame. In B. Broberg, J. Hult, & F. Niordson (Eds.), *Recent Progress in Applied Mechanics: The Folke Odqvist* Volume (pp. 337–354). Almqvist & Wiksell, 1966.
15. D. O. Brush and B. O. Almroth. *Buckling of Bars, Plates and Shells*. McGraw-Hill, New York, 1975.
16. E. Trefftz. Zur Theorie der Stabilität des elastischen Gleichgewichts. *Zeitschrift für Angewandte Mathematik und Mechanik (ZAMM)*, 13(2): 160–165, 1933.
17. D. O. Brush, and B. O. Almroth. *Buckling of Bars, Plates, and Shells*. McGraw-Hill, New York, 1975.
18. M. Beck. Die Knicklast des einseitig eingespannten, tangential gedrückten Stabes. *Zeitschrift für Angewandte Mathematik und Physik (ZAMP)*, 3(3): 225-–228, 1952.
19. H. Ziegler. Die Stabilit´atskriterien der Elastomechanik. *Ingen. Arch.*, 20: 49–56, 1952.
20. S. P. Timoshenko and J. M. Gere. *Theory of Elastic Stability*. Dover, New York, 2009.

10 Discrete Approximation via "Finite Elements"

10.1 INTRODUCTION

In the initial chapters, we noted that the concept of a continuous medium was merely a mathematical model of a material, which is composed of a discrete atomic structure. It has proven to be an effective model, but not without significant limitations. The continuous body is described via differential equations and boundary conditions; often, those conditions defy solutions. Indeed, the annals of mechanics reveal few solutions for bodies of realistic mechanical and structural forms. Prerequisites to the solution are regular coordinate systems and boundaries, which are amenable to the mathematical solutions. One can only imagine an attempt to obtain such a solution for the connecting rod of an engine; one notes but two simple surfaces viz., the cylindrical bearings. For actual bodies, the engineer has always sought effective schemes of approximation. It is the subject of this chapter.

One must not denigrate the theories of the continuous body, since important results have emerged: We cite the models of bending and torsion, particularly Saint-Venant's theory of torsion for an arbitrary cross-section. Of great importance are solutions that identify stress concentrations and the singularities at the root of a notch. These latter results are important to our understanding of fracture (Sections 4.19 and 4.20). Still, most of the precise solutions are given for the simplest model, the Hookean isotropic material. Finally, one must acknowledge that the concepts of a continuous medium, e.g., strain and stress, and related theories provide the foundations for contemporary methods of approximation. Here, we focus our attention on the concepts and methods of 'finite elements.'

In our introduction to the concepts/methods of finite elements, we would be remiss without a few historical comments. Firstly, we note that the notion of discrete approximations to the continuum model is not recent, nor is it profound. For example, just retracing the differential dy/dx to its origins, i.e., $\Delta y/\Delta x$, it is evident that one might simply suppress the limit $(\Delta x, \Delta y \nrightarrow 0)$ and admit small *discrete* increments of dimensions Δy and Δx. One has thereby approximated the differential equations by difference equations, algebraic equations, which govern discrete values of the continuous function $y(x)$. In any practical situation, a discrete approximation leads to very large systems of algebraic equations. That also presented a great challenge prior to the developments of the electronic digital computers. In short, we are much indebted to those who discovered the semiconductive properties of certain solids and to those who devised the means to etch micro-circuits within silicon chips. Of course, as we shall see, classical mechanics and mathematics are essential to convert the continuous model to the discrete. The important tools are the principles of work and energy as described, albeit briefly, in Chapter 9.

Here, we make no pretenses about our introduction to finite elements. Many books[1] are devoted to the various alternative means of devising elements and the procedures for treating the assemblies. Our intent is a lucid description of the basic concepts to effect the discrete approximations of the "continuous" models of structural and mechanical components.

DOI: 10.1201/9781003665861-10

10.2 CONCEPT OF THE FINITE ELEMENT

Although there are various means to approximate continuous functions by discrete values, e.g., finite differences, series, etc., the method of finite elements is most widely employed to model solid bodies. It is particularly appealing to the engineer because it is readily amenable to physical/geometrical interpretations. Therefore, it is the sole focus of our attention.

The body, structural or mechanical component, is subdivided by a network of surfaces, as depicted in Figure 10.1. Intersections, so-called *nodes*, are also the corners of the *finite elements*. If the network of lines and surfaces lies along orthogonal coordinates (e.g., rectangular, cylindrical, spherical), then the elements are rectangular blocks. The elements need not be hexahedral as that at A in Figure 10.1; the element B is quintahedral with one face lying on the exterior surface. Each element is defined by the nodal (corner) values. If we assume a linear interpolation of the displacements, $u(x,y,z)$, $v(x,y,z)$, and $w(x,y,z)$, then element A has 24 degrees of f3reedom: the 3 components of displacement at each of the eight corners; unless subject to constraints, element B has 18 degrees of freedom. The entire deformed configuration is defined by n discrete nodal values (u_n, v_n, w_n) ($n =$ number of unconstrained nodes) which constitute our approximations of the continuous functions (u, v, w).

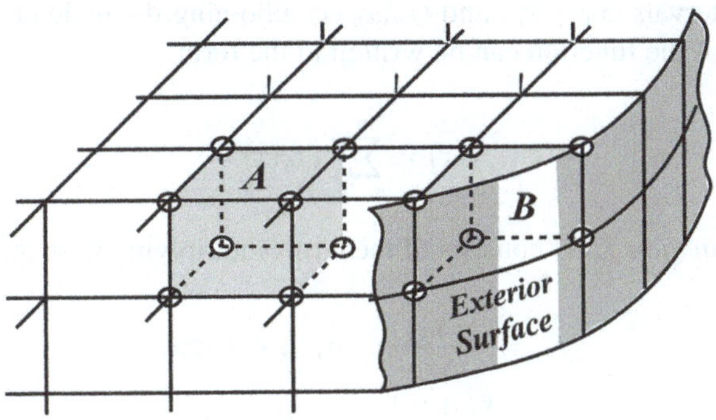

Figure 10.1 Finite Elements

10.3 INTERPOLATION

Within the finite element, the functions of the continuum (displacements, strains, and stresses) are expressed by continuous functions. These are most commonly polynomials, which serve to interpolate between nodes and the contiguous surfaces. It is most desirable to preserve continuity of the displacement throughout the body and, hence, at the interfaces of elements. We begin by examining the simpler approximation of a function $y(x)$:

The simplest approximation is the succession of straight segments as depicted in Figure 10.2. In the segments (x_{n-1}, x_n) and (x_n, x_{n+1}) the function is approximated by the linear terms.

$$y \doteq y_{n-1} + (y_n - y_{n-1})\left(\frac{x - x_{n-1}}{x_n - x_{n-1}}\right), \qquad x_{n-1} < x < x_n \tag{10.1a}$$

$$y \doteq y_n + (y_{n+1} - y_n)\left(\frac{x - x_n}{x_{n+1} - x_n}\right), \qquad x_n < x < x_{n+1} \tag{10.1b}$$

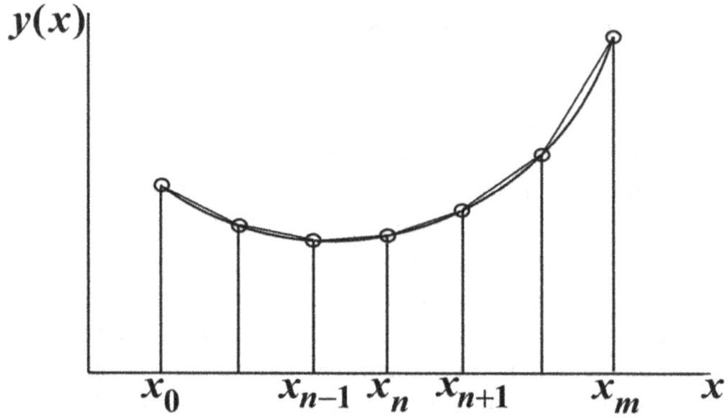

Figure 10.2 Linear/Lagrangean Interpolation

In most cases, the increments are equal, i.e. $x_n - x_{n-1} = x_{n+1} - x_n = \Delta x$. For the sake of brevity, we assume such uniform spacing. Note that the discrete value y_n appears only in these approximations in the two intervals (x_{n-1}, x_n) and (x_n, x_{n+1}) adjoining the node at x_n. Consequently, the entire approximation of the function can be written in the form:

$$y(x) \doteq \sum_{n=0}^{m} y_n f_n(x) \tag{10.2}$$

The so-called *shape function* $f_n(x)$ consists of the terms multiplying y_n in the Equations (10.1a,b):

$$\begin{aligned} f_n &= \frac{x - x_{n-1}}{\Delta x}, \quad x_{n-1} < x < x_n \\ &= \frac{x_{n+1} - x}{\Delta x}, \quad x_n < x < x_{n+1} \end{aligned}$$

We exhibit shape functions in Figure 10.3. Interpolation in the form of (10.2) with the shape functions of Figure 10.3 is termed *Lagrangean*.

Approximation by a cubic polynomial preserves the continuity of the derivatives $(\mathrm{d}y/\mathrm{d}x \equiv y')$, as well as the function (y). Now, the two discrete values (y_n, y'_n) enter the approximations throughout (x_0, x_m), but again these appear only in the two adjoining intervals (x_{n-1}, x_n) and (x_n, x_{n+1}); in place of Equation (10.2), we have

$$y(x) \doteq \sum_{n=0}^{m} \left(y_n f_n(x) + y'_n g_n(x) \right) \tag{10.3}$$

The two shape functions are obtained when the cubic polynomial is expressed in terms of the nodal values (y_n and y'_N):

$$f_n = 1 - \frac{3}{\Delta x^2}(x - x_n)^2 - \frac{2}{\Delta x^3}(x - x_n)^3, \quad x_{n-1} < x < x_n \tag{10.4a}$$

$$= 1 - \frac{3}{\Delta x^2}(x - x_n)^2 + \frac{2}{\Delta x^3}(x - x_n)^3, \quad x_n < x < x_{n+1} \tag{10.4b}$$

and

$$g_n = (x - x_{n+1}) + \frac{2}{\Delta x}(x - x_n)^2 + \frac{1}{\Delta x^2}(x - x_n)^3, \quad x_{n-1} < x < x_n \tag{10.5a}$$

$$= (x - x_n) - \frac{2}{\Delta x}(x - x_n)^2 + \frac{1}{\Delta x^2}(x - x_n)^3, \quad x_n < x < x_{n+1} \tag{10.5b}$$

These are shown graphically in Figure 10.4. Interpolation in the form of Equation (10.3), which preserves continuity of the derivative, is termed *Hermitean*

A function $w(x, y)$ can be interpolated as well by the linear (Lagrangean) or by the cubic (Hermitean) approximations. The shape function for the former appears as shown in Figure 10.5; the approximation of the function appears in the form:

$$f(x, y) \doteq \sum_n \sum_m w_{nn} f_{nm}(x, y) \tag{10.6}$$

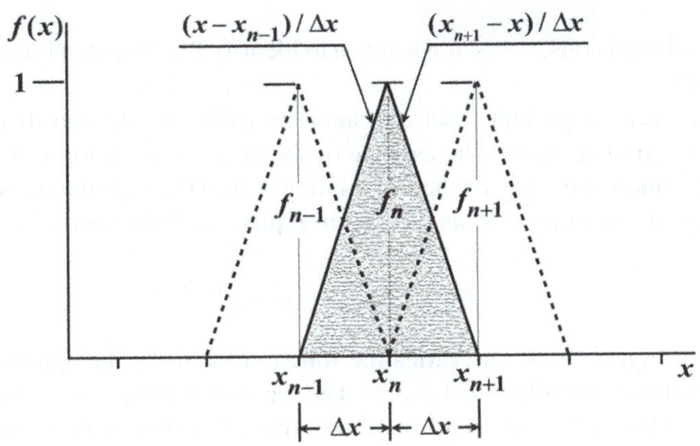

Figure 10.3 Shape Function (Linear Interpolation)

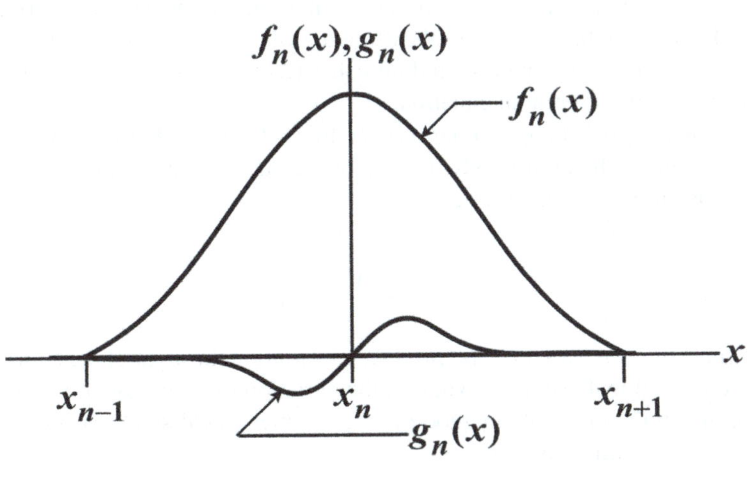

Figure 10.4 Shape Functions (Hermitean Interpolation)

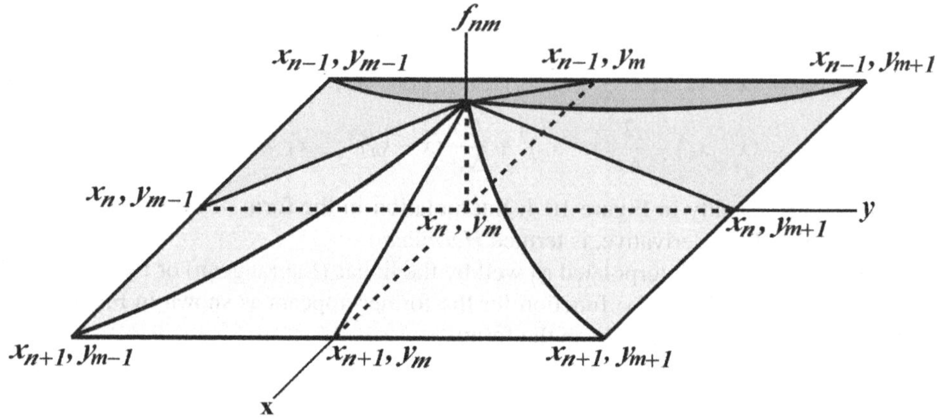

Figure 10.5 Shape Function in Two Dimensions

Here, to suit the two dimensions, we assign n and m to identify the discrete nodal sites. The number of nodes is then $n \times m$.

There is only one very important feature of approximation in terms of discrete nodal values and the internodal interpolation; it is the essence of the *finite element* form of approximation: A variation of a nodal value varies only the approximation within the region of the contiguous nodes: For example, to vary the function w at site (x_2, y_3) in Equation (10.6) results in

$$\delta f(x,y) \doteq \delta w_{23} f_{23}(x,y)$$

As we observe in Figure 10.5, this varies the function (actually the approximation) over the rectangular region with corners (3,4;1,4;1,2;3,2). The region shrinks as the element shrinks (nodal distances diminish). Thus, when we employ the variational method, e.g., a virtual displacement δw_{nm}, the variational process is analogous to that employed in our treatment of the continuum. For example, the derivation of the differential Equation (9.97) hinges upon the argument that δw in Equation (9.93) can be varied over an arbitrarily small region; hence, the factor in parentheses must vanish at all points. In the variation of the discrete value δw_{nm}, the variation extends only over a small adjoining region and, as that region diminishes ($\Delta x \to 0, \Delta y \to 0$), the resulting difference equation converges to the differential equation.

The "finite element" form of approximation also differs from another form wherein the discrete values extend throughout the entire body. For example, one can approximate the deflection of a simply supported beam by a Fourier series, viz

$$w(x) = \sum_{n=1}^{\infty} A_n \sin \frac{n\pi x}{l}$$

A variation of the discrete variable (A_n) produces a variation throughout the entire beam. We say that this variable has "global" support, whereas the variation of y_n in Equation (10.2) has "local" support. Again, the "finite element" method of nodal values (local support) is entirely analogous to our approach to the continuous model.[2]

10.4 APPROXIMATION VIA THE MINIMUM POTENTIAL

A consistent method of discrete approximation of the conservative system is the application of the principle of minimum potential. In a structural application, this requires elastic behavior. The virtual displacement is a variation of the potential. An important feature of the procedure is the assurance of convergence of the difference equation(s) to the differential equation(s) which we strive to approximate. The proof[3] follows:

Let vector $\mathbf{V}_E(x,y,z)$ denote the nodal displacement of the exact solution, i.e. the solution of the differential equations and boundary conditions. This renders the minimum potential P_E among *all* admissible $\mathbf{V}(x,y,z)$. Let $\mathbf{V}_A = \sum f_n \mathbf{V}_n$ denote the approximation which provides the minimum potential for the model as described by the nodal values $\mathbf{V}_n$ and the interpolation (shape functions). The latter must be chosen to meet geometrical and mechanical conditions. Stated otherwise, the exact solution $\mathbf{V}_E$ can be approximated in the form $\sum f_n \mathbf{V}_n$. Indeed, we can define the approximation

$$\mathbf{V}_I \equiv \sum_{n=0}^{m} f_n \mathbf{V}_E(x_n, y_n, z_n) \tag{10.7}$$

Let P_I denote the potential of the approximation (10.7). By the principle,

$$P_I > P_E$$

However, our approximation $\mathbf{V}_A$ was chosen to provide the minimum potential among all displacements of the form (10.7). It follows that

$$P_E < P_A < P_I$$

In the limit $P_I \to P_E$ and therefore the potential $P_A \to P_E$. As the elements diminish in size the potential of the approximation approaches that of the exact solution.

10.5 DEFORMATION OF THE FINITE ELEMENT

Since this is an introduction to the underlying concepts, most aspects are examined in the context of the simplest approximation. Therefore, let us explore the deformation of an element in a state of plane stress and deformed according to bilinear interpolation, within the element of Figure (10.6).

$$u(x,y) = u_o + \bar{\varepsilon}_x x + \gamma_x y + \kappa_x xy \tag{10.8a}$$

$$v(x,y) = v_o + \bar{\varepsilon}_y y + \gamma_y x + \kappa_y xy \tag{10.8b}$$

The square element of Figure 10.6 has nodes 1, 2, 3, and 4. This two-dimensional element has 8 degrees of freedom, two components (u, v) of displacement at each of the 4 nodes; these displacements are designated $(u_n, v_n;\, n = 1,2,3,4)$.

The 8 coefficients in Equations (10.8a,b) are expressed in terms of the nodal displacements as follows:

$$u_0 = \frac{1}{4}(u_1 + u_2 + u_3 + u_4), \quad \bar{\varepsilon}_x = \frac{1}{2h}(u_1 - u_2 + u_4 - u_3) \tag{10.9a}$$

$$\gamma_x = \frac{1}{2h}(u_1 + u_2 - u_3 - u_4), \quad \kappa_x = \frac{1}{h^2}(u_1 - u_2 + u_3 - u_4) \tag{10.9b}$$

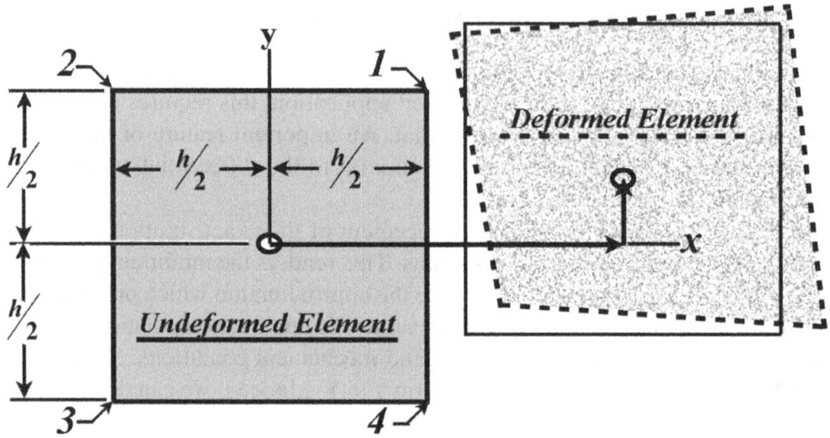

Figure 10.6 Bilinear Interpolation

Coefficients v_0, $\bar{\varepsilon}_y$, γ_y, and κ_x are expressed similarly:

$$v_0 = \frac{1}{4}(v_1 + v_2 + v_3 + v_4), \quad \bar{\varepsilon}_y = \frac{1}{2h}(v_1 + v_2 - v_3 - v_4) \tag{10.9c}$$

$$\gamma_y = \frac{1}{2h}(v_1 - v_2 - v_3 + v_4), \quad \kappa_y = \frac{1}{h^2}(v_1 - v_2 + v_3 - v_4) \tag{10.9d}$$

To achieve an efficient element, viz. accuracy with fewer elements and minimal computation, let us examine the role of the terms that constitute the 8 degrees of freedom. In Figure (10.7), we have decomposed the movement and deformation expressed by the Equations (10.8a,b). We observe that the lead terms, u_0 and v_0, represent the rigid body translations, which carry the element at A to positions B and C, respectively.

In our introduction to the method, we confine ourselves to linear theory; this means that strains and rotations remain small. Then a *small* rigid rotation, as depicted at D in Figure 10.7, has the form:

$$\omega_z(0,0) \equiv \gamma_y - \gamma_x \tag{10.10}$$

With hindsight the coefficients $\bar{\varepsilon}_x$ and $\bar{\varepsilon}_y$ represent uniform extensional strains, as depicted at a and b in Figure 10.7. A simple shear ($\bar{\gamma}_{xy}$), as depicted at c, has the form:

$$\bar{\gamma}_{xy} = \gamma_x + \gamma_y \tag{10.11}$$

Finally, the two remaining modes, κ_x and κ_y, are the higher-order terms (HOT), as depicted at d and e in Figure 10.7. All-together, we observe three rigid-body modes, three dominant deformational modes, and two higher-order modes; these eight modes are expressed in terms of the eight nodal displacements via the Equations (10.9)–(10.11). The decomposition is important to our understanding of the different roles of the dominant and higher-order deformational modes. The following example serves to demonstrate those roles.

In the method of finite elements, each deformational mode must contribute to the internal energy. Otherwise, one or more modes are unconstrained; the system is then mechanically unstable and mathematically indeterminate. On the other hand, the higher-order terms (HOT) increase the

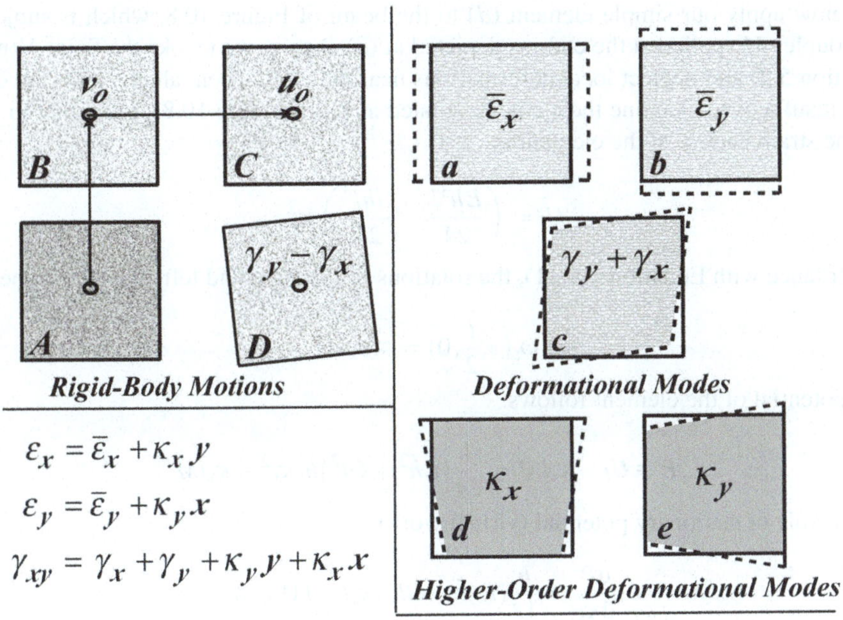

Figure 10.7 Modes of the Bilinear Element

potential and can impede convergence; in some instances, these HOT terms can render the element quite useless. The following example demonstrates the latter contention.

ILLUSTRATIVE EXAMPLE 10.1

Let us apply the foregoing concepts to a simple beam as depicted in Figure 10.8. We begin with an element as depicted in Figures 10.6 and 10.7 and impose the usual assumptions of most theories, viz a state of plane stress ($\sigma_y = \sigma_z = \tau_{xz} = \tau_{yz} = 0$); therefore, the strain energy assumes the form:

$$U_I = \int_{-l/2}^{l/2} \int_{-h/2}^{h/2} \left(\frac{E}{2} \varepsilon_x^2 + \frac{G}{2} \gamma_{xy}{}^2 \right) \mathrm{d}y\mathrm{d}x \tag{10.12}$$

If we adopt the assumption that a cross-section remains plane, then by Equations (10.8a,b)

$$\varepsilon_x = \bar{\varepsilon}_x + \kappa_x y \tag{10.13}$$

Since ε_y does not contribute, $\bar{\varepsilon}_y = \kappa_y = 0$. In simple bending $\bar{\varepsilon}_x = 0$. Then, we have

$$\varepsilon_x = \kappa_x y, \quad \gamma_{xy} = \gamma_x + \gamma_y + \kappa_x x \tag{10.14}$$

Our model is a superposition of the modes *c* and *d* of Figure 10.7. If we also impose the Kirchhoff-Bernoulli assumption, viz $\gamma_{xy}(0,0) = 0$, then Equation (10.14) is replaced by

$$\gamma_{xy} = \kappa_x x \tag{10.15}$$

Our very simple element is *d* of Figure 10.7.

Let us now apply our simple element (*d*) to the beam of Figure 10.8, which is subject to the bending couples M applied at the ends as depicted at (*a*). Again, we invoke the Saint-Venant principle (Section 5.2) and neglect local deformations near the ends. Then, all elements are deformed similarly; it suffices to examine the element isolated at (*c*) of Figure 10.8. According to Equation (10.12), the strain energy of the element is

$$U_I = \left(\frac{Eh^3 l}{24} + \frac{Ghl^3}{24}\right)\kappa_x^2 \tag{10.16}$$

In accordance with Equation (10.11), the rotations of the right and left ends are, respectively,

$$\omega_z(\pm\frac{l}{2},0) = \mp\kappa_x l/2$$

Then the potential of the element follows:

$$P \equiv U_I - \kappa_x lM = \frac{1}{24}(Eh^2 + Gl^2)hl\kappa_x{}^2 - \kappa_x lM$$

By the principle of stationary potential (virtual work)

$$\frac{dP}{d\kappa_x} = \frac{h}{12}(Eh^2 + Gl^2)\kappa_x l - lM = 0$$

The rotation of the right end relative to the left is $l\kappa_x$. Let us look to the analysis of flexure (Section 5.13), and observe that, in the limit $l \to 0$, the curvature κ_x is the "relative" rotation divided by the length l:

$$M = \frac{h}{12}(Eh^2 + Gl^2)\kappa_x$$

We note that the second term dominates since we anticipate $l > h$. However, the result converges to the exact value in the limit, viz[4]

$$M = \frac{Eh^3}{12}\kappa_x, \quad l \to 0$$

Early proponents of the "finite element" observed the very excessive stiffness of the element; it was termed "shear locking." After some numerical experimentation, some found that they obtained

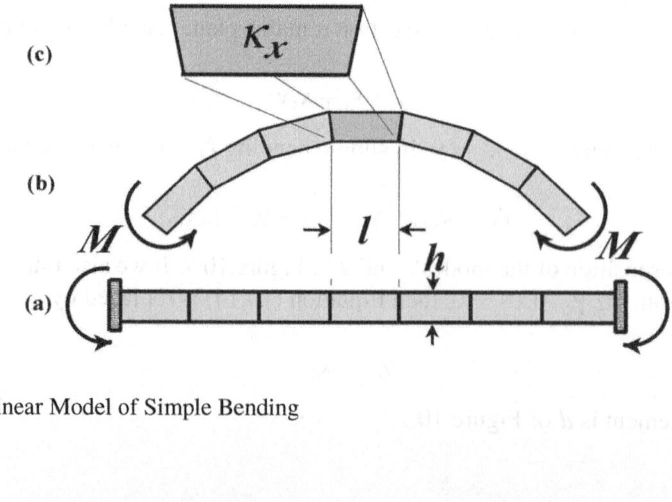

Figure 10.8 Bilinear Model of Simple Bending

a satisfactory result if they replaced the internal energy by its median (midpoint) value within the integral (10.12). It was termed "one point integration." Subsequently, we (see Wempner, G and Talaslidis, D) made our observation: A HOT term need only appear in but one term of the strain energy to ensure resistance to that mode. In the present case, that mode is inhibited by the first term of Equation (10.16). The so-called "one point integration" served to omit the unnecessary shear energy. We should also note that the mode in question, though normally HOT in most bodies, plays a dominant role in the beam. In a *thin* body in which *bending* is the predominant form of deformation; height *h remains finite* whereas *l vanishes* as the discrete approximation (algebraic equation) converges to the continuous model (differential equation). We observe the similar circumstance in thin plates and shells.[5]

ILLUSTRATIVE EXAMPLE 10.2

Consider the cantilever beam of Figure 10.9. It is clamped at the left end ($x = 0$), free at the right end ($x = l$), and subject to a distributed load $p(x)$, symmetric about the $x - z$ plane.The beam has a symmetric rectangular cross-section as shown in Figure 10.9. It is composed of a homogeneous isotropic Hookean material (HI-HO).

Consistent with earlier theories of beams, we assume

$$\sigma_y = \sigma_z = \tau_{yz} = \tau_{yx} = 0$$

Since no work is expended by these components of stress, the following strain components can be neglected:

$$\varepsilon_y = \varepsilon_z = \gamma_{yz} = \gamma_{yx} = 0 \tag{10.17}$$

Then the strain energy density assumes the form of Equation (10.12):

$$u = \frac{E}{2}\varepsilon_x^2 + \frac{G}{2}\gamma_{xz}^2 \tag{10.18}$$

Note: The roles of coordinates y and z are interchanged to conform to our prior treatment of beams.

Let us employ the bilinear interpolation. Then our element undergoes the extensional strain of Figure 10.7a and the shear strain at (c) and (d) (z replaces y). The displaced and deformed element is depicted in Figure 10.10.

Here we define

$w(x)$ = lateral displacement
$\theta(x)$ = rotation of the axis (x)
$\beta(x)$ = rotation of the cross-section

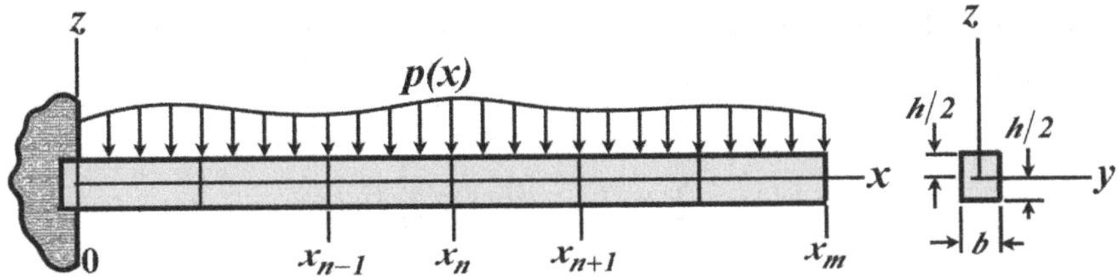

Figure 10.9 Cantilever Beam

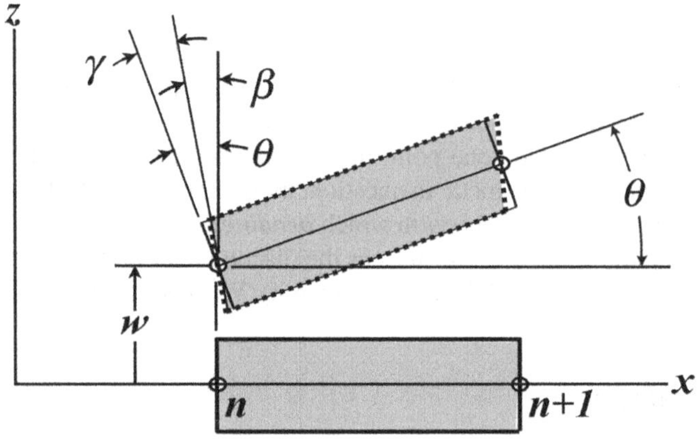

Figure 10.10 Deformed Element

$$\gamma \equiv \gamma_{xz} = \theta(x) - \beta(x)$$

If we were to invoke the Bernoulli hypothesis, then $\beta(x) = \theta(x)$ and $\gamma = \gamma_{xy} = 0$. Then the element would appear as that in Figure 10.8c. Here, the term $\kappa_x y$ in Equation (10.13) is replaced by $\mathrm{d}\beta/\mathrm{d}x$:

$$\varepsilon_x = \frac{\mathrm{d}\beta}{\mathrm{d}x}z \tag{10.19}$$

The transverse shear strain of Equation (10.14) is replaced by

$$\gamma_{xz} = \theta - \beta \doteq \frac{\mathrm{d}w}{\mathrm{d}x} - \beta \tag{10.20}$$

We require the potential P in order to apply the principle of stationary potential; the stationary conditions are the equations governing the discrete nodal variables (β_n and w_n). The potential consists of the internal energy and the potential of the load $p(x)$:

$$P = b\int_0^l \int_{-h/2}^{h/2} u \mathrm{d}z\mathrm{d}x + \int_0^l p(x)w(x)\mathrm{d}x \tag{10.21}$$

Here $p(x)$ denotes the load per unit length. Substituting Equations (10.19) and (10.20) into Equations (10.18) and (10.21), we obtain

$$P = \frac{Ebh^3}{24}\int_0^l \left(\frac{\mathrm{d}\beta}{\mathrm{d}x}\right)^2 \mathrm{d}x + \frac{Gbh}{2}\int_0^l \left(\frac{\mathrm{d}w}{\mathrm{d}x} - \beta\right)^2 + \int_0^l p(x)w(x)\mathrm{d}x \tag{10.22}$$

We turn now to the Lagrangian (linear) interpolation for the functions, $w(x)$, $\beta(x)$, and $p(x)$ as expressed by Equation (10.2):

$$w(x) \doteq \sum_0^m w_n f_n, \quad \beta(x) \doteq \sum_0^m \beta_n f_n \tag{10.23a,b}$$

$$\frac{\mathrm{d}w}{\mathrm{d}x} \doteq w_n f_n', \quad \frac{\mathrm{d}\beta}{\mathrm{d}x} \doteq \sum_0^m \beta_n f_n' \tag{10.23c,d}$$

where $f_n' \equiv \frac{df_n}{dx}$. Then

$$f_n(x) = (x - x_{n-1})/\Delta x, \qquad f_n'(x) = 1/\Delta x, \qquad x_{n-1} \le x \le x_n$$
$$f_n(x) = (x_{n+1} - x)/\Delta x, \qquad f_n'(x) = -1/\Delta x, \qquad x_n \le x \le x_{n+1}$$

We note that the shear strain of Equation (10.20) contains $\beta(x)$, which is approximated by Equation (10.23a); the linear term of $f_n(x)$ introduces the parasitic mode (κ_x) of Figure 10.7d. It must be suppressed in the shear energy of the potential. Then, γ_{xz} has the approximation:

$$\gamma_{xz} \doteq \gamma_{n-} \equiv \left[\frac{w_n - w_{n-1}}{\Delta x} - \frac{(\beta_n + \beta_{n-1})}{2}\right], \quad (x_{n-1}, x) \tag{10.24a}$$

$$\doteq \gamma_{n+} \equiv \left[\frac{w_{n+1} - w_n}{\Delta x} - \frac{(\beta_{n+1} + \beta_n)}{2}\right], \quad (x_n, x_{n+1}) \tag{10.24b}$$

With the introduction of these approximations, the potential, Equation (10.22), is a function of the discrete variables, β_n and w_n. That function is to be stationary with respect to the variations, $\delta\beta_n$ and δw_n (virtual displacements). Recall that the shape functions f_n, also γ_{n-} and γ_{n+}, have "local support"; therefore the integrations of functions f_n, f_n', γ_{n-}, γ_{n+}, extend only over the interval (x_{n-1}, x_{n+1}). The stationary conditions at x_n follow:

$$\frac{\partial P}{\partial \beta_n} = \frac{Ebh^3}{12}\int_{x_{n-1}}^{x_{n+1}} f_n(\beta_{n-1}f_{n-1}' + \beta_n f_n' + \beta_{n+1}f_{n+1}')\mathrm{d}x + \frac{Gbh}{2}\int_{x_{n-1}}^{x_n}\gamma_{n-}\mathrm{d}x + \frac{Gbh}{2}\int_{x_n}^{x_{n+1}}\gamma_{n+}\mathrm{d}x = 0 \tag{10.25a}$$

$$\frac{\partial P}{\partial w_n} = \frac{Gbh}{2}\int_{x_{n-1}}^{x_n}\gamma_{n-}\mathrm{d}x - \frac{Gbh}{2}\int_{x_n}^{x_{n+1}}\gamma_{n+}\mathrm{d}x + \int_{x_{n-1}}^{x_{n+1}} f_n(p_{n-1}f_{n-1} + p_n f_n + p_{n+1}f_{n+1})\mathrm{d}x = 0 \tag{10.25b}$$

We need only evaluate the specific integrals in these equations, e.g.

$$\int_{x_{n-1}}^{x_{n+1}} f_n f_{n-1}'\mathrm{d}x = -\frac{1}{2}, \qquad \int_{x_{n-1}}^{x_{n+1}} f_n f_{n-1}\mathrm{d}x = \frac{\Delta x}{6}$$

etc. Substituting these results into Equations (10.25a and b), we obtain the two equations applicable to node x_n:

$$-\frac{Ebh^3}{12}\left(\frac{\beta_{n+1} - 2\beta_n + \beta_{n-1}}{\Delta x}\right) - \frac{Gbh}{2}\left[(w_{n+1} - w_{n-1}) - \frac{(\beta_{n+1} + 2\beta_n + \beta_{n-1})}{2}\Delta x\right] = 0 \tag{10.26a}$$

$$-Gbh\left[\frac{(w_{n+1} - 2w_n + w_{n-1})}{\Delta x} - \frac{(\beta_{n+1} - \beta_{n-1})}{2}\right] + (p_{n-1}\frac{\Delta x}{6} + \frac{2}{3}p_n\Delta x + p_{n+1}\frac{\Delta x}{6}) = 0 \tag{10.26b}$$

If the Equations (10.26a and b) are to approximate the continuous beam, then they must converge to the differential equations ($\Delta x \to 0$). By definition

$$\lim_{x\to 0}\left(\frac{w_{n+1}-w_{n-1}}{2\Delta x}\right)=\frac{\mathrm{d}w}{\mathrm{d}x} \quad \text{at } w = w_n$$

$$\lim_{x\to 0}\frac{1}{\Delta x}\left[\frac{(\beta_{n+1}-\beta_n)}{\Delta x}-\frac{(\beta_n-\beta_{n-1})}{\Delta x}\right]=\frac{\mathrm{d}^2\beta}{\mathrm{d}x^2} \quad \text{at } \beta=\beta_n$$

$$\lim_{x\to 0}\frac{1}{\Delta x}\left[\frac{(w_{n+1}-w_n)}{\Delta x}-\frac{(w_n-w_{n-1})}{\Delta x}\right]=\frac{\mathrm{d}^2 w}{\mathrm{d}x^2} \quad \text{at } w=w_n,$$

Therefore, if we divide the Equations (10.26a and b) by Δx and pass to the limit ($\Delta x \to 0$), we obtain

$$-\frac{Ebh^3}{12}\frac{\mathrm{d}^2\beta}{\mathrm{d}x^2}-Gbh\left(\frac{\mathrm{d}w}{\mathrm{d}x}-\beta\right)=0 \tag{10.27a}$$

$$-Gbh\left(\frac{\mathrm{d}^2w}{\mathrm{d}x^2}-\frac{\mathrm{d}\beta}{\mathrm{d}x}\right)+p=0 \tag{10.27b}$$

Note that

$$\frac{Ebh^3}{12}\frac{\mathrm{d}^2\beta}{\mathrm{d}x^2}=\frac{\mathrm{d}M}{\mathrm{d}x}, \quad Gbh\left(\frac{\mathrm{d}w}{\mathrm{d}x}-\beta\right)=V$$

where M and V denote the bending couple and transverse shear force, respectively. In these notations we recognize the Equations (10.27a and b) as equations of equilibrium (Section 6.2):

$$\frac{\mathrm{d}M}{\mathrm{d}x}+V=0, \quad \frac{\mathrm{d}V}{\mathrm{d}x}=p$$

If we differentiate Equation (10.27a) and subtract the result from Equation (10.27b), we obtain

$$\frac{Ebh^3}{12}\frac{\mathrm{d}^3\beta}{\mathrm{d}x^3}=-p \tag{10.28}$$

If we now introduce the Bernoulli hypothesis ($\mathrm{d}w/\mathrm{d}x=\beta$), then Equation (10.28) becomes

$$\frac{Ebh^3}{12}\frac{\mathrm{d}^4w}{\mathrm{d}x^4}=-p(x)$$

This is the previous result (Section 6.2, equation 6.52a).

We have established the equations governing the discrete variables (β_n, w_n) at an interior node x_m. However, we also require the equations which govern the end node x_m; here the integrations extend only over the interval (x_{m-1}, x_m); these two equations follow:

$$\frac{Ebh^3}{12}\left(\frac{\beta_m-\beta_{m-1}}{\Delta x}\right)-\frac{Gbh}{2}\left[(w_m-w_{m-1})-\left(\frac{\beta_m+\beta_{m-1}}{2}\right)\Delta x\right]=0 \tag{10.29a}$$

$$Gbh\left[\frac{w_m-w_{m-1}}{\Delta x}-\frac{(\beta_m+\beta_{m-1})}{2}\right]+\left(\frac{1}{6}p_{m-1}+\frac{1}{3}p_m\right)\Delta x=0 \tag{10.29b}$$

In the limit ($\Delta x \to 0$, Equations (10.29 a and b) converge to

$$\frac{Ebh^3}{12}\frac{d\beta}{dx} = M = 0$$

$$Gbh\left(\frac{dw}{dx} - \beta\right) = V = 0$$

It is to be noted that the suppression of the parasitic term, mode (*d*) of Figure 10.7, from the shear energy does not affect convergence, nor the computation. Indeed, the accuracy is improved, because excessive energy and resistance are thereby removed.

The solution to the differential Equations (10.27a and b) with uniform load (p = constant) follows:

$$\beta = \frac{2pl^3}{Ebh^3}\left[\left(1-\frac{x}{l}\right)^3 - 1\right]$$

$$w = -\frac{pl^4}{2Ebh^3}\left[\left(1-\frac{x}{l}\right)^4 + \frac{4x}{l} - 1 + 2(1+\nu)\frac{h^2}{l^2}\left(\frac{2x}{l} - \frac{x^2}{l^2}\right)\right]$$

The final term in parentheses represents the contributions from transverse shear deformations. It is worth comparing that to the deflection given by the Bernoulli-Euler theory, viz the preceding terms. Suppose we set $l/h = 10$, $\nu = 0.30$. Then, at $x = l$

$$w = \frac{pl^4}{2Ebh^3}[3 + \underline{0.026}] \tag{10.30}$$

We note that the contribution caused by transverse shear is less than 1% of the total. However, our primary reason for employing the shear deformable element is not such refinement, but to implement effectively the very simple element, the linear element, as shown in Figure 10.10.

To illustrate the efficacy of the model, we consider the simple cantilever subject to an end load P. According to the Bernoulli-Euler analysis,

$$\frac{w(l)}{h} = \frac{4Pl^3}{Ebh^4}$$

With $l/h = 10, \nu = 0.3$ and $w/h = 0.100$, we obtain our dimensionless loading:

$$\frac{(1+\nu)P}{Ebh} = 3.25 \times 10^{-5}$$

Results for the three and five element models are depicted in Figure 10.11. End deflections of both differ from the anticipated value (0.100) by 2%. The model with five elements exhibits the additional shear deformation.

The beam of Figure 10.12 poses a greater challenge: The additional simple support requires curvatures in both directions. There, the Bernoulli-Euler analysis yields

$$\frac{w(3l)}{l} = 14.69\frac{P}{Ebh}\left(\frac{l}{h}\right)^2$$

With $l/h = 4, \nu = 0.30$ and $w(3l)/l = 0.10$, we obtain a dimensionless loading at node A:

$$\bar{P} \equiv \frac{2(1+\nu)P}{Ebh} = 1.11 \times 10^{-3}$$

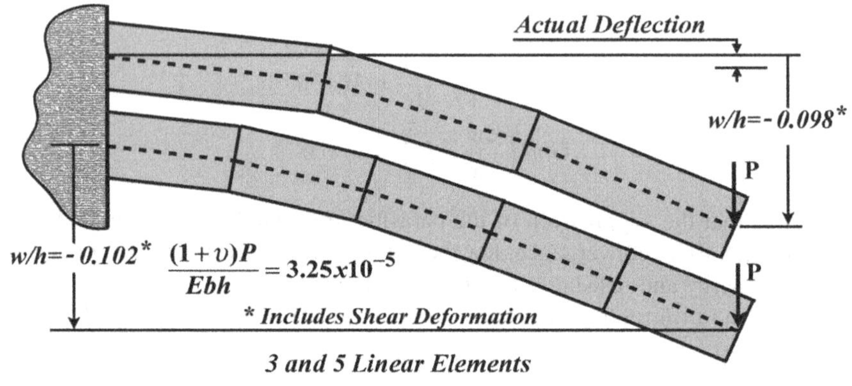

Figure 10.11 Beam Elements

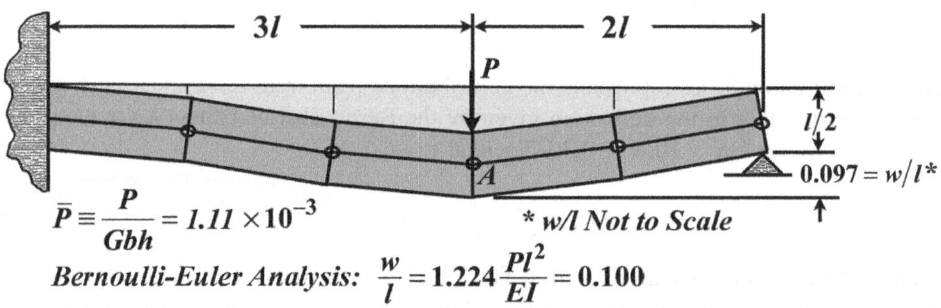

Figure 10.12 Beam Model via 5 Linear Elements

Our model of five elements undergoes a deflection $w(3l)/l = 0.097$ at node A. These approximations of the deformations are remarkably good. However, the models do not provide accurate descriptions of strain and stress distributions. In each linear segment, the "curvature" $(\mathrm{d}\beta/\mathrm{d}x)$ and, therefore, the moment, is constant.

The foregoing examples are offered only to illustrate the concept and method. The following approximation, via the cubic polynomial, is more effective for thin beams. That incorporates the Bernoulli hypothesis, which does not account for transverse shear deformations.

ILLUSTRATIVE EXAMPLE 10.3

Here we develop a discrete approximation of the (HI-HO) beam of Figure 10.9 based on the Bernoulli hypothesis, viz. plane cross-sections remain plane and perpendicular to the centroidal axis. That hypothesis requires that the approximation of the deformed axis have a continuous derivative. Specifically, the displacement $w = \mathrm{d}w/\mathrm{d}x$ must be continuous. Our approach admits the Lagrangean and Newton views of the method. To achieve the requisite continuity, the displacement w must be approximated by a cubic. Let us designate the center of the elements in Figure 10.13a by $'0'$ in (x_{n-1}, x_n) and $'1'$ in (x_n, x_{n+1}). With the origins at the center of each, the displacement is expressed by the cubic equations.

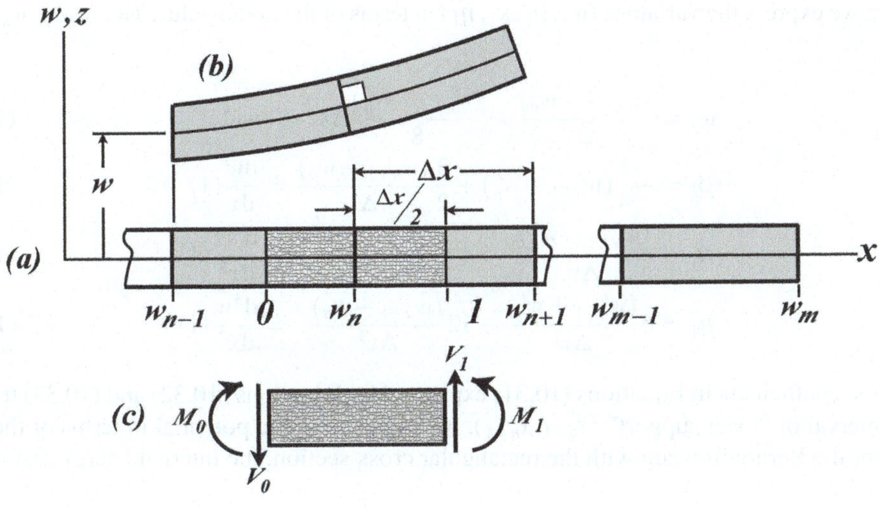

Figure 10.13 Beam Elements

$$w = w_0 + \beta_0 x + \frac{1}{2}\kappa_0 x^2 + \frac{1}{6}\eta_0 x^3, \quad \text{in } (x_{n-1}, n) \tag{10.31a}$$

$$w = w_1 + \beta_1 x + \frac{1}{2}\kappa_1 x^2 + \frac{1}{6}\eta_1 x^3, \quad \text{in } (x_n, x_{n+1}) \tag{10.31b}$$

In each element the variable x is measured from the midsection. In element $'0'$,

$$w_{n-1} = w_0 - \frac{1}{2}\beta_0 \Delta x + \frac{1}{8}\kappa_0 \Delta x^2 - \frac{1}{48}\eta_0 \Delta x^3$$

$$w_n = w_0 + \frac{1}{2}\beta_0 \Delta x + \frac{1}{8}\kappa_0 \Delta x^2 + \frac{1}{48}\eta_0 \Delta x^3$$

$$w'_{n-1} = \beta_0 - \frac{1}{2}\kappa_0 \Delta x + \frac{1}{8}\eta_0 \Delta x^2$$

$$w'_n = \beta_0 + \frac{1}{2}\kappa_0 \Delta x + \frac{1}{8}\eta_0 \Delta x^2$$

Inverting these equations expresses the coefficients $(w_0, \beta_0, \kappa_0, \eta_0)$ in terms of the nodal values $(w_{n-1}, w_n, w'_{n-1}, w'_n)$:

$$w_0 = \frac{(w_n + w_{n+1})}{2} - \frac{(w'_n - w'_{n-1})}{8}\Delta x \doteq w(0) \tag{10.32a}$$

$$\beta_0 = -\frac{1}{4}(w'_n + w'_{n-1}) + \frac{3}{2}\frac{(w_n - w_{n-1})}{\Delta x} \doteq \frac{dw}{dx}(0) \tag{10.32b}$$

$$\kappa_0 = \frac{(w'_n - w'_{n-1})}{\Delta x} \doteq \frac{d^2 w}{dx^2}(0) \tag{10.32c}$$

$$\eta_0 = 6\frac{(w'_n + w'_{n-1})}{\Delta x^2} - 12\frac{(w_n - w_{n-1})}{\Delta x^3} \doteq \frac{d^3 w}{dx^3(0)} \tag{10.32d}$$

Similarly, we express the variables $(w_1, \beta_1, \kappa_1, \eta_1)$ in terms of the nodal values $(w_n, w_{n+1}, w'_n, w'_{n+1})$

$$w_1 = \frac{(w_{n+1} + w_n)}{2} - \frac{(w'_{n+1} - w'_n)}{8}\Delta x \quad \doteq w(1) \tag{10.33a}$$

$$\beta_1 = -\frac{1}{4}(w'_{n+1} + w'_n) + \frac{3}{2}\frac{(w_{n+1} - w_n)}{\Delta x} \doteq \frac{dw}{dx}(1) \tag{10.33b}$$

$$\kappa_1 = \frac{(w'_{n+1} - w'_n)}{\Delta x} \doteq \frac{d^2w}{dx^2}(1) \tag{10.33c}$$

$$\eta_1 = 6\frac{(w'_{n+1} + w'_n)}{\Delta x^2} - 12\frac{(w_{n+1} - w_n)}{\Delta x^3} \doteq \frac{d^3w}{dx^2}(1) \tag{10.33d}$$

With the coefficients in Equations (10.31) expressed by Equations (10.32) and (10.33) throughout the interval of "local support" (x_{n-1}, x_{n+1}), we can express the potential in terms of the nodal values. For the Bernoulli beam with the rectangular cross-section, the internal energy assumes the form:

$$U_I = \frac{Ebh^3}{24}\int_0^l \kappa^2 dx = \frac{Ebh^3}{24}\int_0^l \left(\frac{d^2w}{dx^2}\right)^2 dx$$

The potential of the load takes the form:

$$U_p = \int_0^l p(x)w(x)dx$$

Again, we require that our approximation of the potential be stationary with respect to each of the variables (w_n and w'_n). In each of the elements, 0 and 1 of Figure 10.13,

$$\frac{d^2w}{dx^2} \doteq \kappa_0 + \eta_0 x \quad (x_{n-1}, x_n)$$

$$\frac{d^2w}{dx^2} \doteq \kappa_1 + \eta_1 x \quad (x_n, x_{n+1})$$

Therefore, in the interval (x_{n-1}, x_{n+1})

$$\begin{aligned}\Delta P = \Delta U_I + \Delta U_p = & \frac{Ebh^3}{24}\int_{-\Delta x/2}^{\Delta x/2}({\kappa_0}^2 + 2\kappa_0\eta_0\, x + {\eta_0}^2 x^2)dx \\ & + \frac{Ebh^3}{24}\int_{-\Delta x/2}^{\Delta x/2}({\kappa_1}^2 + 2\kappa_1 n_1 x + {\eta_1}^2 x^2)dx \\ & + \int_{-\Delta x/2}^{\Delta x/2} p(x)(w_0 + \beta_0 x + \frac{1}{2}\kappa_0 x^2 + \frac{1}{6}\eta_0 x^3)dx \\ & + \int_{-\Delta x/2}^{\Delta x/2} p(x)(w_1 + \beta_1 x + \frac{1}{2}\kappa_1 x^2 + \frac{1}{6}\eta_1 x^3)dx\end{aligned}$$

Note: ΔP denotes only that part of the potential which depends on the variables (w_n and w'_n), i.e. the interval of "support." If $p = p_n =$ constant in this interval, then

$$\Delta P = \frac{Ebh^3}{24}\left[\kappa_0{}^2\Delta x + \frac{1}{12}\eta_0^2\Delta x^3 + \kappa_1{}^2\Delta x + \frac{1}{12}\eta_1^2\Delta x^3\right]$$
$$+ p_n\left[w_0\Delta x + \frac{1}{24}\kappa_0\Delta x^4 + w_1\Delta x + \frac{1}{24}\kappa_1\Delta x^4\right]$$
$$\frac{\partial \Delta P}{\partial w_n} = \frac{Ebh^3}{12}\left[\kappa_0\frac{\partial \kappa_0}{\partial w_n}\Delta x + \frac{1}{12}\eta_0\frac{\partial \eta_0}{\partial w_n}\Delta x^3 + \kappa_1\frac{\partial \kappa_1}{\partial w_n}\Delta x + \frac{1}{12}\eta_1\frac{\partial \eta_1}{\partial w_n}\Delta x^3\right] \tag{10.34}$$
$$+ p_n\left[\frac{\partial w_0}{\partial w_n}\Delta x + \frac{1}{24}\frac{\partial \kappa_0}{\partial w_n}\Delta x^3 + \frac{\partial w_1}{\partial w_n}\Delta x + \frac{1}{24}\frac{\partial \kappa_1}{\partial w_n}\Delta x^3\right] = 0$$
$$= Ebh^3\left[\frac{(w'_{n+1} - w'_{n-1})}{2\Delta x^2} - \frac{(w_{n+1} - 2w_n + w_{n-1})}{\Delta x^3}\right] + p_n\Delta x = 0$$

In like manner, we obtain the second equation of equilibrium:

$$\frac{\partial \Delta P}{\partial w'_n} = \frac{Ebh^3}{6}\left[\frac{(w'_{n+1} + 4w'_n + w'_{n-1})}{\Delta x} - \frac{3(w_{n+1} - w_{n-1})}{\Delta x^2}\right] = 0 \tag{10.35}$$

Equations (10.34) and (10.35) are the difference equations (equilibrium conditions) which apply at each interior node x_m. We also require the conditions at the free end, node x_m. There the integrations extend only over the interval (x_{m-1}, x_m). The results follow:

$$\frac{\partial \Delta P}{\partial w_m} = Ebh^3\left[\frac{(w'_m + w'_{m-1})}{2\Delta x^2} - \frac{(w_m - w_{m-1})}{\Delta x^3}\right] + p_n\frac{\Delta x}{2} = 0 \tag{10.36}$$
$$\frac{\partial \Delta P}{\partial w'_m} = \frac{Ebh^3}{12}\left[\frac{(4w'_m + 2w'_{m-1})}{\Delta x} - \frac{6(w_m - w_{m-1})}{\Delta x^2}\right] = 0 \tag{10.37}$$

Let us now correlate the foregoing equations with the physical actions and the equations governing the continuous beam. To that end, we look to the discrete segment of Figure 10.13c, at the left and right ends, respectively.

$$M_0 = \frac{Ebh^3}{12}\frac{d^2w}{dx^2}|_0 \doteq \frac{Ebh^3}{12}\frac{(w'_n - w'_{n-1})}{\Delta x} \tag{10.38a}$$
$$M_1 = \frac{Ebh^3}{12}\frac{d^2w}{dx^2}|_1 \doteq \frac{Ebh^3}{12}\frac{(w'_{n+1} - w'_n)}{\Delta x} \tag{10.38b}$$
$$V_0 = -\frac{Ebh^3}{12}\frac{d^3w}{dx^3}|_0 \doteq \frac{Ebh^3}{12}\left[12\frac{(w_n - w_{n-1})}{\Delta x^3} - 6\frac{(w'_n + w'_{n-1})}{\Delta x^2}\right] \tag{10.38c}$$
$$V_1 = -\frac{Ebh^3}{12}\frac{d^3w}{dx^3}|_1 \doteq \frac{Ebh^3}{12}\left[12\frac{(w_{n+1} - w_n)}{\Delta x^3} - 6\frac{(w'_{n+1} + w'_n)}{\Delta x^2}\right] \tag{10.38d}$$

The equation for equilibrium of forces upon the continuous beam (See Section 6.2) follows:

$$\frac{dV}{dx} - p = 0 \tag{10.39}$$

The counterpart for the finite element of Figure 10.13c follows:

$$\frac{V_1 - V_0}{\Delta x} + \bar{p} = 0 \tag{10.40}$$

Substituting the approximations (10.38c,d) into Equation (10.40), we obtain

$$Ebh^3\left[\frac{(w_{n+1}-2w_n+w_{n-1})}{\Delta x^4}-\frac{(w'_{n+1}+w'_{n-1})}{2\Delta x^3}\right]-p=0$$

This is Equation (10.34) divided by Δx; both converge to the difference Equation (10.39).

The differential equation, which governs the equilibrium of moments, follows:

$$\frac{\mathrm{d}M}{\mathrm{d}x}+V=0 \tag{10.41a}$$

The counterpart for the finite element of Figure 10.13c follows:

$$\frac{M_1-M_0}{\Delta x}+\frac{V_1+V_2}{2}=0 \tag{10.41b}$$

Substitution of the approximations (10.38a–d) into Equation (10.41b) yields the difference equation:

$$\frac{Ebh^3}{6}\left[\frac{(w'_{n+1}-4w'_n+2w'_{n-1})}{\Delta x^2}-\frac{3(w_{n+1}-w_{n-1})}{\Delta x^3}\right]=0$$

Again, this is the difference Equation (10.35) divided by Δx. This converges to the differential Equation (10.41a).

Equations (10.36) and (10.37) converge to the end conditions ($\Delta x \to 0$):

$$V=M=0$$

In the preceding example, we developed a discrete approximation based upon linear interpolation; here we developed the approximation based upon the cubic of Equations (10.31a,b). Both are defined by two nodal quantities at each node. The former is defined by displacement w_n and rotation β_n; it admits a transverse shear strain $\gamma_{xz}=\gamma$. The latter is defined by the displacement w_n and the derivative w'_n (rotation of the tangent line and cross-section). The latter admits no transverse shear strain.

ILLUSTRATIVE EXAMPLE 10.4

The cubic polynomial in our approximations, Equations (10.31a,b), is the exact solution of the homogeneous differential Equation governing the Bernoulli-Euler beam, i.e. the beam with no distributed load (p). That polynomial is written in terms of the end values (w_n, w'_n) via Equations (10.32) and (10.33). It follows that any segment of such beam without a distributed load is "exactly" described by the cubic element. For example, the beam depicted in Figure 10.14 is described "exactly" by five elements (A, B, C, D, and E) of lengths a, b, c, d, and e. That model possesses eight degrees of freedom. If we denote the internal energies of these segments (elements) by $I_A, I_B, I_C, I_D, and I_E$, respectively, then we require the solution of the following eight equations:

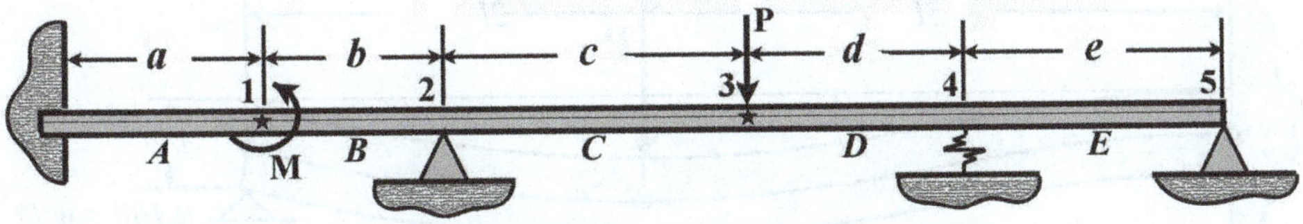

Beam Requiring 5 Elements and 8 Equations

Figure 10.14

$$\frac{\partial(I_A+I_B)}{\partial w_1}=0, \quad \frac{\partial(I_A+I_B)}{\partial w_1'}=M$$

$$\frac{\partial(I_B+I_C)}{\partial w_2'}=0, \quad \frac{\partial(I_C+I_D)}{\partial w_3}=-P, \quad \frac{\partial(I_C+I_D)}{\partial w_3'}=0$$

$$\frac{\partial(I_D+I_E)}{\partial w_4}+Kw_4=0, \quad \frac{\partial(I_D+I_E)}{\partial w_4'}=0$$

$$\frac{\partial(I_E)}{\partial w_5'}=0$$

Here, w_n and w_n' denote the transverse displacement and slope (rotation).

As a simple example, a cantilever beam with only an end load P is essentially one element with two degrees of freedom, viz the end displacement w and rotation w'. By Equations (10.36) and (10.37) we obtain

$$\frac{w'}{2}-\frac{w}{l}=-\frac{Pl^2}{Ebh^3}$$

$$\frac{w'}{3}-\frac{w}{2l}=0$$

The solution is that for the continuous Bernoulli-Euler beam under small rotations:

$$w(l)=\frac{4Pl^3}{Ebh^3}, \quad w'(l)=\frac{3}{2}\frac{w(l)}{l}$$

As another specific example, consider that of Figure 10.15, which was previously treated via the linear element (see Figure 10.12). In this instance, we require only the two elements, viz. CA and AB. This model possesses 3 degrees of freedom, viz w_A, w_A', and w_B', and the three equations:

$$\frac{\partial(I_{CA}+I_{AB})}{\partial w_A}=P, \quad \frac{\partial(I_{CA}+I_{AB})}{\partial w_A'}=0$$

$$\frac{\partial I_{AB}}{\partial w_B'}=0$$

Again, the result is the "exact" solution according to the Bernoulli hypothesis.

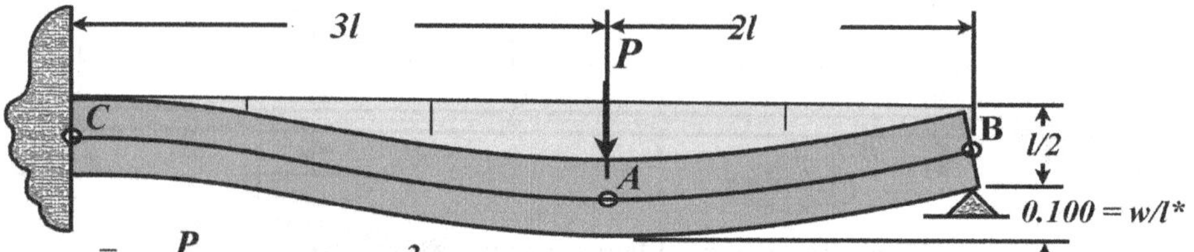

$$\bar{P} \equiv \frac{P}{Gbh} = 1.11 \times 10^{-3}$$

Bernoulli-Euler Analysis: $\frac{w}{l} = 1.224\frac{Pl^2}{EI} = 0.100$

Figure 10.15 Beam Model via 2 Cubic Elements

10.6 EXTENSIONAL DEFORMATIONS OF THIN PLATES

In Chapter 8, we formulated the theory of the continuous HI-HO plate and applied the equations to obtain solutions for simple problems. Such solutions are not possible for most practical situations, wherein boundaries have irregular forms. The methods of finite elements provide useful approximations; digital computers often enable great accuracy. Here, we present some elementary approximations, demonstrate the correlation with the continuous theory, and offer a few illustrative examples. To be specific, we consider a rectangular body, which is subdivided by a rectangular grid, as depicted in Figure 10.16. The corners of the rectangular elements are the nodes identified by numbers, n and m, to signify positions in the directions $\bar{x}$ and $\bar{y}$.

According to the principle of virtual work, or stationary potential, we can employ an approximation of the deformed state, which maintains continuity. A simple approximation is the bilinear displacement within each element. Specifically, with the local coordinates (x, y) originating at the center of an element; the displacements (u, v) are approximated in the form:

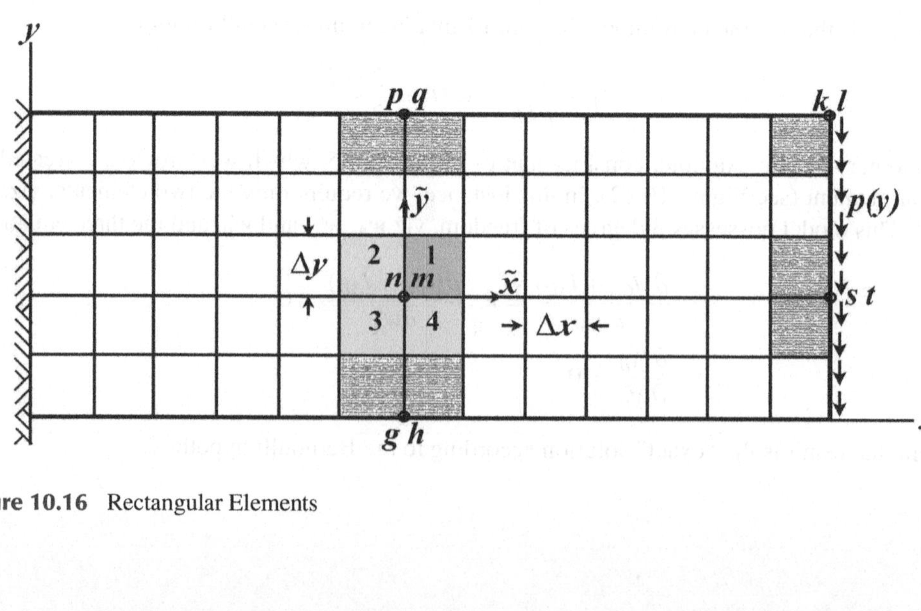

Figure 10.16 Rectangular Elements

$$u = u_0 + \bar{\varepsilon}_x x + \gamma_x y + \kappa_x xy \tag{10.42a}$$

$$v = v_0 + \gamma_y x + \bar{\varepsilon}_y y + \kappa_y xy \tag{10.42b}$$

We note that this description gives the element 8 degrees of freedom, which are associated with the 8 coefficients, our discrete variables. We note too that the components vary linearly along each edge; that variation is determined by the nodal values. If each element is described in this form and nodes remain intact, then the adjoining elements remain contiguous.

To reveal fully the kinematics, we note that the initial constants, u_o and v_o, constitute rigid translation of the element. Also, a small rigid rotation is expressed by

$$\omega_0 \equiv \left(\frac{\partial v}{\partial x} - \frac{\partial u}{\partial y}\right)\Big|_{0,0} = (\gamma_y - \gamma_x)$$

The remaining 5 degrees of freedom comprise the deformational modes:

$$\varepsilon_x \equiv \frac{\partial u}{\partial x} = \bar{\varepsilon}_x + \kappa_x y \tag{10.43a}$$

$$\varepsilon_y \equiv \frac{\partial v}{\partial y} = \bar{\varepsilon}_y + \kappa_y x \tag{10.43b}$$

$$\gamma_{xy} = \left(\frac{\partial v}{\partial x} + \frac{\partial u}{\partial y}\right) = \gamma_y + \gamma_x + \kappa_y y + \kappa_x x \tag{10.43c}$$

Of course, the very accurate approximation of any body requires many small elements. Mathematically, that implies that the dimensions Δx and Δy diminish. In the neighborhood of the node (n, m) the deformation is dominated by the constants $\bar{\varepsilon}_x$, $\bar{\varepsilon}_y$, and $(\gamma_x + \gamma_y)$. The additional deformational modes, κ_x and κ_y, are higher order, but essential to achieve the requisite continuity with adjoining *finite* elements. The rigid-body and deformational modes are depicted in Figure 10.17. Here coordinates (x, y) originate at the center of the element. As noted, the motion and deformation of each element is determined by the nodal displacements; u and v at each of the 4 corners determine the 8 degrees of freedom. This means that the entire configuration of the body is determined by the collective displacements of the nodes. Inversion of Equations (10.42a and b) expresses the coefficients (u_0, $\bar{\varepsilon}_x$, γ_x, κ_x, v_0, $\bar{\varepsilon}_y$, γ_y, κ_y) in terms of the displacements of the four corners. The motion and deformation of the element are thereby expressed in terms of the displacements of the adjacent nodes.

Mathematically, the bilinear approximation can be expressed in the form:

$$u = \sum_{n,m=0}^{N,M} u_{nm} f_{nm}(\tilde{x}, \tilde{y}), \quad v = \sum_{n,m=0}^{N,M} v_{nm} f_{nm}(\tilde{x}, \tilde{y}) \tag{10.44a,b}$$

The "shape function" f_{nm} is dictated by our choice of the linear interpolation; f_{00} appears graphically as the "tent" covering the 4 quadrants adjoining node 00 of Figure 10.18. It is noteworthy that a virtual displacement of a node affects only the strains $(\varepsilon_x, \varepsilon_y, \gamma_{xy})$ in the adjoining four elements; the discrete variable (nodal displacement) has "local support"; at any node nm the function is expressed by the linear equations:

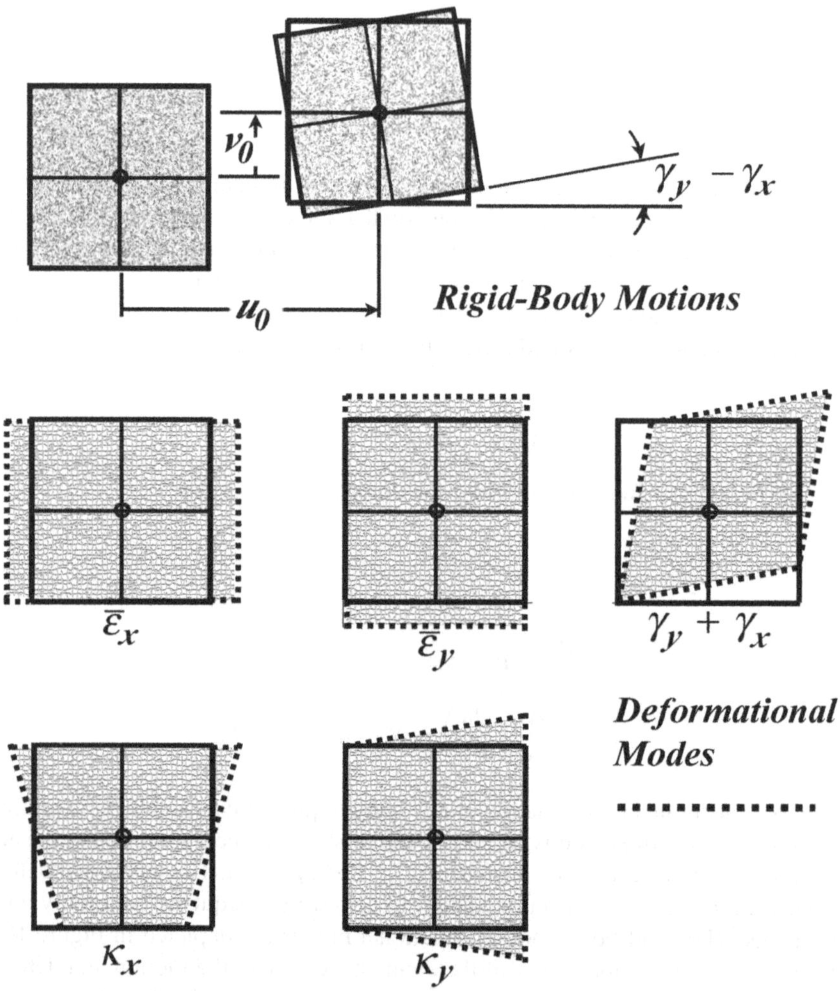

Figure 10.17 Modes of the Bilinear Elements

$$x_n < \tilde{x}, \quad y_m < \tilde{y}, \quad f_{nm} = \left[1 - \frac{(\tilde{x} - x_n)}{\Delta x}\right]\left[1 - \frac{(\tilde{y} - y_m)}{\Delta y}\right]$$

$$x_n > \tilde{x}, \quad y_m < \tilde{y}, \quad f_{nm} = \left[1 + \frac{(\tilde{x} - x_n)}{\Delta x}\right]\left[1 - \frac{(\tilde{y} - y_m)}{\Delta y}\right]$$

$$x_n > \tilde{x}, \quad y_m > \tilde{y}, \quad f_{nm} = \left[1 + \frac{(\tilde{x} - x_n)}{\Delta x}\right]\left[1 + \frac{(\tilde{y} - y_m)}{\Delta y}\right]$$

$$x_n < \tilde{x}, \quad y_m > \tilde{y}, \quad f_{nm} = \left[1 - \frac{(\tilde{x} - x_n)}{\Delta x}\right]\left[1 + \frac{(\tilde{y} - y_m)}{\Delta y}\right]$$

Let us return to an element, specifically, the element in the first quadrant of Figure 10.18, as depicted in Figure 10.19. With the origin at the center, the displacements have the approximations

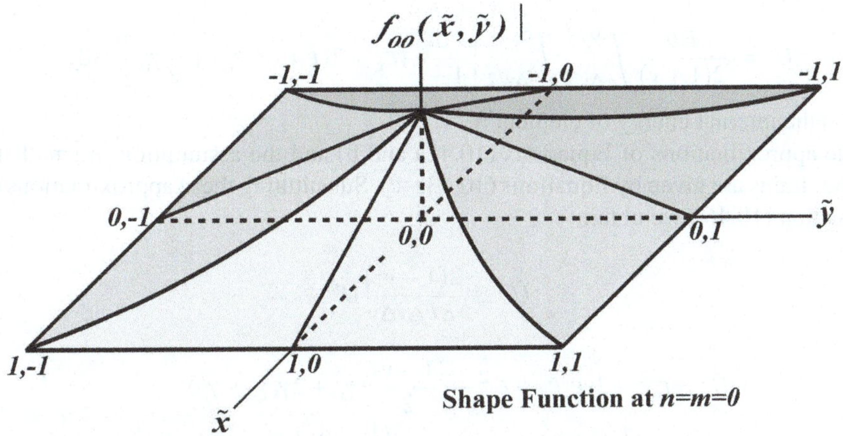

Figure 10.18 Shape Function for the Bilinear Approximation

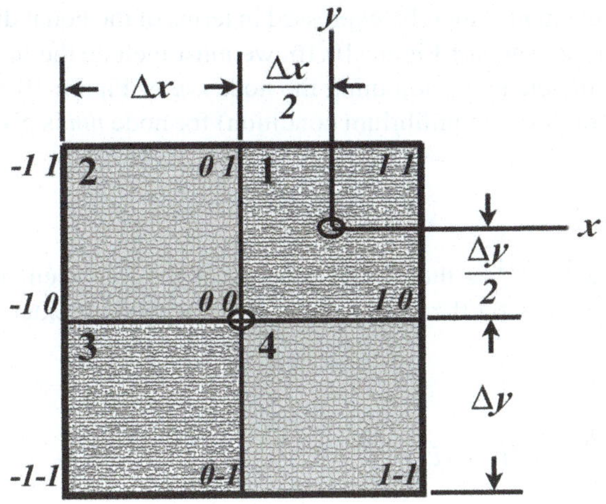

Figure 10.19 Patch of Support for Node 00

(10.42a and b). Inversions of those equations express the coefficients in terms of the nodal values:

$$u_0 = \frac{1}{4}(u_{00} + u_{10} + u_{11} + u_{01}), \quad v_0 = \frac{1}{4}(v_{00} + v_{10} + v_{11} + v_{01}) \tag{10.45a,b}$$

$$\gamma_x = \frac{1}{2\Delta y}(-u_{00} - u_{10} + u_{11} + u_{01}), \quad \bar{\varepsilon}_y = \frac{1}{2\Delta y}(-v_{00} - v_{10} + v_{11} + v_{01}) \tag{10.45c,d}$$

$$\bar{\varepsilon}_x = \frac{1}{2\Delta x}(-u_{00} + u_{10} + u_{11} - u_{01}), \quad \gamma_y = \frac{1}{2\Delta x}(-v_{00} + v_{10} + v_{11} - v_{01}) \tag{10.45e,f}$$

$$\kappa_x = \frac{1}{\Delta x \Delta y}(u_{00} - u_{10} + u_{11} - u_{01}), \quad \kappa_y = \frac{1}{\Delta x \Delta y}(v_{00} - v_{10} + v_{11} - v_{01}) \tag{10.45g,h}$$

To apply the equations of stationary potential (equilibrium conditions), we require the internal energy of the element. Let us consider the most likely circumstance of plane stress (see Section 8.6, Equation 8.8) where b denotes the thickness of the plate:

$$U^n = \frac{Eb}{2(1+\nu)} \int_{-\Delta x/2}^{\Delta x/2} \int_{-\Delta y/2}^{\Delta y/2} \left[\frac{1}{1-\nu} (\varepsilon_x^2 + 2\nu \varepsilon_x \varepsilon_y + \varepsilon_y^2) + \frac{1}{2} \gamma_{xy}^2 \right] dxdy \tag{10.46}$$

where U^n is the internal energy of element n.

With the approximations of Equations (10.42a and b) and the assumption of small strain and rotation, the strains are given by Equations (10.43a–c). Substituting these approximations (10.43a–c) into Equation (10.46), we obtain

$$\tilde{U}^n \equiv \frac{2(1-\nu^2)}{Eb\Delta x \Delta y} U^n \tag{10.47}$$

$$\begin{aligned} \tilde{U}^n =& \bar{\varepsilon}_x{}^2 + 2\nu \bar{\varepsilon}_x \bar{\varepsilon}_y + \bar{\varepsilon}_y{}^2 + \frac{(1-\nu)}{2} (\gamma_x^2 + 2\gamma_x \gamma_y + \gamma_y^2) \\ &+ \frac{1}{12} (\kappa_x^2 \Delta y^2 + \kappa_y^2 \Delta x^2) + \frac{(1-\nu)}{24} (\kappa_x^2 \Delta x^2 + \kappa_y^2 \Delta y^2) \end{aligned} \tag{10.48}$$

With the coefficients ($\bar{\varepsilon}_x$, $\bar{\gamma}_x$, etc.) all expressed in terms of the nodal values (u_{11}, - -, v_{11}, - -) the strain energy of the one element is thereby expressed in terms of the nodal displacements. To apply the stationary criteria to the patch of Figure 10.16, we must include the internal energies of the 4 adjoining elements, i.e. the elements adjoining the node *nm* of Figure 10.16 (node 0,0 of Figure 10.19). One stationary condition (equilibrium condition) for node *nm* is given by

$$\frac{\partial(\bar{U} + \bar{P})}{\partial u_{mn}} \tag{10.49a}$$

where $\bar{U} = \bar{U}^1 + \bar{U}^2 + \bar{U}^3 + \bar{U}^4$ is the internal energy of the four elements, and $\bar{P}$ denotes the potential of any external loads on the patch. If the patch is without external loading, i.e. $\tilde{p} = 0$, then

$$\begin{aligned} \frac{\partial \tilde{U}}{\partial u_{nm}} =& 2 \Bigg[(\bar{\varepsilon}_x + \nu \varepsilon_y) \frac{\partial \bar{\varepsilon}_x}{\partial u_{nm}} + (\bar{\varepsilon}_y + \nu \bar{\varepsilon}_x) \frac{\partial \bar{\varepsilon}_y}{\partial u_{nm}} + \frac{(1-\nu)}{2} (\gamma_x + \gamma_y) \frac{\partial \gamma_x}{\partial u_{nm}} + \frac{(1-\nu)}{2} (\gamma_x + \gamma_y) \frac{\partial \gamma_y}{\partial u_{nm}} \\ &+ \frac{1}{12} \left(\kappa_x \frac{\partial \kappa_x}{\partial u_{nm}} \Delta y^2 + \kappa_y \frac{\partial \kappa_y}{\partial u_{nm}} \Delta x^2 \right) + \frac{(1-\nu)}{24} \left(\underline{\kappa_x \frac{\partial \kappa_x}{\partial u_{nm}} \Delta x^2 + \kappa_y \frac{\partial \kappa_y}{\partial u_{nm}} \Delta y^2} \right) \Bigg] = 0 \end{aligned} \tag{10.49b}$$

The second equation of equilibrium has the same form (u_{nm} is replaced by v_{nm}): Note that the final terms containing the HOT modes, κ_x and κ_y, are higher order (Δx^2 or Δy^2). These are necessary to resist those HOT modes, but need only appear in one term of the energy. For example, we might delete the underlined terms; this could only improve convergence by reducing the internal energy.

To simplify our notations, let us consider the patch as shown in Figure 10.19, merely replacing *nm* by 00. Then in quadrant 1, the derivatives in the integral (10.49b) follow from Equations (10.45):

$$\frac{\partial \bar{\varepsilon}_x^1}{\partial u_{00}} = -\frac{1}{2\Delta x}, \quad \frac{\partial \bar{\varepsilon}_y^1}{\partial u_{00}} = 0, \quad \frac{\partial \kappa^1}{\partial u_{00}} = \frac{1}{\Delta x \Delta y} \tag{10.50a,b,c}$$

$$\frac{\partial \gamma_x^1}{\partial u_{00}} = -\frac{1}{2\Delta y}, \quad \frac{\partial \gamma_y^1}{\partial u_{00}} = 0, \quad \frac{\partial \kappa^1}{\partial u_{00}} = 0 \tag{10.50d,e,f}$$

Similarly, for the quadrants 2, 3, and 4, where the subscript signifies the quadrant,

$$\frac{\partial \bar{\varepsilon}_x^2}{\partial u_{00}} = \frac{1}{2\Delta x}, \quad \frac{\partial \bar{\gamma}_x^2}{\partial u_{00}} = -\frac{1}{2\Delta y}, \quad \frac{\partial \kappa_x^2}{\partial u_{00}} = -\frac{1}{\Delta x \Delta y}$$

$$\frac{\partial \bar{\varepsilon}_x^3}{\partial u_{00}} = \frac{1}{2\Delta x}, \quad \frac{\partial \bar{\gamma}_x^3}{\partial u_{00}} = \frac{1}{2\Delta y}, \quad \frac{\partial \kappa_x^3}{\partial u_{00}} = \frac{1}{\Delta x \Delta y}$$

$$\frac{\partial \bar{\varepsilon}_x^4}{\partial u_{00}} = -\frac{1}{2\Delta x}, \quad \frac{\partial \bar{\gamma}_x^4}{\partial u_{00}} = \frac{1}{2\Delta y}, \quad \frac{\partial \kappa_x^4}{\partial u_{00}} = -\frac{1}{\Delta x \Delta y}$$

Note that $\bar{\varepsilon}_y$, $\bar{\gamma}_y$, and κ_y are dependent only on the displacement v.

10.6.1 CONVERGENCE TO THE DIFFERENTIAL EQUATIONS

With a view toward interpreting the equilibrium equation (10.49b), we recall (see Section 8.3, Equation 8.2) that

$$\sigma_x = \frac{E}{1-\nu^2}(\varepsilon_x + \nu\varepsilon_y), \quad \tau_{xy} = \frac{E}{2(1+\nu)}\gamma_{xy}$$

Let us also recall Equations (10.43a–c) and note that the quantities $\bar{\varepsilon}_x$, $\bar{\varepsilon}_y$ and $(\gamma_x + \gamma_y)$ are not only dominant but also the mean values of the strain components of each quadrant. Again, using superscripts (1,2,3,4) to identify the quadrants, we have the approximation of Equation (10.49):

$$\frac{2(1-\nu^2)}{E}\left[-\frac{\sigma_x^1}{2\Delta x} + \frac{\sigma_x^2}{2\Delta x} + \frac{\sigma_x^3}{2\Delta x} - \frac{\sigma_x^4}{2\Delta x} - \frac{\gamma_{xy}^1}{2\Delta y} - \frac{\gamma_{xy}^2}{2\Delta y} + \frac{\gamma_{xy}^3}{2\Delta y} + \frac{\gamma_{xy}^4}{2\Delta y}\right] = 0$$

where $\bar{\gamma}_{xy} = \gamma_x + \gamma_y$. Rearranging the terms, we obtain

$$-\frac{(\sigma_x^1 + \sigma_x^4)}{2\Delta x} + \frac{(\sigma_x^2 + \sigma_x^3)}{2\Delta x} - \frac{(\tau_{xy}^1 + \tau_{xy}^2)}{2\Delta y} + \frac{(\tau_{xy}^3 + \tau_{xy}^4)}{2\Delta y} = 0 \tag{10.51}$$

Let

$$\frac{(\sigma_x^1 + \sigma_x^4)}{2} \equiv \bar{\sigma}_x\left(\frac{\Delta x}{2}\right), \quad \frac{(\tau_{xy}^1 + \tau_{xy}^2)}{2} \equiv \bar{\tau}_{xy}\left(\frac{\Delta y}{2}\right)$$

$$\frac{(\sigma_x^2 + \sigma_x^3)}{2} \equiv \bar{\sigma}_x\left(-\frac{\Delta x}{2}\right), \quad \frac{(\tau_{xy}^3 + \tau_{xy}^4)}{2} \equiv \bar{\tau}_{xy}\left(-\frac{\Delta y}{2}\right)$$

These express the averages of σ_x and τ_{xy} in quadrants at $(\pm\Delta x/2)$ and $(\pm\Delta y/2)$, respectively. In the limit $(\Delta x \to 0)$, i.e. we converge to the midpoint (n,m) of the "patch".

$$\lim_{x\to 0} \frac{\bar{\sigma}_x\left(\frac{\Delta x}{2}\right) - \bar{\sigma}_x\left(-\frac{\Delta x}{2}\right)}{\Delta x} = \frac{\partial \sigma_x}{\partial x}$$

Likewise

$$\lim_{x\to 0} \frac{\bar{\tau}_{xy}\left(\frac{\Delta y}{2}\right) - \bar{\tau}_{xy}\left(-\frac{\Delta y}{2}\right)}{\Delta y} = \frac{\partial \tau_{xy}}{\partial y}$$

In the limit ($\Delta x \to 0,\ \Delta y \to 0$), Equation (10.51) approaches the differential equation (see Section 8.4, Equation 8.4):

$$\frac{\partial \sigma_x}{\partial x} + \frac{\partial \tau_{xy}}{\partial y} = 0$$

Likewise, the second equation of equilibrium, viz. $\partial \tilde{U}/\partial v_{nm}$, approaches the differential equation:

$$\frac{\partial \sigma_y}{\partial y} + \frac{\partial \tau_{xy}}{\partial x} = 0$$

The term $\tilde{U}$ of Equation (10.49a) must contain the contributions from the 4 quadrants adjoining node nm of Figure 10.16. For our illustration, we use the simpler notations of Figures 10.18 and 10.19 where 00 replaces nm; the first quadrant ($\bar{x} > 0,\ \bar{y} > 0$) is depicted in the latter. The Equations (10.45) apply to the first quadrant of Figure 10.19. When Equations (10.45) and (10.50) are substituted into Equation (10.49b), we obtain

$$\begin{aligned}\frac{\partial \tilde{U}^1}{\partial u_{00}} &= \frac{(u_{00} - u_{10} - u_{11} + u_{01})}{2\Delta x^2} + \nu\frac{(v_{00} + v_{10} - v_{11} - v_{01})}{2\Delta x \Delta y} \\ &+ \frac{(1-\nu)}{4}\left[\frac{(u_{00} + u_{10} - u_{11} - u_{01})}{\Delta y^2} + \frac{(v_{00} - v_{10} - v_{11} + v_{01})}{\Delta x \Delta y}\right] \\ &+ \frac{1}{6}\left[\frac{1}{\Delta x^2} + \frac{(1-\nu)}{2\Delta y^2}\right](u_{00} - u_{10} + u_{11} - u_{01})\end{aligned} \tag{10.52a}$$

Similarly, we can express the contribution from the energies of quadrants 2, 3, and 4. Then, the equilibrium equation takes the form:

$$\begin{aligned}\frac{\partial \tilde{U}}{\partial u_{00}} &= \frac{1}{2\Delta x^2}(4u_{00} + 2u_{01} - 2u_{-10} + 2u_{0-1} - 2u_{10} - u_{11} - u_{-11} - u_{-1-1} - u_{1-1}) \\ &+ \frac{\nu}{2\Delta x \Delta y}(-v_{11} + v_{-11} - v_{-1-1} + v_{1-1}) \\ &+ \frac{(1-\nu)}{4\Delta y^2}(4u_{00} - 2u_{01} + 2u_{-10} - 2u_{0-1} + 2u_{10} - u_{11} - u_{-11} - u_{-1-1} - u_{1-1}) \\ &+ \frac{(1-\nu)}{4\Delta x \Delta y}(-v_{11} + v_{-11} - v_{-1-1} + v_{1-1}) \\ &+ \frac{1}{6}\left[\frac{1}{\Delta x^2} + \frac{(1-\nu)}{2\Delta y^2}\right](4u_{00} - 2u_{01} - 2u_{-10} - 2u_{0-1} - 2u_{10} + u_{11} + u_{-11} + u_{-1-1} + u_{1-1})\end{aligned} \tag{10.52b}$$

This equation represents the net component of force in the x direction associated with the displacement u_{00}. One can regard this as the sum of the actions exerted by the 4 adjacent elements at the node 00. Similarly, we can obtain the component associated with the component of displacement v_{00}, i.e. $\partial \tilde{U}/\partial v_{00}$. Quite simply, the roles of u and v are interchanged, and the indices 1 and 0 are interchanged. In the absence of a body force acting in the plane (x-y), we must enforce the two equations of equilibrium at each interior node:

$$\frac{\partial \tilde{U}}{\partial u_{nm}} = 0, \quad \frac{\partial \tilde{U}}{\partial v_{nm}} = 0 \tag{10.52c,d}$$

In the continuum theory, we have two differential equations in the variables, $u(x,y)$ and $v(x,y)$. If our model converges, then Equation (10.52a) must converge to the differential equation, which enforces equilibrium in the direction x. By rearranging terms in the first parenthetical term, we obtain

$$\lim_{\Delta x\to 0}\left(-\frac{1}{\Delta x}\right)\left[\left(\frac{u_{10}-u_{00}}{\Delta x}-\frac{u_{00}-u_{-10}}{\Delta x}\right)+\frac{1}{2}\left(\frac{u_{11}-u_{01}}{\Delta x}-\frac{u_{01}-u_{-11}}{\Delta x}\right)\right.$$
$$\left.+\frac{1}{2}\left(\frac{u_{1-1}-u_{0-1}}{\Delta x}-\frac{u_{0-1}-u_{-1-1}}{\Delta x}\right)\right]$$
$$=-\left(\frac{\partial^2 u}{\partial x^2}+\frac{1}{2}\frac{\partial^2 u}{\partial x^2}+\frac{1}{2}\frac{\partial^2 u}{\partial x^2}\right)=-2\frac{\partial^2 u}{\partial x^2}$$

Similarly, the line 3 converges to $-(1-\nu)\dfrac{\partial^2 u}{\partial y^2}$. Lines 2 and 4 provide the following:

$$\lim_{\substack{\Delta x\to 0\\ \Delta y\to 0}}-\frac{1+\nu}{4\Delta y}\left(\frac{v_{11}-v_{-11}}{\Delta x}-\frac{v_{1-1}-v_{-1-1}}{\Delta x}\right)=-(1+\nu)\frac{\partial^2 v}{\partial x\partial y}$$

Dividing each by (-2) and adding the limits of these 4 lines of Equation (10.52a), we obtain

$$\frac{\partial^2 u}{\partial x^2}+\frac{(1-\nu)}{2}\frac{\partial^2 u}{\partial y^2}+\frac{(1+\nu)}{2}\frac{\partial^2 v}{\partial x\partial y}=0$$

This is the differential equation:

$$\frac{\partial \sigma_x}{\partial x}+\frac{\partial \tau_{xy}}{\partial y}=0$$

Note that the final line of Equation (10.52a) vanishes in the limit $(\Delta x\to,\Delta y\to 0)$. These are the HOT terms, which derive from the second line of Equation (10.48).

At a boundary node, one or both components of displacement may be constrained, i.e. $u_{nm}=0$ and/or $v_{nm}=0$. If a boundary node is unconstrained, then the force derives from the strains of the adjoining elements. For example, the forces at nodes pq and gh of Figure 10.16 derive from the 2 adjoining elements; $\tilde{U}$ contains the energies of those 2 elements. At the corner kl, the equilibrium condition derives from the internal energy $\tilde{U}$ of the one element *and* the potential of the edge load $p(y)$. To illustrate, let us consider the boundary node st of Figure 10.16. The affected elements are depicted in Figure 10.20. Now, the affected potential consists of the internal energies of elements 2 and 3 and the potential of the edge load $p(y)$; for simplicity, let us assume that p is a constant shear (τ_{xy}). Then the relevant potential P follows:

$$\tilde{P}=\tilde{U}^2+\tilde{U}^3-\tilde{p}\int_{-\Delta y}^{\Delta y} v\,dy \tag{10.53a}$$

where

$$\tilde{p}=\frac{2(1-\nu^2)p}{Eh\Delta x\Delta y}$$

$\tilde{p}$ is force per unit area; Equation (10.53a) is dimensionless. We recall our linear approximation of v, viz.

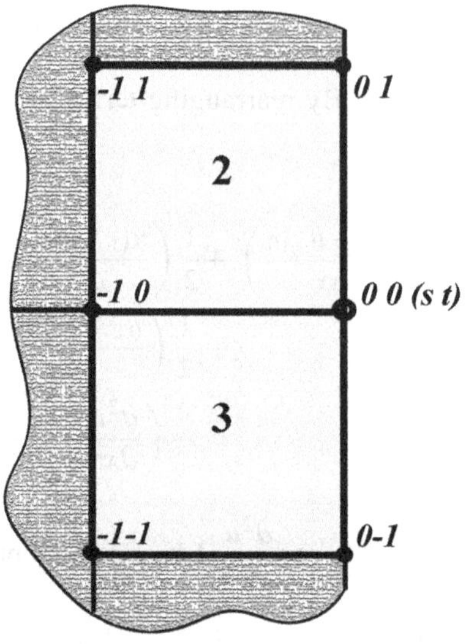

Figure 10.20 Patch of Support for Boundary Node 'st'

$$v = v_{00} + \frac{v_{00} - v_{0-1}}{\Delta y} y \qquad -\Delta y < y < 0$$

$$v = v_{00} + \frac{v_{10} - v_{00}}{\Delta y} y \qquad 0 < y < \Delta y$$

Equation (10.53a) takes the form:

$$\tilde{P} = \tilde{U}^2 + \tilde{U}^3 - \tilde{p}(v_{0-1} + 2v_{00} + v_{01})\frac{\Delta y}{2} \tag{10.53b}$$

The stationary condition follows:

$$\frac{\partial \tilde{P}}{\partial v_{00}} = \frac{\partial}{\partial v_{00}}(\tilde{U}^2 + \tilde{U}^3) - \tilde{p}\Delta y = 0 \tag{10.53c}$$

To identify the physical significance, we turn to Equation (10.48). Also, in quadrants 2 and 3, respectively,

$$\bar{\varepsilon}_y = \frac{1}{2\Delta y}(-v_{-10} - v_{00} + v_{01} + v_{-11}), \qquad \text{quadrant 2}$$

$$\bar{\varepsilon}_y = \frac{1}{2\Delta y}(-v_{-1-1} - v_{0-1} + v_{00} + v_{-10}), \qquad \text{quadrant 3}$$

$$\gamma_y = \frac{1}{2\Delta x}(-v_{-10} + v_{00} + v_{01} - v_{-11}), \qquad \text{quadrant 2}$$

$$\gamma_y = \frac{1}{2\Delta x}(-v_{-1-1} + v_{0-1} + v_{00} - v_{-10}), \qquad \text{quadrant 3}$$

Also, we note that

$$\frac{\partial \bar{\varepsilon}_x}{\partial v_{00}} = \frac{\partial \bar{\gamma}_x}{\partial v_{00}} = 0$$

$$\frac{\partial \bar{\varepsilon}_y}{\partial v_{00}} = -\frac{1}{2\Delta y}, \qquad \frac{\partial \gamma_y}{\partial v_{00}} = \frac{1}{2\Delta x}, \qquad \text{quadrant 2}$$

$$\frac{\partial \bar{\varepsilon}_y}{\partial v_{00}} = \frac{1}{2\Delta y}, \qquad \frac{\partial \gamma_y}{\partial v_{00}} = \frac{1}{2\Delta x}, \qquad \text{quadrant 3}$$

From Equation (10.53c) we obtain via the chain rule, as in Equation (10.49):

$$\begin{aligned}\frac{\partial \tilde{P}}{\partial v_{00}} = &-\frac{1}{\Delta y}(\bar{\varepsilon}_y + v\bar{\varepsilon}_x)_2 + \frac{(1-v)}{2\Delta x}(\gamma_x + \gamma_y)_2 \\ &+ \frac{1}{\Delta y}(\bar{\varepsilon}_y + v\bar{\varepsilon}_x)_3 + \frac{(1-v)}{2\Delta x}(\gamma_x + \gamma_y)_3 - \tilde{p}\Delta y = 0\end{aligned} \tag{10.54}$$

To view the physical quantities, we multiply the foregoing equation by the factor $\dfrac{E\Delta x}{2(1-v^2)}$. The result is as follows:

$$-\frac{1}{2}\frac{\Delta x}{\Delta y}\left[(\bar{\sigma}_y)_2 - (\bar{\sigma}_y)_3\right] + \frac{1}{2}\left[(\bar{\tau}_{xy})_2 + (\bar{\tau}_{xy})_3\right] = p$$

In the limit $(\Delta x \to 0, \Delta y \to 0)$, $(\bar{\sigma}_y)_2 \to (\bar{\sigma}_y)_3$, $(\bar{\tau}_{xy})_2 \to (\bar{\tau}_{xy})_3$; the preceding equation attains the limit:

$$\tau_{xy} = p$$

Similarly, the condition $\partial\tilde{P}/\partial u_{00} = 0$ leads to the result:

$$\sigma_x = 0$$

Note that the HOT terms of (10.48) are not relevant in the previous argument, since these vanish in the limit. Of course, the implementation of the discrete approximation requires the inclusion of the HOT terms, and all terms of the equation must be expressed in terms of the nodal displacement (u_{nm}, v_{nm}).

In summary, at each interior node, two equations of equilibrium are required, viz.

$$\frac{\partial \tilde{U}}{\partial u_{nm}} = 0, \qquad \frac{\partial \tilde{U}}{\partial v_{nm}} = 0$$

The first is explicitly given here [Equation (10.52b)]. A boundary may be subject to a kinematic constraint, e.g. $u_{nm} = 0$ or $v_{nm} = 0$; then those conditions supplant the equilibrium condition. If a traction acts upon a boundary, e.g. p_x or p_y in directions x or y, respectively, then an equilibrium condition applies, as in our example at node *st* of Figure 10.16 and 10.20.

10.6.2 IMPLEMENTATION

Every 2-D model of a solid body composed of finite elements possesses a discrete number, m, of *active* nodes, viz. unconstrained, it has $2m$ degrees of freedom. Therefore, at each node of the two-dimensional model, we must enforce two equations of equilibrium in accordance with the principle of virtual work $(\delta u_m, \delta v_m)$. In the conservative system, e.g. elastic body, the work of the internal force components is the change of internal energy. At an interior node, such as *nm* in

Figure 10.16, that internal force derives from the internal energies of the four adjoining elements. At a surface node, such as *oo* in Figure 10.20, it derives from the internal energies of the two adjoining nodes. Let us label the forces F_n and V_n in the u and v directions; n denotes the position of the element relative to the node. Figure 10.21 depicts a rectangular patch and the adjoining four elements. If U^n is the internal energy of element n, then

$$F_n = \frac{\partial U^n}{\partial u_{00}}, \quad V_n = \frac{\partial U^n}{\partial v_{00}} \tag{10.55a,b}$$

These are the internal forces exerted by one element n at the node 00.

Figure 10.21 depicts the patch of Figure 10.19 and the force components, F_n and V_n, which are exerted by the individual elements $(1,2,3,4)$. In the absence of an external load, the equations of equilibrium at the interior node (00) follow:

$$F_1 + F_2 + F_3 + F_4 = 0, \quad V_1 + V_2 + V_3 + V_4 = 0$$

If the node were at a corner, these equations would contain but one term:

$$F_n = V_n = 0$$

If the node lies on a surface, then the node adjoins two elements, n and m; these equations contain the contributions from both, Viz. Equation

$$F_n + F_m = 0, \quad V_n + V_m = 0$$

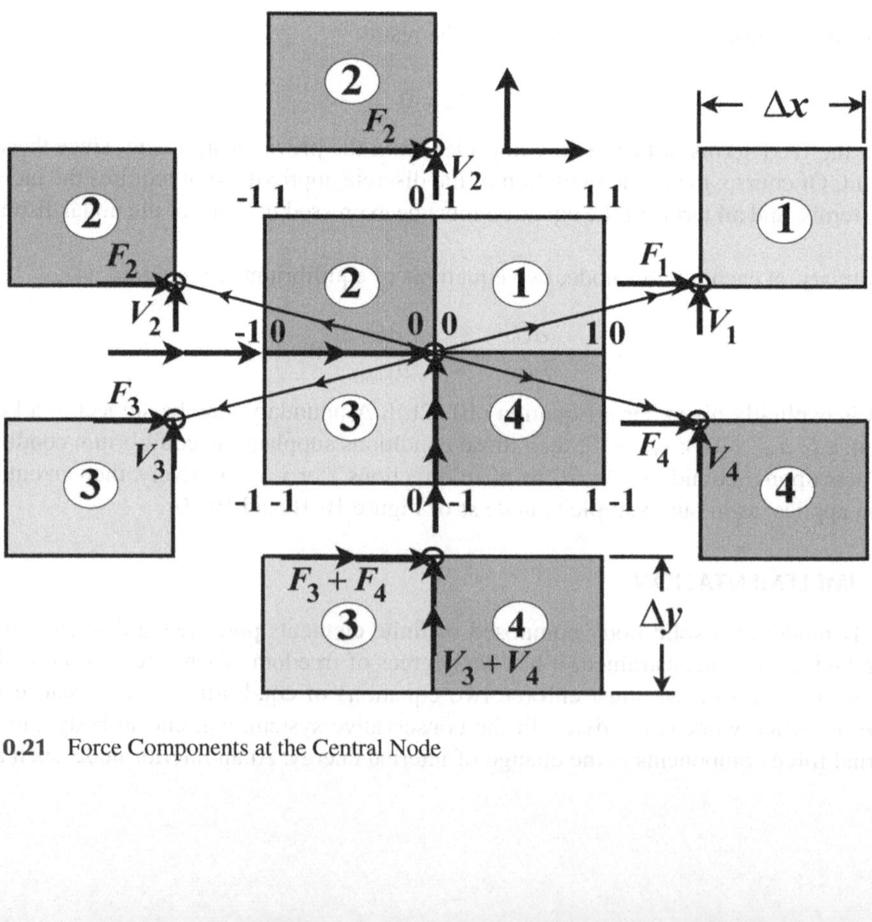

Figure 10.21 Force Components at the Central Node

Let us review the steps whereby we express the forces F_n and V_n in terms of the displacement components, u_{nm} and v_{nm}:

The expression U^n is the integral of Equation (10.46) with strains as given by Equations (10.43). The derivatives in Equation (10.55) are expressed in terms of the strains ($\bar{\varepsilon}_x, \bar{\varepsilon}_y, \bar{\gamma}_x, \bar{\gamma}_y, \kappa_x,$ and κ_y). These are expressed in terms of the displacements by Equations (10.45) and the derivatives by Equations (10.50). With those substitutions, Equation (10.55a) is expressed in terms of the nodal displacements of element n. As mentioned before, we employ the nondimensional variables:

$$\bar{u} \equiv \frac{u}{\Delta y}, \ \bar{v} \equiv \frac{v}{\Delta y}, \ \bar{U} \equiv \frac{2(1-\nu^2)}{Eb\Delta x\Delta y}U \tag{10.56a,b,c}$$

$$\bar{F} \equiv \frac{2(1-\nu^2)}{Eb\Delta x}F, \ \bar{V} \equiv \frac{2(1-\nu^2)}{Eb\Delta x}V \tag{10.56d,e,f}$$

In the dimensionless terms,

$$\begin{aligned}
\bar{F}_1 &= \frac{\partial U^1}{\partial \bar{u}_{00}} \\
&= \left[\frac{2}{3}\left(\frac{\Delta y}{\Delta x}\right)^2 + \frac{(1-\nu)}{4}\right]\bar{u}_{00} - \left[\frac{2}{3}\left(\frac{\Delta y}{\Delta x}\right)^2 - \frac{(1-\nu)}{4}\right]\bar{u}_{10} \\
&\quad - \left[\frac{1}{3}\left(\frac{\Delta y}{\Delta x}\right)^2 + \frac{(1-\nu)}{4}\right]\bar{u}_{11} - \left[\frac{1}{3}\left(\frac{\Delta y}{\Delta x}\right)^2 - \frac{(1-\nu)}{4}\right]\bar{u}_{01} \\
&\quad + \left[\frac{(1+\nu)}{4}\left(\frac{\Delta y}{\Delta x}\right)\right](\bar{v}_{00} - \bar{v}_{11}) + \left[\frac{(1-3\nu)}{4}\left(\frac{\Delta y}{\Delta x}\right)\right](\bar{v}_{01} - \bar{v}_{10})
\end{aligned} \tag{10.57}$$

The remaining force components follow in like manner. It is a matter of "book keeping". The Equation (10.57) applies to any node nm; 00 is replaced by nm, O1 by nm+1, etc. If node nm lies at the corner with no external load, the equilibrium dictates that $\bar{F} = 0$.

The foregoing steps in the derivation of the forces (F_n, V_n) are burdensome when done manually; however, these are robotic and can be committed to our digital data processors, i.e. "software". However, it is important that we explore these procedures on a basic conceptual level. To do so, is to understand the limitations and possible failures. For example, the so-called "locking" which made certain models unusable. This was particularly troublesome until we examined the modes and energies associated with higher-order modes. The locking of thin elements in plates and shells found a partial solution with the introduction of the discrete version of Kirchhoff's approximation.[6] A better solution is illustrated by Illustrative Example 10.1.[7] The "locking" is absent in our present formulation as we have omitted the "parasitic" shear energy, the underlined higher order terms of Equation (10.49b)

ILLUSTRATIVE EXAMPLE 10.5

To illustrate the method of approximation of a body in a state of plane stress via the bilinear element, we turn to the rectangular body of Figure 10.22. Notations are the same as those of Figures 10.6 and 10.21. The force $\bar{F}_1$ is given by Equation (10.57). Forces $\bar{F}_2, \bar{F}_3$ and $\bar{F}_4$ are obtained in like manner, viz., from the internal energies of elements 2, 3, and 4, e.g.

$$\bar{F}_n = \frac{\partial U^n}{\partial \bar{u}_{00}}, \ \bar{V}_n = \frac{\partial U^n}{\partial \bar{v}_{00}}$$

where $\bar{U}^n$ is given by Equation (10.48).

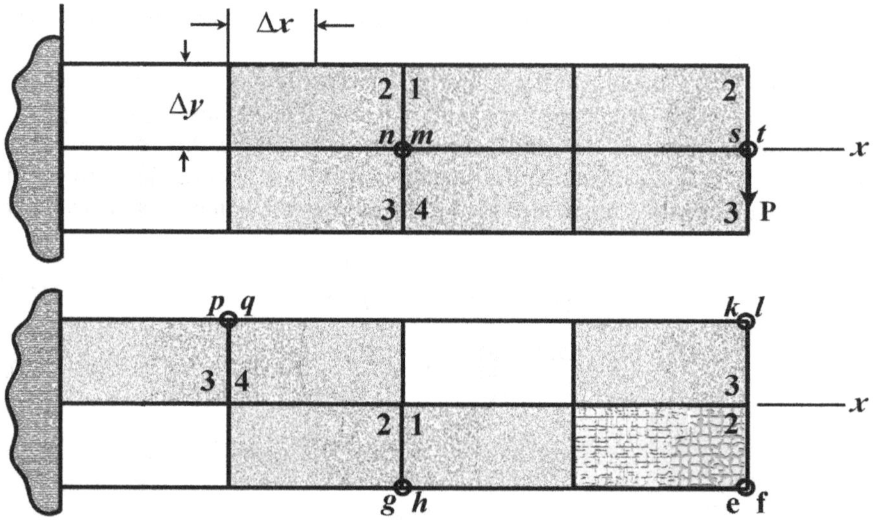

Figure 10.22 Cantilever Beam as a Plane tress Problem

At the interior node *nm*, the total internal force derives from the four adjoining elements. That is Equation (10.52b); here, in terms of the nondimensional displacements $(\bar{u}, \bar{v})$:

$$
\begin{aligned}
\bar{F} &\equiv \frac{\partial(\bar{U}^1 + \bar{U}^2 + \bar{U}^3 + \bar{U}^4)}{\partial \bar{u}_{00}} \\
&= \left[\frac{8}{3}\left(\frac{\Delta y}{\Delta x}\right)^2 + (1-\nu)\right]\bar{u}_{00} + \left[\frac{2}{3}\left(\frac{\Delta y}{\Delta x}\right)^2 - \frac{(1-\nu)}{2}\right](\bar{u}_{0-1} + \bar{u}_{01}) \\
&- \left[\frac{4}{3}\left(\frac{\Delta y}{\Delta x}\right)^2 - \frac{(1-\nu)}{2}\right](\bar{u}_{10} + \bar{u}_{-10} \\
&- \left[\frac{1}{3}\left(\frac{\Delta y}{\Delta x}\right)^2 + \frac{(1-\nu)}{4}\right](\bar{u}_{-1-1} + \bar{u}_{1-1} + \bar{u}_{11} + \bar{u}_{-11}) \\
&+ \left[\frac{(1+\nu)}{4}\left(\frac{\Delta y}{\Delta x}\right)\right](\bar{v}_{1-1} - \bar{v}_{-1-1} + \bar{v}_{-11} - \bar{v}_{11})
\end{aligned}
\tag{10.58}
$$

Let us apply the foregoing equations to the simple cantilever beam of Figure 10.22 with a transverse load P applied at the node s,t.

At each of the three interior nodes, we have the two equations of equilibrium, e.g. *nm*:

$$\bar{F}^1 + \bar{F}^2 + \bar{F}^3 + \bar{F}^4 = 0, \ \bar{V}^1 + \bar{V}^2 + \bar{V}^3 + \bar{V}^4 = 0$$

At the three nodes on the top surface, we have the two equations, e.g. *pq*:

$$\bar{F}^3 + \bar{F}^4 = 0, \ \bar{V}^3 + \bar{V}^4 = 0$$

At the 3 nodes on the lower surface we have the 2 equations, e.g. *gh*:

$$\bar{F}^1 + \bar{F}^2 = 0, \ \bar{V}^1 + \bar{V}^2 = 0$$

At the upper corner (κl) and the lower corner (e, f), we have the equations:

$$\bar{F}^3 = 0,\ \bar{V}^3 = 0,\ \text{and}\ \bar{F}^2 = 0,\ \bar{V}^2 = 0$$

Finally, at the node (s,t) we must include the applied force $\bar{P}$ in the equations of equilibrium:

$$\bar{F}^2 + \bar{F}^3 = 0,\ \bar{V}^2 + \bar{V}^3 = -P$$

Together we have a system of 24 simultaneous linear equations in the 24 variables. The cantilever beam is thereby approximated by the displacements (u_{nm}, v_{nm}) at the 12 active nodes.

The solution of the simultaneous equations, the 24 components of displacement, are distilled in Figure 10.23. Here, we note that the displacements at the end A are approximately 5% less than those given by the Bernoulli beam theory. The longitudinal displacement vanishes at the center line $(\bar{u} = 0)$. At the midpoint, A, the transverse displacement is the same as that at the upper and lower nodes on the section, e.g. at A $\bar{v} = 0.0946$. These results indicate that there is no significant transverse extensional strain, viz. $\varepsilon_y \doteq 0$. Finally, with the displacements at all nodes, we can return to Equations (10.45c,f) to calculate the mean shear strain (γ_{xy} in each of the eight elements and then the consequent deflection caused by the transverse shear: Those computations involve the small differences between the rotation of the lateral (γ_x) and axial (γ_y) axial lines. We lose two decimal places in accuracy; however, those calculations show that the shear causes less than 1% of that deflection.

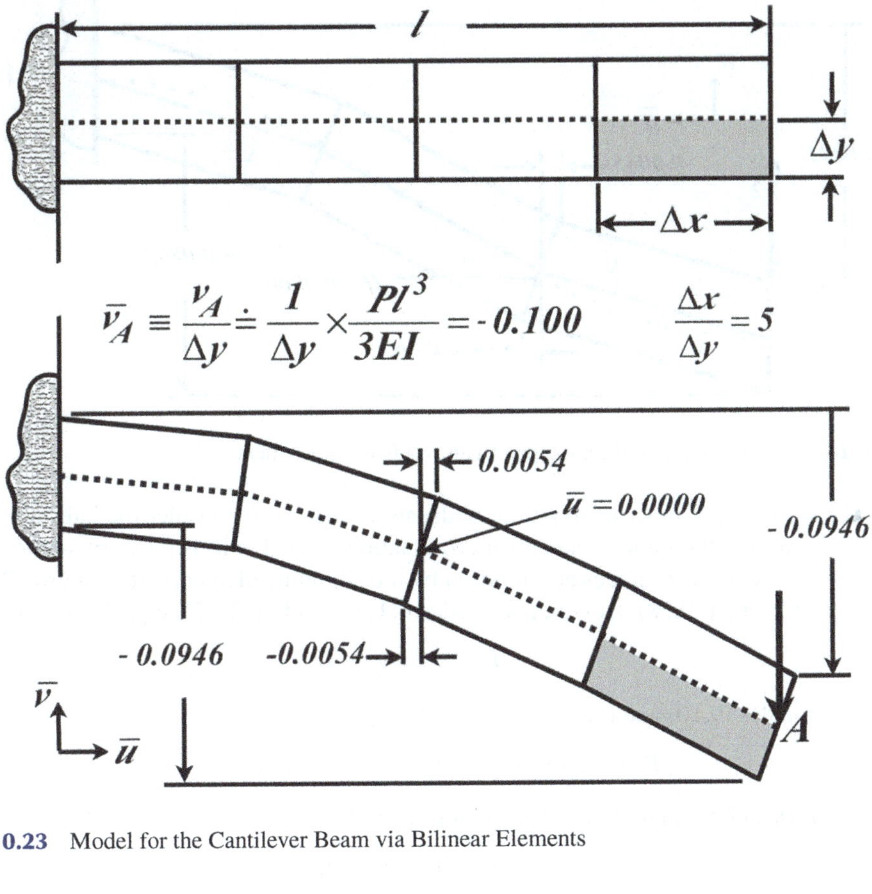

Figure 10.23 Model for the Cantilever Beam via Bilinear Elements

ILLUSTRATIVE EXAMPLE 10.6

The approximation of the beam in Figure 10.24a poses a more challenging problem. Here the deformation exhibits the convex curvature near the clamped ends and concave curvature at the center, the site of the load P.

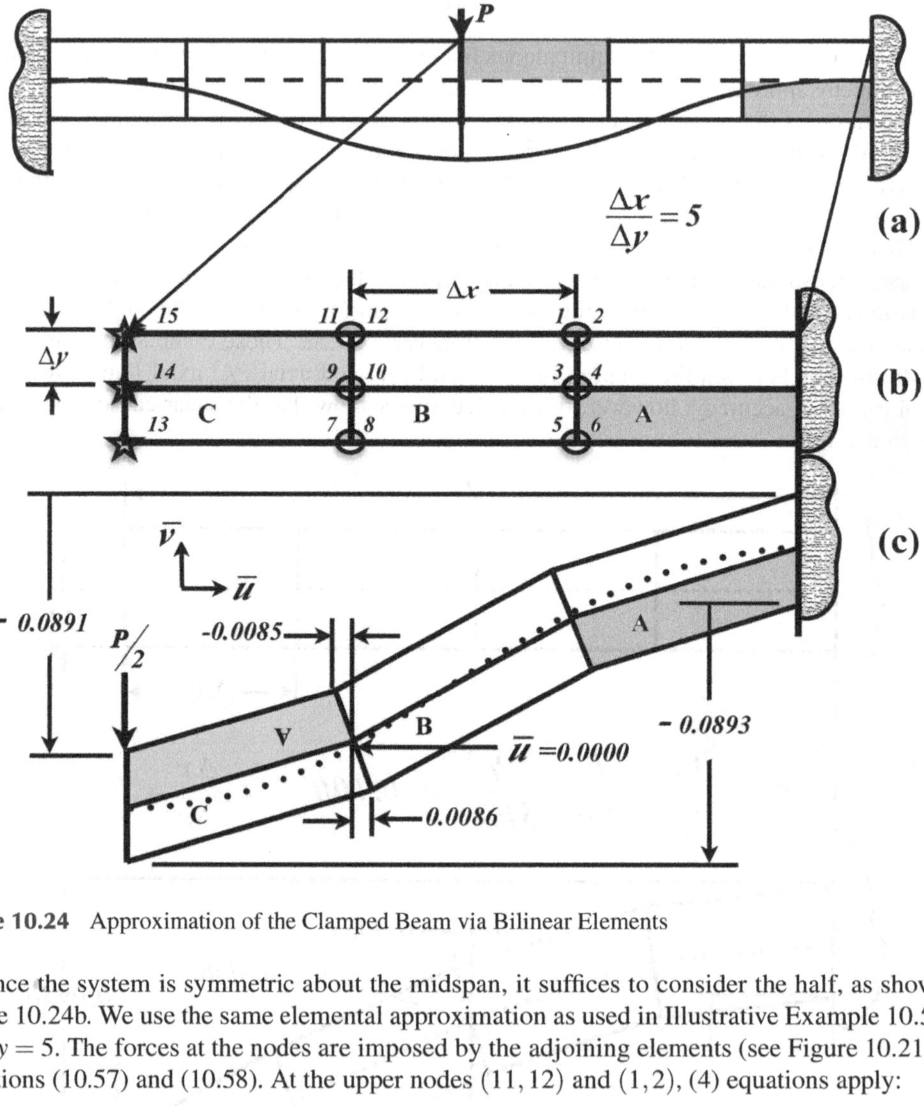

Figure 10.24 Approximation of the Clamped Beam via Bilinear Elements

Since the system is symmetric about the midspan, it suffices to consider the half, as shown in Figure 10.24b. We use the same elemental approximation as used in Illustrative Example 10.5, viz $\Delta x/\Delta y = 5$. The forces at the nodes are imposed by the adjoining elements (see Figure 10.21) and Equations (10.57) and (10.58). At the upper nodes $(11, 12)$ and $(1,2)$, (4) equations apply:

$$F_3 + F_4 = V_3 + V_4 = 0$$

At the interior nodes (9,10) and (3,4), (4) equations apply:

$$F_1 + F_2 + F_3 + F_4 = V_1 + V_2 + V_3 + V_4 = 0$$

At the lower nodes $(7,8)$, and $(5,6)$, (4) equations apply:

$$F_1 + F_2 = V_1 + V_2 = 0$$

At node (15), $V_4 = -P/2$
At node (14), $V_1 + V_4 = 0$
At node (13), $V_1 = 0$

Note that nodes $(13), (14)$, and (15) are inactive in the $\bar{u}$ direction., i.e,

$$\bar{u}_{13} = \bar{u}_{14} = \bar{u}_{15} = 0$$

The result is a system of 15 linear algebraic equations in the 15 displacement components.

The results are displayed in Figure 10.24c. Once again, we can insert the displacements, u_{nm} and v_{nm}, into Equations (10.45c,f) to compute γ_{xy}. The results indicate a shear deformation of less than 2%. Of course, the present beam is much more constrained against flexural deformations.

ILLUSTRATIVE EXAMPLE 10.7

Section 4.19 addressed the matter of stress concentrations, which occur at the ends of an elliptical hole in a plate. Such concentrations of stress (and strain) occur at any re-entrant corner of a machine or structural component. These are accompanied by high gradients of stress (and strain) in the vicinity of the concentration. Similarly, a localized support causes high gradients. An approximation via finite elements requires a refinement of the mesh, i.e. smaller elements, in the neighborhood of the corner or support. This example serves to illustrate one simple means of transition to a finer mesh in such regions.

The notched plate of Figure 10.25a is gripped at the top and bottom, i.e. along lines RE and ST, respectively. These are subjected to the force P acting along the y axis. The specimen and loading are symmetrical with respect to the x and y axes. Therefore, the one quadrant of Figure 10.25b reflects the response of all: Symmetry requires that the displacement in the x direction vanish on the midline AF, $u(0,y) = 0$.

Symmetry also dictates that line AB remain straight. Since a rigid motion is irrelevant, we take line AB to remain on the x axis, $v(x,0) = 0$. The specimen is relatively thin; the material is steel (HI-HO), i.e. linearly elastic to a state of yielding or fracture. Therefore, we adopt the theory of plane stress (see Sections 8.3 ... 8.6). Then the methods and equations of the preceding examples apply:

a) The bilinear approximation of displacements (u,v) in the rectangular elements, viz Equations (10.42) and (10.43).
b) The dimensionless displacements $(\bar{u},\bar{v})$ and forces $(\bar{F},\bar{V})$ and the energy ($\bar{U}$ of Equations (10.56). The explicit expression expression for energy $(\bar{U}^n)$ of an element n, Equation (10.48), and force $(\bar{F}_n)$ of element n. The forces $(\bar{F}_n,\bar{V}_n)$ as given by Equation (10.55a,b).

As before, an interior node, e.g. node R, in Figure 10.25, which is unconstrained, is subject to the two equilibrium equations

$$F_1 + F_2 + F_3 + F_4 = V_1 + V_2 + V_3 + V_4 = 0$$

Elements $(1,2,3,4)$ are labeled as those in Figure 10.21.

In the present example, we have nodes that are partially constrained. Nodes on line FE are clamped: $u = 0, v = K$. Nodes on line AB are partially constrained: $(v = 0)$. Nodes on line AF are partially constrained $(u = 0)$.

Note that specific nodes as identified with a star (*), on lines GD, HC, and PR are "transitional"; these nodes accommodate the transitions to a finer mesh. These are not entirely free; their value is

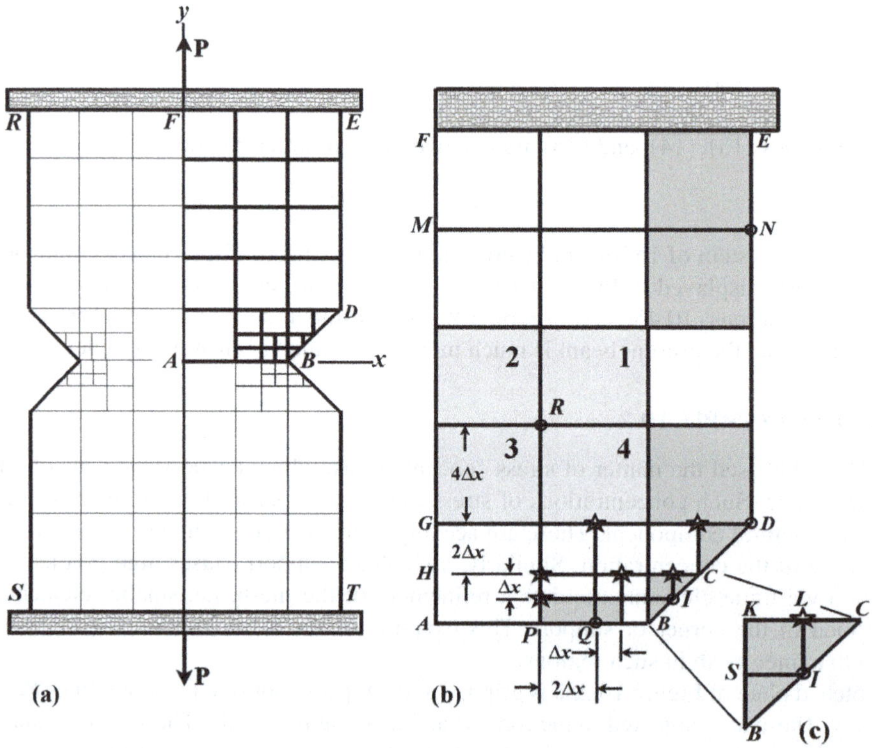

Figure 10.25 Notched Specimen

established by the simple equation, which assigns the median values (u and v) at the two adjoining nodes: Figure 10.25c depicts one such node (L); here, the governing equations are simply,

$$u_L = \frac{1}{2}(u_K + u_C), \quad v_L = \frac{1}{2}(v_K + v_C) \tag{10.59}$$

The enforcement of the equilibrium conditions at all unconstrained nodes provides a system of linear algebraic equations in displacement components (u_{nm} and v_{nm}).

At the nodes B and D, the forces are imposed at the juncture of a rectangular and a triangular element, at C and at I, the forces are imposed by a rectangular and two triangular elements. Therefore, we must now address the approximation of the triangular element as depicted in Figure 10.26. Here, we employ the linear equations:

$$u = u_0 + \bar{\varepsilon}_x x + \gamma_x y, \quad v = v_0 + \bar{\varepsilon}_y y + \gamma_y x$$

The four coefficients $\bar{\varepsilon}_x$, $\bar{\varepsilon}_y$, γ_x, and γ_y, and two central displacements, u_0 and v_0 can be expressed in terms of nodal displacements at the three corners:

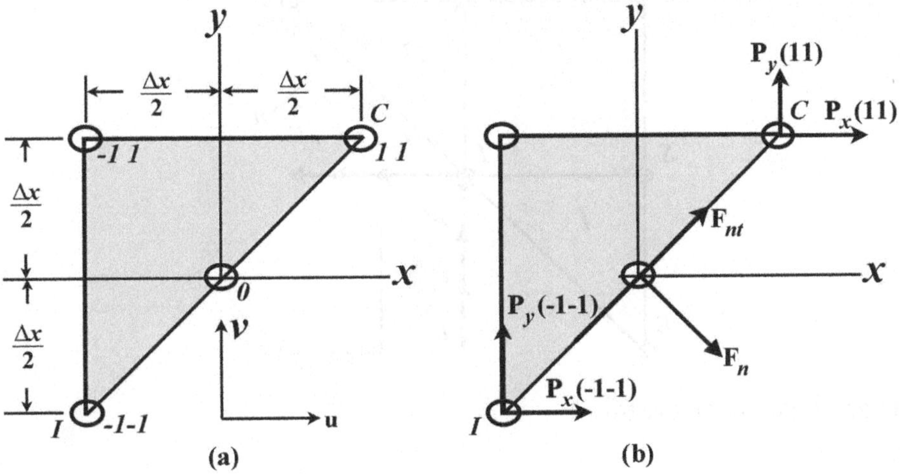

Figure 10.26 Triangular Element

$$
\begin{aligned}
&u_0 = \frac{1}{2}(u_{11} + u_{-1-1}), \quad v_0 = \frac{1}{2}(v_{11} - v_{-1-1}) \\
&\bar{\varepsilon}_x = \frac{u_{11} - u_{-11}}{\Delta x}, \quad \bar{\varepsilon}_y = \frac{v_{-11} - v_{-1-1}}{\Delta y}, \quad \gamma_x = \frac{u_{-11} - u_{-1-1}}{\Delta y}, \quad \gamma_y = \frac{v_{11} - v_{-11}}{\Delta x} \\
&\gamma_{xy} = \gamma_x + \gamma_y = \frac{u_{-11} - u_{-1-1}}{\Delta y} + \frac{v_{11} - v_{-11}}{\Delta x}
\end{aligned}
\tag{10.60}
$$

The force components (P_x, P_y) at the nodes are drawn from the internal energy U^Δ of the triangular element. This is a modified version of the energy for the rectangular element (see Equations 10.46–10.48):

$$\bar{U}^\Delta = \frac{1}{2}\left(\bar{\varepsilon}_x^2 + 2\nu\bar{\varepsilon}_x\bar{\varepsilon}_y + \bar{\varepsilon}_y^2\right) + \frac{1-\nu}{4}(\gamma_x + \gamma_y)^2 \tag{10.61}$$

$$\bar{P}_x(C) = \bar{\varepsilon}_x + \nu\bar{\varepsilon}_y, \quad \bar{P}_y(C) = \frac{1-\nu}{2}(\gamma_x + \gamma_y) \tag{10.62a}$$

$$\bar{P}_x(I) = -\frac{1-\nu}{2}(\gamma_x + \gamma_y), \quad \bar{P}_y(I) = -(\bar{\varepsilon}_y + \bar{\nu}\varepsilon_x) \tag{10.62b}$$

At the node I of Figure 10.25c, the above components must be added to those drawn from the adjoining rectangular element. From Equation (10.49b):

$$\bar{F}_x = (\bar{\varepsilon}_x + \nu\bar{\varepsilon}_y) - \frac{(1-\nu)}{2}(\gamma_x + \gamma_y) - \frac{\kappa_x}{6} \tag{10.63a}$$

$$\bar{V}_y = -(\bar{\varepsilon}_y + \nu\bar{\varepsilon}_x) + \frac{(1-\nu)}{2}(\gamma_x + \gamma_y) - \frac{\kappa_y}{6} \tag{10.63b}$$

Implementation requires that the strains ($\bar{\varepsilon}_x$, $\bar{\varepsilon}_y$, and κ) be expressed in terms of the nodal displacements. Equations (10.45) apply to Equations (10.63a,b). Equations (10.60) apply to Equations (10.62a,b). Bear in mind that the notations of Figure (10.26) and Equation (10.60) only signify the relative positions of the nodes.

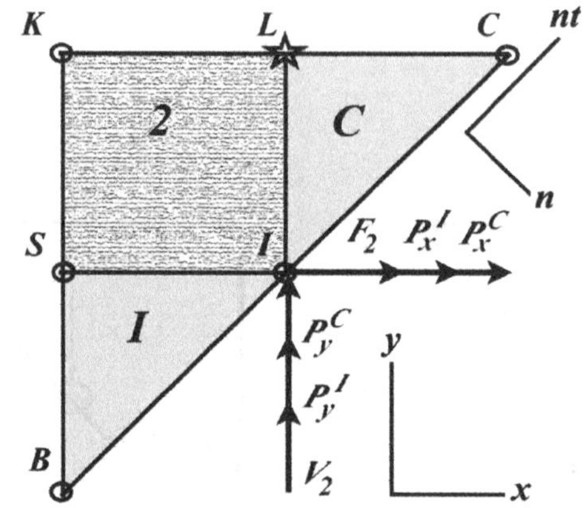

Figure 10.27 Actions on Surface Node I

Let us examine the physical meaning of the action at the node I of Figure 10.27. Firstly, we note that these must equilibrate the actions upon the segment CB with area $2\sqrt{2}b\Delta x$. These forces are the mean stress components when multiplied by the factor

$$\frac{E}{2\sqrt{2}b\,\Delta x\,(1-\nu^2)}$$

Then the sums $(\bar{F}_x + \bar{P}_x^I + \bar{P}_x^C)$ and $(\bar{F}_y + \bar{P}_y^I + \bar{P}_y^C)$ take the respective form:

$$\bar{\sigma}_x \equiv \frac{\sigma_x}{\sqrt{2}} - \frac{\tau_{xy}}{\sqrt{2}}, \quad \bar{\sigma}_y \equiv -\frac{\sigma_y}{\sqrt{2}} + \frac{\tau_{xy}}{\sqrt{2}}$$

When expressed by components in the directions normal $'n'$ and tangent (nt) to the surface BC, we obtain

$$\bar{\sigma}_{nt} = \frac{\sigma_x - \sigma_y}{2}, \quad \bar{\sigma}_n = \frac{\sigma_x + \sigma_y}{2} - \tau_{xy}$$

These are precisely the Equations (3.32) and (3.33) for the transformation of stress components with $\theta = -45°$. Again, the overbar $(^-)$ signifies a mean value. In the limit $(\Delta x \to 0)$, the nodal displacements merge, as do the strain components of the elements adjoining node I.

Finally, we conclude the study of the notched specimen with the results of Figure 10.28. If the mesh were refined further, the maximum stress at the notch (B) would continually grow, since the model presumes unlimited elastic behavior. At some load, the computation would indicate a state of stress sufficient to cause yielding or fracture. Numerical results in Figure 10.28 were obtained by Sergei Khakalo.[8]

10.6.3 BOUNDARIES AND TRIANGLES

Previously, we focused upon assemblies of rectangular elements, which offer the simplest developments and correlations with the continuous body. Although a rectangular grid can be imposed upon any plate, boundaries must be accommodated by triangular elements, e.g. in Figure 10.29, the curved boundary is simulated by a succession of straight lines, the hypotenuse of each successive triangle. At a node, such as B in Figure 10.29, the internal force derives from the internal energies

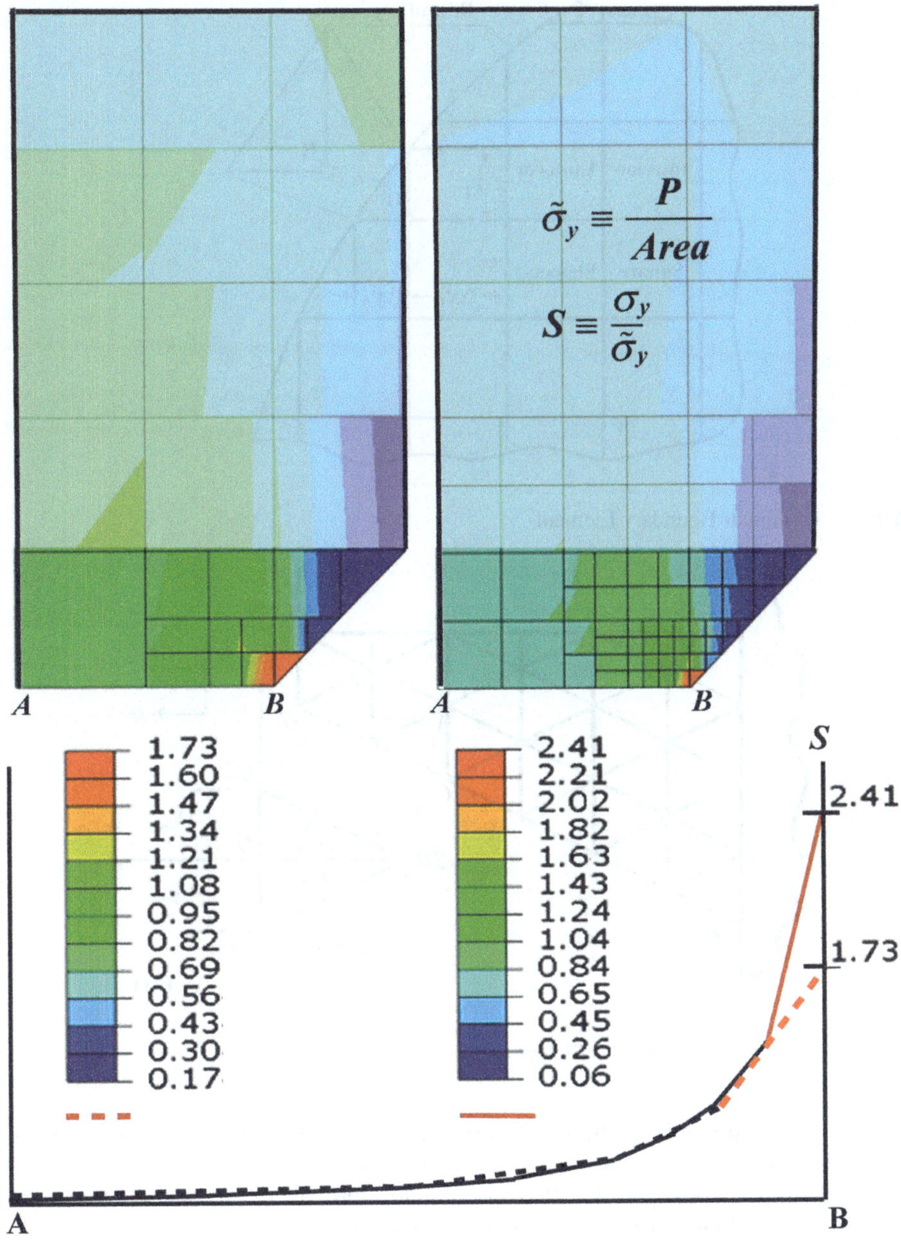

Figure 10.28 Mesh Reduction near a Stress Concentration

of the four adjoining elements, three rectangular and one triangular. In some circumstances, it is advantageous to approximate the entire body via an assemblage of triangular elements. Such is the case of the support depicted in Figure 10.30.

In Chapter 2, in Illustrative Example 2.15, we show that the strains at a point in a surface are entirely described by the extensional strains in three distinct directions (a, b, c in Figure 2.31).

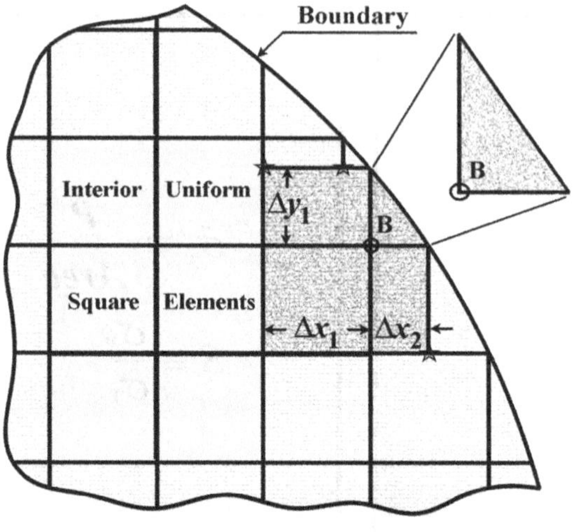

Figure 10.29 Triangular Boundary Element

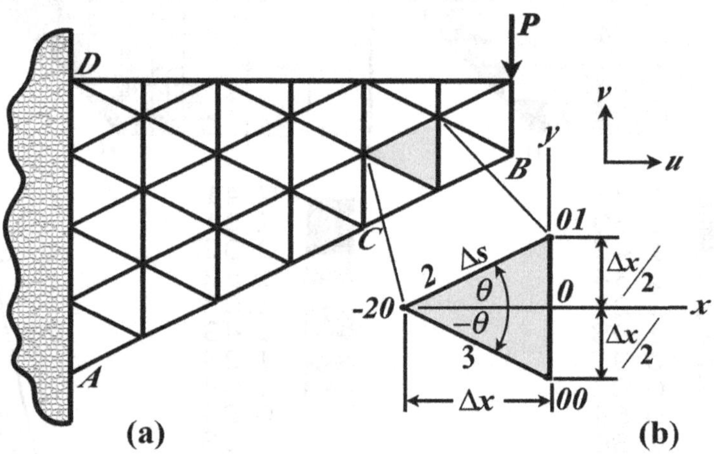

Figure 10.30 Support with Triangular Element

Therefore, it is not surprising that the simplest triangular element is one of uniform strain, and that state can be expressed by the extensional strains of the three sides.

Let us use the support of Figure 10.30 as an example. Here, the mesh consists of lines parallel to the lower edge AB and lines, such as CD, which have the same, but negative slope, and vertical lines passing through the vertices. A typical element is depicted in Figure 10.30b. Since our expressions (Equation 10.61) of internal energy are given in terms of the rectangular components of strain (and displacement 0, let us transform the extensional strains of edges 2 and 3. The extensional strains in directions 2 and 3 are obtained as simple differences, e.g.

$$\varepsilon_2 = \frac{1}{\Delta s}[(u_{01} - u_{-20})\cos\theta + (v_{01} - v_{-20})\sin\theta]$$

$$\varepsilon_3 = \frac{1}{\Delta s}[(u_{00} - u_{-20})\cos\theta - (v_{00} - v_{-20})\sin\theta]$$

Here,

$$\cos\theta = \frac{\Delta x}{\Delta s}, \quad \sin\theta = \frac{\Delta x}{2\Delta s}$$

$$\varepsilon_2 = \frac{\Delta x}{\Delta s^2}\left[(u_{01} - u_{-20} + \frac{1}{2}(v_{01} - v_{-20})\right] \tag{10.64a}$$

$$\varepsilon_3 = \frac{\Delta x}{\Delta s^2}\left[(u_{00} - u_{-20} - \frac{1}{2}(v_{00} - v_{-20})\right] \tag{10.64b}$$

$$\varepsilon_y = \frac{1}{\Delta x}(v_{01} - v_{00}) \tag{10.64c}$$

Let us apply Equation (2.47) to each of the strains, ε_2 and ε_3

$$\varepsilon_2 = \varepsilon_x \cos^2\theta + \varepsilon_y \sin^2\theta + \gamma_{xy}\sin\theta\cos\theta = \frac{\Delta x^2}{\Delta s^2}(\varepsilon_x + \frac{1}{4}\varepsilon_y) + \frac{\Delta x^2}{2\Delta s^2}\gamma_{xy} \tag{10.65a}$$

$$\varepsilon_3 = \varepsilon_x \cos^2\theta + \varepsilon_y \sin^2\theta - \gamma_{xy}\sin\theta\cos\theta = \frac{\Delta x^2}{\Delta s^2}(\varepsilon_x + \frac{1}{4}\varepsilon_y) - \frac{\Delta x^2}{2\Delta s^2}\gamma_{xy} \tag{10.65b}$$

We must solve the two simultaneous equations to express ε_x and γ_{xy} in terms of the strains ε_x, ε_2, ε_3:

$$\varepsilon_x = \frac{1}{\Delta x}\left[\frac{u_{00} + u_{01}}{2} - u_{-20}\right] \tag{10.66a}$$

$$\gamma_{xy} = \left[\frac{(v_{01} + v_{00})}{2\Delta x} - \frac{v_{-20}}{\Delta x}\right] + \left[\frac{u_{01} - u_{00}}{\Delta x}\right] \tag{10.66b}$$

As engineers, we can appreciate a geometric interpretation of these equations: The right side of Equation (10.66a) is the difference of the displacement component u_x at sites (-20) and (0) divided by the distance Δx. This is by definition ε_x in an element of constant strain. The first bracketed term of Equation (10.66b) is the counter-clockwise rotation of line (-20) to (0); the second bracketed term is the clockwise rotation of line (00) to (01). The sum is the change of angle between the x and y lines, τ_{xy} by definition.

In these examples, one line lies parallel to the y axis, hence ε_y is readily obtained [Equation (10.64c)]. For any triangle with sides a, b, and c with arbitrary angles, we could obtain the extensional strains in the manner of Equations (10.64) and then solve 3 equations like Equations (10.65a,b). With the strains ε_x, ε_y, and τ_{xy}, we would proceed to obtain the nodal force as in the previous example, viz

$$F_x = \frac{\partial U}{\partial u_{nm}}, \quad v_y = \frac{\partial U}{\partial v_{nm}}$$

For the equilateral triangles of Figure 10.30, the internal energy of Equation (10.61) takes the symmetric form:

$$U = \frac{1}{3}\left(\frac{5}{3} - \nu\right)(\bar{\varepsilon}_1^2 + \bar{\varepsilon}_2^2 + \bar{\varepsilon}_3^2) - \frac{2}{3}\left(\frac{1}{3} - \nu\right)(\bar{\varepsilon}_1\bar{\varepsilon}_2 + \bar{\varepsilon}_1\bar{\varepsilon}_3 + \bar{\varepsilon}_2\bar{\varepsilon}_3)$$

10.7 FLEXURAL DEFORMATION OF THIN PLATES

In Chapter 8, we formulated theories to describe the small flexural deformations of HI-HO plates. Again, our capabilities to solve those differential equations and to satisfy edge conditions are very limited. The methods of finite elements admit good approximations via linear *algebraic* equations; the enormous capacities of modern digital computers enable the numerical solution of large systems of those linear equations. Our purpose is to introduce the simpler formulations, to cast them in the context of the earlier theory of the continuum and to illustrate them with a few simple examples.

The Kirchhoff theory of Section 8.11 has proven very effective, although the constraint against transverse shear deformations (γ_{xz} and γ_{yz},) limits applicability to thin plates. That theory expresses the flexural strains, hence the internal energy, in terms of the curvatures of the reference surface; $\kappa_x = \partial^2 w/\partial x^2$, $\kappa_y = \partial^2 w/\partial y^2$, $\kappa_{xy} = \partial^2 w/\partial x \partial y x$. An approximation of a curved surface requires continuity of slope; $\partial w/\partial x$, $\partial w/\partial y$ must be continuous. To achieve continuity between contiguous elements requires a high order of approximation, viz., a polynomial of degree six in the interpolation of $w(x,y)$. Consequently, we turn to a simpler approximation, one which admits transverse shear (see Section 8.19). We accept the bilinear approximation of the transverse displacement within the element:

$$w = w_0 + \theta_x x + \theta_y y + \theta_{xy} xy \tag{10.67}$$

This provides for continuity of displacement at the contiguous edges of rectangular elements, but does introduce kinks at those edges. To preserve the continuity of the interfaces, we relax the Kirchhoff approximation; we require only that normals remain straight and admit independent rotation, β_x and β_y (see Section 8.19). Interfacial continuity is achieved by the bilinear approximations:

$$\beta_x = \bar{\beta}_x + \kappa_x x + \gamma_y y + \beta_{xy} xy \tag{10.68a}$$

$$\beta_y = \bar{\beta}_y + \gamma_x x + \kappa_y y + \beta_{yx} yx \tag{10.68b}$$

The rigid-body rotation of the element is identified with the motion of the reference surface as described by Equation (10.67). We suppose that the origin ($x = y = z = 0$) is set at the center of the rectangular element, as shown in Figure 10.31. Then, w_0 is the translation, $\bar{\theta}_x$ and $\bar{\theta}_y$ are the small rotations about the y and x lines, respectively.

Since we anticipate no relevant extensions of the reference surface ($z = 0$), the tangential components (u, v) of displacement are a consequence of the rotations:

$$u = -\beta_x z, \quad v = -\beta_y z \tag{10.69a,b}$$

The strain components follow the approximations (10.67) to (10.69a,b)

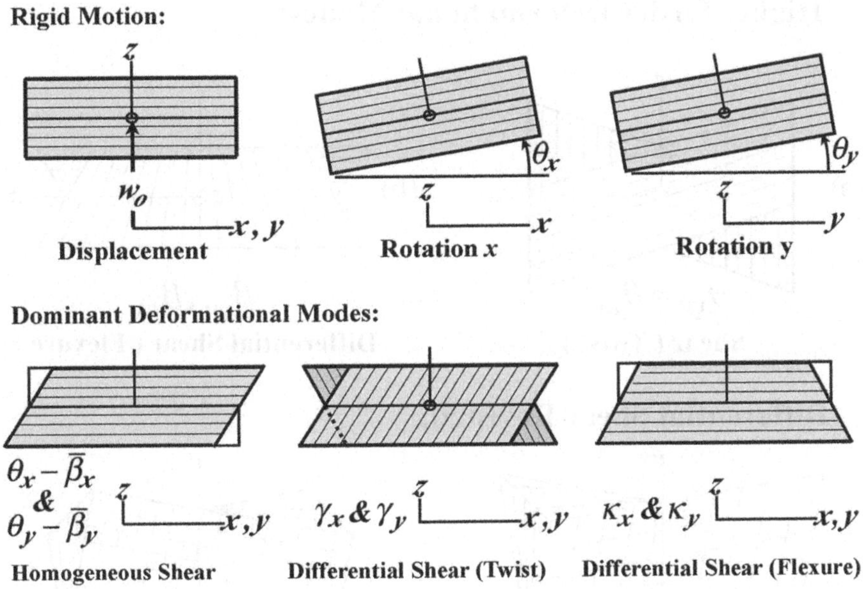

Figure 10.31 Motion and Dominant Deformations of the Bilinear Element

$$\varepsilon_x = \frac{\partial u}{\partial x} = -(\kappa_x + \beta_{xy} y) z, \quad \varepsilon_y = \frac{\partial v}{\partial y} = -(\kappa_y + \beta_{yx} x) z, \tag{10.70a,b}$$

$$\gamma_{xy} = \frac{\partial u}{\partial y} + \frac{\partial v}{\partial x} = -(\gamma_y + \gamma_x + \beta_{xy} x + \beta_{yx} y) z, \tag{10.70c}$$

$$\gamma_{xz} = \frac{\partial w}{\partial x} + \frac{\partial u}{\partial z} = (\bar{\theta}_x - \bar{\beta}_x) + (\theta_{xy} - \gamma_y) y - \underline{(\kappa_x x + \beta_{xy} xy)} \tag{10.70d}$$

$$\gamma_{yz} = \frac{\partial w}{\partial y} + \frac{\partial u}{\partial z} = (\bar{\theta}_y - \bar{\beta}_y) + (\theta_{xy} - \gamma_x) x - \underline{(\kappa_y y + \beta_{yx} yx)} \tag{10.70e}$$

Let us summarize the elemental approximations embodied in the approximations (10.67), (10.68), and (10.69), and now reflected in the strain components (10.70a–e): Collectively the element possesses twelve degrees of freedom as reflected by the coefficients in the three Equations (10.67) and (10.68a,b). Three coefficients describe rigid movement:

$$\textbf{Rigid Body Motions: } w_0, \bar{\theta}_x, \bar{\theta}_y \tag{10.71a}$$

In Equations (10.70a–c), we identify the three dominant flexural modes; the initial terms of equations (10.70d,e) constitute the homogeneous transverse shear modes. These five dominant deformational modes are characterized by the six constants:

$$\textbf{Flexural Modes: } \kappa_x, \kappa_y, \gamma_x + \gamma_y \tag{10.71b}$$

$$\textbf{Transverse Shear Modes: } (\bar{\theta}_x - \bar{\beta}_x), (\bar{\theta}_y - \bar{\beta}_y) \tag{10.71c}$$

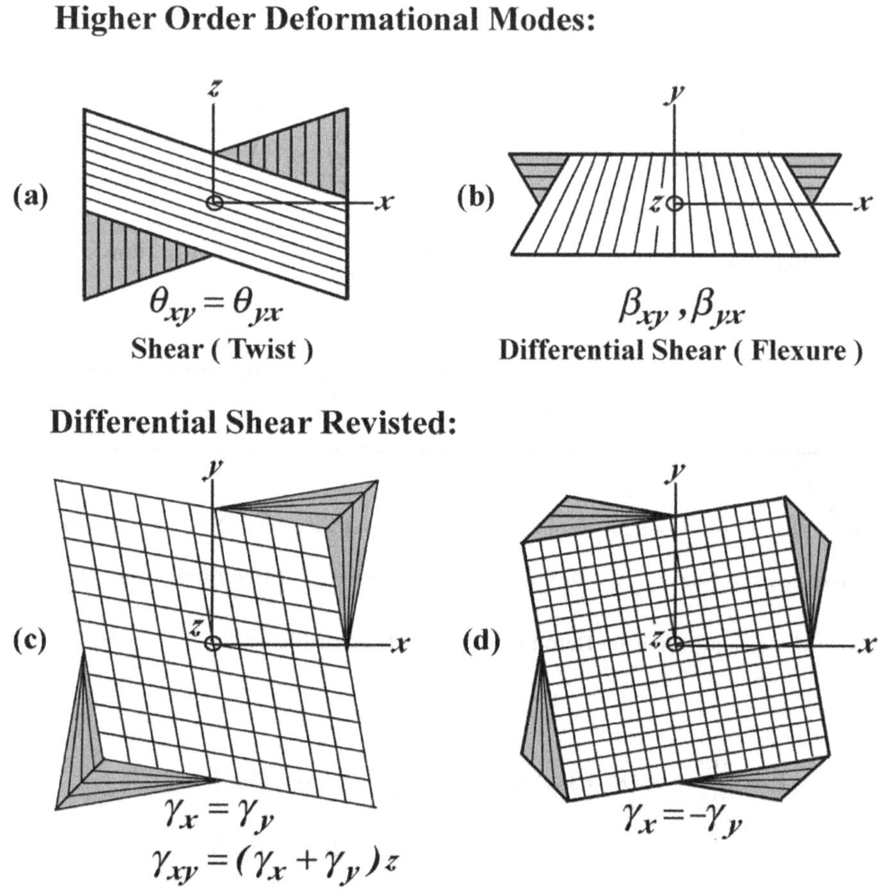

Figure 10.32 Higher Order Deformational Modes

The remaining three coefficients comprise the higher-order modes:

$$\textbf{Higher Order Modes: } \beta_{xy}, \beta_{yx}, \theta_{xy} = \theta_{yx} \tag{10.71d}$$

It is helpful to examine the kinematics of the deformational modes. The dominant deformational modes (10.71b,c) are depicted in Figure 10.31. The three higher-order deformational modes are shown in Figure 10.32. Modes γ_x and γ_y deserve further examination.

It is the sum $(\gamma_x + \gamma_y)(= \gamma_{xy})$ which constitutes the shear strain in an x-y plane. The difference $(\gamma_y - \gamma_x)$ represents a rotation in a plane $x - y$, but produces the transverse shear strains γ_{xz} and γ_{yz} as depicted in Figure 10.32d. Two extremes are illustrated in Figure (10.32c and d). The modes of Figure (10.32a and b) contribute terms of higher order to the internal energy and the nodal forces: Specifically, the internal energy (per unit area $\Delta x \Delta y$) includes terms of higher- order, e.g. $\beta_{xy} \Delta x \Delta y$, which vanish in the limit. They are essential to the *finite* element and the approximation via *finite* elements. However, the contribution of the higher-order terms diminishes with the size of the element (see Illustrative Example 10.1, where we discarded a parasitic and unnecessary higher-order term, viz. Gl^3h).

10.8 HIGHER ORDER DEFORMATIONAL MODE

Let us examine the various deformational modes in the context of their role in the theory of plates. Foremost are the modes κ_x, κ_y, γ_x, and γ_y which account for the "curvature" (rotation of the normal line z) and the twist; they account for resistive moments M_x, M_y, and $M_{xy} = M_{yx}$. The higher-order terms β_{xy}, β_{yx}, and $\theta_{xy} = \theta_{yx}$ are lesser contributions; their associated energies and forces are needed to resist these modes in the *finite* element. Finally, the simple bilinear element includes the transverse shear, e.g. $\theta_x - \beta_x$ of Figure 10.31; this deformation is absent from the Kirchhoff theory.

The modes $\gamma_x = -\gamma_y$ and $\theta_{xy} = \theta_{yx}$ deserve special attention; these modes produce transverse shear $(\gamma_{yz}, \gamma_{xz})$, but no shear strain in the plane $x - y$. It is the latter (γ_{xy}) which accompanies the twist and the resisting moment (M_{xy}) in the Kirchhoff theory. The particular modes $\gamma_x = -\gamma_y$ and θ_{xy} are unique: these modes produce only transverse shear $(\gamma_{xz} = \gamma_{yz})$ and the twist depicted in Figure 10.32(a and d); neither introduces a shear strain γ_{xy} in plane $x - y$.

The mode $\gamma_x = \gamma_y = \theta_{xy} = \gamma_{yx}$ is also unique; it produces only the strain γ_{xy}. This is depicted in Figure 10.33; it can be termed "pure twist" and is resisted solely by a twisting couple $M_{xy} = M_{yx}$. This is the deformation treated in Section 8.16. In the HI-HO Kirchhoff plate, this twist is produced by the shear stresses τ_{xy}. The edges rotate rigidly; hence, those shear stresses can be replaced by a single couple (see Figure 8.14).

To formulate the algebraic equations, we require the internal energy of the element in terms of the twelve nodal values, the displacements and rotations of the corners (w, β^x, β^y). Since this is a thin HI-HO plate, we employ the strain energy based on a state of plane stress, viz., $\sigma_z = 0$:

$$U = \frac{E}{2(1-\nu^2)} \int_{-\Delta y/2}^{\Delta y/2} \int_{-\Delta x/2}^{\Delta x/2} \int_{-h/2}^{h/2}$$

$$[\varepsilon_x^2 + 2\nu\varepsilon_x\varepsilon_y + \varepsilon_y^2 + \frac{(1-\nu)}{2}(\gamma_{xy}^2 + \gamma_{xz}^2 + \gamma_{xz}^2)]dzdxdy$$

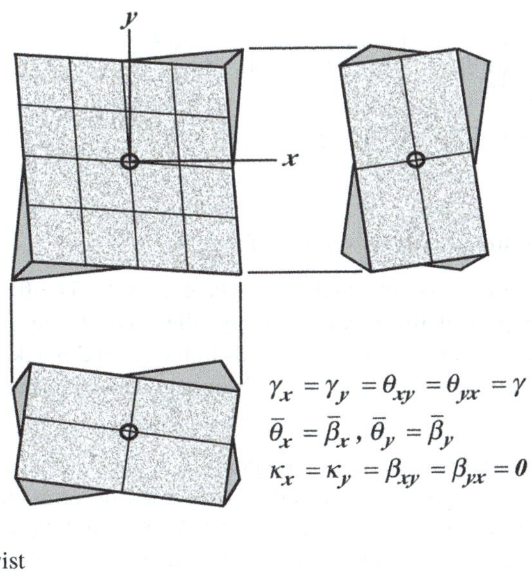

Figure 10.33 "Pure" Twist

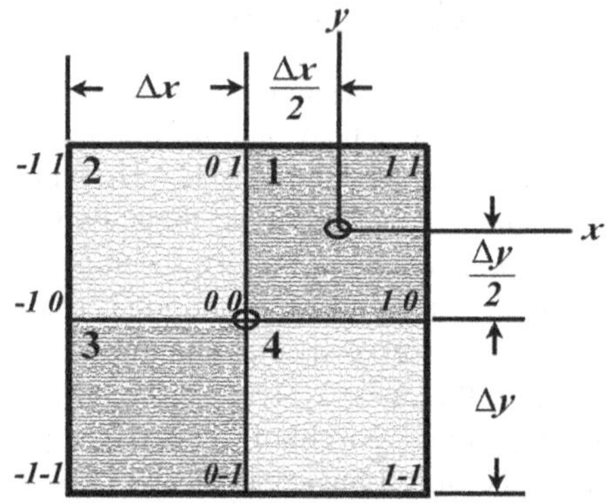

Figure 10.34 Patch of Support for Node 00

From our approximation of a beam via a similar (linear) finite element, we know that the retention of terms, κ_x and κ_y, in the transverse shear $(\gamma_{xz}, \gamma_{yz})$ is unacceptable, but also unwarranted. The associated modes are inhibited by their presence in the flexural modes. The same is true of the terms, β_{xy} and β_{yx}; these too are not desirable in the transverse shear $(\gamma_{xz}, \gamma_{yz})$. Indeed, the internal energy is reduced by their omission and convergence is accelerated.[9] Accordingly, we delete those terms, underlined in Equations (10.70d,e). We insert the expressions of (10.70a–e), delete underlined terms, and integrate to obtain:

$$\tilde{U} \equiv \frac{2(1-\nu^2)}{Eh} U \tag{10.73}$$

$$\begin{aligned}\tilde{U} = \frac{h^2}{12} \Bigg[& (\kappa_x^2 + 2\nu\kappa_x\kappa_y + \kappa_y^2) + \frac{(1-\nu)}{2}(\gamma_x + \gamma_y)^2 \\ & + \frac{\Delta y^2}{12}\beta_{xy}^2 + \frac{\Delta x^2}{12}\beta_{yx}^2 + \frac{(1-\nu)}{2}\left(\frac{\Delta x^2}{12}\beta_{xy}^2 + \frac{\Delta y^2}{12}\beta_{yx}^2\right)\Bigg] \Delta x \Delta y \\ & + \frac{(1-\nu)}{2}\left[(\theta_x - \beta_x)^2 + (\theta_y - \beta_y)^2 + (\theta_{xy} - \gamma_y)^2 \frac{\Delta y^2}{12} + (\theta_{xy} - \gamma_x)^2 \frac{\Delta x^2}{12}\right] \Delta x \Delta y\end{aligned}$$

$\bar{U}$ is merely a convenient dimensionless version of energy U.

It is worth assessing the various contributions to the energy $\tilde{U}$. The first line in the first bracket constitutes the dominant part of the flexural energy; the second line contains the higher-order contributions (Δy^2 and Δx^2). Similarly, the first terms in the second bracket hold the dominant part of the energy attributed to transverse shear; the subsequent terms are the higher-order terms (Δy^2 and Δx^2).

Our element has 12 degrees of freedom corresponding to the 12 coefficients in the approximations, Equations (10.67) and (10.68). These can be expressed in terms of the nodal values of the displacements w and rotations β_x and β_y at the corners: With the notations of Figure 10.34 and the Equations (10.67) the displacements at the corners of element 1 of Figure 10.34 follow:

$$w_{00} = w_0 - \theta_x \frac{\Delta x}{2} - \theta_y \frac{\Delta y}{2} + \theta_{xy} \frac{\Delta x \Delta y}{2} \tag{10.74a}$$

$$w_{10} = w_0 + \theta_x \frac{\Delta x}{2} - \theta_y \frac{\Delta y}{2} - \theta_{xy} \frac{\Delta x \Delta y}{2} \tag{10.74b}$$

$$w_{11} = w_0 + \theta_x \frac{\Delta x}{2} + \theta_y \frac{\Delta y}{2} + \theta_{xy} \frac{\Delta x \Delta y}{2} \tag{10.74c}$$

$$w_{01} = w_0 - \theta_x \frac{\Delta x}{2} + \theta_y \frac{\Delta y}{2} - \theta_{xy} \frac{\Delta x \Delta y}{2} \tag{10.74d}$$

Inversion of the four equations gives us

$$w_0 = \frac{1}{4}(w_{00} + w_{10} + w_{11} + w_{01}) \tag{10.75a}$$

$$\theta_y = \frac{1}{2\Delta y}(-w_{00} - w_{10} + w_{11} + w_{01}) \tag{10.75b}$$

$$\theta_x = \frac{1}{2\Delta x}(-w_{00} + w_{10} + w_{11} - w_{01}) \tag{10.75c}$$

$$\theta_{xy} = \frac{1}{\Delta x \Delta y}(w_{00} - w_{10} + w_{11} - w_{01}) \tag{10.75d}$$

In like manner, we have

$$\left.\begin{aligned} \beta_x &= \tfrac{1}{4}(\beta^x_{00} + \beta^x_{10} + \beta^x_{11} + \beta^x_{01}), \\ \gamma_y &= \frac{1}{2\Delta y}(-\beta^x_{00} - \beta^x_{10} + \beta^x_{11} + \beta^x_{01}), \\ \kappa_x &= \frac{1}{2\Delta x}(-\beta^x_{00} + \beta^x_{10} + \beta^x_{11} - \beta^x_{01}), \\ \beta_{xy} &= \frac{1}{\Delta x \Delta y}(\beta^x_{00} - \beta^x_{10} + \beta^x_{11} - \beta^x_{01}) \end{aligned}\right\} \left.\begin{aligned} \beta_y &= \tfrac{1}{4}(\beta^y_{00} + \beta^y_{10} + \beta^y_{11} + \beta^y_{01}) \\ \kappa_y &= \frac{1}{2\Delta y}(-\beta^y_{00} - \beta^y_{10} + \beta^y_{11} + \beta^y_{01}) \\ \gamma_x &= \frac{1}{2\Delta x}(-\beta^y_{00} + \beta^y_{10} + \beta^y_{11} - \beta^y_{01}) \\ \beta_{yx} &= \frac{1}{\Delta x \Delta y}(\beta^y_{00} - \beta^y_{10} + \beta^y_{11} - \beta^y_{01}) \end{aligned}\right\} \tag{10.75e,f}$$

Finally, we are in a position to write the equations governing the equilibrium at a node; at an interior node, we require the contributions from each of the adjoining elements, the four elements (quadrants of the patch depicted in Figure (10.34). We denote the internal energies of these elements by the indicated numbers. The conditions of equilibrium are expressed by the sums:

$$\frac{\partial}{\partial w_{00}}(U_1 + U_2 + U_3 + U_4) = -\frac{\partial \bar{V}}{\partial w_{00}} \tag{10.76a}$$

$$\frac{\partial}{\partial \beta^x_{00}}(U_1 + U_2 + U_3 + U_4) = -\frac{\partial \bar{V}}{\partial \beta^x_{00}} \tag{10.76b}$$

$$\frac{\partial}{\partial \beta^y_{00}}(U_1 + U_2 + U_3 + U_4) = -\frac{\partial \bar{V}}{\partial \beta^y_{00}} \tag{10.76c}$$

Here, $\bar{V}$ denotes the potential of external forces acting upon the patch; at an interior node that typically derives from a transverse load applied to the patch. The derivatives in Equations (10.76a–c) can be obtained from (10.75a–f) via the chain rule. Of course, the explicit forms of equations (10.75a–f) must be supplanted by the corresponding forms for each quadrant; Equations (10.75) are the forms applicable to quadrant 1. The others are readily formed by the appropriate interchanges of the notations, e.g., in quadrant two corner 00 replaces 01, etc.: We record the results in brief form:

$$\begin{aligned}\frac{\partial \tilde{U}}{\partial w_{00}} &= \frac{(1-\nu)}{2}[(\theta_y-\beta_y)_3+(\theta_y-\beta_y)_4-(\theta_y-\beta_y)_1-(\theta_y-\beta_y)_2]\Delta x\\ &\quad+\frac{(1-\nu)}{2}[(\theta_x-\beta_x)_2+(\theta_x-\beta_x)_3-(\theta_x-\beta_x)_4-(\theta_x-\beta_x)_1]\Delta y\\ &\quad+\frac{(1-\nu)}{12}[(\theta_{xy}-\gamma_y)_1-(\theta_{xy}-\gamma_y)_2+(\theta_{xy}-\gamma_y)_3-(\theta_{xy}-\gamma_y)_4]\Delta y^2\\ &\quad+\frac{(1-\nu)}{12}[(\theta_{yx}-\gamma_x)_1-(\theta_{yx}-\gamma_x)_2+(\theta_{yx}-\gamma_x)_3-(\theta_{yx}-\gamma_x)_4]\Delta x^2\\ &= -\tilde{p}\Delta x\Delta y\end{aligned} \tag{10.77a}$$

Here, the subscripts, 1, 2, 3, and 4, denote the quadrant, e.g., 1 signifies the first quadrant of Figure 10.34; the specific forms of Equations (10.75) give the coefficients,

$$\theta_y, \theta_x, \beta_x, \beta_y, \theta_{xy} = \theta_{yx}$$

in terms of the displacements and rotations at the corners of quadrant 1. We have given the external force derived from a uniform downward ($-z$) load $\tilde{p} = [2(1-\nu^2)/E]p$, where p is actual load (force/area). The derivatives with respect to nodal rotations provide the other two equations of equilibrium:

$$\begin{aligned}\frac{\partial \tilde{U}}{\partial \beta^x_{00}} &= \frac{h^2}{12}[-(\kappa_x+\nu\kappa_y)_1+(\kappa_x+\nu\kappa_y)_2+(\kappa_x+\nu\kappa_y)_3-(\kappa_x+\nu\kappa_y)_4]\Delta y\\ &\quad+\frac{(1-\nu)h^2}{24}[-(\gamma_y+\gamma_x)_1-(\gamma_y+\gamma_x)_2+(\gamma_y+\gamma_x)_3+(\gamma_y+\gamma_x)_4]\Delta x\\ &\quad-\frac{(1-\nu)}{4}[(\theta_x-\beta_x)_1+(\theta_x-\beta_x)_2+(\theta_x-\beta_x)_3+(\theta_x-\beta_x)_4]\Delta x\Delta y\\ &\quad+\frac{h^2}{72}\left[(\beta_{xy})_1-(\beta_{xy})_2+(\beta_{xy})_3-(\beta_{xy})_4\right]\left[\Delta y^2+\frac{(1-\nu)}{2}\Delta x^2\right]\\ &\quad+\frac{(1-\nu)}{24}[(\theta_{xy}-\gamma_y)_1+(\theta_{xy}-\gamma_y)_2-(\theta_{xy}-\gamma_y)_3-(\theta_{xy}-\gamma_y)_4]\Delta y^2\end{aligned} \tag{10.77b}$$

$$\begin{aligned}\frac{\partial \tilde{U}}{\partial \beta^y_{00}} &= \frac{h^2}{12}[-(\kappa_y+\nu\kappa_x)_1-(\kappa_y+\nu\kappa_x)_2+(\kappa_y+\nu\kappa_x)_3+(\kappa_y+\nu\kappa_x)_4]\Delta x\\ &\quad+\frac{(1-\nu)h^2}{24}[-(\gamma_y+\gamma_x)_1+(\gamma_y+\gamma_x)_2+(\gamma_y+\gamma_x)_3-(\gamma_y+\gamma_x)_4]\Delta y\\ &\quad-\frac{(1-\nu)}{4}[(\theta_y-\beta_y)_1+(\theta_y-\beta_y)_2+(\theta_y-\beta_y)_3+(\theta_y-\beta_y)_4]\Delta x\Delta y\\ &\quad+\frac{h^2}{72}[(\beta_{yx})_1-(\beta_{yx})_2+(\beta_{yx})_3-(\beta_{yx})_4]\left[\Delta x^2+\frac{(1-\nu)}{2}\Delta y^2\right]\\ &\quad+\frac{(1-\nu)}{24}[(\theta_{yx}-\gamma_x)_1-(\theta_{yx}-\gamma_x)_2-(\theta_{yx}-\gamma_x)_3+(\theta_{yx}-\gamma_x)_4]\Delta x^2\end{aligned} \tag{10.77c}$$

The equilibrium equations (10.77a,b,c) can be put in terms of the nodal values of displacements (w_{nm}) and rotations ($\beta^x_{nm}, \beta^y_{nm}$) via equations (10.75a-f); the coefficients, e.g., $(\kappa_x + \nu\kappa_y)_n$, are the constants for the element n, in accordance with the approximations of Equations (10.67) and (10.68a,b). Equation (10.77) is in a form, which readily admits mechanical interpretation. Firstly, we cite actions, bending couples (M_x, M_y), twisting couples ($M_{xy} = M_{yx}$), and transverse shear forces (q_x, q_y) as given by the continuum theory:

$$M_x = \frac{Eh^3}{12(1-\nu^2)}(\kappa_x + \nu\kappa_y), \quad M_y = \frac{Eh^3}{12(1-\nu^2)}(\kappa_y + \nu\kappa_x) \tag{10.78a,b}$$

$$M_{xy} = \frac{Eh^3}{24(1-\nu^2)}(\gamma_x + \gamma_y) \tag{10.78c}$$

$$q_x = \frac{Eh}{2(1+\nu)}(\theta_x - \beta_x), \quad q_y = \frac{Eh}{2(1+\nu)}(\theta_y - \beta_y) \tag{10.78d,e}$$

These actions are depicted in Figure 10.35. They include only the dominant terms in the strains of Equations (10.70); higher order terms vanish in the limit ($\Delta x \to 0, \Delta y \to 0$). We multiply the Equations (10.77) by the factor $Eh/2(1-\nu^2)$; they are then expressed in terms of forces (10.78d,e) and moments (10.78a–c). We employ the notation of Equations (10.78) and omit higher-order terms in Equations (10.77a–c) to obtain

$$\begin{aligned}\frac{\partial U}{\partial w_{00}} =& \frac{1}{2\Delta y}[(q_y)_4 + (q_y)_3 - (q_y)_1 - (q_y)_2]\Delta x\Delta y \\ &+ \frac{1}{2\Delta x}[(q_x)_2 + (q_x)_3 - (q_x)_4 - (q_x)_1]\Delta x\Delta y = -p\Delta x\Delta y\end{aligned} \tag{10.79a}$$

$$\begin{aligned}\frac{\partial \tilde{U}}{\partial \beta^x_{00}} =& \frac{1}{2\Delta x}[-(M_x)_1 + (M_x)_2 + (M_x)_3 - (M_x)_4]\Delta x\Delta y \\ &+ \frac{1}{2\Delta y}[-(M_{xy})_1 - (M_{xy})_2 + (M_{xy})_3 + (M_{xy})_4]\Delta x\Delta y \\ &- \frac{1}{4}[(q_x)_1 + (q_x)_2 + (q_x)_3 + (q_x)_4]\Delta x\Delta y = 0\end{aligned} \tag{10.79b}$$

$$\begin{aligned}\frac{\partial \tilde{U}}{\partial \beta^y_{00}} =& \frac{1}{2\Delta y}[-(M_y)_1 - (M_y)_2 + (M_y)_3 + (M_y)_4]\Delta x\Delta y \\ &+ \frac{1}{2\Delta x}[-(M_{xy})_1 + (M_{xy})_2 + (M_{xy})_3 - (M_{xy})_4]\Delta x\Delta y \\ &- \frac{1}{4}[(q_y)_1 + (q_y)_2 + (q_y)_3 + (q_y)_4]\Delta x\Delta y = 0\end{aligned} \tag{10.79c}$$

We have set the right sides of Equations (10.79b,c) to zero, assuming no external couples. The forces $[(q_x)_n, (q_y)_n]$ and couples $[(M_x)_n, (M_y)_n, (M_{xy})_n]$ act upon the patch in Figure 10.35. The actions, as shown on the positive edges of the patch, are only to indicate the positive senses; couples are assigned directions in accordance with the right-hand screw.

Dividing Equation (10.79a) by ($\Delta x\Delta y$) and regrouping the terms, we have

$$\left[-\frac{(q_y)_1 - (q_y)_4}{2\Delta y} - \frac{(q_y)_2 - (q_y)_3}{2\Delta y} - \frac{(q_y)_1 - (q_y)_2}{2\Delta x} - \frac{(q_y)_4 - (q_y)_3}{2\Delta x}\right] = -p.$$

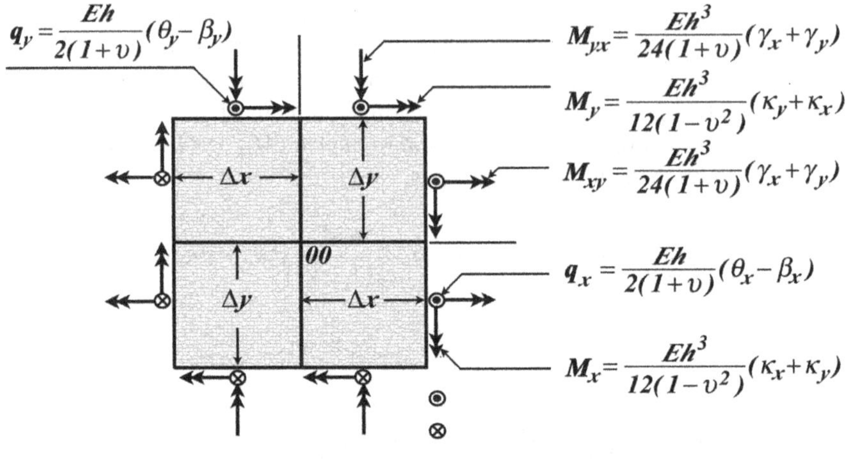

Figure 10.35 Forces and Couples on the Patch

The first term converges to $-\frac{1}{2}\frac{\partial q_y}{\partial y}$ as $\Delta y \to 0$; the second to $-\frac{1}{2}\frac{\partial q_y}{\partial y}$ as $\Delta y \to 0$; the sum is $\frac{\partial q_y}{\partial y}$. By similar analyses of the third and fourth terms, we obtain the differential equation

$$\frac{\partial q_y}{\partial y} + \frac{\partial q_x}{\partial x} = p \tag{10.80a}$$

By similar arguments, the Equations (10.79b,c) lead to the differential equations:

$$\frac{\partial M_x}{\partial x} + \frac{\partial M_{xy}}{\partial y} + q_x = 0, \quad \frac{\partial M_y}{\partial y} + \frac{\partial M_{xy}}{\partial x} + q_y = 0 \tag{10.80b,c}$$

Our mission is the system of algebraic equations, which governs the discrete nodal values. At every interior node, we have the three equations, which follow from (10.77a–c) upon substituting the equations of the form (10.75) for each of the elements adjoining the node. For example, Equation (10.77a) assumes the form:

$$\begin{aligned}
\frac{\partial \tilde{U}}{\partial w_{00}} = (1-\nu)&\left[-\left(\frac{w_{0-1}-2w_{00}+w_{01}}{2\Delta y^2}\right)-\left(\frac{w_{1-1}-2w_{10}+w_{11}}{4\Delta y^2}\right)-\left(\frac{w_{-1-1}-2w_{-10}-w_{-11}}{4\Delta y^2}\right)\right]\\
&+(1-\nu)\left[\left(\frac{\beta^y_{01}-\beta^y_{0-1}}{4\Delta y}\right)+\left(\frac{\beta^y_{-11}-\beta^y_{-1-1}}{8\Delta y}\right)+\left(\frac{\beta^y_{11}-\beta^y_{1-1}}{8\Delta y}\right)\right]\\
+(1-\nu)&\left[\left(-\frac{w_{-10}-2w_{00}+w_{10}}{2\Delta x^2}\right)-\left(\frac{w_{-1-1}-2w_{0-1}+w_{1-1}}{4\Delta x^2}\right)-\left(\frac{w_{-11}-2w_{01}+w_{11}}{4\Delta x^2}\right)\right]\\
&+(1-\nu)\left[\left(\frac{\beta^x_{10}-\beta^x_{-10}}{4\Delta x}\right)+\left(\frac{\beta^x_{1-1}-\beta^x_{-1-1}}{8\Delta x}\right)+\left(\frac{\beta^x_{11}-\beta^x_{-11}}{8\Delta x}\right)\right]
\end{aligned}$$

$$
\begin{aligned}
&+\frac{(1-\nu)}{6}\left[-\left(\frac{w_{10}-2w_{00}+w_{-10}}{\Delta x^2}\right)+\left(\frac{w_{11}-2w_{01}+w_{-11}}{2\Delta x^2}\right)+\left(\frac{w_{1-1}-2w_{0-1}+w_{-1-1}}{2\Delta x^2}\right)\right] \\
&\qquad\qquad +\frac{(1-\nu)}{6}\left[\left(\frac{\beta^x_{10}-\beta^x_{-10}}{2\Delta x}\right)-\left(\frac{\beta^x_{1-1}-\beta^x_{-1-1}}{4\Delta x}\right)-\left(\frac{\beta^x_{11}-\beta^x_{-11}}{4\Delta x}\right)\right] \\
&+\frac{(1-\nu)}{6}\left[-\left(\frac{w_{0-1}-2w_{00}+w_{01}}{\Delta y^2}\right)+\left(\frac{w_{-1-1}-2w_{-10}+w_{-11}}{2\Delta y^2}\right)+\left(\frac{w_{1-1}-2w_{10}+w_{11}}{2\Delta y^2}\right)\right] \\
&\qquad\qquad +\frac{(1-\nu)}{6}\left[\left(\frac{\beta^y_{01}-\beta^y_{0-1}}{2\Delta y}\right)-\left(\frac{\beta^y_{-11}-\beta^y_{-1-1}}{4\Delta y}\right)-\left(\frac{\beta^y_{11}-\beta^y_{1-1}}{4\Delta y}\right)\right]=-\tilde{p}
\end{aligned}
\tag{10.81}
$$

This is the final version of the first equation of equilibrium; it asserts the vanishing of the transverse nodal force (z direction). The first four lines are the dominant terms; these are the terms that converge to (10.80a), now expressed in terms of nodal displacements (w_{nm}) and rotations ($\beta^x_{nm}, \beta^y_{nm}$). The final four lines are higher-order terms, which are essential to the discrete (algebraic) system.

Again, we can interpret each parenthetical expression: In the first and third brackets, the parenthetical terms are difference equations, which converge to the second derivatives; $-\partial^2 w/\partial y^2$ and $-\partial^2 w/\partial x^2$, respectively. The second and fourth brackets contain the difference equations, which converge to the first derivatives, $\partial\beta^y/\partial y$ and $\partial\beta^x/\partial x$. Together the sum of the first and second lines converges to

$$
-(1-\nu)\left(\frac{\partial^2 w}{\partial y^2}-\frac{\partial\beta^y}{\partial y}\right)
$$

Likewise, the third and fourth lines converge to

$$
-(1-\nu)\left(\frac{\partial^2 w}{\partial x^2}-\frac{\partial\beta^x}{\partial x}\right)
$$

When multiplied by $Eh/2(1-\nu^2)$, we recognize the terms $\partial q_y/\partial y$ and $\partial q_x/\partial x$ in Equations (10.80a).

Finally, it is most revealing to examine the higher-order terms of equation (10.81), viz., the final four lines: The first parenthetical term of the fifth line converges to $-\partial^2 w/\partial x^2$;the second and third converge to $+\partial^2 w/2\partial x^2$. Thus, the bracketed expression vanishes in the limit ($\Delta x \to 0, \Delta y \to 0$). In like manner, each of the higher-order terms vanishes in the limit. The remaining two equations of equilibrium, Equations (10.77b,e), must also be expressed in terms of nodal values. In their expanded forms they too display the differences of the nodal displacements and rotations ($w_{nm}, \beta^x_{nm}, \beta^y_{nm}$); they too converge to the differential equations governing the continuous functions: $w(x,y)$, $\beta^x(x,y)$, and $\beta^y(x,y)$.

10.9 BOUNDARIES AND TRIANGLES

Although a rectangular mesh can always be imposed on the plate, nonrectangular boundaries require elements of triangular form. One or more elements adjoining a node may be triangular as the nodes 00, 10, or 01 in Figure 10.36. This requires suitable approximations of the variables (w, β^x, β^y), the internal energy of the element and the force and couples derived from that energy; the latter are combined with the contributions of the adjoining elements and the external actions to ensure equilibrium.

The approximation of displacement and rotations in the triangle must be continuous with the adjoining elements. The following linear approximation in element 1 of Figure 10.36 provides continuity along the adjoining edges.

$$w = w_{00} + \theta_x x + \theta_y y \tag{10.82a}$$

$$\beta_x = \bar{\beta}^x_{00} + \kappa_x x + \gamma_y y \tag{10.82b}$$

$$\beta_y = \bar{\beta}^y_{00} + \kappa_y y + \gamma_x x \tag{10.82c}$$

These equations are an abbreviated version of Equations (10.67) and (10.68) for the rectangular element. As before,

$$u = -\beta_x z, \quad v = -\beta_y z \tag{10.82d,e}$$

The expression of strains $(\varepsilon_x, \varepsilon_y, \gamma_{xy}, \gamma_{xz}, \gamma_{yz})$ is an amended version of Equations (10.70):

$$\varepsilon_x = -\kappa_x z, \quad \varepsilon_y = -\kappa_y z, \quad \gamma_{xy} = -(\gamma_x + \gamma_y) z \tag{10.83a,b,c}$$

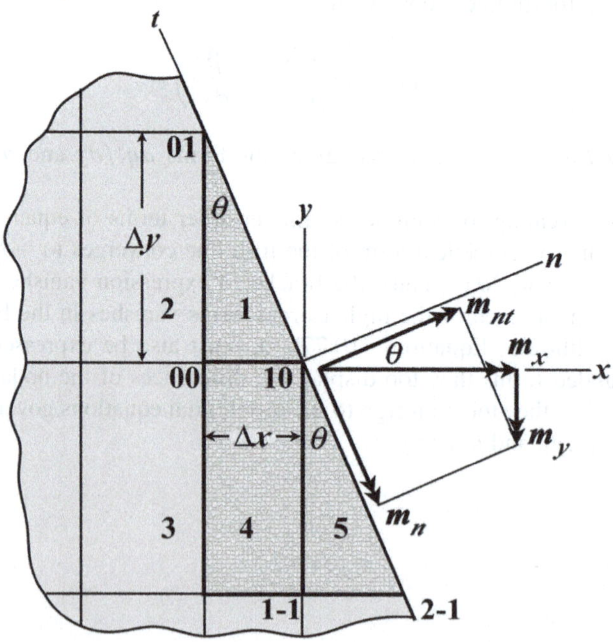

Figure 10.36 Moment at an Edge

$$\gamma_{xz} = (\theta_x - \beta^x_{00}) - \gamma_y y - \underline{\kappa_x x} \tag{10.832d}$$

$$\gamma_{yz} = (\theta_y - \beta^y_{00}) - \gamma_x x - \underline{\kappa_y y} \tag{10.83e}$$

Our triangle has 9 degrees of freedom as expressed by the 9 coefficients in Equations (10.82). Once again, we can express the nodal values $(w_{nm}, \beta^x_{nm}, \beta^y_{nm})$ in terms of the nine coefficients; the element exhibits four rigid body motions $(w_0, \theta_x, \theta_y)$ and the 5 strains of Equations (10.83).

Inversion of those equations provides the expression of the strains in terms of the nodal displacements. For element 1 of Figure 10.35, we obtain

$$\varepsilon_x = -\frac{1}{\Delta x}(\beta^x_{10} - \beta^x_{00})z, \quad \varepsilon_y = -\frac{1}{\Delta y}(\beta^y_{01} - \beta^y_{00})z \tag{10.84a,b}$$

$$\gamma_{xy} = -\left[\frac{1}{\Delta y}(\beta^x_{01} - \beta^x_{00}) + \frac{1}{\Delta x}(\beta^y_{10} - \beta^y_{00})\right] \tag{10.84c}$$

$$\gamma_{xz} = \frac{1}{\Delta x}(w_{10} - w_{00}) - \beta^x_{00} - \underline{\frac{1}{\Delta x}(\beta^x_{10} - \beta^x_{00})x} - \frac{1}{\Delta y}(\beta^x_{01} - \beta^x_{00})y \tag{10.84d}$$

$$\gamma_{yz} = \frac{1}{\Delta y}(w_{01} - w_{00}) - \beta^y_{00} - \underline{\frac{1}{\Delta y}(\beta^y_{01} - \beta^y_{00})y} - \frac{1}{\Delta x}(\beta^y_{10} - \beta^y_{00})x \tag{10.84e}$$

Note: The underlined terms in the transverse shear, γ_{xz} and γ_{yz}, represent the modes which cause excessive energy, i.e. the so-called "locking" (see Illustrative Example 10.1). These modes are the dominant terms in strains ε_x, ε_y, and γ_{xy}. The underlined terms are suppressed in our subsequent analysis.

To obtain the three equations governing the equilibrium at node 10 of Figure 10.36, we turn to the internal energy, viz. Equation (10.72). Using Equations (10.84), we express the energy in terms of the coefficients. For triangle 1, we obtain

$$U^1 = \frac{Eh}{2(1-\nu^2)}\left\{\frac{h^2}{24}\left(\kappa_x^2 + \nu\kappa_x\kappa_y + \kappa_y^2\right) + \frac{(1-\nu)h^2}{48}(\gamma_x + \gamma_y)^2 + \frac{(1-\nu)}{4}\left[(\theta_x - \beta^x_{00})^2 + (\theta_y - \beta^y_{00})^2\right]\right.$$
$$\left. - \frac{(1-\nu)}{6}\left[(\theta_x - \beta^x_{00})\gamma_y\Delta y + (\theta_y - \beta^y_{00})\gamma_x\Delta x\right] + \frac{(1-\nu)}{24}\left[\gamma_y^2\Delta y^2 + \gamma_x^2\Delta x^2\right]\right\}\Delta x\Delta y \tag{10.85}$$

At the node 10 of Figure 10.36, we require the contribution from each adjoining element, i.e. the elements 1, 4, and 5: The transverse force follows:

$$\frac{\partial(U^1 + U^4 + U^5)}{\partial w_{10}} \equiv Q_{10}$$
$$= \frac{Eh}{2(1+\nu)}\left[(\theta_y - \beta_y)\Delta x + (\theta_x - \beta_x)\Delta y\right] = Q_{10}$$

In view of Equations (10.78d,e), this assumes the form

$$q_x\Delta x + q_y\Delta y = Q_{10} = q\Delta s$$

where q denotes the edge shear force per unit length. In the limit, we obtain the edge condition for the continuous plate, viz.

$$q_x \sin\theta + q_y \cos\theta = q$$

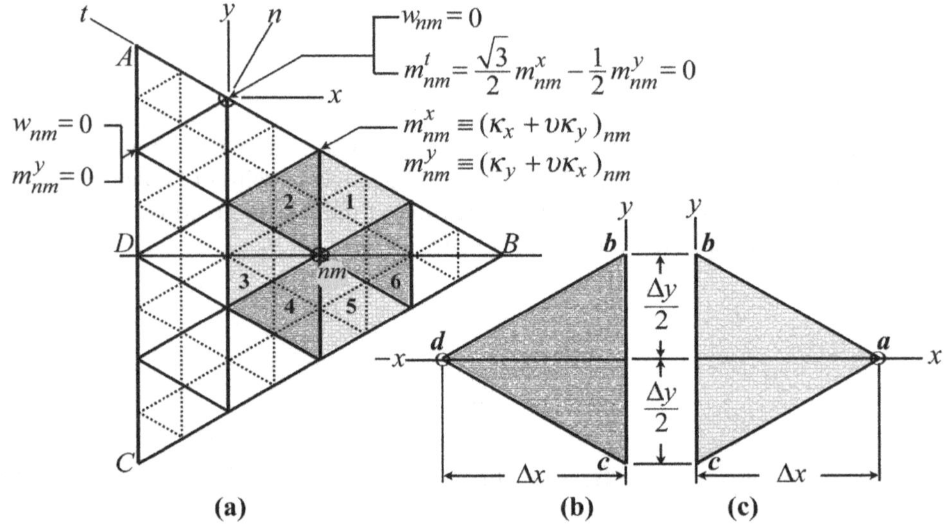

Figure 10.37 Patch of 6 Elements

In the discrete model, Q_{10} is the nodal value in the approximation of the shear force distribution on the edge. Similarly, we require two equilibrium conditions of moment:

$$\frac{\partial(U^1+U^4+U^5)}{\partial\beta^x_{10}}=M^x_{10},\quad \frac{\partial(U^1+U^4+U^5)}{\partial\beta^y_{10}}=M^y_{10}$$

Here too, the moments can be represented by components, normal (M_{tn}) and tangent (M_t) to the edge

$$M_n=M^y_{10}\cos\theta+M^x_{10}\sin\theta,\quad M_{nt}=-M^y_{10}\sin\theta+M^x_{10}\cos\theta$$

ILLUSTRATIVE EXAMPLE 10.8

In some instances, a triangular element is well suited to the geometry of the plate. The triangular plate of Illustrative Example 8.18 is a case in point. Figure 10.37 depicts the plate with a regular mesh of equilateral triangles.

Since our expression of internal energy, Equation (10.72), is given in terms of rectangular components of strain, $\varepsilon_x,\varepsilon_y,\gamma_{xy},\gamma_{xz},\gamma_{yz}$, we turn to Figure 10.37c in order to express those strains in terms of the nodal displacements (w_{nm}) and rotations ($\beta^x_{nm},\beta^y_{nm},\theta^x_{nm},\theta^y_{nm}$). Again we adopt the simple linear approximations of Equations (10.82a–e). The strains are then expressed in the form of Equations (10.83a–e). Again, the useful approximation of the flexural deformations requires that we omit the "parasitic" modes (κ_x,κ_y) in the shear strains (γ_{xz},γ_{yz}). Then,

$$\varepsilon_x\approx-\kappa_x z,\quad \varepsilon_y\approx-\kappa_y z,\quad \gamma_{xy}=-(\gamma_x+\gamma_y)z \qquad (10.86\text{a,b,c})$$

$$\gamma_{xz}\approx(\theta_x-\bar{\beta}_x)-\gamma_y y,\quad \gamma_{yz}\approx(\theta_y-\bar{\beta}_y)-\gamma_x x \qquad (10.86\text{d,e})$$

For the triangular element *abc* of Figure 10.37c, the coefficients (modes) are expressed in terms of the nodal displacements and rotations, as follows:

$$\kappa_x = \frac{1}{2\Delta x}(2\beta_a^x - \beta_b^x - \beta_c^x), \quad \kappa_y = \frac{1}{\Delta y}(\beta_b^y - \beta_c^y) \tag{10.87a,b}$$

$$\theta_x = \frac{1}{2\Delta x}(2w_a - w_b - w_c), \quad \theta_y = \frac{1}{\Delta y}(w_b - w_c) \tag{10.87c,d}$$

$$\bar{\beta}_x = \frac{1}{2}(\beta_b^x + \beta_c^x), \quad \bar{\beta}_y = \frac{1}{2}(\beta_b^y + \beta_c^y) \tag{10.87f,g}$$

$$\gamma_x = \frac{1}{2\Delta x}(2\beta_a^y - \beta_b^y - \beta_c^y), \quad \gamma_y = \frac{1}{\Delta y}(\beta_b^x - \beta_c^x) \tag{10.87h,i}$$

For the triangular element *bcd* the subscript *d* replaces *a* and the signs are reversed on the coefficients κ_x, θ_x, and γ_x; e.g.

$$\kappa_x = \frac{1}{2\Delta x}(\beta_b^x + \beta_c^x - 2\theta_d^x)$$

Again, the internal energy can be integrated through the element with respect to the coordinates (z, x, y); then,

$$U = \frac{Eh\Delta x\Delta y}{4(1-\nu^2)}\left\{\frac{h^2}{12}(\kappa_x^2 + \nu\kappa_x\kappa_y + \kappa_y^2) + \frac{h^2}{24}(1-\nu)(\gamma_x + \gamma_y)^2 + \frac{(1-\nu)}{2}\left[(\theta_x - \bar{\beta}_x)^2 + (\theta_y - \bar{\beta}_y)^2\right] + \frac{(1-\nu)}{2}\left(\frac{\Delta y^2}{24}\gamma_y^2 + \frac{\Delta x^2}{6}\gamma_x^2\right)\right\} \tag{10.88}$$

At each interior node *nm* (see Figure 10.37), we have three conditions of equilibrium:

$$M_x = \frac{\partial U^T}{\partial \beta_{nm}^x} = 0, \quad M_y = \frac{\partial U^T}{\partial \beta_{nm}^y} = 0, \quad \frac{\partial U^T}{\partial w_{nm}} = p \tag{10.89}$$

Here U^T is the sum of the internal energies of the 6 adjoining elements, and p is the discrete external load at the node. As an example, let us examine M_x. Contributions stem from the coefficients κ_x, γ_y, $\bar{\beta}_x$. We designate the individual contributions from the adjoining elements by superscripts 1, 2, 3, 4, 5, 6, in concert with the labels on the shaded patch in Figure 10.37a, e.g.

$$M_x^1 = -\frac{Eh\Delta x\Delta y}{2(1-\nu^2)}\left[\frac{h^2}{12}(\kappa_x + \nu\kappa_y)^1\left(\frac{1}{2\Delta x}\right) + \frac{h^2}{24}(1-\nu)(\gamma_x + \gamma_y)^1\left(\frac{1}{\Delta y}\right) + \frac{(1-\nu))}{2}(\theta_x - \bar{\beta}_x)^1(\frac{1}{2})\right] \tag{10.90}$$

Recall the moments and forces per unit length, in the continuum theory (see Section 8.12):

$$M_x = \frac{Eh^3}{12(1-\nu^2)}(\kappa_x + \nu\kappa_y)$$

$$M_{xy} = \frac{Eh^3}{12(1+\nu)}\kappa_{xy} \approx \frac{Eh^3}{24(1+\nu)}(\gamma_x + \gamma_y)$$

$$Q_x \approx \frac{Eh}{2(1+\nu)}(\theta_x - \bar{\beta}_x)$$

Therefore, Equation (10.90) takes the form

$$M_x^1 = -\frac{\Delta x \Delta y}{2}\left(\frac{M_x^1}{2\Delta x} + \frac{M_{xy}^1}{\Delta y} + \frac{Q_x^1}{2}\right)$$

Similar contributions derive from the adjoining elements (2,3,4,5,6). Collecting the six "moments", we obtain

$$\frac{M_{nm}^x}{\Delta x \Delta y} = -\left[\frac{(M_x^1 - M_x^2)}{4\Delta x} + \frac{(M_x^6 - M_x^3)}{2\Delta x} + \frac{(M_x^5 - M_x^4)}{4\Delta x} + \frac{(M_{xy}^1 - M_{xy}^5)}{2\Delta y} + \frac{(M_{xy}^2 - M_{xy}^4)}{2\Delta y} + Q_x\right] = 0$$

The sum must vanish if no external couple acts at a node nm. Note the relative positions of the elements, as shown in Figure 10.37a. For example, elements 1 and 2 lie along the same y coordinate, at some distance Δx. Similarly, elements 6 and 3, 5 and 4, are separated only by positions along an x line. Accordingly, we identify $M_x^1 - M_x^2$ as an incremental change with respect to coordinate x. In the limit, $\Delta x \to 0, \Delta y \to 0$, the bracketed term approaches the differential equation:

$$\frac{\partial M_x}{\partial x} + \frac{\partial M_{xy}}{\partial y} + Q_x = 0$$

Note that M_{nm}^x is moment (force $\times$ length) whereas the differential equation is moment/length^2 (force/length).

Implementation of the computation for the triangular plate requires the system of linear algebraic equations in terms of the nodal displacements (w_{nm}, β_{nm}^x, β_{nm}^y). These consist of the three Equations (10.89) at each interior node, and 1 equation at each boundary node ($\partial U^T/\partial \beta_{nm}^t = m_{nm}^t = 0$); exceptions are the corners (A, B, C) where $w = \beta^x = \beta^y = 0$.

Note that symmetry admits an analysis of one half, above or below the line DB. In this case, we would set $\beta_{nm}^y = 0$ along the line DB; 2 equations would apply at those nodes, viz

$$\beta_{nm}^y = 0, \quad \frac{\partial U^T}{\partial w_{nm}} = Q_y = 0$$

The results of the computation with the 16 elements of Figure 10.37 and the solution of Illustrative Example 8.18 are shown in Figure 10.38. Results with 36 elements are shown in Figure 10.39. These numerical results were provided by Sergei Khakalalo.[10]

The 16 elements are the minimum that can simulate the deformed configuration: The edge constraint ($w = 0$) prevents any transverse displacement of the 3 elements at the corners. Twelve of the elements can only rotate about one edge. Only the central element experiences the translation;, $w/h = 0.01198$. The principle of minimum potential tells us that this is the best approximation by this mechanism. Note that the bending of these elements is simulated by the relative rotations of the edges (β). Despite the limited freedom, the maximum (central) deflection (w) is 83% of that for the continuous model (Kirchhoff theory). Figure 10.39 depicts the deformation of the model with 36 elements. Here the maximum deflection (of the central element) is in good agreement with the continuous model: $w/h = 0.01440$ vs. 0.01444.

The triangular plate is subject to severe kinematic constraint, as it possesses symmetry with respect to three lines, $01, 23$ and AB in Figure 10.38. A normal line in these sections can only rotate within the section; at a node on those lines, we have but one component (β),e.g. at nodes $1, 2, 3$ of Figure 10.38. Therefore, the 16 element model has only 6 degrees of freedom. The 36 element model is a mechanism with 12 degrees of freedom.

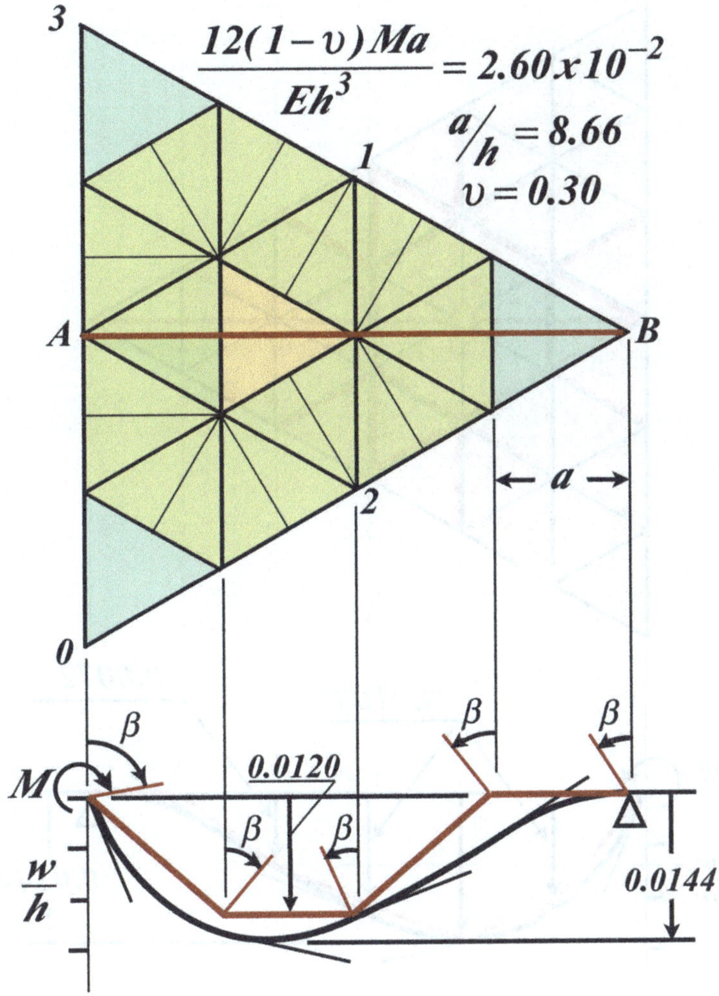

Figure 10.38 Plate with 16 Triangular Elements

10.10 THE KIRCHHOFF PLATE VIA FINITE ELEMENTS

10.10.1 INTRODUCTION

Chapter 8 recounts the theory of a continuous plate according to the Kirchhoff hypothesis. That theory rests upon the hypothesis (assumption) that a normal to a reference surface (e.g. mid-surface) remains perpendicular upon deformation of the plate/surface. In effect, the hypothesis acts as a constraint against transverse shear deformation; median values of the shear strains vanish ($\bar{\gamma}_{xz} = \bar{\gamma}_{yz} = 0$). Let us recall two essential features of the continuum theory: The displacement w and derivatives are continuous.

We confine our attention to small strains and rotations, wherein the rotations (of normals *and*

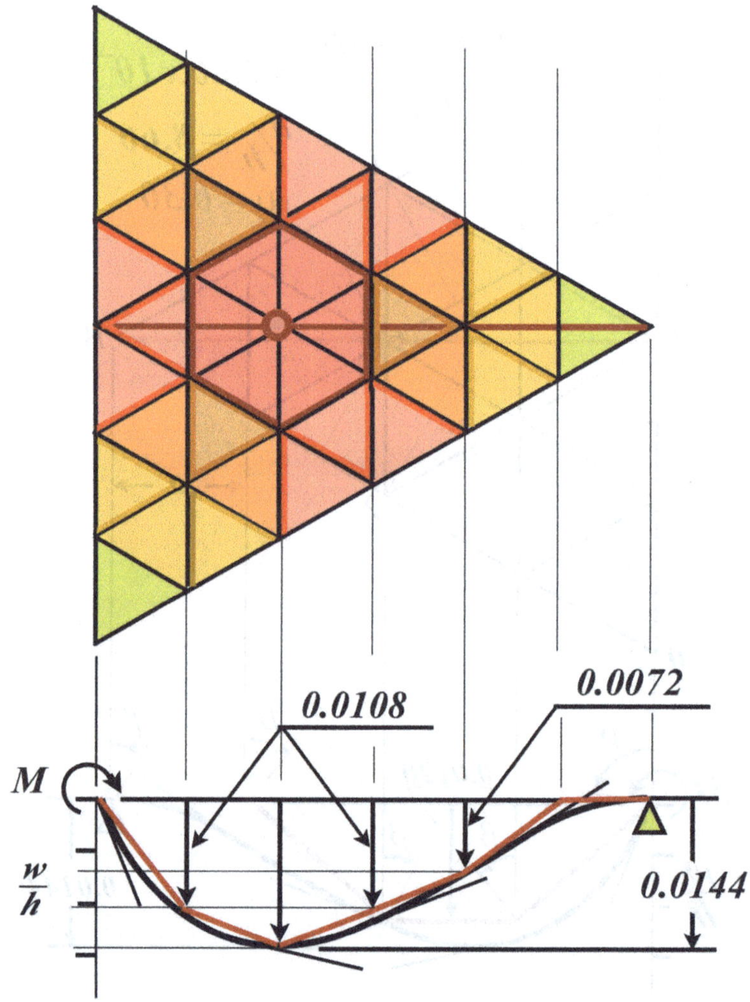

Figure 10.39 Plate with 36 Triangular Elements

tangents) are given by the derivatives:

$$\theta_x \equiv \frac{\partial w}{\partial y}, \quad \theta_y \equiv \frac{\partial w}{\partial x} \tag{10.91a,b}$$

The strains are determined by the curvatures (κ_x, κ_y) and twist ($\kappa_{xy} = \kappa_{yx}$) (see Section 8.11):

$$\varepsilon_x = -\kappa_x z, \quad \varepsilon_y = -\kappa_y z, \quad \gamma_{xy} = -2\kappa_{xy} z$$

The curvatures and twist are determined by the rotations (θ_x, θ_y) which are determined by the displacement (w) according to Equations (10.99a,b). The curvatures (κ_x, κ_y) and twist (κ_{xy}) follow:

$$\kappa_x = \frac{\partial \theta_y}{\partial x} = \frac{\partial^2 w}{\partial x^2}, \quad \kappa_y = \frac{\partial \theta_x}{\partial y} = \frac{\partial^2 w}{\partial y^2} \tag{10.92a,b}$$

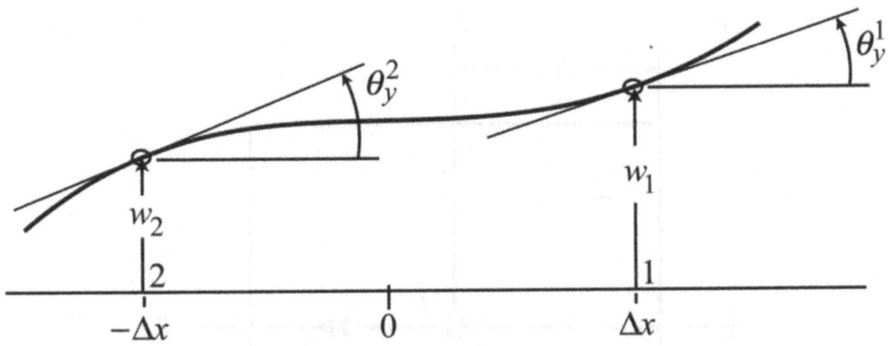

Figure 10.40 Hermititean Interpolation

$$\kappa_{xy} = \frac{1}{2}\left(\frac{\partial \theta_y}{\partial y} + \frac{\partial \theta_x}{\partial x}\right) = \frac{\partial^2 w}{\partial x \partial y} \tag{10.92c}$$

An approximation via finite elements must describe a surface, which *converges* to a smooth surface, i.e., the deformed surface should exhibit no kinks. The displacement must be approximated by a polynomial of higher degree; cubic terms are essential to ensure interelement continuity along the curved interfaces. Other features are noteworthy:

The smooth curve along an interelement line, for example an x line ($y = 0$) requires a cubic approximation of displacement; e.g.,

$$w = \bar{w} + \bar{\theta}_y x + \frac{1}{2}\bar{\kappa}_x x^2 + \psi_x x^3 \tag{10.93}$$

Such approximation provides 4 degrees of freedom, as embodied in the 4 coefficients. These constitute four generalized coordinates of the discrete model. The polynomial (10.93) admits independent approximation of two nodal values, e.g., sites 1 and 2 of Figure 10.40. Each of the four nodal displacements (w) and rotations (θ_y) are expressed in terms of the four coefficients ($\bar{w}, \bar{\theta}_y, \bar{\kappa}_x, \psi_x$) via Equations (10.91a,b): those equations are readily inverted to express those generalized coordinates in terms of the nodal coordinates ($w_1, w_2, \theta_y^1, \theta_y^2$):

$$\bar{w} = \frac{1}{2}(w_1 + w_2) - \frac{\Delta x}{4}(\theta_y^1 - \theta_y^2), \tag{10.64a}$$

$$\bar{\theta}_y = \frac{3}{4\Delta x}(w_1 + w_2) - \frac{1}{4}(\theta_y^1 - \theta_y^2) \tag{10.94a,b}$$

$$\bar{\kappa}_x = \frac{1}{2\Delta x}(\theta_y^1 - \theta_y^2), \tag{10.64b}$$

$$\psi_x = \frac{1}{4\Delta x^2}(\theta_y^1 + \theta_y^2) - \frac{1}{4\Delta x^3}(w_1 - w_2) \tag{10.94c,d}$$

The foregoing polynomial (10.93) is but a one-dimensional counterpart, i.e., $w(x)$, of the polynomial approximation required for the plate, i.e., $w(x, y)$. Equation (10.93) is an acceptable approximation for an element of the Bernoulli beam (see Illustrative Example 10.3), the one-dimensional counterpart of the Kirchhoff plate.

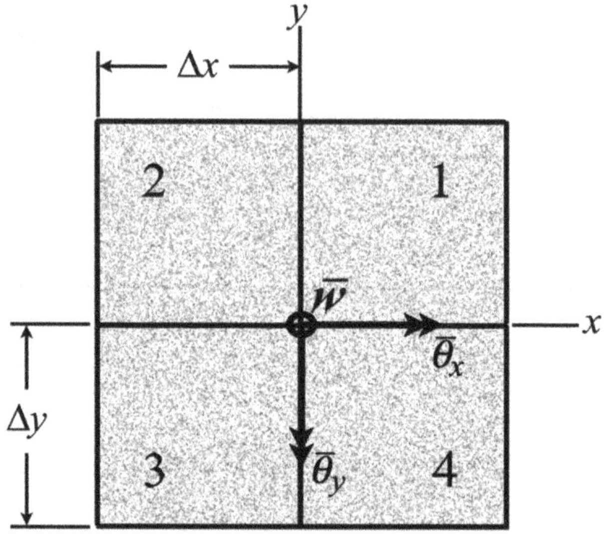

Figure 10.41 One Element

10.10.2 APPROXIMATION OF THE PLATE ELEMENT

A polynomial, which meets the minimal requirements in the two dimensions (x,y) follows[11].

$$
\begin{aligned}
w \doteq \bar{w} + \bar{\theta}_x y + \bar{\theta}_y x + \frac{1}{2}\bar{\kappa}_x x^2 + \frac{1}{2}\bar{\kappa}_y y^2 + \bar{\kappa}_{xy} xy \\
+\psi_{xy} x^2 y + \psi_{yx} y^2 x + \psi_x x^3 + \psi_y y^3 \\
+\phi_{xy} x^3 y + \phi_{yx} y^3 x
\end{aligned}
\tag{10.95}
$$

This approximation of the displacement within the rectangular element of Figure 10.41 provides 12 degrees of freedom embodied in the 12 coefficients; these are generalized coordinates, which fully describe the geometrical/kinematic state of the deformed element. The corresponding rotations are given by (10.91a,b):

$$\frac{\partial w}{\partial y} = \bar{\theta}_x + \bar{\kappa}_y y + \bar{\kappa}_{xy} x + \psi_{xy} x^2 + 2\psi_{yx} yx + 3\psi_y y^2 + \phi_{xy} x^3 + 3\phi_{yx} y^2 x \tag{10.96a}$$

$$\frac{\partial w}{\partial x} = \bar{\theta}_y + \bar{\kappa}_x x + \bar{\kappa}_{xy} y + \psi_{yx} y^2 + 2\psi_{xy} yx + 3\psi_x x^2 + 3\phi_{xy} x^2 y + \phi_{yx} y^3 \tag{10.96b}$$

Our choices of notation become evident: $\bar{w}$, $\bar{\theta}_x$, and $\bar{\theta}_y$ are the displacement and rotations at the midpoint of the element (see Figure 10.41).

To serve our purpose, the 12 generalized coordinates (coefficients) must be expressed in terms of the 4 nodal displacements (w_1, w_2, w_3, w_4) and 8 rotations $(\theta_x^1, \theta_x^2, \theta_x^3, \theta_x^4, \theta_y^1, \theta_y^2, \theta_y^3, \theta_y^4)$. Twelve equations express the latter in terms of the former via relationships (10.95) and (10.96a,b). The inversion of those equations provides the following:

$$\bar{w} = \frac{1}{4}(w_1 + w_2 + w_3 + w_4) - \frac{\Delta x}{8}(\theta_y^1 - \theta_y^2 - \theta_y^3 + \theta_y^4) - \frac{\Delta y}{8}(\theta_x^1 + \theta_x^2 - \theta_x^3 - \theta_x^4) \quad (10.97a)$$

$$\bar{\theta}_x = \frac{3}{8\Delta y}(w_1 + w_2 - w_3 - w_4) - \frac{\Delta x}{8\Delta y}(\theta_y^1 - \theta_y^2 + \theta_y^3 - \theta_y^4) - \frac{1}{8}(\theta_x^1 + \theta_x^2 + \theta_x^3 + \theta_x^4) \quad (10.97b)$$

$$\bar{\theta}_y = \frac{3}{8\Delta x}(w_1 - w_2 - w_3 + w_4) - \frac{\Delta y}{8\Delta x}(\theta_x^1 - \theta_x^2 + \theta_x^3 - \theta_x^4) - \frac{1}{8}(\theta_y^1 + \theta_y^2 + \theta_y^3 + \theta_y^4) \quad (10.97c)$$

$$\bar{\kappa}_y = \frac{1}{4\Delta y}(\theta_x^1 + \theta_x^2 - \theta_x^3 - \theta_x^4) \quad (10.97d)$$

$$\bar{\kappa}_x = \frac{1}{4\Delta x}(\theta_y^1 - \theta_y^2 - \theta_y^3 + \theta_y^4) \quad (10.97e)$$

$$\kappa_{xy} = \frac{1}{2\Delta x\Delta y}(w_1 - w_2 + w_3 - w_4) - \frac{1}{8\Delta y}(\theta_y^1 + \theta_y^2 - \theta_y^3 - \theta_y^4) - \frac{1}{8\Delta x}(\theta_x^1 - \theta_x^2 - \theta_x^3 + \theta_x^4) \quad (10.97f)$$

$$\psi_{yx} = \frac{1}{8\Delta x\Delta y}(\theta_x^1 - \theta_x^2 + \theta_x^3 - \theta_x^4) \quad (10.97g)$$

$$\psi_{xy} = \frac{1}{8\Delta x\Delta y}(\theta_y^1 - \theta_y^2 + \theta_y^3 - \theta_y^4) \quad (10.97h)$$

$$\psi_y = \frac{1}{8\Delta y^2}(\theta_x^1 + \theta_x^2 + \theta_x^3 + \theta_x^4) - \frac{1}{8\Delta y^3}(w_1 + w_2 - w_3 - w_4) \quad (10.97i)$$

$$\psi_x = \frac{1}{8\Delta x^2}(\theta_y^1 + \theta_y^2 + \theta_y^3 + \theta_y^4) - \frac{1}{8\Delta x^3}(w_1 - w_2 - w_3 + w_4) \quad (10.97j)$$

$$\phi_{yx} = \frac{1}{8\Delta x\Delta y^2}(\theta_x^1 - \theta_x^2 - \theta_x^3 + \theta_x^4) - \frac{1}{8\Delta x\Delta y^3}(w_1 - w_2 + w_3 - w_4) \quad (10.97k)$$

$$\phi_{xy} = \frac{1}{8\Delta y\Delta x^2}(\theta_y^1 + \theta_y^2 - \theta_y^3 - \theta_y^4) - \frac{1}{8\Delta y\Delta x^3}(w_1 - w_2 + w_3 - w_4) \quad (10.97l)$$

As noted previously, the generalized coordinates, $\bar{w}$, $\bar{\theta}_x$, and $\bar{\theta}_y$, represent displacement and rotations at the midpoint; these terms describe rigid motion of our element. The remaining generalized coordinates describe deformational modes; the dominant terms are $\bar{\kappa}_x$, $\bar{\kappa}_y$, and $\bar{\kappa}_{xy}$, which describe the central values of the curvatures and twist. Other deformational modes are expressed by the remaining terms: ψ_{yx}, ψ_{xy}, ψ_y, ψ_x, ϕ_{yx}, and ϕ_{xy}; these are higher-order terms (HOT), which are needed in the *finite* element to inhibit those modes. The contributions of the HOT terms diminish as the element shrinks ($\Delta x \to 0, \Delta y \to 0$).

To be effective, our model must simulate the simplest states of deformation, viz., simple bending and simple twist. The simple bending of a HI-HO Kirchhoff plate is characterized by a constant curvature κ_x in one direction x and an anticlastic curvature $-\nu\kappa_x$ in the transverse direction y (see Section 8.15). Such deformation requires no nodal displacements ($w_1 = w_2 = w_3 = w_4 = 0$), no central rotations ($\bar{\theta}_x = \bar{\theta}_y = 0$) and no central twist ($\bar{\kappa}_{xy} = 0$). Symmetry dictates that

$$\theta_y^1 = \theta_y^4 = -\theta_y^2 = -\theta_y^3, \quad \theta_x^1 = \theta_x^2 = -\theta_x^3 = -\theta_x^4$$

According to (10.97d,e),

$$\bar{\kappa}_y = -\nu\bar{\kappa}_x = \frac{\theta_x^1}{\Delta y}, \qquad \bar{\kappa}_x = \frac{\theta_y^1}{\Delta x}$$

$$\theta_x^1 = -\nu\bar{\kappa}_x\Delta y, \qquad \theta_y^1 = \bar{\kappa}_x\Delta x$$

From (10.97f),

$$\kappa_{xy} = 0$$

According to (10.97g,h) $\psi_{yx} = \psi_{xy} = 0$; Equations (10.97i,j) yield $\psi_y = \psi_x = 0$ and Equations (10.97k,l) yield $\phi_{yx} = \phi_{xy} = 0$. Our element provides an accurate description of simple bending. Note that the bending produces a central deflection, as expected:

$$\bar{w} = -\frac{1}{2}\kappa_x(\Delta x - \nu \Delta y)$$

A simple state of twist is characterized by the constant $\bar{\kappa}_{xy} = \bar{\kappa}_{yx} = \bar{\kappa}$ (see Section 8.16),

$$w_1 = -w_2 = w_3 = -w_4, \quad \theta_x^1 = \theta_x^4 = -\theta_x^2 = -\theta_x^3, \quad \theta_y^1 = \theta_y^2 = -\theta_y^3 = -\theta_y^4$$

This state is depicted in Figure 10.42 in views from the x and y axes. It follows from the foregoing equations that

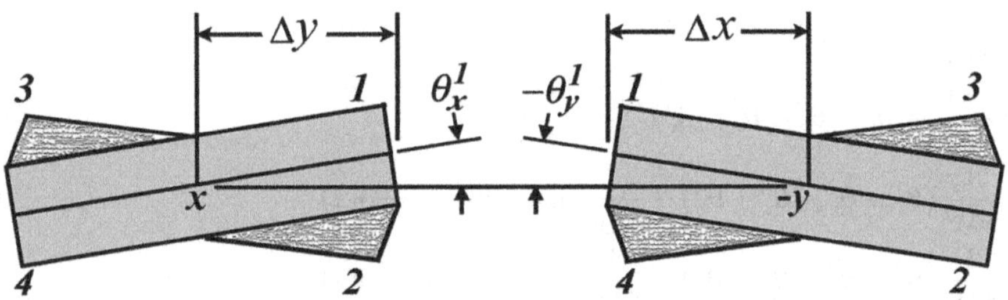

Figure 10.42 Simple Twist

$$\bar{w} = 0, \qquad \bar{\theta}_x = \bar{\theta}_y = 0, \qquad \bar{\kappa}_x = \bar{\kappa}_y = 0$$

$$\psi_{yx} = \psi_{xy} = 0, \qquad \psi_y = \psi_x = 0, \qquad \phi_{yx} = \phi_{xy} = 0$$

The latter follows from the relations:

$$\theta_x^1 = \frac{w_1}{\Delta y}, \quad \theta_y^1 = -\frac{w_1}{\Delta x}$$

10.10.3 DEFORMATION OF THE ELEMENT

The deformation of the element is embodied in the curvatures and twist in accordance with Equations (10.92a–c), (10.95):

$$\kappa_x = \bar{\kappa}_x + \underline{\underline{2\psi_{xy}y}} + \underline{6\psi_x x} + \underline{\underline{6\phi_{xy}xy}} \tag{10.98a}$$

$$\kappa_y = \bar{\kappa}_y + \underline{\underline{2\psi_{yx}x}} + \underline{6\psi_y y} + \underline{\underline{6\phi_{yx}yx}} \tag{10.98b}$$

$$\kappa_{xy} = \bar{\kappa}_{xy} + \underline{\underline{2\psi_{xy}x}} + \underline{\underline{2\psi_{yx}y}} + \underline{\underline{3\phi_{xy}x^2}} + \underline{\underline{3\phi_{yx}y^2}} \tag{10.98c}$$

Higher-order terms (HOT) are underlined in Equations (10.98). As previously noted, these play a lesser role; their contribution diminishes as the element shrinks ($\Delta x \to 0, \Delta y \to 0$). They play an essential role in the finite element: These HOT terms must contribute to the strain energy to resist the

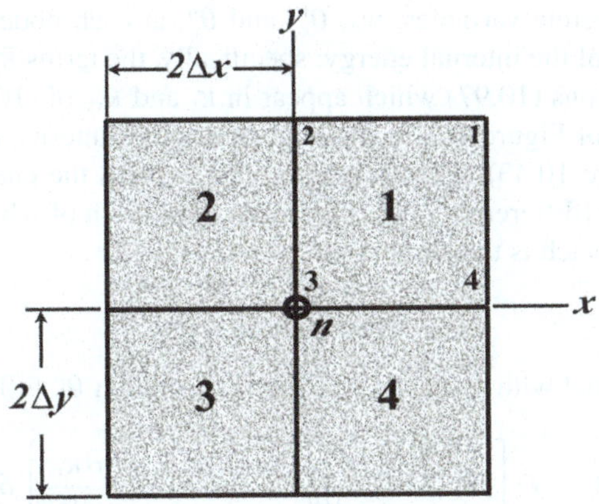

Figure 10.43 Patch of 4 Elements

associated deformational modes. We note that each of four higher-order modes ($\psi_{xy}, \psi_{yx}, \theta_{xy}, \theta_{yx}$) appears in a curvature (flexural mode) and a twist (torsional mode); these underlined terms are repeated in Equations (10.98). For example, the mode ψ_{xy} appears in κ_x and also in κ_{xy}; that mode is inhibited if only the term is retained in one. For example, the term $2\psi_{xy}x$ could be omitted from κ_{xy}; the omission would serve to reduce the strain energy (in the total potential) and improve the convergence to the continuum solution (to the minimum of the potential). Let us now examine the potential energy and the variation which provide the algebraic (difference) equations and the convergence to the differential equations of the Kirchhoff theory.

10.10.4 POTENTIAL AND VARIATION (VIRTUAL WORK)

The potential P consists of the strain energy (U_I) and the potential of the load (U_E). For purposes of illustration, we assume a uniform transverse load p (per unit area) and suppose that it acts downward ($-z$ direction); then

$$U_E = \iint_{\text{Area}} pw\text{d}x\text{d}y \tag{10.99a}$$

The strain energy in the HI-HO Kirchhoff plate is

$$U_I = \iint_{\text{Area}} \frac{D}{2}\left[\kappa_x^2 + 2\nu\kappa_x\kappa_y + \kappa_y^2 + 2(1-\nu)\kappa_{xy}^2\right] \tag{10.99b}$$

where

$$D \equiv \frac{Eh^3}{12(1-\nu^2)}$$

We recall that the motion/deformation of each element is expressed entirely in terms of the nodal motions, viz., w^n, θ_x^n, and θ_y^n, in accordance with Equations (10.95) and (10.97). The variations (δw^n, $\delta\theta_x^n$, and $\delta\theta_y^n$) at node n cause variations of potential, but are put only in the four adjoining elements, i.e., in the patch of Figure 10.43.

In the analysis of the continuous plate, the deformations, curvatures (κ_x, κ_y) and twist (κ_{xy}) are all functions of the one variable, $w(x,y)$. In our discrete approximation, the deformation is approximated by three discrete variables, w_n, θ_x^n, and θ_y^n, at each node. A variation of one, e.g. θ_y^n, introduces a variation of the internal energy; specifically, the terms $\bar{\kappa}_x$, $\bar{\kappa}_{xy}$, ψ_{xy}, ψ_x and ϕ_{xy} are varied according to Equations (10.97) which appear in κ_x and κ_{xy} of (10.98a,c). That variation of θ_y^n at node n in the patch of Figure 10.43 causes a variation of internal energy in the 4 adjoining elements (1,2,3,4 of Figure 10.43). Therefore, we must express the energy of those elements in terms of those coefficients (different modes of deformation), each of which is a function of θ_y^n.

The potential P of the patch is the sum:

$$P = U_I + U_E$$

The variation of the potential with respect to the discrete rotation θ_y^n follows:

$$\delta P = \iint_{\text{Patch}} D\left[(\kappa_x + \nu\kappa_y)\frac{\partial \kappa_x}{\partial \theta_y^n} + 2(1-\nu)\kappa_{xy}\frac{\partial \kappa_{xy}}{\partial \theta_y^n}\right]\delta\theta_y^n \mathrm{d}x\mathrm{d}y$$

From Equations (10.97) and (10.98) we obtain the required derivatives for element 1 of Figure 10.43:

$$\frac{\partial \kappa_x}{\partial \theta_y^3} = -\frac{1}{4\Delta x} + \frac{y}{4\Delta x\Delta y} + \frac{3x}{4\Delta x^2} - \frac{3yx}{4\Delta y\Delta x^2}$$

$$\frac{\partial \kappa_{xy}}{\partial \theta_y^3} = +\frac{1}{8\Delta y} + \frac{1}{4\Delta x\Delta y}x - \frac{3x^2}{8\Delta y\Delta x^2}$$

The superscript 3 identifies the corner of element 1, which lies at the center n of the patch in Figure 10.43. For elements 2, 3, and 4, the corresponding corners (and superscripts) are 4, 1, and 2, respectively.

For element 1 in the patch, we have the following virtual work:

$$\begin{aligned}\delta\theta_y^n \int_{-\Delta x}^{\Delta x}\int_{-\Delta y}^{\Delta y} D\Bigg[(\kappa_x + \nu\kappa_y)^1\left(\frac{-1}{4\Delta x} + \underline{\frac{y}{4\Delta x\Delta y}} + \frac{3x}{4\Delta x^2} - \underline{\frac{3}{4}\frac{yx}{\Delta y\Delta x^2}}\right)\\ +2(1-\nu)\kappa_{xy}^1\left(\frac{1}{8\Delta y} + \underline{\frac{1}{4\Delta x\Delta y}} - \underline{\frac{3}{8}\frac{x^2}{\Delta y\Delta x^2}}\right)\Bigg]\mathrm{d}y\mathrm{d}x\end{aligned} \tag{10.100}$$

Note that the underlined terms derive from the higher-order (HOT) terms which vanish in the limit. Note too that Equation (10.100) is the variation of energy in element 1 as signified by the superscripts, i.e. $(\kappa_x + \nu\kappa_y)^1$, κ_{xy}^1.

If we omit the HOT terms, then the variations of θ_y^n θ_x^n, and w_n provide the following variation δP for the entire patch, the four elements of Figure 10.43:

$$
\begin{aligned}
\delta P = &\Big\{ D\left[-(\bar{\kappa}_y^1 + \nu\bar{\kappa}_x^1) - (\bar{\kappa}_y^2 + \nu\bar{\kappa}_x^2) + (\bar{\kappa}_y^3 + \nu\bar{\kappa}_x^3) + (\bar{\kappa}_y^4 + \nu\bar{\kappa}_x^4)\right] \Delta x \\
&+ (1+\nu) D\left[\bar{\kappa}_{xy}^1 - \bar{\kappa}_{xy}^2 - \bar{\kappa}_{xy}^3 + \bar{\kappa}_{xy}^4\right] \Delta y + HOT \Big\} 4\delta\theta_x^n \\
&+ \Big\{ D\left[-(\bar{\kappa}_x^1 + \nu\bar{\kappa}_y^1) + (\bar{\kappa}_x^2 + \nu\bar{\kappa}_y^2) + (\bar{\kappa}_x^3 + \nu\bar{\kappa}_y^3) - (\bar{\kappa}_x^4 + \nu\bar{\kappa}_y^4)\right] \Delta y \\
&+ (1-\nu) D\left[\bar{\kappa}_{xy}^1 + \bar{\kappa}_{xy}^2 - \bar{\kappa}_{xy}^3 - \bar{\kappa}_{xy}^4\right] \Delta x + HOT \Big\} 4\delta\theta_y^n \\
&+ \Big\{ (1-\nu) D\left[\bar{\kappa}_{xy}^1 - \bar{\kappa}_{xy}^2 + \bar{\kappa}_{xy}^3 - \bar{\kappa}_{xy}^4\right] + \left[p\Delta x\Delta y + HOT\right] \Big\} 4\delta w^n
\end{aligned}
\tag{10.101}
$$

By the conditions of stationary potential, the variation (virtual work) $\delta P = 0$ for arbitrary (but admissible) variations; stated otherwise, each of the bracketed terms of (10.101) must vanish. These represent three conditions of equilibrium: The first (coefficient of $\delta\theta_x^n$) asserts the vanishing of moment acting in the direction of rotation θ_x. The second (coefficient of $\delta\theta_y^n$) asserts the vanishing of moment acting in the direction of rotation θ_y. The final condition (coefficient of δw^n) expresses the vanishing of force in the direction of the displacement w. Implementation also requires that the generalized strains, e.g., $\bar{\kappa}_x^n$, $\bar{\kappa}_y^n$, and $\bar{\kappa}_{xy}^n$, be expressed in terms of the nodal rotations and displacement, viz., θ_x^n, θ_y^n, and w^n. The result is a system of $3M$ algebraic equations for the entire plate with M nodes, in terms of the $3M$ quantities (θ_x^n, θ_y^n, w^n, $n = 1,\ldots,M$). The number M is reduced by nodal constraints, specifically, at edges. For example, at node q on a *fixed* edge: $\delta\theta_x^q = \delta\theta_y^q = \delta w^q = 0$; the system of equations is reduced by three; the number of unknowns is also reduced by three, since $\theta_x^q = \theta_y^q = w_q = 0$. At a simply supported edge, say an x line: $\delta\theta_y^q = \delta w^q = 0$; θ_x^q in the transverse direction is not constrained.

10.10.5 INTERPRETATION AND CONVERGENCE

To give additional meaning to the equations, we recall the moment-curvature and moment-twist relations of the Kirchhoff theory (see Equation 8.26):

$$M_x = D(\kappa_x + \nu\kappa_y), \quad M_y = D(\kappa_y + \nu\kappa_x), \quad M_{xy} = (1-\nu)D\kappa_{xy}$$

As previously noted, the coefficients, $\bar{\kappa}_x$, $\bar{\kappa}_y$, and $\bar{\kappa}_{xy}$, of (10.98a,b,c) are the central values of the curvatures and twist; these dominate, as the HOT decay ($\Delta x \to 0, \Delta y \to 0$). Accordingly, we identify the central moments of each element, i.e. 1, 2, 3, and 4 in the patch of Figure 10.44.

$$\bar{M}_x^n \equiv D(\bar{\kappa}_x^n + \nu\bar{\kappa}_y^n), \quad \bar{M}_y^n \equiv D(\bar{\kappa}_y^n + \nu\bar{\kappa}_x^n) \tag{10.102a,b}$$

$$\bar{M}_{xy}^n \equiv (1-\nu)D\left(\bar{\kappa}_{xy}^n\right). \quad n = 1,2,3,4 \tag{10.102c}$$

With these notations, the three conditions of equilibrium, as drawn from Equation (10.101), take the forms:

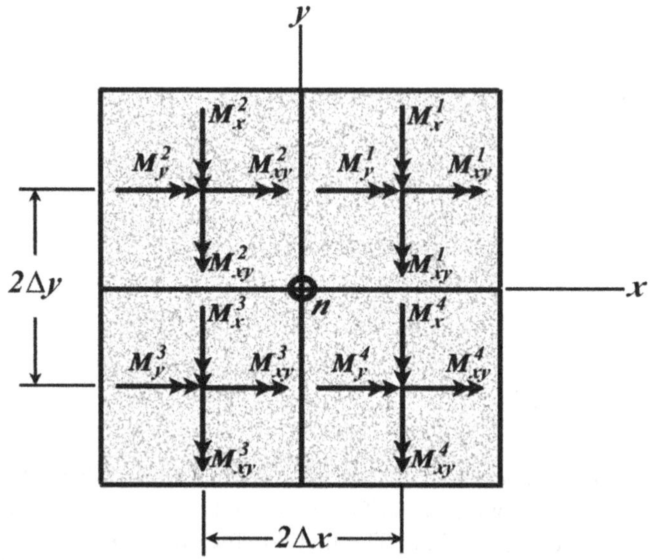

Figure 10.44 Moments on the Elements of Patch

$$\frac{-\bar{M}_y^1 - \bar{M}_y^2 + \bar{M}_y^3 + \bar{M}_y^4}{4\Delta y} + \frac{\bar{M}_{xy}^1 - \bar{M}_{xy}^2 - \bar{M}_{xy}^3 + \bar{M}_{xy}^4}{4\Delta x} = 0 \tag{10.103a}$$

$$\frac{-\bar{M}_x^1 + \bar{M}_x^2 + \bar{M}_x^3 - \bar{M}_x^4}{4\Delta x} + \frac{\bar{M}_{xy}^1 + \bar{M}_{xy}^2 - \bar{M}_{xy}^3 - \bar{M}_{xy}^4}{4\Delta y} = 0 \tag{10.103b}$$

$$+\frac{\bar{M}_{xy}^1 - \bar{M}_{xy}^2 + \bar{M}_{xy}^3 - \bar{M}_{xy}^4}{\Delta x \Delta y} + p = 0 \tag{10.103c}$$

Here, each expression has been divided by the area of the patch ($16\Delta x\Delta y$); that serves to express the energy per unit area and the moments per unit length. The moments M_x^n, M_y^n, and M_{xy}^n are those acting at the midpoints of the elements, as shown in Figure 10.45. Let us define the following average values:

$$\frac{\bar{M}_y^1 + \bar{M}_y^2}{2} \equiv \bar{M}_y(\Delta y), \quad \frac{\bar{M}_y^3 + \bar{M}_y^4}{2} \equiv \bar{M}_y(-\Delta y)$$

$$\frac{\bar{M}_{xy}^1 + \bar{M}_{xy}^4}{2} \equiv \bar{M}_{xy}(\Delta x), \quad \frac{\bar{M}_{xy}^2 + \bar{M}_{xy}^3}{2} \equiv \bar{M}_{xy}(-\Delta x),$$

Note that the averages $\bar{M}_y$ act along the lines $y = \pm\Delta y$ of Figure 10.44; averages $\bar{M}_{xy}$ act along the lines $x = \pm\Delta x$. The Equation (10.103a) takes the form:

$$-\frac{\bar{M}_y(\Delta y) - \bar{M}_y(-\Delta y)}{2\Delta y} + \frac{\bar{M}_{xy}(\Delta x) - \bar{M}_{xy}(-\Delta x)}{2\Delta x} = 0$$

This equation converges to the differential equation in the limit ($\Delta x \to 0, \Delta y \to 0$):

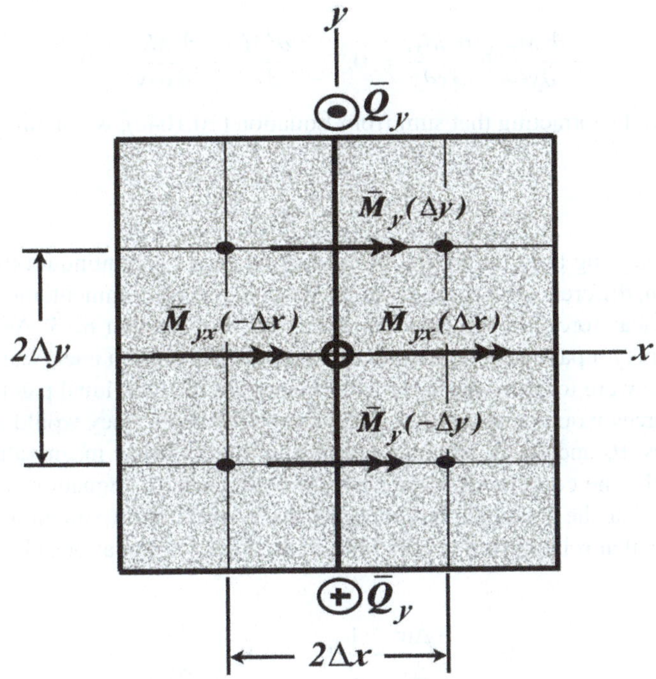

Figure 10.45 Moment about the x-axis

$$-\frac{\partial M_y}{\partial y} + \frac{\partial M_{xy}}{\partial x} = 0 \tag{10.104a}$$

Similarly, the Equations (10.103b) converge to

$$-\frac{\partial M_x}{\partial x} + \frac{\partial M_{xy}}{\partial y} = 0 \tag{10.104b}$$

Equation (10.103c) converges as follows:

$$\lim_{\substack{\Delta x \to 0 \\ \Delta y \to 0}} \frac{4}{2\Delta x}\left[\frac{\bar{M}^1_{xy}(\Delta x, \Delta y) - \bar{M}^4_{xy}(\Delta x, -\Delta y)}{2\Delta y} - \frac{\bar{M}^2_{xy}(-\Delta x, \Delta y) - \bar{M}^3_{xy}(-\Delta x, -\Delta y)}{2\Delta y}\right] + p = 0$$

$$= 4\frac{\partial^2 M_{xy}}{\partial y \partial x} + p = 0 \tag{10.104c}$$

The three equations appear different from what one might anticipate (see Equations 8.27 and 8.28). We note the absence of transverse shear forces, which produce moments and transverse forces upon an *individual* element. Here, we address the work upon a *patch*. Transverse shear forces upon interfaces are internal actions and, in the absence of shear strains, these forces contribute no internal (strain) energy.

In a traditional treatment of equilibrium, the reactive shear forces are eliminated[12] in order to obtain one equation in terms of moments. Here, too, the three Equations (10.104a–c) can be combined as follows: First, we differentiate (10.104a) and (10.104b):

$$-\frac{\partial^2 M_y}{\partial y^2}+\frac{\partial^2 M_{xy}}{\partial x \partial y}=0, \qquad -\frac{\partial^2 M_x}{\partial x^2}+\frac{\partial^2 M_{xy}}{\partial x \partial y}=0$$

Adding the above and subtracting that sum from Equation (10.104c), we obtain

$$\frac{\partial^2 M_y}{\partial y^2}+\frac{\partial^2 M_x}{\partial x^2}+2\frac{\partial^2 M_{xy}}{\partial x \partial y}=-p \tag{10.105}$$

Although the foregoing account provides a bridge between the continuum theory and the discrete approximation, differences in the derivations deserve further comment, most notably the role of the transverse shear forces in the Newtonian approach of Section 8.13. As noted, the transverse shear forces play a passive role; neither deformations nor strain energy are associated with those actions. If one were to apply the variational method to the individual patch of Figure 10.35, then those shear forces would appear as edge load upon the patch; they would do work upon the rigid-body rotations, $\bar{\theta}_x$ and $\bar{\theta}_y$, in amounts $\bar{q}_y\delta\theta_x$ and $\bar{q}_x\delta\theta_y$. These mean values would appear in Equation (10.101); the continuous counterparts would appear in Equations (10.104a,b). Also, it should be noted that the rotations θ_x and θ_y in the discrete approximation are not linked to the displacement w; that relationship occurs only in the limit via Equations (10.97b,c), e.g., from Equation (10.97b)

$$\bar{\theta}_x \Rightarrow \frac{3}{2}\frac{\Delta w}{\Delta y}-\frac{1}{2}\bar{\theta}_x, \qquad \bar{\theta}_x=\frac{\partial w}{\partial y}$$

Convergence is ultimately assured, since the approximation (10.95) achieves a smooth surface, i.e. function w and derivatives are continuous in the limit ($\Delta x \to 0, \Delta y \to 0$).

10.11 APPROXIMATIONS OF THE 3-D ELEMENT

Of course, all bodies, structural or mechanical components, are three-dimensional. It is only the slender form of a beam and the thinness of the plate or shell, which admit the approximations and simplifications to one or two dimensions. The concepts of finite elements apply as well in three dimensions. Here, we introduce the simplest form of approximation, viz., the tri-linear approximation in the rectangular Cartesian format. Moreover, we again employ the principle of stationary potential to illustrate the concepts. As previously noted, the continuous *approximation* of a deformed configuration always possesses excessive energy; it is too stiff! Again some terms in the polynomial approximations of the elements are higher-order (H.O.); they are needed only to inhibit all deformational modes. Judicious suppression of repeated H.O. terms reduces excessive stiffness and improves convergence toward the "exact" solution of the continuum theory. In our applications, we invoke that feature to achieve the more practical approximation. Features are drawn from the authors early work.[13]

10.11.1 THE SIMPLE TRI-LINEAR ELEMENT

To illustrate the concepts and formulation, we turn to the tri-linear approximation of the displacement **V** within the rectangular element shown in Figure 10.46. The displacement within the element is expressed by the vector

$$\begin{aligned}\mathbf{V} =& \bar{\mathbf{V}}+\mathbf{V}_x x+\mathbf{V}_y y+\mathbf{V}_z z\\ &+\mathbf{V}_{xy}xy+\mathbf{V}_{zx}zx+\mathbf{V}_{yz}yz+\mathbf{V}_{xyz}xyz\end{aligned} \tag{10.106}$$

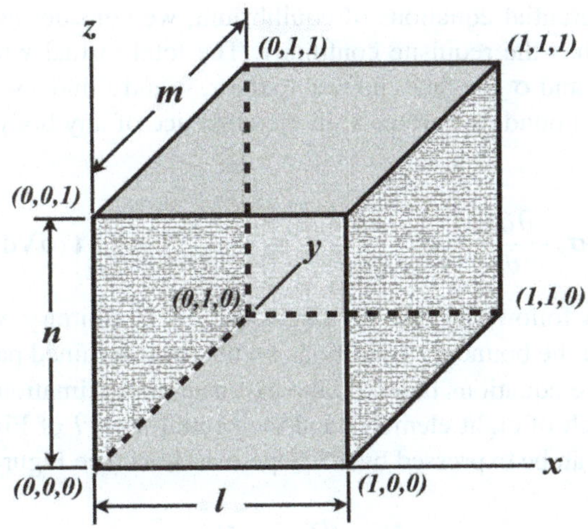

Figure 10.46 Rectangular Element

Each coefficient has three components, e.g.

$$\bar{\mathbf{V}} = u\hat{\mathbf{i}} + v\hat{\mathbf{j}} + w\hat{\mathbf{k}}, \text{ etc} \tag{10.107}$$

This provides continuity between contiguous elements. An element possesses 24 (3 × 8) degrees of freedom, corresponding to the three components of displacement at each of the 8 corners. The linearity along each of the contiguous edges signals the complete conformity/continuity of the assemblage.

The formulation requires the system of algebraic equations governing the nodal displacements, i.e. the displacements of the corners. Our treatment is limited to the small displacements of a linearly elastic (HI-HO) body, and, therefore, a system of linear algebraic equations. However, the concepts and methods would apply as well to the large displacements of an elastic body.

An element is depicted in Figure 10.46 with the local rectangular coordinates (x, y, z) originating at the corner $(0,0,0)$. For brevity, we define the dimensionless local coordinates:

$$\bar{x} = \frac{x}{l}, \quad \bar{y} = \frac{y}{m}, \quad \bar{z} = \frac{z}{n} \tag{10.108a,b,c}$$

where l, m, and n are dimensions of the element as depicted in Figure 10.46. The eight nodal values of the element lie at corners. In Section 10.10.3, we express the strain components in terms of the coefficients in Equation (10.106) and those coefficients in terms of the 24 nodal values of the displacement components. The lead term $\bar{V}$ represents the translation of the origin $(0,0,0)$. The nine components of the linear terms (3×3) are dominant as the element shrinks $(m, n, l \to 0)$. These represent the three rigid rotations and the six strains $(\varepsilon_x, \varepsilon_y, \varepsilon_z, \gamma_{xy}, \gamma_{xz}, \gamma_{yz}$ at the origin $(0,0,0)$. These are explicitly explored in Section 10.10.3. The remaining 12 components are higher-order terms (HOT).

10.11.2 THE VARIATIONAL METHOD OF APPROXIMATION

As mentioned previously (see Section 10.3), the approximation via finite elements is analogous to the variational method of the continuum theories, specifically, in the application of virtual work. To derive the differential equations of equilibrium, we consider an incremental displacement $\delta \mathbf{V}(x,y,z)$, which has the requisite continuity. The total virtual work includes that of the internal stresses ($\boldsymbol{\sigma}_x$, $\boldsymbol{\sigma}_y$, and $\boldsymbol{\sigma}_z$ on faces normal to the x, y, and z lines) within the body v and the external traction $\mathbf{t}$ on the bounding surface s; in the abscence of any body force the virtual work follows:

$$\iiint_v \left(\boldsymbol{\sigma}_x \cdot \frac{\partial \delta \mathbf{V}}{\partial x} + \boldsymbol{\sigma}_y \cdot \frac{\partial \delta \mathbf{V}}{\partial y} + \boldsymbol{\sigma}_z \cdot \frac{\partial \delta \mathbf{V}}{\partial z} \right) - \iint_s \mathbf{t} \cdot \delta \mathbf{V} ds = 0 \qquad (10.109)$$

An integration by parts is followed by the argument that $\delta \mathbf{V}$ is arbitrary, which leads to the three differential equations and the boundary conditions on the unconstrained parts of the boundary s.

Let us seek the discrete equations based on the tri-linear approximation of the rectangular element. We focus on a patch of eight elements and the central node I of Figure 10.47. Within that patch, the displacement can be expressed by a "shape function" (see Figure 10.3 and 10.5).

$$\mathbf{V} = f^I(x,y,z)\mathbf{V}_I$$

Here the origin is placed at the central node I. Dimensions of each element are l, m, and n in directions x, y, and z, respectively. The "shape function" for the polynomial (10.106) is expressed as follows in the first octant ($x \geq 0, y \geq 0, z \geq 0$):

$$\begin{aligned} f^I &= 1 - \frac{x}{l} - \frac{y}{m} - \frac{z}{n} + \frac{xy}{lm} + \frac{xz}{ln} + \frac{yz}{mn} - \frac{xyz}{lmn} \\ &0 \leq x \leq l, \quad 0 \leq y \leq m, \quad 0 \leq z \leq n \end{aligned} \qquad (10.110)$$

Corresponding forms for the remaining elements are obtained by altering the signs. For example, in the second octant ($-l \leq x \leq 0$)

$$f^I = 1 + \frac{x}{l} - \frac{y}{m} - \frac{z}{n} - \frac{xy}{lm} - \frac{xz}{ln} + \frac{yz}{mn} + \frac{xyz}{lmn}$$

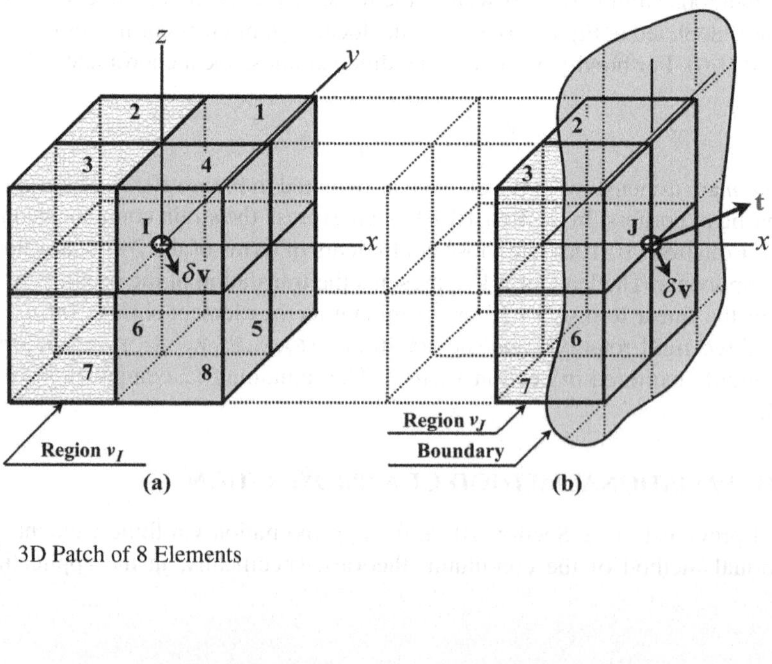

Figure 10.47 3D Patch of 8 Elements

The function f^I vanishes at the boundaries ($x = \pm l, y = \pm m, z = \pm n$ where the $\pm$ signs apply depending on the octant [element]). The function has the following properties.

$$\iiint_{V_I} f^I \mathrm{d}x\mathrm{d}y\mathrm{d}z = 1, \quad \iint_{s_x} f^I \mathrm{d}y\mathrm{d}z = 1 \tag{10.111a,b}$$

where V_I denotes the space of the patch of eight elements and s_x denotes the x surface through node I.

If we apply the variational concept to the assemblage of elements, we are only free to vary a nodal displacement $\delta \mathbf{V}_I$. Instead of the increment work of Equation (10.109), we have

$$\delta \mathbf{V}_I \cdot \iiint_{V_I} \left[\boldsymbol{\sigma}_x \frac{\partial f^I}{\partial x} + \boldsymbol{\sigma}_y \frac{\partial f^I}{\partial y} + \boldsymbol{\sigma}_z \frac{\partial f^I}{\partial z} \right] = 0 \tag{10.112}$$

Note: like the variation $\delta V(x,y,z)$ in the continuum theory, the increment $\delta \mathbf{V}_I$ at an interior node I is independent of the increment $\delta \mathbf{V}_J$ at a surface node J. Our principle asserts that both vanish: One provides the equilibrium equations at an interior node I; the other ensures the equilibrium at the surface node J.

Let us examine one equation of equilibrium, corresponding to a displacement increment $\delta u_I = \delta \mathbf{V} \cdot \hat{\mathbf{i}}$, specifically, the associated virtual work must vanish:

$$\iiint_{V_I} \left(\sigma_x \frac{\partial f^I}{\partial x} + \tau_{xy} \frac{\partial f^I}{\partial y} + \tau_{xz} \frac{\partial f^I}{\partial z} \right) \delta u_I \,\mathrm{d}x\,\mathrm{d}y\,\mathrm{d}z = 0 \tag{10.113}$$

It is the integral that must vanish for equilibrium. If we restrict our attention to small deformations, then

$$\frac{\partial}{\partial x}(f^I \delta u_I) \approx \frac{\partial \delta u}{\partial x} = \delta \varepsilon_x, \quad \text{etc.}$$

When we substitute this into Equation (10.113), we recognize the integral as a weighted average of the increment of internal work within the patch V_I.

To evaluate the integral and obtain the algebraic equations, viz the 3 components of force, we require the derivatives, e.g. $\partial f^I / \partial x$, for each octant of the patch. In the first octant ($x \geq 0, y \geq 0, z \geq 0$),

$$\frac{\partial f^I}{\partial x} = -\frac{1}{l} + \frac{y}{lm} + \frac{z}{ln} - \frac{yz}{lmn} \tag{10.1114}$$

All terms play a role in the algebraic equations governing the approximation of the body. However, our intent is an interpretation and a demonstration of convergence to the differential equations of the continuum theory. To that end, we require only the dominant terms; higher-order terms vanish in the limit ($l \to 0, m \to 0, n \to 0$). For each of the adjoining elements, those dominant terms follow:

$$\frac{\partial f_1^I}{\partial x} = -\frac{1}{l}, \quad \frac{\partial f_2^I}{\partial x} = \frac{1}{l}, \quad \frac{\partial f_3^I}{\partial x} = \frac{1}{l}, \quad \frac{\partial f_4^I}{\partial x} = -\frac{1}{l}$$
$$\frac{\partial f_5^I}{\partial x} = -\frac{1}{l}, \quad \frac{\partial f_6^I}{\partial x} = \frac{1}{l}, \quad \frac{\partial f_7^I}{\partial x} = \frac{1}{l}, \quad \frac{\partial f_8^I}{\partial x} = -\frac{1}{l}$$
$$\frac{\partial f_1^I}{\partial y} = -\frac{1}{m}, \quad \frac{\partial f_2^I}{\partial y} = -\frac{1}{m}, \quad \cdots \text{etc.}$$
$$\frac{\partial f_1^I}{\partial z} = -\frac{1}{n}, \quad \frac{\partial f_2^I}{\partial z} = -\frac{1}{n}, \quad \cdots \text{etc.}$$

The consistent approximation also requires that the stresses in Equations (10.112) and (10.113) be consistent with our approximation of the deformation. For example, if the material is HI-HO, then

$$\sigma_x = \frac{E}{(1-2\nu)(1+\nu)}[(1-\nu)\varepsilon_x + \nu\varepsilon_y + \nu\varepsilon_z]$$
$$\tau_{xy} = G\gamma_{xy}, \text{ etc.}$$
$$\text{and} \quad \varepsilon_x \approx \frac{\partial f^I}{\partial x}u_I, \quad \gamma_{xy} \approx \frac{\partial f^I}{\partial x}v_I + \frac{\partial f^I}{\partial y}u_I, \text{ etc.}$$

For our present purposes, we define the integrals of those stresses in each octant. For example, in octant n,

$$\sigma_x^n = \frac{1}{lmn}\iiint_{V_n} \sigma_x \mathrm{d}x\mathrm{d}y\mathrm{d}z \tag{10.115}$$

Our approximation of the integral in (10.113) then contains only the mean values $\sigma_x^n, \tau_{yx}^n, \tau_{zx}^n$ and the dominant terms of the derivatives, viz.

$$\frac{\partial f^I}{\partial x}, \frac{\partial f^I}{\partial y}, \frac{\partial f^I}{\partial z}$$

With the foregoing approximations, the integral of (10.112), i.e. the equilibrium equation, assumes the form:

$$\begin{aligned}
&-\frac{4}{l}\left[\left(\frac{\sigma_x^1+\sigma_x^4+\sigma_x^5+\sigma_x^8}{4}\right) - \left(\frac{\sigma_x^2+\sigma_x^3+\sigma_x^6+\sigma_x^7}{4}\right)\right] \\
&-\frac{4}{m}\left[\left(\frac{\tau_{xy}^1+\tau_{xy}^2+\tau_{xy}^5+\tau_{xy}^6}{4}\right) - \left(\frac{\tau_{xy}^3+\tau_{xy}^4+\tau_{xy}^7+\tau_{xy}^8}{4}\right)\right] \\
&-\frac{4}{n}\left[\left(\frac{\tau_{xz}^1+\tau_{xz}^2+\tau_{xz}^3+\tau_{xz}^4}{4}\right) - \left(\frac{\tau_{xz}^5+\tau_{xz}^6+\tau_{xz}^7+\tau_{xz}^8}{4}\right)\right] = 0
\end{aligned} \tag{10.116}$$

The first parenthetical term in the first bracket is the average of the mean stress components (σ_x) in the four elements ($x \geq 0$); the second parenthetical term is the average of the components (σ_x) in the four elements ($x \leq 0$). The difference in that first bracket is divided by the median distance between the 2 layers, (elements $x \geq 0$) – (elements $x \leq 0$). That difference converges ($l \to 0, m \to 0, n \to 0$) to the derivative $\partial\sigma_x/\partial x$. By the same logic, the second and third brackets converge to $\partial\tau_{xy}/\partial y$ and $\partial\tau_{xz}/\partial z$, respectively. Our difference equation converges to the differential equation (see Equation 3.41), viz.

$$\frac{\partial \sigma_x}{\partial x} + \frac{\partial \tau_{xy}}{\partial y} + \frac{\partial \tau_{xz}}{\partial z} = 0$$

Neglecting higher-order contributions to the derivatives of f^I and introducing the mean values of stress components $(\sigma_x^n, \tau_{xy}^n, \tau_{xz}^n)$ does not invalidate our result, since all converge to those values in the limit $(l \to 0, m \to 0, n \to 0)$.

At the surface node J of Figure 10.47b, the increment $\delta \mathbf{V}_J$ is arbitrary. The variation of the internal work not only extends through the adjoining patch V_j of four elements but also includes the external work of traction $\mathbf{t}$ on the boundary surface s_j:

$$\delta \mathbf{V}_j \cdot \left[\iiint_{V_J} \left(\boldsymbol{\sigma}_x \frac{\partial f^I}{\partial x} + \boldsymbol{\sigma}_y \frac{\partial f^I}{\partial x} + \boldsymbol{\sigma}_z \frac{\partial f^I}{\partial x} \right) \mathrm{d}x\,\mathrm{d}y\,\mathrm{d}z - \iint_{s_J} \mathbf{t} f^J \mathrm{d}x\,\mathrm{d}z \right] = 0 \qquad (10.117)$$

Again, the bracketed term must vanish for equilibrium.

For our purposes, we require only the initial (dominant) terms in the function f^I and the derivatives. Since we are concerned only with the convergence $(l \to 0, m \to 0, n \to 0)$, we define the mean values

$$\frac{1}{lmn} \iiint_{V_n} \boldsymbol{\sigma}_x \,\mathrm{d}x\,\mathrm{d}y\,\mathrm{d}z \equiv \boldsymbol{\sigma}_x^n$$

where n denotes the patch of four elements $(2,3,6,7)$. Similarly, we define the mean values:

$$\frac{1}{mn} \iint_{S_n} \mathbf{t}\,\mathrm{d}y\,\mathrm{d}z \equiv \mathbf{t}^n$$

With these notations, the bracketed term of Equation (10.117) assumes the form

$$(\boldsymbol{\sigma}_x^2 + \boldsymbol{\sigma}_x^3 + \boldsymbol{\sigma}_x^6 + \boldsymbol{\sigma}_x^7) + (-\boldsymbol{\sigma}_y^2 + \boldsymbol{\sigma}_y^3 - \boldsymbol{\sigma}_y^6 + \boldsymbol{\sigma}_y^7)$$
$$+ (-\boldsymbol{\sigma}_z^2 - \boldsymbol{\sigma}_z^3 + \boldsymbol{\sigma}_z^6 + \boldsymbol{\sigma}_z^7) - (\mathbf{t}^2 + \mathbf{t}^3 + \mathbf{t}^6 + \mathbf{t}^7) = 0$$

In the limit, the first parenthetical term converges to $4\boldsymbol{\sigma}_x$, the second and third vanish, and the last converges to $\mathbf{4t}$. This is the boundary condition:

$$\boldsymbol{\sigma}_x = \mathbf{t}$$

10.11.3 DEFORMATION OF THE TRILINEAR ELEMENT

It is worth examining the geometrical implications of the coefficients in the polynomial (10.106). Firstly, we have the constant term $\bar{\mathbf{V}}$ which represents displacement of the center $(0,0,0)$, with the 3 components:

$$\bar{u}, \bar{v}, \bar{w} \qquad (10.118\text{a,b,c})$$

Secondly, we have the linear terms, which can be decomposed as follows:

$$\omega_x \equiv \frac{1}{2}\left(\hat{\mathbf{k}}\cdot\frac{\partial \mathbf{V}}{\partial y} - \hat{\mathbf{j}}\cdot\frac{\partial \mathbf{V}}{\partial z}\right)\bigg|_{000} \equiv \frac{1}{2}(w_y - v_z) \tag{10.119a}$$

$$\omega_y \equiv \frac{1}{2}\left(\hat{\mathbf{i}}\cdot\frac{\partial \mathbf{V}}{\partial z} - \hat{\mathbf{k}}\cdot\frac{\partial \mathbf{V}}{\partial x}\right)\bigg|_{000} \equiv \frac{1}{2}(u_z - w_x) \tag{10.119b}$$

$$\omega_z \equiv \frac{1}{2}\left(\hat{\mathbf{j}}\cdot\frac{\partial \mathbf{V}}{\partial x} - \hat{\mathbf{i}}\cdot\frac{\partial \mathbf{V}}{\partial y}\right)\bigg|_{000} \equiv \frac{1}{2}(v_x - u_y) \tag{10.119c}$$

These represent three rigid-body rotations about axes x, y, and z, respectively, of lines at the origin $(0,0,0)$. Here the subscripts signify the derivatives, e.g. $u_x = \mathbf{V_x}\cdot\hat{\mathbf{i}}$

Let us evaluate the strains at the origin.

$$\bar{\varepsilon}_x \equiv \frac{\partial \mathbf{V}}{\partial x}\cdot\hat{\mathbf{i}}\bigg|_{000} = u_x, \quad \bar{\varepsilon}_z \equiv \frac{\partial \mathbf{V}}{\partial z}\cdot\hat{\mathbf{k}}\bigg|_{000} = w_z, \quad \bar{\varepsilon}_y \equiv \frac{\partial \mathbf{V}}{\partial y}\cdot\hat{\mathbf{j}}\bigg|_{000} = v_y \tag{10.120a,b,c}$$

$$\bar{\gamma}_{xy} \equiv \left(\frac{\partial \mathbf{V}}{\partial x}\cdot\hat{\mathbf{j}} + \frac{\partial \mathbf{V}}{\partial y}\cdot\hat{\mathbf{i}}\right)\bigg|_{000} = v_x + u_y \tag{10.120d}$$

$$\bar{\gamma}_{xz} \equiv \left(\frac{\partial \mathbf{V}}{\partial x}\cdot\hat{\mathbf{k}} + \frac{\partial \mathbf{V}}{\partial z}\cdot\hat{\mathbf{i}}\right)\bigg|_{000} = w_x + u_z \tag{10.120e}$$

$$\bar{\gamma}_{yz} \equiv \left(\frac{\partial \mathbf{V}}{\partial y}\cdot\hat{\mathbf{k}} + \frac{\partial \mathbf{V}}{\partial z}\cdot\hat{\mathbf{j}}\right)\bigg|_{000} = w_y + v_z \tag{10.120f}$$

Equations (10.120) are the six components of strain at the origin. These are the dominant strains which prevail in the limit $(\Delta x \to 0, \Delta y \to 0, \Delta z \to 0,)$, i.e. as the element shrinks.

Equations (10.118), (10.119), and (10.120) are the 12 dominant modes. The remaining terms of Equation (10.106) are higher-order terms. These are HOT terms, which are essential to the discrete model. These HOT terms contribute to the energy and stiffness, but can be suppressed where they appear in more than one strain component (see Illustrative Example 10.1). Six of the HOT terms represent "extensional modes":

$$u_{xy}, u_{xz}, v_{yx}, v_{yz}, w_{zx}, w_{zy}$$

Three HOT terms represent "shear modes":

$$u_{yz} \equiv \gamma_x, \quad v_{xz} \equiv \gamma_y, \quad w_{xy} \equiv \gamma_z$$

Three cubic terms represent "flexural modes":

$$u_{xyz} \equiv \kappa_x, \quad v_{xyz} \equiv \kappa_y, \quad w_{xyz} \equiv \kappa_z$$

Our descriptive terminology can be appreciated by viewing the geometrical features shown in Figure 10.48. Note that modes u_{xy} and u_{xz} exhibit extension in the x direction; $u_{yz} \equiv \gamma_x$ exhibits a variable shear in the x-y plane and $u_{xyz} \equiv \kappa_x$ exhibits a variable flexure in the x-y plane.

10.11.4 FORMULATION OF THE TRILINEAR ELEMENT

Let us return to the trilinear approximation of Equation (10.106) and examine the roles of the various coefficients as they appear in the following strain components:

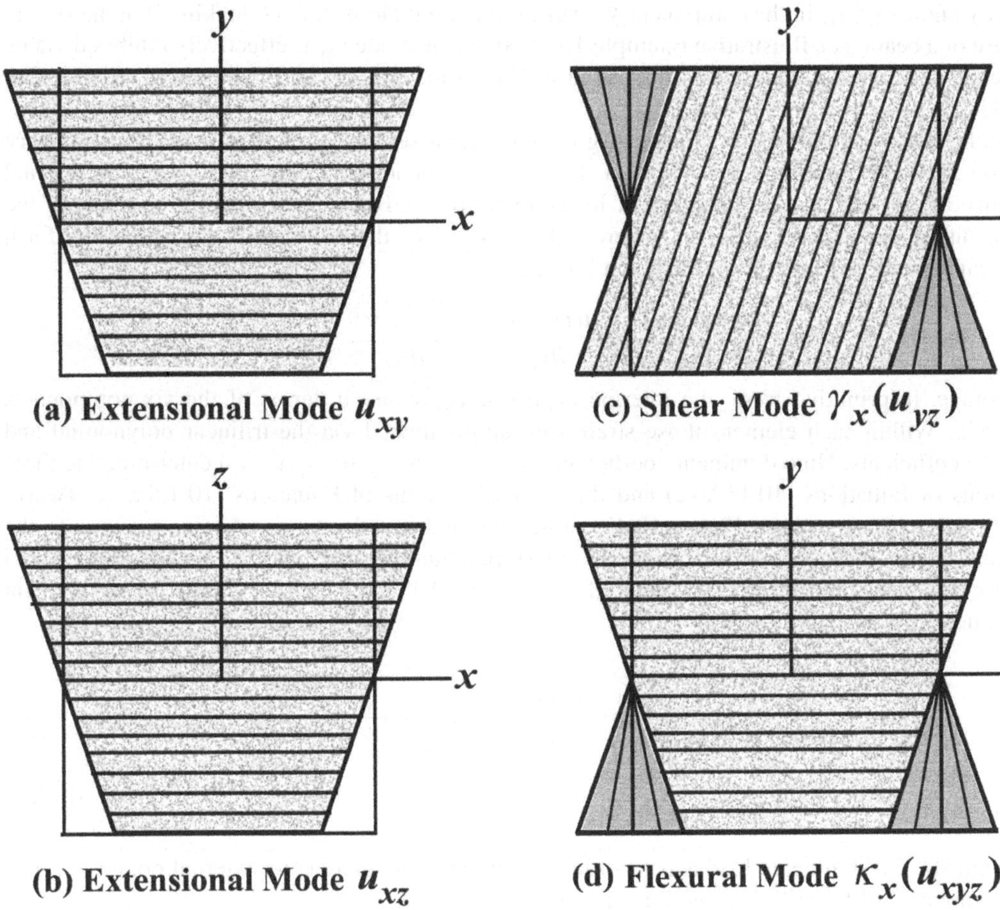

Figure 10.48

$$
\begin{aligned}
\varepsilon_x &= \hat{\mathbf{i}} \cdot \frac{\partial \mathbf{V}}{\partial x} \\
&= u_x + u_{xy}y + u_{xz}z + u_{xyz}yz \\
\gamma_{xy} &= \hat{\mathbf{j}} \cdot \frac{\partial \mathbf{V}}{\partial x} + \hat{\mathbf{i}} \cdot \frac{\partial \mathbf{V}}{\partial y} \\
&= (v_x + u_y) + (v_{xy}y + u_{xy}x) + (v_{xz} + u_{yz} + v_{xyz}y + u_{xyz}x)z \\
\gamma_{xz} &= \hat{\kappa} \cdot \frac{\partial \mathbf{V}}{\partial x} + \hat{\mathbf{i}} \cdot \frac{\partial \mathbf{V}}{\partial z} \\
&= (w_x + u_z) + (w_{xz}z + u_{xz}x) + (w_{xy} + u_{yz} + w_{xyz}z + u_{xyz}x)y
\end{aligned}
$$

In these expressions, we note that the extensional mode (u_{xy}) could be inhibited by retaining the term in ε_x *or* in γ_{xy}; it could be deleted in one to improve the approximation. Similarly, the term u_{xz} can be omitted from ε_x or γ_{xz}; the term u_{yz} from γ_{xy} or γ_{xz}; u_{xyz} from ε_x, γ_{xy}, or γ_{xz}. The choices depend upon the problem, i.e. the role that the particular strain plays. For example, we already know that u_{xy} in the component γ_{xy} causes unacceptable stiffness ("locking") in the simple flexure of a beam (see Illustrative Example 10.1). Since the mode u_{xy} is effectively inhibited via its presence in ε_x, it can be omitted from γ_{xy}; indeed, that is the only practical solution to the so-called "locking".

As before, the equilibrium equations, governing the assembly of elements, are the stationary criteria imposed upon the total potential. Let U and V denote the contributions of the internal (strain) energy and external potential of loads, respectively. Both are expressible in terms of the nodal displacements, viz. u_I, v_I, w_I where I denotes any of the m nodes in the assembly. Each algebraic equation derives from the potential, e.g.

$$\frac{\partial(U+V)}{\partial u_I} = \frac{\partial(U+V)}{\partial v_I} = \frac{\partial(U+V)}{\partial w_I} = 0$$

Of course, in primitive form, the internal energy is expressed in terms of the six components of strain. Within each element those strains are approximated via the trilinear polynomial and the 21 coefficients: Nine dominant coefficients $(u_x, v_x, w_x, u_y, v_y, w_y, u_z, v_z, w_z)$ determine the three rotations of Equations (10.119a–c) and the six mean strains of Equations (10.120a–f). Twelve additional coefficients (see Figure 10.47) appear as higher-order terms. Again, we turn to the internal energy U and the principle of stationary potential (virtual work) to establish the nodal force components. For the HI-HO material (see Section 4.11), the internal energy for the element of Figure 10.46 is the integral

$$U^n = lmn \int_0^{+1}\int_0^{+1}\int_0^{+1} \left\{ \frac{E}{2(1-2\nu)(1+\nu)} \left[(1-\nu)(\varepsilon_x^2+\varepsilon_y^2+\varepsilon_z^2) + 2\nu(\varepsilon_x\varepsilon_y+\varepsilon_y\varepsilon_z+\varepsilon_x\varepsilon_z)\right] + \frac{G}{2}(\gamma_{xy}^2+\gamma_{xz}^2+\gamma_{yz}^2) \right\} d\bar{x}\, d\bar{y}\, d\bar{z}$$

Typically, no significant load acts at an internal node I, and there is no external potential. Then the component of force F_x in direction x is given by

$$F_x = \frac{\partial U}{\partial u_I} = \sum_{n=1}^{8} \frac{\partial U^n}{\partial u_I}$$

Note that the energy encompasses the 8 adjoining elements. By the chain rule for differentiation we obtain the strains in terms of the nodal displacements, viz. u_m, v_m, and w_m. For example, from Equations (10.106) and (10.107):

$$\varepsilon_x = \frac{\partial u}{\partial x} = u_x + u_{xy}y + u_{xz}z + u_{xyz}yz$$

This in turn requires the coefficients expressed in terms of the nodal displacement. The Equations (10.106) and (10.107) serve to express the displacement components u^n, v^n, w^n at each of the eight corners of the element in terms of the coefficients. These can be inverted to express the 24 coefficients in terms of the displacements. For example, with the corners labeled as shown in Figure 10.49, we obtain

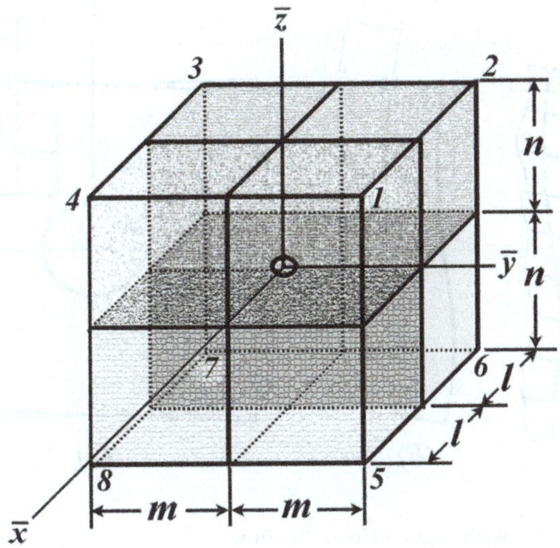

Figure 10.49 Square Element

$$\bar{u} = \frac{1}{8}(u^1 + u^2 + u^3 + u^4 + u^5 + u^6 + u^7 + u^8)$$
$$u_x = \frac{1}{8}(u^1 - u^2 - u^3 + u^4 + u^5 - u^6 - u^7 + u^8)$$
$$u_y = \frac{1}{8}(u^1 + u^2 - u^3 - u^4 + u^5 + u^6 - u^7 - u^8)$$
$$u_z = \frac{1}{8}(u^1 + u^2 + u^3 + u^4 - u^5 - u^6 - u^7 - u^8)$$
$$u_{xy} = \frac{1}{8}(u^1 - u^2 + u^3 - u^4 + u^5 - u^6 + u^7 - u^8)$$
$$u_{xz} = \frac{1}{8}(u^1 - u^2 - u^3 + u^4 - u^5 + u^6 + u^7 - u^8)$$
$$u_{yz} = \frac{1}{8}(u^1 + u^2 - u^3 - u^4 - u^5 - u^6 + u^7 + u^8)$$
$$u_{xyz} = \frac{1}{8}(u^1 - u^2 + u^3 - u^4 - u^5 + u^6 - u^7 + u^8)$$

To convert to dimensions, multiply by l, m, n accordingly as the subscripts are designated x, y, z, respectively; then x, y, z are the actual dimensions of an element of Figure 10.46 and 10.49.

ILLUSTRATIVE EXAMPLE 10.9

Let us revisit the torsion of a HI-HO bar with a rectangular cross-section (see Illustrative Example 5.17). Figure 10.50 depicts the twisted bar; (a) and (b) exhibit the shear strain, which must vanish at the corners since $\tau_{xz} = \tau_{zy} = 0$ on the surfaces. We assume that cross-sections are unconstrained against axial displacement (w), i.e. a cross-section is free to warp:

$$w = \psi(x, y)$$

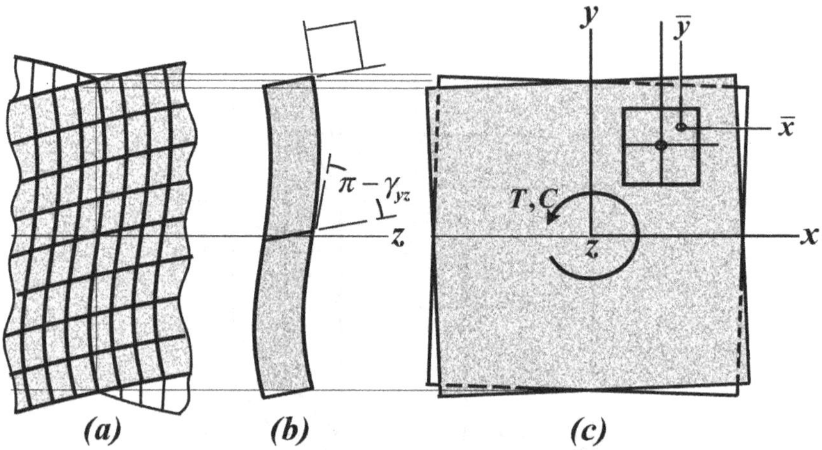

Figure 10.50 Torsion of a Bar with Square Cross-Section

We can adopt the well-established assumptions of St. Venant: There is no deformation *within* a cross-section, that is

$$\varepsilon_x = \varepsilon_y = \gamma_{xy} = 0$$

Moreover

$$\varepsilon_z = \frac{\partial w}{\partial z} = 0$$

Then the sole strain components are

$$\tau_{xz}(x, y), \quad \tau_{yz}(x, y)$$

Any cross-sectional slice, as depicted in Figure 10.50b, exhibits the same deformation. Since there is no deformation within a cross-section, the displacements u and v result from the rotation θ/per unit length about the z axis. The displacement is

$$\mathbf{V} = u\hat{\mathbf{i}} + v\hat{\mathbf{j}} + w\hat{\mathbf{k}}$$
$$u = -yz\theta, \quad v = xz\theta$$

where $z\theta$ is the rotation at distance z from the origin.

We can focus our attention on the deformation of the slice of Figure 10.50b. The strains follow:

$$\gamma_{xz} = -y\theta + \frac{\partial \psi}{\partial x}, \quad \gamma_{yz} = x\theta + \frac{\partial \psi}{\partial y}$$

Mathematically, our problem is two-dimensional. We require the function $\psi(x, y)$. Then, for any given twist θ, we can compute the strain/stress distribution and resultant torque T.

For element 1 of Figure 10.51, the bilinear approximation of the function ψ has the form:

$$w \approx \psi = \bar{\psi} + \psi_x \bar{x} + \psi_y \bar{y} + \psi_{xy} \bar{x}\bar{y} \tag{10.121}$$

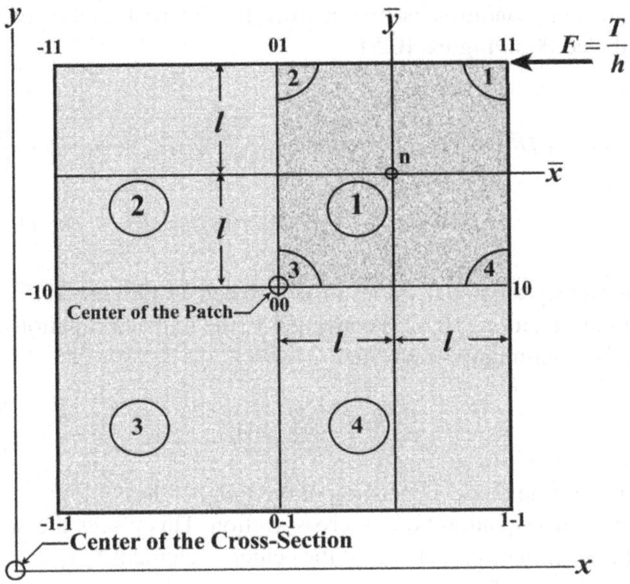

Figure 10.51 Patch of 4 Square Elements

The coefficients are expressed in terms of nodal values as follows:

$$\bar{\psi} = \frac{1}{4}(\psi_1 + \psi_2 + \psi_3 + \psi_4) \tag{10.122a}$$

$$\psi_x = \frac{1}{4l}(\psi_1 + \psi_4 - \psi_2 - \psi_3) \tag{10.122b}$$

$$\psi_y = \frac{1}{4l}(\psi_1 + \psi_2 - \psi_3 - \psi_4) \tag{10.122c}$$

$$\psi_{xy} = \frac{1}{4l^2}(\psi_1 + \psi_3 - \psi_2 - \psi_4) \tag{10.122d}$$

Note that coordinates (x, y) originate at the center of the cross-section; coordinates $(\bar{x}, \bar{y})$ originate at the center n of the element. Then,

$$x = x_n + \bar{x}, \quad y = y_n + \bar{y} \tag{10.23a,b}$$

$$\gamma_{xz} = -(y_n + \bar{y})\theta + \psi_x + \psi_{xy}\bar{y} \tag{10.124a}$$

$$\gamma_{yz} = (x_n + \bar{x})\theta + \psi_y + \psi_{xy}\bar{x} \tag{10.124b}$$

We require the internal energy within the patch. The internal energy of one element follows:

$$U^n = \frac{Gl}{2}\int_{-l}^{l}\int_{-l}^{l}({\gamma_{xz}}^2 + {\gamma_{yz}}^2)\, d\bar{x}\, d\bar{y} \tag{10.125}$$

As a matter of convenience, we take the slice of thickness l and a square element $(2l \times 2l)$.

$$U^n = 2Gl^3\left[\theta^2\left({x_n}^2 + {y_n}^2 + \frac{2}{3}l^2\right) + 2\theta(x_n\psi_y + y_n\psi_x) + {\psi_x}^2 + {\psi_y}^2 + \frac{2}{3}l^2{\psi_{xy}}^2\right] \tag{10.126}$$

The required equilibrium condition is drawn from the internal energy of the four elements adjoining the node, labeled 00 in Figure 10.51.

$$\frac{\partial}{\partial \psi_{00}}(U^1+U^2+U^3+U^4)=Gl\left[-\frac{1}{3}(\psi_{11}+\psi_{-11}+\psi_{-1-1}+\psi_{1-1})+\frac{8}{3}\psi_{00}\right.$$
$$\left.-\frac{1}{3}(\psi_{10}+\psi_{-10}+\psi_{01}+\psi_{0-1})\right]=0 \quad (10.127)$$

Here the subscripts signify the nodal values of ψ as labeled in Figure 10.51. It is worth noting that the bracketed term in Equation (10.127) converges to the Laplace equation, which governs the axial displacement in the continuum theory, viz.

$$\frac{\partial^2 \psi}{\partial x^2}+\frac{\partial^2 \psi}{\partial y^2}=0$$

Let us formulate the discrete model consisting of the four patches of Figure 10.51. The patch of Figure 10.51 occupies the first quadrant of the cross-section. This places the node $(-1,-1)$ at the center, the origin of the coordinates (x,y), e.g. at the center of element 1: $x_1=y_1=3l$. Symmetries dictate that displacement w vanish on the axes:

$$\psi_{-11}=\psi_{-10}=\psi_{-11}=\psi_{0-1}=\psi_{1-1}=0$$

Also, the warping function $w(x,y)$ and our approximation must be antisymmetric with respect to the diagonal $x=y$. Therefore,

$$\psi_{00}=\psi_{11}=0$$

Then our discrete model has but three degrees of freedom, viz.

$$\psi_{10},\ \psi_{01},\ \text{and}\ \theta$$

These are governed by the equations for a stationary (minimum) potential:

$$\frac{\partial U}{\partial \psi_{10}}=0,\quad \frac{\partial U}{\partial \psi_{01}}=0,\quad \frac{\partial U}{\partial \theta}=\frac{T}{4} \quad (10.128a,b,c)$$

Here, U denotes the internal energy of the four elements, $\ell\theta$ is the rotation of the slice, and T is the torque upon the entire (4 quadrants) of the cross-section.

$$\frac{4}{3}\psi_{10}-\frac{1}{3}\psi_{01}=4\ell^2\theta$$
$$-\frac{1}{3}\psi_{10}+\frac{4}{3}\psi_{01}=-4\ell^2\theta$$

These provide the two nodal values and our approximation of the warping function.

$$\psi_{10}=+\frac{12}{5}\ell^2\theta,\quad \psi_{01}=-\frac{12}{5}\ell^2\theta \quad (10.129a,b)$$

The internal energy is now rendered as a function of the twist θ. The Equation (10.128c) provides the stiffness, i.e. the torque-twist relation:

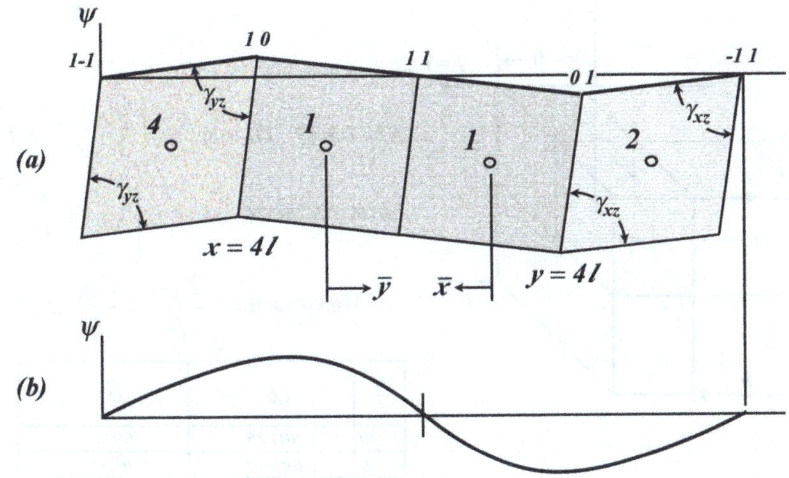

Figure 10.52 Approximation of Warping

$$T = 606G\ell^4\theta$$

A more accurate result is based upon a much better approximation[14]:

$$T = 575G\ell^4\theta$$

If we neglect warping, i.e. assume

$$\gamma_{xz} = -y\theta, \quad \gamma_{yz} = x\theta$$

then

$$T = 683G\ell^4\theta$$

This violates the boundary conditions: Most notably, $\tau_{yz} = -\tau_{xz} = 4\ell\theta$ at the corner $x = y = 4\ell$.

It is instructive to view the geometry of our simple approximation as seen in Figure 10.52a. The elements 4 and 1 are viewed on the side $x = 4\ell$; elements 1 and 2 are viewed on the side $y = 4\ell$. Note the antisymmetry, the greater shear strains in elements 4 (γ_{yz}) and 2 (γ_{xz}). Our simple approximation achieves the minimum energy consistent with the approximations. For comparison, Figure 10.52b depicts the plot of our approximation (5.88). That continuous function $w = \psi(x,y)$ provides the better result:

$$T = 545G\ell^4\theta$$

ILLUSTRATIVE EXAMPLE 10.10

The simple torsion of a square plate, as depicted in Figure 10.53, was previously treated [Section 8.16] in accordance with the Kirchhoff theory. With our current notations, the twist (θ) versus torque (T) is expressed by the equation:

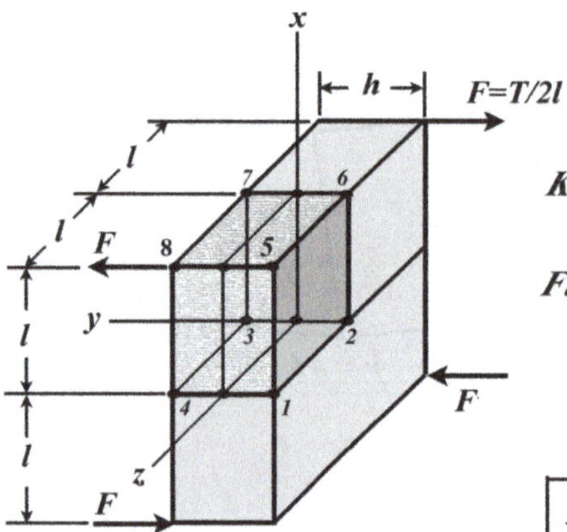

Kirchhoff Theory: $\alpha = \frac{2}{3}\frac{Gh^3\theta}{T} = 1$

FiniteElement : $\alpha * = \alpha\left[1 + \frac{1}{2}\left(\frac{h}{l}\right)^2\right]$

Maximum : $\frac{\gamma_{yz}}{\theta} = \frac{\gamma_{yx}}{\theta} \equiv \beta$

$\frac{h}{l}$	α^*	β
1/20	1.00125	0.00125
1/10	1.00500	0.00500
1/5	1.02000	0.02000
1/2	1.12500	0.12500

Figure 10.53 Torsion of a Square Plate

$$\frac{Gh^3\theta}{T} = \frac{3}{2}$$

Here we approach the same problem via the methods of finite elements.

Firstly, we recognize that a good approximation of the deformation is embodied in the trilinear approximation of Equation (10.106), e.g.

$$v = \bar{v} + v_y\frac{2y}{h} + v_z\frac{z}{l} + v_x\frac{x}{l} + v_{xy}\frac{2yx}{hl} + v_{yz}\frac{2yz}{hl} + v_{xz}\frac{xz}{l^2} + v_{xyz}\frac{2yxz}{hl^2}$$

We adopt the similar trilinear approximations of the displacement components u and w (directions x and z). Here the displacement w represents the warping of the previous example.

Note that the four quadrants of the plate must deform the same. Therefore, we need only focus on the first quadrant, as delineated in Figure 10.53. Rigid-body motion is ruled out by fixing the lines x, y, and z at the midsection. As in the earlier analyses, we discount the extensional strains,

$$\varepsilon_x = \varepsilon_y = \varepsilon_z = 0$$

By our approximation, all coordinate lines, e.g. edges of the element, remain straight but do rotate. The z lines of the upper face and x lines of the front face rotate through the angle θ. The displacement v follows:

$$v = \theta\frac{xz}{l}$$

After enforcing the conditions of symmetry and inextensionality ($\varepsilon_x = \varepsilon_y = \varepsilon_z = 0$), the entirety of coefficients is reduced to two, viz.

$$u = u_{yz}\frac{2yz}{hl}, \quad w = w_{yx}\frac{2yx}{hl}$$

Note that $u_{yz} = u_8$, $w_{yz} = w_8$ (see node 8 of Figure 10.53). The nonvanishing strains follow:

$$\gamma_{xz} = (u_8 + w_8)\frac{2y}{hl}, \quad \gamma_{yz} = \left(\theta + \frac{2}{h}w_8\right)\frac{x}{l}, \quad \gamma_{yx} = \left(\theta + \frac{2}{h}u_8\right)\frac{z}{l} \qquad (10.130a,b,c)$$

Again, to determine the unknowns, u_8 and w_8, we turn to the principle of stationary potential. Since there are no normal forces acting on the faces:

$$\frac{\partial U}{\partial u_8} = 0 = G\iiint_v \left[\gamma_{xz}\frac{\partial \gamma_{xz}}{\partial u_8} + \gamma_{yx}\frac{\partial \gamma_{yx}}{\partial u_8}\right] dxdydz \qquad (10.131a)$$

$$\frac{\partial U}{\partial w_8} = 0 = G\iiint_v \left[\gamma_{xz}\frac{\partial \gamma_{xz}}{\partial w_8} + \gamma_{yz}\frac{\partial \gamma_{yz}}{\partial w_8}\right] dxdydz \qquad (10.131b)$$

When Equations (10.130a–c) are inserted into the integrands of Equations (10.131a,b) and the latter are integrated over the volume of the element, i.e. the quadrant, we obtain

$$u_8 + w_8\left(1 + 4\frac{l^2}{h^2}\right) = -\frac{2l^2\theta}{h}$$

$$\left(1 + 4\frac{l^2}{h^2}\right)u_8 + w_8 = -\frac{2l^2}{h}\theta$$

The simultaneous solution follows:

$$u_8 = w_8 = -\frac{-l^2\theta}{h\left(1 + 2\frac{l^2}{h^2}\right)}$$

Now the shear stresses are expressed in terms of the rotation θ.

It remains to determine the force F exerted upon node 8. According to the principle of stationary potential:

$$F = \frac{\partial U}{l\partial \theta} = G\iiint_v \left[\gamma_{xz}\frac{\partial \gamma_{xz}}{\partial \theta} + \gamma_{yz}\frac{\partial \gamma_{yz}}{\partial \theta} + \gamma_{yx}\frac{\partial \gamma_{yx}}{\partial \theta}\right] dxdydz$$

Omitting terms of order h^4/l^4 as compared to unity, we obtain our result.

$$\frac{2Gh^3\theta}{3T} = \left(1 + \frac{1}{2}\frac{h^2}{l^2}\right) \equiv \alpha^* \qquad (10.140)$$

These values of the "twist" α^* are tabulated in Figure 10.53. Note that the twist of our element increases slightly with the ratio h/l. This can be attributed to the transverse shear strains γ_{xy} and γ_{zy} which are suppressed by the Kirchhoff hypothesis. Note too that the twist agrees with the result (See Illustrative Example 8.16, Figure 8.15) via the Kirchhoff theory with $h/l = 0$.

ILLUSTRATIVE EXAMPLE 10.11

Consider the square plate of Figure 10.54, which is fixed at one end and subject to the couple $2\ell F$ at the other. This circumstance is quite different than that depicted in Figure 10.53. The fixed

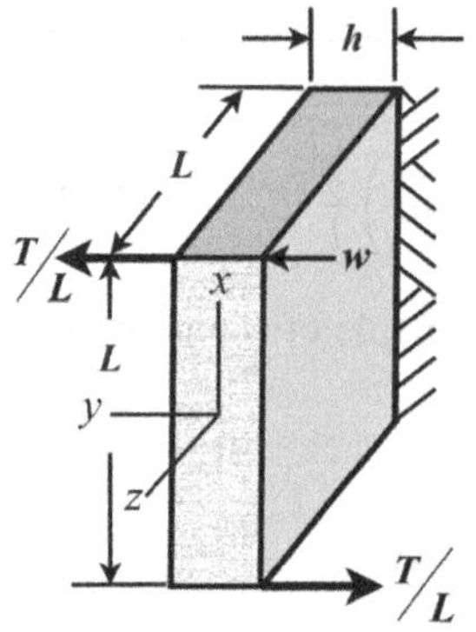

Twist $\beta \equiv 16wh^3G/LT(\equiv 12\alpha)$

Grid	1 x 1		2 x 2			3 x 3	
L/h	All Terms	Omit H-O-T	All Terms	Omit H-O-T	*	All Terms	Omit H-O-T
20	0.46	10.4	1.61	9.60	12.1	2.99	9.65
10	0.84	5.23	2.19	4.91	6.12	3.15	4.95
5	1.24	2.74	2.04	2.66	3.24	2.40	2.74

* *Asterisk signifies the twist of the unconstrained plate*

Figure 10.54 Torsion of a Clamped Plate

end is constrained; most importantly, $w = 0$ at the fixed end. Then $w = w(x, y, z)$. We can however accept the inextensionality in the y direction, i.e. $\varepsilon_y = 0$. Then four terms can be dismissed in the trilinear approximation of displacement (Equation 10.106); v: v_y, v_{xy}, v_{zy} and v_{xyz} are eliminated. Each element possesses 20 degrees of freedom.

Let us examine the strains ε_x, ε_z, γ_{yz}, γ_{yx}, and γ_{xz}.

$$\varepsilon_x = u_x + u_{xy}y + \underline{u_{xz}z} + \underline{u_{xyz}yz}$$

$$\varepsilon_z = w_z + \underline{w_{zx}x} + \underline{w_{zy}y} + \underline{w_{xyz}xy}$$

$$\gamma_{yz} = v_z + v_{xz}x + w_y + \underline{w_{yz}y} + \underline{w_{xz}x} + \underline{w_{xyz}xz}$$

$$\gamma_{yx} = v_x + v_{xz}z + u_y + \underline{u_{yx}x} + \underline{u_{xz}z} + \underline{u_{xyz}xz}$$

$$\gamma_{xz} = w_x + w_{yx}y + \underline{w_{xz}z} + \underline{w_{xyz}yz} + u_z + \underline{u_{xz}x} + \underline{u_{yz}y} + \underline{u_{xyz}yz}$$

Observe the repetition of higher order terms (underlined). Note that the present situation imposes flexure in the zy plane. The repetition of terms w_{zx}, w_{zy}, w_{xyz}, u_{xy}, u_{xz}, and u_{xyz} in the shear strains is unnecessary and, indeed, will only increase the internal energy and cause excessive stiffness ("locking").

Figure 10.54 exhibits the results obtained with different "grids" (one, four, and nine elements). Note the striking difference when the higher-order (HOT) terms are omitted. The differences between the result with one and nine elements is only 8% when HOT terms are omitted. For the thicker plate $l/h = 5$ the differences are less. On the other hand, the best results (nine elements) with all terms retained indicate a discrepancy of 10%. One must acknowledge the role of higher order terms; omitting HOT terms when duplicated in an unnecessary role does not affect convergence. As a matter of interest, the values for the unconstrained torsion are also noted (signified by the asterisk *).

10.12 ALTERNATIVE ELEMENTS

In Sections 10.7 and 10.8, we introduced the triangular element, with a linear variation of the displacement, that element possesses 6 degrees of freedom, 3 rigid motions, and 3 deformational modes (3 strain components). The counterpart in three dimensions is a tetrahedron, as depicted

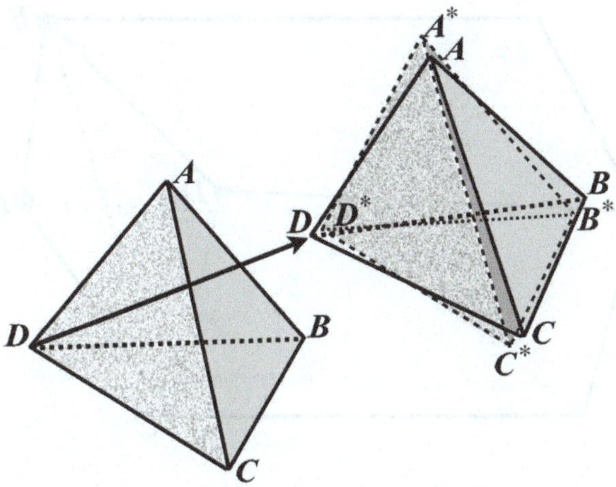

Figure 10.55 Motion+Deformation of the Tetrahedron

in Figure 10.55. With a linear variation of displacement, that element possesses 12 degrees of freedom. These can be expressed by the three components of motion at each of the four corners and account for the six rigid motions (three translations and three rotations) and six homogeneous strains. The latter can be described by the extensional strains in the directions of the six edges or, via the transformations of Sections 2.11 and 2.12, these can be described by the three extensional and three shear strains in orthogonal directions. Indeed, the reader will note that many of the computations related to finite elements are transformations between alternative systems.

A subdivision of tetrahedral elements is not so apparent as one of hexahedral elements. One can more readily perceive an assemblage of the hexahedral blocks, as shown in Figure 10.56. The block needs not be rectilinear. We can set an arbitrary central node A; radial lines from A to each of the eight corners delineate eight pyramids like that with base DBEC. The bisector CB completes the plane ABC, the common sides of the two tetrahedrons ABCD and ABCE. Similarly, each of the remaining six pyramids can be divided such that the original block is composed of eight tetrahedrons. If each is described via linear displacements, then the entire hexahedral element possesses 27 degrees of freedom. A variation at the central node A of the elastic block produces a variation of internal energy within that block only; the hexahedral block is the region of local support (see Section 10.3).

Finally, the reader should be advised of other nonorthogonal elements as those depicted in Figure 10.57. The elements of (b) and (d) depict curvilinear coordinates. Typically, the analyst employs a system of local coordinates with origin at the center. The central lines bisect the boundaries, and the coordinates are normalized such that the coordinates η, ξ, ζ take values ± 1 at the corners. Such elements are usually designated as "isoparametric." Practitioners have reported poor results when a dihedral angle (α in Figure 10.57c), is significantly less than 90°. Comparisons with established results indicate excessive stiffness of an elastic body. We can only suppose that the distorted element acts much as a constraint upon the physical system.

10.13 EPILOGUE

The "method of finite elements" can be described as a procedure, which serves to approximate the continuous variables, which describe a continuous body, by a system of discrete vari-

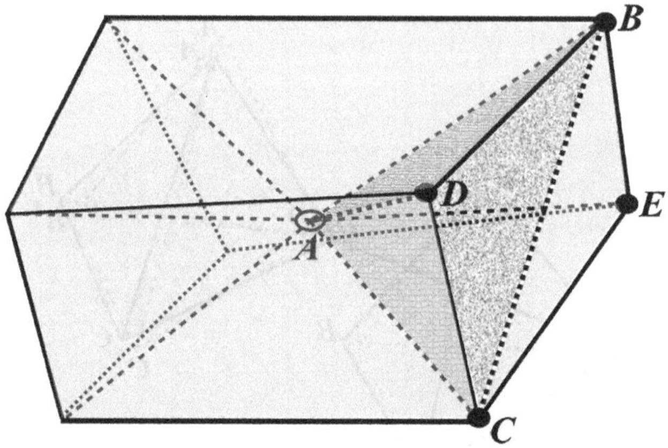

Figure 10.56 Hexahedral Element (12 Tetrahedrons)

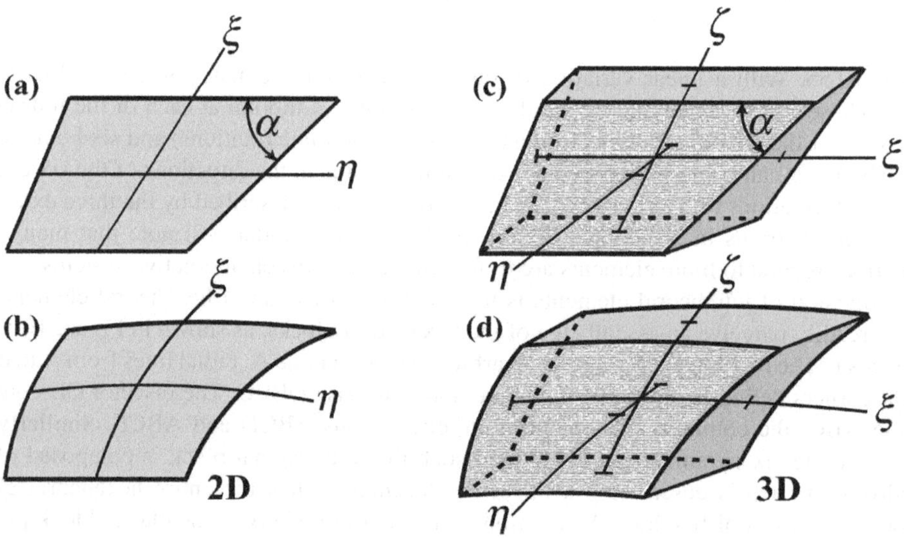

Figure 10.57 Nonorthogonal Elements

ables (nodal values). The differential equations and boundary conditions, which govern the continuous model, are replaced by a system of algebraic equations, which determine the discrete values.

Many writings suggest that the method of finite elements is distinct from the methods of finite differences. In the preceding examples, we have shown that the finite elements lead to algebraic difference equations which, in turn, converge to the differential equations *and* boundary conditions. There is one noteworthy difference: The algebraic equations, as derived via finite elements, converge to the differential equations *and* the consistent boundary values. The latter arise via the conditions imposed on the boundary nodes.

Many articles and books adopt an entirely mathematical perspective. Certainly, the effective implementation of the method(s) of finite elements requires facility with various mathematical tools: For example, one requires the transformation of coordinate systems, which we have largely avoided in the foregoing presentation. The reader who intends to use the method, to model mechanical and structural bodies, will need the additional mathematical procedures. Today the reader will find a plethora of books devoted to formulation and computation.[15],[16]

A purely mathematical perspective deprives us of the physical implications and can be an impediment to an effective implementation. A case in point is our Illustrative Example 10.1: There the simple bilinear approximation is mathematically valid: Indeed, that approximation does converge to the "exact" solution, *but* it is useless until we recognize the physical origins of the difficulty, i.e. the so-called "locking". Upon closer scrutiny, we recognize the impediment, a nonessential higher-order term. The recognition of the role of such higher-order terms, i.e. the contributions to the internal energy, is founded on physical insight.[17] Often it is the rudimentary properties of an approximation, which lead to the practical consequences: For example, in the application of the bi-linear element to the beam, we acknowledge the approximation of a smooth curve by a sequence of straight lines. Similarly, in Section 10.7 the bi-linear element of Figure 10.31 introduces kinks in the deformed surface of a plate; yet, the simple element is effective when we suppress the nonessential higher-order terms, i.e. so-called "parasitic" shear.

The method is applicable to nonlinear and to nonconservative systems.[18] Successful applications to such systems are again founded on the notion of successive linear steps. The evolution of motion and deformation traces a *smooth* path in time and/or loading. That path is typically nonlinear; we are replacing it with a succession of linear steps. The general procedure was developed in 1969 by the author who introduced the "generalized arc-length".[19]

Our introduction to finite element is intended to set forth the essential concepts, mathematical and physical, and to provide a bridge between the discrete and continuous formulations. Mathematically, we have the fundamental arguments of the variational calculus (see Figures 10.3 and 10.4). Physically, we have the correlation of the discrete nodal forces and the continuous stresses (see Equation 10.115). We hope that these analogies prove beneficial to the engineer who requires assurance that the underlying differential equations and their discrete counterparts lead to a sound structure.

NOTES

1. See for example:
   R. H. Gallagher. *Finite Element Analysis; Fundamentals.* Prentice Hall, Upper Saddle River, 1975;
   O. C. Zienkiewicz and R. C. Taylor. *The Finite Element Method, Volume 1: The Basis, Volume 2: Solid Mechanics*, fifth edition. Butterwoth Heinemann, Oxford, 2000.
2. See G. Wempner. Discrete approximations related to nonlinear theories of solids. *Int. J. Solids Struct.*, 7:1581–1599, 1971.
3. M. W. Johnson and R. W. McLay. Convergence of the finite element method in the theory of elasticity. *J. Appl. Mech.*, 35:274–278, 1968.
4. Note: M is moment per unit width.
5. G. Wempner, D. Talaslidis, and C.-M. Hwang. A simple and efficient approximation of shells via finite quadrilateral elements. *J. Appl. Mech.*, 49:115–120, 1982.
6. G. A. Wempner. New concepts for finite elements of shells. *Zeit. Ang. Math. u. Mech.*, 1968.
7. G. A. Wempner and D. A. Talaslidis. Simple and efficient approximation of shells via finite quadrilateral elements. *J. Appl. Mech.*, 1982.
8. Khakalo, Sergei, DSc, Integrated Computational Materials Engineering, VTT, Espoo, Finland.
9. C. D. Pionke. *Convergence of Finite Elements Based on the Hu-Washizu Variational Theorem with Minimal Compatibility*. Doctoral Thesis, Georgia Institute of Technology, Georgia, Atlanta, 1993.
10. Khakalo, Sergei, DSc, Integrated Computational Materials Engineering, VTT, Espoo, Finland, 1993.

11. This polynomial was employed by A. Adini and R. W. Clough. "Analysis of Plate Bending by the Finite Element Method", NSF Report, Grant G7337, 1960. Numerical results can be found in "Finite Element Stiffness Matrices for Analysis of Plate Bending", Proceedings First Conference on Matrix Methods in Structural Mechanics, AFFDL, TR 66-80, 1965.
12. see S. Timoshenko and S. Woinwsky-Krieger. *Theory of Plates and Shells*, second edition, p. 8. McGraw-Hill, 1951.
13. G. Wempner. Finite element modeling of solids via the Hu-Washizu functional. *Computers in Mechanical Engineering*, 2(1): 57–62, 1973.
14. S. P. Timoshenko. *Theory of Elasticity*, second edition, p. 278. McGraw-Hill, Columbus, 1951.
15. R. D. Cook, M. E. Plesha, and D. S. Malkus. *Concepts and Applications of Finite Element Analysis*. John Wiley & Sons, Hoboken, 1989.
16. Introduction to Finite Element Methods, Univ. of Colorado, 2016, Internet Note: Analysis is confined to linear analysis, i.e. small deflections of Hookean bodies.
17. G. Wempner, D. Talaslidis, and C. M. Hwang. A simple and efficient approximation of shells via finite quadrilateral elements. *J. Appl. Mech.*, 49(1): 115–120, 1982.
18. G. Wempner, and D. Talaslidis. A simple finite element for elastic-plastic deformations of shells. *Comp. Meth. In Appl. Mech. & Eng.*, 34:1051–1064, 1982.
19. G. A. Wempner. Discrete approximations related to nonlinear theories of solids. *Int. J. Solids & Struct.*, 7:1581–1599, 1971.

Appendix A: Centroid of an Area

In the study of HI-HO prismatic members under extension and flexure, the centroid of the cross-section plays a useful role. Specifically, it is convenient to represent the resultant by a force at the centroid and a couple (moment about the centroid); consequently, the centroid is a convenient origin of the coordinates. If the origin of a Cartesian system (y, z) is at the centroid, then

$$\int\int_A y\,dA = \int\int_A z\,dA = 0 \qquad \text{(A.1)}$$

Equations A.1 *define* the centroid and enable us to locate the centroid.

Consider an arbitrary section A, depicted in Figure A.1, and the arbitrary coordinates (y, z). We require the location of the centroid C. It is located by the two unknowns, Y and Z. Let $\bar{y}$ and $\bar{z}$ be Cartesian coordinates with axes parallel to y and z, originating at C. By definition (A.1)

$$\int\int_A \bar{y}dA = \int\int_A = \bar{z}dA = 0$$

or

$$\int\int_A (y - Y)dA = \iint_A (z - Z)dA = 0$$

It follows that

$$Y = \frac{1}{A}\iint_A ydA, \quad Z = \frac{1}{A}\iint_A zdA \qquad \text{(A.2a,b)}$$

To illustrate the application of Equations (A.2) for locating the centroid, let us apply them to the section depicted in Figure A.2. The region consists of the rectangular strip A_1 and the semicircular region A_2. We choose the apparently convenient system (y, z) with its origin at the center of the circle. This lies on a line of symmetry; centroid C lies on the line of symmetry by definition (A.1). Then, the unknown Z is determined by (A.2b); the integral over the area consists of the integrals over areas A_1 and A_2.

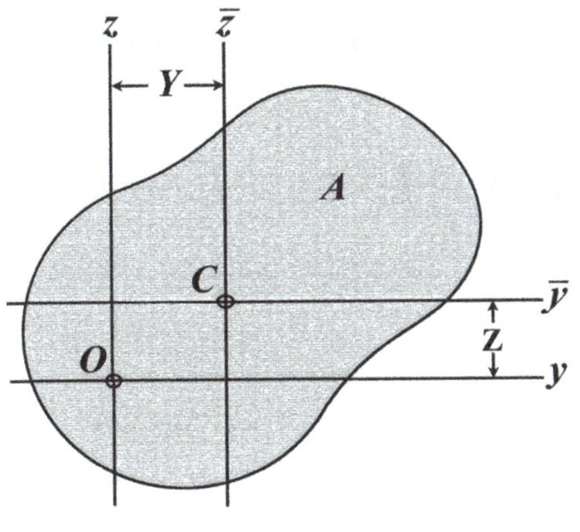

Figure A1

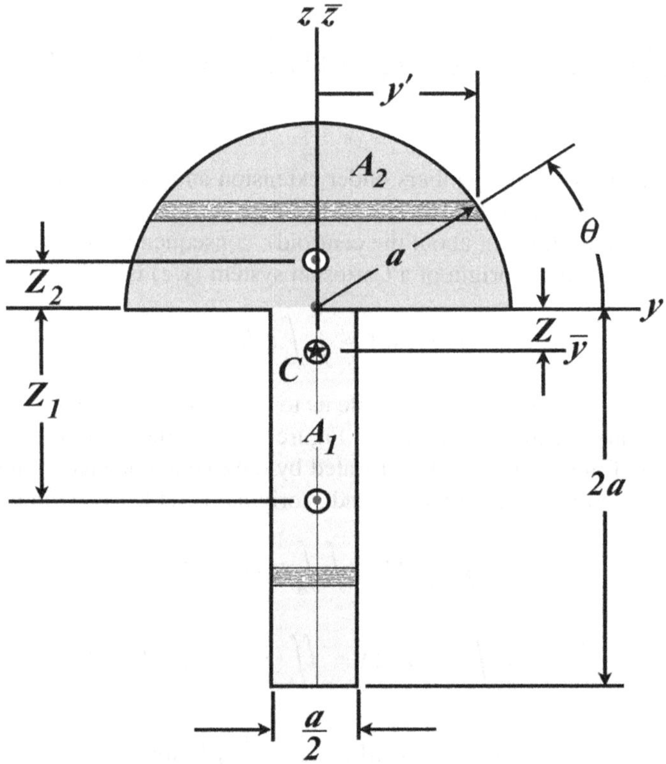

Figure A2

The differential areas are depicted by the shaded strips. In A_1 and A_2

$$dA = \frac{A}{2}dz, \quad dA = 2y'dz$$

respectively. We note that on the circular boundary,

$$z = a\sin\theta, \quad y = y' = a\cos\theta \tag{A.3}$$

Therefore,

$$\iint_A z dA = \frac{a}{2}\int_{-2a}^{0} z dz + 2a^3\int_0^{\pi/2} \sin\theta\cos^2\theta d\theta$$
$$= -a^3 + 2\frac{a^3}{3} = -\frac{a^3}{3}$$

By Equation (A.2b)

$$Z = \frac{-\frac{1}{3}a^3}{\left(\frac{\pi}{2}a^2 + a^2\right)} = -\frac{2a}{3(\pi+2)}$$

Note that Equations (A.2b) and (A.3) give

$$Z = \frac{1}{A}\iint_{A_1} z dA + \frac{1}{A}\iint_{A_2} z dA \tag{A.4}$$

If Z_1 and Z_2 are coordinates to the centroids of A_1 and A_2, respectively, then by Equations (A.2) and (A.4)

$$Z = \frac{Z_1 A_1 + Z_2 A_2}{A} \tag{A.5}$$

In short, if we know the positions Z_1 and Z_2 of the centroids for the individual parts, we can obtain the result by the algebraic equation A.5. Values have been tabulated for simple geometric forms. The centroid of the rectangle (or any doubly symmetric area) is at the midpoint; $Z_1 = -a$. The centroid of the semicircular region is at $Z_2 = 4a/3\pi$. Then, by Equation (A.5)

$$Z = \frac{-a^3 + \frac{4a}{3\pi} \times \frac{\pi a^2}{2}}{a^2 + \frac{\pi}{2} a^2} = -\frac{2a}{3(2+\pi)}$$

Appendix B: Integrals I_{yy}, I_{zz}, and I_{yz}

B.1 ROTATION OF AXES

In the Bernoulli-Euler theory of beams, we encounter the integrals (in alternative notations):

$$I_y \equiv I_{yy} \equiv \int\int_A y^2 dA \tag{B.1a}$$

$$I_z \equiv I_{zz} \equiv \int\int_A z^2 dA \tag{B.1b}$$

$$I_{yz} \equiv I_{zy} \equiv \int\int_A yz dA \tag{B.1c}$$

The reader must be forewarned that many books use another notation: $I_y = I_{yy}$ is often denoted by I_z. Their notation can be traced to a physical analog, namely, the "moment of inertia," which arises in the dynamics of rotating rigid bodies. A body rotating about the axis z has a moment of inertia about that axis:

$$I_z \equiv \iiint_v (x^2 + y^2)\rho dv$$

where ρ denotes mass density, and v denotes volume. If the body is very thin in the x direction (thickness t) and the density ρ is constant, then

$$I_z \approx t\rho \iint_A y^2 dA$$

Hence, the term *area moment of inertia* is attached to the integral, and the subscript z persists. We adopt the notation of Equation (B.l) for two reasons. (1) The subscript signals the integrand: $I_y y$ has an integrand y^2, I_{yz} has an integrand yz, and I_{zz} has an integrand z^2. (2) More importantly, the notations of Equation (B.l) conform completely to those employed previously for the *mathematically* similar components of stress (σ) and strain (ε). Let us show that these "components" $(I_{yy}, I_{yz} I_{zz})$ transform via the same linear equations.

Figure B.1 shows an arbitrary cross section of area A and rectangular coordinates (y, z) and (n, t). Both originate at the common point O, which is arbitrary. From the geometry of the figure

$$n = a + b = y\cos\theta + z\sin\theta$$
$$t = c - d = z\cos\theta - y\sin\theta$$

From the definitions:

$$\begin{aligned} I_{nn} &= \iint_A n^2 dA \\ &= \cos^2\theta \iint_A y^2 dA + 2\cos\theta\sin\theta \iint_A yz dA + \sin^2\theta \iint z^2 dA \\ &= I_{yy}\cos^2\theta + 2I_{yz}\cos\theta\sin\theta + I_{zz}\sin^2\theta \end{aligned} \tag{B.2a}$$

Likewise,

$$I_{nt} = -(I_{yy} - I_{zz})\cos\theta\sin\theta + I_{yz}(\cos^2\theta - \sin^2\theta) \tag{B.2b}$$

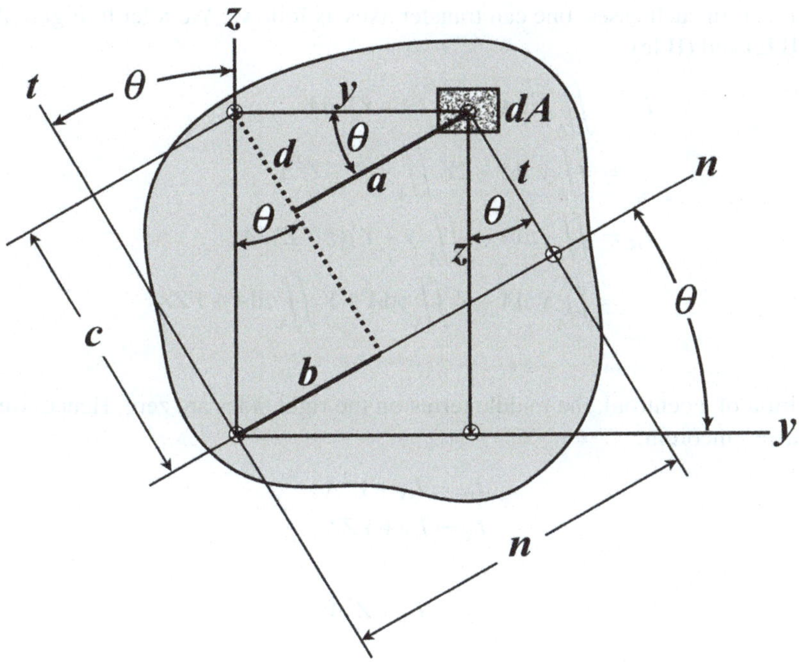

Figure B1

In the notation of Sections (2.18) and (3.12),

$$l_n = \cos\theta, \quad m_n = \sin\theta$$
$$l_t = -\sin\theta, \quad m_t = \cos\theta$$

$$I_{nn} = l_n l_n I_{yy} + l_n m_n I_{yz} + m_n l_n I_{yz} + m_n m_n I_{zz} \tag{B.3a}$$
$$I_{nt} = l_n l_t I_{yy} + l_n m_t I_{yz} + m_n l_t I_{yz} + m_n m_t I_{zz} \tag{B.3b}$$

Of course, here the coordinates y and z replace x and y in keeping with our notations for the cross section of a beam.

with the notations of Equations (B.1), the linear transformations of Equations (B.2) or (B.3) are the same as those for the components of stress (σ) and strain (ε). It follows by the same arguments that principal directions $(\bar{y}, \bar{z})$ always exist, such that $I_{\bar{y}\bar{z}} = 0$ and $I_{\bar{y}\bar{y}}$ and $I_{\bar{z}\bar{z}}$ are extrema (maximum or minimum). As before, their orientation is determined by setting Equation (B.2b) to zero. In terms of the double angle 2θ,

$$\tan 2\theta = \frac{2I_{yz}}{I_{yy} - I_{zz}}$$

B.2 TRANSLATION OF AXES

In the Bernoulli-Euler theory of beams, the centroid of the cross section is often the convenient origin of coordinates (y, z). Then we need the integrals I_{yy}, I_{yz}, and I_{zz} with respect to the centroidal axes $(\bar{y}, \bar{z})$. In such cases, one can transfer axes as follows: We refer to Figure A.1 and the definitions (B.la) and (B.lc):

$$\begin{aligned} I_{yy} &= \iint_A y^2 dA = \iint_A (\bar{y}+Y)^2 dA \\ &= \iint_A \bar{y}^2 dA + 2Y \iint_A y dA + Y^2 A \\ I_{yz} &= \iint_A yz dA = \iint_A (\bar{y}+Y)(\bar{z}+Z) dA \\ &= \iint_A \bar{y}\bar{z} dA + Z \iint_A \bar{y} dA + Y \iint_A \bar{z} dA + YZA \end{aligned}$$

By the definition of a centroid, the middle terms on the right sides are zero. Hence, we obtain the "transfer of axes" theorem:

$$I_{yy} = I_{\bar{y}\bar{y}} + Y^2 A \quad \text{(B.4a)}$$

$$I_{yz} = I_{\bar{y}\bar{z}} + YZA \quad \text{(B.4b)}$$

Likewise,

$$I_{zz} = I_{\bar{z}\bar{z}} + Z^2 A \quad \text{(B.4c)}$$

ILLUSTRATIVE EXAMPLE B.1

Consider again the section shown in Figure A.2. Recall that the centroid of the semicircular part is at $z = 4a/3pi$. We know that the integral $''_{I}zz$ (the double prime signifies part 2) for this part is

$$I''_{zz} = \frac{pi}{8} a^4$$

With respect to the centroid of A_2,

$$\begin{aligned} \bar{I}''_{zz} = I''_{zz} &= -\left(\frac{4a}{3\pi}^2 A_2\right) \\ &= \pi a^4 8 - \frac{4a}{3\pi}^2 \left(\frac{\pi a^2}{2}\right) \\ &= \left(\frac{\pi}{8} - \frac{8}{p\pi}\right) a^4 \end{aligned}$$

The integral I'_{zz} (the prime signifies part 1) with respect to the centroid of A_1 is

$$\bar{I}'_{zz} = \frac{a^4}{3}$$

The integral for the cross section with respect to the centroid of the entire section is

$$I_{\bar{z}\bar{z}} = I''_{\bar{z}\bar{z}} + I'_{\bar{z}\bar{z}} \quad \text{(B.5)}$$

The two integrals on the right can be obtained by transfer of axes according to Equation (B.4c)

$$I_{\bar{z}\bar{z}} = I''_{zz} + I'_{zz} + (\bar{z}'')^2 A_2 \quad \text{(B.6a)}$$

$$I_{\bar{z}\bar{z}} = I'_{zz} + I'_{zz} + (\bar{z}')^2 A_1 \quad \text{(B.6b)}$$

Note: The distances $\bar{z}'$ and $\bar{z}''$ are the distances between the centroidal axis of the cross section and the centroidal axes of parts 1 and 2, respectively. The result follows from Equations (B.5) and (8.6):

$$I_{\bar{z}\bar{z}} = I''_{zz} + I'_{zz} + (\bar{z}'')^2 A_2 + (\bar{z}')A_1$$

$$I_{\bar{z}\bar{z}} = \left(\frac{\pi}{8} - \frac{8}{9\pi}\right) a^4 + \left(\frac{4a}{3\pi} - Z\right) \frac{\pi a^2}{2} + \frac{a^4}{3} + (a+Z)^2 a^2 \quad \text{(B.7a)}$$

By direct integration

$$I_{\bar{z}\bar{z}} = 2a^2 \int_0^{\pi/2} (Z + a\sin\theta)^2 \cos^2\theta \, d\theta + \frac{a}{2} \int_{-} 2a + z^{-Z} \bar{z}^2 dA \quad \text{(B.7b)}$$

where $Z = -2a/3(\pi+2)$, as previously determined. Both Equations (B.7a) and (B.7b) give $I_{\bar{z}\bar{z}} = 1.683a^4$, the integral throughout the cross section with respect to the centroidal axes.

Index

Note: *Italic* page numerals refer to figures and those followed by "n" refer to end notes.

For Product Safety Concerns and Information please contact our EU representative GPSR@taylorandfrancis.com Taylor & Francis Verlag GmbH, Kaufingerstraße 24, 80331 München, Germany

Batch number: 10733697

Printed by Printforce, the Netherlands